UNSOLVED PROBLEMS OF NOISE AND FLUCTUATIONS

Previous Proceedings in the Series of International Conferences on Unsolved Problems of Noise and Fluctuations

	Year	Held in	Publisher	ISBN
3rd	2002	Washington, USA	AIP Conf. Proceedings Vol. 665	0-7354-0127-6
2nd	1999	Adelaide, Australia	AIP Conf. Proceedings Vol. 511	1-56396-826-6
1st	1996	Szeged, Hungary	World Scientific	981-02-3199-7

To learn more about these titles, or the AIP Conference Proceedings Series, please visit the webpage **http://proceedings.aip.org**

UNSOLVED PROBLEMS OF NOISE AND FLUCTUATIONS

UPoN 2005: Fourth International Conference on Unsolved Problems of Noise and Fluctuations in Physics, Biology, and High Technology

Gallipoli, Lecce, Italy 6 - 10 June 2005

EDITORS

L. Reggiani
C. Pennetta
V. Akimov
E. Alfinito
M. Rosini

Università di Lecce
Lecce, Italy

SPONSORING ORGANIZATIONS

Lecce University and the "Dipartimento di Ingegneria dell' Innovazione"
Office of Naval Research Global
Institute of Pure and Applied Physics (IUPAP)
Italian Ministry of University and Research MIUR
European Commission
Nanotechnology National Laboratory of CNR-INFM
"Provincia di Lecce"
"Comune di Gallipoli"
Hewlett Packard Invent, the INFOTEL, and the INTEL
"Associazione degli Industriali" of Lecce,
Industrial Liason Office of Lecce
European Physical Society
Institute of Electrical and Electronics Engineering (IEEE)

AMERICAN INSTITUTE OF PHYSICS

Melville, New York, 2005

AIP CONFERENCE PROCEEDINGS ■ VOLUME 800

Editors:

L. Reggiani
C. Pennetta
V. Akimov
E. Alfinito
M. Rosini

Dipartimento di Ingegneria dell'Innovazione
Università di Lecce
Via Arnesano s/n
73100 Lecce
Italy

E-mail: lino.reggiani@unile.it
cecilia.pennetta@unile.it
vladimir.akimov@unile.it
eleonora.alfinito@unile.it
marcello.rosini@unile.it

L.C. Catalog Card No. 2005935381
ISBN 0-7354-0289-2
ISSN 0094-243X
Printed in the United States of America

CONTENTS

ENHANCED AND SUPPRESSED SHOT NOISE

NOISE AND COHERENCE

NOISE AND CHAOS

CONSTRUCTIVE ROLE OF NOISE

NOISE IN BIOLOGICAL SYSTEMS

NOISE IN DEVICES

NOISE IN COMPLEX SYSTEMS AND NON-GAUSSIAN FLUCTUATIONS

Preface

The 4th International Conference on Unsolved Problems of Noise (UPoN) and Fluctuations in Physics, Biology and High Technology was held in Italy at the Conference Centre of the Hotel Costa Brada located in Gallipoli (Lecce) and chaired by Lino Reggiani. According to the tradition, the Conference follows a three years term since its first edition held in 1996 in Szeged (Hungary). The first conference, organized by Laszlo Kish (former Kiss), was mostly devoted to high technology devices. Conference participants recall its warm and productive atmosphere. The total number of talks was 57, the number of attendees was 83 and the Proceedings were published by Singapore Worldwide publishing. The second one, organized by Derek Abbott, was hosted by the University of Adelaide, Australia in 1999 and focused mainly on mathematical aspects and paradoxes in noise and fluctuation research. The total number of talks was 73, the number of attendees was 74, and the Proceedings were published by the American Institute of Physics (AIP). The conference received coverage in Nature (Random fluctuations - Unsolved problems of noise, Nature 401, p.23, Sept. 2, 1999). The third one, organized by Sergey Bezrukov, was held in the National Institutes of Health campus in Bethesda (Washington, DC, USA), and focused mainly on biology, biophysics, and biomedical engineering aspects. The total number of talks was 73, the number of attendees was 106, and the Proceedings were published by AIP. This fourth edition focused mainly on noise and fluctuations at the nanometric scale-length in electron-devices, bio-materials, and mesoscopic systems. The total number of contributions, including a poster session, was around 90, the number of attendees was slightly above 110 coming from 4 continents and 27 countries.

Since its first edition, the aim of the UPoN conference is to provide a forum for researchers working on noise and fluctuations, where they can present and discuss their scientific problems which do resist solutions. These unsolved problems need a more thorough, more open-minded, and less biased refereeing process than the presentation of solved problems typical in the publications of the standard literature. The submitted contributions were reviewed by the members of the Scientific Committee.

To favour cross-fertilization from different disciplines and knowledge transfer no parallel sessions were foreseen. The conference focused on the following main topics. 1 - Theoretical frontiers of noise and fluctuations, 2 - Experimental frontiers of noise and fluctuations, 3 - Enhanced and suppressed shot noise, 4 - Noise and coherence, 5 - Noise and chaos, 6 - Constructive role of noise, 7 - Noise in complex systems and Non-Gaussian fluctuations, 8 - Noise in biological systems, 9 - Noise in devices, 10 - Noise and fluctuations in complex systems.

To honor and remember the memory of Prof. Michele Costato, a physicist at the Modena University (Italy) who addressed the Chairman of this Conference to the study of electrical noise in solids, an award of 2000 Euro was assigned by the Scientific Committee to Dr. Maria King from Texas University (USA) who was selected as the author of the contribution presenting the most intriguing unsolved problem at the UPoN-4 2005 Conference. The motivation of the award addressed the issue: NOISE FOR HEALTH: Can noise signify the health of a living organism?

The partnership and the financial support which helped the Conference organization and the researchers with budget difficulties came from several Institutions which are below summarized and warmly acknowledged. The Lecce University and the "Dipartimento di Ingegneria dell' Innovazione" of the same University, the Office of Naval Research Global (Grant Number N00014-05-1-1020), the Institute of Pure and Applied Physics (IUPAP), the Italian Ministry of University and Research (PRIN prot. 2003029008), the European Commission (Grant IST-2001-38899-SPOT-NOSED) the Nanotechnology National Laboratory of CNR-INFM, the "Provincia di Lecce", the "Comune di Gallipoli", the Hewlett Packard Invent and INFOTEL, the INTEL, the "Associazione degli Industriali" of Lecce, the Industrial Liason Office of Lecce, the European Physical Society, the Institute of Electrical and Electronics Engineering (IEEE).

A warm acknowledgement is finally addressed to the technical staff who provided assistance during and before the Conference: Dr. M. Leuzzi, and Dr. E. Caputo of the Advanced Computing Laboratory of the "Dipartimento di Ingegneria dell' Innovazione".

L. Reggiani, C. Pennetta, V. Akimov, E. Alfinito and M. Rosini
Editors

Editors' address:

Dipartimento di Ingegneria dell'Innovazione
Universita' di Lecce
Via Arnesano s/n
73100 Lecce
Italy

E-mail: lino.reggiani@unile.it

SCIENTIFIC COMMITTEE

SPONSORS

Lecce University and the "Dipartimento di Ingegneria dell' Innovazione"
Office of Naval Research Global (Grant Number N00014-05-1-1020)
Institute of Pure and Applied Physics (IUPAP),
Italian Ministry of University and Research MIUR (PRIN prot. 2003029008)
European Commission (Grant IST-2001-38899-SPOT-NOSED)
Nanotechnology National Laboratory of CNR-INFM
"Provincia di Lecce"
"Comune di Gallipoli"
Hewlett Packard Invent, the INFOTEL, and the INTEL
"Associazione degli Industriali" of Lecce,
Industrial Liason Office of Lecce
European Physical Society
Institute of Electrical and Electronics Engineering (IEEE)

THEORETICAL FRONTIERS ON NOISE AND FLUCTUATIONS

Mean Extinction Time for Birth-Death Processes & Failure of the Fokker-Planck Approximation

Charles R. Doering[1,2], Khachik V. Sargsyan[1,2] and Leonard M. Sander[1,3]

[1]*Department of Mathematics, University of Michigan, Ann Arbor, MI 48109-1043 USA*
[2]*Michigan Center for Theoretical Physics, Ann Arbor, MI 48109-1040 USA*
[3]*Department of Physics, University of Michigan, Ann Arbor, MI 48109-1040 USA*

Abstract. We consider extinction times for a class of birth-death processes commonly found in applications where there is a control parameter that defines a threshold. Below the threshold the population quickly becomes extinct and above the threshold it persists for an exponentially long time. We derive an asymptotic expression for the mean time to extinction in the discrete population case for large values of the population scale, i.e., in the continuous population limit. The Fokker-Planck (Markov diffusion process) approximation does not generally predict the true behavior of the extinction time for large values of the population scale. This is surprising because the continuum limit is precisely where it could be anticipated that the Fokker-Planck approximation should be most accurate. Rather, it is valid only near the threshold. This constitutes an interesting example of the delicate relationship between discrete and continuum treatments of the same problem.

Keywords: Markov processes, master equations, Fokker-Planck equations, first passage time.
PACS: 02.50.-r, 82.20.Uv, 87.32.-n

INTRODUCTION

Many problems in nature are fundamentally *discrete*, involving the actions and interactions of individual elements. For some such systems this feature simplifies modeling and simulation. In some cases, however, it complicates the problem. For situations where there are many individual elements, the models are complex, and direct simulations are difficult, we might hope to exploit averaging effects and derive a *continuum* description that is more susceptible to theoretical analysis and numerical solution. This is one of the fundamental and classical challenges for physics.

For example, continuous time/discrete state space Markov processes, described by Master Equations, will mimic the solutions of deterministic rate equations in the full continuum limit. And they should, in appropriate limits, also be well approximated by continuous time/continuous state space Markov diffusion processes, described by Fokker-Planck equations. This strategy works in many cases, but here we point out that it is not always possible to get *all* the relevant aspects right in the process of making this transition. The unsolved problem (really a question) we propose is precisely this: is there a consistent and quantitatively accurate way to model discrete stochastic processes by continuous processes that capture both the deterministic limiting dynamics as well as all the important features of the fluctuations?

CP800, *Unsolved Problems of Noise and Fluctuations: UPoN 2005*
edited by L. Reggiani, C. Pennetta, V. Akimov, E. Alfinito, and M. Rosini

BIRTH-DEATH PROCESSES

Let $\mathcal{N}(t) \in \{0, 1, ..., R \leq \infty\}$ be a continuous time Markov processes defined by the birth and death rates:

$$\begin{aligned} \lambda_n &= \text{ rate for } n \to n+1 \\ \mu_n &= \text{ rate for } n \to n-1. \end{aligned} \tag{1}$$

The evolution of $P_n(t) = \text{Prob}\{\mathcal{N}(t) = n\}$ is then defined by the master equation:

$$\frac{dP_n}{dt} = \lambda_{n-1}P_{n-1} - (\lambda_n + \mu_n)P_n + \mu_{n+1}P_{n+1}. \tag{2}$$

As a specific example throughout this report we will focus on the Susceptible-Infected-Susceptible (SIS) model from theoretical epidemiology. In this system there are n infected individuals within total population of N, and "birth" events denote the infection of a new individual while "death" means recovery and instantaneous recovery of susceptibility to re-infection. Assuming a well-mixed population and the usual hypothesis for a Markov process, we may take

$$\lambda_n = \Lambda n\left(1 - \frac{n}{N}\right) \quad \text{and} \quad \mu_n = n \tag{3}$$

where the time units are set by taking recovery at unit rate and Λ (a.k.a. R_0) is the individual infection rate per encounter. When $\Lambda > 1$ there is a non-vanishing equilibrium population (endemic infection level):

$$n_e \approx N\left(1 - \frac{1}{\Lambda}\right). \tag{4}$$

Above threshold the infected segment of the population persists for a long time before extinction. Precisely at threshold the infection dies out "slowly", while below threshold the infected segment of the population is expected to die out very quickly (see the appendix for more discussion of the threshold and sub-threshold cases).

We are interested in the *continuum limit* for the state space where an appropriate population scale N is very large. Let $x = n/N$ be the continuum population variable and suppose that the continuum rate functions defined by

$$\lambda_n = N\bar{\lambda}(x) \quad \text{and} \quad \mu_n = N\bar{\mu}(x) \tag{5}$$

are smooth. Denoting the expected population measured in units of N as $X(t)$, in the continuum $N \to \infty$ limit we recover the deterministic rate equation

$$\frac{dX}{dt} = \bar{\lambda}(X) - \bar{\mu}(X). \tag{6}$$

For example in the SIS model, the continuum rate functions ($x = n/N$) are

$$\bar{\lambda}(x) = \Lambda x(1-x) \quad \text{and} \quad \bar{\mu}(x) = x. \tag{7}$$

Then the continuum limit for $X(t)$ satisfies the *logistic equation*:

$$\frac{dX}{dt} = (\Lambda - 1)X - \Lambda X^2. \tag{8}$$

As is evident, the threshold in the SIS model is at $\Lambda = 1$. That is, if $\Lambda \leq 1$, $X(t) \rightarrow 0$, and if $\Lambda > 1$ and $X(0) > 0$, $X(t) \rightarrow x_e = (1-1/\Lambda) > 0$.

EXTINCTION TIMES

For this class of Birth-Death processes with $\lambda_0 = 0$, the *extinct* state $\mathcal{N}(t) = 0$ is an absorbing state. That is, starting from any initial level the system will achieve extinction at a finite time with probability one. This is a feature of the discrete stochastic model that is absent in the deterministic continuum description. Let τ_n (or $\tau(x)$) be the mean extinction time starting from n (or x). The question of interest here is, how does $\tau(x)$ behave as $N \rightarrow \infty$? Above threshold it will diverge and the challenge is to accurately estimate the rate of divergence.

We might expect that there are two alternative and equivalent approaches to this question. For one we could perform asymptotic analysis on the exact solution of the discrete mean first passage time problem,

$$-1 = \lambda_n \tau_{n+1} - (\lambda_n + \mu_n)\tau_n + \mu_n \tau_{n-1}, \tag{9}$$

or else we could study the exact solution of the "near-continuum" Fokker-Planck approximation

$$-1 = f(x)\frac{dT(x)}{dx} + \frac{g(x)^2}{2N}\frac{d^2T(x)}{dx^2} \tag{10}$$

where

$$f(x) = \bar{\lambda}(x) - \bar{\mu}(x) \quad \text{and} \quad g(x)^2 = \bar{\lambda}(x) + \bar{\mu}(x) \tag{11}$$

and $T(x) = \tau_n$ is assumed to be a smooth function of x. The Fokker-Planck approximation is simply a truncated Taylor expansion: with $x = n/N$ and $\Delta x = 1/N$,

$$\tau_{n\pm 1} = T(x) \pm \frac{dT(x)}{dx}\Delta x + \frac{1}{2}\frac{d^2T(x)}{dx^2}\Delta x^2 + \cdots, \tag{12}$$

and inserting this into Equation (10) followed by truncation at the lowest order in $1/N$ produces Equations (11) and (12).

The exact solution of the discrete problem is [2]

$$\tau_n = \sum_{m=1}^{n}\left[\frac{1}{\mu_m} + \prod_{i=1}^{m-1}\frac{1}{\rho_i}\sum_{j=m+1}^{R}\frac{1}{\mu_j}\prod_{k=1}^{j-1}\rho_k\right] \tag{13}$$

where

$$\rho_n = \frac{\lambda_n}{\mu_n} = \overline{\rho}(x) = \frac{\overline{\lambda}(x)}{\overline{\mu}(x)}. \tag{14}$$

In principle this solves the problem once and for all. However, this formula is neither theoretically transparent nor practical to evaluate numerically at large N. In order to discern the behavior as $N\to\infty$ it is necessary to perform a careful asymptotic analysis. This has recently been carried out in [1]. Not unexpectedly, the asymptotic $N\to\infty$ behavior depends crucially on whether the system is super-threshold, at threshold, or sub-threshold.

For the super-threshold case we find

$$\begin{aligned}\tau(x) = \left(1 - e^{-N\log\overline{\rho}(0)x}\right) \times \sqrt{\frac{2\pi\overline{\rho}(0)}{N[\overline{\lambda}(x_e)\overline{\mu}'(x_e) - \overline{\lambda}'(x_e)\overline{\mu}(x_e)]}} \times \\ \times \frac{e^{N[\Phi(0)-\Phi(x_e)]}}{\overline{\rho}(0)-1} \times \left(1 + \mathcal{O}(\frac{1}{N})\right).\end{aligned} \tag{15}$$

This diverges exponentially $\sim \mathcal{O}(e^{N\Delta\Phi})$ for large N where the *effective potential* is

$$\Phi(x) = -\int^{x}\log\overline{\rho}(\xi)d\xi. \tag{16}$$

Figure 1 illustrates the structure of $\tau_n = \tau(x)$ for the SIS model, and the exact results are compared with the asymptotic large N approximation. The boundary layer structure is apparent: when N is "large", $\tau(x)$ rapidly achieves and maintains the value $\tau(x_e)$ once $n = Nx > \mathcal{O}(1)$. While some discrepancy between the exact data and the approximation curve is evident, they actually differ by less 2% uniformly over the entire range of n. See [1] for explicit numerical verification that the formula in Equation (15) truly is within a factor of $1+\mathcal{O}(N^{-1})$ uniformly in x for the SIS model.

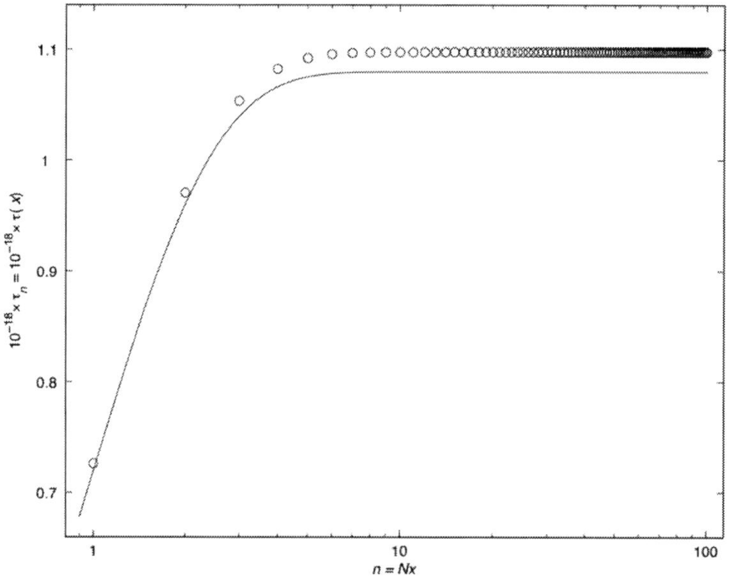

Figure 1. Mean extinction times $\tau_n = \tau(x)$ vs. $n = Nx$ for the SIS model with $N = 100$ and $\Lambda = 3$. The circles are the exact values from Equations (13) and (14) while the solid line is its large-N asymptotic approximation from Equations (15) and (16). The Fokker-Planck approximation (not shown) falls a factor of about 50 below the exact values, well off the scale of this plot.

It is an elementary exercise to solve the Fokker-Planck model in Equation (11) for $T(x)$, resulting in a double integral expression. For the super-threshold case where there is only one deterministic equilibrium point x_e, a straightforward and standard steepest descents analysis at large N yields,

$$T(x_e) = \frac{2}{|V'(0)|}\sqrt{\frac{2\pi}{NV''(0)}}\frac{e^{N[V(0)-V(x_e)]}}{g(x_e)^2} \times \left(1 + \mathcal{O}(\frac{1}{N})\right) \tag{17}$$

with effective potential

$$V(x) = -2\int^x \frac{f(\xi)}{g(\xi)^2}d\xi = 2\int^x \frac{\bar{\mu}(\xi) - \bar{\lambda}(\xi)}{\bar{\mu}(\xi) + \bar{\lambda}(\xi)}d\xi. \tag{18}$$

The effective potential $V(x)$ from the Fokker-Planck equation is *different* from the effective potential $\Phi(x)$ emerging from the asymptotic limit of discrete solution. This means that the Fokker-Planck approximation has a systematic error in the exponent; it does not produce the same large N asymptotics for the extinction time as the discrete process it is supposed to mimic. For the case illustrated in Figure 1 with $N = 100$, for example, the Fokker-Planck approximation underestimates the actual values by a factor of about 50. That error factor subsequently grows exponentially for larger N.

The existence of the discrete effective potential has been recognized for a long time. We mention in particular the early works of Weiss and Dishon [3] and Hänggi, Grabert, Talkner & Thomas [4]. More recently this class of problems has received a lot of attention in the population biology literature; see, for example, the paper of Nasell [5]. It is discussed as well in the classic monograph of van Kampen [6]. There it is noted and exploited that the effective potentials do in fact coincide in two very important ways: $V(x)$ and $\Phi(x)$ have minima at the deterministic equilibrium point x_e (i.e., $V'(x_e) = \Phi'(x_e) = 0$) and they have the same curvature there ($V''(x_e) = \Phi''(x_e)$). This means that both potentials may have the equivalent "local" Gaussian approximations. The mismatch of the tails in the effective potentials, however, dominates resulting in the discrepancy for the kinetics of rare events.

SUMMARY AND CONCLUSIONS

For this class of Birth-Death processes with $\lambda_0 = 0$, $\mathcal{N}(t) = 0$ is an absorbing state and every realization starting from any initial level will go extinct in a finite time. The population may live for *exponentially long* times in super-threshold situations. This extinction phenomena is *absent* deterministic continuum models, but the Markov diffusion process approximation to the discrete state space process (i.e., the Fokker-Planck approximation to the Master Equation, or equivalently the appropriate Langevin stochastic differential equation generalization of the deterministic model) *does* predict finite time extinction. Equilibrium and quasi-steady-state statistics, i.e., some small fluctuation moments, may also be correctly predicted by Fokker-Planck approximation, but we have seen here that there is generally a *residual asymptotic error* (in the exponent!) for rare events such as the mean extinction time.

As discussed in detail in [1], this residual asymptotic error vanishes when the drift $f(x)$ is sufficiently small. That is, the Fokker-Planck approximation will produce the correct asymptotic result when $f(x) = o(N^{-1/3})$ uniformly over the relevant x-interval as $N \rightarrow \infty$. This means that the system must be very close to the threshold for the Fokker-Planck approximation to give an accurate extinction time. Note that we can modify the drift $f(x)$ and diffusion $g(x)^2$ "by hand" to transform the effective potential for the diffusion process approximation into the "proper" effective potential $\Phi(x)$. This may make both (quasi) steady-state statistics *and* the exponential factor in the extinction time correct, but there will still necessarily be an error in the pre-factor for $\tau(x)$. The implication is that it is generally impossible for a Markov diffusion process approximation to capture *all* these features of the discrete kinetics simultaneously.

Fokker-Planck, and diffusion-process, and stochastic differential equation, approximations to discrete state processes are useful for analysis and estimation. Moreover, such approximations to discrete state processes are extremely useful for simulations where in a single time step one can often include the effects of very many individual fluctuation events. The (unsolved) problem that we point out here is that there may be unavoidable systematic errors even in the "natural" $N \to \infty$ asymptotic continuum limit. The open question is whether there is any foolproof way to simplify such multi-scale problems conveniently *and* with uniform quantitative accuracy.

APPENDIX: THRESHOLD AND SUBTHRESHOLD EXTINXTION

Extinction *is* predicted by the deterministic continuum rate equations for this class of Birth-Death processes at and below threshold, albeit in infinite time. Because the discreteness of the population will become apparent when $X(t)$ reaches the elemental level $1/N$, we may consider the deterministic extinction time scale to be the time when this level is achieved. That is, at and below threshold we simply define an effective deterministic extinction time $T_d(x)$ by the relation $1/N = X(T_d(x))$ when $X(0) = x$.

Generically the deterministic limit precisely at threshold is a differential equation of the form $dX/dt \sim -X^2$ for small X, predicting a t^{-1} decay. This implies that the time for the population to reach $n = 1$ satisfies $T_d(x) = \mathcal{O}(N)$ with a weak dependence on x. The *actual* (mean) extinction time, however, is much smaller and $\sim N^{1/2}$. In [1] we derive

$$\tau(x) = \frac{(\pi/2)^{3/2}}{\sqrt{\Phi''(0)\bar{\lambda}'(0)\bar{\mu}'(0)}} \times \sqrt{N} + \frac{\log Nx}{\bar{\mu}'(0)} + \mathcal{O}(1). \tag{19}$$

In the sub-threshold case the deterministic rate equation is of the form $dX/dt \sim -X$ as $X \to 0$, predicting exponential decay and $T_d(x) = \mathcal{O}(\log[Nx])$ as $N \to \infty$ at fixed x. In this situation the deterministic approximation predicts not only the proper scaling, but the correct pre-factor as well. In [1] it is shown for this case that

$$\tau(x) = \frac{1}{\bar{\mu}'(0)[1 - \bar{\rho}(0)]} \log Nx + \mathcal{O}(1). \tag{20}$$

ACKNOWLEDGMENTS

This work supported in part by National Science Foundation award DMS-0244419: *Fronts, Fluctuations and Growth.*

REFERENCES

1. C. R. Doering, K. V. Sargsyan and L. M. Sander, *SIAM J. Multiscale Modeling & Sim.* **3**, 283–299 (2005).
2. S. Karlin and H. M. Taylor, *A Second Course in Stochastic Processes*, New York: Academic Press, 1981.
3. G. H. Weiss and M. Dishon, *Math. Biosci.* **11**, 261-265 (1971).
4. P. Hänggi, H. Grabert, P. Talkner and H. Thomas, *Phys. Rev. A* **29**, 371-378 (1984).
5. I. Nasell, *Math. Biosci.* **156**, 21-40 (1999).
6. N. van Kampen, *Stochastic Processes in Physics and Chemistry*, Amsterdam: North Holland, 1981.

How Good Is Langevin?

N. G. van Kampen

Institute for Theoretical Physics of the University, Utrecht, Netherlands

Abstract. The Langevin equation is universally applied to describe noise in physical systems of all kinds. In order to derive it, rather than postulating it, one has to introduce a noise source such as an environment or "reservoir". The solution requires that at some initial time the reservoir is in equilibrium and uncorrelated with the system. This is unphysical and moreover, in the quantum mechanical case leads to inconsistencies. The way to overcome this anomaly is by considering the equilibrium of system and noise source combined. The resulting equations for the time dependent correlations between various operators can be computed. In order to find out whether the Langevin equation is able to reproduce the result we consider the explicitly solvable system of a harmonic oscillator in a harmonic bath. Conclusion: Langevin cannot describe quantum noise but belongs to the realm of mesoscopic physics.

Keywords: Langevin approach, quantum noise
PACS: 02.50.Ey, 05.10.Gg, 05.40.Ca

INTRODUCTION

In 1908 Langevin published an equation for the treatment of a Brownian particle in a fluid. The velocity of the particle was assumed to obey

$$\frac{dv}{dt} = -\gamma v + \xi(t), \tag{1}$$

where the first term is friction and the second term is due to the collisions of the surrounding fluid molecules, described by a rapidly varying force $\xi(t)$. Later the theory was formalised mathematically by describing the force as a stochastic function, assumed Gaussian, with zero average over small time intervals Δt, and

$$\langle \xi(t)\xi(t')\rangle = 2D\delta(t-t'). \tag{2}$$

This provided a way of introducing both fluctuations and irreversibility in a simple way, not only in physics, but also in biology and even economics. One merely adds a "Langevin force" accompanied by a term for the corresponding dissipation. The question whether the underlying physics justifies these assumptions is usually answered by an appeal to interaction with a surrounding world. A more conscientious treatment introduces the surroundings in the shape of a reservoir or bath, endowed with the required properties, in particular the ones needed for irreversiblity [1, 2]. The question whether this reservoir is not itself a physical system and therefore also subject to the fundamental property of irreversibility is answered by an appeal to the entire universe.

CP800, *Unsolved Problems of Noise and Fluctuations: UPoN 2005*
edited by L. Reggiani, C. Pennetta, V. Akimov, E. Alfinito, and M. Rosini

THEORETICAL CALCULATIONS

The standard way of deriving the equations from basic physics regards the system S as interacting with an outside bath B, both quantum mechanical. Together they constitute a total system T, described by a density matrix $\rho_{\rm T}$ obeying the evolution equation

$$\frac{d}{dt}\rho_{\rm T} = -i[H_{\rm T},\rho_{\rm T}] , \quad (3)$$

$$H_{\rm T} = H_{\rm S}+H_{\rm B}+\lambda H_{\rm I} , \quad (4)$$

where λ measures the interaction. This equation is formally solved for given initial $\rho_{\rm T}$ at time $t=0$, by

$$\rho_{\rm T}(t) = e^{-itH_{\rm T}}\rho_{\rm T}(0)e^{itH_{\rm T}} . \quad (5)$$

The density matrix of S itself is $\rho_{\rm S}(t) = {\rm Tr}_{\rm B}\rho_{\rm T}(t)$. However $\rho_{\rm S}(t)$ does not obey an equation by itself and cannot be found without solving the total evolution equation(5). Accordingly it is not sufficient to choose an initial state $\rho_{\rm S}(0)$, but one needs to know the initial state $\rho_{\rm T}(0)$ of the total system.

In the standard treatment these problems are met by assuming that the bath B is near equilibrium at all times. If it were exactly in equilibrium its effect on the motion of system S would average out. The presence of S modifies the state of B, to order λ. This modification acts back on S, with the result that S is affected to order λ^2. The initial-value problem is dismissed by assuming that initially the bath is in equilibrium and uncorrelated with the system,

$$\rho_{\rm T}(0) = \rho_{\rm S}(0)\times\rho_{\rm B}^{\rm eq}. \quad (6)$$

One assumes for the bath a set of harmonic oscillators, which makes it possible to determine explicitly the operator $e^{-itH_{\rm B}}$. One may then expand $e^{-itH_{\rm T}}$ to second order in λ and take the trace over $\rho_{\rm B}$ so as to obtain $\rho_{\rm S}(\Delta t)$ for small time Δt. Subsequently one interprets $[\rho_{\rm S}(\Delta t)-\rho_{\rm S}(0)]/\Delta t$ as a derivative $\frac{d}{dt}\rho_{\rm S}(t)$. The result is

$$\frac{d}{dt}\rho_{\rm S}(t) = -i[H_{\rm S},\rho_{\rm S}]-\lambda^2\int_0^{\Delta t} dt'{\rm Tr}_{\rm B}\left\{[H_{\rm I},[e^{-it'(H_{\rm S}+H_{\rm B})}H_{\rm I}{\rm e}^{it'(H_{\rm S}+H_{\rm B})}\rho_{\rm B}^{\rm eq}]]\right\}\rho_{\rm S}(t) . \quad (7)$$

Although the upper limit of integration is Δt, one usually argues that the integrand is rapidly varying so that the integral converges in a time even shorter than Δt so that the upper limit may be taken $+\infty$. For instance, in the case of a Brownian particle Δt should cover a number of successive individual collisions, but on the other hand it must still be small enough that the velocity of the heavy particle is barely changed [3, 4].

Unfortunately (7) is unacceptable because it does not obey the criterion for the density matrix to remain positive definite [6, 7]. This flaw is caused by the absence of correlations in (6) [8]. There is no valid reason for this initial condition: nobody believes that it is necessary for understanding Brownian motion to assume that at some time in the past the Brownian particle was uncorrelated to the surrounding fluid. Moreover, if one postulates this condition at $t=0$, after Δt a correlation between S and B builds up and the same formula (7) no longer holds. The conclusion is that $\rho_{\rm S}(t)$ does not really obey a differential equation in time, that is, it does not constitute a semi-group with (7) as infinitesimal generator. Hence there does not exist a Langevin equation (1) with delta-correlated force (2).

The new approach eliminates assumption (6). Brownian motion, rather than starting at one initial moment, is a persistent phenomenon in the total system while it is in equilibrium. We therefore start from the total equilibrium distribution

$$\rho_{\mathrm{T}}^{\mathrm{eq}} = Z^{-1}e^{-\beta H_{\mathrm{T}}} \tag{8}$$

$$Z = \mathrm{Tr}_{\mathrm{T}}e^{-\beta H_{\mathrm{T}}} \tag{9}$$

and compute fluctuations and their correlations occurring in this equilibrium state. When G, F are two operators acting on the system, the correlation between their values at two times t_1, t_2 is

$$\langle G(t_1)F(t_2)\rangle^{\mathrm{eq}} = Z^{-1}\mathrm{Tr}_{\mathrm{T}}e^{-\beta H_{\mathrm{T}}}e^{it_1H_{\mathrm{T}}}Ge^{-i(t_1-t_2)H_{\mathrm{T}}}Fe^{-it_2H_{\mathrm{T}}}. \tag{10}$$

This can be computed to second order in λ with the result given in [9, 10].

Comparison with the standard result shows a number of correction terms. They disappear for low β, high kT, that is, $|\lambda H_I| \ll kT$. The question about the initial condition and the lack of positivity has been eliminated.

A soluble model obtains when S is a harmonic oscillator coupled bilinearly to the bath [11, 12, 13].

$$H_{\mathrm{T}} = \frac{1}{2}(P^2+\Omega^2Q^2)+\frac{1}{2}\sum_k(P_k^2+k^2Q_k^2)+Q\sum_k\lambda_kQ_k\,. \tag{11}$$

The quadratic form in the Q can be diagonalized with an orthogonal transformation X, with the result [14]

$$H_{\mathrm{T}} = \frac{1}{2}\sum_\nu(p_\nu^2+\omega_\nu^2q_\nu^2)\,. \tag{12}$$

The eigenvalues ω_ν are the zeros of the analytic function

$$G(\omega) = \omega^2-\Omega^2-\sum_k\frac{\lambda_k^2}{\omega^2-k^2}\,. \tag{13}$$

In this way one gets the *exact* formula for the autocorrelation function of Q

$$\langle Q(0)Q(\tau)\rangle = \sum_\nu X_{0\nu}^2\left(\frac{1}{2\omega_\nu}\coth\frac{\beta\omega_\nu}{2}\cos\omega_\nu\tau+\frac{i}{2}\frac{\sin\omega_\nu\tau}{\omega_\nu}\right), \tag{14}$$

where $\tau = t_2-t_1$. The sum may be written as a path integral

$$\langle Q(0)Q(\tau)\rangle = \frac{1}{2\pi i}\oint\frac{d\omega}{G(\omega)}\left\{\coth(\frac{1}{2}\beta\omega)\cos\omega\tau+i\sin\omega\tau\right\} \tag{15}$$

the path extending over a loop around the positive real axis in the complex ω-plane. Suppposing that the bath frequencies k lie dense with density $g(k)$, this may be evaluated to give

$$\langle Q(0)Q(\tau)\rangle = \frac{1}{2\Omega_1} e^{-\Gamma\tau} \left\{ \coth(\frac{1}{2}\beta\Omega_1)\cos\Omega_1\tau + i\sin\Omega_1\tau \right\} , \tag{16}$$

where $\Gamma = \pi g(\Omega)/4\Omega^2$ and $\Omega_1^2 = \Omega^2 - \Gamma^2$. The coth in (15) has poles $\pm i/\beta$, which give contributions $e^{-kT/\hbar\Omega}$; they must be omitted if Langevin is to be obtained; this is the quantum condition $kT \gg \hbar\Omega$, which expresses the fact that white noise does not exist in quantum mechanics.

Classical Langevin gives for our harmonic oscillator with damping coefficient γ,

$$\langle Q(t)Q(t+\tau)\rangle = \frac{D}{\gamma\Omega^2} e^{-\gamma\tau/2} \{\cos\Omega_1\tau + \frac{\gamma}{2\Omega_1}\sin\Omega_1\tau\} . \tag{17}$$

Comparison with (16) shows $\gamma = 2\Gamma$ and Ω_1 is the same as above. Agreement with of (16) with Langevin is obtained if one sets $\coth\frac{1}{2}\beta\Omega_1 \approx 2/\beta\Omega_1$ and $D/\gamma\Omega^2 = 1/\beta\Omega_1^2$. This is close enough to the classical fluctuation-dissipation theorem $D = \gamma/\beta = \gamma kT$.

CONCLUSIONS

Numerous efforts have been made to modify the Langevin equation so as to take quantum mechanics into account, with various results [15, 16, 17, 18]. The second term in (16), however, is imaginary and hence of purely quantum mechanical origin. This is often overcome by the *ad hoc* excuse that for comparing with the classical case one should look at the symmetrized expression:

$$\frac{1}{2}\{\langle Q(0)Q(\tau)\rangle + \langle Q(\tau)Q(0)\rangle\} = \frac{1}{\beta\Omega_1^2} e^{-\Gamma\tau}\cos\Omega_1\tau , \tag{18}$$

This does coincide with the first term of (16). The second term in (17), however, is not reproduced by our exact result (16) for the equilibrium correlation.

My conclusion is that the Langevin approach is essentially classical since it deals with probabilities. More precisely, it is *mesoscopic,* in the original sense of that word: it describes quantum physics on the coarse-grained level, on which interference of complex probability amplitudes is negligible and only probabilities are relevant. Therefore it is unable to reproduce the quantum mechanical expressions for the correlations in equilibrium.

REFERENCES

1. I.R. Senitzky, Phys Rev. **119,** 670 (1960)
2. J. Luczka, Physica A **167,** 919 (1990)
3. N.G. van Kampen and I. Oppenheim, Physica A **138,** 231 (1986)
4. N.G. van Kampen and I. Oppenheim, J. Stat. Phys. **87,** 1325 (1997)

5. I. Oppenheim and V. Romero-Rochin, Physica A **147,** 184 (1987)
6. A. Kossakowski, Bull. Acad. Pol. Sci., Série Math. Astr. Phys. **20,** 1021 (1971); Rep. Math. Phys. **3,** 247 (1972)
7. G. Lindblad, Comm. Math. Phys. **40,** 147 (1975)
8. A. Suàrez, R. Silbey, and I. Oppenheim, J. Chem. Phys. **97,** 5101 (1992)
9. N.G. van Kampen, Fluct. Noise Lett. **1,** C7-C13 (2001).
10. Zero temperature has been considered by A.N. Jordan and M. Büttiker, Phys. Rev. Lett. **92,** 247901 (2005)
11. P. Ullersma, Physica **32,** 27,56,74,90 (1966)
12. H. Grabert, U. Weiss, and P. Talkner, Z. Phys. B **55,** 87 (1984)
13. F. Haake and R. Reibold, Phys. Rev A **32,** 2462 (1985)
14. N.G. van Kampen, J. Stat. Phys. **115,** 1057 (2004)
15. R. Benguria and M Kač, Phys. Rev. Lett. **46,** (1981)
16. C. W. Gardiner, IBM J. Res. Develop. **32,** 127 (1988)
17. H. Kleinert and S. V. Shabanov, Phys. Rev. Lett. A **200,** 224 (1995)
18. P. Shikritov et al., Semicond. Sci. Technol. **19,** S232 (2004).

Applications Of Stochastic Differential Equations In Electronics

Andrew Allison[1]and Derek Abbot

Centre for Biomedical Engineering (CBME) and School of Electrical & Electronic Engineering, The University of Adelaide, SA 5005, Australia.

Abstract. We argue that the stochastic differential calculus of Itô is the best tool for the analysis of noise in electronic circuits. We begin by showing how the nodal and mesh equations of ordinary circuit analysis can be extended to model the effects of thermal fluctuations. This leads to a systematic method for formulating Langevin equations for electronic circuits. These can then be transformed into ordinary differential equations, allowing the calculation of noise power without the need to explicitly solve the stochastic differential equations.

We have encountered a difficult unsolved problem of noise, which arises when our technique is applied to the standard noise model for a MOSFET. The resulting equations are singular and do not yield to the usual algebraic manipulations. We do not appear to have enough independent equations for the number of (apparently) independent variables. Why is it so?

Keywords: Circuit analysis, Electronics, Noise, Langevin equations, Stochastic Differential Equations

PACS: 72.70.+m, 07.50.Hp

INTRODUCTION

The traditional approach to thermal noise in circuits can be traced back to Johnson [1]. Over the years, a number of empirical techniques have been developed to estimate noise in circuits that filter the thermal noise. Our aim is to extend these results in a more systematic manner.

White noise cannot be constructed as an ordinary function of time, say $Z(t)$. It should be regarded a generalised function, or "distribution" of the type described by Schwartz [2, 3]. Since noise is a generalised function, it must appear inside an integration. Itô's differential equation

$$dX_t = \sigma(X_t)dB_t + \mu(X_t)dt \tag{1}$$

is really short hand for an integral equation

$$X_t - X_{t_0} = \int_{t_0}^{t} \sigma(X_\tau)dB_\tau + \int_{t_0}^{t} \mu(X_\tau)d\tau. \tag{2}$$

We can make a finite difference approximation for a very short time interval $\Delta t = t - t_0$:

$$\Delta X_t \approx \sigma(X_t)\Delta B_t + \mu(X_t)\Delta t. \tag{3}$$

---

[1] Further author information: (Send correspondence to A. Allison, School of Electrical and Electronic Engineering, The University of Adelaide, SA 5005, Australia.) E-mail: aallison@eleceng.adelaide.edu.au, Telephone:+61 8 8303 5283

CP800, *Unsolved Problems of Noise and Fluctuations: UPoN 2005*
edited by L. Reggiani, C. Pennetta, V. Akimov, E. Alfinito, and M. Rosini

We can consider the behavior of a circuit over a short but finite time interval, Δt. This interval needs to be large enough to avoid problems with physical representation, such as infinite bandwidth, and yet short enough to represent the changing behaviour of the system under investigation. We shall work with infinitesimal increments, of the form of Equation 1, knowing that these are really short hand for integrals over finite time intervals of the form of Equation 2 and we can approximate the process in computer simulations using Equation 3.

We can re-write Kirchhoff's laws for infinitesimal intervals of time to provide a single simple systematic basis for combining noise signals and deterministic signals.

KIRCHHOFF'S CURRENT LAW

The infinitesimal version of Kirchhoff's Current Law (KCL) is written in terms of infinitesimal increments of electric flux (electric charge) flowing out of a node:

$$\sum_{\forall k} dQ_k = 0. \tag{4}$$

If all the contributions, dQ_k are deterministic, then we can integrate and get $\sum_{\forall k} I_k = 0$, which is the more usual form of the law. The proof of the infinitesimal form follows from one of Maxwell's equations [4], called Gauss's electric law: $\oint_A \underline{D} \cdot d\underline{A} = \int_v \rho \cdot dv = Q$. This can be reduced to: $\sum_{k=1}^{N} dQ_k = dQ$. The correct modelling of dQ depends the capacitance, C, of the surface, A, with respect to earth. The stored energy of the total charge in the node is: $U = \frac{1}{2C}Q^2$ If we assume that the system is at thermodynamic equilibrium then it is possible to show that $E\left[\frac{1}{2C}Q^2\right] = \frac{1}{2}kT$, which is simply the mean energy for each degree of freedom, in a classical thermodynamic system at temperature T. This implies that $E\left[Q^2\right] = kT \cdot C$. We regard the charge enclosed in a node as being a normally distributed random process, $Q \sim N(0, kT \cdot C)$ and we can write $\sum_{k=1}^{N} dQ_k = \sqrt{kTC} \cdot dB$. This differs from the homogeneous form of Kirchhoff's current law, in Equation 4. Kirchhoff's laws are ultimately statistical and there will be small fluctuations that can usually be ignored, as long as the circuit is well constructed and the capacitance of the node to earth, C is small. If this is not the case, then we would need to include the parasitic capacitance, C, in our model of the circuit. In the remainder of this paper, we will use the homogeneous, differential form of Kirchhoff's current law, Equation 4, where infinitesimal increments of charge are represented in the form, $dQ = I dt$ for deterministic currents, and also in the form, $dQ = i_n dB$ for noise currents.

KIRCHHOFF'S VOLTAGE LAW

The infinitesimal version of Kirchhoff's Voltage Law (KVL) is written in terms of infinitesimal increments of magnetic flux through a mesh:

$$\sum_{\forall k} d\Phi_k = 0. \tag{5}$$

If all the contributions, $d\Phi_k$, are deterministic then we can integrate and get $\sum_{\forall k} V_k = 0$, which is the more usual form of the law. This is equivalent to saying that electric fields are the product of a static potential field, $\underline{E} = -\nabla V$. The proof of the infinitesimal form of Kirchhoff's voltage law follows from one of Maxwell's equations [4], called Faraday's law: $\oint_\lambda \underline{E} \cdot d\underline{\lambda} = -\frac{\partial}{\partial t} \int_S \underline{B} \cdot d\underline{A} = -\frac{\partial \Phi}{\partial t}$. This can be reduced to: $\sum_{k=1}^N V_k dt = \sum_{k=1}^N d\Phi_k = -d\Phi$. The correct modelling of $d\Phi$ depends on the physical geometry of the current path. Specifically, we need to know the inductance, L, of the current path. The stored energy of the total charge in the node is: $U = \frac{1}{2L}\Phi^2$. If we assume that the system is at thermodynamic equilibrium with the surroundings, then we can show that $E\left[\Phi^2\right] = kT{\cdot}L$. We regard the magnetic flux linked by the contour, λ, as being a normally distributed random process, $\Phi{\sim}N(0, kT{\cdot}L)$ and we can write $\sum_{k=1}^N d\Phi_k = -\sqrt{kTL}{\cdot}dB$. This differs from the homogeneous from of Kirchhoff's voltage law, in Equation 5. Kirchhoff's laws are ultimately statistical and there will be small fluctuations, which can usually be ignored. If this is not the case, then we would need to include the parasitic inductance, L, in our model of the circuit. In the remainder of this paper, we will use the homogeneous, differential form of Kirchhoff's voltage law, Equation 5, where infinitesimal increments of magnetic flux are represented in the form $d\Phi = V dt$ for deterministic currents, and also in the form, $d\Phi = v_n dB$ for noise currents.

MODELS FOR RESISTORS

There are two possible linear models for thermal noise in a resistor, a Thévenin and a Norton model, which are shown in Figure 1. The Thévenin model, Figure 1(a), places

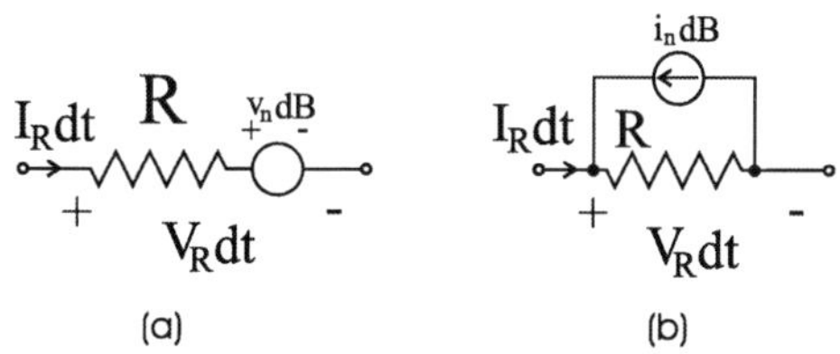

FIGURE 1. The Thévenin model, (a), and Norton model, (b), noise models for a resistor.

a noise voltage source in series with an ideal noiseless resistor. During an infinitesimal time interval, dt, the source contributes a magnetic flux of $v_n dB$ to all circuit meshes in which it sits. We can use the results of Johnson and Nyquist, $\Delta\langle V^2 \rangle = 4kTR{\cdot}\Delta f$, to estimate intensity of the noise voltage, $v_n = \sqrt{2kTR}$. The units of noise voltage are $\mathrm{V}/\sqrt{\mathrm{Hz}}$.

The use of the constant "2" rather than "4" is due to the convention that we use for frequency. It is consistent with widely used definitions of power spectral density and autocorrelation functions [5, 6, 7, 8, 9] and is sometimes called the two-sided power spectrum [9]. An interesting summary of the technicalities and history of spectral noise density can be found in [10].

The Norton model, Figure 1(b), places a noise current source in parallel with an ideal noiseless resistor. During an infinitesimal time interval, dt, the source contributes an electric flux, or charge, of $dQ_1 = i_n dB$ to the node into which the terminal flows. A corresponding opposite charge of $dQ_2 = -i_n dB$ is added at the other terminal. The corresponding intensity of noise current is $i_n = \sqrt{2kT/R}$. The units of noise current are $\mathrm{A}/\sqrt{\mathrm{Hz}}$.

We can write the increments of electric flux associated with a resistor as $dQ = i_n dB$. These increments of electric flux can be used directly in the differential form of KCL, Equation 4. We can write the increments of magnetic flux associated with a resistor as $d\Phi = v_n dB$. These increments of magnetic flux can be used directly in the differential form of KVL, Equation 5. We can also imagine other noise models for other devices. These will generally have different expressions for current and voltage noise intensities, v_n and i_n, but the calculations will generally be of the same form as for the resistor.

A functional mapping between electrical and mechanical systems

There is a widely used functional mapping between quantities in electrical and mechanical systems. These can be expressed in terms of the Hamiltonian function, $\mathscr{H}$, using Hamilton's canonical equations [11].

Variable	Electrical Quantity	Mechanical Quantity
Effort	Voltage, $-V = +\frac{\partial\Phi}{\partial t} = -\frac{\partial\mathscr{H}}{\partial Q}$	Force, $+F = +\frac{dp}{dt} = -\frac{\partial\mathscr{H}}{\partial x}$
Flow	Current, $I = \frac{+\partial Q}{\partial t} = +\frac{\partial\mathscr{H}}{\partial\Phi}$	Velocity, $v = +\frac{dx}{dt} = +\frac{\partial\mathscr{H}}{\partial p}$
Momentum	Magnetic flux,Φ	Momentum, p
Displacement	Electric flux, Q	Displacement, x

Further details of the mapping are described by Karnopp et al. [12]. There are differences of sign in the definitions of the effort variables. These are purely due to historical convention. The mapping applies to all equations of motion, including the Langevin equation.

Variable	Electrical Quantity	Mechanical Quantity
Viscous damping force	$-V = -R\cdot I$	$+F = -\alpha\cdot v$
Noise intensity	$E[{v_n}^2] = 2kT\cdot R$	$E[{f_n}^2] = 2kT\cdot\alpha$
Langevin Equation	$LdI + RIdt = v_n dB_t$	$mdv + \alpha vdt = f_n dB_t$

This demonstrates that the resulting equations for electrical circuits are Langevin equations in the literal sense.Established results for mechanical systems can be applied directly to electrical systems, by mapping solutions to the mechanical problems into the electrical domain. These considerations also suggest that we should use the generalised coordinates, Φ and Q to describe the dynamics of electrical systems, as an analogy of the use of p and x for mechanical systems.

CAPACITORS AND INDUCTORS

For a fixed capacitor, C, with terminal voltage, V, and terminal current, I, we require the following equivalent forms for increments of electric flux $dQ = CdV = Idt$. This allows the capacitor model to be used in Norton and Thévenin models. These increments of electric flux can be used directly in the differential form of KCL, Equation 4.

For a fixed inductor, L, with terminal voltage, V, and terminal current, I, we require the following equivalent forms for increments of magnetic flux $d\Phi = LdI = Vdt$. This allows the inductor model to be used in Norton and Thévenin models. These increments of magnetic flux can be used directly in the differential form of KVL, Equation 5.

A ONE-DIMENSIONAL EXAMPLE OF STOCHASTIC CIRCUIT ANALYSIS

We consider the well known case of a resistor in parallel with a capacitor [13]. This is shown in Figure 2. We expect to find a mean-square voltage across the capacitor of,

$$\langle V^2 \rangle = \frac{kT}{C}. \tag{6}$$

This is clear if we regard the voltage across the capacitor as a single degree of freedom and apply the principle of equipartition of energy, $\langle \frac{1}{2}CV^2 \rangle = \frac{1}{2}kT$. We represent the resistor using a Norton model of an noiseless resistor, R, in parallel with a noise source, denoted by $i_0 dB$ where $i_0 = \sqrt{2kT/R}$. We consider the nodal equation for the circuit. We require $\sum dQ = 0$. For the capacitor, we have $dQ = CdV$, where C is the capacitance and dV is an infinitesimal increment of voltage. For the resistor, we have $dQ = \frac{V}{R}dt$ and for the equivalent noise source in the resistor, we have $dQ = i_0 dB$, as discussed above. The nodal equation then becomes

$$CdV + \frac{1}{R}Vdt = i_0 dB, \tag{7}$$

which is the SDE for this system. It has the same form and plays exactly the same role as the Langevin equation in statistical physics [5, 14, 15, 16].

It is possible to evaluate the infinitesimal moments implied by a Langevin equation and to formulate a Partial Differential Equation (PDE), describing the probability density of an ensemble of solutions to the SDE in Equation 7. This PDE is known as the Fokker-Planck equation [5, 17, 16]. If we could solve the Fokker-Planck equation then we could apply operators to the solutions to calculate quantities of interest, such as the noise power as a function of time, $w(t)$. Unfortunately, this approach is cumbersome. Instead, we suggest the application of a method described by Gubner [18], where the SDE is transformed into a much simpler ODE in the noise power.

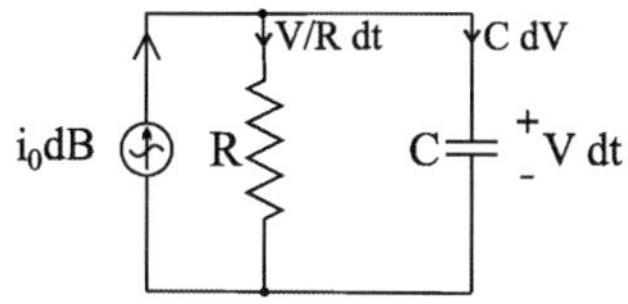

FIGURE 2. *RC* parallel circuit, with resistor, R, noise current source, $i_o dB$, and capacitor, C.

Equation 7 can be re-written as an SDE in the narrow sense: $dV = \frac{-1}{RC} V\,dt + \frac{i_0}{C} dB$. This is of the form $dV = \alpha(t)V\,dt + \beta(t)dB$ where $\alpha(t) = \alpha = -1/(RC)$ and $\beta(t) = \beta = +i_0/C = \sqrt{2kT/(RC^2)}$. If we define $\mu = E[V]$ and the variance as $w = E\left[(V-\mu)^2\right]$ then it can be shown [18] that $\frac{dw(t)}{dt} = 2\alpha(t)w(t) + \beta(t)^2$. We have now derived an Ordinary Differential Equation (ODE) in the variable of interest, noise power. If we consider the steady-state situation after all transients have decayed then we have $\frac{dw(t)}{dt} = 0$, which implies that $w = E\left[(V-\mu)^2\right] = -\frac{\beta(t)^2}{2\alpha(t)} = \frac{kT}{C}$, which is the classical result in Equation 6. We did not solve the SDE directly. We only used it to derive an ODE. We did not solve the ODE but only used it to derive an algebraic equation, which we then used to solve for the steady state value of the noise power. We believe that this simple and systematic method is general and should find wide application in the analysis of noise in circuits.

NOISE MODELS FOR THE JFET

The noise model that we use here, shown in Figure 3 is the one used by Abbott *et al.* [19] and is essentially a van der Ziel model [20, 9] with all noise sources referred to the input. For a JFET, the gate currents are limited by reverse biased PN junctions. We regard the gate leakage current as negligible and have not included it in the model.

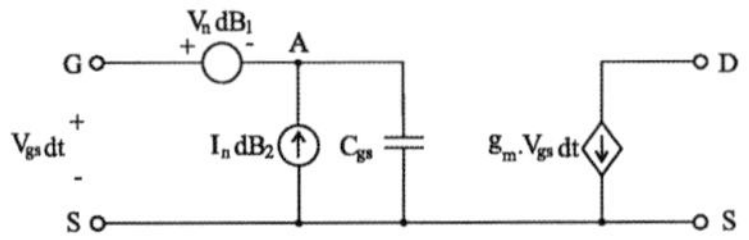

FIGURE 3. Noise model for a JFET. This is basically a van der Ziel model with all noise sources referred to the input. The gate to source capacitance is represented by C_{gs}. This model includes the standard noise-free small signal model for a JFET. The amplifying effect of the JFET is represented by the dependent current source, $g_m{\cdot}V_{gs}$. The noise is represented by two sources at the input, $V_n{\cdot}dB_1$ and $I_n{\cdot}dB_2$ where dB_1 and dB_2 are infinitesimal increments of Brownian motion.

ANALYSIS OF A SIMPLE JFET CIRCUIT

We use a very simple version of the Colpitts oscillator with a FET as the amplifying element. This is shown in Figure 4. In the Colpitts topology, the chain of capacitors, C_1 and C_2, allows for a feedback path with high impedance and voltage amplification. If we analyse the circuit in Figure 4, using small-signal technique, and insert the noise model from Figure 3, then we obtain the complete small signal noise model for the Colpitts oscillator, shown in Figure 5. If the circuit did not have noise sources then ordinary mesh and nodal analysis could be performed and we obtain a standard state-variable model [21]. With the presence of noise sources, we can still write down mesh equations. Kirchhoff's voltage law takes the form: $\sum d\Phi = \sum V\,dt = 0$ where the contribution from a noise voltage source would be $d\Phi = V_n dB$. We obtain a mesh equation for the mesh

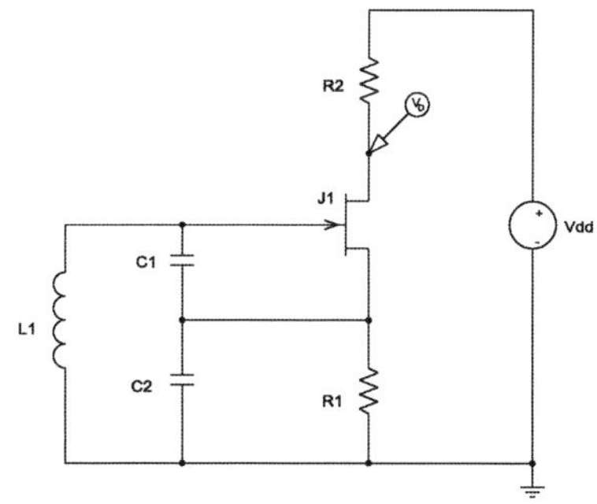

FIGURE 4. Standard, large-signal, schematic circuit diagram for a Colpitts oscillator using a JFET as the amplifying element. There are noise sources in R_1, R_2 and J_1.

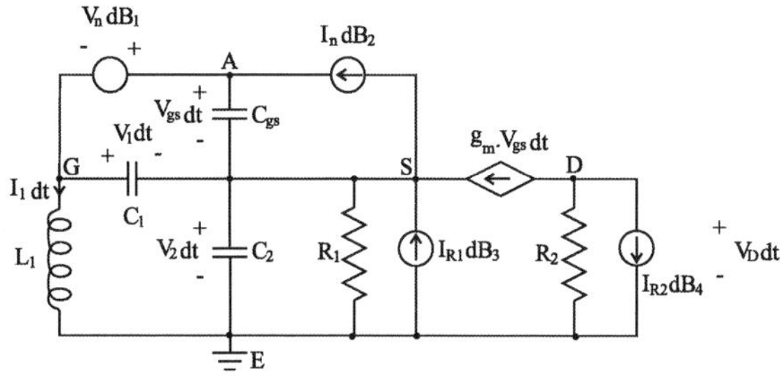

FIGURE 5. Small signal equivalent circuit of the Colpitts oscillator, including the van der Ziel noise model. The FET is mapped into the equivalent circuit with Source, S, Gate, G, Drain D. The earth is represented by E. The circuit has four energy storage elements, C_{gs}, C_1, C_2 and L_1. There are four corresponding state variables, V_{gs}, V_1, V_2 and I_1. Further analysis shows that only three of these are independent. There are four noise sources, $V_n{\cdot}dB_1$, $I_n{\cdot}dB_2$, $I_{R1}{\cdot}dB_3$ and $I_{R2}{\cdot}dB_4$. The nominal output is the drain voltage, V_D.

including the Gate G and the node A,

$$V_1\,dt - V_{gs}\,dt + V_n\,dB_1 = 0, \tag{8}$$

which indicates that the state-variables, V_1 and V_{gs} are not independent. We can obtain a mesh equation including L_1, C_1 and C_2,

$$dI_1 = \frac{1}{L_1}{\cdot}V_1\,dt + \frac{1}{L_1}{\cdot}V_2\,dt. \tag{9}$$

The current through the noise source, $V_n\,dB_1$ is undetermined but we can regard G and A as an enlarged node and write

$$I_1\,dt + C_1\,dV_1 + C_{gs}\,dV_{gs} - I_n\,dB_2 = 0. \tag{10}$$

Finally, the source is simply a very large node

$$C_1\,dV_1 - C_2\,dV_2 + C_{gs}\,dV_{gs} = \frac{V_2}{R_1}\,dt - g_m\,V_{gs}\,dt + I_n\,dB_2 - I_{R1}\,dB_3. \tag{11}$$

Equations 8, 9, 10 and 11 are the equations of state for this system.

The nominal output of the circuit is the drain voltage , V_D. This can be expressed with an output equation,

$$V_D\,dt = -R_2\,g_m\,V_{gs}\,dt - I_{R2}\,R_2\,dB_4. \tag{12}$$

This completes the formulation of the circuit. We could solve the resulting SDEs exactly if we could write these equations in matrix form:

$$d\underline{X_t} = \mathbf{A}\cdot\underline{X_t}dt + \mathbf{K}d\underline{B_t} \tag{13}$$

where $\mathbf{A}$ is a square state-transition matrix and $\exp(+\mathbf{A}t)$ is defined, $\underline{X_t}$ is a state vector containing the state-variables, $\{\,V_{gs},\,V_1,\,V_2,\,I_1\,\}$ and $\mathbf{K}$ is a matrix that combines the independent noise sources,$\{dB_1,dB_2,dB_3,dB_4\}$. In this case the matrix, $\mathbf{K}$ is square but this does not have to be the case. There will be an explicit solution of the form:

$$\underline{X_t} = \exp(+\mathbf{A}t)\cdot\left[\underline{X_0} + \exp(-\mathbf{A}t)\cdot\mathbf{K}\cdot\underline{B_t} + \int_0^t \exp(-\mathbf{A}s)\cdot\mathbf{A}\cdot\mathbf{K}\underline{B_s}ds\right]. \tag{14}$$

The derivation of this solution relies on the multi-dimensional form of Itô's lemma and may be found in Øksendal [22]. Of course, these equations can also be solved using numerical methods described in the literature, see for example [3, 23].

SUMMARY AND OPEN QUESTIONS

So far, we have constructed models for L,R and C. We have devised practical and consistent forms for Kirchhoff's laws. We have shown that the noise equations for electrical circuits are completely equivalent to the Langevin equations of statistical mechanics. We have shown that explicit solution of the SDEs is not always necessary, since we can often formulate much simpler ODEs in the variables of interest. Finally, we have made some progress on the formulation of a practical and consistent noise models for a JFET.

A cursory examination of the JFET in Figure 5 circuit would suggest that the state variables should be: $\{V_1,V_2,V_{gs},I_1\}$. There "ought" to be 4 state variables. There seem to be 4 independent equations, 8, 9, 10 and 11, so we might expect that these equations could be written directly in the form of Equation 13. This turns out not to be the case, because Equation 8 is degenerate. It does not contain any explicit reference to dV_1 or dV_{gs} and we might expect that the noise components of these variables are not independent. We do not appear to have enough independent equations for the number of (apparently) independent variables.

Our unsolved problem of noise is that we cannot yet implement the noise model for the JFET until we can put Equations 8, 9, 10 and 11 into the form of Equation 13.

It seems that the solution must ultimately depend on a reduction of the number of state variables required to describe the circuit. We must somehow reduce V_1 and V_{gs} to a single variable. We would then have three truly independent variables in three truly independent equations, in the whole circuit, and the equations could be solved.

At first sight, this problem looks like it might be a simple oversight. Perhaps we "forgot to carry the one?" If the noise source, $v_n{\cdot}dB_1$ were a deterministic voltage source, say $V_3 \cdot dt$ then there would be no problem. Equation 8 would reduce to an ordinary mesh equation of the usual type, $V_1 - V_{gs} + V_3 = 0$, and the elimination of one of the state-variables, V_1 or V_{gs} would be trivial. Unfortunately, the circuit *does* contain a noise source, $v_n \cdot dB_1$, and the mesh equation does not reduce in the usual way. We cannot simply "cancel by dt," as it were.

We suspect that the required reduction will depend on the unique decomposition of semi-martingales as expounded by Doob [24]. The state-variables V_1 and V_{gs} cannot be independent because they do not have independent decompositions. There can only be "one noise current" in that mesh, or can there? We would welcome correspondence on this problem.

REFERENCES

1. J. B. Johnson, *Phys. Rev.*, **32** (1928).
2. B. B. Mandelbrot, *The Fractal Geometry of Nature*, W. H. Freeman, New York, 1977.
3. P. E. Kloeden, and E. Platen, *Numerical Solution of Stochastic Differential Equations*, Springer, Berlin, 1999.
4. W. H. Hayt, *Engineering Electromagnetics*, McGraw-Hill Book Co., New York, 1988.
5. F. Reif, *Fundamentals of Statistical and Thermal Physics*, McGraw Hill, Singapore, 1965, international ed. 1985 edn.
6. R. N. Bracewell, *The Fourier Transform and its Applications*, McGraw-Hill, 1965, 3rd ed. 2000, third edn.
7. R. P. Yates, and D. J. Goodman, *Probability and Stochastic Processes*, John Wiley & Sons inc., New York, 1999.
8. P. Z. Peebles, *Probability, Random Variables, and Random Signal Principles*, McGraw Hill, Singapore, 2001.
9. B. Razavi, *Design of Analog CMOS Integrated Circuits*, McGraw Hill, Boston, 2001.
10. D. Abbott, B. R. Davis, N. J. Phillips, and K. Eshraghian, *IEEE Transaction on Education*, **39**, 1–13 (1999).
11. C. Lanczos, *The Variational Principles of Mechanics*, Dover Publications Inc., 1949, dover 1986 edn.
12. D. C. Karnopp, D. C. Margolis, and R. C. Rosenberg, *System Dynamics, Modelling and Simulation of Mechatronic Systems*, John Wiley and Sons inc., New York, 2000.
13. R. P. Feynman, R. B. Leighton, and M. Sands, *The Feynman Lectures on Physics*, vol. 1, Addison-Wesley, Reading, Massachusetts, 1963.
14. C. E. Hecht, *Statistical Thermodynamics and Kinetic Theory*, Dover, New York, 1990, dover 1990 edn.
15. G. H. Wannier, *Statistical Physics*, Dover Publications Inc., New York, 1966, dover 1987 edn.
16. H. Risken, *The Fokker-Planck Equation*, Springer, Berlin, 1996.
17. H. C. Tuckwell, *Elementary Applications of Probability Theory*, Chapman & Hall, London, 1988.
18. J. A. Gubner, *The Control Handbook*, CRC Press and IEEE Press, 1996, chap. 60, Stochastic Differential Equations, pp. 1067–1078.
19. D. Abbott, B. R. Davis, and K. Eshraghian, "Output circuit noise analysis of the XY array imager," in *Noise in Physical Systems and $1/f$ Fluctuations*, edited by C. Claeys, and E. Simoen, World Scientific, Leuver, Belgium, 1997, pp. 482–487.
20. B. G. Streetman, *Solid State Electronic Devices*, Prentice Hall, Englewood Cliffs, New Jersey, 1995, fourth edn.
21. P. M. DeRusso, R. J. Roy, and C. M. Close, *State Variables for Engineers*, John Wiley & Sons inc., New York, 1965.
22. B. Øksendal, *Stochastic Differential Equations*, Springer, Berlin, 1998, sixth ed. edn.
23. C. Penski, *Journal of Computational and Applied Mathematics*, **115**, 461–70 (2000).
24. P. Protter, *Stochastic Integration and Differential Equations*, Springer Verlag, Berlin, 1990.

Full Counting Statistics And Shot Noise In Diffusive Mesoscopic Superconductors

Wolfgang Belzig

Department of Physics und Astronomy, University of Basel, Klingelbergstrasse 82, CH-4056 Basel, Switzerland

Abstract. We study shot noise in mesoscopic diffusive wires between a normal and a superconducting terminal. We particularly focus on the regime, in which the proximity-induced reentrance effect is important. We will examine the difference between a simple Boltzmann-Langevin description, which neglects induced correlations beyond the conductivity correction, and a full quantum calculation. In the latter approach, it turns out that two Andreev pairs propagating coherently into the normal metal are anti-correlated for $E \lesssim E_c$, where $E_c = \hbar D/L^2$ is the Thouless energy.

Keywords: Andreev reflection, shot noise, quantum transport
PACS: 74.50.+r

INTRODUCTION

Fluctuations of the current in mesoscopic conductors originate from the quantum scattering of the charge carriers, and are sensitive to their interference, statistics and interaction. This makes the theoretical and experimental study of noise in small electronic circuits interesting and challenging (for recent reviews, see Refs. [1, 2]). Statistical correlations in the transport of fermions have led to a number of interesting predictions. For example, the noise of a single-channel quantum contact of transparency T at zero temperature has the form $S_I = 2|eI|(1-T)$ [3, 4]. The noise is thus suppressed below the Schottky value $2|eI|$ of uncorrelated charge transfer. The suppression is a direct consequence of the Pauli principle, which prevents two electrons from tunneling together. It is convenient to measure the deviation from the Schottky result by the so-called Fano factor $F = S_I/2|eI|$. For a number of generic conductors, it turns out that the suppression of the Fano factor is universal, i. e. it does not depend on details of the conductor like geometry or impurity concentration. In particular, a diffusive metal with elastic scattering leads to $F^N_{\text{diff}} = 1/3$ [5], which is independent on the concrete shape of the conductor [6] and has been confirmed experimentally [7].

If superconductivity comes into play the fundamental charge transport mechanism at energies below the superconducting gap is Andreev reflection. Two electrons enter the superconductor simultaneously and form a Cooper pair, which can propagate in the superconductor. Thus, in this process a charge transfer of $2e$ occurs, but with a reduced probability, since two particles have to tunnel. The shot noise is proportional to the charge of the elementary processes, and one thus naivly expects a doubling of the shot noise, which was indeed found theoretically [8] and experimentally [9] for diffusive conductors. It is remarkable that this doubling occurs for diffusive conductors, whereas it is not found for other conductors like, e. g., single-channel contacts [3, 8, 10, 11],

CP800, *Unsolved Problems of Noise and Fluctuations: UPoN 2005*
edited by L. Reggiani, C. Pennetta, V. Akimov, E. Alfinito, and M. Rosini

double tunnel junctions [12], or diffusive junctions with a tunneling barrier [13]. A doubling of the full Schottky noise is recovered for low transmission contacts and was recently observed experimentally [14].

CURRENT STATISTICS

Consider the following Gedanken experiment. A constant voltage bias is applied to a mesoscopic conductor for a certain time interval t_0. During this time interval we count the number of charges N passing the conductor. The outcome of the experiment is described by a probability $P_{t_0}(N)$ that N charges have passed the conductor. This is the so-called *full counting statistics* (FCS), which is the most natural description of quantum transport. Alternatively we may study the *characteristic function* $\Phi(\chi) = \sum_N P_{t_0}(N) e^{iN\chi}$. The kth coefficient in the expansion of the characteristic function in powers of χ yields the moments of the counted charge $\langle N^k \rangle$. An equivalent description is to obtain the *cumulant generating function* (CGF) $S(\chi) = \ln \Phi(\chi)$, which defines the cumulants. The cumulants are directly connected to the (measureable) zero-frequency current correlations functions. One route to counting statistics is the Keldysh-Green's function approach in combination with the circuit theory of mesoscopic transport developed by Nazarov [15]. For details, we refer to several articles in [2]. In this approach, terminals are described by 4×4 Green's function matrices. For a normal terminal (N) with occupation f we introduce the 2×2 matrix $\hat{f}$ =diag$(f(E), f(-E))$ and a superconducting terminal (S) at equilibrium and for $E \ll \Delta$ we have

$$\check{G}_N(\chi) = \begin{pmatrix} \hat{\tau}_3(1-2\hat{f}) & -2\hat{\tau}_3 e^{i\chi\hat{\tau}_3}\hat{f} \\ -2\hat{\tau}_3 e^{-i\chi\hat{\tau}_3}(1-\hat{f}) & \hat{\tau}_3(2\hat{f}-1) \end{pmatrix}, \check{G}_S = \begin{pmatrix} \hat{\tau}_1 & 0 \\ 0 & \hat{\tau}_1 \end{pmatrix}. \tag{1}$$

For a general contact described by a scattering matrix with corresponding transmission eigenvalues $\{T_n\}$ the counting statistics is obtained as [16]

$$S(\chi) = \frac{t}{h} \int dE \sum_n \mathrm{Tr} \ln \left[1 + \frac{T_n}{4} \left(\{\check{G}_L, \check{G}_R\} - 2 \right) \right]. \tag{2}$$

To give a simple example, we evaluate the counting statistics of a quantum contact between two normal terminals and obtain in agreement with the scattering matrix approach [17]

$$S(\chi) = \frac{2t_0}{h} \int dE \sum_n \ln[1 + T_n f_L(1-f_R)(e^{i\chi}-1) + T_n f_R(1-f_L)(e^{-i\chi}-1)]. \tag{3}$$

Thus, the statistics is a simple multinomial form of the two possible events of left and right transfer of charges. The formula (2) includes in addition the statistics of SN- and SS-contacts [10, 16].

In diffusive conductors we have to find the general counting statistics by a different method. The quantum kinetic equation for the diffusive wire is the so-called Usadel equation [18]

$$\hbar \frac{\partial}{\partial x} D(x) \check{G}(x) \frac{\partial}{\partial x} \check{G}(x) = -i \left[E\hat{\tau}_3, \check{G}(x) \right], \tag{4}$$

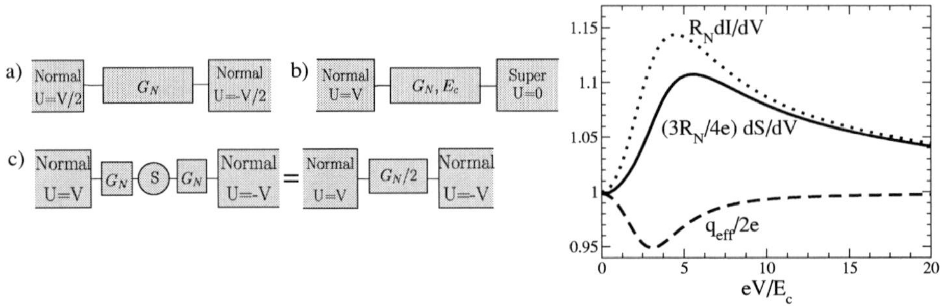

FIGURE 1. Left: a) diffusive wire with conductance G_N between two normal terminals. b) diffusive Andreev wire, characterized by the energy scale $E_c = \hbar D/L^2$. c) in the incoherent regime the wire b) is mapped onto a series of two wires between normal terminals. For diffusive connectors the system is equivalent to one wire with conductance $G_N/2$. Right: Shot noise of a diffusive proximity wire. Both the differential conductance and the noise show a reentrant behaviour. The effective charge reveals that the correlated Andreev pair transport suppresses the noise below the uncorrelated Boltzmann-Langevin result.

with continuous boundary conditions at both ends of the conductor. The right-hand side of Eq. (4) accounts for the decoherence of electrons and holes during the propagation in the normal metal at finite energies E. A full solution has so far been only obtained numerically, and we will discuss the implications of the decoherence on the shot noise later. For diffusive wires between normal terminals or with one superconducting terminal (see Fig. 1a and Fig. 1b for smalle energies $eV, k_BT \ll E_c$) the right hand side of Eq. (4) can be neglected. We obtain the general solution [19, 20, 21]

$$S(\chi) = \frac{t_0 G_N}{8e^2} \int dE \, \mathrm{Tr} \, \mathrm{acosh}^2 \left(\frac{1}{2} \{ \check{G}_L, \check{G}_R \} \right) . \tag{5}$$

The same result can be obtained by averaging Eq. (2) over the transmission eigenvalue distribution of a diffusive connector, i.e. the bimodal distribution $\rho(T) = G_N/2G_Q T \sqrt{1-T}$ [22].

Another drastic simplification can be made if the proximity effect is negligible , i. e. the right hand side of Eq. (4) is dominant. In Ref. [23] it was shown, that the diffusive wire can be mapped onto a series of two diffusive wires contacted by normal terminals, which constitute the electron and hole propagation (see Fig. 1c). For the special case of diffusive connectors the counting statistics is independent of the geometry. As a consequence, the counting statistics is exactly given by that of a *normal* conductor, with *halved* conductance and a *negative* counting field for the hole terminal. The energy dependence resulting from the right hand side of Eq. (4) leads to interesting effects related to the quantum propagation of electron-hole pairs [24]. Below, we will address in detail the dependence of the shot noise on voltage and temperature, and compare our theoretical predictions, based on (4) with available experimental results.

SHOT NOISE IN DIFFUSIVE CONDUCTORS

The shot noise is obtained via $S_I = -(2e^2/t_0)(\partial^2/\partial\chi^2)S(\chi)|_{\chi=0}$. From Eq. (2) we obtain for a general scatterer between normal terminals

$$S_I^{NN} = \frac{2e^2}{h}\sum_n \int dE \Big[T_n(1-T_n)(f_L(E)-f_R(E))^2 + T_n\big[f_L(E)(1-f_L(E))+f_R(E)(1-f_R(E))\big]\Big]. \tag{6}$$

Averaging (6) over the Dorokovs distribution [22] we find

$$S_I^{NN} = 2G_N \int dE \left[f_L(E)(1-f_L(E))+f_R(E)(1-f_R(E))+\frac{1}{3}(f_L(E)-f_R(E))^2\right]. \tag{7}$$

The energy integration can be done and we obtain

$$S_I^{NN}(T,V) = \frac{8}{3}G_N k_B T + \frac{2}{3}eG_N V \coth\left(\frac{eV}{2k_BT}\right). \tag{8}$$

Using the result (5) we can obtain the general formula for the noise

$$S_I = -\frac{G}{4e^2}\int dE\,\mathrm{Tr}\left[\frac{\partial^2}{\partial\chi^2}\{\check{G}_L,\check{G}_R\} - \frac{2}{3}\left(\frac{\partial}{\partial\chi}\{\check{G}_L,\check{G}_R\}\right)^2\right]_{\chi\to 0}. \tag{9}$$

For the Andreev wire at $eV, k_BT \ll E_c$ we obtain

$$S_I^{NS}(V) = 2G_N\int dE\left[2f(E)(1-f(E))+\frac{1}{3}(f(E)-f(-E))^2\right]. \tag{10}$$

Evaluating the energy integration we find

$$S_I^{NS}(V) = \frac{8}{3}G_N k_B T + \frac{4}{3}eG_N V\coth\left(\frac{eV}{k_BT}\right). \tag{11}$$

This result can, indeed, be infered from the normal-state result Eq. (9) by the replacement $S_I \to 2S_I$, $G_N \to G_N/2$ and $V \to 2V$, or equivalently $e \to 2e$ in this case.

ENERGY-DEPENDENT CURRENT NOISE

The full quantum-mechanical description requires the solution of the Usadel equation (4) to first order in the counting field. We will discuss these results later. Let us first note, that one can obtain an approximate expression for the energy dependence of the shot noise by a generalized Boltzmann-Langevin approach. We recall, that the kinetic equation for the average distribution function has the form

$$\frac{\partial}{\partial x}\sigma(E,x)\frac{\partial}{\partial x}(1-f(E,x)-f(-E,x)) = 0. \tag{12}$$

The local energy-dependent conductivity incorporates the effect of the proximity induced coherence and is given by $\sigma(E,x) = \sigma_N \cosh^2(\mathrm{Re}\theta(E,x))$, where the spectral angle θ obeys $\hbar D(\partial^2/\partial x^2)\theta(E,x) = -i2E\sin(\theta(E,x))$ with appropriate boundary conditions (see Ref. [25] for details). Solving Eq. (12) the current (per unit area of the contact) is

$$I(V,T) = \frac{1}{e}\int dE G(E)\left[1 - f_N(E) - f_N(-E)\right], \frac{1}{G(E)} \equiv \int_0^L \frac{dx}{L}\frac{1}{\sigma(E,x)}, \qquad (13)$$

Here we have defined the spectral conductance $G(E)$ for for a one-dimensional wire of length L with a uniform cross section. For an arbitrary geometry the spectral conductance has to be found from the solution of a diffusion equation.

Guided by the kinetic equation for the average transport, we may try to find the current noise by generalizing the Boltzmann-Langevin approach [26] to include the energy- and space-dependent conductivity. In fact, this task has already been performed. In Ref. [20] it was shown, that the counting statistics of a diffusive normal wire only depends on the conductance, defined in exactly the same way as (13). In this proof an energy-independent conductivity $\sigma(x)$, but an arbitrary form of the diffusive metal was taken into account. However, it is obvious, that the same result is obtained if a spectral conductivity $\sigma(E,x)$ is assumed. As a result the current would be given by Eq. (13). We combine this observation with the results for incoherent Andreev transport obtained in Ref. [23]. In this work it was shown that the counting statistics of a diffusive wire between a normal and a superconducting terminal is mapped onto a series of two diffusive wires. Again, this holds equally well if we assume *ad hoc* an energy-dependent conductance of both diffusive wires. Due to electron-hole symmetry the spectral conductance of both wires and therefore of the total wire is the same. Therefore, we obtain the current noise from Eq. (7) by inserting the spectral conductance (13) *inside* the energy integral and multiplying by a factor of 2. In this way we obtain the result of the modified Boltzmann-Langevin approach

$$S_I^{BL} = 2\int dE G(E)\left[2f(E)(1-f(E)) + \frac{1}{3}(f(E)-f(-E))^2\right]. \qquad (14)$$

This result my be written as

$$S_I^{BL} = \frac{8}{3}G(V,T)k_BT + \frac{4}{3}eI(V,T)\coth\left(\frac{eV}{k_BT}\right). \qquad (15)$$

Here we introduced the temperature-dependent differential conductance $G(V,T) = dI(V,T)/dV$. The same result was obtained by a different method recently in Ref. [27]. At zero temperature, we obtain the result [28]

$$S_I^{BL}(V) = \frac{4}{3}eI(V). \qquad (16)$$

This result was derived neglecting correlations of scattering events *between electrons and holes* in the normal metal.

The full quantum-mechanical calculation of the energy-dependent shot noise was performed in Ref. [24] and the results are shown in Fig. 1. A direct comparison of the differential shot noise and the differential conductance (for zero temperature) shows the difference in the energy dependence. The effective charge is defined as deviation of the quantum noise from the Boltzmann-Langevin result (16) of $2e$. At energies below the Thouless energy $E_c = \hbar D/L^2$ the effective charge is suppressed below $2e$. This shows that the correlated Andreev pair transport suppresses the noise below the uncorrelated Boltzmann-Langevin result.

CONCLUSIONS

In conclusion, we have shown, that shot noise in diffusive heterostructures between normal and superconducting terminal provides valuable information on the correlated Andreev pair transport. We have examined the difference between a simple Boltzmann-Langevin description, which neglects these correlations, and a full quantum calculation. Examining the effective charge we have shown that two Andreev pairs propagating coherently into the normal metal are anti-correlated.

ACKNOWLEDGMENTS

We thank C. Bruder, Yu. V. Nazarov, D. Prober, B. Reulet, and P. Samuelsson for discussions and the Swiss SNF and NCCR Nanoscience for financial support.

REFERENCES

1. Ya. M. Blanter and M. Büttiker, Phys. Rep. **336**, 1 (2000).
2. *Quantum Noise in Mesoscopic Physics*, edited by Yu. V. Nazarov (Kluwer, Dordrecht, 2003).
3. V. A. Khlus, Sov. Phys. JETP **66**, 1243 (1987).
4. G. B. Lesovik, Pis'ma Zh. Eksp. Teor. Fiz. **49**, 513 (1989); M. Büttiker, Phys. Rev. Lett. **65**, 2901 (1990); Phys. Rev. B **46**, 12485 (1992).
5. C. W. J. Beenakker and M. Büttiker, Phys. Rev. B **46**, 1889 (1992).
6. Yu. V. Nazarov, Phys. Rev. Lett. **73**, 1420 (1994).
7. P. Dieleman *et al.*, Phys. Rev. Lett. **79**, 3486 (1997); T. Hoss *et al.*, Phys. Rev. B **62**, 4079 (2000).
8. M. J. M. de Jong and C. W. J. Beenakker, Phys. Rev. B **49**, 16070 (1994); K. E. Nagaev and M. Büttiker, Phys. Rev. B **63**, 081301(R) (2001).
9. X. Jehl *et al.*, Phys. Rev. Lett. **83**, 1660 (1999); X. Jehl *et al.*, Nature **405**, 50 (2000); A. A. Kozhevnikov, R. J. Schoelkopf, and D. E. Prober, Phys. Rev. Lett **83**, 1660 (2000).
10. B. A. Muzykantskii and D. E. Khmelnitzkii, Phys. Rev. B **50**, 3982 (1994).
11. T. Martin, Phys. Lett. A **220**, 137 (1996).
12. S. V. Naidenov and V. A. Khlus, Fiz. Nizk. Temp. **21**, 594 (1995); A. L. Fauchere, G. B. Lesovik, and G. Blatter, Phys. Rev. B **58**, 11177 (1998); K. M. Schep and G. E. W. Bauer, Phys. Rev. B **56**, 15860 (1997); P. Samuelsson, Phys. Rev. B **67**, 054508 (2003).
13. M. P. V. Stenberg and T T. Heikkilä, Phys. Rev. B **66**, 144504 (2002); F. Pistolesi, G. Bignon, and F. W. J. Hekking, Phys. Rev. B **69**, 214518 (2004); M. Houzet and F. Pistolesi, Phys. Rev. Lett. **92**, 107004 (2004).
14. F. Lefloch, C. Hoffmann, M. Sanquer, and D. Quirion, Phys. Rev. Lett. **90**, 067002 (2003).
15. Yu. V. Nazarov, Phys. Rev. Lett. **73**, 134 (1994); Superlattices Microst. **25**, 1221 (1999).
16. W. Belzig and Yu. V. Nazarov, Phys. Rev. Lett. **87**, 197006 (2001).

17. L. S. Levitov and G. B. Lesovik, JETP Lett. **58**, 230 (1993); L. S. Levitov, H. W. Lee, and G. B. Lesovik, J. Math. Phys. **37**, 4845 (1996).
18. K. D. Usadel, Phys. Rev. Lett. **25**, 507 (1970).
19. H. Lee, L. S. Levitov, and A. Yu. Yakovets, Phys. Rev. B **51**, 4079 (1996).
20. Yu. V. Nazarov, Ann. Phys. (Leipzig) **8**, SI-193 (1999).
21. W. Belzig in [2].
22. O. N. Dorokhov, Solid State Comm. **51**, 381 (1984).
23. W. Belzig and P. Samuelsson, Europhys. Lett. **64**, 253 (2003).
24. W. Belzig and Yu. V. Nazarov, Phys. Rev. Lett. **87**, 067006 (2001).
25. Yu. V. Nazarov and T. H. Stoof, Phys. Rev. Lett. **76**, 823 (1996).
26. K. E. Nagaev, Phys. Lett. A **169**, 103 (1992); Phys. Rev. B **57**, 4628 (1998).
27. M. Houzet and V. P. Mineev, Phys. Rev. B **67**, 184524 (2003).
28. B. Reulet, D.E. Prober, and W. Belzig, in [2].
29. B. Reulet *et al.*, Phys. Rev. Lett. **90**, 066601 (2003).

The Role Of Noise In Quantum Communication

János A. Bergou

Department of Physics and Astronomy, Hunter College of the City University of New York, 695 Park Avenue, New York, NY 10021

Abstract. In quantum information and quantum computing, the carrier of information is a quantum system and information is encoded in its state. Secure communication schemes employ non-orthogonal states that can not be discriminated with 100% probability of success. We briefly review various possible optimized state discrimination strategies since they provide a quantitative figure of merit for the acceptable noise level in quantum communication. We also report on recent results in two special areas. First, we consider discrimination between mixed quantum states. Second, we show that communication via completely unknown quantum states is possible and discuss programmable quantum state discriminators that accomplish this task optimally.

Keywords: state discrimination, measurement optimization, POVM, detection noise
PACS: 03.67.-a,03.65.Bz,42.50.-p

INTRODUCTION

In quantum information and quantum computing the carrier of information is a quantum system and information is encoded in its state [1]. The state, however, is not an observable [2] and, thus, a fundamental problem arises: after processing the information it has to be read out or, in other words, the state of the system has to be determined. When the possible target states are orthogonal, this is a relatively simple task. But when the possible target states are not orthogonal they cannot be discriminated perfectly, and optimum discrimination with respect to some appropriately chosen criteria is far from being trivial even if the set of the possible nonorthogonal states is known. The two most frequently used discrimination strategies are discrimination with minimum error [3] and unambiguous discrimination (for recent reviews see [4, 5]). Here we focus on the latter. The problem appears in its cleanest form in the B92 quantum cryptographic protocol [6] where a sender (Alice) prepares a system in one of two known nonorthogonal quantum states and sends it to a receiver (Bob). Without further processing Bob has to determine which of the two states the system he receives is in.

Unambiguous discrimination started with the works of Ivanovic, Dieks and Peres [7] who studied the following problem. A collection of quantum systems is prepared so that each single system is equally likely to be prepared in one of two known states, $|\psi_1\rangle$ or $\psi_2\rangle$, that are not orthogonal, $\langle\psi_1|\psi_2\rangle \neq 0$. The sender then sends the systems one by one to the receiver whose task is to determine which of the two states has been prepared in each case. All the receiver can do is to perform a single measurement or perhaps a series of measurements on the individual system. Ivanovic came to the conclusion that if one allows inconclusive detection results to occur then in the remaining cases the receiver can conclusively determine the state of the individual system.

It is rather easy to see that a simple von Neumann measurement can accomplish this

CP800, *Unsolved Problems of Noise and Fluctuations: UPoN 2005*
edited by L. Reggiani, C. Pennetta, V. Akimov, E. Alfinito, and M. Rosini

task. Let us denote the Hilbert space of the two given states by $\mathscr{H}$ and introduce the projector P_1 for $|\psi_1\rangle$ and $\bar{P}_1$ for the orthogonal subspace, such that $P_1 + \bar{P}_1 = 1$, the unity in $\mathscr{H}$. Then we know for sure that $|\psi_2\rangle$ was prepared if in the measurement of $\{P_1, \bar{P}_1\}$ a click in the $\bar{P}_1$ detector occurs. A similar conclusion for $|\psi_1\rangle$ can be reached with the roles of $|\psi_1\rangle$ and $|\psi_2\rangle$ reversed. Of course, when a click along P_1 (or P_2) occurs then we learn nothing about which state was prepared thus corresponding to inconclusive results. Ivanovic's startling observation was that a sequence of measurements can sometimes do better than the single von Neumann measurement described here. Dieks then found that this sequence of measurements can be realized with a single generalized measurement (POVM) and Peres subsequently showed that this POVM is optimal in the sense that its failure probability, the probability that an inconclusive outcome occurs, is minimum. The probability of the inconclusive outcome, or failure, is $Q_{IDP} = |\langle\psi_1|\psi_2\rangle|$ and the probability of success is then given by what is called the Ivanovic-Dieks-Peres (IDP) limit,

$$P_{IDP} = 1 - Q_{IDP} = 1 - |\langle\psi_1|\psi_2\rangle| \,. \tag{1}$$

This result can be generalized for the case when the preparation probabilities of the states, η_1 and η_2, are different, $\eta_1 \neq \eta_2$. The preparation probabilities are also called a priori probabilities. The IDP result thus corresponds to the case of equal a priori probabilities for the two states, $\eta_1 = \eta_2 = 1/2$, and the generalization for arbitrary a priori probabilities is due to Jaeger and Shimony [8].

Clearly, communication via noisy channels or, in general, interaction with the environment will turn an initially pure state into a mixed one. Therefore, in the next section we give a brief overview of recent progress in the area of unambiguous discrimination of two mixed quantum states and we point out optimum strategies for the receiver and eavesdropper. Surprisingly, they turn out to be different. Then, in the subsequent section, we describe a scheme that can be used for communication via unknown quantum states which appears to the unsuspecting observer as communiaction via pure noise. To this end we shall consider programmable state discriminators that can be programmed to optimally discriminate between two unknown quantum states. We conclude with a brief summary of the positive role noise can play in quantum communication. First, it can be used to identify the presence of an eavesdropper. Second, communication via pure noise can provide complete security against eavesdropping.

DISCRIMINATION OF TWO MIXED STATES

The problem of unambiguously discriminating mixed quantum states arises for instance when given pure states undergo a specified decoherence process during transmission through a noisy quantum channel. It will be useful to introduce some terminology first. The support S_j of the density operator ρ_j, describing a quantum state, is the subspace of the entire Hilbert space $\mathscr{H}$ spanned by the eigenvectors of ρ_j belonging to nonzero eigenvalues (for $j = 1,2$). The rank of the density operator is equal to the dimension of its support. The subspace orthogonal to its support, $\bar{S}_j$, is the kernel of the density operator ρ_j such that $\mathscr{H} = S_j \oplus \bar{S}_j$. We shall denote the projector to S_j by P_j and the projector to $\bar{S}_j$ by $\bar{P}_j$. Equipped with these definitions we can now make the following general statement. For two mixed quantum states unambiguous discrimination is possible with a finite probability of success if and only if the supports of their density operators are not identical. Indeed, in such a case the kernel of at least one of them is not empty and a projective measurement along this kernel unambiguously identifies the other state. On the other hand, if the supports are identical, then so are their kernels, and there is no direction in $\mathscr{H}$ that could unambiguously identify at least one of the density operators.

While for two pure states the minimum failure probability has been known since long [7, 8], the study of unambiguous discrimination among mixed states has begun only relatively recently [9]-[17]. A complete solution, determining the minimum achievable failure probability for arbitrary prior probabilities of the states, has been obtained for the special cases of discriminating a pure and a mixed state [9, 10], and of two mixed states of rank d in a $(d+1)$-dimensional joint Hilbert space [11]. For discriminating two arbitrary mixed states, bounds have been derived for the failure probability [11], in terms of the fidelity of the states. Here we give a brief overview of recent progress, based mainly on our recent work [16].

A key quantity that plays an important role in unambiguous discrimination is the fidelity, F, which for two density operators, ρ_1 and ρ_2, is given by

$$F \equiv \mathrm{Tr}(|\sqrt{\rho_1}\sqrt{\rho_2}|)| \,. \tag{2}$$

We assume that the two density operators are prepared with the prior probabilities η_1 and $\eta_2 = 1 - \eta_1$, respectively.

In terms of these quantities the optimum failure probability Q, i.e. the minimum probability of obtaining an inconclusive outcome in our optimized measurement strategy saisfies a lower bound. This bound for the various regimes of the prior probability η was established in Ref. [16] and here we just give the final expressions without going into details of the derivation:

$$Q \geq \begin{cases} 2\sqrt{\eta_1\eta_2}F = Q_0 & \text{if } \frac{\mathrm{Tr}(P_2\rho_1)}{F} \leq \sqrt{\frac{\eta_2}{\eta_1}} \leq \frac{F}{\mathrm{Tr}(P_1\rho_2)}, \\ \eta_1\mathrm{Tr}(P_2\rho_1) + \eta_2\frac{F^2}{\mathrm{Tr}(P_2\rho_1)} & \text{if } \sqrt{\frac{\eta_2}{\eta_1}} \leq \frac{\mathrm{Tr}(P_2\rho_1)}{F}, \\ \eta_1\frac{F^2}{\mathrm{Tr}(P_1\rho_2)} + \eta_2\mathrm{Tr}(P_1\rho_2) & \text{if } \frac{F}{\mathrm{Tr}(P_1\rho_2)} \leq \sqrt{\frac{\eta_2}{\eta_1}}. \end{cases} \tag{3}$$

For density operators of the form $\rho_1 = \sum_{i=1}^{d} r_i|r_i\rangle\langle r_i|$ and $\rho_2 = \sum_{i=1}^{d} r_i|s_i\rangle\langle s_i|$, with $\langle r_i|s_j\rangle = b\,\delta_{ij}$, the fidelity is found to be $F = |b|$. For this case the equality sign holds in Eq. (3) and it reprsents the full solution to the problem. In the intermediate range of η_2/η_1 the top line holds in Eq. (3) and there exists a generalized measurement (POVM) that attains this failure probabilty. At the boundaries of this interval the POVM crosses over continuously into von Neumann measurements with the failure probability of the middle line at the lower boundary and the bottom line at the upper boundary, in (3), respectively. This case, however, is very special. Eq. (3), in general, does not provide a solution. But the solution has to obey these general inequalities and therefore they are useful in eliminating nonphysical solutions. For further details see [16].

Next, we consider two density operators of rank 2 in a 4-dimensional joint Hilbert space, with the corresponding spectral representations

$$\rho_1 = \sum_{l=1}^{2} r_l|r_l\rangle\langle r_l|, \qquad \rho_2 = \sum_{m=1}^{2} s_m|s_m\rangle\langle s_m|, \tag{4}$$

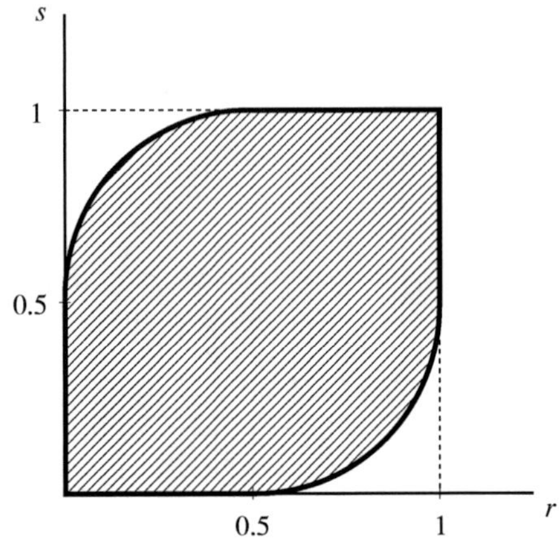

FIGURE 1. The parameter space of two rank 2 density operators from Eq. (4), with $d = 2$. For this case each of the density operators can be characterized by a single parameter, $r \equiv r_1 = 1 - r_2$ for ρ_1 and $s \equiv s_1 = 1 - s_2$ for ρ_2, with $0 \leq r, s \leq 1$. In addition, we assume $\langle r_i | s_j \rangle = \frac{1}{\sqrt{2}} \delta_{i,j}$ $(i, j = 1, 2)$. For this case $F = \frac{1}{\sqrt{2}}[\sqrt{rs} + \sqrt{(1-r)(1-s)}]$ and $\mathrm{Tr}(P_1\rho_2) = \mathrm{Tr}(P_2\rho_1) = 1/2$. The shaded area is where the fidelity bound can be reached for some value of the prior probabilities, and the white corners represent the part of the parameter space where it can not be reached for any value of the prior probabilities. The boundaries of the shaded area are given by $s = 0$ for $0 \leq r \leq 1/2$, by $(r - 1/2)^2 + (s - 1/2)^2 = 1/4$ for $1/2 \leq r \leq 1,\ 0 \leq s \leq 1/2$ and for $0 \leq r \leq 1/2, 1/2 \leq s \leq 1$, and by $s = 1$ for $1/2 \leq r \leq 1$.

where $r_l, s_m \neq 0$, and $\langle r_l | r_m \rangle = \delta_{l,m} = \langle s_l | s_m \rangle$. We notice that in this case each density operator can be characterized by a single parameter, $r \equiv r_1 = 1 - r_2$ for ρ_1 and $s \equiv s_1 = 1 - s_2$ for ρ_2, with $0 \leq r, s \leq 1$. Thus the parameter space is the unit square in the positive quarter of the r, s plane with the origin as its lower left corner and the axes as its sides. In certain regions of the parameter space the fidelity bound can be reached and in others it can not. The boundaries between these regions are shown in Fig. 1.

With increasing dimensionality d, the region where the fidelity bound cannot be reached also increases. We also see that the case $r = s$ is the symmetry axis of the allowed region, as far away from the boundaries as possible so, in the following, we restrict ourselves to this case. More specifically, we consider the special case that $r_i = s_i$ for $i = 1, \ldots, d$, for which $F = 1/\sqrt{2}$. For this case the results can be summarized as

$$Q = \begin{cases} Q_0 = 2\sqrt{\eta_1\eta_2}F = \sqrt{2\eta_1\eta_2} & \text{if } \frac{\mathrm{Tr}(P_2\rho_1)}{F} = \frac{1}{\sqrt{2}} \leq \sqrt{\frac{\eta_2}{\eta_1}} \leq \frac{F}{\mathrm{Tr}(P_1\rho_2)} = \sqrt{2}, \\ Q_2 = \eta_1 \mathrm{Tr}(P_2\rho_1) + \eta_2 = \frac{1}{2}\eta_1 + \eta_2 & \text{if } \sqrt{\frac{\eta_2}{\eta_1}} \leq \frac{1}{\sqrt{2}}, \\ Q_1 = \eta_1 + \eta_2 \mathrm{Tr}(P_1\rho_2) = \eta_1 + \frac{1}{2}\eta_2 & \text{if } \sqrt{2} \leq \sqrt{\frac{\eta_2}{\eta_1}}. \end{cases} \tag{5}$$

For a complete derivation see again Ref. [16]. Thus we have the complete otimum solution, as opposed to bounds only, for this case. The optimum measurement strategy yields the equality sign in Eq. (3).

In closing this section we note that for $\eta_1 = \eta_2 = 1/2$ the upper line holds in Eq.

(5) yielding a failure probability of about 71% which is the optimum the receiver can have. If an eavesdropper follows this strategy she, too, will fail about 71% of the time. But in the cases when she fails she still has to send a signal to the receiver. The best she can do is guess and if half the time she guesses right she will introduce an error rate of 35.5%. If, on the other hand, she follows a different strategy which is called the minimum error strategy [15], she will only have a 14.5% error rate so clearly this second strategy is advantageous for eavesdropping. Nevertheless, her presence can be detected if the sender and receiver compare a part of their respective key strings and find that they disagree about 14.5% of the time, or more.

PROGRAMMABLE STATE DISCRIMINATORS

In this section we will revisit the simpler problem of discriminating between two pure quantum states, $|\psi_1\rangle$ and $|\psi_2\rangle$, that we discussed in the Introduction. However, we add a new twist: we do not assume that we know anything about the states. In other words, we want to construct a device that will unambiguously discriminate between two unknown quantum states. The input states can be completely arbitrary, with the constraint that for unambiguous discrimination to be possible they must be linearly independent [18] but not necessarily orthogonal. We expect that the success probability for *unknown* states will be smaller than the IDP limit and in this section we present the optimum strategy.

The actual state-distinguishing device for two *known* states depends on the two states, $|\psi_1\rangle$ and $|\psi_2\rangle$, i. e. these two states are "hard wired" into the machine. If this device is given a system in one of these two states, it will produce one of three outputs, 1, 2, or 0. If the output is 1, then the input was $|\psi_1\rangle$, if the output is 2, then the input was $|\psi_2\rangle$, and if the output is 0, which we call failure, then we learn nothing about the input. The device for unambiguous discrimination will not make an error, it will never give an output of 2 if the input was $|\psi_1\rangle$, and vice versa. We show that it is possible to construct a machine in which the information about $|\psi_1\rangle$ and $|\psi_2\rangle$ is supplied in the form of a program. This machine is then capable, with the correct program, of distinguishing any two quantum states.

We refer to this device as a programmable state discriminator [19] (see this reference also for many of the details omitted in this section). The program consists of the two qubit states that we wish to distinguish. In other words, we are given two qubits, one in the state $|\psi_1\rangle$ and another in the state $|\psi_2\rangle$. We have no knowledge of the states $|\psi_1\rangle$ and $|\psi_2\rangle$. Then we are given a third qubit that is guaranteed to be in one of these two program states, and our task is to determine, as best we can, in which one. We are allowed to fail, but not to make a mistake and we want to determine the optimum success probability. We shall consider the first two qubits we are given as a program. They are fed into the program register of the programmable state discriminator, and the third, unknown qubit is fed into the data register of this device. The device then tells us, with optimal probability of success, which one of the two program states the unknown state of the qubit in the data register matches. We can design such a device by viewing our problem as a task in measurement optimization. Our measurement is allowed to return an inconclusive result but never an erroneous one. Thus, it is a generalized measurement that returns 1 (the unknown state stored in the data register matches $|\psi_1\rangle$), 2 (the unknown state stored in the data register matches $|\psi_2\rangle$), or 0 (we do not learn anything about the unknown state stored in the data register).

Our task is then reduced to the following measurement optimization problem. One has two input states

$$\begin{aligned} |\Psi_1^{in}\rangle &= |\psi_1\rangle_A|\psi_2\rangle_B|\psi_1\rangle_C , \\ |\Psi_2^{in}\rangle &= |\psi_1\rangle_A|\psi_2\rangle_B|\psi_2\rangle_C, \end{aligned} \tag{6}$$

where the subscripts A and B refer to the program registers (A contains $|\psi_1\rangle$ and B contains $|\psi_2\rangle$), and the subscript C refers to the data register. Our goal is to unambiguously distinguish between these inputs, keeping in mind that one has no knowledge of $|\psi_1\rangle$ and $|\psi_2\rangle$, i. e. we want to find the measurement that will accomplish this.

The fact that we know nothing about $|\psi_1\rangle$ and $|\psi_2\rangle$ means that the only way we can guarantee satisfying the above conditions is to take advantage of the symmetry properties of the states, i.e. that $|\Psi_1^{in}\rangle$ is symmetric in AC and $|\Psi_2^{in}\rangle$ is symmetric in BC. A click in the detector that projects onto the antisymmetric subspace of AC can come only from the second entry and a click in the detector that projects onto the antisymmetric subspace of BC can come only from the first entry. These detectors will unamiguously tell us which entry we were given, with the respective success probabilities

$$P_1 = \frac{1}{2}\eta_1(1-|\langle\psi_1|\psi_2\rangle|^2) , \tag{7}$$

for the first entry and

$$P_2 = \frac{1}{2}\eta_2(1-|\langle\psi_1|\psi_2\rangle|^2) , \tag{8}$$

for the second.

Unfortunately, the corresponding projectors are not orthogonal and, therefore we can perform either the first measurement or the the second but not both. Again, a generalized measurement (POVM) comes to the rescue. It performs a little bit of both measurements simultaneously with the success probability [19]

$$P_{\mathrm{POVM}} = \frac{2}{3}(1-\sqrt{\eta_1\eta_2})(1-|\langle\psi_1|\psi_2\rangle|^2) . \tag{9}$$

The POVM that yields this success probability can be implemented only when

$$\frac{1}{5} \leq \eta_1, \eta_2 \leq \frac{4}{5} . \tag{10}$$

Outside this interval, we have to perform one or the other projective von Neumann measurement.Our findings can be summarized as follows

$$P^{opt} = \begin{cases} P_{\mathrm{POVM}} & \text{if } \frac{1}{5} \leq \eta_1 \leq \frac{4}{5} , \\ P_2 & \text{if } \eta_1 < \frac{1}{5} , \\ P_1 & \text{if } \frac{4}{5} < \eta_1 . \end{cases} \tag{11}$$

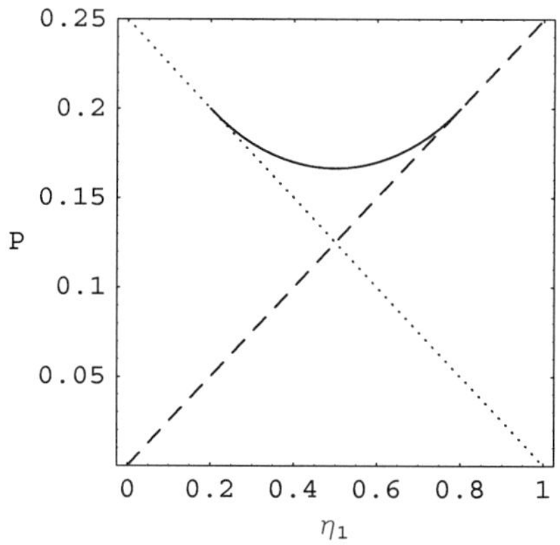

FIGURE 2. Optimal average success probability, P, vs. the *a priori* probability, η_1. Dashed line: P_1 from Eq. (7), dotted line: P_2 from Eq. (8), and solid line: P_{POVM} from Eq. (9). For the figure we replaced $1-|\langle\psi_1|\psi_2\rangle|^2$ by its average, $\frac{1}{2}$. The optimal P is given by P_2 for $\eta_1 < 0.2$, by P_{POVM} for $0.2 \leq \eta_1 \leq 0.8$ and by P_1 for $0.8 < \eta_1$.

Equation (11) represents the main result of this section. In the intermediate range of the *a priori* probability the optimal failure probability, Eq. (9), is achieved by a generalized measurement or POVM. Outside this region, for very small *a priori* probability, $\eta_1 \leq 1/5$, when the preparation is dominated by the second input, or very large *a priori* probability, $\eta_1 \geq 4/5$, when the preparation is dominated by the first input, the optimal failure probabilities, Eqs. (7) and (8), are realized by standard von Neumann measurements. For very small η_1 the optimal von Neumann measurement is a projection onto the antisymmetric subspace of the A and C qubits. For very large η_1 the optimal von Neumann measurement is a projection onto the antisymmetric subspace of the B and C qubits. At the boundaries of their respective regions of validity, the optimal measurements transform into one another continuously. The results depend on the overlap of the unknown states only. If we do not know the states but we know their overlap then Eqs. (7)-(9) immediately give the optimal solutions. If we know nothing about the states, not even their overlap, then we average these expressions over all input states, which results in the factor, $1-|\langle\psi_1|\psi_2\rangle|^2$, being replaced by its average value of $\frac{1}{2}$. Then we have the optimum average probabilities of success in the various regions.

This situation is depicted in Fiq. 2. In its range of validity the POVM performs better than either one of the two unambiguously discriminating von Neumann measurements. The difference between the performance of the POVM and that of the von Neumann projective measurements is largest for $\eta_1 = \eta_2 = \frac{1}{2}$. For these values $P_{POVM}^{ave} = \frac{1}{6}$ while $P_1^{ave} = P_2^{ave} = \frac{1}{8}$ so the POVM represents a 33% improvement over the standard quantum measurement. A striking feature of the programmable state discriminator is that it is universal. Neither the optimal detection operators nor the boundaries for their region of validity depend on the unknown states. This device, then, realizes the best strategy for the discrimination of unknown states and it is very flexible. All of the information about the states is carried by a quantum program, which means that it works for any two states.

Consequently, it can be used as part of a larger device that produces quantum states that need to be unambiguously identified.

SUMMARY AND OUTLOOK

In the first part of this paper we gave an overview of recent advances in the area of unambiguous discrimination of mixed quantum states. The results have wide ranging applications in quantum information processing. Some examples are communication via noisy channels and probabilistic quantum algorithms. The optimal measurements (POVMs) can be implemented experimentally by standard techniques of linear optics via generalized interferometers. In the second part we presented a scheme for communication via completely unknown quantum states and discussed programmable quantum state discriminators that are universal, i.e. perform optimally on average, independently of the actual states used for the communication scheme.

Acknowledgments. This research was partially supported by a PSC-CUNY Grant. The author is grateful for discussions of various aspects of this project with Ulrike Herzog, Mark Hillery, Masoud Mohseni, Aephraim Steinberg, and Yuqing Sun.

REFERENCES

1. M. A. Nielsen and I. L. Chuang, *Quantum Computation and Information*, Cambridge University Press, 2000.
2. A. Peres, *Quantum Theory: Concepts and Methods*, Kluwer Academic, Dordrecht, 1995.
3. C. W. Helstrom, *Quantum Detection and Estimation Theory*, Academic Press, New York, 1976.
4. A. Chefles, *Contemp. Phys.* **41**, 401, 2000; quant-ph/0010114.
5. J. A. Bergou, U. Herzog and M. Hillery, "Discrimination of quantum states," in *Lecture Notes in Physics*, J. Řeháček and M. Paris, eds., pp. 419–467, Springer, Berlin 2004.
6. C. H. Bennett, *Phys. Rev. Lett.* **68**, 3121 1992.
7. I. D. Ivanovic, *Phys. Lett. A* **123**, 257 (1987), D. Dieks, *Phys. Lett. A* **126**, 303 (1988), A. Peres, *Phys. Lett. A* **128**, 19 (1988).
8. G. Jaeger and A. Shimony, *Phys. Lett. A* **197**, 83, 1995.
9. Y. Sun, J. A. Bergou, and M. Hillery, *Phys. Rev. A* **66**, 032315, 2002.
10. J. A Bergou, U. Herzog, and M. Hillery, *Phys. Rev. Lett.* **90**, 257901 (2003); and *Phys. Rev. A* **72**, 012302 (2005).
11. T. Rudolph, R. W. Spekkens, and P. S. Turner, *Phys. Rev. A* **68**, 010301(R) (2003).
12. Ph. Raynal, N. Lütkenhaus, and S. van Enk, *Phys. Rev. A* **68**, 022308 (2003).
13. Y. C. Eldar, *Phys. Rev. A* **67**, 042309 (2003), Y. C. Eldar, M. Stojnic, and B. Hassibi, *Phys. Rev. A* **69**, 062318 (2004).
14. Y. Feng, R. Duan, and M. Ying, *Phys. Rev. A* **70**, 012308 (2004).
15. U. Herzog and J. A. Bergou, *Phys. Rev. A* **70**, 022302 (2004).
16. U. Herzog and J. A. Bergou, Phys. Rev. A **71**, 050301(R) (2005).
17. Ph. Raynal and N. Lütkenhaus, *quant-ph/0502165* (2005).
18. A. Chefles, *Phys. Lett. A* **239**, 339 (1998).
19. J. Bergou and M. Hillery, Phys. Rev. Lett. **94**, 160501 (2005).

Nucleation Time Distribution In The Two-Dimensional Ising Model With Spin-Flip Dynamics

Kevin Brendel*, G.T. Barkema* and Henk van Beijeren*

*Theoretical Physics, Utrecht University, Leuvenlaan 4, 3584 CE Utrecht, the Netherlands

Abstract. The distribution of nucleation times is studied for the two-dimensional Ising model with spin-flip dynamics in an external magnetic field. Direct simulation results for the nucleation times, spanning more than four orders of magnitude, are compared with theoretical predictions. Cluster free energies and effective transition rates for growth and shrinkage of clusters are determined directly from our simulations. The free energies so determined are well described by the classical Becker-Döring expression, provided one uses an effective surface tension that exceeds the macroscopic surface tension by up to 20%. Within this framework there is good agreement between simulation results and theoretical predictions for the mean nucleation time.

Keywords: Ising model, Monte Carlo, nucleation time
PACS: 64.60.ab

INTRODUCTION

Homogeneous nucleation is a prototypic example of escape from a metastable state through thermal activation. In this process a stable nucleus spontaneously forms in a metastable environment. It has been studied extensively, and excellent books and reviews exist [1, 2]. A most convenient tool for studying this phenomenon, both analytically and numerically, is the well-known Ising model. In equilibrium, above the critical temperature and in the absence of an external magnetic field, up- and down-pointing spins are roughly equally abundant. Below the critical temperature, the system resides in either of two states: one with a positive magnetization, in which most spins are pointing up, and another one with a negative magnetization. In the presence of a small external field one of these states is metastable and will decay to the stable equilibrium state through nucleation.

In *classical nucleation theory* (CNT), developed by Volmer and Weber [3], Farkas [4], Becker and Döring [5] and Zeldovich [6], among others, this phenomenon is described as the stochastic evolution of a droplet, which may grow or shrink by the attachment or detachment of single molecules. The rates for these elementary processes are assumed to satisfy a detailed balance condition determined by the free energy of the droplet. In its simplest form this free energy consists of a negative bulk contribution $C\Delta\mu$ and a positive surface contribution σA. Here C is the number of particles in the cluster, $\Delta\mu$ is the chemical potential difference between the stable and the metastable state (also called the supersaturation), A is the surface area and σ is the surface tension, which is supposed to be the same as that for coexisting states in equilibrium. The shape of the droplet is assumed to be isotropic (circular in two, spherical in three dimensions). Since for small

CP800, *Unsolved Problems of Noise and Fluctuations: UPoN 2005*
edited by L. Reggiani, C. Pennetta, V. Akimov, E. Alfinito, and M. Rosini

droplets the surface term dominates and for large droplets the bulk term, the free energy as a function of cluster size C goes through a maximum at a critical cluster size C_x. For small supersaturation, the free energy difference $\Delta F \equiv F(C_x) - F(0)$ is typically large compared to $k_B T$, with T temperature and k_B Boltzmann's constant. Therefore nucleation is a thermally activated process and the *nucleation rate*, i.e. the average number of nucleations per units of volume and time is suppressed by an Arrhenius factor $\exp[-\Delta F/(k_B T)]$.

>From our simulations we find that for the two-dimensional Ising model under non-conserving dynamics, CNT and specifically the Becker-Döring equations, work very well, provided one uses realistic values for the cluster free energies and for the effective transition rates between clusters of different sizes. We find nucleation rates that differ never more than 20% from their theoretical values, although the range of values found spans over four decades.

In addition we calculate the effects of an initial quench from a high temperature state into the metastable state. Right after the quench only very small droplets of the stable phase may be found. This first of all gives rise to an asymptotic time shift in the nucleation time distribution, which can be calculated in terms of the parameters of the Becker-Döring equations. It is independent of system size and therefore becomes relatively more important as the system gets larger. Secondly, the Becker-Döring equations also predict the short-time behavior of the probability distribution function for nucleation after the initial quench. It starts out at zero and differs strongly from the Poisson distribution that holds asymptotically for large times (but still small enough that the clusters that have nucleated already, cover only a negligible fraction of the total surface area). Both the asymptotic time shift and the short-time nucleation probabilities thus obtained show very good agreement with simulation results.

The organization of our manuscript is as follows. In section , we describe the Ising model with spin-flip dynamics and we outline our theoretical framework and apply it to our model. In section we compare the theoretical predictions with the results of high-accuracy computations, and in the final section we discuss our results.

THEORY

We consider the Ising model on a square lattice with lateral dimension L, periodic (helical) boundary conditions, and an external magnetic field h which favors downward-pointing spins. The Hamiltonian of this model reads

$$H = -J \sum_{\langle i,j \rangle} s_i s_j + h \sum_i s_i, \tag{1}$$

in which $s_i = \pm 1$ is the spin at site i, and J is the coupling constant. The first summation runs over all pairs of nearest-neighbor sites; under our helical boundary conditions the neighbors of site i are $j = i \pm 1$ modulo N and $j = i \pm L$ modulo N, with $N = L^2$. Note that in the sum each pair is counted only once. The magnetization is defined as $M \equiv \sum_i s_i$; it can take values $M = -N, -N+2, \ldots, N$. We restrict ourselves to systems in which L is even. As a consequence, M takes only even values, and summations over a range of possible magnetizations only run over even numbers, with an increment of 2.

The system evolves in time according to single-spin-flip dynamics with Metropolis acceptance probabilities [7]. If S_i is the configuration after i proposed spin flips, a trial configuration S'_{i+1} is generated by flipping a single spin at a random site. This trial configuration is then either accepted ($S_{i+1} = S'_{i+1}$) or rejected ($S_{i+1} = S_i$); the acceptance probability is given by

$$P_a = \min\left[1, \exp(-\beta\left\{E(S'_{i+1}) - E(S_i)\right\})\right], \tag{2}$$

in which $\beta = 1/(k_B T)$ with k_B Boltzmann's constant and T temperature. Our unit of time is one Monte Carlo step per site, so in one unit of time *on average* one spin flip is tried for each site.

To study the behavior of nucleation times at temperatures below the critical one we consider an ensemble of systems prepared in configurations with no large negative clusters present (typically by quenching from a positive field equilibrium state). For each system in the ensemble we keep track of the sizes of all of its clusters, defined as the numbers of aligned spins, connected to each other by nearest-neighbor bonds, contained in them. We define the nucleation time of a system as the first time at which one of its clusters grows beyond a given size A, which is comparable to, but larger than the critical cluster size for reasons that will be made clear below. For a theoretical description we will derive an expression for the probability distribution of the nucleation time under the stochastic dynamics of the system.

The spin-flip dynamics described in this section may be represented by a master equation for the probability distribution $P(\mathbf{S},t)$ of finding a system in the configuration $\mathbf{S}$ at time t. Due to the huge number of possible configurations this master equation cannot be solved analytically or even numerically for system sizes of practical interest. Therefore we resort to a couple of approximations that are made standardly in classical nucleation theory. First, we assume that we may treat the clusters of negative spins in the system as independent. In this approximation, the probability that none of the N_c clusters has grown beyond the critical size, is simply the N_c^{th} power of the probability that a single cluster has not grown beyond the critical size. Secondly, following the Becker-Döring approach [5], we assume that the dynamics of a single cluster may be modeled by a master equation for the probability $P(C,t)$ that this cluster contains C spins at time t:

$$\begin{aligned}\frac{dP(C,t)}{dt} &= \Gamma_{C,C+1}P(C+1,t) + \Gamma_{C,C-1}P(C-1,t) \\ &\quad - (\Gamma_{C+1,C} + \Gamma_{C-1,C})P(C,t),\end{aligned} \tag{3}$$

By adopting these Becker-Döring equations we make several approximations: we suppress all dependence of the transition rates on the shape of the clusters and we neglect the possibilities of merging or splitting of clusters. In sections and we will come back to this and argue how these effects can be accounted for to a large extent by choosing the transition rates in Eq. (3) in an appropriate effective way.

In order that the equilibrium distribution of cluster sizes be a stationary solution of the master equation we impose the condition of detailed balance

$$\frac{\Gamma_{C+1,C}}{\Gamma_{C,C+1}} = \exp\left[\beta(F(C) - F(C+1))\right], \tag{4}$$

where $\beta F(C) = -\ln[P_{eq}(C)/N]$, with $P_{eq}(C)$ the average number of clusters of size C for a system forced to be in equilibrium in the metastable state (see section). For C large enough that it is very improbable to find more than one cluster of size C simultaneously in the metastable system, $P_{eq}(C)$ may likewise be interpreted as the probability of finding a cluster of size C in it.

The long-time nucleation rate predicted by the master equation (3) follows as the largest eigenvalue (with a minus sign) of this equation, supplemented with an absorbing boundary at $C = A$. Here A is an integer larger than the critical cluster size C_x, chosen such that $P_{eq}(A) \gg P_{eq}(C_x)$, and clusters with size A are almost certain to nucleate. The absorbing boundary condition is implemented by setting $\Gamma_{A-1,A}$ equal to zero.

The largest eigenvalue $-\nu$ of $\Gamma_{C,C'}$ in Eq. (3), as well as the corresponding eigenvector $P_0(C)$, may be found by requiring that the net current away from cluster size C assumes the value $\nu P_0(C)$. Using conservation of probability one easily shows that this may be expressed as

$$\begin{aligned} j_{C+1,C} &\equiv \Gamma_{C+1,C}P_0(C) - \Gamma_{C,C+1}P_0(C+1) \\ &= \nu \sum_{c \leq C} P_0(c), \end{aligned} \tag{5}$$

$$\nu = \frac{\Gamma_{A,A-1}P_0(A-1)}{\sum_{c\leq A-1} P_0(c)}, \tag{6}$$

where $j_{C+1,C}$ is defined as the net current flowing from C to $C+1$. This current may be approximated by

$$j_{C+1,C} = \begin{cases} \nu \dfrac{\sum_{c=1}^{C} \exp[-\beta F(c)]}{\sum_{c=1}^{C_x} \exp[-\beta F(c)]}, & c \leq C_x \\ \nu, & c \geq C_x \end{cases} \tag{7}$$

because the sum on the right-hand side of Eq. (5) is dominated by the terms with small c-values, for which $P_0(c)$ is approximately proportional to $\exp[-\beta F(c)]$. This may be checked in hindsight against the solution obtained. With this approximation the equation may be solved recursively for $P_0(c)$ in terms of $P_0(A-1)$ for $c = A-2,\ A-3,\ \ldots$, with the result

$$\frac{P_0(c)}{P_0(A-1)} = \frac{\Gamma_{A,A-1}}{\nu} \sum_{m=c}^{A-1} j_{m+1,m} \frac{\exp[\beta(F(m) - F(c))]}{\Gamma_{m+1,m}}. \tag{8}$$

Identifying $j_{m+1,m}$ with ν, which is an excellent approximation for m near C_x, we arrive at the result

$$\nu = \left(\sum_{m=1}^{A-1} \frac{\exp[\beta F(m)]}{\Gamma_{m+1,m}} \sum_{c=1}^{C_x} \exp[-\beta F(c)] \right)^{-1}. \tag{9}$$

This well-known expression is usually derived by considering a state with a stationary current in which mass is inserted at a constant rate on one side (e.g. at $C = 0$ in our case) and taken out as soon as it reaches the absorbing boundary (see e.g. Ref. [8], section IV E). Under those conditions the replacement of $j_{m+1,m}$ by a constant is exact.

For arbitrary initial distributions $P(C,0)$ that are concentrated near the origin the probability $S(t)$ that the system has not yet nucleated at time t for large t behaves as

$$S(t) = \sum_{C=1}^{A-1} P(C,t) = \exp[-\nu(t-t_d)] \text{ for } t \to \infty. \tag{10}$$

where t_d is called the delay time. In Ref. [9] we showed that it can be expressed as

$$t_d = \frac{\sum_{n=1}^{A-1}\sum_{m=1}^{n-1}\left(\frac{W(n)}{\Gamma_{m+1,m}W(m)}\sum_{k=1}^{\min[m,C_x]} W(k)\right)\tilde{\psi}^l(n)}{\sum_{n=1}^{A-1}\left(W(n)\tilde{\psi}^l(n)\right)}, \tag{11}$$

with $W(n) \equiv \exp(-\beta F(n))$.

SIMULATIONS AND RESULTS

For applying the master equation approach discussed above to the Ising model one needs the appropriate values of the cluster free energies $F(C)$ and the transition rates $\Gamma_{C',C}$ from cluster sizes C to C'. We calculated both from Monte Carlo simulations employing a technique known as multispin coding [10]. This technique is very fast, hence allows for obtaining good statistics.

Cluster free energies

We measured the distribution of cluster sizes for various values of β and h, in a system with 64×64 spins. Since we specifically want to know the quasi-equilibrium distribution of cluster sizes in the metastable state, we defined a cut-off cluster size C_{max}, which is typically chosen to be 300. To prevent clusters from growing beyond this size we modified our algorithm in the following way: starting with a configuration S_i perform a fixed number M of Monte Carlo steps, and measure the sizes of all the clusters of down-spins in the system. If there is a cluster with more than C_{max} spins reject the new configuration and choose $S_{i+1} = S_i$. Otherwise accept the new configuration as S_{i+1}. In either case, add the numbers of clusters of given size in S_{i+1} to a histogram. After many repetitions of this loop the cluster free energy $F(C)$ is obtained, up to some additive constant, from

$$\beta F(C) = -\ln\frac{\langle N(C)\rangle}{N} \tag{12}$$

where $\langle N(C)\rangle$ is the average number of clusters of size C in a system with N sites.

If C_{max} is chosen too large, the results are no longer representative for the metastable system. In particular, excluded volume effects inside and around large clusters suppress larger clusters in the rest of the system more strongly than smaller ones. As a consequence, the distribution of cluster sizes depends on C_{max}. To determine which values of

C_{max} are allowed, we check that the free energy curve obtained for some given value, coincides (up to a small constant) with the same curve, obtained with a lower value of C_{max}. Figure 1 shows our measurements for the free energy according to Eq. (12) as a function of cluster size, for some combinations of temperature and external field.

Large clusters at not too low temperature on average are nearly circular in shape. For these Becker-Döring theory predicts a free energy [5] of the form

$$F(C) \approx F_0 + 2\sigma\sqrt{\pi C} - 2hC, \tag{13}$$

with σ the interfacial free energy per unit length, and h the strength of the external field.

Strictly speaking, in order to account correctly for the difference between the free energies of the metastable and the stable phase, one should replace $2hC$ by $2hCm_s$, with m_s the average spin per lattice site. At the temperatures considered here, however, m_s is very close to unity, so this correction is very small. In addition there are correction terms such as the ones resulting from the Laplace effect of compression under the influence of surface tension, and Tolman, or curvature corrections [11] to the surface tension. We fitted F_0 and σ to the data in figure 1, and added the corresponding curves as lines in the same figure. The measurements are well fitted by the curves, as long as the cluster size is not too small. However, as noted before by Shneidman et al. [12], the surface tensions obtained in this way are larger than those given by the Onsager expression [14]

$$\sigma = 2J + \beta^{-1} \ln \tanh \beta J, \tag{14}$$

valid for long horizontal or vertical interfaces, at zero external field (see figure 2). At the temperatures we considered the surface tension is known to be almost isotropic [13], so anisotropy effects cannot explain this difference. As suggested by Fig. 2 both field dependence of the surface tension and corrections due to the finite size of the clusters may play a role here. Note that similar corrections for the surface tension on a finite cylinder are present already in Onsager's exact solution [14]. In Ref. [15] an explicit plot of this was given. For a precise determination of nucleation rates these corrections are quite important. They may easily cause a change by a few orders of magnitude in the values of average nucleation rates. We will address these issues in more detail in a separate paper [16].

Interface diffusion coefficient

To estimate the rates $\Gamma_{C+1,C}$ of cluster growth, we study the diffusion of a single interface in a system with anti-periodic boundary conditions in the absence of an external field, as described in [15]. The location of the interface is obtained from the magnetization M. Its diffusion coefficient D is defined as:

$$D = \lim_{t\to\infty} \left[\frac{\langle (M(t) - M(0))^2 \rangle}{2t} \right]. \tag{15}$$

It is found, to a good approximation, to be proportional to the length of the interface B:

$$D(B, L, \beta J) = g(\beta J)B. \tag{16}$$

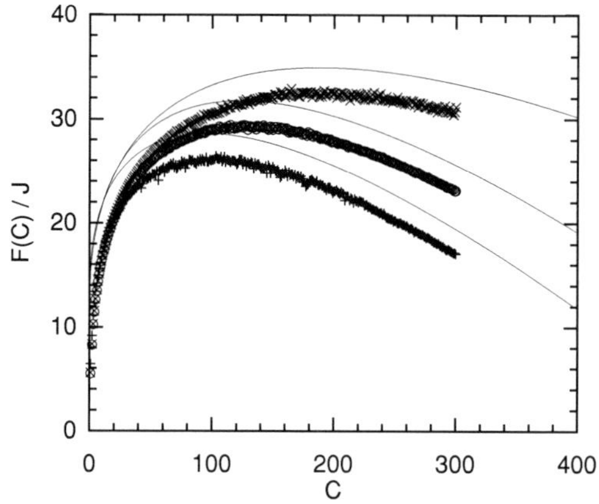

FIGURE 1. Free energy in units of k_BT as a function of cluster size in the 64×64 system, at $\beta J = 0.56$ and $\beta h = 0.06$ ($\times$), $\beta J = 0.58$ and $\beta h = 0.08$ ($\circ$), and $\beta J = 0.53$ and $\beta h = 0.08$ ($+$). The lines represent the Becker-Döring expression Eq. (13), with fitted values $\beta\sigma = 0.93$, 1.00 and 0.88, and $\beta F_0 = 12.3$, 11.9 and 13.4.

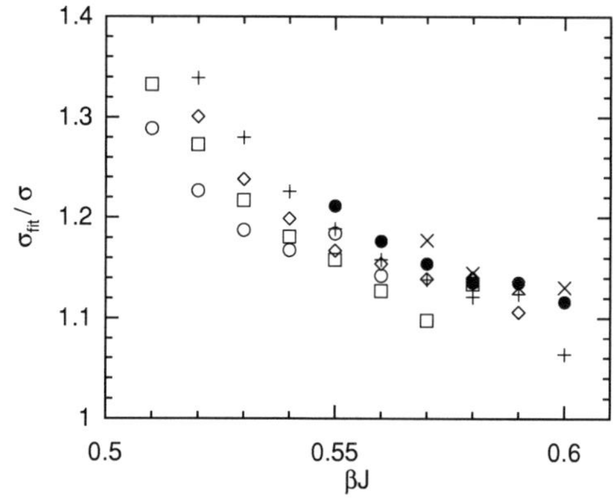

FIGURE 2. Ratio of the surface tension obtained by fitting the free energy curve and the theoretical value given in Eq. (14), as a function of βJ. Different symbols denote different strengths of the external field: $\beta h = 0.05(\circ)$, $\beta h = 0.06(\square)$, $\beta h = 0.07(\diamond)$, $\beta h = 0.08(+)$, $\beta h = 0.09(\bullet)$ and $\beta h = 0.10(\times)$.

The results for $g(\beta, J)$ as function of β are plotted in figure 3.

To arrive at an estimate for the rate $\Gamma_{C+1,C}$ for cluster growth and shrinkage, we assume that the diffusion coefficient neither depends on the external field nor on the shape of the interface (straight or circular), but only on the length of the cluster boundary, for which we use $2\sqrt{\pi C}$, assuming that the shape of the cluster is almost circular and that the increase of area due to the presence of a few vacancies inside the cluster is negligible. This then gives rise to a jump rate

$$\Gamma_{C+1,C} = \frac{g(\beta J)}{4} 2\sqrt{\pi C}. \tag{17}$$

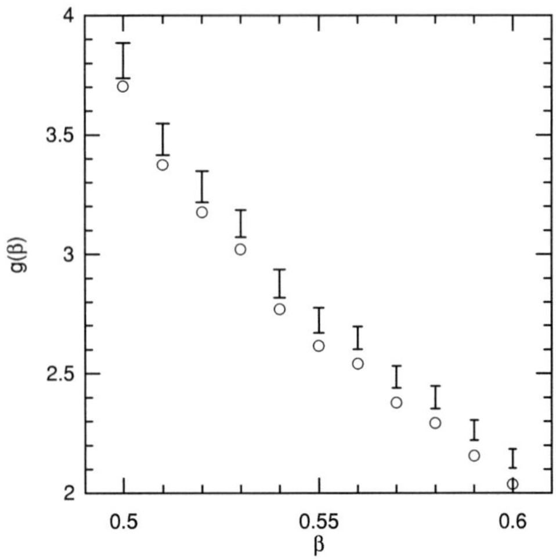

FIGURE 3. Monte Carlo measurements of the diffusion coefficient per interface length $g(\beta J)$, as a function of inverse temperature βJ.

Nucleation rates

To measure nucleation rates, we first bring the system into equilibrium in the presence of a magnetic field. At time $t = 0$, we then reverse all the spins (equivalent to an instantaneous reversal of the magnetic field) so that the system is near its metastable quasi-equilibrium. We measure the time the system needs to reach the magnetization corresponding to the stable equilibrium.

Nucleation has taken place as soon as a cluster has grown well beyond the critical nucleus size. Once that happens, the relatively small systems in our simulations will quickly equilibrate in the state in which most spins align with the external field. Since monitoring the size of the biggest cluster is computationally demanding, we monitor instead the total magnetization, and conclude that nucleation has taken place as soon as it has reached a value corresponding to the stable equilibrium. Figure 4 shows a semi-logarithmic plot of a histogram of all the measured times.

DISCUSSION

We may conclude that for two-dimensional Ising models with spin-flip dynamics classical nucleation theory provides an excellent description of nucleation time distributions, provided a realistic description is used for the free energy of the growing droplets. We determined this free energy from cluster size distributions in equilibrium Monte Carlo simulations and found that it may be fitted well by the Becker-Döring expression, provided one uses surface tensions that are 10 to 20% higher than the surface tension of a bulk interface at zero field.

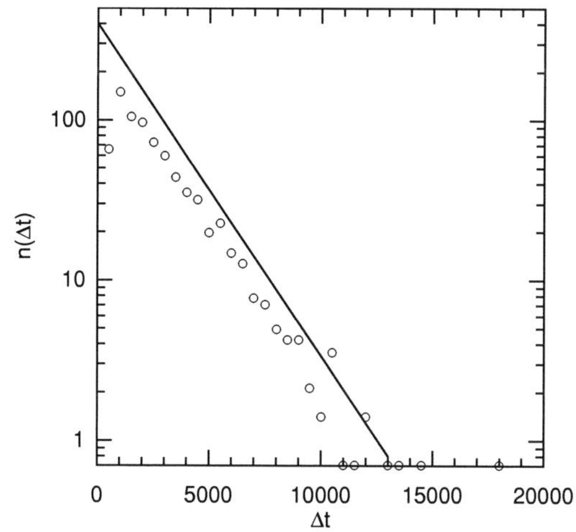

FIGURE 4. Histogram of the time Δt elapsed before nucleation occurs, at inverse temperature $\beta J = 0.54$ and external field strength $h = 0.08$, in a system with 64×64 sites. The straight line is a fit, given by $n(\Delta t) \sim \exp(\Delta t/1990)$.

A point one may question is whether the assumption of independent noninteracting clusters holds even for clusters of small size. This may indeed be doubted, but fortunately, at least for the asymptotic nucleation frequency $\langle \nu N_c \rangle$ this is not really relevant. One may choose to define as clusters only those clusters that have a size $c \geq m_0$, with m_0 chosen such that clusters of this size already are very rare, but still $\exp(\beta F(m_0)) \ll \exp(\beta F(C_x))$. From Eqs. (9) and (12) it then follows that

$$\nu_{m_0} = \frac{\sum_{c=1}^{C_x} N_c}{\sum_{c=m_0}^{C_x} N_c} \nu,$$

since the terms with $m < m_0$ in the first sum in Eq. (9) basically do not contribute. As a consequence of this the asymptotic nucleation rate in the system is independent of the choice of m_0.

In our estimations we have been using a number of assumptions, whose validity is not guaranteed under all conditions:

- Strong fields should modify the diffusion coefficient; this effect is neglected. The freedom to modify the field strength within the metastable region is limited though, and long nucleation times, as seen mostly in real experiments, require weak fields.
- The diffusion coefficient is assumed to be determined by the size of the cluster alone, and is calculated on the assumption that its shape is strictly spherical. This requires that the temperature is not too low, because at very low temperatures the equilibrium shape of the cluster is more square than circular [13] (this however, is an effect that may easily be corrected for without making any basic changes in the theory). On the other hand the temperature should not be too close to the critical temperature for shape fluctuations to be reasonably small. To some extent these

fluctuations are taken into account, since our calculation of the diffusion coefficient is done for a fluctuating interface around a cylinder. But it is by no means certain that the fluctuations of a circular interface are in all aspects comparable to those of the interface around the cylinder.

- It may happen that a cluster splits up or that two clusters merge, corresponding in our theoretical framework to non-zero transition rates $\Gamma_{C,C+i}$ and $\Gamma_{C+i,C}$ with $i > 1$. This effect also is partly accounted for through the numerical determination of the diffusion constant on a cylinder, but especially for larger clusters the difference in geometry may cause additional effects. However, these will only become important in systems that contain a sizable density of clusters of spins aligned with the external field. Hence, also this approximation can be trusted least near the critical point.
- No memory effects are accounted for explicitly. For spin-flip dynamics, memory effects will chiefly be due to the influence of shape fluctuations on the transition rates. Since shape fluctuations on larger length scales will decay only slowly, these may be fairly long lasting effects. Again, our way of determining the diffusion coefficient will take many of these effects into account implicitly, but memory effects involving large shape fluctuations may be different for the present cluster geometry. For magnetization conserving dynamics (consisting e.g. of local spin exchanges) much stronger memory effects exist due e.g. to the effect that a spin that evaporates from a large cluster has high probability of condensing back to it soon.

Under conditions in which the effects above are negligible, our theoretical framework is capable of estimating nucleation rates with an accuracy in the range of 20%. The small systematic overestimation by about 10% of the nucleation time by theory may have several causes. The radius of a cluster will be slightly larger than our estimate because especially a large cluster will typically contain a few holes in its interior. Since the equilibrium magnetization is always larger than 0.92 for the temperatures studied, the density of vacancies in the cluster is less than 4%, and hence the increase in interface length due to holes is less than 2%. Also the assumptions that the diffusion coefficient is independent of the magnetic field and of the orientation of the interface may be not entirely correct. At low temperatures a diagonal interface is much more mobile than a straight one, but at the fairly elevated temperatures studied here one would not expect a large orientation dependence. Further there could be effects from the possibility of cluster splittings and mergings, though some of these are accounted for through our numerical determination of the diffusion constant on a cylinder.

The results of this paper have been reported before in [9].

REFERENCES

1. P. G. Debenedetti, *"Metastable Liquids"*, (Princeton University Press, Princeton, 1996).
2. P. A. Rikvold and B. M. Gorman, Annual Reviews of Computational Physics I, edited by D. Stauffer (World Scientific, Singapore, 1994), pp. 149-191.
3. M. Volmer and A. Weber, Z. Phys. Chem. **119**, 277 (1926).
4. Z. Farkas, Z. Phys. Chem. A **125**, 236 (1927).

5. R. Becker and W. Döring, Ann. Phys. (Leipzig) **24**, 719 (1935).
6. J. B. Zeldovich, Acta Physicochim. **18**, 1 (1943).
7. N. Metropolis, A. W. Rosenbluth, M. N. Rosenbluth, A. H. Teller, and E. Teller, J. Chem. Phys. **21**, 1087 (1953).
8. P. Hänggi, P. Talkner and M. Borkovec, Rev. Mod. Phys. **62**, 251 (1990).
9. K. Brendel, G.T. Barkema and H. van Beijeren, Phys. Rev. E **71** 031601 (2005). Phys. **62**, 251 (1990).
10. M. E. J. Newman and G. T. Barkema, *"Monte Carlo Methods in Statistical Physics"*, Oxford University Press, Oxford, 1999.
11. R. C. Tolman, J. Chem. Phys. **17**, 333 (1949).
12. V. A. Shneidman, K. A. Jackson and K. M. Beatty, J. Chem. Phys. **111**, 6932 (1999).
13. J. E. Avron, H. van Beijeren, L. S. Schulman and R. K. P. Zia, J. Phys. A **15**, L81 (1982).
14. L. Onsager, Phys. Rev. **65**, 117 (1944).
15. K. Brendel, G. T. Barkema and H. van Beijeren, Phys. Rev. E **67**, 026119.
16. K. Brendel, G.T. Barkema and H. van Beijeren, to be published.

Unsolved Problems of Surface Diffusion at Low Friction

Katja Lindenberg*, A. M. Lacasta†, J. M. Sancho** and A. H. Romero‡

*Department of Chemistry and Biochemistry 0340, and Institute for Nonlinear Science, University of California, San Diego, La Jolla, California 92093-0340, USA
†Departament de Física Aplicada, Universitat Politècnica de Catalunya, Avinguda Doctor Marañon 44, E-08028 Barcelona, Spain
**Departament d'Estructura i Constituents de la Matèria, Facultat de Física, Universitat de Barcelona, Diagonal 647, E-08028 Barcelona, Spain
‡Cinvestav-Queretaro, Libramiento Norponiente No 2000, Real de Juriquilla, 76230 Queretaro, Qro, Mexico

Abstract. Surface transport and diffusion at low damping reveals a number of unexplained or at least not systematically explained behaviors. Here we present two problems that require further study for a full understanding. One involves motion on a random surface, the unresolved issue being the parameter regimes leading to subdiffusive, diffusive, and superdiffusive motions at intermediate times. The other involves the temperature dependence of maximal diffusion on a periodic surface in the presence of a constant external force.

Keywords: Friction, Surfaces, Langevin equation, Superdiffusion
PACS: 05.40.-a,68.35.Fx,68.43.Jk

INTRODUCTION

Transport and diffusion of atoms, molecules, molecular clusters, and colloidal particles on periodic or random surfaces continues to be a problem of ubiquitous interest because there are constantly new and technologically important experiments that involve such motions, and because there are still many unanswered theoretical issues associated with these problems in spite of their long history ([1, 2, 3, 4, 5, 6] and extensive references therein). This being a volume associated with a conference on unsolved problems involving noise, we focus on some transport and diffusion problems on surfaces that lead to as yet unresolved issues.

Theoretical analyses of the motion of particles on surfaces would ideally rely on *ab initio* or molecular dynamics models in which the particles and the surfaces are represented in their full microscopic detail. This is as yet impossible over experimentally relevant time scales. A compromise is to rely on mesoscopic phenomenological paradigms in which one focuses on a few dynamical variables that describe the particle of interest. Its interaction with the surface is represented by a surface potential, a frictional force, and a stochastic component to represent thermal effects and other degrees of freedom not microscopically considered. External forces for the study of transport processes can also be included.

Among the mesoscopic models of choice in the longevous history of the subject are *overdamped* Langevin equations. The generic form for such models to describe the

CP800, *Unsolved Problems of Noise and Fluctuations: UPoN 2005*
edited by L. Reggiani, C. Pennetta, V. Akimov, E. Alfinito, and M. Rosini

evolution of a dynamical variable $x(t)$ (e.g. the displacement of a particle in a one-dimensional system) would be $\gamma\dot{x} = -V'(x) + \xi(t) + F(x,t)$, where a dot denotes a derivative with respect to time t, a prime is a derivative with respect to the argument, $\xi(t)$ is δ-correlated Gaussian noise, $\langle\xi_i(t)\xi_j(t')\rangle = 2\gamma k_B T\delta_{ij}\delta(t-t')$ where T is the temperature, and $F(x,t)$ is an external force. The damping coefficient γ is usually not written explicitly as we have done because one can of course simply divide through by it. The equation represents an overdamped system because it is missing the inertial contribution proportional to $\ddot{x}$ that would appear in Newton's Law. The associated model of choice on a discrete lattice is a nearest neighbor random walk, and the probabilistic description based on the density $P(x,t)$ that the particle is at x at time t is the Fokker-Planck equation. The equation with full inertial contributions from which the overdamped equation emerges in the large γ limit is $m\ddot{x} = -\gamma\dot{x} - V'(x) + \xi(t) + F(x,t)$, where m is the particle mass (set to unity henceforth). This equation is not analyzed nearly as often as is the overdamped case because the second time derivative seriously complicates the mathematics.

The overdamped model and its discrete counterpart (and its higher dimensional versions) are of course not able to reproduce all observations, especially those that can not be described as one would a nearest neighbor random walk or simple diffusion in a potential. This has long been recognized, and extensions of the models have been the subject of an extensive literature. A few examples involve the extension of random walk models to include jumps to neighbors beyond the nearest (and even to very distant ones), persistent behavior that causes a particle to preferentially continue jumping along a previously chosen direction, and the generalization of $\xi(t)$ to non-white and/or non-Gaussian noise. In recent years there have been a number of experimental observations of particles on surfaces that seem to move over long distances in one direction before moving in another (see references in [3]), and a number of these generalizations have been invoked in an attempt to explain the observations. It is our thesis that many of these observations do not require special modeling of the noise or other memory effects, but can be explained by including the inertial contribution and exploring the small-γ regime.

In the course of these investigations, we have encountered some results that lead to open questions to be explored, and here we exhibit two of them. One is related to the nature of the diffusive process on random surfaces in the absence of external forces, and will be presented in the next Section. The other concerns a temperature dependence in the diffusion on periodic surfaces in the presence of a constant external force. This will be presented in the subsequent Section. We close with a brief summary.

SUPERDIFFUSION?

We consider a two-dimensional surface in the absence of external forces. In conveniently scaled variables the equations are (with independent noises in different directions)

$$
\begin{aligned}
\ddot{x} &= -\gamma\dot{x} - \frac{\partial}{\partial x}V(x,y) + \xi_x(t), \\
\ddot{y} &= -\gamma\dot{y} - \frac{\partial}{\partial y}V(x,y) + \xi_y(t),
\end{aligned} \tag{1}
$$

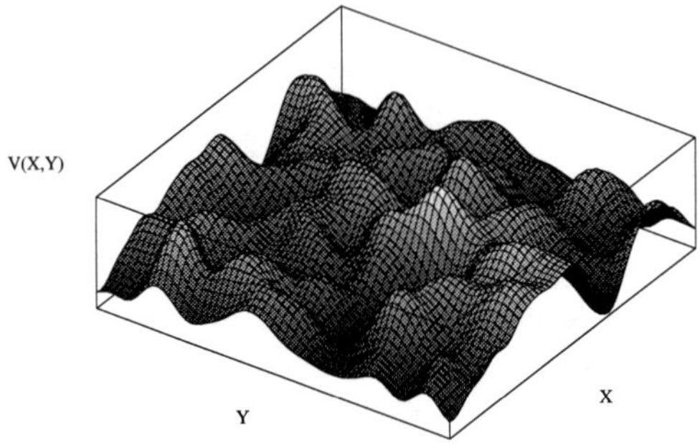

FIGURE 1. Typical random potential.

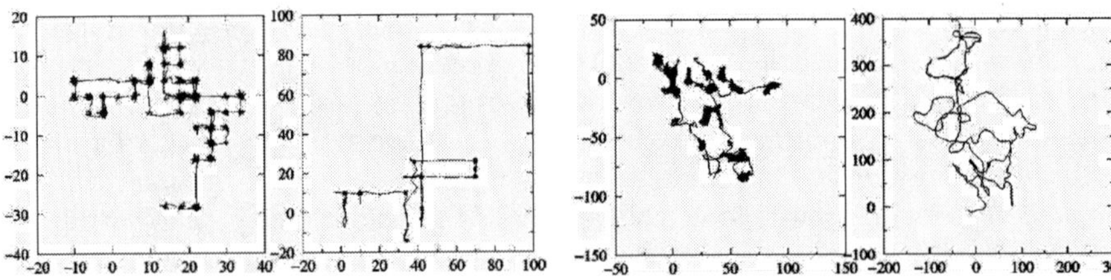

FIGURE 2. Left two panels: typical trajectories for high damping (far left) and low damping (second from left) on a periodic lattice. Right two panels: typical trajectories for high damping (second from right) and low damping (far right) on a random lattice. The average potential heights is the same for both lattices. For detailed parameter values see [3].

where the potential $V(x,y)$ is either periodic or random [2, 3]. If random, it is characterized by a distribution, which we choose to be a Gaussian, with a correlation function that we choose to be exponential, $\langle V(\mathbf{x})V(\mathbf{x}')\rangle = g(\mathbf{x}-\mathbf{x}')$ with

$$g(\mathbf{x}-\mathbf{x}') = \frac{\varepsilon}{2\pi\lambda^2} e^{-|\mathbf{x}-\mathbf{x}'|^2/2\lambda^2}, \tag{2}$$

and $\mathbf{x} = (x,y)$, $\mathbf{x}' = (x',y')$. The correlation function is parameterized by the intensity ε (specified in terms of the temperature) and the characteristic length λ (we set $\lambda = 4$ in our simulations). In Fig. 1 we show a small portion of such a random potential. In Fig. 2 we show typical trajectories for high and low damping, both for a periodic potential (for comparison) and for a random potential. The situation of interest here is that of the rightmost panel in the figure.

In a periodic potential we have shown that, for a fixed set of potential parameters and temperature, the motion ranges from diffusive to superdiffusive over many intermediate decades of time as one lowers the damping [2, 3]. That is, the motion over many decades of time at sufficiently low damping is characterized by a mean square displacement

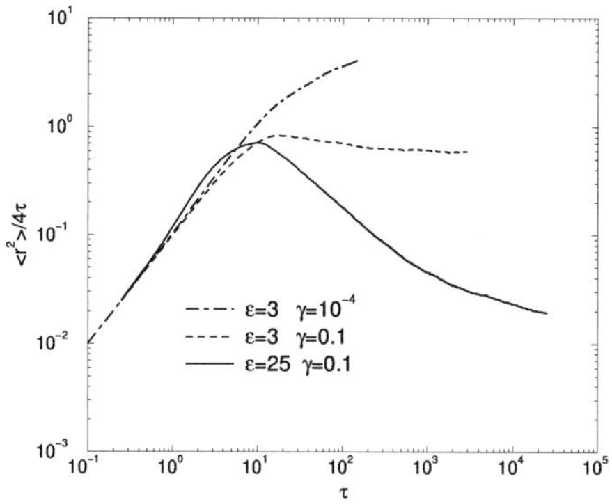

FIGURE 3. $\langle r^2 \rangle / 4\tau$ for three sets of values of the random potential intensity parameter ε and the damping coefficient γ as defined in [2, 3, 7].

proportional to t^{α} with $1 < \alpha \leq 2$ (at asymptotic times the motion becomes diffusive, that is, $\alpha \to 1$). We analyzed this motion in further detail and showed that the probability distribution function $P(r, \tau)$ of particle displacements r at time τ in the intermediate time regime shows many characteristics typical of Lévy walks. We stress that this behavior is caused entirely by inertial effects that become increasingly important as damping decreases, since the fluctuations in the model are white and Gaussian.

In a random potential we conjecture that the possible range of behaviors extends all the way from subdiffusive ($\alpha < 1$) to diffusive and superdiffusive [2, 3, 7]. It is fairly easy and not surprising to find parameter values that lead to subdiffusive behavior, since particles may get trapped for very long time intervals in extremely deep potential wells from which it is difficult to emerge [7, 8]. It is also fairly easy to find parameters that lead to diffusive behavior. The uncertain issue (*unsolved problem*) is the question of occurrence, parameter range, and time range of superdiffusion, which is associated with long open pathways that can occur with some probability in a random system. While we have not found a systematic set of parameter values such that the variation of a single parameter (e.g. damping) takes one through the subdiffusive-diffusive-superdiffusive behaviors, we have observed all three behaviors [7], as shown in Fig. 3.

TEMPERATURE DEPENDENCE

Our second *unsolved problem* deals with a particular temperature dependence in the case of transport over a *periodic* surface subject to a constant external force. The potential in suitably scaled units now is (again, in our simulations we set $\lambda = 4$)

$$V(x,y) = \frac{1}{2}\left[\cos\left(\frac{2\pi x}{\lambda}\right) + \cos\left(\frac{2\pi y}{\lambda}\right)\right]. \tag{3}$$

A constant force F_0 is applied along the x direction. We have explored the dependence of the mean velocity $v \equiv \lim_{t\to\infty}(\langle x(t)\rangle / t)$ and the diffusion coefficient $D \equiv$

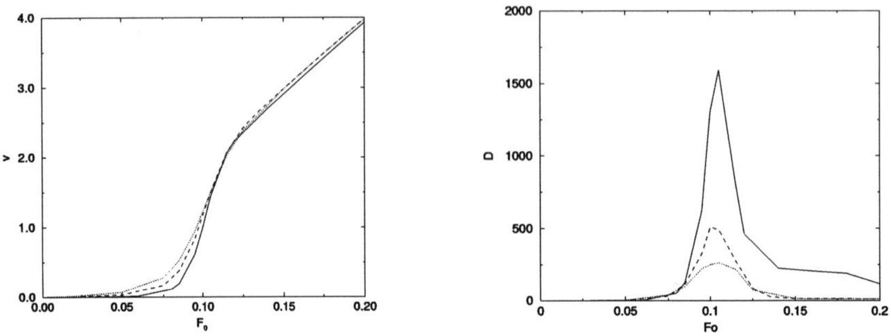

FIGURE 4. Mean velocity (left panel) and diffusion coefficient (right panel) vs F_0 for three values of the temperature: $\varepsilon = 0.15$ (solid curve), $\varepsilon = 0.20$ (dashed curve), and $\varepsilon = 0.25$ (dotted curve). The friction coefficient is fixed at $\gamma = 0.05$.

$\lim_{t\to\infty}(\langle[x(t) - \langle x(t)\rangle]^2/2t)$ on damping, force, and temperature [4]. The unexplained behavior concerns the behavior observed in Fig. 4.

In the left panel we present the velocity vs F_0 curves for three values of the temperature for a system with friction coefficient $\gamma = 0.05$. The panel on the right shows the associated diffusion coefficient for each of the three temperatures. The transition behavior observed in the velocity has been discussed in [4], and the principal effect of lowering the temperature is to sharpen the transition region, essentially without moving its location F_c ($F_c \sim 0.1$ for $\gamma = 0.05$). However, the effect on the diffusion coefficient is far more pronounced. While the location F_c of the peak is essentially temperature independent, the peak *grows* with *decreasing* temperature, that is, diffusion is *stronger* as temperature decreases. This behavior is consistent with a transition between locked and running states, as described in [4].

A plot of the maximum diffusion as a function of the temperature reveals the strongly nonlinear dependence

$$D_{max} \sim T^{-3.5}, \tag{4}$$

that is, the diffusion coefficient seems to grow without bound as temperature is lowered (in the overdamped system one finds that $D_{max} \sim T^{1/3}$ [4]). An interesting scaling behavior is observed if we plot the product $\varepsilon^{3.5}D$ as a function of $(F_0 - F_c)/\varepsilon^{1.5}$, as done in Fig. 5. With this scaling, all the curves in the right panel of Fig. 4 collapse onto the same curve. We stress that the exponents 3.5 and 1.5 of ε are simply numerical fits (and not necessarily the very best ones at that), although the divergent increase of D with decreasing temperature is supported by the simulations for the temperatures studied. In any case, the result (4) as well as the scaling behavior in Fig. 5 (or ones with even more accurate exponents) are the theoretically *unsolved problems* of this section.

SUMMARY

We have presented two unsolved problems involving noise. Both have to do with diffusion on surfaces when damping is low. One concerns motion over a random surface:

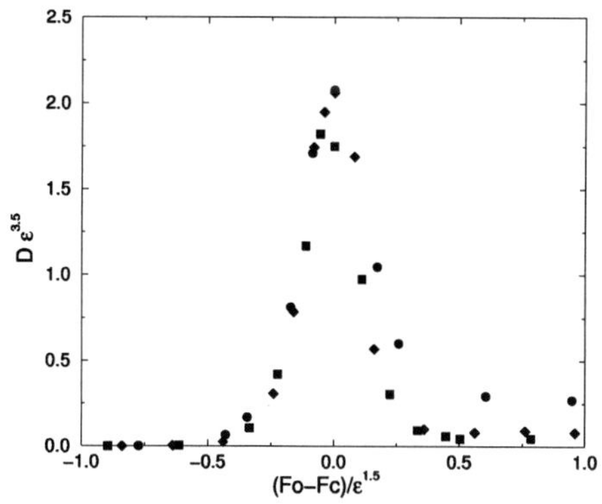

FIGURE 5. Scaled diffusion for $\gamma = 0.05$ for various temperatures. Dots: $\varepsilon = 0.15$; squares: $\varepsilon = 0.21$; diamonds: $\varepsilon = 0.25$.

while it is apparent that the motion may be subdiffusive, diffusive, or superdiffusive at intermediate times for appropriate parameter values, there is as yet no systematic understanding of the conditions that lead to each of these behaviors. The other concerns the temperature dependence of the maximum of the diffusion coefficient on a periodic lattice when there is an external force along a crystallographic axis. This temperature dependence, as well as an apparent scaling law associated with it, remain to be explained.

ACKNOWLEDGMENTS

This work was supported by the MCyT (Spain) under project BFM2003-07850, by the Engineering Research Program of the Office of Basic Energy Sciences at the U. S. Department of Energy under Grant No. DE-FG02-04ER46179, and by a grant from the University of California Institute for México and the United States (UC MEXUS) and the Consejo Nacional de Ciencia y Tecnología de México (CoNaCyT). A.H.R. acknowledges support from CONACyT-Mexico through Grant J-42647-F.

REFERENCES

1. H. Risken, 1989 *The Fokker–Planck Equation*, Springer Verlag, New York, 1989, Chapter 11.
2. J. M. Sancho, A. M. Lacasta, K. Lindenberg, I. M. Sokolov, and A. H. Romero, *Phys. Rev. Lett.* **92**, 250601 (2004).
3. A. M. Lacasta, J. M. Sancho, A. H. Romero, I. M. Sokolov, and K. Lindenberg, *Phys. Rev. E* **70**, 051104 (2004).
4. K. Lindenberg, A. M. Lacasta, J. M. Sancho, and A. H. Romero, *New J. Phys.* **7**, 29 (2005).
5. A. M. Lacasta, J. M. Sancho, A. H. Romero, and K. Lindenberg, *Phys. Rev. Lett.* **94**, 160601 (2005).
6. K. Lindenberg, A. M. Lacasta, J. M. Sancho, and A. H. Romero, *Proceedings of SPIE* **5845**, XXX (2005).
7. J. M. Sancho, A. M. Lacasta, K. Lindenberg, I. M. Sokolov, and A. H. Romero, *Phys. Rev. Lett.* **94**, 188902 (2005).
8. J-D. Bao and Y. Zhou, *Phys. Rev. Lett.* **94**, 188901 (2005).

Open Problems In Noise In NEMS

Antti-Pekka Jauho*, Andrea Donarini†, Christian Flindt** and Tomáš Novotný‡,§

*NanoDTU, MIC – Department of Micro and Nanotechnology, Technical University of Denmark, DTU - Building 345 East, DK - 2800 Kongens Lyngby, Denmark
†Institut I – Theoretische Physik, Universität Regensburg, D-93040 Regensburg, Germany
**NanoDTU, MIC - Department of Micro and Nanotechnology, Technical University of Denmark, DTU - Building 345 East, DK - 2800 Kongens Lyngby, Denmark
‡Nano-Science Center, University of Copenhagen, Universitetsparken 5, DK - 2100 Copenhagen Ø, Denmark
§Department of Electronic Structures, Faculty of Mathematics and Physics, Charles University, Ke Karlovu 5, 121 16 Prague, Czech Republic

Abstract. Nanoelectromechanical systems (NEMS) is a new technology which achieves its functionality by combining electrical transport with mechanical motion of nanoscopic bodies. The measured current-voltage characteristics does not always yield sufficient information to uniquely identify the underlying charge transport mechanism, and thus different theoretical models may result in indistinguishable predictions for the IV-characteristics. Therefore, the measurement of noise, and even the full counting statistics is viewed as an important step towards a more complete understanding of the dynamics of NEMS devices. While significant progress has been made in recent years in the theory of noise in NEMS, several open issues and challenges remain, both concerning the appropriate quantum mechanical equations to be used, their simplified semiclassical counterparts, and, finally, their numerical evaluation. We illustrate some of these issues using a quantum mechanical charge shuttle as a bench-marking system.

Keywords: Micro- and nano-electromechanical systems and devices, Noise processes and phenomena, Coulomb blockade, single-electron tunneling
PACS: 85.85.+j, 72.70.+m,73.23.Hk

INTRODUCTION

A generic example of a nanoelectromechancial (NEMS) device is given by the *charge shuttle* (originally proposed by Gorelik *et al.* [1]): a movable single-electron device, working in the Coulomb blockade regime, which can exhibit regular charge transport, where one electron within each mechanical oscillation cycle is transported from the source to the drain – see Figure 1 for a schematic illustration. The shuttling transition can be induced by varying a control parameter, such as the applied bias, or the mechanical damping. Since its introduction, the charge shuttle has inspired a large number of theoretical papers (see, *e.g.*, Refs. [2, 3, 4, 5, 6, 7, 8, 9]). To the best of our knowledge, an experimental verification of these ideas is not yet available, though significant progress has been made, such as the *driven* shuttle of Erbe *et al.* [10], or the nanopillars studied by Scheible and Blick [11]. In the present communication, we use the quantum shuttle, *i.e.* a device where also the mechanical motion needs to be quantized (the physical condition for this to happen is $\lambda \simeq x_{zp}$, where λ is the tunneling length describing the exponential decay of the wave-functions into vacuum, and x_{zp} is the quantum-mechanical zero-point

CP800, *Unsolved Problems of Noise and Fluctuations: UPoN 2005*
edited by L. Reggiani, C. Pennetta, V. Akimov, E. Alfinito, and M. Rosini

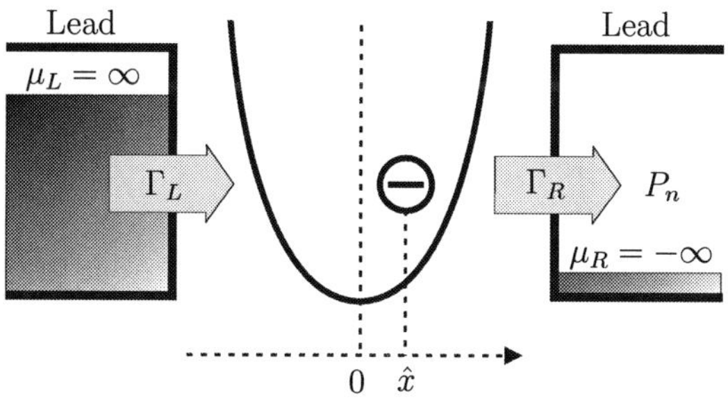

FIGURE 1. Schematic picture of the quantum shuttle. The shuttle consists of a nanosized grain moving in a harmonic potential between two leads. A high bias between the leads drives electrons through the grain. Strong Coulomb blockade prohibits double occupancy of the grain.

amplitude), to illustrate how far the theory has progressed, and where difficulties begin. While our examples are given in the context of one specific physical realization, we believe that they possess certain generic features which *any* theory addressing noise in NEMS must consider, at least in the quantum limit.

GENERALIZED MASTER EQUATION

We first state the starting point of the most general calculations reported so far. The language is that of quantum optics, where the central object is the "system" density matrix, which is obtained from the "total" density matrix by projecting out the environmental degrees of freedom. The system density matrix evolves in time according to the Liouville equation, $\dot{\hat{\rho}} = \mathscr{L}\hat{\rho}$, where the Liouvillean superoperator $\mathscr{L}$ contains the physics. Thus, in the case of a quantum shuttle, the system degrees of freedom are the charge states of the movable nanoscopic device (in the simplest case of spinless electrons there are just two: the empty state (denoted by $|0\rangle$), and the filled state (denoted by $|1\rangle$)), and the center-of-mass coordinate of the movable body, denoted by $\hat{x}$, and which will be quantized in due course. Following the ideas of Gurvitz and Prager [12], it is advantageous to work with the *charge-resolved* density matrix $\hat{\rho}_{ii}^{(n)}$ $(i = 0, 1)$, where n is the number of electrons that have passed through the system in time-span t. This object satisfies

$$\begin{aligned}\dot{\hat{\rho}}_{00}^{(n)}(t) =& \frac{1}{i\hbar}[\hat{H}_{\text{osc}}, \hat{\rho}_{00}^{(n)}(t)] - \frac{\Gamma_L}{2}\{e^{-\frac{2\hat{x}}{\lambda}}, \hat{\rho}_{00}^{(n)}(t)\} + \Gamma_R e^{\frac{\hat{x}}{\lambda}} \hat{\rho}_{11}^{(n-1)}(t) e^{\frac{\hat{x}}{\lambda}} + \mathscr{L}_{\text{damp}} \hat{\rho}_{00}^{(n)}(t),\\ \dot{\hat{\rho}}_{11}^{(n)}(t) =& \frac{1}{i\hbar}[\hat{H}_{\text{osc}} - eE\hat{x}, \hat{\rho}_{11}^{(n)}(t)] - \frac{\Gamma_R}{2}\{e^{\frac{2\hat{x}}{\lambda}}, \hat{\rho}_{11}^{(n)}(t)\} + \Gamma_L e^{-\frac{\hat{x}}{\lambda}} \hat{\rho}_{00}^{(n)}(t) e^{-\frac{\hat{x}}{\lambda}} + \mathscr{L}_{\text{damp}} \hat{\rho}_{11}^{(n)}(t),\end{aligned} \tag{1}$$

where the commutator terms describe the coherent time evolution, the Γ-terms describe the coupling between the movable part and the contacts with a characteristic exponential position-dependence, and, finally, the damping of the oscillating body (with oscillation frequency ω) is embodied in a term having the generic structure

$$\mathscr{L}_{\text{damp}}\hat{\rho} = -\frac{i\gamma}{2\hbar}[\hat{x},\{\hat{p},\hat{\rho}\}] - \frac{\gamma m\omega}{\hbar}(N+1/2)[\hat{x},[\hat{x},\hat{\rho}]]. \tag{2}$$

The reason for studying $\hat{\rho}^{(n)}$ is twofold. First, *if* it can be computed, the central physical quantities, such as the mean current and the zero-frequency shot noise spectrum follow immediately [13]:

$$I = e\frac{d}{dt}\sum_n nP_n(t)\Big|_{t\to\infty} = e\sum_n n\dot{P}_n(t)\Big|_{t\to\infty}, \tag{3}$$

$$S(0) = 2e^2\frac{d}{dt}\left[\sum_n n^2 P_n(t) - \left(\sum_n nP_n(t)\right)^2\right]\Big|_{t\to\infty}, \tag{4}$$

where $P_n(t) = \mathrm{Tr}_{\mathrm{osc}}[\hat{\rho}_{00}^{(n)}(t) + \hat{\rho}_{11}^{(n)}(t)]$. Second, efficient numerical techniques have become available and many results have been reported in the recent literature [3, 6, 7, 8, 14]. Among the physical results obtained, here we want to point out the giant enhancement of noise (Fano factor around 600) that was found in a so-called coexistence regime, where the system slowly switches between an orderly mode of transport (the shuttling regime), and an irregular mode of transport (incoherent tunneling). An interpretation of these results in terms of a general theory for switching in bistable systems [15, 16], and a discussion of the rates at which the switching takes place will soon be reported elsewhere [17]. The numerical issues will be addressed below, here we first discuss the inherent limitations of Eq. (1).

Problem (1): The Markov approximations. At present, we are not aware of numerical techniques that would allow an efficient solution of a non-Markovian time-evolution of the density-matrix with the degree of complexity that can be expected in NEMS. As long as this situation prevails, the Markov limit appears unavoidable. To obtain the Markovian form of Eq. (1) several important approximations had to be made. For example, it is *not* possible to derive a Markovian damping term, which simultaneously leads to the correct equations of motion for the coordinate and the momentum, guarantees a positive semidefinite density matrix, and leads to a solution approaching thermal equilibrium state for asymptotic times [18]. In our numerical runs we experimented with various possible damping terms and found that they do not yield very different results, nor are the results ever unphysical. Nevertheless, an element of uncontrollability has entered the theory, and ideally should be eliminated. Next, the damping term is evaluated within the weak-coupling approximation. In the present case several issues should merit further investigation: (i) The additivity of the thermal baths - one bosonic (causing the external dissipation controlled by γ), the other fermionic (the two contact Fermi seas), becomes an issue for multi-dot arrays [2, 7]. (ii) In reality, the confining potential is not harmonic, which would lead to an unequally spaced energy level spectrum for the oscillator, which may complicate enormously the algebra associated with the weak-coupling procedure. Yet another assumption is the high-bias limit (as indicated in Fig. 1): without this the Markov property would again be lost. This leads to the unfortunate consequence that no true IV-curves can be evaluated based on Eq. (1): the electric field parameter E is not connected to the chemical potential difference (which is largest remaining energy scale of the problem, after elimination of the infinite Hubbard U forcing the strong Coulomb blockade).

NUMERICAL EVALUATION

We next outline the numerical solution of Eq.(1). All relevant quantities are evaluated in the harmonic oscillator basis, and then the basis is truncated at some suitable, sufficiently large number N. The central numerical task is to evaluate the null eigenvector of the Liouvillean superoperator, because this object equals the steady-state density matrix, $\mathscr{L}\hat{\rho}^{\text{stat}} = 0$, from which physical quantities follow. The cut-off is determined empirically: once the physical quantities have converged, there is no need for further increase in N. Unfortunately, the resulting matrices sizes scale as $(m \times N^2) \times (m \times N^2)$, where m is the dimension of the electronic subspace (for example, $m = 2$ for a spinless single level, or $m = 10$ for the triple dot studied in [2, 7]), and since in practice one often requires $N > 10^2$, one very soon ends up dealing with very large matrices. By using a technique based on Arnoldi iteration (a detailed account is given in Ref. [7]) we were able to routinely solve systems with N of the order of few tens, but going to even larger systems (and this is necessary if one wants to study the experimentally most relevant semiclassical limit) posed serious problems. We thus formulate the next open problem.

Problem (2): Numerics of large matrices. The convergence of the Arnoldi scheme crucially depends on finding a suitable preconditioner. For certain subtasks, such as the calculation of the stationary current, we believe to have found it (the Sylvester part of the Liouvillean [7]). However, for the calculation of noise (in particular, the frequency dependent noise [14]), this preconditioner did not turn out to be very helpful, and convergence was tedious. Thus: can one find better preconditioners? Or, should one develop alternative numerical schemes - such as directly propagating the formal solution of the Liouville equation, $\hat{\rho}(t) = \exp[i\mathscr{L}t]\hat{\rho}(0)$? Here we strongly believe that a close collaboration with numerical analysts would be helpful.

KRAMERS EQUATION

We have in our papers [3, 6, 7] promoted the use of the Wigner function, defined as

$$W_i(q,p,t) = \frac{1}{2\pi\hbar}\int_{-\infty}^{+\infty} d\xi \left\langle q - \frac{\xi}{2}\right| \hat{\rho}_{ii}(t) \left| q + \frac{\xi}{2}\right\rangle \exp\left(\frac{ip\xi}{\hbar}\right), \tag{5}$$

as a strong interpretational tool to distinguish between the various charge transport regimes the shuttle may exhibit. For example, the Wigner function W_0 gives the empty-dot "probability density" in the phase-space (q,p) with the caveat that it may obtain negative values. Instead of studying the Generalized Master equation Eq.(1) one can also to study directly the equation of motion for the Wigner function itself. This reads

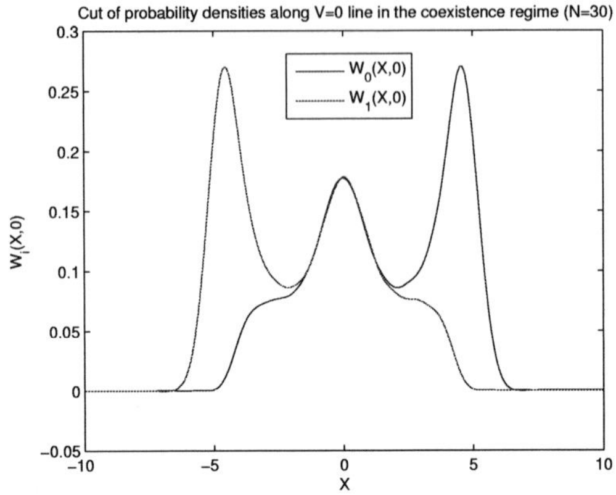

FIGURE 2. The cut through the two probability densities (occupied $W_1(x,0)$ and empty $W_0(x,0)$) along the line $v=0$ for the parameters $\Lambda \equiv \lambda\omega/v_0 = 1$, ($v_0$ is the thermal velocity) $\gamma = 0.0035\omega, \Gamma_{L,R} = 0.015\omega, D \equiv d\omega/v_0 = 0.05$ which corresponds to the coexistence regime. The size of the retained basis is $N=30$.

[19]

$$
\begin{aligned}
\frac{\partial W_0}{\partial t} &= \left[m\omega^2 q\frac{\partial}{\partial p} - \frac{p}{m}\frac{\partial}{\partial q} + \gamma\frac{\partial}{\partial p}p + \gamma m\hbar\omega\left(n_B+\frac{1}{2}\right)\frac{\partial^2}{\partial p^2}\right]W_0 \\
&\quad + \Gamma_R e^{2q/\lambda}W_1 - \Gamma_L e^{-2q/\lambda}\sum_{n=0}^{\infty}\frac{(-1)^n}{(2n)!}\left(\frac{\hbar}{\lambda}\right)^{2n}\frac{\partial^{2n}W_0}{\partial p^{2n}} \qquad (6)\\
\frac{\partial W_1}{\partial t} &= \left[m\omega^2 (q-d)\frac{\partial}{\partial p} - \frac{p}{m}\frac{\partial}{\partial q} + \gamma\frac{\partial}{\partial p}p + \gamma m\hbar\omega\left(n_B+\frac{1}{2}\right)\frac{\partial^2}{\partial p^2}\right]W_{11} \\
&\quad + \Gamma_L e^{-2q/\lambda}W_0 - \Gamma_R e^{2q/\lambda}\sum_{n=0}^{\infty}\frac{(-1)^n}{(2n)!}\left(\frac{\hbar}{\lambda}\right)^{2n}\frac{\partial^{2n}W_1}{\partial p^{2n}}, \qquad (7)
\end{aligned}
$$

where the explicit appearance of Planck's constant allows one to consider the semiclassical limit. In the limit $\hbar \to 0$ one finds (here we prefer to use the variables $x = q + d/2$ ($d = eE/m\omega^2$ is the equilibrium position of the charged shuttle), and $v = p/m$)

$$
\begin{aligned}
\partial_t W_0(x,v;t) &= -v\,\partial_x W_0(x,v;t) + \omega^2(x+d/2)\partial_v W_0(x,v;t) + \gamma\partial_v v W_0(x,v,t) + \\
&\quad + \gamma k_B T/m\,\partial_v^2 W_0(x,v,t) - \Gamma_L e^{-\frac{2x}{\lambda}}W_0(x,v;t) + \Gamma_R e^{\frac{2x}{\lambda}}W_1(x,v;t)\,, \qquad (8)\\
\partial_t W_1(x,v;t) &= -v\,\partial_x W_1(x,v;t) + \omega^2(x-d/2)\partial_v W_1(x,v;t) + \gamma\partial_v v W_1(x,v,t) + \\
&\quad + \gamma k_B T/m\,\partial_v^2 W_1(x,v,t) - \Gamma_R e^{\frac{2x}{\lambda}}W_1(x,v;t) + \Gamma_L e^{-\frac{2x}{\lambda}}W_0(x,v;t)\,. \qquad (9)
\end{aligned}
$$

Here, $W_i(x,v,t)$ is the probability density of the shuttle being either charged ($i=1$) or empty ($i=0$) at the position x with the velocity v. Equation (9) can be interpreted as a matrix generalization of the scalar Kramers equation, which for a Brownian particle in

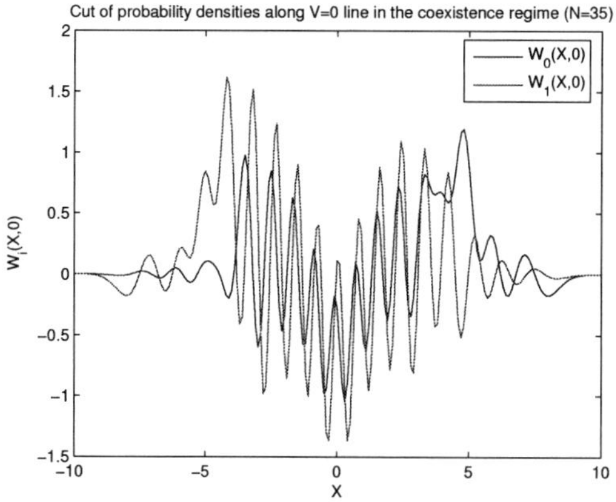

FIGURE 3. The cut through the two probability densities (occupied $W_1(x,0)$ and empty $W_0(x,0)$) along the line $v = 0$ for the same parameters as in Fig. 2 but for $N = 35$.

a potential $V(x)$ reads [20]

$$\partial_t W(x,v;t) = -v\,\partial_x W(x,v;t) + \frac{V'(x)}{m}\partial_v W(x,v;t) + \gamma\partial_v v W(x,v,t) + \gamma\frac{k_B T}{m}\partial_v^2 W(x,v,t). \tag{10}$$

Making use of the fact that v and ∂_v occur only in linear or quadratic form, it is possible to recast the solution of this equation to a continued fraction form, with an expected rapid convergence (the details can be found in Ref.[20]). It is then natural to ask whether the procedure can be generalized to a matrix problem, such as Eq.(9). We have attempted this procedure, but only with limited success, as we shall now discuss. Figure 2 shows an example of the computed probability distributions in a so-called co-existence regime [6], where the shuttle simultaneously exhibits shuttling transport (corresponding to the maxima at finite x), and incoherent tunneling (corresponding to the maximum at $x = 0$). In this figure matrix size entering the continued fraction scheme was (arbitrarily) truncated at $N = 30$.

The seemingly tranquil picture is strongly disturbed if one moves just a bit further, and attempts $N = 35$. The results are wildly oscillating, as evidenced in Fig. 3, and clearly unphysical, since the classical probability distribution cannot obtain negative values. We have attempted to go beyond the most naive implementation, which led to Figures 2 and 3, but have never succeeded in going beyond $N \simeq 100$, which is needed for low values of the damping parameter γ. An example of the results obtained for the physical quantities is given in Fig. 4, which compares the results obtained with matrix continued fraction (dots) and Kramers rate theory [19, 21]. As can be seen from the figure, the lower the damping becomes, the more unstable the matrix continued fraction becomes. This is highly undesirable, because the physically most interesting region, which is the cross-over of the extreme quantum limit (where the Generalized Master Equation works well), and the semiclassical limit (where Kramers rate equation works well), cannot be reached. We thus pose the third open problem.

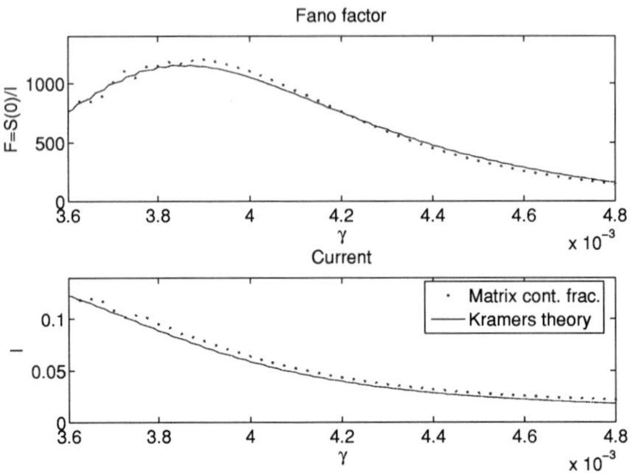

FIGURE 4. Comparison of the results for the Fano factor $F = S(0)/I$ (upper panel) and mean current I (lower panel) around the shuttling transition calculated by the (modified version of the) matrix continued fraction method (dots) and by the semi-analytical Kramers rate theory (line). The number of states used was $N = 100$ and, apparently, the numerical method stops working reliably for low enough damping.

Problem (3): Applicability of Matrix Continued Fraction method. Is it possible to solve (numerically) coupled Kramers equations with the continued fraction method? If so, a large number of applications could be addressed.

CONCLUSION

We have analyzed the noise properties of NEMS using the quantum shuttle as a specific example. We identify three separate open problems which we believe are generic for all noise calculations for nanoscale systems involving many degrees of freedom: (i) Reliability and generalizations of the Markovian Generalized Master equation which underlies all state-of-the art calculations; (ii) Ability to efficiently deal with large matrix equations resulting from the truncation of infinite matrices; and (iii) Solution methods for matrix Krarmers equations.

REFERENCES

1. L. Y. Gorelik, A. Isacsson, M. V. Voinova, B. Kasemo, R. I. Shekther, and M. Jonson, *Phys. Rev. Lett.*, **80**, 4526 (1998).
2. A. D. Armour, and A. MacKinnon, *Phys. Rev. B*, **66**, 245409 (2002).
3. T. Novotný, A. Donarini, and A. P. Jauho, *Phys. Rev. Lett.*, **90**, 256801 (2003).
4. D. Fedorets, L. Y. Gorelik, R. I. Shekther, and M. Jonson, *Phys. Rev. Lett.*, **92**, 166801 (2004).
5. F. Pistolesi, *Phys. Rev. B*, **69**, 245409 (2004).
6. T. Novotný, A. Donarini, C. Flindt, and A. P. Jauho, *Phys. Rev. Lett.*, **92**, 248302 (2004).
7. C. Flindt, T. Novotný, and A. P. Jauho, *Phys. Rev. B*, **70**, 205334 (2004).
8. C. Flindt, T. Novotný, and A. P. Jauho, *Europhys. Lett.* , **69**, 475 (2005).
9. F. Pistolesi, and R. Fazio, *Phys. Rev. Lett.*, **94**, 036806 (2005).

10. A. Erbe, C. Weiss, W. Zwerger, and R. H. Blick, *Phys. Rev. Lett.*, **87**, 096106 (2001).
11. D. Scheible, and R. H. Blick, *Appl. Phys. Lett.*, **84**, 4632 (2004).
12. S. A. Gurvitz, and Y. S. Prager, *Phys. Rev. B*, **53**, 15932 (1996).
13. B. Elattari, and S. A. Gurvitz, *Phys. Lett. A*, **292**, 289 (2002).
14. C. Flindt, T. Novotný, and A. P. Jauho, to appear in *Physica E*, (2005); cond-mat/0412425.
15. A. N. Jordan, and E. V. Sukhorukhov, *Phys. Rev. Lett.*, **93**, 260604 (2004).
16. A. N. Jordan, and E. V. Sukhorukhov, cond-mat/0503751.
17. C. Flindt, T. Novotný, and A. P. Jauho, submitted to *Proceedings of ICNF*, Salamanca, Spain (2005); cond-mat/0504555.
18. D. Kohen, C. C. Marston, and D. J. Tanner, *J. Chem. Phys.*, **107**, 5236 (1997).
19. A. Donarini, *Dynamics of Shuttle Devices*, Ph. D. Thesis, MIC, Technical University of Denmark (2004); cond-mat/0501242.
20. H. Risken, *The Fokker-Planck equation: Methods of solution and applications, 2nd ed.*, Springer, Berlin, 1996
21. A. Donarini, T. Novotný, and A. P. Jauho, submitted to *New Journal of Physics*, (2005).

Fluctuations In Electrohydrodynamic Instability

Francesco Bianco,[1] Mauro Lucchesi,[1] Simone Capaccioli,[3] Leone Fronzoni[1,2,3] and Paolo Allegrini[1]

[1]*Department of Physics, University of Pisa, Pisa, Italy*
[2] *Centro Interdisciplinare per lo Studio dei Sistemi Complessi (CSSC).*
[3]*CNR-CRS.SOFT*

Abstract. Electrohydrodynamic Convection in Liquid Crystals (EHC) is a good system for the experimental study of spatio-temporal chaos. Particularly interesting is the behavior of the Nematic in presence of weak turbulence where ordered and disordered states are mixed. In this case, the fluctuations of velocity and electric current, for instance, are typical fluctuations of a system far from equilibrium. Recently some authors have analyzed the amplitude of the fluctuations as function of the applied electric field [1-2] and they present interesting interpretations provided by some theories. Although important results have been obtained by these authors, many aspects of the dynamical behavior have to be further analyzed as the role of some localized coherences inside the turbulence regions. The direct optical observation allows us to make a correspondence between fluctuations and patterns, providing important information for a theoretical interpretation.

Keywords: Electroconvection, turbulence, entropy.
PACS: 47.65.+a

INTRODUCTION

Recently, the study of fluctuation in non-equilibrium thermodynamic has attracted the interest of the scientific community. In literature there are some theoretical studies that making prediction on the statistics of the fluctuation as the Probabilistic Distribution Functions (PDF) [3]. Particularly interesting are the fluctuations that appear in proximity of the threshold of formations of patterns and spatial structures [4]. In many cases these instabilities come from the formation of defects on the structures generating fluctuations of, for instance, density or velocity. This kind of phenomenon is known as turbulence mediated by defects and can be observed in the thermo-convective instabilities (Rayleigh-Benard instability) or in electro-convection that appears in samples of Nematic Liquid Crystals with the presence of an applied electric field [5-6]. This last phenomenon shows many advantages in comparison with the thermo-convective instability. This is due to the fact that its experimental set is easier to realize and shows a dynamics fast enough.

NEMATIC LIQUID CRYSTAL SAMPLE

Nematic Liquid Crystals are constituted of long molecules with a preferential orientation named "director". This material is sandwiched between two plates. By means of a special treatment of the surfaces it is possible to induce a molecular

CP800, *Unsolved Problems of Noise and Fluctuations: UPoN 2005*
edited by L. Reggiani, C. Pennetta, V. Akimov, E. Alfinito, and M. Rosini

orientation parallel to the surfaces as shown in Figure1 (left). If an electric field is applied perpendicularly to the "director", because of the strong anisotropy of the medium, a periodic fluctuation of the molecular orientation induces a perpendicular electric current to the electric field. This current generates periodic spatial charges. These charges interact with the field and move the fluid. This generates a cylindrical configuration of the velocity (rods). The pattern appears over a threshold of the applied field in correspondence of which the drive forces exceed the viscosity forces. This motion is very similar to that generated by a temperature gradient in thermo-convection. The diameter of the rods are of the order of the thickness of the nematic layer that in our experiment is 50 micrometers, it means that the observation of the samples needs a polarized microscope.

FIGURE 1. Left: Cell of Liquid Crystal with an electric field applied. Right: Top view of a cell of nematic liquid crystals in presence of pattern. A defect is present in the center of the sample

In Figure 1 (right) a top view of a sample in proximity of the threshold of the instability is shown. It is important to observe the defect presents in the center of the photo. The number of these defects increases as function the electric field applied and, as later shown, these defects are responsible of the wide fluctuations of the current that flows in the sample.

DEFECT DYNAMICS AND CURRENTS

In proximity of the threshold it is possible to describe the formation of the pattern and defects by means of the Ginzburg-Landau Equation (GLE) [7]. In a recent paper [8]we have shown how the motion of the defects is driven by the field of the pattern phase and a simple formula was obtained that provides the value of the defect velocity as function of phase gradient. Increasing the electric field the validity of this theory falls and the defect motion becomes more complex including the presence of periodic trajectories. Erratic motion of the defects and fluctuations of the pattern phase modulate the current that flows through the sample, so that a measurement of this current gives global information about the dynamics that are present in the nematic cell. Goldburg and co-workers have measured this current as function the intensity of the electric field applied [1]. They found that this current shows a maximum of fluctuation in proximity of the threshold. In that paper they made interesting considerations on the Gallavotti-Choen theorem, but important questions remain open as about the role and the contribution of the defects at the fluctuations of the current. The spectral analysis of the current shows peaks that indicate the presence of an oscillations, in highly turbulent regime also. These oscillations of the current come from a self-oscillation of the defects as indicated by a direct observation of the sample. The frequency of the oscillation increases quasi-linearly as a function of the voltage applied [9].

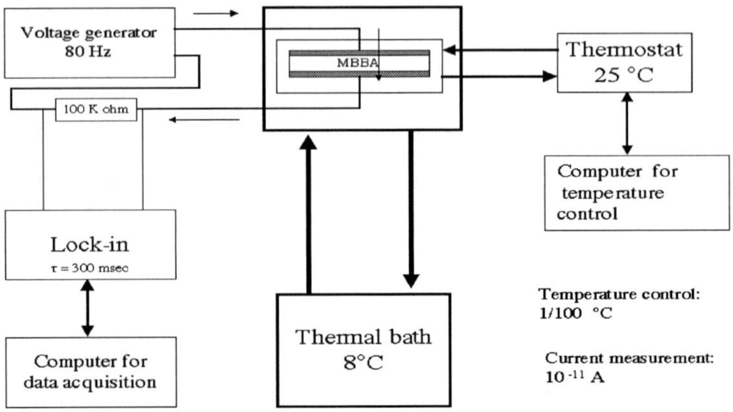

FIGURE 2. Experimental apparatus: A periodic voltage generator of amplitude V is applied to the conducting plates of the sample of MBBA. A lock-in reads the voltage between the caps of a resistor where the cell current is flowing.

We argue that these oscillations are the main cause of the increasing of the fluctuations in proximity of the threshold of the instability. We decided to repeat the experiment of Goldburg *et al,* but with our attention focused on the motion of the defects. The scheme of our apparatus is shown in Figure 2 and the nematic used was MBBA. In order to have a high accuracy of the temperature control we used two thermostats, one with a cooled liquid and another with an electric thermostat controlled by a computer. The current that flow through Liquid Crystal was measured my means a locking technique (STANFORD RESEARCH SYSTEMS Mod SR830 DPS LocK-In Amplifier) that allowed to measure current fluctuations of the order of 10^{-11} Amperes. A function generator at the frequency of 80 Hz gives an amplitude voltage V to the plates of the sample. A computer made a sampling of the current with a integration time of 300 msec. The presence of a bias that affects the data gives serious difficulties in order to compute the Probability Density Functions (PDF) and to estimate the standard deviation of the fluctuations. In spite of this disadvantage we found results similar to have obtained in Ref.[1] using the same method in removing the bias from the data. Of course, this provides an arbitrariness that must be removed in order to get a correct statistics. However the variance of the fluctuations of the current as function of the sample voltage is shown in Figure 3, where a maximum of the fluctuations is presents in proximity of the threshold.

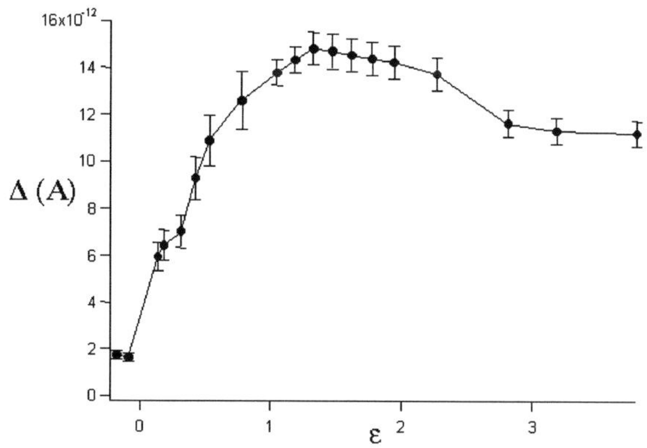

FIGURE 3. Standard deviation as function a control parameter ε= (V/V_0-1).

THE DIFFUSION ENTROPY METHOD

In order to characterize the dynamic properties of this instability, although in the presence of bias, we have used a method of Diffusion Entropy [10]. On the basis of this technique, the experimental data are transformed in a boolean sequence, therefore a Brownian-like diffusion is generated using this sequence. For instance, we suppose to assign the value $\xi = 1$ every time the experimental value is increasing and $\xi = -1$ when this is decreasing (Symmetric case). We obtain a sequence of the following form:

1 -1 -1 1 1 1 1 -1 -1 -1 -1 1 1 -1 1 1 -1 -1 -1 -1 -1 -1 1 1 1 1 1 1 -1

Then, let consider a window of lengh t and define a new variable

$$y(t) = \sum^{t}_{i=1} \xi_I \tag{1}$$

Moving step by step the window along the sequence one finds a collection of y(t). Let be P(y,t) the probability to find the result y. The Shannon entropy as function of t assumes the following form

$$S(t) = -\int P(y,t) \ln P(y,t) \tag{2}$$

If the probability has rescaling properties, this reads

$$P(y,t) = t^{-\delta} F(y / t^{\delta}), \tag{3}$$

and inserting (3) into (2) we obtain

$$S(t) = A + \delta \ln t, \tag{4}$$

where δ is the scaling parameter. The plot of S(t) as function of ln(t) provides the rescaling given by the slope of the straight line. Of course, generally, a transient precedes the straight line.

RESULTS

In Figure 4 (left) we show the Diffusion Entropy for four different values of voltage applied to the sample. We can observe three regions: the first part represents a transient. In the central region the data show some oscillations and, for high values of time, the data converge in a straight line. The trend for long time values gives us the information about the scaling properties of the nonperiodic fluctuations. The line under the data indicated by full circle symbols shows a slope $\delta = 0.67$. This means that the fluctuations generate superdiffusive processes. The oscillations observed on the entropy diffusion are related to the periodic behavior observed in the spectrum. Because of the presence of these oscillations we define an average slope obtained drawing a line that pass through points with change of curvature sign. The slope of the line drawn over the data indicated by full square symbols, shows an average slope $\delta_{av} < 0.5$. It is important to observe that δ_{av} has low values just slightly over of the instability threshold, V= 33 volts (V_0 = 22 volts). This behavior can be interpreted assuming a dynamics characterized by an alternation of two processes, one with random fluctuations with $\delta = 0.5$ and the other with periodic oscillations due to the presence of the defects which corresponds $\delta = 0$ (localization), so that the δ_{av} assumes value smaller than 0.5

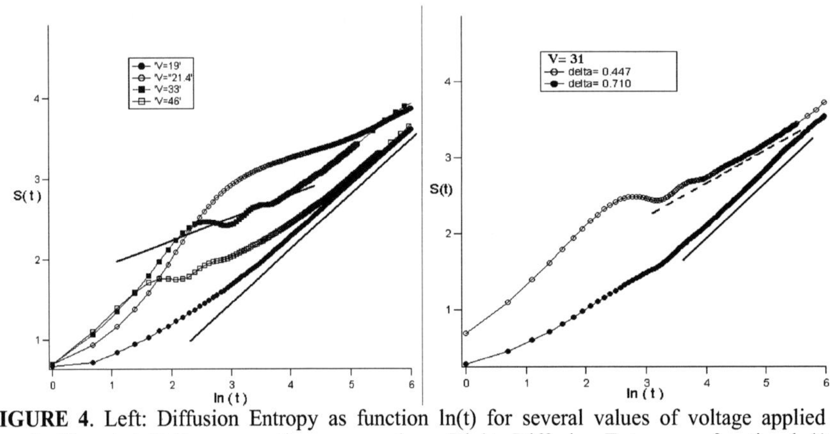

FIGURE 4. Left: Diffusion Entropy as function ln(t) for several values of voltage applied to the sample. The straight lines mean the average slope. Right: Diffusion Entropy as function ln(t) in the asymmetric (full circle) and symmetric (white circle) cases. The straight lines gives $\delta_{av} = 0.447$ and $\delta_{av} = 0.71$.

We tested again this model using another procedure to get the Diffusion Entropy. We consider a sequence obtained from the experimental data, taking 1 when the data is at the maximum and zero for the others values (asymmetric case). This sequence must generate a balistic diffusion, with $\delta = 1$, when the current behaves periodically and $\delta = 0.5$ in case of random data. In presence of the alternation between the two modes, one must observe a values δ higher than 0.5. It is important to observe that this prediction is valid for times of the order of the duration of the two phases. Figure 4 (right) shows the Diffusion Entropy corresponding to the symmetric and asymmetric sequences extracted from the same experimental data (V=31 Volts). The dotted line and the outlined line give, respectively, $\delta_{av} = 0.447$ and $\delta_{av} = 0.71$, in the same interval of time, which is in agreement with the model. It is important to observe that for very long times the two curves reach the same slopes with values higher than 0.5 and with absence of oscillations. This means that the behavior for long times is not affect by the defects oscillations and the behavior is essentially superdiffusive. Moreover, as indicated by figure 4 (left) for high voltages (open square data) the oscillations in the diffusion entropy reduce, that means that the defects disappear as indicated by direct observations of the nematic.

CONCLUSIONS

The dynamics of weak turbulence can be studied using a method defined Diffusion Entropy. We have studied, by means of this technique, the turbulence mediated by defects in electro-convection, realized in a sample of Nematic Liquid Crystals. We performed this analysis looking at the current that flow through the sample. This current shows a maximum of the fluctuations when defects are present in the sample. The analysis shows that the dynamics is characterized by the alternation of two behaviors, one dominated by fluctuation of phase and the other by periodic motion of the defects. By means of the Diffusion Entropy we are able to recognize the scaling properties of the weak turbulence although in the presence of the defects which induce localization in the diffusion. This localization generates a scaling corresponding to a subdiffusive dynamics, whereas, for long times and high values of voltage the process appears superdiffusive.

REFERENCES

1 W.I. Goldburg ,Yadin Y. Golschmidt, and H. Kellay,. *Phys.Rev.Lett.* , **87**, 245502 (2001).
2 Tibor Tóth-Katona et al. *Electronic-Liquid Communications*, May **15** (2003).
3 G.Gallavotti and E.G.D. Cohen, *Phys.Rev.Lett.* **74,** 2694 (1995).
4 S.Ciliberto and C.Laroche, "An experimental test of the Gallavotti-Cohen fluctuation Theorem. "*J.Phys.IV France* **8,** 215, 219 (1998).
5 P.G de Gennes and J.Prost, *The Physics of Liquid Crystals*. Claredon Press, (Oxford, 1993)
6 L.Kramer, I.S Aranson, *Rev. Mod. Phys.* **74,** 99 (2002).
7 M,.C.Cross, P.Hohenberg, " Pattern formation outside of equilibrium", *Rev. Mod. Phys.*, **65**, 851(1993).
8 P. D' Amico, L.Fronzoni, I.A. Maione , L. Palatella. "Experimental defect dynamics driven by Phase in liquid-crystal electro-convective patterns". *Physica D*, **157**, 353,374 (2004).
9 S.Kai and K.Hirakawa, *Prog.Theor.Phys.Suppl.* **64,** 212 (1978).
10 P. Allegrini, V.Benci, P.Grigolini, P.Hamilton, M. Ignaccolo, G.Menconi, L.Pallatela, G.Raffaeli, N.Scafetta, M.Virgilio, J.Yang. "Compression and Diffusion: a joint approach to detect complexity". *Chaos, Solitons and Fratals*, **15**, 517-585 (2003).

Effective Temperature In Sheared Binary Systems

F. Corberi*, G. Gonnella†, E. Lippiello* and M. Zannetti*

*Istituto Nazionale per la Fisica della Materia, Unità di Salerno and Dipartimento di Fisica "E.R.Caianiello", Università di Salerno, 84081 Baronissi (Salerno), Italy
†Istituto Nazionale per la Fisica della Materia, Unità di Bari and Dipartimento di Fisica, Università di Bari, and Istituto Nazionale di Fisica Nucleare, Sezione di Bari, via Amendola 173, 70126 Bari, Italy

Abstract. We study the dynamics of a binary system with a non conserved order parameter under the action of a plain shear flow in the large-N limit. A phase transition is observed at a critical temperature $T_c(\gamma)$. After a quench from a high temperature equilibrium state to a lower temperature T a non-equilibrium stationary state is entered when $T > T_c(\gamma)$, while aging dynamics characterizes the phases with $T \leq T_c(\gamma)$. The behavior of the effective temperature T_{eff} is discussed. For $T > T_c(\gamma)$ a non monotonous behavior is observed while, below $T_C(\gamma)$, the usual result $T_{eff} = \infty$ of coarsening systems is found in the aging regime.

Keywords: fluctuation-dissipation Theorem, glasses, order parameters
PACS: 05.70.Ln; 64.75.+g; 83.50.Ax

INTRODUCTION

Understanding the general properties of systems out of equilibrium is one of the central topics of modern statistical mechanics. Many relevant systems, among which glasses and coarsening systems, are intrinsically far from equilibrium and, therefore, cannot be described by theories formulated as extensions of Gibbs statistical mechanics. Instead, specific concepts relative to the non-equilibrium state must be developed. A promising subject in this field is the attempt to build an out of equilibrium fluctuation-dissipation theorem (FDT). These studies investigate the relation between the response function $R(t,t')$ and the autocorrelation function $D(t,t')$. In equilibrium these two quantities are time translational invariant (TTI) and are related by means of the standard FDT

$$TR(t-t') = \frac{\partial D(t-t')}{\partial t'}. \tag{1}$$

In Eq. (1) a fundamental quantity, the temperature of the system, appears as the coefficient of this relation. For off-equilibrium systems we do not have nowadays an analogous theorem, of such a wide generality. Nevertheless there are indications [1, 2, 3, 4] that, in restricted contexts, the relation between R and D can be used to define an *effective* temperature of the system, different from that of the thermal bath, through

$$T_{eff}(t,t')R(t,t') = \frac{\partial D(t,t')}{\partial t'} \tag{2}$$

CP800, *Unsolved Problems of Noise and Fluctuations: UPoN 2005*
edited by L. Reggiani, C. Pennetta, V. Akimov, E. Alfinito, and M. Rosini

This concept, which was originally developed in the study of turbulence [5] has been successively applied to the glassy state [6, 1] and, more generally, to aging systems [7] where the non-equilibrium state is generally determined by the change of a thermodynamic control parameter, as in a temperature quench.

A different non-equilibrium situation is very often considered because of its physical relevance: The case of driven systems. These systems are subjected to an external forcing that prevents them to attain equilibrium. Examples range from systems subjected to a force which does not derive from a potential, as sheared fluids or conductor wires with a potential difference applied at their hands, to systems with externally imposed thermal gradients or magnetic materials under the action of an oscillatory magnetic field and tapped materials. Despite many studies on this subject [8] much less is known on the effect of the drive on the fluctuation-dissipation relation, although some progress has been made recently [1, 9].

In this paper we focus on driven phase ordering systems which, besides being physically interesting of their own, can be regarded as simple paradigms of aging behaviour. We consider a coarse-grained system described by the usual double-well φ^4 hamiltonian where φ is a scalar order parameter field. A Langevin dynamics is considered and the order parameter is not conserved. The driving field is a planar shear or Couette flow. This model, without the shear flow, is referred to in the literature as model A. The choice of this particular flow, besides analytical tractability, is also motivated by its practical relevance, which appears clearly in the physics of complex fluids and binary mixtures, where the behaviour of the stress as a function of the shear rate is one of the main characterizations of fluid systems. Moreover, the structure of binary and complex fluids is strongly affected by the flow and this is relevant in many applications [10] and interesting for its theoretical implications. In phase separation of binary mixtures, for instance, it was found that not only the morphology but also the laws of dynamical scaling are affected by the shear-induced anisotropy [11, 12] and new phenomena related to the presence of stress in segregating systems, as logarithmic-time periodic oscillations [13] and violation of dynamical scaling [14] have been observed.

In this paper we focus on a non-conserved order parameter; however we believe that our study could be also useful as a preliminary step for understanding more complex systems with conserved ordered parameter. By its own the model of this paper can describe the behaviour of liquid crystals undergoing the isotropic-nematic transition where the order parameter is not conserved [15]. The phase separation properties of this model have been studied in [16] where only the behaviour of one-time correlations was considered.

The role of the non-linear term in the φ^4 hamiltonian is crucial for quenching below and at the critical temperature. However its presence makes the analytical study of the model impossible unless one considers approximate theories. Among these the so-called large-N limit has played a prominent role in the study of phase separation. The approximation takes into account a vectorial system with an infinite number of components N. This model, which can be solved by a self-consistent closure, is one of the few cases where it is possible to illustrate the out-of-equilibrium behaviour of a coarsening system by exact calculations [15].

In the case without shear the large-N approximation captures the essence of the phenomenon at a semi-quantitative level [17]. In this work we use this approximation to discuss the behavior of the effective temperature (2).

The plan of the paper is the following. In Sec. the model is defined. In Sec. we sketch the solution of the model and introduce the basic observables. The autocorrelation function and the response function, are studied in Sec. , and the behavior of the effective temperature is discussed. The conclusions are contained in Sec. ().

THE MODEL

We consider a system with dynamics described by the equations

$$\frac{\partial \varphi^\alpha(\vec{x},t)}{\partial t} + \vec{\nabla}\cdot(\varphi^\alpha(\vec{x},t)\vec{v}) = -\Gamma \frac{\delta \mathscr{H}[\vec{\varphi}]}{\delta \varphi^\alpha(\vec{x},t)} + \eta^\alpha(\vec{x},t) \quad , \qquad \alpha = 1,\ldots,N \tag{3}$$

where $\{\varphi^\alpha\}$ are the N components of the vectorial order parameter. For instance, in the case of magnetic systems, $\vec{\varphi}$ is the local magnetization, Γ is a transport coefficient and the gaussian white noise η, representing thermal fluctuations, has expectations $\langle \eta^\alpha(\vec{x},t)\rangle = 0$, $\langle \eta^\alpha(\vec{x},t)\eta^\beta(\vec{x}',t')\rangle = 2T\delta_{\alpha\beta}\Gamma\delta(\vec{x}-\vec{x}')\delta(t-t')$, where T is the temperature of the thermal bath. Without the convective term on the left hand side, Eq. (3) would be the usual Langevin equation which governs the purely relaxational dynamics of a system with non-conserved order parameter and hamiltonian $\mathscr{H}[\vec{\varphi}]$. The velocity $\vec{v}$ in Eq. (3) is chosen to be the steady planar shear flow $\vec{v} = \gamma y \vec{e}_x$ where γ is the spatially homogeneous shear rate and $\vec{e}_x$ is a unit vector in the flow direction. The hamiltonian has the Ginzburg-Landau form

$$\mathscr{H}[\vec{\varphi}] = \int_V d^d x \left[\frac{1}{2} \mid \nabla\vec{\varphi} \mid^2 + \frac{r}{2}\vec{\varphi}^2 + \frac{g}{4N}(\vec{\varphi}^2)^2 \right] \tag{4}$$

where $r < 0$, $g > 0$ and V is the volume of the system. We will be interested in processes where the system is initially in the equilibrium ($\gamma = 0$) infinite temperature state and then is suddenly put in contact with a heat bath at temperature T, and subjected to the shear flow.

In the large-N limit [15] the equation of motion for the Fourier transform $\varphi^\alpha(\vec{k},t) = \int_V d^d x \varphi^\alpha(\vec{x},t)\exp(i\vec{k}\cdot\vec{x})$ takes the linear form

$$\frac{\partial \varphi^\alpha(\vec{k},t)}{\partial t} - \gamma k_x \frac{\partial \varphi^\alpha(\vec{k},t)}{\partial k_y} = -\Gamma[k^2 + I(t)]\varphi^\alpha(\vec{k},t) + \eta^\alpha(\vec{k},t) \tag{5}$$

where the function $I(t) = r + \frac{g}{N} < \vec{\varphi}^2(\vec{x},t) >$ has to be calculated self-consistently. The quantity $< \vec{\varphi}^2(\vec{x},t) >$ can be expressed as

$$\frac{1}{N} < \vec{\varphi}^2(\vec{x},t) > = \frac{1}{V}\sum_{\vec{k}} C(\vec{k},t) \tag{6}$$

where the structure factor $C(\vec{k},t) = \frac{1}{NV} < \vec{\varphi}(\vec{k},t)\cdot\vec{\varphi}(-\vec{k},t) >$ is solution of the equation

$$\frac{\partial C(\vec{k},t)}{\partial t} - \gamma k_x \frac{\partial C(\vec{k},t)}{\partial k_y} = -2\Gamma[k^2 + I(t)]C(\vec{k},t) + 2\Gamma T. \tag{7}$$

The momentum sum in Eq. (6) extends up to a phenomenological ultraviolet cut-off Λ.

DEFINITION OF THE MAIN OBSERVABLES AND SOLUTION OF THE MODEL

In the following we briefly sketch the method used to solve the model, further details are discussed in [18]. >From now on we will take the infinite volume limit. The formal solution of Eq. (7) is given by

$$C(\vec{k},t) = C(\vec{k},0)y(t)e^{-2\int_0^t du \mathscr{K}^2(u)}e^{-k^2/\Lambda^2} + \frac{2T}{y(t)}\int_0^t e^{-2\int_0^{t-z} du \mathscr{K}^2(u)}e^{-k^2/\Lambda^2}y(z)dz \quad (8)$$

where $\vec{\mathscr{K}}(s) = \vec{k} + \gamma s k_x \vec{e}_y$,

$$y(t) = \exp\left(2\int_0^t [r + gS(t')]dt'\right) \quad (9)$$

and, from Eq. (6)

$$S(t) = \lim_{V\to\infty}\frac{1}{N}\langle\vec{\varphi}^2(\vec{x},t)\rangle = \lim_{V\to\infty}\frac{1}{V}\sum_{\vec{k}} C(\vec{k},t) = \int \frac{d^dk}{(2\pi)^d}e^{-k^2/\Lambda^2}C(\vec{k},t) \quad . \quad (10)$$

Eqs. (8,9,10) are a set of coupled self-consistent equations whose solution can be found by means of Laplace transform techniques. The solution of the model shows [19, 18] that a *critical* temperature $T_c(\gamma)$ exists such that for $T > T_c(\gamma)$ the system is in disordered state with a finite correlation length $\xi_\perp$ in the direction trasverse to the flow. $\xi_\perp$ diverges as $T_c(\gamma)$ is approached from above and remains infinite below $T_c(\gamma)$ where an ordered phase is found.

The autocorrelation function is defined as $D(t,t') \equiv <\varphi_\alpha(\vec{r},t)\varphi_\alpha(\vec{r},t')>$ and the response function as $R(t,t') \equiv \left.\frac{\delta<\varphi_\alpha(\vec{r},t)>}{\delta h_\alpha(\vec{r},t')}\right|_{h=0}$. These quantities do not depend on the component index due to rotational symmetry.

Since $D(t,t')$ for a given t' is a monotonously decreasing function of t, the time t in Eq. (2) can be reparametrized in terms of D obtaining $T_{eff}(t,t') = \widehat{T}_{eff}(D,t')$ and, keeping D fixed, the limit $\mathscr{T}(D) = \lim_{t'\to\infty}\widehat{T}_{eff}(D,t')$ defines the effective temperature in the time sector corresponding to the chosen value of D. This quantity is important also because, when appropriate conditions are satisfied, provides the connection between dynamic and static properties [20] through the relation $P(q) = T\frac{d}{dD}\mathscr{T}(D)^{-1}|_{D=q}$ where $P(q)$ is the overlap probability function of the equilibrium state. Therefore, different shapes of $\mathscr{T}(D)$ are associated to different degrees of complexity of the equilibrium state. A first broad classification of systems has been made [21, 22] drawing the distinction between coarsening systems, structural glasses and the infinite range spin glass model on the basis of the patterns displayed by $\mathscr{T}(D)$, or related quantities such as the fluctuation-dissipation ratio $X(t,t') = T/T_{eff}(t,t')$. The limiting value $X_\infty = \lim_{t'\to\infty}\lim_{t\to\infty}X(t,t')$ of this quantity has been recently intensively studied. In systems without drive X_∞ takes the value $X_\infty = 1$ for equilibrium systems while $X_\infty = 0$ in the low temperature phase of coarsening systems with a non-vanishing critical temperature [23, 24, 25]. For critical systems the value of X_∞ has been computed for the random walk and the Gaussian model where the non trivial value $X_\infty = 1/2$ is found [26]. This quantity has been also computed for some coarsening systems at the lower critical dimension d_L, such as the X-Y model in $d = 2$ [26] and the Ising chain [27], where the same result $X_\infty = 1/2$ is recovered. However, concerning systems quenched to the critical temperature above d_L, the value of X_∞ turns out to possess a different value. Numerical simulations of the Ising model with heat bath dynamics in $d = 2$ indicate that $X_\infty \sim 0.26$ [28] and the exact solution [28] of the spherical model (that is equivalent to the large-N model) gives

$$X_\infty = \frac{d-2}{4} \qquad 2 < d < 4 \tag{11}$$

$$X_\infty = \frac{1}{2} \qquad d > 4. \tag{12}$$

In [28] it is proposed that this quantity is a novel universal quantity of non-equilibrium critical dynamics.

EFFECTIVE TEMPERATURE

In the following we will discuss the behavior of the effective temperature in $d = 3$ in the three cases $T > T_c(\gamma)$, $T = T_c(\gamma)$, and $T < T_c(\gamma)$ separately, since the model behaves in a qualitativly different way.

$\mathbf{T > T_c(\gamma)}$

For large t' the autocorrelation reads

$$D(t,t') \simeq D_{st}(\tau,\xi_\perp) = \frac{4T}{(8\pi)^{d/2}} \int_{\tau/2}^{\infty} e^{-2\xi_\perp^{-2} y} (y + 1/2\Lambda^2)^{-d/2} \times \\ \times \frac{1}{\sqrt{4 + \frac{1}{3}\gamma^2 y^2 + \frac{\gamma^2}{2}\tau^2 \left(1 - \frac{\tau^2}{8y^2}\right)}} dy, \tag{13}$$

with $\tau = t - t'$, and the response function

$$R(t,t') \simeq R_{st}(\tau,\xi_\perp) = \frac{2^{1-d/2}}{(4\pi)^{d/2}} e^{-\xi_\perp \tau} \left(\frac{\tau}{2} + \frac{1}{2\Lambda^2}\right)^{-d/2} \frac{1}{\sqrt{4 + \frac{1}{3}\gamma^2\tau^2}}. \tag{14}$$

This implies that i) D and R are TTI quantities, and ii) D and R decay from one to zero for $\tau \simeq \xi_\perp^2$ meaning that $\xi_\perp^2$ is the characteristic relaxation time of fluctuations above the critical temperature.

With expressions (13,14) $\mathscr{T}(D)$ can be easily obtained. This quantity is plotted in Fig. 1 where, for comparison, the case without shear is also plotted. Without shear the

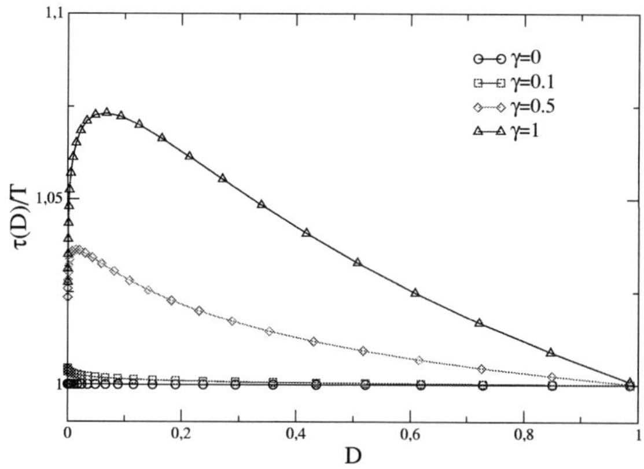

FIGURE 1. $\mathscr{T}(D)$ for $T = 1.5T_c(\gamma)$, $t' = 10^3$.

system attains the equilibrium state, FDT holds and $\mathscr{T}(D) \equiv T$. The situation is radically different when the shear is present since a non-equilibrium stationary state is found. For $D \simeq 1$ the fast degrees of freedom are probed, which are equilibrated and $\mathscr{T}(D) = T$. Moving towards smaller values of D, however, slow degrees of freedom are probed, FDT does not hold and $\mathscr{T}(D)$ is a non monotonous function.

$\mathbf{T = T_c(\gamma)}$

At the critical temperature, for large t' one has $D(t,t') = D_{st}(\tau,\infty)$ and $R(t,t') = R_{st}(\tau,\infty)$. In other words, $D(t,t')$ and $R(t,t')$ take the same forms (13) and (14), with $\xi_\perp = \infty$ since the system is at criticality. Notice that one has TTI again but, for large τ, these functions decays as power laws, implying the absence of a characteristic relaxation time, as usual at criticality.

With the expressions for D and R the effective temperature can be computed. $\mathscr{T}(D)$ is plotted in Fig. 2. For $D \simeq 1$ we again find $\mathscr{T}(D) = T$ for the reason discussed above. As D is decreased the effective temperature increases and, at $D = 0$ a discontinuity occurs. In order to discuss this behavior it is useful to consider the quantity X_∞ introduced in Sec.. It can be shown that $X_\infty = 1/2$ both for $d = 3$ (this can be checked from the value of $\mathscr{T}(0)$ in Fig. 2) and for $d = 2$. This can be understood as follows: The mathematical structure of the solution of the model with applied flow in dimension d is similar to the structure of the solution without shear in an increased dimensionality $d+2$, as discussed in [18]. Then, with respect to the case $\gamma = 0$ where Eqs. (11,12) hold, the effect of the flow is to raise the *effective* dimensionality at or above $d_U = 4$ when $d \geq 2$, and the result (12) is found for both $d = 2$ and $d = 3$.

$\mathbf{T < T_c(\gamma)}$

In this case there are two time regimes of interest:
i) short time separation: $t' \to \infty$, $\frac{\tau}{t'} \to 0$ (Quasi-stationary regime)
ii) large time separation: $t' \to \infty$, $\frac{\tau}{t'} \to \infty$ (Aging regime).

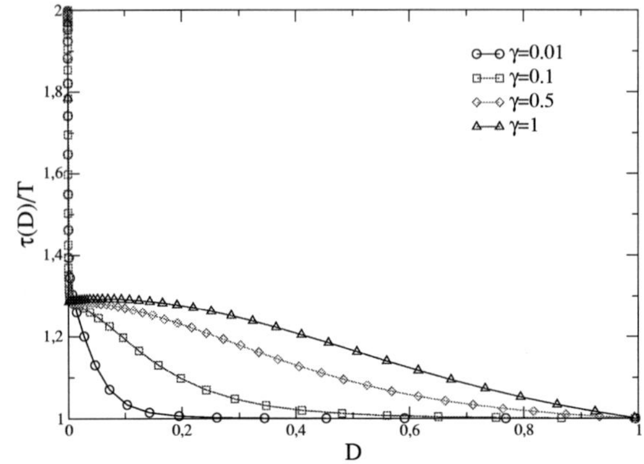

FIGURE 2. $\mathscr{T}(D)$ for $T = T_c(\gamma)$, $t' = 10^3$.

In the time sector *i*), one finds

$$D(t,t') = M^2 + D_{st}(\tau,\infty) \tag{15}$$

where $M_0^2 = -r/g$ and $M = M_0\sqrt{1 - \frac{T}{T_c(\gamma)}}$ is the analogue of the spontaneous magnetization in equilibrium.

In the other limit *ii*), one gets

$$D(t,t') = D_{ag}(t/t') = M^2 \left(\frac{t'}{t}\right)^{\frac{d+2}{4}} (1+t'/t)^{-\frac{d+2}{2}} \frac{2^{\frac{d+2}{2}}}{\sqrt{4\frac{2-(1-t'/t)^3}{1+t'/t} - 3\frac{(2-(1-t'/t)^2)^2}{(1+t'/t)^2}}}. \tag{16}$$

Notice the dependence of D_{ag} on the ratio t/t' alone, as usual in slowly relaxing aging systems. Furthermore, for $t/t' \gg 1$, a generalization of the Fisher-Huse exponent λ defined by $D(t,t') \propto (t'/t)^\lambda$ gives $\lambda = \frac{d+2}{4}$.

The structure of the autocorrelation function provided in Eqs. (16,15) for large t' allows its splitting into the sum of two distinct contributions

$$D(t,t') = D_{st}(\tau,\infty) + D_{ag}(t/t'). \tag{17}$$

In the regime *i*) $D_{ag}(t/t')$ remains fixed to its initial value $D_{ag}(1) = M^2$, while $D_{st}(\tau,\infty)$ decays to zero. This makes Eq. (17) consistent with Eq. (15). Conversely, in the regime *ii*), $D_{st}(\tau,\infty)$ has already decayed to zero and the whole time dependence is carried by $D_{ag}(t/t')$, providing again consistence between Eqs. (17) and (15).

This whole pattern of behaviours of $D(t,t')$ is closely reminiscent of what is known for the large-*N* model without shear [23] and for systems with a scalar order parameter [24].

The response function is given by

$$R(t,t') = \frac{2^{1-d/2}}{(4\pi)^{d/2}} \left(\frac{t'}{t}\right)^{-\frac{d+2}{4}} \left(\frac{t-t'}{2} + \frac{1}{2\Lambda^2}\right)^{-d/2} \frac{1}{\sqrt{4 + \frac{1}{3}\gamma^2(t-t')^2}}. \tag{18}$$

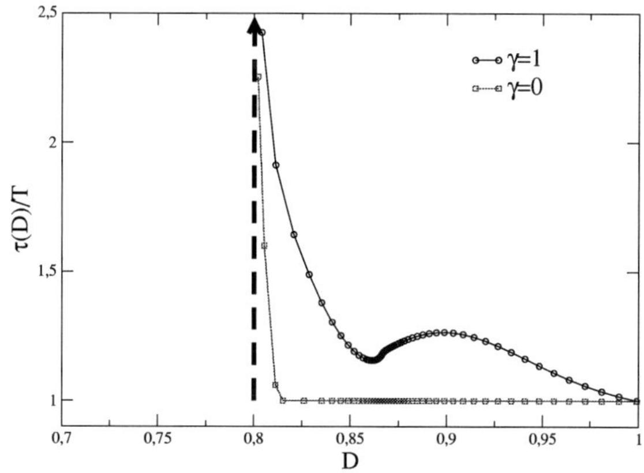

FIGURE 3. $\mathscr{T}(D)$ for $T = 0.5T_c(\gamma)$, $t' = 10^4$.

The response function can be written [18] as $R(t,t') = R_{st}(\tau,\infty) + R_{ag}(t,t')$ and, for large t', $R_{ag}(t,t')$ can be neglected with respect to $R_{st}(\tau,\infty)$ for every choice of t and t'. Before discussing the behaviour of $\mathscr{T}(D)$ under the action of the flow it is useful to overview the behaviour without shear. In this case a structure like (17) is also found. In the short time separation regime, or stationary regime, $D_{st}(\tau,\infty)$ decays from $D_{st}(0,\infty) = M_0^2 - M^2$, to $D_{st}(\tau \simeq t_w,\infty) \simeq 0$ while $D_{ag}(t/t_w) \simeq D_{ag}(1) = M^2$ remains constant. Therefore, in this regime $D(t,t_w)$ ranges from M_0^2 down to M^2. In the same range of times $R_{st}(\tau,\infty)$ decays from the initial value $R_{st}(0,\infty)$ to zero, and is related to $D_{st}(\tau,\infty)$ by the equilibrium FDT, Eq. (1). As a consequence the effective temperature equals the temperature of the thermal bath. This gives rise to the constant part of $\mathscr{T}(D)$ on the right of Fig. 3.

In the large time separation sector, or aging regime, $D_{st}(\tau,\infty) \simeq 0$ whereas $D_{ag}(t/t_w)$ decays from $D_{ag}(1) = M^2$ to zero for t/t_w large. Hence the aging regime corresponds to the region $0 \leq D(t,t_w) \leq M^2$ in the plot of Fig. 3. The response function is zero in this regime because $R_{st}(\tau,\infty)$ is already decayed to zero. Then $\mathscr{T}(D) \equiv \infty$, which accounts for behavior on the left side of Fig. 3.

Next we turn to the case with shear. In this situation, most of the considerations discussed for $\gamma = 0$ still apply. In particular in the short time separation regime $D_{st}(\tau,\infty)$ decays from $D_{st}(0,\infty) = M_0^2 - M^2$ to zero while $D_{ag}(t/t') \simeq M^2$ keeps staying constant. In this regime $D(t,t')$ ranges from M_0^2 down to M^2 and $R_{st}(\tau,t')$ decays to zero. However, when shear is applied, the relation between $R_{st}(\tau,\infty)$ and $D_{st}(\tau,\infty)$ is not the equilibrium FDT (1) but, instead, is analogous to the one found in the case $T \geq T_c(\gamma)$ discussed above. Therefore, one finds $\mathscr{T}(D) = T$ for $D = 1$ and a non trivial function for $M^2 < D < 1$, as can be seen on the right of Fig. 3. For $\tau > t'$ (namely $D < M^2(\gamma)$), the aging regime is explored and, for the same reasons discussed for $\gamma = 0$, $\mathscr{T}(D)$ is infinite.

CONCLUSIONS

In this paper we have studied analytically the out of equilibrium kinetics of a solvable model for phase ordering systems subjected to a uniform shear flow. After solving the model, we have discussed the behavior of the effective temperature. The pattern of $\mathscr{T}(D)$ found for coarsening under shear flow can be compared with the behaviour of driven mean field models for glassy kinetics [1]. In both cases the system without drive exhibits a phase transition at a critical temperature T_c characterized by an aging dynamics below T_c, and a qualitatively similar behaviour of the autocorrelation function. However several differences occur between these systems when the drive is switched on, some of which are revealed by $\mathscr{T}(D)$. The most important difference is due to the fact that mean field glass models always attain asymptotically a stationary state, regardless of the temperature, while the model considered in this paper exibits an aging kinetics below a critical temperature even in the presence of shear. Furthermore, in glassy models, $\mathscr{T}(D)$ is piecewise constant for every temperature and takes only two values. This can be associated to the existence of two well defined temperatures in the system, the bath temperature T and the effective temperature T_{eff} of the slow degrees of freedom. A similar behaviour is found in molecular dynamics simulations of binary Lennard-Jones mixture under shear flow [29]. On the other hand, at least in the stationary state observed above $T_c(\gamma)$, the only one that can be compared with the time translational invariant states of the glassy case, the model considered here is characterized by a much more complex structure of $\mathscr{T}(D)$. This suggests that, despite some qualitative similarities, coarsening systems under the action of an external drive behave quite differently from glassy systems and that effective temperature is an efficient tool to detect it. Let us notice, in conclusion, that in the high temperature phase $T > T_c(\gamma)$ the large-N approximation discussed in this paper is believed to provide reliable results. Therefore we expect the non monotonous behavior of the effective temperature to be observable in real experiments with systems subjected to shear flow.

ACKNOWLEDGMENTS

We acknowledge support by MURST(PRIN 2005).

†corberi@na.infn.it ; *gonnella@ba.infn.it
‡lippiello@sa.infn.it ; §zannetti@na.infn.it

REFERENCES

1. L.F. Cugliandolo, J. Kurchan, and L. Peliti Phys. Rev. E **55**, 3898 (1997); L. Berthier, J.-L. Barrat and J. Kurchan, Phys. Rev. E **61**, 5464 (2000).
2. Sollich P, Fielding S and Mayer P, Fluctuation-dissipation relations and effective temperatures in simple non-mean field systems, 2002 J.Phys:Cond.Matt. 14 1683 [cond-mat/0111241].
3. Calabrese P and Gambassi A, On the definition of a unique effective temperature for non-equilibrium critical systems [cond-mat/0406289].
4. F. Corberi, E. Lippiello and M. Zannetti, Jstat, P12007 (2004).
5. P.C. Hohenberg and B.I. Shraiman, Physica D **37**, 109 (1989).

6. L.F. Cugliandolo and J. Kurchan, Phys.Rev.Lett. **71**, 173 (1993); Philos.Mag. **71**, 501 (1995); J.Phys. A **27**, 5749 (1994).
7. S. Fielding and P. Sollich, Phys. Rev. Lett. **88**, 050603 (2002); cond-mat/0209645.
8. B. Schmittmann and R.K.P. Zia, "Statistical Mechanics of Driven Diffusive Systems" in "Phase Transitions and Critical Phenomena", Vol. 17, eds. C. Domb and J.L. Lebowitz, (Academic 1995).
9. G. Gallavotti and E.G.D Cohen, Phys. Rev. Lett. **74**, 2694 (1995); J. Stat. Phys. **80**, 931 (1995).
10. See, e.g., R.G. Larson, *The structure and Rheology of Complex fluids* (Oxford University Press, New York, 1999).
11. For a review, see A. Onuki, J. Phys.: Condens. Matter **9**, 6119 (1997).
12. F. Corberi, G. Gonnella, and A. Lamura, Phys. Rev. Lett. **83**, 4057 (1999).
13. F. Corberi, G. Gonnella, and A. Lamura, Phys. Rev. Lett. **81**, 3852 (1998).
14. F. Corberi, G. Gonnella and A. Lamura, Phys. Rev. E **66**, 016114 (2002).
15. A.J. Bray, Adv. in Phys. **43**, 357 (1994).
16. A.J. Bray and A. Cavagna, J. Phys A: Math. Gen. **33**, L305 (2000).
17. A. Coniglio, P. Ruggiero, and M. Zannetti, Phys. Rev. E **50**, 1046 (1994).
18. F. Corberi, G. Gonnella, E. Lippiello and M. Zannetti, J.Phys. A **36**, 4729 (2003).
19. G. Gonnella and M. Pellicoro, J. Phys. A, 7043 (2000). C. Chan and L. Lin, Europhys. Lett. **11**, 13 (1990).
20. Franz S, Mézard M, Parisi G and Peliti L, Measuring equilibrium properties in aging systems, 1998 Phys.Rev.Lett. 81 1758 [cond-mat/9803108]; The response of glassy systems to random perturbations: a bridge between equilibrium and off-equilibrium, 1999 J.Stat.Phys. 97 459 [cond-mat/9903370].
21. Cugliandolo L F, Effective temperatures out of equilibrium [cond-mat/9903250].
22. Parisi G, Ricci-Tersenghi F and Ruiz-Lorenzo J J, Generalized off-equilibrium fluctuation-dissipation relations in random Ising systems, 1999 Eur.Phys.J. B 11 317 [cond-mat/9811374].
23. F. Corberi, E. Lippiello, and M. Zannetti, Phys. Rev. E **65**, 046136 (2002).
24. F. Corberi, E. Lippiello and M. Zannetti, Phys.Rev. E **63**, 061506 (2001); Eur.Phys.J. B. **24**, 359 (2001).
25. A. Barrat, Phys.Rev.E **57**, 3629 (1998).
26. L.F. Cugliandolo, J. Kurchan and G. Parisi, J.Phys.I France **4**, 1641 (1994).
27. C.Godreche and J.M.Luck, J.Phys. A **33**, 1151 (2000); E.Lippiello and M.Zannetti, Phys.Rev. E **61**, 3369 (2000).
28. C. Godreche and J.M. Luck, J. Phys A **33**, 9141 (2000).
29. J.-L. Barrat and L. Berthier, Phys. Rev. E **63**, 012503 (2001); cond-mat/0111312.

Shot Noise Spectrum Cut-off in Schottky-Barrier Diodes

P. Shiktorov*, E. Starikov* and V. Gružinskis*

*Semiconductor Physics Institute, A. Goštauto 11, LT 01108 Vilnius, Lithuania

Abstract. We develop an analytical model for the shot-noise spectral density of n^+n-metal Schottky-barrier diodes (SBDs). It is shown that the cut-off of shot-noise takes place in the frequency region corresponding to the resonance of carriers reflected from the Schottky-barrier. The model is validated by comparison with Monte Carlo simulations of GaAs SBDs.

Keywords: Shot noise, Schottky-barrier diodes
PACS: 72.20.Ht, 72.30.+q, 72.70.+m

INTRODUCTION

The analysis of the intrinsic current noise of Schottky-Barrier Diodes (SBD) under barrier-limited transport conditions is usually performed by subdividing the noise spectrum into low- (LF) and high-frequency (HF) components which are then considered separately. Such a separation is dictated by the different physical nature of the LF and HF noise. The LF noise spectrum follows the full shot noise law: $S_{II}(0) = 2eI$ which describes fluctuations of the *number* of free carriers which are able to cross the depletion region. The HF noise spectrum is originated by the fluctuations of the *velocity* of free carriers which cannot overcome the depletion region due to reflection from the Schottky barrier.

The analytical model of SBD noise spectrum in both LF and HF regions proposed in [1] is based on the analysis of only individual motions of carriers in the static electric field of the depletion region when the boundary between neutral and depletion regions is assumed as a cathode terminal of SBD. In essence, it is a model for a vacuum diode rather than for a semiconductor structure. The carriers reflected from the Schottky barrier form the HF spectrum, while the LF spectrum is determined by carriers which cross the depletion region. As a result, the model involves only one characteristic time responsible for the frequency of the HF cut-off of the shot noise and LF cut-off of the reflected carrier noise. This time is determined by the donor concentration in the depletion region and is independent of the depletion region length, what contradicts a more rigorous approach as given by the Monte Carlo particle (MCP) simulation of the SBD. To overcome these discrepancies with MCP in [2] there was proposed a model of HF noise of SBD which assumes that the formation of the boundary between the neutral and depletion regions occurs dynamically as a consequence of the carrier motion in the self-consistent electric field. Such a model was found [2] to describe reasonably well the HF region of the SBD noise spectrum. However, since the model neglected carrier transport through the depletion region, it could not describe the LF behavior of the shot-noise. The aim of the

CP800, *Unsolved Problems of Noise and Fluctuations: UPoN 2005*
edited by L. Reggiani, C. Pennetta, V. Akimov, E. Alfinito, and M. Rosini

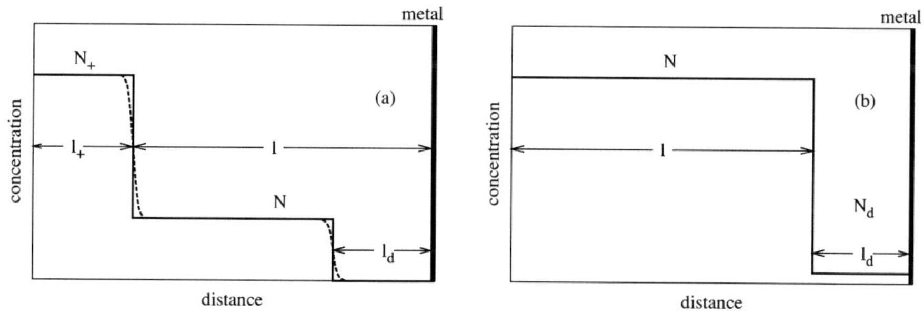

FIGURE 1. Schematic representation of the n^+n–metal SBD with the depletion region near to the metal contact: (a) without and (b) with carrier transport through the SBD

present work is to propose and to discuss the extension of this model able to include the case of LF shot noise.

THEORETICAL MODEL

Let us briefly summarize the main assumptions and results of the model developed for the SBD HF noise [2]. It is assumed that: (i) carriers move in the self-consistent electric field of the whole n^+n-metal SBD structure, as sketched in Fig. 1 (a); (ii) all the carriers in the n-region are reflected when hitting the Schottky barrier so that in the presence of a depletion region no static current is flowing in the structure (i.e. shot noise is not considered); (iii) the potential drop at the n^+n-homojunction is negligible so that the electric field is zero outside the depletion region. Accordingly, we take $E(x) = 0$ for $x < L - l_d$ and $E(x) = (e/\varepsilon\varepsilon_0)N(x - L + l_d)$, for $L - l_d < x < L$, where $L = l_+ + l$ is the total length of the SBD and the length of the depletion region, $l_d = \sqrt{2\varepsilon\varepsilon_0(\phi_0 - U)/eN}$, is determined by the barrier height ϕ_0 at the semiconductor-metal boundary and the voltage U applied to the SBD.

Under constant voltage conditions, the instantaneous response to any local perturbation originated by a scattering event of a free carrier leads to the appearance inside the whole SBD of the electric field fluctuation, $\delta E(x)$, which results in a self-consistent motion of all the free carriers in the SBD volume, described by spatial shifts δx_+ and δx_- in the n^+ and n-regions respectively. These collective motions in each region (labelled below as $i = +, -$) are described by the Langevin equations as:

$$\frac{d^2}{dt^2}\delta x_i - v_i \frac{d}{dt}\delta x_i = \frac{e}{m}\delta E_i + f_i \tag{1}$$

where

$$\delta E_i = \frac{e}{\varepsilon\varepsilon_0}\begin{cases} -(1 - r_+)N_+\delta x_+ + r_- N\delta x_- & , i = + \\ r_+ N_+ \delta x_+ - (1 - r_-)N\delta x_- & , i = - \end{cases} \tag{2}$$

is the electric field perturbation, v_i the carrier momentum relaxation rate, f_i the Langevin force describing thermal fluctuations in n^+ and n regions, $r_i = l_i/L$ the relative length of

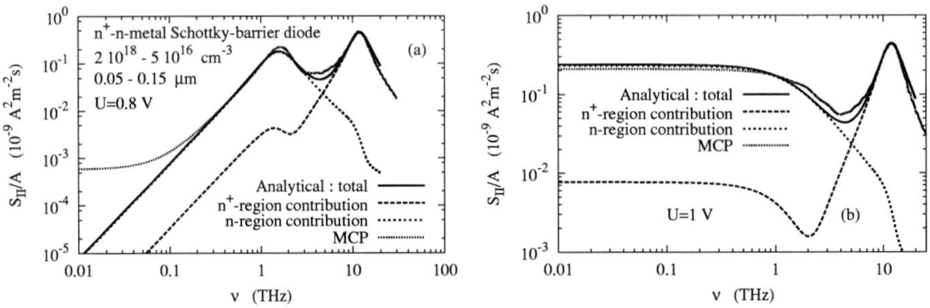

FIGURE 2. Spectra of current fluctuations per unit surface (A) in GaAs $0.05 - 0.15\ \mu m$ n^+n–metal SBDs at room temperature calculated under (a) barrier-limited and (b) flat-band conditions.

the i-region, $l_- = l - l_d$ the length of the neutral part of the n-region. To be consistent with Nyquist theorem, the spectral density of the thermal Langevin force, $S^i_{ff}(\omega) = \frac{1}{AN_i l_i} 4k_B T \frac{\nu_i}{m}$, must be normalized to the total number of free carriers inside the i-region.

By solving Eq. (1) in the spectral representation, one obtains the spectral density of thermal fluctuations of the SBD current as:

$$S_{II}(\omega) = 4k_B T C \omega^2 \frac{\nu_+ r_+ \omega_+^2 o(\omega_-) + \nu_- r_- \omega_-^2 o(\omega_+)}{o(\Omega_+) o(\Omega_-)} \tag{3}$$

where $C = \varepsilon\varepsilon_0 A/L$ is the geometrical capacitance of the SBD, $\omega_i^2 = (e^2/\varepsilon\varepsilon_0 m)N_i$ is the plasma frequency of the free carriers in the i-region, $o(\omega_i) = (\omega^2 - \omega_i^2)^2 + \nu_i^2\omega^2$ and Ω_i $(i = +, -)$ are the resonant frequencies determined by the roots of the characteristic equation $\Delta(\omega) = 0$ at $\nu_i = 0$. For $l_d < l$, one obtains the following approximation for the resonant frequencies: $\Omega_1^2 = \omega_h^2 - \beta\omega_-^2 r_d$, $\quad \Omega_2^2 = (1+\beta)\omega_-^2 r_d$ where $\omega_h^2 = \omega_+^2(r_- + r_d) + \omega_-^2 r_+$, $\beta = r_+(\omega_+^2 - \omega_-^2)/\omega_h^2$, $r_d = l_d/L$.

Figure 2 reports the comparison of the spectra of current fluctuations for a GaAs n^+n SBD with $N = 5 \times 10^{16}\ cm^{-3}$, $N_+ = 2 \times 10^{18}\ cm^{-3}$, $l = 0.15\ \mu m$ and $l_+ = 0.05\ \mu m$ calculated at 300 K with the MCP technique and with the analytical model described above. For completeness, the figure also presents the separate contributions coming from the thermal noise sources of n^+ and n-regions (long- and short-dashed curves, respectively). The results refer to the two operation regimes corresponding to barrier-limited Fig. 2(a) and flat-band Fig. 2(b) conditions. Analytical calculations are performed in accordance with Eq. (3) by taking $\nu_+ = 4 \times 10^{13}$ and $\nu_- = 10^{13}\ s^{-1}$ as estimated from MC simulations of bulk GaAs with the same carrier concentrations.

Under the barrier limited transport (see Fig. 2(a)), the HF noise spectrum of the SBD contains two resonant peaks, namely: (i) the low-frequency peak centered at $\omega = \Omega_- \approx \omega_-\sqrt{r_d}$ which describes the resonant reflection of carriers pertaining to the n-region from the Schottky barrier, and (ii) the high-frequency peak at $\omega = \Omega_+ \approx \sqrt{\omega_+^2(r_- + r_d) + \omega_-^2 r_+}$ caused by the resonant reflection of carriers pertaining to the n^+-region from the n^+n-homojunction, and which practically is independent of the depletion region length r_d. As follows from Fig. 2(a), the MCP and analytical model give

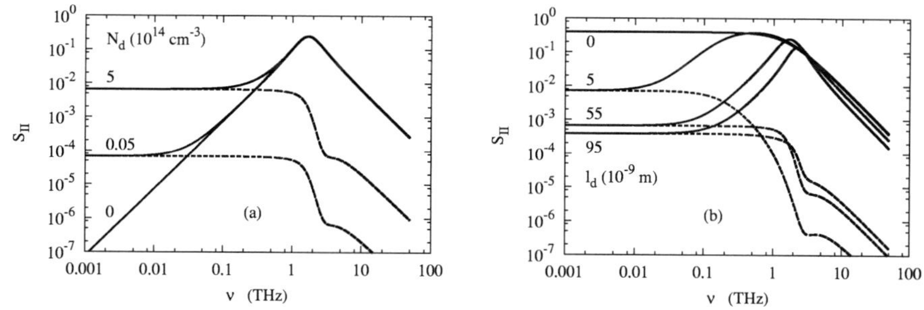

FIGURE 3. Total spectra of current fluctuations and transport-induced contribution (solid and dashed lines, respectively) in GaAs 0.15 μm nn_d−metal SBDs at room temperature calculated by the analytical model at different values of: (a) depletion region doping level N_d ($l_d = 0.055$ μm) and (b) depletion region length l_d ($N_d = 5 \times 10^{13}$ cm^{-3}).

full qualitative and good quantitative agreement in the HF region. In the LF region, in contrast to the MCP results, the model predicts the decrease of the intensity of fluctuations as ω^2 what corresponds to the absence of carrier flow through the depletion region.

In the limit of flat-band conditions (see Fig. 2(b)), corresponding to the disappearance of the empty depletion region at $r_d \to 0$, the considered SBD transforms into an inhomogeneous resistor with n^+n-homojunction and the noise spectrum predicted by the analytical model fully agrees with MCP calculations in the whole frequency range. There appears a plateau in the LF region while the resonant peak related with the presence of the n^+n-homojunction remains practically the same. As follows from Eq. (3), under the limit of flat-band condition with $\Omega_- = 0$ the LF behavior of the noise spectrum as ω^2 disappears, i.e. $S_{II}(0) \neq 0$. This means that there is the opening of the channel responsible for carrier transport through the SBD. From the formal point of view, such a transport becomes possible when one of the characteristic frequencies of collective self-consistent motions of carriers inside the structure under test becomes equal to zero.

To overcome the problem of the absence of carrier transport through the SBD under barrier limited conditions, and thus open a channel for such a transport through the depletion region, we assume the presence of some small concentration of carriers, N_d in this region. In other words, the Schottky barrier will be treated as a nn_d-homojunction similarly to the n^+n-homojunction considered above. For simplicity, in further considerations the latter will be omitted. In this case, in the absence of the depletion region ($l_d = 0$), the n^+ and neutral n regions of the above described model (see Fig 1(a)) will play the role of the n and n_d regions of the extended model (see Fig. 1(b)). Indeed, in this case $\Omega_- = 0$, that is there takes place carrier transport through the structure, and, under the condition $\omega_+ >> \omega_-$, the characteristic frequency $\Omega_+ \approx \omega_+\sqrt{r_-}$ coincides with the resonant frequency of carriers reflected from the Schottky barrier.

Figure 3 illustrates the spectrum of fluctuations with a channel open in such a way for carrier transport through the SBD. The modifications of the spectral density of current fluctuations $S_{II}(\omega)$ with the variation of the depletion region concentration N_d at the same l_d/l and of the depletion region length l_d at the same N_d are presented in Fig. 3 (a) and (b), respectively. The solid lines correspond to the total $S_{II}(\omega)$ while the noise induced by carrier transport through the n_d-region is shown by the dashed lines. As follows from Fig. 3, the approximation of the SBD by a nonhomogenous resistor with nn_d-homojunction leads to the following results: (i) it well reproduces the SBD noise spectrum in the whole frequency region, and (ii) it allows to extract the low-frequency (shot-noise-like) component induced by carrier transport through the structure the cut-off of which takes place at the same frequency as the resonance of carriers reflected from the Schottky-barrier.

DISCUSSION AND OPEN QUESTIONS

The above-used approximation model of SBD by n^+n-homojunction means that we replace the shot-noise description related with fluctuations of the number of carriers which cross the depletion region by usual diffusion noise related with velocity fluctuations. Indeed, by determining from Eq. (3) a value of the spectral density at the zero frequency as

$$S_{II}(0) = 4kT\left[\frac{1}{C}\left(\frac{r_+\nu_+}{\omega_+^2} + \frac{r_-\nu_-}{\omega_-^2}\right)\right]^{-1} \tag{4}$$

and calculating then the voltage drop at the structure in the presence of the flowing current I

$$U = \left[\frac{1}{C}\left(\frac{r_+\nu_+}{\omega_+^2} + \frac{r_-\nu_-}{\omega_-^2}\right)\right]I \tag{5}$$

it is easy to see that the noise described by the model is the diffusion noise in full agreement with Nyquist theorem (the quantity in square brackets is the Ohmic resistance of the structure). Here, there appear a question: is such a replacement physically meaningful or it is merely a game based on the identical form of the noise spectrum?

As shown in Refs. [3,4], in the presence of long-range Coulomb interactions in a linear homogeneous resistor with diffusive current flow there is a crossover from the usual diffusion noise to the shot noise when:

$$2eI = 4kT/R \tag{6}$$

where I is the current flowing through the linear resistor R. Such a crossover to appear requires that the transit time due to drift be shorter than that due to diffusion processes. The latter conditions is just fulfilled in the depletion region. In essence, this justifies, from the physical point of view, the possibility to use the resistive model for the description of the noise of carriers flowing through the depletion region, i.e. the shot noise.

The crossover condition given by Eq. (6) allows us to estimate an effective concentration of carriers in the depletion region approximated by a linear diffusive resistor:

$$R_d = \frac{1}{C}\frac{r_-\nu_-}{\omega_-^2} \tag{7}$$

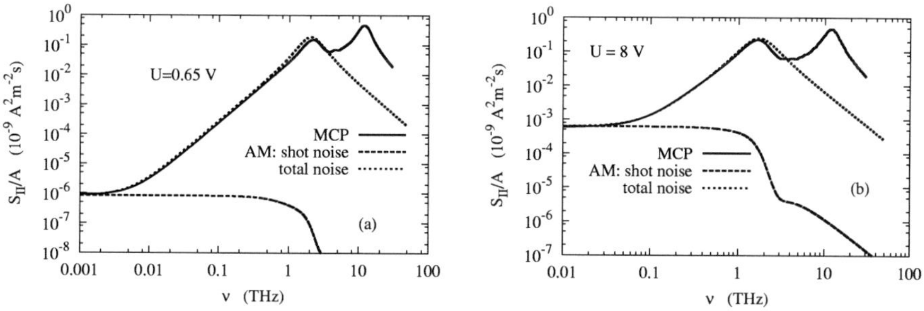

FIGURE 4. Spectra of current fluctuations per unit surface (A) calculated by MCP method in GaAs $0.05 - 0.15\ \mu m\ n^+n-$metal SBD at room temperature and by analytical model in nn_d region of the same SBD.

In the exponential region of current-voltage relation of the SBD the flowing current can be represented as:

$$I = AT^2 exp[-\frac{1}{2}\frac{l_d^2}{\lambda_d^2}] \tag{8}$$

where $\lambda_d = \sqrt{kT\varepsilon\varepsilon_0/e^2N}$ is the Debye length and A the Richardson constant. For simplicity we neglect in Eq. (8) the carrier flow from the metal to the semiconductor. By combining Eqs. (6)-(8) one obtains the estimation for the plasma frequency ω_- (concentration) in the depletion region as function of l_d, thus defining all the parameters necessary for the analytical model. The results of such calculations for the SBD described above are presented in Fig. 4 where short- and long-dashed curves show the total and transport-induced spectral densities of current fluctuations obtained by the analytical model (AM) with l_d as the only fitting parameter. For comparison, the solid line presents MCP results which fully coincides with analytical model in the low-frequency region including the reflected carrier resonance.

Thus, the same analytical model based on Eq. (3) is able to describe both the low- and high-frequency parts of the SBD noise spectrum.

ACKNOWLEDGMENTS

This work is supported by NATO Grant PST.EAP.CLG 980629.

REFERENCES

1. M. Trippe, G. Bosman, and A. van der Ziel, IEEE Trans. Microwave Theory Techn. **MTT-34**, 1183 (1986).
2. P. Shiktorov, E. Starikov, V. Gružinskis, L. Reggiani, L. Varani, J.C. Vaissière, IEEE Electr. Dev. Lett. **26**, 2 (2005).
3. G. Gomila and L. Reggiani, Phys. Rev. B **62**, 8068 (2000).
4. G. Gomila, C. Pennetta, L. Reggiani, M. Sampietro, G. Ferrari and G. Bertuccio, Phys. Rev. Lett. **92**, 226601 (2004).

How Does One Describe A Time-Varying Statistical Spectrum: Transforming Stochastic Differential Equations Into Phase-Space

Lorenzo Galleani* and Leon Cohen†

*Politecnico di Torino, Corso Duca degli Abruzzi 24, 10129 Torino, Italy
†Hunter College, City University of New York, 695 Park Avenue, New York

Abstract. We address some fundamental issues regarding nonstationary stochastic processes. Among the questions we discuss are how one describes a nonstationary process and how one obtains a governing differential equation for it. We argue that a simplification occurs when one studies stochastic process in the time-frequency phase space. This leads to a number of interesting questions. How can one obtain the equations of motion for reduced quantities such as the mean and variance of a nonstationary stochastic process? How can one define local stationarity? How can nonstationary stochastic systems be modeled from experimental data? Among other questions.

Keywords: nonstationary noise, time-frequency analysis, Wigner spectrum
PACS: 05.40+j

INTRODUCTION

Nonstationary noise is common in nature and there are many ways it is generated. Systems whose input is stationary noise can produce as outputs nonstationary noise in a variety of ways. Commonly, the transient solution after the system is turned on is certainly nonstationary. Also, the system itself can have random time dependent coefficients. In these cases the statistical properties of the output noise vary with time and hence the common statistical methods to analyze these stochastic processes are not appropriate. In approaching the issue of nonstationary noise the following fundamental question arises:

a) How does one describe nonstationary noise from a mathematical point of view? For the stationary case one uses the power spectrum and the autocorrelation function and associated ideas. What one needs for the nonstationary case is a generalization of the power spectrum. That is one needs a definition of a time dependent power spectrum. One also needs a generalization of the associated ideas, such as the generalization of the autocorrelation function for the time dependent case.

b) What is the equation of motion for the time dependent spectrum?

c) How can one obtain the equations of motion for reduced quantities such as the mean and variance of a nonstationary stochastic process?

d) How can one define local stationarity? Or approximate stationarity? Or find the places where a time-varying stochastic process seems to be locally stationary?

e) There are a number of representations that can be used to define a time dependent spectrum. They are all equivalent in some sense but is one representation better

CP800, *Unsolved Problems of Noise and Fluctuations: UPoN 2005*
edited by L. Reggiani, C. Pennetta, V. Akimov, E. Alfinito, and M. Rosini

than another?

f) How can nonstationary stochastic systems be modeled from experimental data?

In a series of recent papers [11, 12] we have addressed some of these issues, but the fundamental questions remain unsolved and we discuss these issues here.

RANDOM DIFFERENTIAL EQUATIONS AND THEIR TRANSFORMATION INTO PHASE SPACE

To keep things concrete we will deal with a stochastic process, $X(t)$, that is the solution of a linear stochastic differential equation

$$a_n(t)\frac{d^nX(t)}{dt^n}+a_{n-1}(t)\frac{d^{n-1}X(t)}{dt^{n-1}}\cdots+a_1(t)\frac{dX(t)}{dt}+a_0(t)X(t)=F(t) \qquad (1)$$

where $F(t)$ is a wide sense stationary random process. The coefficients can be time dependent and random but we here consider the case where the coefficients are deterministic functions. What is the output $X(t)$ and how can one define the appropriate time dependent statistical properties? It is clear that $X(t)$ is nonstationary and that is the case even if the coefficients are not time dependent. We have argued that one way to study $X(t)$ is by way of the Wigner spectrum[1] [6, 7]

$$\overline{W}_X(t,\omega)=\frac{1}{2\pi}\int E\left[X^*(t-\tfrac{1}{2}\tau)X(t+\tfrac{1}{2}\tau)\right]e^{-i\tau\omega}\,d\tau \qquad (2)$$

which is the ensemble average of the Wigner distribution [1]. This can be thought of as a generalization of the power spectrum. We give an example in Fig. 1. It is an approximation of the solution to the following time-varying random differential equation

$$\frac{d^2X(t)}{dt^2}+2\mu\frac{dX(t)}{dt}+(\omega_0^2+\varepsilon t)X(t)=F(t) \qquad (3)$$

where $F(t)$ is white Gaussian noise. The coefficients μ and ω_0 have been chosen so that the oscillator is underdamped and also we have chosen $\varepsilon \ll 1$. The figure clearly shows that the stochastic process is nonstationary. The nonstationarity is due to the time-varying coefficient and has the effect of making the resonant frequency of the harmonic oscillator to be a function of time. This effect cannot be observed if one evaluates the ordinary frequency spectrum.

We have argued that a proper way to study such systems is directly in phase space, and we have shown how to obtain the equation of motion for $\overline{W}_X(t,\omega)$. We consider the

---

[1] All integrals without limits imply integration from $-\infty$ to $+\infty$,

$$\int \equiv \int_{-\infty}^{+\infty}$$

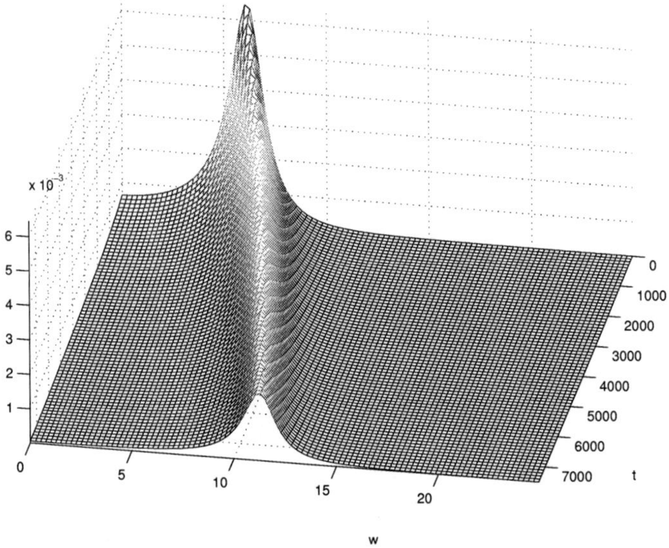

FIGURE 1. Approximated Wigner spectrum $\overline{W}_X(t,\omega)$ of the nonstationary random process $X(t)$ defined in Eq. (3). The instantaneous spectrum shows a bandpass behavior with a central frequency that varies with time.

class of processes defined by Eq. (1), which we rewrite in polynomial form

$$P(D,t)X(t) = F(t) \tag{4}$$

where

$$P(D,t) = a_n(t)D^n + a_{n-1}(t)D^{n-1}\cdots + a_1(t)D + a_0(t) \tag{5}$$

and

$$D = \frac{d}{dt} \tag{6}$$

The Wigner spectrum $\overline{W}_X(t,\omega)$ of $X(t)$ satisfies

$$P^*(A,\mathscr{E})P(B,\mathscr{F})\overline{W}_X(t,\omega) = \overline{W}_F(t,\omega) \tag{7}$$

where the operators are

$$A = \frac{1}{2}\frac{\partial}{\partial t} - i\omega \qquad\qquad B = \frac{1}{2}\frac{\partial}{\partial t} + i\omega \tag{8}$$

$$\mathscr{E} = t + \frac{1}{2i}\frac{\partial}{\partial\omega} \qquad\qquad \mathscr{F} = t - \frac{1}{2i}\frac{\partial}{\partial\omega} \tag{9}$$

and the complex conjugation of $P^*(A,\mathscr{E})$ means that the coefficients used in the expansion of the functions $a_k(t)$ have to be conjugated. Eq. (7) is an equation in time and frequency, and can now be used to ascertain the dynamic of the system in the time-frequency domain.

EQUATION OF MOTION FOR THE MEAN AND SPREAD

Is it possible to obtain an equation of motion for the instantaneous mean value of the Wigner spectrum $\overline{W}_X(t,\omega)$? Can one obtain an equation for the instantaneous conditional spread or variance? Both of these questions are of fundamental interest in the study of systems governed by Eq. (3). There are several reasons why that is true. First, the instantaneous mean and variance can summarize the behavior of the instantaneous spectrum at any given time with much less information than the entire spectrum itself. Also, it is much easier to estimate the mean and variance from experimental data than to estimate the entire process. Finally, having an equation for the instantaneous mean and variance could clarify the dynamics of the system. In terms of the Wigner distribution the mean $m(t)$ and instantaneous conditional spread $\sigma^2_{\omega|t}(t)$ are defined by

$$m(t) = \int \omega \overline{W}_X(t,\omega) d\omega \tag{10}$$

$$\sigma^2_{\omega|t}(t) = \int (\omega - m(t))^2 \overline{W}_X(t,\omega) d\omega \tag{11}$$

LOCAL STATIONARITY

Suppose we are given $\overline{W}_X(t,\omega)$, can one find where the process is locally stationary and more fundamentally how can one define local stationarity? The reason this is important is that often stochastic systems appear to be stationary for a small amount of time. But how can one find where that happens? This is perhaps analogous to asking where a function is approximately constant and the answer is of course the region around a point where the derivative is zero. Can a similar criterion be given for a nonstationary process? The concept of local stationarity has a long history [8, 9] but now we have an equation of motion for the process and hence that may provide a handle towards the problem.

OTHER REPRESENTATIONS

There have been many representations that have been studied besides the Wigner distribution [2, 4, 3, 5, 10] and their properties investigated for many years. In 1966 Cohen showed how an infinite number of distributions may be generated. All distributions can be studied by way of the kernel method and in particular, they may be obtained from [4]

$$C(t,\omega) = \frac{1}{4\pi^2} \iiint E[X^*(u-\tau/2)X(u+\tau/2)]\, \phi(\theta,\tau)\, e^{-j\theta t - j\tau\omega + j\theta u} du\, d\tau\, d\theta \tag{12}$$

where $\phi(\theta,\tau)$ is the kernel which characterizes the distribution and its properties. One can express the general class, $C(t,\omega)$, in terms of the Wigner distribution

$$C(t,\omega) = \iint g_{12}(t'-t, \omega'-\omega) W(t',\omega')\, dt'\, d\omega' \tag{13}$$

with

$$g(t,\omega) = \frac{1}{4\pi^2}\iint \phi(\theta,\tau)\, e^{j\theta t + j\tau\omega}\, d\theta\, d\tau \tag{14}$$

We have been able to obtain the equations of motion for an arbitrary $C(t,\omega)$. It is an interesting question to ask whether the equation for $C(t,\omega)$ is better, in both the conceptual sense and possible solution.

HOW CAN NONSTATIONARY STOCHASTIC SYSTEMS BE MODELED

Suppose we have a nonstationary stochastic process: can it be modeled as the output of an ordinary stochastic differential equation, as per Eq. (1)? That is, how can one find the functional form and the coefficients of the equation? Since we proved that there is an immense simplification when the equation is transformed into phase space, we believe that it may be easier to obtain the functional form and the coefficients by devising the estimation problem in phase space. But we do not know how to do it.

CONCLUSION

The characterization of nonstationary random processes is a subject of great interest, bus this subject has not been developed thoroughly, and there are numerous unanswered fundamental questions. We have addressed some of these questions and have described an approach that is based on the time-frequency analysis of the random process defined as the solution of a stochastic differential equation. Of particular interest is to find an equation of motion for the instantaneous mean and conditional variance of a random process defined by a time-varying stochastic differential equation. The solution of this problem could lead to a deeper understanding of the system dynamics and could improve estimation algorithms for nonstationary random processes.

ACKNOWLEDGMENTS

The research of Leon Cohen was supported by the Air Force Information Institute Research Program (Rome, NY) and the NSA HBCU/MI program.

REFERENCES

1. E. P. Wigner, "On the quantum correction for thermodynamic equilibrium," *Physical Review*, vol. 40, pp. 749–759, 1932.
2. L. Cohen, "Time-Frequency Distributions – A Review," *Proc. of the IEEE*, vol. 77, pp. 941–981, 1989.
3. L. Cohen, *Time-Frequency Analysis*, Prentice-Hall, 1995.
4. L. Cohen, "Generalized phase–space distribution functions," *Jour. Math. Phys.*, vol. 7, pp. 781–786, 1966.

5. M. G. Amin, "Time–varying spectrum estimation of a general class of nonstationary processes," *Proceedings of the IEEE,* vol. 74, pp. 1800–1802, 1986.
6. W. D. Mark, "Spectral analysis of the convolution and filtering of nonstationary stochastic processes," *J. Sound Vib.*, vol. 11, pp. 19-63, 1970.
7. W. D. Mark, "Power spectrum representation for nonstationary random vibration," in: *Random Vibration – Status and Recent Developments*, D. H. Ielishakoff (Ed.), Elsevier, 1986.
8. R. A. Hedges and B. W. Suter, "The Diagonally Optimized Spread: An Optimized Spread for Quantifying Local Stationarity," *Proc. SPIE*, vol. 4391, pp. 102-108, 2001.
9. R. A. Hedges and B. W. Suter, "Improved radon-transform-based method to quantify local stationarity," Proc. SPIE, vol. 4116, pp. 17-24, 2000.
10. Special Issue on Applications of Time-Frequency Analysis, P. Loughlin (ed.), *Proc. of the IEEE*, vol. 84, no. 9, 1996.
11. L. Galleani, L. Cohen, "Approximation of the Wigner Distribution for Dynamical Systems Governed by Differential Equations," *EURASIP Journal on Applied Signal Processing*, vol. 2002, no. 1, pp. 67 72, 2002.
12. L. Galleani, L. Cohen, "Time Frequency Wigner Distribution Approach To Differential Equations," Invited chapter in K. Barner and G. Arce (eds.), *Nonlinear Signal and Image Processing: Theory, Methods, and Applications*, Nov. 2003, CRC Press.

Renewal, Modulation and Blinking Quantum Dots

P. Paradisi*, P. Allegrini†, F. Barbi†, S. Bianco** and P. Grigolini‡,**

*ISAC-CNR, Sezione di Lecce, Str. Prov. Lecce-Monteroni Km 1.2, 73100 Lecce, Italy
†Dipartimento di Fisica "E. Fermi", Università di Pisa, Largo Pontecorvo 3, 56127 Pisa, Italy
**Center for Nonlinear Sciences, University of North Texas, P.O. Box 311427, Denton, Texas 76203-1427, USA
‡IPCF-CNR, Area della Ricerca di Pisa, Via G. Moruzzi, 56124 Pisa, Italy

Abstract. A method of time series analysis based on the aging properties of renewal processes is introduced. The method compute the amount of *renewal aging* in a time series. It is shown that the method allows to distinguish between systems with a renewal dynamics from those driven by slow modulation of the parameters. The application of the method to Blinking Quantum Dots shows a maximum of renewal aging. This excludes the possibility of interpreting the dynamics as a slow modulation process and indicates a possible modelling approach in the renewal perspective.

Keywords: Renewal Aging, Renewal Theory, Modulation Processes, Time Series Analysis, Stochastic Models
PACS: 78.67.Hc, 05.40.Fb, 05.45.Tp, 78.20.Bh

Introduction

In the last few years there has been an increasing interest in the intermittent fluorescence of new nanomaterials [1, 2], known as *Blinking Quantum Dots* (BQDs) [3]. The fluorescence of BQDs is essentially described in terms of a dichotomous dynamics with random jumps between a *light-on* (+) and a *light-off* (−) state. The Probability Density Function (PDF) $\psi(\tau)$ of Waiting Times (WTs) in a certain state shows an inverse power-law tail, with exponent between 1.5 and 2. The interest towards BQDs in this paper is related to this last feature, which allows to consider BQDs as a prototype of a complex system. The term *complexity* is related to the breakdown of the conditions on which Boltzmann's view is based, i.e., short-range interactions and short memory. A typical feature of complex systems is related to the power-law behaviour of some dynamical characteristics such as the distribution of Poincaré recurrence times and of WTs or the decay of correlations. The origin of power-law behaviour is an open question and different approaches have been proposed. Here we limit ourselves to considering two different proposals, generating the same PDF of WTs, which will be referred to as *renewal* [4, 5, 6] and *modulation* [7]. A modulation process is defined as a stochastic process satisfying standard statistical mechanics, but with parameters randomly fluctuating in time. Conversely, a renewal process is related to the presence of recurrent *critical events* or jumps, originated from the dynamics, whose time distances are random independent variables. After each critical event, memory of the past is erased and the system restarts from a new initial condition. The behaviour of the PDF of WTs is not sufficient to discriminate between these two approaches. In order to address this point, a new method of time series analysis, based on the aging properties of renewal processes [8], will be

CP800, *Unsolved Problems of Noise and Fluctuations: UPoN 2005*
edited by L. Reggiani, C. Pennetta, V. Akimov, E. Alfinito, and M. Rosini

introduced. This method, which will be referred to as *renewal aging analysis*, allows to evaluate a sort of *renewal aging amount* from the WT sequence. This sequence is usually obtained from a generic experimental time series, such as the emission intensity of BQDs, after some definition of critical event is given. The results of this analysis can be used to reveal if the dynamical origin of the time series can be described by means of a *renewal* or a *slow modulation* process, known also as *superstatistics* [9, 10]. In fact, it is shown that the two approaches are characterized by different aging properties, i.e., a maximum of aging for renewal processes and absence of aging for slow modulation processes. The application of the method to BQD data shows that the underlying dynamics should be reasonably modelled with a renewal perspective.

Renewal processes and aging

In the renewal approach, the system behaves as it restarts from a new initial condition after each critical event. If the relaxation time to an equilibrium distribution (if it exists) is much longer than the mean WT, then the system shows aging during the time interval between two events, while it shows a sudden rejuvination at the time instant of the critical event. Considering that observation of the system starts randomly at some (*aging*) time t_a between two such events, then the observed (*aged*) PDF of WTs depend on the starting time and can be used as an indicator of the aging properties of the system. Aging properties of renewal processes have been extensively studied in recent years [11, 8], also using a simplified stochastic model of the Manneville map [12]. This is a physical model, used as a prototype of turbulent intermittency, showing the occurrence of critical events [13]. The stochastic model can be used as a generator of WTs whose PDF is a Pareto-Nutting distribution: $\psi(\tau) = (\mu - 1)T^{\mu-1}/(\tau + T)^{\mu}$ with $T > 0$ and $\mu > 1$. The limiting case $\mu = T = \infty$ corresponds to an exponential PDF for the WTs: $\psi(\tau) = r\exp(-r\tau)$. This case corresponds to having a Poisson process, i.e., a stochastic process with Poisson distribution of the number of events in a certain time interval. Now, a useful approximated expression relating the *aged* PDF $\psi_{t_a}(\tau)$ to the *non-aged* PDF $\psi(\tau)$ was already derived under the renewal assumption [12, 8]. This is given by:

$$\psi^{\text{ren}}_{t_a}(\tau) = \frac{1}{K_{t_a}} \int_0^{t_a} dy \psi(\tau + y), \tag{1}$$

where K_{t_a} is the normalization constant. In the case of a renewal process, a difference between aged and non-aged PDFs is the indication of the presence of aging effects. It must be stressed that the renewal Poisson process (defined above) shows no aging. On the contrary, renewal processes with a Pareto-Nutting distribution of the WTs show aging effects.

Now, let us consider a time series obtained from the experimental measurement of the observable of a physical system, as the emission intensity of BQDs. Suppose that a set of recurrent events can be defined and let us extract the WTs sequence. Note that, as regards BQDs, WTs in $+$ and $-$ states must be distinguished. Given the experimental sequence of WTs, it is suitable to set up a technique of statistical analysis being able to specify the renewal character of the dynamics generating the time series. First of all, a stationary assumption must be introduced. Then, the first analysis performed on the WT sequence is the check of stochastic independence, obtained by the evaluation of the standard correlation function. As we will see in next subsection, this is not enough to say that the process is renewal. For this reason, let us define a sort of measure of the *renewal aging amount* in the time series. In order to do this, let us first introduce the *experimental aged PDF* $\psi_{t_a}^{\exp}(\tau)$. Given the experimental WT sequence, this is computed from the direct definition of the aged PDF without the renewal assumption. Practically, this is done as follows. Firstly, a Gibbs ensemble is generated from the WT sequence by relating the first WT of the n-th trajectory to the n-th WT in the sequence: the first trajectory begins from the first WT, the second one from the second WT and so on. Then, given the *aging time* t_a, a truncated WT is defined for each trajectory as the First Exit Time obtained imposing that the observation begins at $t = t_a$. The non-aged PDF $\psi(\tau) = \psi_0^{\exp}(\tau)$ is simply obtained by imposing $t_a = 0$. From Eq. (1) it is then possible to compute also the *renewal aged PDF* $\psi_{t_a}^{\mathrm{ren}}(\tau)$. Let us now introduce the Survival Probability (SP), defined as the probability for the WT being longer of a given value τ. This is easily obtained starting from the related PDF:

$$\Psi_{t_a}^{\mathrm{ren}}(\tau) = \int_\tau^\infty \psi_{t_a}^{\mathrm{ren}}(\tau')d\tau' = 1 - \int_0^\tau \psi_{t_a}^{\mathrm{ren}}(\tau')d\tau'. \tag{2}$$

A similar definition is valid for $\Psi_{t_a}^{\exp}(\tau)$. Note that, as for the non-aged PDF, the non-aged SP is simply given by the aged one with $t_a = 0$: $\Psi(\tau) = \Psi_0^{\mathrm{ren}}(\tau) = \Psi_0^{\exp}(\tau)$. It is now possible to define the Aging Intensity Function (AIF) as:

$$I_a(\tau) = \frac{\Psi_{t_a}^{exp}(\tau) - \Psi_0^{exp}(\tau)}{\Psi_{t_a}^{ren}(\tau) - \Psi_0^{exp}(\tau)}. \tag{3}$$

If the analysis of the WT sequence shows $\Psi_{t_a}^{\exp}(\tau) = \Psi_{t_a}^{\mathrm{ren}}(\tau)$ and, consequently, $I_a(\tau) = 1$, the time series has a maximum of *renewal aging*. Conversely, if it results $\Psi_{t_a}^{\exp}(\tau) = \Psi_0^{\exp}(\tau)$, so that $I_a(\tau) = 0$, there's *lack of aging* in the time series. In the former case it is reasonable to assume that the dynamics has a renewal character. In the latter case, it is possible that the system is described by a renewal Poisson process. This can be verified checking if the non-aged PDF of WT is an exponential function. If not, the time series cannot be generated by a renewal process. It must be stressed that, due to approximation errors in Eq.(1), the AIF is usually overestimated. Random generation of WT sequences with Pareto-Nutting distributions have been performed using an inversion method [14]. The values $T = 1$ and, alternatively, $\mu = 1.65$ or $\mu = 1.8$ were used. Applying the renewal aging analysis, i.e., computation of AIF, to these WT sequences, we found, as expected, asymptotic values of AIF, $I_a(\infty)$, which will be referred as Aging Indicator (AI). From this analysis, we should obtain $I_a(\infty) = 1$, but overestimation errors of about $20 - 30\%$ were found. As an example, with $t_a = 220$ the values $I_a(\infty) = 1.28$ for $\mu = 1.65$ and $I_a(\infty) = 1.23$ for $\mu = 1.8$ were found. As expected, the approximation errors decrease as t_a increases.

Modulation processes

Following the approach given in [7], a fundamental example of superstatistics (or slow modulation) is given by a Orstein-Ulenbeck process [15] with a slowly changing relaxation rate. This process was studied by Beck [7], who found that, if the fluctuating rate has an equilibrium distribution given by a Γ distribution, then the distribution of velocities is a Tsallis distribution [16]. The physical interpretation of this approach is that of a Brownian particle in equilibrium with the thermal bath on short times, but with slowly fluctuating properties of the thermal bath itself. In the following, the same approach is applied to WT statistics. Let us remind that a renewal Poisson process describes the sequence of WTs between two collisions in standard Brownian motion. The PDF of WTs is an exponential function, whose rate r depends on the mean free path. If the rate slowly fluctuates in time and has an equilibrium distribution $p_{\text{eq}}(r)$, then the modulated PDF can be computed as:

$$\psi(\tau) = \int_0^\infty dr\, p_{\text{eq}}(r)\, r e^{-r\tau}. \tag{4}$$

If $p_{\text{eq}}(r)$ is given by the Γ distribution [7] the resulting PDF of WTs is given by the Pareto-Nutting distribution. In order to find the aging amount of this process, numerical generation of random WT sequences have been performed. Numerically, this could be obtained with the following approach. Firstly, random draws from the distribution $p_{\text{eq}}(r)$ are obtained using a rejection method [14]. Then, for each given rate r, N_m random WTs are drawn from the related exponential PDF using an inversion method [14]. The resulting PDF was verified to be given by the Pareto-Nutting distribution. The slow modulation case is obtained in the limit $N_m \gg 1$. The *renewal aging analysis*, described above, was applied to the simulated WT sequences. The values $T = 1$ and $\mu = 1.8$ were chosen. In order to enhance the effect of increasing N_m, the results for $N_m = 1$ and $N_m = 100$ are shown in Fig. 1. As regards aging properties, a fast modulation process ($N_m = 1$) is virtually equivalent to a renewal process, while it is clear that the increase of N_m produces a considerable aging reduction. Fig. (2) shows the decay of the AI $I_a(\infty)$ as

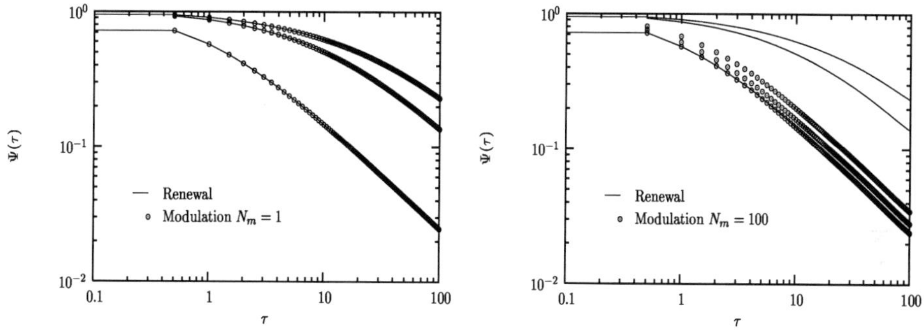

FIGURE 1. Comparison between renewal and modulation processes with $N_m = 1$ (left) and $N_m = 100$ (right). Both cases refer to a Pareto-Nutting distribution of WTs, Eq. (??). $T = 1$ and $\mu = 1.8$. From bottom to top: $t_a = 0, 20, 60$.

a function of N_m, which was found to be an inverse power-law,i.e., $I_a(\infty) \sim (N_m)^{-\alpha}$ with $\alpha = 0.70 \pm 0.02$. From these results and remembering that a Poisson process does not yield aging, it can be argued that, in this approach, the amount of renewal aging in the modulation process decreases as the number of Poisson events (relative to the number of modulation events) increases.

Renewal aging analysis of BQD data

We are now in a position to discuss the experimental results on real BQD data sequences. The experimental data discussed in this Section have been obtained by Prof. M. Kuno

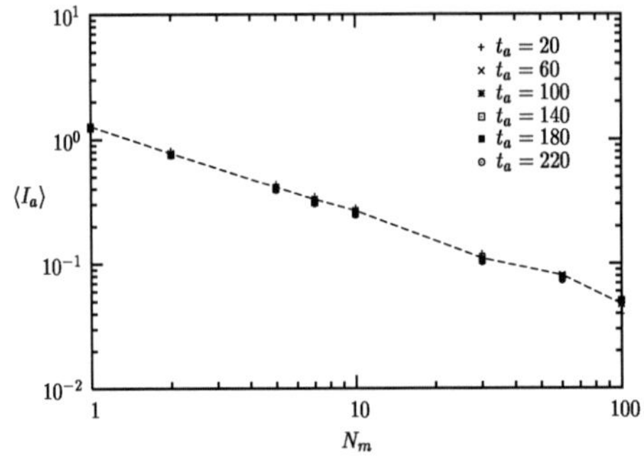

FIGURE 2. $I_a(\infty)$ as a function of N_m, for different aging times. $\mu = 1.8$, $T = 1$.

and V. Protasenko, Dept. of Chemistry and Biochemistry, University of Notre Dame. Our data set consists of 32 sequences of BQD fluorescence intensities. Each sequence contains 1 hour of records, sampled every ms, for a total amount of 360000 data per sequence. In order to separate the two states, it is necessary to define a threshold intensity, above which the signal corresponds to $+$ and below to $-$ state. This separation was established following the approach adopted in [17]. Denoting with σ the Root Mean Square of the $-$ state, the threshold turns out to be located at the value 2σ over the $-$ state itself. With this definition a sequence of symbols $+$ and $-$ is obtained and it is straightforward to derive the related WT sequence. In this case, it is also suitable to distinguish between the WTs referring to $+$ and $-$ states. Renewal aging analysis was performed on these sequences. Asymptotics values of AIF were found to be well-defined, so that a non-ambiguous Aging Indicator can be obtained for given N_m and t_a. The results are shown in Fig. 3. It is worth noting the difference between the WTs in $+$ and $-$ states. In the former case, the decay is faster and it seems to have a sort of passage between an intermediate power-law behaviour and an asymptotic exponential decay. In the latter case, the decay is much slower and the tail is a well-defined inverse power-law function with exponent ranging from 1.65 ± 0.02 to 1.80 ± 0.05. This result confirms the earlier observation of Chung and Bawendi [18]. The computation of $I_a(\infty)$ for $t_a = 220$, averaged over all WT sequences, gives 0.96 for the $-$ state and 0.87 for the $+$ state. Considering the overestimation error discussed above and looking at Fig. 2, it can be argued that the most reasonable modelling approaches should be given by the renewal or by fast modulation perspectives. In any case, this analysis shows that slow modulation perspective cannot be the right modelling approach. A possible interpretation is that the dynamics of BQD intermittent fluoroscence gives rise to a sequence of critical main events with WTs satisfying renewal condition with a non-exponential PDF. Then, the time distances between these main events are filled with secondary Poisson events reducing the total aging amount.

Conclusions

A new method of statistical analysis, based on the concept of aging properties of renewal process and referred to as *Renewal Aging Analysis* has been introduced. This method was shown to be able to distinguish between different processes leading to the same PDF of

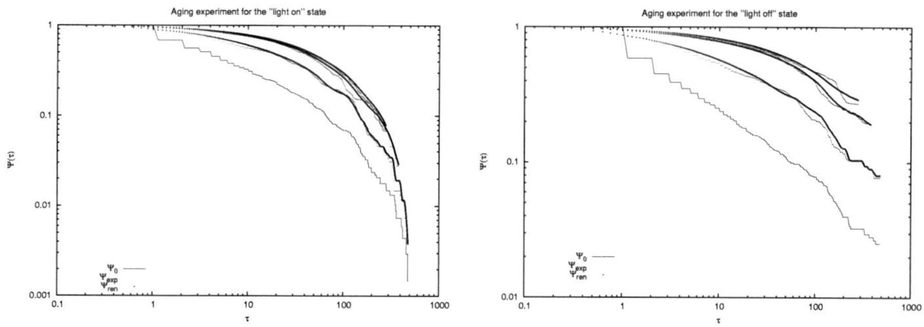

FIGURE 3. Aging analysis on WTs in the + (left) and − (right) state. (a) thin continuous line: $\Psi_0^{exp}(\tau)$, (b) dotted line: $\Psi_{t_a}^{exp}(\tau)$, (c) thick continuous line: $\Psi_{t_a}^{ren}(\tau)$. From bottom to top: $t_a = 20, 120, 220$.

WTs, with particular attention to renewal and (fast or slow) modulation processes. The method has been applied to BQD fluoroscence data, finding that BQD dynamics should have a renewal character.

We thankfully acknowledge Prof. M. Kuno and V. Protasenko for kindly making their experimental data available to us. S.B. and P.G. thankfully acknowledge financial support from Welch through grant # 70525. P.P. thankfully acknowledges financial support from MIUR through grant "bando 1105/2002", project #245 and CNR 2005 short term mobility program.

REFERENCES

1. R.G. Neuhauser, K. T. Shimizu, W.K. Woo, S.A. Empedocles, and M.G. Bawendi, Phys. Rev. Lett. **85**, 3301 (2000).
2. M. Kuno, D.P. Fromm, H.F. Hamann, A. Gallagher, and D.J. Nesbitt, J. Chem. Phys. **112**, 3177 (2000).
3. Y.-J. Jung, E. Barkai, and R.J. Silbey, Chem. Phys. **284**, 181 (2002).
4. E.W. Montroll and G. E. Weiss, J. Math. Phys. **6**, 167 (1965).
5. D.R. Cox, *Renewal Theory*, Chapman and Hall, London (1962).
6. I.M. Sokolov, Phys. Rev. E **68**, 041101 (2002).
7. C. Beck, Phys. Rev. Lett. **87**, 180601 (2001).
8. P. Allegrini, G. Aquino, P. Grigolini, L. Palatella, and A. Rosa, Phys. Rev. E **68**, 056123 (2003).
9. C. Beck, E.G.D. Cohen, Physica A **322**, 267 (2003).
10. E.G.D. Cohen, Physica D 193 (2004) 35.
11. G. Aquino, L. Palatella, and P. Grigolini, Phys. Rev. Lett. **93**, 050601 (2004).
12. G. Aquino, M. Bologna, P. Grigolini, B.J. West, Phys. Rev. E **70**, 036105 (2004).
13. P. Manneville, J. Physique **41**, 1235 (1980).
14. W.H. Press *et al.* , *Numerical recipes in C++ : the art of scientific computing* Cambridge University Press, Cambridge (2002).
15. C.W. Gardiner, *Handbook of stochastic methods*, Springer-Verlag, New York, II edition, 1983.
16. C. Tsallis, J. Stat. Phys. **52**, 479 (1988).
17. M. Kuno, D.P. Fromm, H.F. Hamann, A. Gallagher, and D. J. Nesbitt, J. Chem. Phys. **115**, 1028 (2001).
18. I. Chung and M. G. Bawendi, Phys. Rev. B **70**, 165304 (2004).

Non-Exotic Theory Of 1/f Noise As A Trace Of Infralow-Frequency Fluctuations

A. Ya. Shul'man

Institute of Radio Engineering and Electronics, Russian Academy of Sciences, Moscow 125009, Russia. E-mail: ash@cplire.ru

Abstract. This report is aimed at reviving the explanation of flicker-noise observations as the result of spectral measurement of very low-frequency but stationary narrow-band fluctuations named as infralow-frequency noise (ILF noise). Such a kind of the spectrum analyzer output takes place when the ILF-noise correlation time is much longer than the analyzer reciprocal bandwidth. This result is valid for both analog and digital spectral measurements. The measured signal is proportional in this case to the mean square and not to the spectral density of the noise. The equilibrium temperature fluctuations and the defect-motion as mechanisms of $1/f$ noise in metal films are reconsidered from this point of view. It is shown that the ILF-noise approach allows to remove the main objection against temperature fluctuation model and difficulties within the defect-motion model widely discussed in literature.

Keywords: 1/f noise, spectral methods in noise spectroscopy
PACS: 02.70.Hm, 05.40.Ca, 07.50.Ek, 72.70+m

INTRODUCTION

It was shown in [1, 2] that flicker-noise observations can be explained as the result of spectral measurement of very low-frequency but stationary narrow-band fluctuations named as infralow-frequency noise (ILF noise). Such a kind of the spectrum analyzer output takes place when the measured noise spectrum concentrates on frequencies far lower than the operation range of the spectrum analyzer and the ILF-noise correlation time is much longer than the analyzer reciprocal bandwidth. This result is valid for both analog and digital spectral measurements. The measured signal is proportional in this case to the mean square rather than the spectral density of the noise, and does not reflect in any way the dependence of the latter on the frequency. Within the scope of this model the universality of the $1/f$ noise follows from the universality of inevitable fluctuations of intensive thermodynamic variables like the temperature and a power-low falling of narrow-band filter spectral distributions in the wings. It is of importance here the more isolated is the sample under study from the environment the more low-frequency are corresponding fluctuations. Thus, it is impossible to reach the low-frequency saturation region of the noise spectral density if the duration of the measurements is less than response time of the sample - environment coupling.

The necessity to find not only slow fluctuation process but also with $1/f$ -spectrum serves as the reason to call in question some physical mechanisms that by their qualitative attributes and quantitative estimations could be a cause of the excess current noise. Example of this is the temperature model of $1/f$ noise in metal films suggested by Voss and Clarke on the basis of convincing experiments [3]. Taken initially with hopes and

CP800, *Unsolved Problems of Noise and Fluctuations: UPoN 2005*
edited by L. Reggiani, C. Pennetta, V. Akimov, E. Alfinito, and M. Rosini

an enthusiasm [4, 5] this theory has later been rejected as an explanation of $1/f$ noise in metal films [6, 7]. In addition to the proof that the heat conductivity equation does not lead under no circumstances to the $1/f$ -spectrum of temperature fluctuations [8], another weighty argument against this model was the absence of spatial correlations of the current fluctuations [9, 10].

On the contrary, there is no need in the scope of the ILF-theory for exponentially-wide distribution of relaxation times of any kind [2]. The interpretation of the spatial correlation measurements is also to be reconsidered since the mean square of the conterminous fluctuations is measured in the case of ILF-noise rather than the spectral density of them. It will be shown here that all abovementioned objections against temperature fluctuations are withdrawn from the ILF-noise point of view.

THE NOISE SPECTROSCOPY AND INFRALOW-FREQUENCY FLUCTUATIONS

Analogue measurements

It was shown in [1, 2] that the output $G^{out}(\omega_0)$ of an analogue spectrum analyzer as a function of the resonant frequency ω_0 can be described in a good approximation by the convolution integral

$$G^{out}(\omega_0) = \frac{1}{\Delta}\int_{-\infty}^{\infty} \frac{d\omega}{2\pi}\mathscr{K}\left(\frac{\omega_0 - \omega}{\Delta}\right) G_y(\omega), \Delta/\omega_0 \ll 1, 0 < |\omega_0| < \infty, \tag{1}$$

(the notation is slightly changed). Here $G_y(\omega)$ is the spectral density of the stationary random input process $y(t)$ with $\langle y^2 \rangle < \infty$, $\mathscr{K}\left(\frac{\omega_0-\omega}{\Delta}\right)$ plays the role of the instrumental function of the spectrum analyzer normalized by the condition $\mathscr{K}(0) = 1$, Δ is the effective bandwidth of the analyzer defined by

$$\frac{1}{\Delta}\int_{-\infty}^{\infty} \frac{d\omega}{2\pi}\mathscr{K}\left(\frac{\omega}{\Delta}\right) = 1. \tag{2}$$

An important point is that $\mathscr{K}(\omega)$ and $G_y(\omega)$ are nonnegative and normalizable functions of the frequency and can be approximated by δ-function each at proper interrelation between the effective bandwidths. If $\mathscr{K}(\omega)$ is noticeably more narrow-band function then

$$G^{out}(\omega_0) = G_y(\omega_0)$$

and the result of measurement $G^{out}(\omega_0)$ is the common-adopted estimation of the true spectral density $G_y(\omega)$. On the other hand if the investigated spectrum has a narrower contour than the instrumental function, then the result of the measurements will be the contour of the instrumental function

$$G^{out}(\omega_0) = \frac{1}{\Delta}\mathscr{K}\left(\frac{\omega_0 - \omega_m}{\Delta}\right)\langle y^2 \rangle, \tag{3}$$

where ω_m is the frequency at which the maximum of $G_y(\omega)$ takes place. Spectra are always measured with a narrow-band filter, when the quality factor or the figure of merit

$$Q \equiv f_0/\Delta \gg 1, (f_0 = \omega_0/2\pi) \tag{4}$$

. It is now obvious from Eq. (3) that when the spectral density of the noise is narrower than $\mathscr{K}(\omega)$, and the noise frequency is low enough $\omega_m \ll \omega_0$ [this is the definition of the term "infralow-frequency" noise], then in the usual $Q = const(\omega_0)$ measurements the output should have a dependence of the $1/f$ type:

$$G^{out}(\omega_0) = \frac{Q}{f_0}\mathscr{K}(2\pi Q)\left\langle y^2\right\rangle. \tag{5}$$

On the other hand, if the measurements are carried at $\Delta = const(\omega_0)$, then the output signal has the frequency dependence of the far wing of $\mathscr{K}(\omega)$, which likewise has as a rule the form $1/f^n$. In either of these case the measured values are proportional to the mean square $\left\langle y^2\right\rangle$ of the input random process.

Digital spectral analysis

In some respect the analysis of digital methods, from our point of view, is even simpler, since the formula similar to Eq. (1) is already known. There is the exact expression of the form

$$G^{out}(\omega_0) = T_0 \int_{-\infty}^{\infty} \frac{d\omega}{2\pi}\mathscr{W}(T_0(\omega_0 - \omega))\, G_y(\omega), \quad -\infty < \omega_0 < \infty, \tag{6}$$

where T_0 is the time slice of the one realization of the random process and $\mathscr{W}(\omega)$ is the spectral window. Widely used windows can be written in the form

$$\mathscr{W}(\omega) \propto \left[\frac{\sin(\omega T_0/n)}{\omega T_0/n}\right]^n \tag{7}$$

where $n = 2,4$ corresponds to the Bartlett and Parzen window, respectively.

The account for finiteness of the time step τ_0 and application of discrete Fourier transformation lead to a little bit changed expressions for spectral windows. For Bartlett window one has [2]

$$\mathscr{W}_d(\omega) = \frac{\sin^2(\omega N\tau_0/2)}{N^2\sin^2(\omega\tau_0/2)} \tag{8}$$

and $T_0 = N\tau_0$, N is the number of data points in the measured realization. So long as the frequency ω is small compared with the Nyquist frequency, defined by the condition $\omega_N = 2\pi f_N = \pi/\tau_0$ the continuous and discrete Fourier transformations yield practically identical expressions for the instrumental function (spectral window). However, since $\mathscr{W}_d$ is the periodical function of ω, the formula (6) can give rise to the aliasing effect.

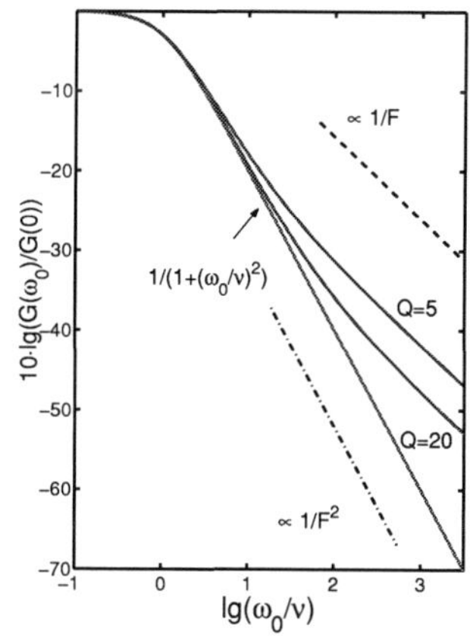

FIGURE 1. Dependence of the spectrum analyzer output on the tuning frequency, Eq. (11), under condition of the constant quality factor $Q = f_0/\Delta$. Two typical values of Q are presented corresponding to the equivalent noise bandwidth of 20% and 5%, respectively. The spectral density of the noise at the analyzer input, Eq. (10), is also shown for comparison.

To avoid this, the input process $G_y(\omega)$ should be filtered by low-pass filter with cut-off frequency of order of f_N. With this condition in mind, the comparison of Eqs.(1)-(2) to Eqs.(6)-(8) shows that the effective bandwidth is $\Delta = 1/T_0$ in the digital spectral analysis with the Barttlet window.

Comparing Eqs.(6)-(7) with Eqs.(1) and (3) we can easily verify that in the case of digital processing of ILF noise measurements, an erroneous observation of the spectrum of the $1/f^n$ type will also take place. The digital measurements are always carried out with a constant bandwidth ($\sim 1/T_0$), at least in each of several frequency intervals into which the entire investigated band is subdivided. As a result of effective "averaging" of the resultant curve over all the sub-bands, a spectrum of the type $1/f^n$ can be obtained with $1 < n < 2$ for Barttlet window under conditions of ILF-noise measurements.

An illustrative example

The observable signature of ILF noise measurements is shown in the figure 1. It was calculated in the case of Lorentzian form of both the noise spectral density and the narrow-band input filter of the spectrum analyzer. The following expressions were substituted in Eq. (1) as the $\mathscr{K}(\omega)$, Δ and $G_y(\omega)$

$$\mathscr{K}(\omega_1 - \omega) = \frac{\gamma^2/4}{(\omega_1-\omega)^2 + \gamma^2/4}, \; \omega_1^2 = \omega_0^2 - \gamma^2/4, \Delta = \gamma/4 \tag{9}$$

$$G_y(\omega) = \langle y^2 \rangle \frac{2\nu}{\omega^2 + \nu^2} \tag{10}$$

to obtain the explicit formula for the spectrum analyzer output

$$G^{out}(\omega_1) = \langle y^2 \rangle \frac{2(\nu+\gamma/2)}{\omega_1^2 + (\nu+\gamma/2)^2}. \tag{11}$$

The parameters ω_0 and γ are defined by the equation of the simple resonant filter[1]

$$\ddot{x} + \gamma\dot{x} + \omega_0^2 x = y(t).$$

It is easily seen that the condition (4) for the analyzer to be narrow-band (in our case it takes the form $\pi\gamma/2\omega_0 \ll 1$) still does not determine uniquely the $G^{out}(\omega_1)$ dependence. It is also important to know the relation between γ and ν, i.e., between the widths of the input spectrum and the resonant filter. At $\gamma/\nu \ll 1$, i.e., in the case of a relatively narrower filter, we have the usual result $G^{out}(\omega_1) = G_y(\omega_1)$. In the other limiting case, however, when $\gamma/\nu \gg 1$, we obtain

$$G^{out}(\omega_1) = \langle y^2 \rangle \frac{\gamma}{\omega_1^2} \equiv \langle y^2 \rangle \frac{4\Delta}{\omega_1^2} \tag{12}$$

From Eq. (12) it follows directly that if the measurements are carried out in this situation at a constant Q, i.e., under the condition $\gamma/\omega_1 = const(\omega_1)$, then the output signal will imitate the presence of noise with a spectrum $1/f$ at the input of the analyzer. It is clearly seen in Fig.1 that although at high frequencies ($\omega_1 \gg \nu$) the true spectral density of the noise (10) is everywhere proportional to $1/\omega^2$ and the condition (4) is satisfied, all this is still not enough to let the region of the maximum $\mathscr{K}(\omega_1 - \omega)$ to make the main contribution to the integral in Eq. (1). Conversely, as soon as the filter bandwidth γ becomes larger than the width of the input spectrum ν, a transition from the expression (10) to (12) takes place quite rapidly. In other words, the output signal of the analyzer ceases to yield information on the form of the spectrum of the input and begins to vary with frequency in accordance with the $1/f$ law.

The measured curves of such a kind can be found in literature without a discussion of possible ILF-noise interpretation. As a result, some speculations take place in order to explain the observed transition of measured spectral density from $1/f^2$ to $1/f$ either by means of an unusual frequency dependence of the sample properties (see, for instance [11]) or a hypothesis is offered based on a decomposition of measured noise spectrum on the sum of Lorentzian and $1/f$ components originated by different causes [12].

$1/F$ NOISE IN METAL FILMS

Investigations of the excess current noise in metal films purposed to shift the study of unclear $1/f$ noise onto the objects that should be physically more defined than non-uniform semiconductors or thermionic emission. If refer a start of that researches into the

[1] To compare Eqs. (11) and (12) with corresponding Eqs. (10) and (12) in [2] one has to correct the misprint there and change $\gamma \rightarrow \gamma/2$ taking into account the condition (4).

beginning of 1970th, then it is to be ascertained that more 30 elapsed years and numerous new obtained data have not led to the decision of a problem of origin of such a spectrum. Although observed dependencies of spectral density of resistance fluctuations at any fixed frequency on various parameters of a sample material or measurement conditions point out as if these variables are related with $1/f$ noise, it is failed to deduce from here such its statistical properties as the spectral form or the behavior of spatial correlations. The approach from the point of view of ILF-noise allows to remove an acuteness of some contradictions and restate the problem for additional experimental studies.

The temperature fluctuations

Temperature fluctuations as the cause of $1/f$-noise have been rejected by three principal arguments [6, 7]: 1. The impossibility to deduce the $1/f$ spectral density for fluctuations of the volume-averaged film temperature. 2. The absence of correlations of current fluctuations in spatially dividual parts of the film. 3. An effect of the substrate on the measured resistance fluctuations. Let us consider these conclusions from ILF-noise point of view.

To analyze the Voss and Clarke experiments it is necessary to address to Eqs.(6) and (8). Assuming $G_y(\omega)$ is ILF-noise and ω is not too close to Nyquist frequency ω_N we have for the relative voltage fluctuations [2]

$$\frac{G_U^{+out}(f)}{U^2} = \frac{2\sin^2(N\pi\tau_0 f)}{\pi^2(f/\Delta)}\frac{\beta^2\langle\delta T^2\rangle}{f} = \frac{1}{\pi^2(f/\Delta)}\frac{\beta^2 T^2}{3N_a f}. \tag{13}$$

Here $G^+(f) = 2G(f)$ is the spectral density in $f > 0$ frequency semi-axis, $\beta = d\ln R/dT$, R is the resistance of the film, $\langle\delta T^2\rangle = k_B T^2/C$ is the mean square of the equilibrium temperature fluctuations averaged over the film volume, C is the film heat capacity. In the case of the room temperatures and monatomic metals $C = 3k_B N_a$ and N_a is the number of atoms in the film. The substitution of experimental data [3] combined with the hand-book material parameters for the Bi and Au films in Eq. (13) and hand-made estimations of $\Delta_{Bi} = 2Hz$ and $\Delta_{Au} = 1Hz$ results in[2]

$$G_{Bi}^{+out}(10Hz) = 15\cdot 10^{-17}Hz^{-1},\ G_{Au}^{+out}(10Hz) = 0.58\cdot 10^{-17}Hz^{-1}$$

The measured values are $13\cdot 10^{-16}Hz^{-1}$ (Bi) and $0.6\cdot 10^{-16}Hz^{-1}$ (Au). The difference just in 10 times for both cases suggests that there might be some problem with conversion of digital Fourier-transform coefficients in $1/Hz$ spectral density. Anyway, the obtained of-order-of-magnitude coincidence is worthy of notice, taking into account the lack of fitting parameters and indeterminacy of the film dimensions mentioned in [3]. The $1/f^\alpha$ dependence of the analyzer output with $1 < \alpha < 2$ follows from Eq. (13) as it is explained in Sec.2.2.

As for the lack of spatial correlations of the volume-averaged variables, it is necessary to note, first of all, that the dependence of the relative correlation function like in Eq. (13) on the sample volume Ω in the form of $1/\Omega$ may occur only if the characteristic scale r_c of the spatial correlation function $\langle\delta T(r,t)\delta T(r_1,t_1)\rangle$ is much less than the dimensions of the sample. The discussion of this point can be found in [13] from general point of view and in [6] with reference to $1/f$-noise problems. Therefore, the spectral density of the frequency Fourier components should be spatially correlated if they are formed from the spatial Fourier components with characteristic wavelength of order of the film dimensions. Thus there is no sense to look for the spatial correlations between different parts of the film if the $1/\Omega$ dependence of correlation function is observed and vice versa.

This receipt regards to the case of the usual spectral measurements. In the case of ILF-noise the output signal is proportional to the mean square of temperature fluctuations taken in the same instant time. In an equilibrium macroscopic body out of phase transition region such fluctuations are always uncorrelated on macroscopic scales in accordance with the principle of additivity of extensive variables in thermodynamics. The explicit prove of this statement for the spatial-averaged temperature fluctuations can be found in [8]. So in the case of ILF-noise the absence of spatial correlations can not be an argument against of the temperature-fluctuation origin of $1/f$ -noise.

To consider the dependence of $1/f$ noise on the substrate supporting the film and on a degree of the film-substrate thermal contact one need to account for the last stage of the time relaxation of the averaged temperature fluctuations, i.e. a heat interchange with the environment. The Newton boundary condition for the thermoconductivity equation gives rise to the temperature relaxation with one characteristic time τ, resulting in the simple expression for the spectral density of the form

$$G_T(\omega) = \langle \delta T^2 \rangle \frac{2\tau}{1+\omega^2\tau^2} \tag{14}$$

As τ is very large in comparison with the reciprocal lowest measured frequency the ILF noise should take place and the dependence on the substrate in Eq. (13) comes from the heat capacitance C, which is determined by all structure under study including an involved part of the substrate. The latter depends also on the heat-transfer factor of the film-substrate interface and can increase as τ decreases, giving rise to the decrease of $G^{out}(\omega)$. As $1/\tau$ gets into the measured frequency range the usual spectral measurements takes effect and $1/f$ behavior of the spectral density can give place to an ordinary one.

So both characteristic dependencies of spectral measurements on the substrate, which were discussed in [5] without definite conclusions, can be understood on the base of ILF-noise model within the scope of the temperature fluctuations' theory as well as two other problems considered in this section. One can believe it is premature to discard the temperature fluctuations as the source of $1/f$ noise in metal films.

The defect-motion model

After Eberhard and Horn have found out the activation dependence of the $1/f$ -noise on the metal film temperature [14] the opportunity of $1/f$-noise generation by defects of the lattice was taken into consideration. As usual in the field of flicker noise there is a contradiction between the observable well-defined activation energy and the necessity to have a wide-spread range of these energies to produce $1/f$ spectral dependence over an extended frequency range. It was note in [14] there is no possibility to avoid this contradiction using only one activation process. It was suggested to divide the defect contribution into the current noise by two parts connected with the fluctuations of the defect number due to a vacancy creation/annihilation and the fluctuation of the free-carrier-scattering cross-section due to a defect motion with different activation energy. The last process should be giving the bread distribution of the activation energy. However, the observed noise activation energy in various metal films are less than known vacancy formation energy approximately to 10 times [14]. Hence the observed activation energies should be related with the motion of defects, leaving no place for the usual construction of $1/f$-noise model.

Kogan and Nagaev [15] suggested to connect the motion-defect contribution to the flicker noise with some characteristics of the internal friction process. The reasonable estimation of possible noise level was obtained for that mechanism (see also [7]) but the origin of the wide distribution of the activation energy remained unexplained [6]. The ILF-noise model does not required any real $1/f$ spectrum of fluctuations. It should be sufficiently low-frequency only. The internal friction processes rather meet this requirement having characteristic relaxation frequencies in the diapasons of $10^{-12}, 10^{-8}, 10^{-5}$ Hz with magnitudes of reciprocal quality $Q_{\max}^{-1}$ in the $10^{-1} \div 10^{-2}$ range [16]. One has to recall that the value of $Q^{-1} \approx 10^{-4}$ for the background level of the internal friction was used in [15] with a positive conclusion.

CONCLUDING REMARKS

It is worthy to mention the work [17] where the magnitude of $1/f$ noise in SQUIDs was evaluated as a result of ILF-fluctuations of the current in superconducting or normal metal shield surrounding the SQUID. The results obtained there keep their significance for high-temperature SQUIDs, too [18].

It seems evidently that the ILF-noise model could remove contradictions in existing theories of $1/f$ noise and should be involved in the analysis of experimental results in the field of flicker-noise investigations. A necessary change in the experimental technique has been shortly discussed in [2].

ACKNOWLEDGMENTS

This work was partially supported by Russian Foundation for Basic Researches.

REFERENCES

1. A. Ya. Shul'man. *Pis'ma Zh. Tekh. Fiz.* **7**, 783 (1981) [Sov. Tech. Phys. Lett. **7**, 337 (1981)]
2. A. Ya. Shul'man. *Zh. Eksp. Teor. Fiz.* **81**, 784 (1981) [Sov. Phys. JETP **54**, 420 (1981)]
3. R.F. Voss and J. Clarke. *Phys. Rev.* **B13**, 556 (1976)
4. Sh.M. Kogan. *Usp. Fiz. Nauk*, **123**, 131 (1977) [Sov. Phys. Usp. **20**, 763 (1977)]
5. P. Dutta and P.M. Horn. *Rev. Mod. Phys.* **53**, 497 (1981)
6. Sh.M. Kogan. *Usp. Fiz. Nauk,* **145**, 285 (1985)

7. M. B. Weissman. *Rev. Mod. Phys.* **60**, 537 (1988)
8. L.A. Vainstein. *Zh. Eksp. Teor. Fiz.* **83**, 1841 (1982)
9. J.H. Scofield, D.H. Darling, and W.W. Webb. *Phys. Rev.* **B24**, 7450 (1981)
10. R.D. Black, M.B. Weissman, and F.M. Fliegel. *Phys. Rev.* **B24**, 7454 (1981)
11. S.K. Kim, A. van der Ziel. *Physica* **B98**, 303 (1980)
12. A. Bid, A. Guha, and A.K. Raychaudhuri. *Phys. Rev.* **B67**, 174415 (2003)
13. E.M. Lifshits and L.P. Pitaevskii. *Statistical Physics,* P.2., (Fizmatlit, M., 2001), Ch. IX
14. J.W. Eberhard and P.M. Horn. *Phys. Rev.* **B18**, 6681 (1978)
15. Sh.M. Kogan and K.E. Nagaev. *Sol. State Comm.* **49**, 387 (1984)
16. B.N. Finkelshtein. Internal friction. In *Encyclopedia of Physics,* v.1 ("Sov. Encyclopedia", M. 1988), p.289
17. S.S. Tinchev. *Physics Letters* **108A**, 357 (1985)
18. D. Koelle, R. Kleiner, F. Ludwig, E. Dantsker, and John Clarke. *Rev. Mod. Phys.* **71**, 631 (1999)

EXPERIMENTAL FRONTIERS ON NOISE AND FLUCTUATIONS

Noise In Nonohmic Regimes Of Disordered Systems

K. K. Bardhan*, C. D. Mukherjee* and U. N. Nandi†

*Saha Institute of Nuclear Physics, 1/AF Bidhannagar, Kolkata 700 064, India
†Scottish Church College, 1 & 3 Urquhart Square, Kolkata 700 006, India

Abstract. We present here a short review of mainly experimental properties of noise as disordered systems are driven into non-ohmic regimes by applying voltages of few volts only. It is found that the noise does not simply follow the resistance in that the direction of change of noise could be opposite to that of resistance. It is discussed how this and other properties make the noise a complementary and incisive tool for studying complex systems, particularly its dynamic properties. Study of noise in non-ohmic regimes in physical systems is rather in a nascent stage. Some of the open issues are highlighted.

Keywords: Noise, Non-ohmic/Nonlinear state, Disorder
PACS: 72.70.+m, 72.60.+g, 72.80.Ng

INTRODUCTION

The ubiquitous low frequency resistance fluctuations in conductors have proved in recent times to be an unique and increasingly useful tool for probing various condensed matter systems[1, 2]. For this purpose, fluctuations have been often studied also as a function of other relevent variables such as temperature. But, studies of fluctuations as a function of bias are rather scarce. In the early stage of noise study, the focus was on its fundamental properties and simple systems such as pure metals or semiconductors[3] were used. These systems remain ohmic except at very high electric fields and hence, were not particularly suitable for probing non-ohmic regimes. However, with advent of semiconducting devices that contain abundantly non-ohmic junctions such as metal-semiconductor and semiconductor-semiconductor ones, noise-related performance issues assumed much importance. Study of the noise in these non-ohmic junctions in late seventies (see Hooge et al.[4] for references) constitutes one of the earliest example of investigation of noise in non-ohmic regime. About the same time, noise measurements in a charge-density system of $NbSe_3$ as a function of electric field [5] revealed intricacies of the pinning dynamics, thereby proving its usefulness for the first time in a non-device physical system. In recent times, there are signs of increased level of activities mainly upon realisation that such studies would be uniquely useful especially in complex systems such as composites[6], mangenites[7] and others[8, 9] where usual transport measurements of average quantities prove to be inadequate. These systems have varying degree of disorder and are easily driven into non-ohmic states by application of bias of only few volts (thus, fluctuations in hot-electrons[10] remain beyond the scope of this review). Noise, being proportional to the fourth moment of the current distribution, is more sensitive to the microstructures (i.e. disorder) or onset of non-ohmic behaviour (i.e.

CP800, *Unsolved Problems of Noise and Fluctuations: UPoN 2005*
edited by L. Reggiani, C. Pennetta, V. Akimov, E. Alfinito, and M. Rosini

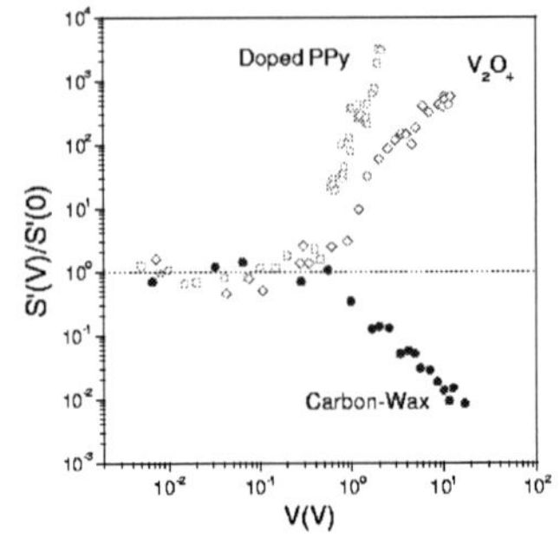

FIGURE 1: Suitably normalised relative noise as a function of bias in three systems with different conduction mechanisms. Resistances in all three cases decrease with bias. Yet, the relative noise behaves differently by decreasing in composites (carbon-wax), and increasing in other two cases. (Figure from Ref. [12]).

change in conduction mechanism) than the resistance which is proportional only to the second moment of the current distribution. Recently, nonlinearity has been used by Vilar and Rubi[11] to suggest an interesting possibility of suppressing intrinsic noise in a nonlinear system by adding external noise. Let us note that Hooge's empirical formula[13] can be generalised in the following manner using chordal resistance $R = V/I$:

$$\mathscr{S} = \frac{S_R}{R^2} = \frac{S_V}{V^2} = \frac{S_I}{I^2} = \frac{V^\gamma}{f^\alpha}\mathscr{R}(R) \tag{1}$$

where $S_X = < \delta X^2 >$ is the spectral density when $X = R,V,I$ is the fluctuating variable. I is the current through sample when V is the applied dc voltage. The value of the exponent γ is normally zero but can assume a non-zero value in disordered systems (see next section). The function $\mathscr{R}$ depends upon the particular system under consideration. In homogeneous samples, $\mathscr{S} \sim \mathscr{R} \sim R$. The relative noise power, $\mathscr{S}$ is analogous to conductance or resistance. Naturally, one would like to know how $\mathscr{S}$ behaves vis-a-vis R as a function of bias. For example, does it increase with R as it does *always* in ohmic states? As seen in Fig. 1, the relative noise does not necessarily follow the resistance in the nonohmic regime. The figure shows bias-dependent relative noise in three different systems with different conduction mechanisms. Resistances of all the samples *decrease* with bias. Yet, the relative noise behaves differently, decreasing in one case and increasing in others, obviously depending upon the underlying conduction dynamics. This amply illustrates the complementary role that the noise studies can play in unravelling complex systems. Interestingly, such behaviour of noise can be seen even from very general consideration of the bias-dependent resistance

$$R(I) = R_o f(I, R_o, c_1, c_2, ...) \tag{2}$$

where R_o is the resistance at zero bias, f is a function such that $f(0,..) = 1$. c_i's are various relevant parameters characterising the non-ohmic regime. Its fluctuation $\mathscr{S} = < \delta R^2 > /R^2$ is then given by

$$\mathscr{S} = s_1 + \sum_i \left(\frac{\partial \ln f}{\partial c_i}\right)^2 < \delta c_i{}^2 > + \sum_{i \neq j} \frac{\partial \ln f}{\partial c_i}\frac{\partial \ln f}{\partial c_j} < \delta c_i \delta c_j > + \sum_i \frac{\partial \ln f}{\partial c_i}\frac{\partial \ln f}{\partial R_o} < \delta c_i \delta R_o > \tag{3}$$

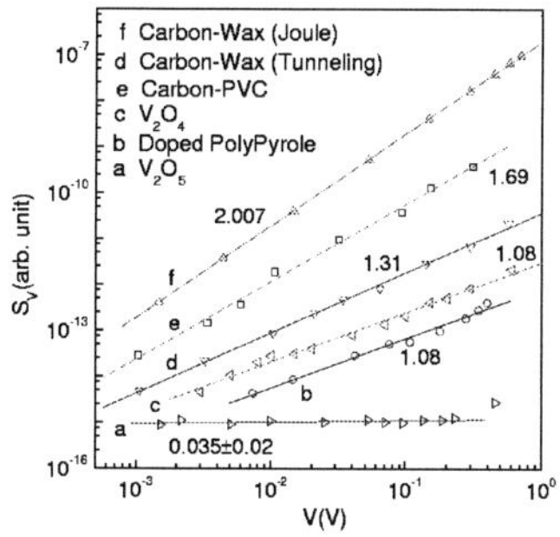

FIGURE 2: Noise voltage power vs. voltage in ohmic regimes of several disordered systems. The bias exponent β in each system is given by the slope of the corresponding curve as indicated. Notice that β values range from almost zero (V_2O_5, curve *a*) to 2 (carbon-wax composite, curve *f*). (Data of curves b,c,d from Ref. [12] and those of f from Ref. [15]).

where $s_1 = \mathscr{S}_o(1+(\partial \ln f/\partial \ln R_o)^2)$ and $\mathscr{S}_o =< \delta R_o{}^2 > /R_o{}^2$ is the relative noise in the ohmic state. If there is no correlation among c_i's and R_o it follows from above that the noise in non-ohmic regimes is *always* greater than that in ohmic state irrespective of whether f increases or decreases with bias. On the other hand, if any correlation does exist the noise may increase or decrease depending on the signs of the correlation terms on right hand side of Eq. (3). In practice, it may be a challenge to identify the appropriate set of parameters c_i's in physical systems. If f is known the corresponding noise could be calculated using Eq. (3). Such a method, for example, was used by Carbone et al.[14] to calculate the photocurrent noise.

We discuss below behaviour of noise in nonlinear regimes of various disordered systems and how the latter provides physical information. However, we first describe an anamolous behaviour in bias exponent ($\gamma \neq 0$ in Eq. (1)) that seems to be observed *only* in the ohmic regimes of disordered systems.

OHMIC REGIME REVISITED

In normal conductors like metals or semiconductors, numerous experiments [13] have verified that the bias exponent $\beta (= 2+\gamma = \partial \ln S_V/\partial \ln V)$ is equal to 2. In the ohmic state, it is an identity resulting from the assumption that resistance fluctuations exist *independent* of any current but become 'visible' as voltage fluctuations when a constant current I is passed: $< \delta V^2 > \; = I^2 < \delta R^2 > \; \sim V^2$. But in case of disordered systems, $\beta(\gamma)$ even in the ohmic states are often found to differ from the value of 2(0) in the ordered systems as seen in the Fig. 2. The systems in the figure range from polaronic (V_2O_5) to Mott (Doped Polypyrol,V_2O_4) to composites (carbon-wax, carbon-PVC). Carbon-wax samples in the figure differ only in carbon fraction. One with $\beta = 2$ is much less disordered than the one with $\beta = 1.3$. In fact in composites, there is a clear trend of β decreasing with increasing disorder (or decreasing conducting fraction) [12, 16]. The lower bound for $\beta(\gamma)$ is obviously 0(-2):

$$0 \leq \beta \leq ? \tag{4}$$

There is no theoretical upper bound of β. A value as high as 4 has been reported in discontinuous Pt films[17]. As per available data, there seems to be a preponderance

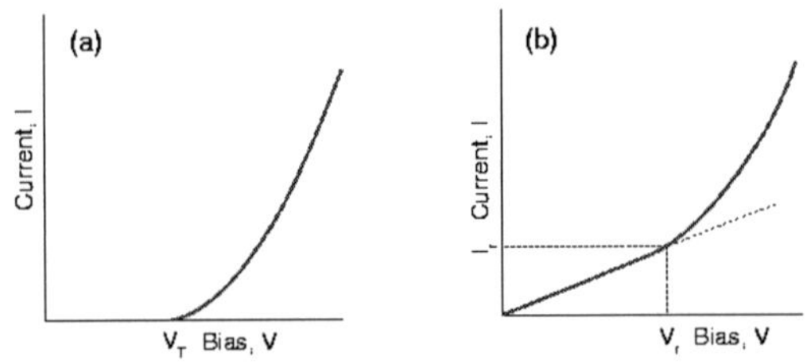

FIGURE 3. Schematic I-V curves corresponding to two possible modes of onset of nonlinearity: one (a) with a threshold, and another (b) without.

of β less than or equal to 2 (i.e. $\gamma \leq 0$)[18, 12, 16, 15]. However, there are several systems[19, 17, 20] where γ varies from negetive to positive values depending on sample structures. Till now, a proper understanding of such anamolous values of β is lacking. Interestingly, similar anamolous dependence of shot noise on bias has also been predicted in mesoscopic systems[21] in that $S/2eI$ is no longer a constant but a function of bias. Here, the electron-electron interaction has been shown not to disturb the ohmicity but affect the current fluctuations.

To take into account generally non-zero γ in disordered systems it is useful to generalise the definition of relative noise as used in Fig. 1: $\mathscr{S}' = \mathscr{S}/\mathscr{V}^{\gamma}$. Some authors (e.g. C. Parman et al.[20]) termed the noise as nonlinear in case of $\gamma \neq 0$. However, the term 'nonlinear noise' is used here to describe, in analogy with nonlinear conductance, bias-dependent noise.

ONSET OF NONLINEARITY

One of the important characteristics of nonlinearity is the mode of its onset. Accordingly, systems could be divided broadly into two groups (Fig. 3). One group of systems requires application of a finite bias V_T before any current can flow through samples (Fig. 3a) whereas in others, there is no apparent threshold for conduction (Fig. 3b). In the latter, one can still define a bias scale V_r (or alternatively, a current scale I_r) for onset of nonlinearity or simply, a scale of nonlinearity. Noise invariably increases rapidly with the onset of current in systems with finite thresholds. But, in systems without thresholds, the relative noise can either increase or decrease with bias as illustrated in Fig. 1.

Another aspect of the onset of nonlinearity involves the question whether noise and resistance start deviating from their respective ohmic behaviour at the *same* value of bias or not. Let V_n and V_r denote suitably defined values of bias for onset of nonlinearity for noise and resistance respectively in a system. Considering the fact that noise is more sensitive to disorder than resistance one would expect that V_n will be always less than or equal to V_r. Thus, the onset ratio $b = V_r/V_n$ is given by

$$b \geq 1 \tag{5}$$

This is seen in Fig. 4 which presents experimental $I-V$ and $S_V - V$ characteristics of a space-charge-limited-current(SCLC) diode.

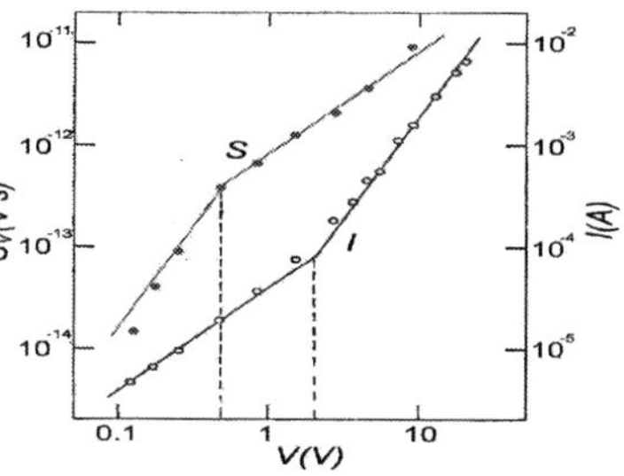

FIGURE 4: Experimental evidence that the onset ratio $b = V_r/V_n$ (about 5) is greater than 1. Curves are $I - V$ and $S_V - V$ characteristics of a SCLC diode. Figure from Ref. [4].

NOISE CHARACTERISTICS: $\mathscr{S}' - V/I/R$

While presentation of noise power as a function of either current or bias is straightforward, that as a function of (current-dependent) resistance requires furthur clarification as $\mathscr{S}' - R$ is not unique. Noise depends strongly on the methods used for changing the resistance. This is seen in Fig. 5 where the composite samples have their resistances varied in two different ways - by varying static disorder (i.e. fraction of conducting component) (a) and by Joule heating (b). The change in (non-ohmic) resistance of a sample in case of Joule heating is much less than the change in ohmic resistance caused by varying disorder but is accompanied by a huge increase in the noise compared to the other case.

Figs. 1 (composite) and 4 (SCLC diode) provide two different examples of monotonically varying noise characteristics in absence of any phase change. However, if there is any field-induced phase change the noise instead usually goes through a peak as in Fig. 7. Phases being referred to are not the usual thermodynamic ones, but rather ones characterised by different transport properties.

Normal systems (without any phase transition)

A typical characteristics posesses an initial linear part, then departs from linearity at some voltage V_n. Like nonlinear conductivity, the nature of the noise characteristics is determined by the physics of the given system. To our knowledge, composites are the most extensively studied system in non-ohmic regime[12, 15]. Limited experimental data that are presently available tend to suggest that there could be general relations

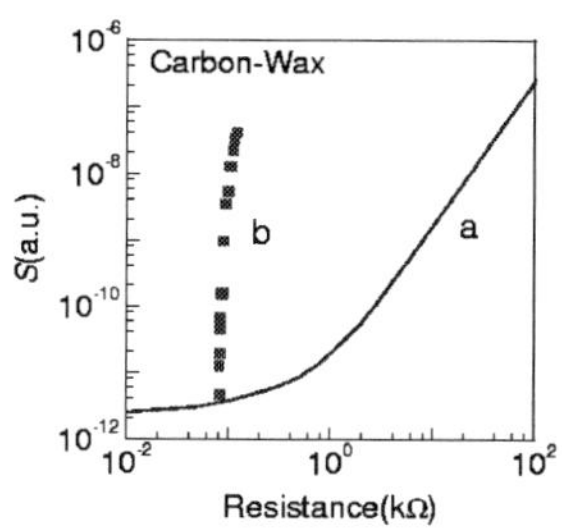

FIGURE 5: Comparision of noise in a composite system as a function of resistance when the latter is varied in two different ways. Curve a is obtained from samples with different degree of static disorder (ohmic resistance) while curve b was obtained from a sample subjected to Joule heating (non-ohmic resistance). (Figure from Ref. [15]).

TABLE 1. Four types of relative behaviour of resistance and noise in non-ohmic regimes of various systems. ΔR and $\Delta\mathscr{S}$ represent changes from ohmic values of resistance and relative noise respectively. $\mathscr{R}$ is the function as defined in Eq. (1). $\uparrow$ and $\downarrow$ indicate increase and decrease respectively. See text for other symbols.

ΔR	$\Delta\mathscr{S}$	**System**	$\mathscr{R}$	**Ref.**
$\downarrow$	$\downarrow$	Composites ($p \geq p_c$)	R^{w_V}	[12]
		Planar lipid membrane		[22]
		SCLC diode		[4]
		1D Charge-transfer salts	?	[23]
$\downarrow$	$\uparrow$	Variable Range Hopping	$\exp(R_o/R)^{w_V}$	[12]
		Solid lipid layers		[24]
		ZnO (varistar)	?	[25]
$\uparrow$	$\uparrow$	Composites ($p \gg p_c$),	Polynomial	[15]
$\uparrow$	$\downarrow$	?	?	

between the generalised relative noise power $\mathscr{S}'$ and the bias-dependent resistance R. When both are considered togeather one comes across four different scenerios according to the relative directions of changes as given in Table I.

R decreasing, $\mathscr{S}'$ decreasing. The first three systems exhibiting this type of variations are quite diverse in nature, yet appear to possess the same power-law dependence (see Fig. 6):

$$\mathscr{S}' \sim R^{w_V} \tag{6}$$

Here w_V is an exponent characteristic of the system in question. It is 1 in SCLC diode[4], about 3 in carbon-wax composites (low conducting fraction, $p \geq p_c$, p_c being the percolation threshold) [12] and 1.5 in planar lipid membrane. The value is well explained within the theoretical modelling of SCLC diode[26]. But there is no theoretical understanding for other systems. A question naturally arises here regarding any general feature that is common to all the three systems, and that could explain the reduction of noise? It turns out that in all these systems, increasing bias leads to progressive increase in number of fluctuators (injected charges in diode, opening of new conducting channels in composites or pores in membranes). The increase in number of fluctuators, in turn, leads to decrease in noise. However, this does not explain the power-law in (6).

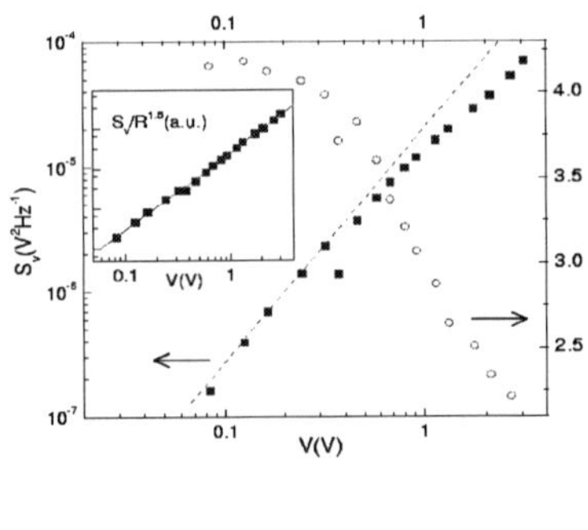

FIGURE 6: Illustration of power-law function in planar lipid membrane. The main panel shows voltage noise power and resistance as a function of voltage. S_V deviates from the initial V^2 variation in the non-ohmic regime but $S_V/R^{1.5}$ becomes a straight line proportional to V^2 (inset). (Figure adapted from Ref. [22]).

R decreasing, $\mathscr{S}'$ *increasing.* This type of variation is observed in Mott systems (see Fig. 1) and others. Comparision with composites is particularly striking since the percolative picture is invoked in both type of systems. An emperical $\mathscr{R}$ is given by

$$\mathscr{S}' \sim \exp\left(R_o/R\right)^{w_V} \tag{7}$$

R increasing, $\mathscr{S}'$ *increasing.* The Joule regime in composites with high $p \gg p_c$[15] is an example of this type of systems. Nonlinearity in both resistance and noise here is well understood. The noise in this case is a polynomial of R in contrast to having a power-law dependence in composites near p_c. Recently, Pennetta and her coworkers[27] have carried out extensive simulation of noise in the Joule regime of a random resistor network under two competing processes, driven by bias and thermally activated.

R increasing, $\mathscr{S}'$ *decreasing.* No system exhibiting this type of variation is known.

Phase Transitions: Noise peaks

Systems with threshold. A topic that has seen much activities in recent times is one of collective motion in disordered systems. Such a collective motion is usually characterised by a finite threshold value of the driving force. Examples of systems exhibiting collective motions include charge density waves (CDW's) [8], flux line lattice in type II superconductors[9], Wigner crystals in semiconductors. A finite threshold is essentially due to presence of pinnings in the system that inhibit the motion until the applied bias is sufficient to overcome the pinning force. As expected, noise measurements in these systems[28, 30, 31] have either confirmed earlier conclusions or yielded new information on dynamic quantities such as coherence lengths of moving entities. For example, the narrow-band noise in CDW's reinforces the picture of sliding CDW's by touching on the details of its periodic interaction with pinning potential. Fig. 7a shows a typical broad-band noise curve in a CDW system. Noise increases rapidly beyond the threshold

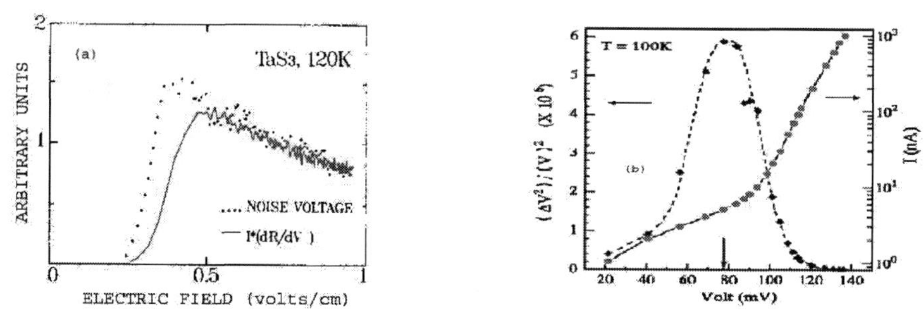

FIGURE 7. a) Noise in charge density wave system of $TaSe_3$ as a function of bias. The increase of noise coincides with appearance of current flow in the sample. (Figure from Ref. [28]); b) Noise in charge-ordered phase of $Nd_{0.5}Ca_{0.5}MnO_3$ as a function of bias. The corresponding $I-V$ curve is also shown. (Figure from Ref. [29]).

field and then goes through a maximum or peak. The authors[28] ascribed the noise to fluctuations between metastable CDW states with random pinning forces. The peak in the figure appears to represent a rather generic feature. It may indicate a transition from the initial noisy state due to jerky motion at the onset to a quieter phase. Indeed, at high field the transport should be less sensitive to the pinnings and consequently, less noisy. In type II superconductors, such structures[30] have been interpreted as a field-induced transition from a quiet elastic flow state of the flux flow lattice to a noisy plastic flow regime corresponding to peak region and finally, to a (molten) fluid flow regime again with low noise.

Transport in a new system of metallic dot arrays[32] exhibit the same threshold phenomena due to Coulomb blockade. Although no noise measurement is known in this system, numerical simulations[33] have brought out many interesting features of noise in collective transport. The current near the threshold is given by $I \sim (V - V_T)^\zeta$ where ζ is an exponent. At larger bias $I - V$ becomes linear. It was found that the crossover to ohmic behaviour in 2D is accompanied by a change in the flow from 2D meandering to straight 1D flow. Interestinly, at low bias noise shows a normal 1/f-type feature but in ohmic regime, it shows a characteristic narrow-band type peak. Such signatures would be easily observable in experimental noise spectra.

Systems without threshold. Noise peaks are also found in field-induced phase transitions which are accompanied by large nonlinearity in transport (see Fig. 7b). Recent examples include melting of charge-ordered states in manganites[29], crossover from ohmic regime to SCLC regime in organic semiconductors[34]. It is seen in Fig. 7b that unlike in the above, the noise starts increasing much earlier than the onset of nonlinearity in $I - V$ curve indicating that $b > 1$.

DISCUSSION AND OPEN ISSUES

Much of the phenomenalogical description of noise in non-ohmic regimes given in sections above lack theoretical support. For example, it will be nice to have some basis for Eqns. (6) and (7) or, for the fact that $\gamma \leq 0$ in many disordered systems. Apart from these, there are several fundamental problems associated with the noise in what is basically field-driven nonequilibrium regimes. One of the earliest theoretical discussion of fluctuation in nonlinear systems was by Van Kampen[35]. Some conceptual issues are as follow:

i) Fluctuation-dissipation theorem: The theorem is known to be valid only for equilibrium systems. However, in last decade there have been attempts to extend it in nonequilibrium conditions. See Ref. [36] and references therein. According to these, asymmetry of fluctuations of energies can be related to the dissipation energies required to maintain a nonequlibrium steady state. *ii) Fluctuation distribution:* Many work[20, 27, 37] have pointed out that fluctuations tend to be nongaussian particularly in media with large disorder. *iii) Coherence length:* The coherence length of fluctuations is normally assumed to be of the order of microscopic lengths. But, in fluids in in nonequilibrium states[38] and in CDW states[28] the length could be of macroscopic order. The same was invoked to explain the huge increase in noise in the Joule regime of composites[15].

ACKNOWLEDGMENTS

We thankfully acknowledge assistance of Arindam Chakrabarti in sample preparation and data analysis. We also thank our collaborators S. De and P. Nandy, and A. Carbone for sending us a preprint in advance.

REFERENCES

1. M. B. Weissman, *Adv. Rev. Mater. Sci.*, **26**, 395 (1996).
2. A. K. Roychaudhuri, *Cur. Opinion. Sol. State Mat. Sci.*, **6**, 67 (2002).
3. P. Dutta, and P. M. Horn, *Rev. Mod. Phys.*, **53**, 497 (1981).
4. F. N. Hooge, T. G. M. Kleinpenning, and L. K. J. Vandamme, *Rep. Prog. Phys.*, **44**, 479 (1981).
5. R. M. Fleming, and C. C. Grimes, *Phys. Rev. Lett.*, **42**, 1423 (1979).
6. D. J. Bergman, and D. Stroud, Solid State Physics **46**, Academic Press, New York, 1992, p. 147.
7. Y. Tokura, in *Colossal-Magnetoresistive Oxides,* Ed. Y. Tokura, Gordon & Breach Sci. Pub., New York, 1999.
8. G. Grunner, *Rev. Mod. Phys.*, **60**, 1129 (1988).
9. S. Bhattacharya, and M. J. Higgins, *Phys. Rev. B*, **49**, 10005 (1994).
10. *Hot-electron transport in semiconductors,* Ed. L. Reggiani, Springer-Verlag, Berlin, 1985.
11. J. M. G. Vilar, and J. M. Rubí, *Phys. Rev. Lett.*, **67**, 2529 (1991).
12. U. N. Nandi, C. D. Mukherjee, and K. K. Bardhan, *Phys. Rev. B*, **54**, 12903 (1996).
13. F. N. Hooge, *Phys. Lett. A.*, **29**, 139 (1969).
14. A. Carbone, P. Mazzetti, and F. Rossi, *Appl. Phys. Lett.*, **78**, 2518 (2001).
15. K. K. Bardhan, and C. D. Mukherjee, *Phys. Rev. B*, **65**, 212302 (2002).
16. C. Chitemed, D. S. McLachlan, and I. Balberg, *Phys. Rev. B*, **67**, 024207 (2003).
17. J. L. Williams, and I. L. Stone, *J. Phys. C*, **5**, 2105 (1972).
18. D. A. Bell, *Electrical Noise*, Van Nostrand, London, 1960.
19. J. L. Williams, and R. K. Burdett, *Brit. J. Appl. Phys.*, **17**, 977 (1966).
20. C. Parman, and J. Kakalios, *Phys. Rev. Lett.*, **67**, 2529 (1991).
21. E. G. Mishchenko, *Phys. Rev. Lett*, **85**, 4144 (2000).
22. S. De, R. Basu, and P. Nandy, *Phys. Rev. B*, **61**, 6689 (2000).
23. H. Rommelmann, A. J. Epstein, J. S. Miller, P. J. Restle, R. D. Black, and M. B. weissman, *Phys. Rev. B*, **32**, 1257 (1985).
24. S. De, R. Basu, U. N. Nandi, C. D. Mukherjee, K. K. Bardhan, and P. Nandy, *Phys. Rev. B*, **56**, 13627 (1997).
25. A. Kusy, and T. G. M. Kleinpenning, *J. Appl. Phys.*, **54**, 2900 (1983).
26. T. G. M. Kleinpenning, *Physica B, C*, **94**, 141 (1978).
27. C. Pennetta, *Fluctuation Noise Lett.*, **2**, R29 (2002).
28. S. Bhattacharya, J. P. Stokes, M. O. Robbins, and R. A. Klemm, *Phys. Rev. Lett.*, **54**, 2453 (1985).
29. A. Guha, A. Ghosh, and A. K. Raychaudhuri, *Appl. Phys. Lett*, **75**, 3381 (1999).
30. A. C. Marley, M. J. Higgins, and S. Bhattacharya, *Phys. Rev. Lett.*, **74**, 3029 (1995).
31. Y. P. Li, T. Sajoto, L. W. Engel, D. C. Tsui, and M. Shayegan, *Phys. Rev. Lett.*, **67**, 1630 (1991).
32. R. Parthasarathy, X.-M. Lin, and H. M. Jaeger, *Phys. Rev. Lett.*, **87**, 186807 (2001).
33. C. Reichhardt, and C. J. O. Reichhardt, *Phys. Rev. Lett.*, **90**, 046802 (2003).
34. A. Carbone, B. K. Kotowska, and D. Kotowski, *cond-matt/0505533* (2005).
35. N. G. van Kampen, in *Fluctuation phenomena in solids,* Ed. R. E. Burgess, Academic Press, New York, 1965, p. 139.
36. N. Garnier, and S. Ciliberto, *Phys. Rev. E*, **71**, 060101(R) (2005).
37. G. T. Seidler, S. A. Solin, and A. C. Marley, *Phys. Rev. Lett.*, **76**, 3049 (1996).
38. R. Schmitz, *Phys. Rep.*, **171**, 1 (1988).

Noise Activated Switching Between Dynamical States In A Nonlinear Micromechanical Oscillator

H. B. Chan and C. Stambaugh

Department of Physics, University of Florida, Gainesville, FL 32608, USA.

Abstract. We explore fluctuation induced switching for a system that is far from equilibrium. The device under study is an underdamped micromechanical torsional oscillator driven into the nonlinear regime. Under a sufficiently strong periodic driving field, the oscillator possesses two stable dynamical states with different oscillation amplitudes within a certain range of driving frequencies. The oscillator is induced to switch from one dynamical state into the other by noise in the excitation. We demonstrate the activated behavior of escape from the high-amplitude state to the low-amplitude state by measuring the rate of transitions as a function of noise intensity.

Keywords: activated escape, fluctuations, nonlinear oscillator, hysteresis.
PACS: 05.40.-a, 05.40.Ca, 05.45.-a, 89.75.Da

INTRODUCTION

Nonlinear systems often display more than one coexisting dynamical states when driven sufficiently strongly. In the presence of fluctuations, the nonlinear system can be induced to escape from one metastable dynamical state into the other. Unlike equilibrium systems where the escape rate can be calculated from the height of the free-energy barrier [1], driven systems are, generally, far from thermal equilibrium and cannot be characterized by free energy [2-5]. Consequently, the escape rate in such non-equilibrium systems must be derived from system dynamics [6-9]. Experiments on noise induced switching has been performed in a number of driven nonlinear systems, including parametrically driven electrons in a Penning trap [10], doubly clamped nanomechanical beams [11] and radio frequency driven Josephson junctions [12]. Calculation of the escape rate in such non-equilibrium systems is a non-trivial problem and has attracted much interest [7,8,13,14].

In this paper, we investigate noise-activated switching in an underdamped micromechanical torsional oscillator periodically driven into nonlinear oscillations. Within a certain range of driving frequencies, the oscillator has two stable dynamical states with different oscillation amplitudes. We induce the oscillator to escape from one state into the other by injecting noise in the driving force. By measuring the frequency of random transitions as a function of noise intensity, we extract the transition rate and demonstrate the activated behavior for switching.

CP800, *Unsolved Problems of Noise and Fluctuations: UPoN 2005*
edited by L. Reggiani, C. Pennetta, V. Akimov, E. Alfinito, and M. Rosini

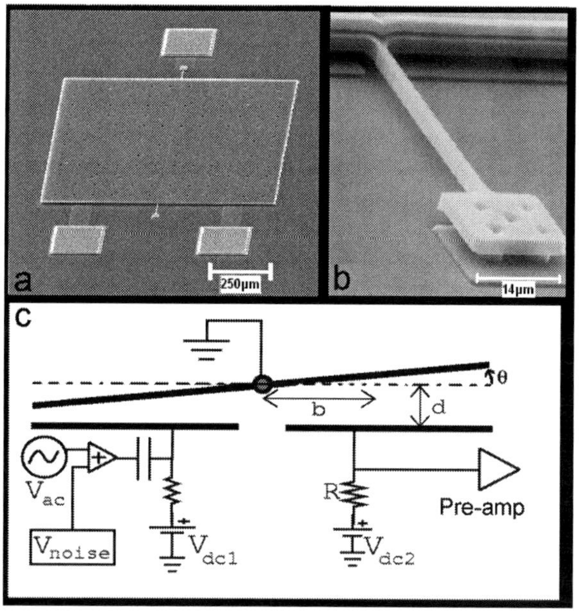

FIGURE 1. (a) Scanning electron micrograph of the micromechanical torsional oscillator. The large square in the middle is a movable polysilicon top plate. The small squares are bond pads that provide electrical connections to the top plate and the two underlying electrodes. (b) Close up on one of the torsional springs. (c) A cross sectional schematic of the torsional oscillator with electrical connections and measurement circuitry.

EXPERIMENTAL SET-UP

In our experiment, the micromechanical oscillator is fabricated using a surface micromachining process. The oscillator consists of a polysilicon plate (500 μm by 500 μm by 3.5 μm) that is supported by two torsional rods. After a 2-μm-thick sacrificial silicon oxide layer beneath the plate is etched away, the plate becomes free to rotate about the torsional springs. Figures 1a and 1b are scanning electron micrographs of a typical oscillator. The moment of inertia I for the plate is 4.3×10^{-17} $kg\ m^2$, while the spring constants are each $9\times10^{-7}\ N\ m\ rad^{-1}$. The other ends of the springs are anchored to the silicon substrate. Beneath the top plate, there are two fixed electrodes (500 μm by 250 μm) on each side of the torsional springs. The two electrodes are used for exciting the torsional oscillations and measuring the response respectively.

Figure 1c shows a cross sectional schematic of the oscillator with electrical connections and measurement circuitry. The application of a periodic voltage with DC bias V_{dc1} to one of the electrodes leads to an electrostatic attraction between the grounded top plate and the electrode. Torsional oscillations of the top plate are excited by the periodic component of the electrostatic torque. The detection electrode is connected to a dc voltage V_{dc2} through a resistor R. As the plate oscillates, the capacitance between the plate and the detection electrode changes. The detection electrode is connected to a charge sensitive preamplifier followed by a lock-in amplifier that measures the signal at the excitation frequency. Measurements were performed at liquid helium temperature and at pressure of less than 2×10^{-7} *torr*. The quality factor Q of the oscillator is about 16,000.

ORIGIN OF NON-LINEARITY IN THE OSCILLATOR

The nonlinear behavior in our torsional oscillator originates mainly from the strongly distance dependent electrostatic interaction between the top plate and the excitation electrode. The equation of motion of the oscillator is given by [15]:

$$\ddot{\theta} + 2\gamma\dot{\theta} + \omega_o^{\ 2}\theta = \tau / I \tag{1}$$

where θ is the angular rotation of the top plate, γ is the damping coefficient, ω_o is the natural frequency of the oscillator and τ is the driving torque. τ arises from the electrostatic interaction between the top plate and the driving electrode. When the rotation angle θ is small, τ can be written as:

$$\tau = bF \tag{2}$$

where b is half the distance from the axis of rotation to the edge of the top plate and the electrostatic force F is evaluated with separation $d\text{-}b\theta$.

$$F = \frac{\varepsilon_o A V^2}{2(d - b\theta)^2} \tag{3}$$

As shown in Fig. 1c, d is the separation between the top plate and the electrode when no electrostatic force is applied, A is the area of the electrode and ε_o is the permittivity of free space. The excitation voltage V is a sum of three components:

$$V = V_{dc1} + V_{ac}\sin(\omega t) + V_{noise}(t) \tag{4}$$

The three terms on the right side of Eq. (4) represent the dc voltage, periodic ac voltage and random noise voltage respectively. V_{dc1} is chosen to be much larger than V_{ac} and V_{noise} to linearize the dependence of F on V_{ac} and V_{noise}. The strong spatial dependence of the electrostatic force leads to nonlinear contributions to the restoring torque. A Taylor expansion of the electrostatic force F about d gives:

$$F(d - b\theta) = F(d) - F'(d)b\theta + \frac{F''(d)(b\theta)^2}{2} - \frac{F'''(d)(b\theta)^3}{6} \tag{5}$$

where F', F'' and F''' denotes the first, second and third spatial derivative of F respectively. Substituting $F(d - b\theta)$ in Eqs. (1) and (2) leads to:

$$\ddot{\theta} + 2\gamma\dot{\theta} + [\omega_0^2 - \eta]\theta + \beta\theta^3 + C = E\sin(\omega t) + N(t) \quad (6)$$

where $\eta = \left(b^2\varepsilon_o A / Id^3\right)V_{dc1}^2$, $\beta = -\left(2b^4\varepsilon_o A / Id^5\right)V_{dc1}^2$, $C = -\left(b\varepsilon_o A / 2Id^2\right)V_{dc1}^2$, $E = \left(b\varepsilon_o A V_{dc1} / 2Id^2\right)V_{ac}$ is the effective amplitude of the periodic excitation, and $N(t) = \left(b\varepsilon_o A V_{dc1} / 2Id^2\right)V_{noise}(t)$ is the effective noise in the excitation. The contribution of the quadratic term to the nonlinearity proves to be negligible compared to the cubic term, while the linear term produces a shift in the natural frequency.

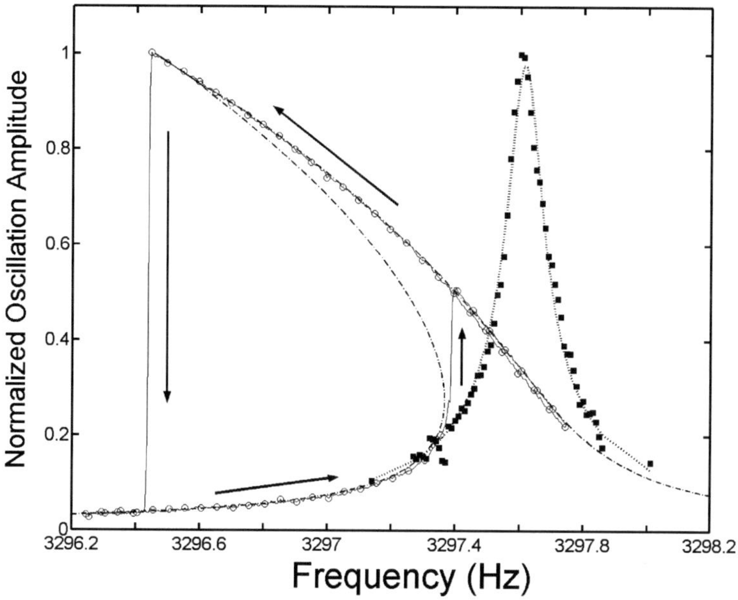

FIGURE 2. Normalized frequency response of the oscillator for excitation voltages amplitudes of 23μV (solid squares) and 91 μV (hollow circles). The dotted line represents a fit to the data at smaller driving force using the response of a damped harmonic oscillator. For the large driving force, two dynamical states coexist from 3296.45*Hz* to 3297.40 *Hz*. The dashed line fits the data to a damped oscillator with cubic non-linearity [15], yielding $\beta = -1.3\text{x}10^{14}$ $rad^{-2}s^{-2}$.

EXPERIMENTAL PROCEDURE AND RESULTS

We first focus on the response of the oscillator with no injected noise in the excitation. Figure 2 shows the frequency response of the oscillator at two different excitation amplitudes. The peak oscillation amplitudes are normalized to unity for both cases. The squares represent the response of the oscillator when the excitation is small. The resonance peak is fitted well by the dotted line that corresponds to the response of a damped harmonic oscillator. As the periodic excitation is increased, contributions from the cubic term in Eq. (6) become more important and lead to nonlinear behavior in the oscillations. The resonance curve becomes asymmetric, with the peak shifting to lower frequencies. At a high enough excitation, the frequency response becomes hysteretic, as shown by the circles in Fig. 2. Within a certain range of driving frequencies, there are two stable dynamical states with different oscillation amplitudes. The system resides in the high-amplitude state or low-amplitude state depending on the history of the oscillator. In the absence of fluctuations, the oscillator remains in one of the stable states indefinitely.

In the presence of noise in the excitation, the oscillator could be induced to escape from one state into the other. Since this driven, bistable system is far from thermal equilibrium and cannot be characterized by free energy, calculation of the escape rate is a non-trivial problem. Theoretical analysis [6,7] suggests that the rate of escape Γ at a particular driving frequency depends exponentially on the ratio of an activation energy E_a to the noise intensity I_N:

$$\Gamma = \Gamma_0 \exp(-E_a / \mathrm{I}_N) \tag{7}$$

Close to the bifurcation frequency where the high-amplitude state disappears the activation energy is expected to display system-independent scaling. For cubic nonlinearity, the activation energy is given by:

$$E_a \propto \Delta\omega^{\alpha} \tag{8}$$

where the frequency detuning $\Delta\omega$ is the difference between the driving frequency and the bifurcation frequency. The activation energy is predicted [6,7] to increase with frequency detuning with critical exponent $\alpha = 3/2$ for all dynamical systems with cubic nonlinearity. We will describe below our comprehensive experimental investigation of activated switching from the high-amplitude to the low-amplitude state of the micromechanical oscillator.

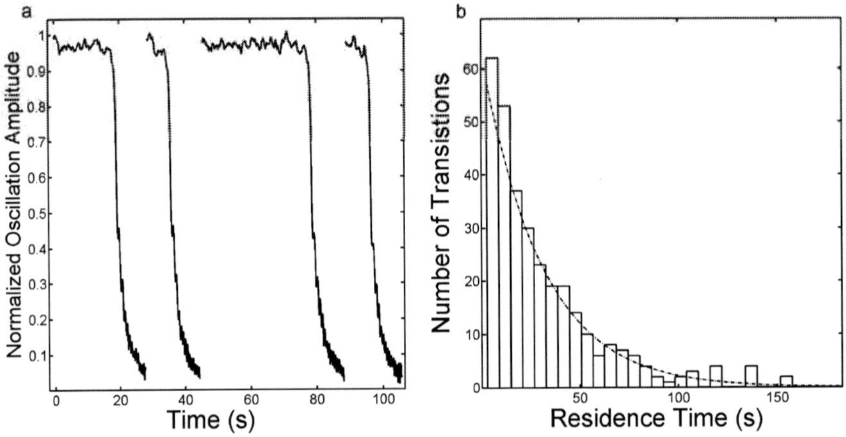

FIGURE 3. (a) In the presence of noise in the excitation, the oscillator switches from the high amplitude state to the low amplitude state at different time intervals. The system is reset to the upper amplitude state between switching events. (b) Histogram of the residence time in the upper state before switching occurs, at detuning frequency of 0.086 *Hz*. The dotted line is an exponential fit.

In our experiment, transitions from the high-amplitude state to the low-amplitude state are induced by injecting noise in the excitation with a bandwidth of $100 Hz$ centered about the resonant frequency. The bandwidth of the noise is much larger than the width of the resonance peak. Figure 3a shows typical switching events where the oscillator resides in the high amplitude state for various durations before escaping to the low amplitude state. Due to the random nature of the transitions, a large number of switching events must be recorded to determine the transition rate accurately. During the time interval between switching events in Fig. 3a, the oscillator is reset to the high amplitude state using the following procedure. First, the noise is turned off and the driving frequency is increased beyond the range of frequencies where bistability occurs (>3297.4 *Hz* as shown in Fig. 2). The driving frequency is then decreased slowly towards the target frequency so that the oscillator remains in the high-amplitude state. Once the target frequency is reached, the noise is turned back on and the time for the oscillator to escape from the high-amplitude state is recorded. This process is then repeated multiple times to accumulate the statistics for the switching. This procedure is necessary because the energy barrier for transitions from the low-amplitude state back to the high-amplitude state is much larger than the barrier for transition in the opposite direction. Thus, noise induced transitions from the low-amplitude to the high-amplitude state will fail to occur in the duration of the experiment and the oscillator must be reset to the high-amplitude state using the steps described above. Figure 3b shows a histogram of the residence time T_R in the high-amplitude state before a transition occurs. The exponential dependence on the residence time indicates that the transitions are random and follow Poisson statistics as expected.

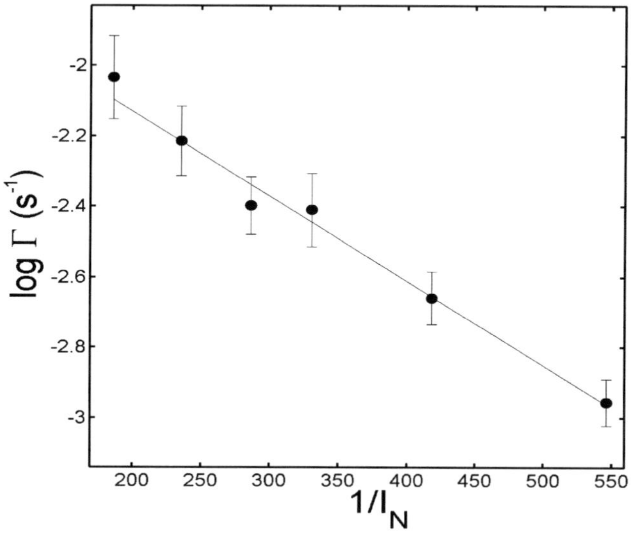

FIGURE 4. Logarithm of the transition rate from the high amplitude state to the low amplitude state as a function of inverse noise intensity at detuning frequency of 0.39 *Hz*. The slope of the linear fit yields the activation energy.

The activation energy at a particular detuning frequency is determined by recording a large number of transitions for multiple noise intensities (I_N). The average residence time at each noise intensity is extracted from the exponential fit to the corresponding histograms. Figure 4 plots the logarithm of the transition rate as a function of inverse noise intensity. The transition rate varies exponentially with inverse noise intensity, demonstrating that escape from the high-amplitude state is activated in nature. According to Eq. (7), the slope in Fig. 4 yields the activation energy for escaping from the high-amplitude state at the particular detuning frequency.

Measurement of the activation barrier at different detuning frequencies is currently in progress. Such data will yield the critical exponent for noise activated switching between dynamical states in a driven, nonlinear system and allow experimental verification of the system-independent scaling of activation energy near the bifurcation point. Micromechanical torsional oscillators provide well-characterized systems with tunable characteristics, allowing quantitative comparisons of experimental results with theoretical predictions of activated escape.

ACKNOWLEDGMENTS

We have benefited from useful discussions with M. I. Dykman and D. Ryvkine

REFERENCES

1. H. A. Kramers, Physica (Utrecht) **7**, 284 (1940).
2. R. Graham and T. Tel, *Existence of a Potential for Dissipative Dynamical-Systems*, Physical Review Letters **52**, 9 (1984).
3. J. Lehmann, P. Reimann and P. Hanggi, *Surmounting oscillating barriers: Path-integral approach for weak noise*, Physical Review E **62**, 6282 (2000).
4. J. Lehmann, P. Reimann and P. Hanggi, *Surmounting oscillating barriers*, Physical Review Letters **84**, 1639 (2000).
5. R. S. Maier and D. L. Stein, *Noise-activated escape from a sloshing potential well*, Physical Review Letters **86**, 3942 (2001).
6. A. P. Dmitriev and M. I. Dyakaonov, *Activated and tunneling transitions between two forced-oscillation regimes of an anharmonic oscillator*, Sov. Phys. JETP **63**, 838 (1986).
7. M. I. Dykman and M. A. Krivoglaz, *Theory of fluctuational transitions between stable states of a nonlinear oscillator*, Sov. Phys. JETP **50**, 30 (1979).
8. M. I. Dykman and M. A. Krivoglaz, *Fluctuations in Non-Linear Systems near Bifurcations Corresponding to the Appearance of New Stable States*, Physica A **104**, 480 (1980).
9. M. I. Dykman, C. M. Maloney, V. N. Smelyanskiy and M. Silverstein, *Fluctuational phase-flip transitions in parametrically driven oscillators*, Physical Review E **57**, 5202 (1998).
10. L. J. Lapidus, D. Enzer and G. Gabrielse, *Stochastic phase switching of a parametrically driven electron in a Penning trap*, Physical Review Letters 83, 899 (1999).
11. J. S. Aldridge and A. N. Cleland, *Noise-enabled precision measurements of a Duffing nanomechanical resonator*, cond-mat, 0312623 (2004).
12. I. Siddiqi, R. Vijay, F. Pierre, C. M. Wilson, L. Frunzio, M. Metcalfe, C. Rigetti, R. J. Schoelkopf, M. H. Devoret, D. Vion and D. Esteve, *Direct observation of dynamical bifurcation between two driven oscillation states of a Josephson junction*, Physical Review Letters 94, 027005 (2005).
13. M. I. Dykman, R. Mannella, P. V. E. Mcclintock, S. M. Soskin and N. G. Stocks, *Noise-Induced Narrowing of Peaks in the Power Spectra of Underdamped Nonlinear Oscillators*, Physical Review A 42, 7041 (1990).
14. D. Ryvkine, M. I. Dykman and B. Golding, *Scaling and crossovers in activated escape near a bifurcation point*, Physical Review E 69, 061102 (2004).
15. D. Landau and E. M. Lifshitz. *Mechanics* (Pergamon, London, 1976).

Random Telegraph Signal In Si n-MOSFETs: A Way Towards Single Spin Resonance Detection

Marco Fanciulli[1], Enrico Prati[1], Giorgio Ferrari[2], Marco Sampietro[2]

[1] *Laboratorio Nazionale Materiali e Dispositivi per la Microelettronica, Istituto Nazionale per la Fisica della Materia, Via Olivetti 2, I-20041 Agrate Brianza,Italy*
[2] *Dipartimento di Elettronica e Informazione, Politecnico di Milano, P.za Leonardo da Vinci 32, I-20133 Milano, Italy*

Abstract. Single spin detection is one of the most challenging tasks towards the realization of a solid-state-based quantum information processor. Spin-dependent (SD) processes in the random telegraph signal (RTS) observed in silicon MOSFETs may lead to quantum bit read-out. In addition, if successful, SD-RTS will allow a direct identification and microscopic characterization of the trap responsible for the RTS. The experiment and its interpretation present a number of technical difficulties and open issues.. We discuss our experimental results towards a deep understanding of spin-dependent RTS aiming at shading light on several open questions related to the influence of the microwave field, the static magnetic field, and the effective temperature of the electron gas.

Keywords: electron spin resonance, interface defects, silicon MOSFETs, quantum computation.
PACS: 82.20.Xr, 73.50.Td, 76.30.-v, 85.30.-z

INTRODUCTION

The electron spin is a natural candidate for the realization of a quantum bit (qubit) [1]. Several schemes have been proposed during the last years to exploit the ingenuities of modern microelectronics envisaging semiconductor-based quantum information processing (QIP) devices where nuclear and/or electronic spins are used as qubits in silicon or silicon-germanium structures [2,3]. The technological challenges are enormous. Manipulation, entanglement, and read-out are key issues to be addressed in any device capable of coherently process the quantum information carried out by the spin. Single-spin detection is a challenging and necessary task in the effort towards the solid state implementation of QIP. Spin-dependent phenomena in the random telegraph noise observed in the channel current of silicon MOSFET devices are among the schemes proposed to achieve single-spin read out [3-4].

For a Si/SiO_2 MOSFET it is now accepted that, when the Fermi level of the conducting channel is in the proximity of the energy level of a defect at the semiconductor/oxide interface, an electron from the channel can tunnel into and out of the trapping center [5,6]. As the RTS is related to a single trap, spin-dependent phenomena in the capture and emission, if observed, will allow single-spin detection, an intriguing objective for condensed matter physicists with potential for application in QIP and spintronics. So far single spin detection has been achieved with scanning

CP800, *Unsolved Problems of Noise and Fluctuations: UPoN 2005*
edited by L. Reggiani, C. Pennetta, V. Akimov, E. Alfinito, and M. Rosini

probe techniques [7], with optical techniques [8], and only recently with electrical detection via a spin-to-charge swap for an electron confined in a quantum dot [9] or via the RTS in a silicon *n*-MOSFET [10].

Spin dependent effects in the RTS can be described considering a defect, acting as the trapping center, that, for example, can be occupied by 1 or 2 electrons (Figure 1). A static external magnetic field B_0 removes the spin degeneracy and splits the energy level E_T in two Zeeman energy levels $E_Z = E_{\downarrow}-E_{\uparrow} = g\ \mu_B B_0$, where g is the electron g factor of the trap, and μ_B is the Bohr magneton. A microwave radiation with an oscillating magnetic field $B_{rf}(t)$ perpendicular to B_0 couples the spin-split levels. In a simplified model, when the conduction channel chemical potential E_F is located between the two Zeeman levels and at sufficiently low temperature only the lower energy level is accessible. In this case the trapping probability of a conduction electron is very low, as the trap state is already occupied. Under microwave irradiation the spin in the trap is excited to the upper level and the trapping is now allowed. In reality one has to consider the spin polarization of the conduction electron (p_e) and of the trap (p_t), $p_{e,t}= \tanh(g_{e,t}\ \mu_B B_0/2kT)$ at the working temperature and field, and the fact that states within few kT from E_F are accessible.

Recently the first experimental evidence of a single electron spin resonance detected via RTS has been reported [10]. The difficulties in the interpretation of the results motivates the need for a complete characterization of the effect and, as a first step, a detailed analysis of the interaction of the device with the environment. In this work we present an overview of the state of the art, the difficulties related to such detection, and the intriguing effects expected to have a relevant role in the design of the experiment and in the interpretation of the results. In particular we have systematically investigated the microwave dependence of the source-drain current and of the random telegraph signal [11,12], the effect of the static magnetic field, and of the low temperatures.

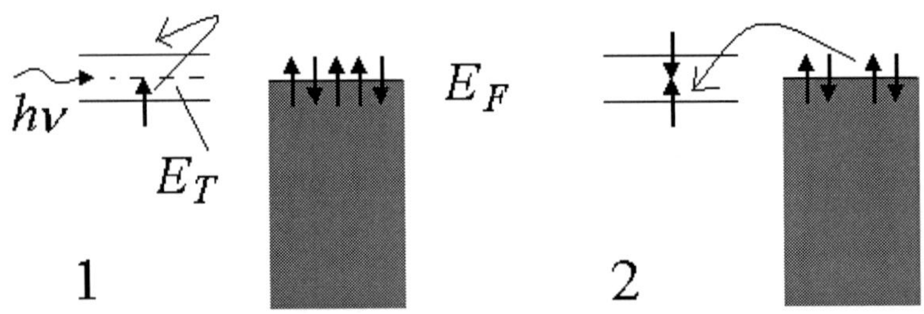

FIGURE 1. Simplified scheme for the single spin detection based on the modification of the population of the accessible trap state using electron spin resonance. The photon $h\nu$ flips the spin (1) making the ground state of the Zeeman split doublet available to a conduction electron (2).

EXPERIMENTAL SETUP

The experimental setup is displayed in Figure 2. The transistors tested are *n*-channel MOS devices realized on a p-well, with a length of 0.18 μm, width of 0.28

μm, and a 3.5 nm thick gate oxide. The threshold voltage V_T was in the range 200 – 500 mV. The source, drain, gate, and well contacts are directly accessible through the bonding pads and connected to wires to exit the cryo-magnetic system or the microwave cavity. The current flowing through drain and source I_{DS} has been measured by a transimpedance amplifier whose output is sampled and digitized for off-line processing. The bandwidth of the amplifier extends from DC to 100 kHz. The transistor and the electronic are powered by independent batteries to avoid power-line pick-up and interferences. The sample was placed into a microwave cavity or into a cryo-magnetic system, to perform experiments related to microwave field induced effects, or to a static magnetic field, respectively.

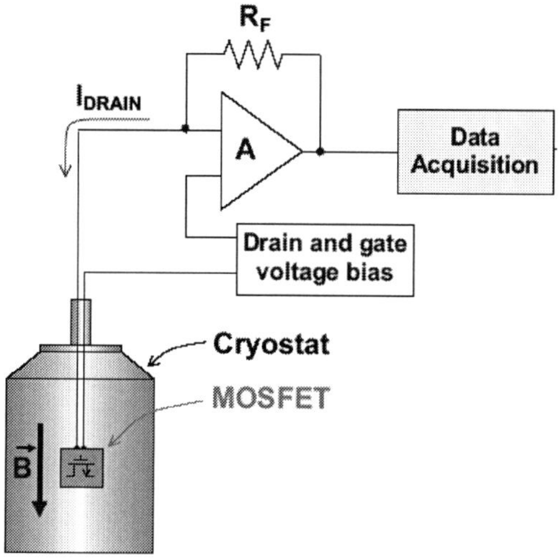

FIGURE 2. Schematic of the set-up for the characterization of the RTS inside a cryo-magnetic system.

MICROWAVE FIELD – MOSFET INTERACTION

We studied the response of the RTS to microwave fields in a MOFSET by introducing the sample into a microwave cavity. Two different cavities, having resonant frequency at 9.5 GHz (X-band) and 34 GHz (Q-band), have been used. The minimum temperature in the X-band cavity was 5 K, while in Q-band 1.6 K could be reached. We did not observe any dependence of the results on the frequency. The irradiation of the sample with a microwave field generates two significant effects. First, the generation of a stationary current that may be detected by a current meter; second a change of the characteristic times of the RTS. In Figure 3 we show the intensity of the current generated by the application of the microwave field. This effect, fully described in a previous publication [11], has been explained by using the non linear response of the MOSFET to the microwave frequency AC current induced into the loop (including the MOSFET) by the harmonic electromagnetic field oscillation. The intensity of the current depends linearly on the nominal microwave power. The effect must be carefully controlled when single spin detection via RTS is based on the measurement of the average dc current: further studies are required to verify that a change of the absorption in the environment do not modify significantly the microwave field coupled to the loop including the MOSFET. Otherwise a change in the dc current may be interpreted as a single spin resonance instead of a spurious effect.

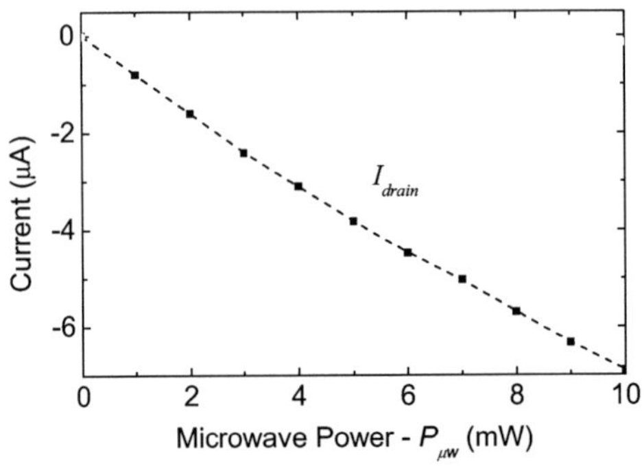

FIGURE 3. Linear dependence of the intensity of the stationary current induced by the microwave irradiation on the MOSFET, and the microwave power.

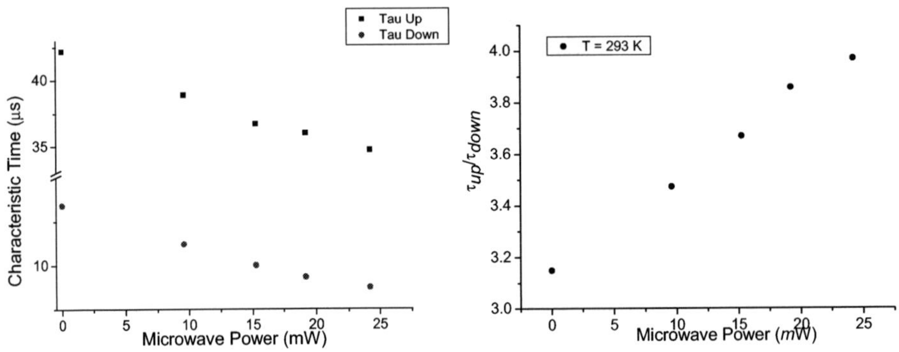

FIGURE 4. Room temperature measurement of the sample n101 in Q band. The MOSFET was operated at V_g = 600 mV and V_d = 582 mV. Identical results have been obtained at cryogenic temperature.

The second effect relates directly to the RTS change induced by the microwave field. We report the experimental evidence (Figure 4) of a reduction of the characteristic times τ_e and τ_c by increasing the microwave power. This has been observed at all the temperatures from 1.6 K to 293 K in all our samples [12]. The ratio $r = \frac{\tau_c}{\tau_e}$ generally increases by increasing the microwave power, but there are some experimental evidences also of an opposite behavior. These behavior is also related to the AC current induced by the microwave radiation and generating the stationary current. Further experiments are in progress to reach a full understanding of the phenomenon.

INFLUENCE OF A STATIC MAGNETIC FIELD ON THE RTS

In order to understand the change of the RTS in presence of a high static magnetic field, we measured the characteristic capture and emission times of some traps operating in a He3 cryostat, nominal T = 250 mK, and as function of the magnetic field B in the range between 0 and 8 T. The sample was inserted into the vacuum jacket of the cryomagnet and cooled by a cold-finger, with the conductive channel parallel to the magnetic field. The relatively high intensity of the current in our experiment did not allowed the temperature to reach the base temperature of the He3 cryostat. The two dimensional electron system (2DES) temperature was determined to be in the range between 1.4 K and 2 K.

In Figure 5 we show the ratio r between the high (capture) and the low (emission) states of the current at V_g = 561 mV and V_d = 15.7 mV, with a current I_d = 37.3 nA. Since the low state occupation increased by increasing the gate voltage (so that the Fermi energy is raised with respect of the trap energy), we deduce that the trap was filled when the current was in the low state.

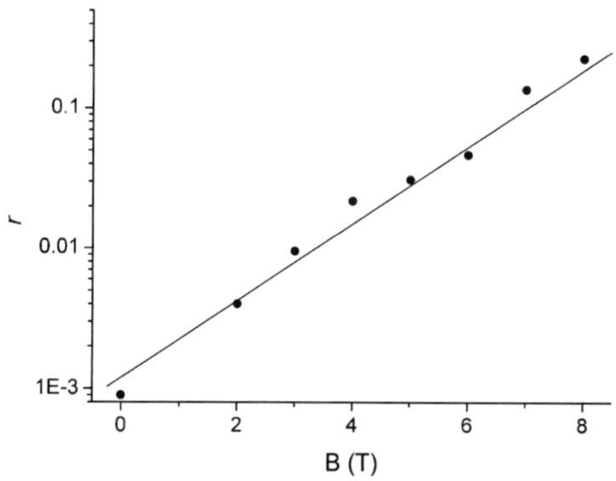

FIGURE 5. Ratio r between the capture and emission times of the trap versus the magnetic field B_0. The values scale by three orders of magnitude, two more than what previously reported [13].

A similar experiment has been reported by Xiao *et al.* [13], however in that case a lower drain-to-source current was used, allowing a lower temperature of the electrons in the 2DES. The change of r we observed is much larger than previously reported [13], even if the actual 2DES temperature in our case is higher. The trapping/emission rates we observed are also faster than those observed by Xiao *et al.* [13]. Using the grand partition function method, assuming the trap in contact with the channel bath having chemical potential E_F, r can be calculated. The experimental data reported by Xiao *et al.* [13] for temperatures below 2 K reveal a dependence of r as function of B weaker than predicted. Recently this discrepancy has been attributed to Kondo screening of the spin in the singly occupied trap which efficiently suppresses the effect of the magnetic field [14,15]. In our case the change in *r* is significantly larger, but still below the prediction given in Ref.13 indicating that Kondo screening could be active also in our system, a possibility which is also consistent with the faster rates, but that additional processes could be responsible for the observed influence of *B* on *r*.

Further work is in progress to expand the experimental information as well as the model used for the interpretation of the data.

CONCLUSION

Spin dependent phenomena in the random telegraph signals in silicon MOSFETs may allow single spin detection paving the way towards semiconductor based QIP. In addition the RTS detection of a spin resonance will allow the identification of the trapping center responsible for the RTS. Presently only one experiment of RTS detected single spin resonance has been reported in Si-MOSFETs. The g-factor of the observed resonance does not correspond to any known defect at the Si/SiO_2 interface. The influence of the microwave and of the static magnetic field on the emission and capture rates is still not fully understood and more work is necessary to achieve a deeper understanding the experimental data.

ACKNOWLEDGMENTS

The authors would like to thank Paolo Fantini and STMicroelectronics for providing the samples. The authors are grateful to Vittorio Pellegrini, Vincenzo Piazza, Pasqualantonio Pingue for the cryomagnet time at the Laboratorio Polvani of NEST-INFM and SNS, Pisa, Italy.

REFERENCES

1. D. Loss and D. P. DiVincenzo, *Phys. Rev. A* 57, 120 (1998)
2. B. E. Kane, *Nature* 393, 133 (1998)
3. R. Vrijen et al., *Phys. Rev. A* 62, 12306 (2000)
4. I. Martin, D. Mozyrsky, and H.W. Jiang, *Phys. Rev. Lett.* 90, 018301 (2003)
5. K. S. Ralls, W. J. Skocpol, L. D. Jackel, R. E. Howard, L. A. Fetter, R. W. Epworth, and D. M. Tennant *Phys. Rev. Lett.* 52, 228 (1984)
6. K. Kandiah, M.O. Deighton, and F.B. Whiting, *J. Appl. Phys.* 66, 93 (1989)
7. D. Rugar, R. Budakian, H. J. Mamin, and B. W. Chui, *Nature* 430, 329 (2004)
8. F. Jelezko, I. Popa, A. Gruber, C. Tietz, and J. Wrachtrup, A. Nizovtsev, and S. Kilin, *Appl. Phys. Lett.* 81, 2160 (2002)
9. J. M. Elzerman, R. Hanson, L. H. Willems van Beveren, B. Witkamp, L. M. K. Vandersypen, and L. P. Kouwenhoven, *Nature* 430, 431 (2004)
10. M. Xiao, I. Martin, E. Yablonovitch, and H. W. Jiang, *Nature*, 430, 435 (2004)
11. G. Ferrari, L. Fumagalli, M. Sampietro, E. Prati, and M. Fanciulli, *J. Appl. Phys.*, to be published
12. E. Prati, G. Ferrari, M. Sampietro, M. Fanciulli, and P. Fantini, unpublished
13. M. Xiao, I. Martin, and H. W. Jiang, *Phys. Rev. Lett.* 91, 078301 (2003)
14. D. Mozyrsky and I. Martin, *IEEE Trans. Nanotech.* 4, 1536 (2005)
15. I. Martin and D. Mozyrsky, *Phys. Rev.B* 71, 165115 (2005)

High-Frequency Noise In AlGaN/GaN Heterostructures

S.A.Vitusevich[1], M.V.Petrychuk[2], S.V.Danylyuk[1], A.M.Kurakin[1], A.E.Belyaev[3] and N.Klein[1]

[1] *Institut für Schichten und Grenzfächen and CNI- Center of Nanoelectronic Systems for Information Technology, Forschungszentrum Jülich, Jülich D-52425, Germany*
[2] *Taras Shevchenko National University, Kiev 01033, Ukraine*
[3] *V.Lashkaryov Institute of Semiconductor Physics, NASU, Kiev 03028, Ukraine*

Abstract. High-frequency noise (10-100MHz) was studied in undoped AlGaN/GaN high electron mobility transistor heterostructures. Two equivalent approaches, noise temperature and resistance fluctuations, were considered for the data analysis. Three possible mechanisms of noise origin were considered and investigated in detail. These include thermal noise, diffusive noise of hot electron and generation-recombination (G-R) noise. The results of our analysis show a dominant influence of G-R noise, which contains two components with different temperature dependences demonstrating the different physical origins of the noise. It is concluded that the concentration fluctuations of electrons on the first quantum level of the quantum well and scattering of the electrons in the barrier layer play a very important role in the noise phenomena.

Keywords: high-frequency noise, AlGaN, GaN, high electron mobility transistor.
PACS: 72.70.+m, 73.40.Kp , 73.50.Gr , 73.50.Mx

INTRODUCTION

The investigation of fluctuation phenomena in a wide range of applied electrical fields is an important research area of modern semiconductor physics. Due to their fundamental properties, III-nitride materials and heterostructures are excellent candidates for high power and therefore for extremely low-phase noise high-frequency oscillator development. Investigations of low-frequency fluctuations have proven to be useful for studying "slow" fluctuations at low electric fields (E), while with increasing E the analysis of "fast" processes, connected with hot electron effects, energy and impulse relaxation mechanisms, become essential.

In this communication, we report the results of investigations of voltage fluctuations in AlGaN/GaN heterostructures in the frequency range from 10 to 100MHz and in a wide range of applied biases. We used two equivalent approaches to the analysis of high-frequency noise: expressing noise through noise temperature as well as through the resistance fluctuation model.

EXPERIMENTAL DETAILS

All measurements were performed at room temperature. The structures were grown by MOCVD on SiC substrate. Two-dimensional electron gas (2DEG) was formed by

CP800, *Unsolved Problems of Noise and Fluctuations: UPoN 2005*
edited by L. Reggiani, C. Pennetta, V. Akimov, E. Alfinito, and M. Rosini

undoped AlGaN/GaN layers. As an object of investigation, a set of ohmic contacts in the form of transmission line model (TLM) structures was used with a channel length between the contacts of L = 5, 10, 15, 20 and 25 μm. The noise was measured in a so-called "longitudinal" configuration, where the same contacts are used for applying current and measuring voltage fluctuations. The noise and high-frequency conductivity were measured with a Hewlett-Packard noise figure meter HP 8970B.

Typical current-voltage dependences of the structures are the same as in our investigations reported earlier [1]. The dependences are linear up to bias voltages $U = 1\text{-}2V$ (depends on length of the structure). At higher voltages current growth decreases and eventually saturates. As was shown earlier [1], dependence of relative change of current on dissipated power $P = IU$ is linear and coincides for samples with different lengths. Assuming that the increase of temperature is proportional to the dissipated power we can conclude that the change of resistance of the sample is proportional to the change of temperature. This fact will be used in our later analysis.

NOISE MEASUREMENT RESULTS AND THEIR ANALYSIS

Typical noise-voltage dependences are shown in Fig.1. The observed noise behaviour correlates with the non-linearity of current-voltage characteristics [1].

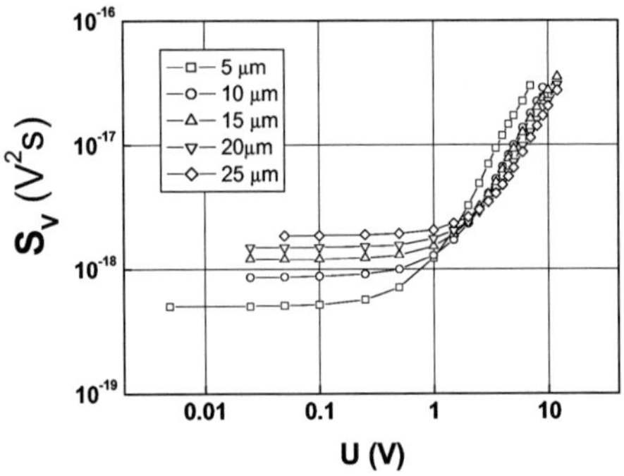

FIGURE 1. Voltage noise, measured at frequency f=90 MHz for TLM structures of different lengths versus applied voltage.

To analyse the noise results obtained we considered three possible mechanisms of noise origin. High-frequency noise f = 10-100 MHz can contain an equilibrium thermal (Nyquist) noise, determined by the formula:

$$S_V = 4kT\,\mathrm{Re}(Z) \tag{1}$$

where S_V –spectral density of noise fluctuations, k –Boltzmann constant, T – temperature, Z – full impedance of the object. It may also contain diffusive noise of hot electrons and excessive current noise, caused by generation-recombination processes.

To evaluate possible contributions from each of the three above-mentioned mechanisms of noise origin, we used two equivalent approaches to the analysis of high-frequency noise: expressing noise through noise temperature as well as through the resistance fluctuation model.

Investigations on the assumption that measured noise is of thermal origin were not confirmed by our calculations. Even if we assume that dissipation of 1W of power leads to overheating $\Delta T = 2200\,\text{K}$ (which is definitely much larger than any experimental evaluations), the value of measured noise is higher than calculated. This leads us to the conclusion that the observed noise cannot be of thermal origin.

The analysis of noise results within the model of diffusive hot electron noise, driven by scattering of hot electrons on phonons [2] also gives unsatisfactory results. We only obtained a fit of the model curve with our experimental points on the assumption that the phonon relaxation time in our structures is $\tau_{ph} = 50\,\text{ns}$, which is much larger than commonly obtained $\tau_{ph} = 350\,\text{fs}$ [3]. Therefore, hot electron noise cannot cause the noise observed in our experiment.

Let us now consider the third possibility: measured noise is generation-recombination (G-R) noise. This is determined by the following formula [4]:

$$S_I = 4\frac{\overline{\Delta N^2}}{N^2}\frac{I^2\tau}{1+(\omega\tau)^2}, \tag{2}$$

where $S_I = S_V/R^2$ – spectral density of current fluctuations, N – mean value of full number of electrons in a sample, $\overline{\Delta N^2}$ – mean-square deviation of number of electrons, ω – frequency, τ – time constant of G-R process. The value $\overline{\Delta N^2}/N = g$ depends only on the statistics of electrons and not on N. Let us assume that the investigated frequency $f = 90\,\text{MHz}$ lies within a region where $\omega\tau \ll 1$. Then expression (2) can be written as:

$$\frac{S_I N}{I^2} = 4g\tau \tag{3}$$

The right side of (3) does not depend on sample size but can depend on temperature (dissipated power). Therefore we shall build a dependence of the left side of (3) on dissipated power, taking into account that a number of electrons, N, are proportional to length of a sample L. Since, as stated earlier, the temperature of the sample is determined solely by dissipated power, the curves obtained for different lengths should coincide (due to coincidence of corresponding resistance-power characteristics for different sample lengths).

As the thermal noise is always present at the no zero temperature, we should subtract value $S_{V0} = 4kT_0R$ from the analyzed spectra. As a result we obtain dependence of value $L(S_V - S_{V0})/I^2R^2$ on dissipated power $P = IU$ (Fig.2).

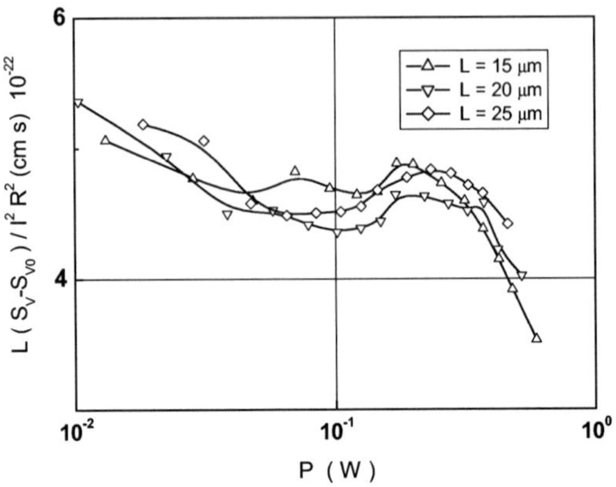

FIGURE 2. Dependence of normalized noise of samples of different lengths on dissipated power at $f = 90\,\text{MHz}$.

We can indeed see that curves coincide quite well. Slight differences at maximal currents are most likely caused by increasing thermal noise due to the rise in temperature with power which we did not take into account in the normalization procedure. Estimations show that this additional noise is of the order of the observed differences between the curves in Fig.2.

It is seen from Fig.2 that not only do the curves for $L = 25, 20, 15\,\mu\text{m}$ coincide well at low power, but also the position and shape of the special feature observed at high power in the range from 0.1W to 0.6 W coincide with the maximum at $P \approx 0.25\,\text{W}$. This feature confirms our assumption of the G-R character of observed noise.

Two kinds of transport mechanisms in two-dimensional electron channels may be responsible for generation-recombination noise. The first is due to the presence of electrons on more than one quantum level with different mobility. In this case, spontaneous transitions between the levels lead to fluctuations of the total conductance of the channel. This model was used in [5] to explain excessive noise at $f = 10\,\text{GHz}$. The second reason for the G-R noise can be as a result of the penetration of the 2D electron wavefunction to the barrier region of the structure, where concentration of traps is generally higher. The mobility of these electrons is therefore reduced due to interaction with the traps.

To analyse the influence of the first suggested mechanism of GR noise we self-consistently calculated the positions of the ground and first excited quantum levels in 2DEG relative to the Fermi energy (Fig.3a). It is interesting to note that at $T \approx 600\,\text{K}$ the position of the first excited level coincides with the Fermi energy. According to the theory of GR noise, the temperature dependence of equation (3) should have a maximum at this point. Indeed, in Fig.2 we observe such a maximum at $W \approx 0.25\,\text{W}$, which allows us to ascribe this noise feature to the fluctuation of the number of electrons at the first excited quantum level. In this case, the estimation of overheating with dissipated power gives a 1200K temperature rise with 1W of dissipated power.

This value is much smaller than that used in our analysis of the influence of thermal noise and can serve as further evidence of the negligible role of equilibrium thermal noise in the total measured noise spectra.

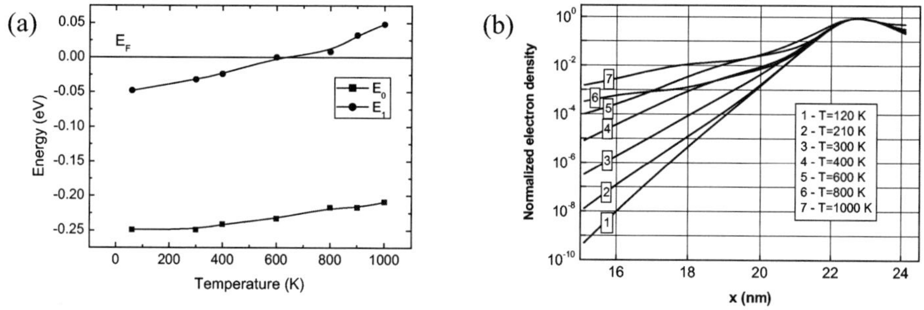

FIGURE 3. Results of self-consistent calculation of Poisson and Schroedinger equations. a) dependence of position of ground and first exited quantum level in 2DEG relative to Fermi energy. b) distribution of density of 2DEG in the direction perpendicular to the interface at different temperatures. The interface position is at 22nm in the calculation.

To evaluate the role of the second above-mentioned reason for excessive noise in the structure we plotted electron density normalized to a maximum (in the ground level of 2DEG) as a function of coordinate x, perpendicular to the 2DEG layer (Fig.3b). Increasing the temperature leads to a rise of electron density within the barrier layer ($x < 22\ nm$). This rise should lead to an increasing of number of electrons with lower mobility and, consequently, to an increase in the noise caused by an exchange between fast and slow electron populations. But, as is seen from Fig.2, the total noise of the structure does not increase with temperature (dissipated power). Therefore this model cannot be used to explain the results of noise measurements at $f = 90\ \text{MHz}$.

A similar analysis was carried for noise results obtained at 20 MHz frequency. The temperature dependence of equation (3) was also found to coincide for samples of different lengths (Fig.4a), but in this case the character of the dependence differs from the results obtained at $f = 90\ \text{MHz}$. Our analysis shows that the noise also has a generation recombination character, but of different origin. The frequency dependence of noise measured at a dissipated power of 0.3W (Fig.4b) shows non-zero frequency slope, which can be analysed as a spectrum combined from two noise components. The first one is a high-frequency component described above for $f = 90\ \text{MHz}$. The second one is at lower frequencies, with a corner frequency of around $f_0 = 50\ \text{MHz}$, which corresponds to the time constant $\tau = 3.2\ \text{ns}$. This component can originate from the increase in the population of electrons in the barrier region of the structure with temperature (Fig.3b).

This model can explain the effect of increasing noise at 20MHz with a rise of dissipated power and, correspondingly, with an increase of temperature (Fig.4a).

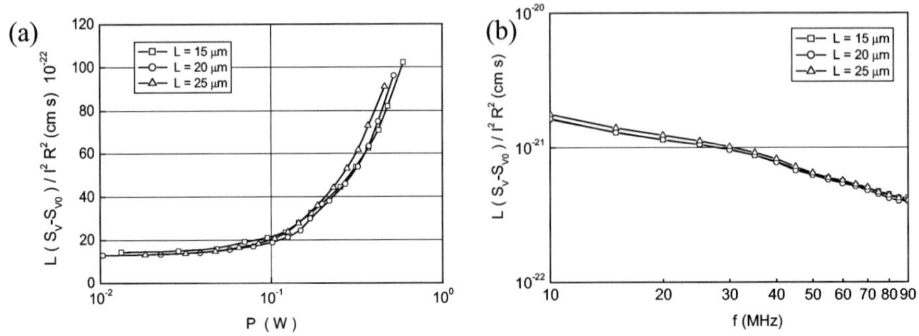

FIGURE 4. a) dependence of normalized noise of samples of different lengths on dissipated power at $f = 20\,\text{MHz}$; b) normalized noise spectra of TLM samples of different lengths measured at P = 0.3W.

CONCLUSION

The analysis of the high-frequency noise results obtained in the frequency range $f = 10\text{-}100\ MHz$ revealed the generation-recombination origin of the noise. Two different components of the noise with different temperature behaviour were observed and investigated. It was determined that the maximum observed in the temperature dependence of noise is caused by GR fluctuations at points of coincidence of the quantum level with Fermi energy. Therefore the noise, measured at $f = 90\,\text{MHz}$, originates from fluctuations of concentration at the first excited quantum level of 2DEG. Another GR component was observed at lower frequencies with the time constant $\tau = 3.2\,\text{ns}$. This demonstrates a different temperature dependence. The noise mechanism can be explained by increasing penetration of electron wavefunctions into the barrier region with increasing self-heating.

ACKNOWLEDGMENTS

This work is supported by the Deutsche Forschungsgemeinschaft (Project KL 1342).

REFERENCES

1. S.A.Vitusevich, S.V. Danyluk, N. Klein, M.V. Petrychuk, and A.E.Belyaev, *Journ. Appl. Phys.* **96**, 5625-5630 (2004).
2. J.Liberis and A.Matulionis. *"Hot-electron noise in nitride 2DEG Channels"* in *Noise and Fluctuations,* edited by Josef Sikula, Brno: CNRL Publisher, 2003, pp. 241-246.
3. A.Van der Ziel. *Noise in Solid State Devices and Circuits*, New York:John Wiley&Sons Publisher,1986, pp. 120-145.
4. A.Matulionis, J.Liberis, I.Matulioniene, M.Ramonas, L.F.Eastman, J.R.Shealy, V.Tilak and A.Vertiatchikh., *Phys.Rev.B,* **68**, 35338-1-7 (2003).
5. L. Ardaravicius, J.liberis, A.Matulionis, L.F.Eastman, J.R.Shealy and A. Vertiatchikh, *Phys.Stat.Sol (a)*, **201**,203-206 (2004).

ENHANCED AND SUPPRESSED SHOT NOISE

On Bunching Of Fractionally-Charged Quasiparticles*

E. Comforti, Y.C. Chung, M. Heiblum, V. Umansky, and D. Mahalu

Braun Center for Submicron Research, Dept. of Condensed Matter Physics, Weizmann Institute of Science, Rehovot 76100, Israel

Abstract: Shot noise measurements are an important tool for characterizing the charge of current carriers and the correlation among them. Most notable recently was the verification that quasiparticles in the Fractional Quantum Hall (FQH) regime have fractional charge $q=e/n$, with e the charge of the electron, n an odd integer,:and $\nu=p/n$ the *fractional filling* of the first Landau level (such as 1/3, 2/5, 3/7, etc.) [1-3]. We investigate here the interaction of quasiparticles with a strong back scatterer and find unexpected results. When a weak back scatterer (with transmission $t\approx1$) is introduced in the path of an otherwise noiseless current of quasiparticles (a fully occupied beam), stochastic *partitioning* and noise proportional to the quasiparticles charge appear. Specifically, at $\nu=1/3$ noise corresponding to a partitioned charge $q=e^*=e/3$ is measured. However, this charge increases monotonically as backscattering becomes stronger – approaching $q=e$ for $t\to0$ [4,5]. In other words, only *electrons* (or, alternatively, three *bunched* quasiparticles) are found to tunnel through high potential barriers when impinged by a noiseless current of quasiparticles. Here we show that the partitioned charge, or the bunching of quasiparticles, after a strong back scatterer, depends on the average state occupation of the *impinging* current (the *dilution* of the quasiparticles beam) [6]. In the limit of a very dilute impinging current, when quasiparticles arrive *one by one*, bunching ceases altogether and the transferred charge approaches $q=e^*=e/3$. These surprising results, not yet understood, prove that sparse quasiparticles, with charge $e^*=e/3$, can tunnel through high potential barriers, originally thought to be opaque.

Keywords: Fractional charge, shot noise, tunneling
PACS: 73.43.Fj, 71.10.Pm, 73.50.Td

INTRODUCTION

As early as in 1918 W. Schottky was able to determine the elementary charge of the electron by measuring the current fluctuations associated with the stochastic emission of electrons in a vacuum tube [7]. His famous expression for the average square of the current fluctuations $\langle i^2 \rangle$ at low frequencies, $S=2qI$, represents the basic characteristics of independent random events obeying Poisson distribution. Here, S is the spectral density of the fluctuations (so called *shot noise* – measured in units of A^2/Hz), q is the charge of the particle, and I is the average current. A similar experiment can be implemented in the FQH regime [8], aiming at verifying the existence of fractionally charged quasiparticles [9]. In this case, a partially-transmitting constriction, a *Quantum Point Contact* (QPC), serves as an adjustable potential barrier in the path of the current, hence partitioning the impinging charges and inducing stochastic charge

CP800, *Unsolved Problems of Noise and Fluctuations: UPoN 2005*
edited by L. Reggiani, C. Pennetta, V. Akimov, E. Alfinito, and M. Rosini

transfer. At zero temperature S=$2qI(1\text{-}t)$, with $t \in [0,1]$ being the transmission of the QPC [4,10,11]. The random process is now described by a Binomial distribution with a noise suppression factor (1-t) added to the original Schottky's formula. In the weak backscattering regime ($t \approx 1$), where the charge-carrying quasiparticles traverse the potential barrier independently, q is found indeed to represent the quasiparticle charge, *e.g.* $q=e^*=e/3$ and $q=e^*=e/5$ for filling factors $\nu=1/3$ and $\nu=2/5$, respectively [1-3]. At $\nu=1/3$, as the transmission of the QPC gets smaller, the tunneling of individual quasiparticles becomes correlated, and in the limit of $t \rightarrow 0$ (*pinched* QPC) three quasiparticles must group in order to tunnel through the barrier. Obviously, Schottky's formula is inapplicable for *correlated quasiparticles*, but it can still be used to characterize the system with the effect of *bunching* of the quasiparticles being incorporated into an effective charge that depends on transmission $q(t)$. Hence, the noise at this strong backscattering limit becomes *electronic like* [4], with an effective charge $q=e$. For a general t of the QPC at $\nu=1/3$, and a noiseless current impinging on the QPC, a universal behavior was found for the effective partitioned charge $q(t)$ [5], starting at $q(1)=e/3$ and monotonically increases towards $q(0)=e$.

In the following work we explore the *bunching properties* of a pinched QPC when a *sparse* beam of $e/3$ quasiparticles impinges upon it. In other words, we ask the following question: When there are not enough quasiparticles in the impinging beam in order to bunch into an 'electron' (a diluted occupation of each propagating state) – will the barrier reflect the impinging quasiparticles, thus reducing the transmission below the 'bare' value? Or will it 'borrow' quasiparticles from lower-energy states in the Fermi sea to compensate for the missing ones? And perhaps the bunching condition will relax and individual $e/3$ quasiparticles will somehow transverse such an opaque barrier? Theory does not provide yet an answer to this question.

RESULTS AND DISCUSSION

To study the previous question samples were fabricated in a high-mobility, low-density, two-dimensional electron gas (2×10^6 cm^2/Vs and 1.1×10^6 cm^{-2}, respectively, both measured at 4.2 K in the dark). Measurements were done in a strong magnetic field $B \approx 13$ T, with electrons in the FQH regime at a filling factor $\nu=1/3$. Vanishing of the longitudinal resistivity ρ_{xx} assures that the (net) current is flowing chiraly along the edges of the sample in the form of *edge states*. This allows measurements in *multi-terminal geometries* shown in Fig. 1. Two techniques are employed in order to dilute

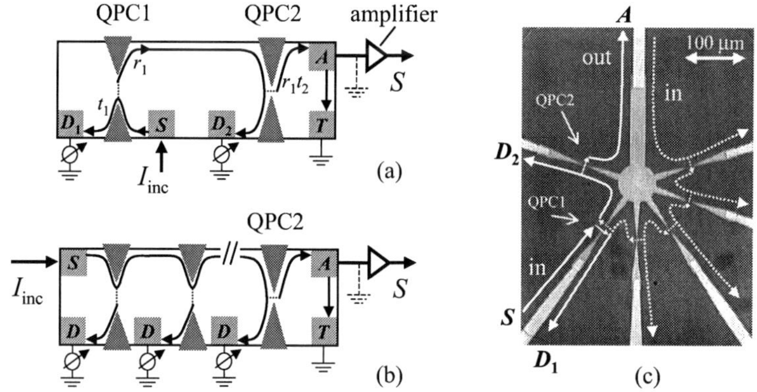

FIGURE 1. **(a)** The *reflected* current, resulting from partitioning a dc (noiseless) incident current (injected from terminal S) by an open QPC1, is composed of a dilute beam of quasiparticles. This dilute current impinges on a pinched QPC2, with the resulting noise measured by a cooled, low-noise, amplifier at terminal A. The intermediate drain D_2 prohibits multiple-reflections and enables determination of the individual transmission of each QPC. The grounded terminal T is used to fix the *output* impedance of the sample, thus setting both the sample's equilibrium noise and the amplifier's sample-dependent noise to a constant background level. **(b)** An alternative scheme, suitable for producing moderately-dilute current, invokes *transmitting* via a cascade of weakly backscattering QPCs. **(c)** Photograph of the vicinity of the QPCs in an actual device, with the direction of the current along the edges. The solid arrows correspond to the direction of current in configuration (a), while the other QPCs on the right are used when configuration (b) is employed (the current in that case is denoted by dotted arrows). The light gray regions are metal gates, deposited on top of the GaAs-AlGaAs heterostructure, which form the QPCs when biased negatively. The 2DEG is embedded some 0.1 µm under the surface, with the Ohmic contacts being outside the picture frame.

the impinging beam of quasiparticles before it arrives at the pinched QPC2: A straightforward scheme, shown in Fig. 1(a), utilizes a noiseless current I_{inc} that arrives at a weakly backscattering QPC1 (its transmission t_1 towards drain $\boldsymbol{D_1}$ approaches unity), resulting with a highly dilute *reflected* beam (reflection r_1) of quasiparticles with charge $q_1=e/3$ [6]. Most of the dilute beam reflects back from QPC2 and is collected by drain $\boldsymbol{D_2}$ while a small part is transmitted (fraction r_1t_2 of the incident current) and reaches terminal $\boldsymbol{A}$, to be amplified and subsequently measured. This simple method cannot be applied, however, to achieve a moderately dilute current of quasiparticles since a *partially-pinched* QPC1 would lead to partial bunching of quasiparticles [5] and to the reflection of an *effective charge* $q_1>e/3$. Hence, we employ also a geometry shown schematically in Fig. 1(b). Here, the incident current from $\boldsymbol{S}$ is being partitioned (gradually) by transmitting through a cascade of weakly backscattering QPCs (for each $t_i\rightarrow 1$), feeding finally QPC2 with a beam of quasiparticles with arbitrary dilution (depending on the number of QPCs) [12]. We also tested carefully whether the dilute beam of quasiparticles, during its transport along the edges of the device, suffers *intra-edge scattering* and a subsequent equilibration. We find that such scattering, that might modify the dilution and the nature of the shot

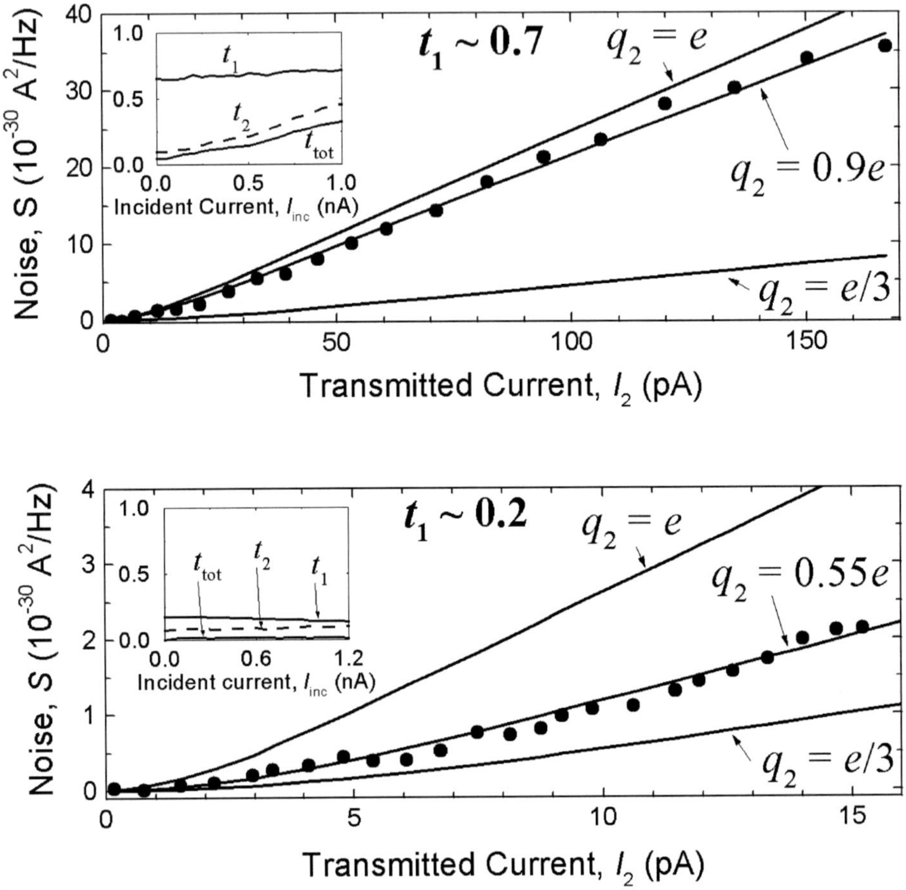

FIGURE 2. Noise and transmission measurements of a pinched QPC (transmission ~0.1 at zero bias) for two different values of dilution of the impinging current, 0.7 (**top**) and 0.2 (**bottom**). In the main graphs the measured noise is plotted against the transmitted current, together with the theoretical prediction of the independent particles model at a finite temperature. The intermediate curve in each graph represents the best fit to an arbitrary charge q_2. Various transmission coefficients, measured simultaneously with the noise, are shown in the insets against the incident (noiseless) current. Each inset shows the dilution level generated by the 'injector' (t_1), the transmission t_2 of the pinched QPC *in response to the dilute impinging current*, and the total transmission t_{tot}. Notice the sensitivity of the *non-linearity* of t_2 to the dilution level of the impinging current (for similar incident currents).

noise altogether, does not take place to any observable measure in our devices (see Ref. [6] for details).

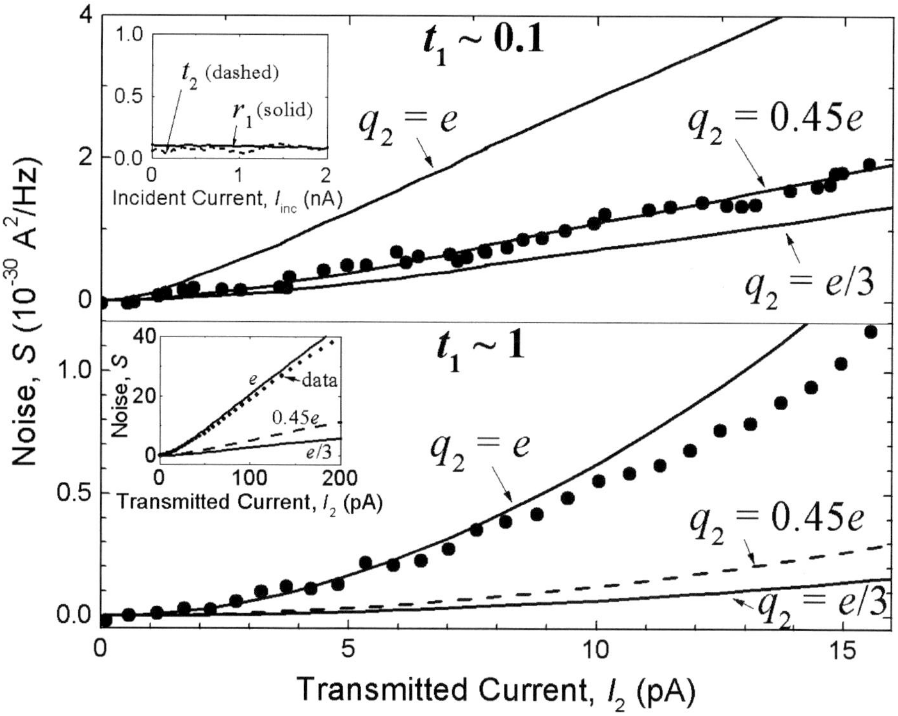

FIGURE 3. Comparison of the charge characterizing a pinched QPC for two extreme cases of impinging current: undilute (full Fermi sea) and highly dilute. **Top:** The noise produced by the pinched QPC2 fed by a highly dilute impinging current (dilution ~0.1) corresponds to a charge $q_2 \approx 0.45e$. The inset shows the current-dependence of the dilution ($r_1 \approx 0.1$) and that of the QPC2 transmission ($t_2 \approx 0.1$, being fairly current independent for such a dilute beam). **Bottom:** The noise produced by QPC2, *keeping the same transmitted current* but using an undilute impinging current, corresponds to almost an *electronic* charge. The inset verifies the charge $q_2 = e$ by measuring the noise over considerably wider range of transmitted current. Clearly, the charge characterizing a QPC depends not only on the potential barrier height but also on the average occupation of the states of the impinging current.

Using one of the two methods depicted in Fig. 1 we create a beam of quasiparticles (of charge $e/3$) with varied dilution. This dilute current is partitioned again by the pinched QPC2, generating noise that is measured with a spectrum analyzer after being amplified by a homemade cooled amplifier. The amplifier, being placed near the sample, has a very low current noise at its input, $\langle i^2_{amp} \rangle = 1.5 \times 10^{-28}$ A^2/Hz, when operating at bandwidth of 30 kHz around a center frequency $f_0 \approx 1.5$ MHz. The value of f_0, chosen well above the cutoff of the ubiquitous $1/f$ noise, is determined by a resonance of a LC circuit, where C is dominated by the unavoidable capacitance of the coaxial cable connecting the sample and the amplifier and L is set by an added

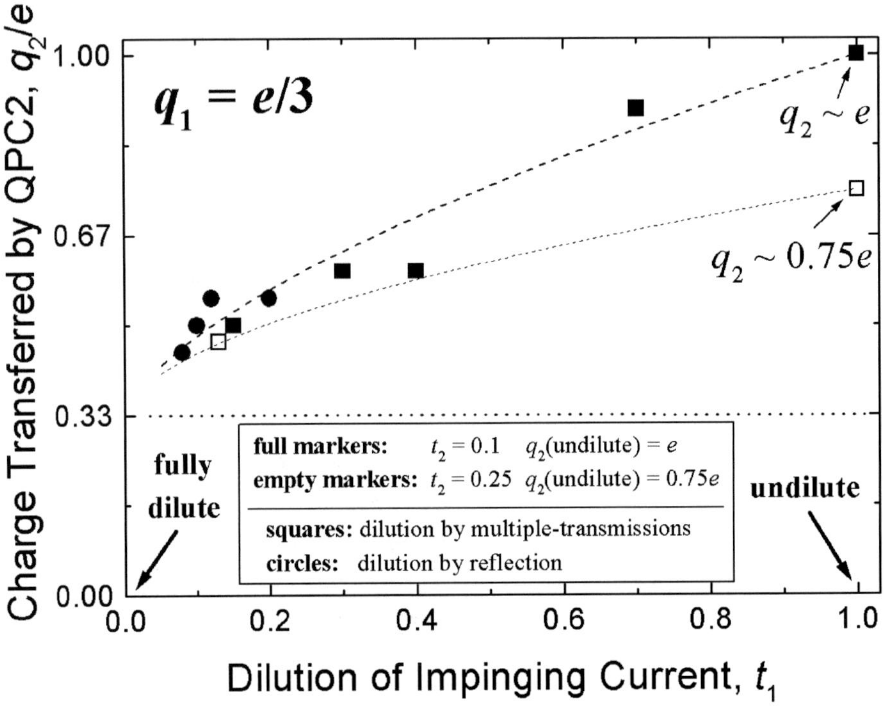

FIGURE 4. Summary of the charge characterizing the pinched QPC2 in response to different values of dilution of the impinging current (extracted form curves similar to the top graph in Fig. 3). Three samples are shown – two complementary sets of data for t_2=0.1, with dilution produced by reflection by a single QPC and by transmission through five weakly backscattering QPCs, and one with t_2=0.25. In the case of t_2=0.25 only the extreme points are shown in order to simplify the graph. Evidently, as the current that impinges on the pinched QPC becomes more dilute, the charge drops from its original value towards ~$e/3$ in the limit of very high dilution – the dashed lines are only a guide to the eye, demonstrating this tendency of charge variation. We conclude that the sparse, individual, quasiparticles, composing the current in the high dilution limit, tunnel through the pinched QPC – originally thought to be opaque for them.

superconducting coil (see Ref. [1] for details). Due to the multi-terminal geometry, reflected currents flow into the grounded terminals ***D*** and ***T***, leading to a constant *input* (at ***S***) and *output* (at ***A***) impedances $G=e^2/3h$ – independent of the transmission of the QPCs. This makes both the sample's equilibrium thermal noise ($4k_BTG$) and the amplifier's sample-dependent back-fed noise ($\langle i^2_{\text{amp}}\rangle / G^2$) independent of the transmission of the QPCs; hence it can be easily subtracted [13]

The system shown in Fig. 1(a), composed of two sources of noise (the two QPCs), can be analyzed by means of *superposition* [14]. Consider first the 'injector',

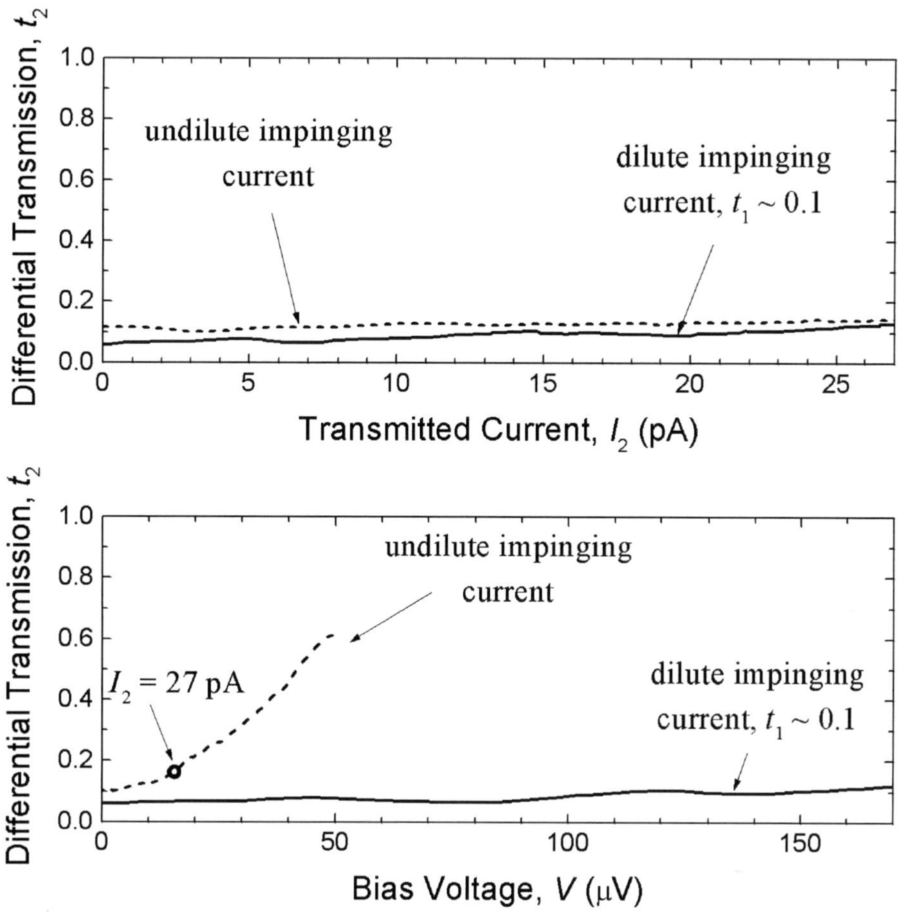

FIGURE 5. The effect of diluting the impinging current on the (differential) transmission of a pinched QPC. **Top:** The transmission of a highly dilute current (solid curve) and that of an undilute current (full Fermi sea, dashed curve) through the pinched QPC2 with the same *transmitted current* in both cases. The bias, on the other hand, reaches a maximal value of ~170 μV in the dilute case but only ~16 μV in the undilute case. Nevertheless the transmissions are similar. **Bottom:** Measurements in the non-linear regime, with the same source voltage applied in both cases (the undilute current is some ten times larger for the same voltage). The non-linearity in the undilute case is much stronger than the dilute case at the larger voltage range.

characterized by reflection r_1 and reflected charge $e/3$: Being a stochastic element, it generates (at zero temperature) noise $2(e/3)I_{inc}r_1(1\text{-}r_1)$, where $I_{inc}r_1$ is the averagreflected current. This noise, however, is transferred via QPC2 (having transmission t_2), thus attenuated by factor t_2^2, resulting with a contribution of the injector to the total noise:

$$S_1 = 2(e/3)I_{\text{inc}}r_1(1-r_1)\cdot t_2^2 \tag{1}$$

Consider now QPC2, characterized by transmission t_2 and charge q_2, when impinged by an undilute current I_{imp} of quasiparticles $e/3$ (a *full Fermi sea*). It produces noise $2q_2 I_{imp} t_2 (1-\tilde{t}_2)$, where $I_{imp}t_2$ is the average transmitted current and $\tilde{t}_2 = t_2 \frac{e/3}{q_2}$ denotes the effective transmission for charge q_2 quasiparticles. This transmission has to be determined self consistently with the charge q_2 in order to maintain the measured conductance of QPC2 [5]. Even though the current impinging on QPC2, $I_{imp}=I_{inc}r_1$ for a partially-reflecting QPC1, is dilute, we still assume in the above written expression that it does not carry any shot noise, since that part was already taken into account in Eq. (1) (the superposition principle). The added contribution of QPC2 to the total noise is therefore

$$S_2 = 2q_2 I_{inc} r_1 t_2 (1-\tilde{t}_2) \quad . \tag{2}$$

The total noise measured in ***A***, at zero temperature, is then given by S_1+S_2. The correctness of this analysis can be validated in the limit of a fixed partitioned charge (say, $e/3$). Then, the total noise equals $S_1+S_2=2(e/3)I_{inc}r_1t_2(1-r_1t_2)$, with r_1t_2 being the total transmission from ***S*** to ***A*** – as expected. When r_1 and t_2 are small, as needed in our experiment, Eqs. (1) and (2) are of order of $O(r_1 t_2^2)$ and $O(r_1 t_2)$, respectively, with the first term much smaller than the second. Hence, the measured noise is dominated by the contribution of the pinched QPC2:

$$S \cong 2q_2 I_{inc} r_1 t_2 = 2q_2 I_2 \quad , \tag{3}$$

with I_2 being the average current traversing QPC2. The orthodox understanding, that models the high potential barrier of pinched QPC as a *bunching filter*, suggests that only three quasiparticles together, or 'electrons', can traverse the barrier with $q_2=e$, thus predicting an *electronic* shot noise. However, as described below, the experimental results are surprisingly different.

We verify first that the noise produced by QPC2, when fed by an non dilute beam (a full Fermi sea of quasiparticles) corresponds, as expected, to $q_2 \approx e$ (not shown here but the results are similar to these in Ref. [5]). Then we allow the dilute beam to impinge on a relatively pinched QPC2, with $t_2 \approx 0.1$. A term t_1 (instead of r_1 employed when the scheme in Fig. 1(a) is used) indicates the level of dilution. As seen in the insets of Fig. 2 t_1 is almost current independent. The top graph in Fig. 2 corresponds to an average state-occupation in the impinging current $t_1 \approx 0.7$ while the bottom graph corresponds to $t_1 \approx 0.2$ (another example can be seen in Fig. 3). The finite temperature of the electrons (65 mK) modifies the linear dependence of the spectral density on current at small currents to a general expression derived within the independent particle model [5]. We also take into account the energy, or current, dependence of the total transmission $t_{tot}=t_1t_2$ when finding the partitioned effective charge q_2. Moreover, the current is being scanned over a large enough range (see Fig. 2, $q_2V>>k_BT$, with V the applied voltage at the source) to allow the noise to reach the linear regime, thus verify further the nature of the noise. A nice agreement is found between the data and the independent particle model, with q_2=0.9e for slightly dilute (t_1=0.7) beam and with q_2=0.55e for highly dilute (t_1=0.2) beam.

A more striking example of the effect of dilution is demonstrated by keeping the average current passing QPC2, I_2, *fixed* while changing its average state-occupation, namely, its dilution. Obviously, a higher source voltage is required to obtain the same current in the dilute case. The top part of Fig. 3 shows a small charge, q_2=0.45e, extracted from the shot noise measurement when the current that impinges on QPC2 is highly dilute, namely, $t_1\approx$0.1. For comparison, the bottom part shows the noise measured at the same current range when an non dilute current (t_1=1) impinges on QPC2 (note that the voltage range here is some ten times smaller). As expected, we find an *electronic* charge, $q_2\approx e$. This result is further verified comparing the expected electronic noise with the data over a much wider range of current in the lower inset (approaching at high current the linear regime of Eq. (2)). A good agreement is observed, with $q_2\approx e$.

Figure 4 summarizes the dependence of q_2 on dilution of the impinging current on QPC2 for two values of transmission, t_2=0.1 and t_2=0.25. As noted in the figure, the partitioned charge by such barriers of non dilute impinging beam is e and 0.75e, respectively. As the current of the quasiparticles is more dilute the effective charge q_2 drops monotonically, approaching ~e/3 for all values of t_2. Due to the scattering of the data, only the most two extreme points are shown for t_2=0.25 in order not to confuse the graph. This leads to the unavoidable conclusion that bunching of quasiparticles is not an essential mechanism for charge transfer through high potential barriers! Apparently, it is only relevant when the states in the incoming current are fully occupied.

We may also ask how is the transmission of a pinched QPC affected by the dilution of the probing quasiparticles. Present theory assumes only an non dilute beam (a full Fermi sea) approaching a one-dimensional constriction within the framework of the Luttinger model. We show in Fig. 5 experimental results of the effect of dilution on transmission. In the top panel we look at the linear regime, where the source voltage is small enough to keep the transmission almost energy independent. Although the applied source voltage in the dilute case is some ten times larger than in the non dilute case (to maintain the same currents), the transmission is almost the same in both cases. In the bottom panel we plot the transmission in the two cases over the same voltage range (the range of the data for the dilute case is essentially equal in the two panels), with the transmitted current in the non dilute case being some ten times larger than in the dilute case (to maintain the same voltages). The transmission of an non dilute beam strongly depends on current and approaches unity at V>50 μV. Since equilibration of the quasiparticles had been ruled out [6], the results suggest strong sensitivity to the *current* flowing through the pinched QPC and much less to the *energy* of the quasiparticles. This rules out that the potential profile of the barrier in the QPC is responsible for non-linearity. Moreover, the insensitivity of the transmission in the linear regime to the dilution of the quasiparticles beam agrees well with our noise results. In other words, noise and transmission measurements show that quasiparticles can transfer, with the same ease, either one by one or bunched in groups. Their bunching depends not only on the transparency of the barrier but also on the preparation of the probing quasiparticles beam. It is now for theory to explain such a bizarre effect.

ACKNOWLEDGMENTS

The work was partly supported by the Israeli Academy of Science and by the German-Israel Foundation (GIF) grant. We thank A. Yacoby, Ady Stern and Y. Levinson for helpful discussions.

REFERENCES

* a similar version was published in nature, Vol. 416, 515 (2002).

1. de Picciotto, R., Reznikov, M., Heiblum, M., Umansky, V., Bunin, G. & Mahalu, D. *Nature* **389**, 162 (1997).
2. Saminadayar, L., Glattli, D. C., Jin, Y. & Etienne, B. *Phys. Rev. Lett.* **79**, 2526 (1997).
3. Reznikov, M., de-Picciotto, R., Griffiths, T. G., Heiblum, M. & Umansky, V. O *Nature* **399**, 238 (1999).
4. Kane, C. L. & Fisher, M. P. A. *Phys. Rev. Lett.* **72**, 724 (1994).
5. Griffiths, T. G., Comforti, E., Heiblum, M., Stern, Ady & Umansky, V. *Phys. Rev. Lett.* **85**, 3918 (2000).
6. Comforti, E., Chung, Y. C., Heiblum, M. & Umansky, V. Phys. Rev. Lett. **89**, 066803 (2002).
7. Schottky, W. Ü *Ann. Phys.* (Leipzig) **57**, 541 (1918).
8. Prange, R. E. & Girvin, S. M. (Eds.) *The Quantum Hall Effect.* Springer, New York, 1987.
9. Laughlin, R. B.. *Phys. Rev. Lett.* **50**, 1395 (1982).
10. Khlus, V. K. *Sov. Phys. JETP* **66**, 1243 (1987).
11. Lesovik, G. B. *JETP lett.* **49**, 592 (1989).
12. This is not an obvious outcome of this configuration. A detailed study of it had been performed [6], proving indeed that a dilute beam of quasiparticles is being produced.
13. The sample's equilibrium thermal noise (5×10^{-29} A^2/Hz) and the amplifier's current noise (1.5×10^{-28} A^2/Hz) sum up to a background level of the order 2×10^{-28} A^2/Hz, while the value of the shot noise is around 10^{-30} A^2/Hz. In a two-terminal geometry current-dependent variations of the background would have overwhelm the minute signal.
14. Oliver, W. D., Kim, J., Liu, R. C. & Yamamoto, Y. *Science* **284**, 299 (1999).

High Frequency Quantum Shot Noise: Can Quiet Electrons Generate Sub-Poissonian Photons?

D.C. Glattli

Laboratoire Pierre Aigrain, Departement de Physique de l'Ecole Normale Superieure, 24 rue Lhomond, 75231 Paris Cedex 05, France
and
Service de Physique de l'Etat Condense, CEA Saclay, F-91191 Gif-sur-Yvette, France

Abstract. Quantum conductors show very low shot noise [1,2]. This is due to the Fermi statistics which regulates the flow of electrons at a frequency eV /h. The electronic shot noise power detected in a measuring circuit can be viewed as the flux of microwave photons emitted by the noisy conductor. An interesting question is whether the flux of photons inherits low noise properties from that of electrons.

Keywords: shot-noise, Fano factor, photon noise
PACS: 73.23.-b, 42.50.Ar, 73.50.Td

INTRODUCTION

A recent answer to the question posed in the abstract has been proposed [3] and quantitatively confirmed in a theoretical calculation [4]: low noise (antibunched) photon emission may occur in very narrow quantum conductors, called Quantum Point contacts (QPC). The photon frequency must however be restricted to the range $[eV/h, eV/2h]$ where V is the bias voltage and the number of transmitted mode is equal to one. This ensures that only one photon is produced when an incoming electron is scattered by the QPC, so allowing to transfer the electron statistics to that of photons.

We discuss the physics of this remarkable effect and the possible tools needed for its observation as the recent Hanbury-Brown and Twiss correlation experiment realized with GHz photons [3]. We also discuss the intimate relation between the photon noise generated by the current fluctuations of a quantum conductor and the Full Counting Statistics of the electronic shot noise.

RESULTS AND DISCUSSION

Single photon sources are the prototype of low noise photons sources. They are usually based on a single two level system which can be either an atom, an optical quantum dot, an ion or a molecule. The upper level is first populated by an electron

CP800, *Unsolved Problems of Noise and Fluctuations: UPoN 2005*
edited by L. Reggiani, C. Pennetta, V. Akimov, E. Alfinito, and M. Rosini

using electrical or optical pumping. Then the transition to the lower level is accompanied by the emission of a single photon. The Pauli principle and the delay necessary to refill the upper level ensures that single photons are well separated in time and that never two photons occupy the same electromagnetic mode. Perfect photon antibunching is thus observed. Its observation requires photon correlations measurements, for example a Hanbury-Brown Twiss experiment.

Instead of starting from discrete initial and final electronic states emitting a photon at a defined frequency within a continuum of frequencies, we propose here a reverse scheme. We start with a continuum of initial and final electronic states and restrict the continuum of photons to a narrow frequency using filtering. According to these unconventional conditions, we will show that the restrictions imposed by the frequency statistics allow nevertheless to generate antibunched photons.

It is well known that the low frequency shot noise of a single mode quantum conductor of transmission D is given by

$$S_I = 2eI(1-D)$$

where S_I is the low frequency spectral density of the current fluctuations, and I is the current in response to a chemical potential difference eV applied between the left and right contact. This is simply derived by considering that the left contact injects regularly electron wavepackets at a rate eV/h toward the conductor. Because of the Fermi statistics each wave packet is occupied by a single electron (here we disregard spin degeneracy). The incoming current is thus $I_O = e(eV/h)$ and the current flowing through the conductor is $I = De^2V/h$. But, more important, the regular injection makes the incoming current noiseless and the only noise comes from the quantum partitioning of electrons between transmitted and reflected wavepackets. The binomial statistics of these events gives rise to a variance of the current fluctuations per unit bandwidth

$$S_I = 2eI_0 D(1-D) = 2eI(1-D)$$

as announced. In the following, we are going to consider the high frequency current noise detected as a flux of photons in the external circuit. This idea is not new and can be found in a seminal paper by Nyquist written in 1928. The shot noise at finite frequency ν is given by

$$S_I(\nu) = S_I(0)\frac{eV - h\nu}{eV}.$$

The interpretation is very simple : the available energy for electrons participating to the current fluctuations is no longer eV but $eV - h\nu$. Indeed, the detectable fluctuations correspond to incoming electrons which have enough energy to both loose a quantum energy $h\nu$ in the external circuit and find an empty outgoing state. For $h\nu > eV$ the current noise vanishes as no such electrons are available. If the external circuit is an electromagnetic transmission line (a coax) of characteristic impedance Z_C (see Fig. 1), the power emitted in the circuit in a bandwidth $\delta\nu$ is

$$P=\frac{R^2 Z_C}{(R+Z_C)^2} S_I(\nu)\delta\nu = N(\nu)h\nu\,\delta\nu$$

where $R=(De^2/h)^{-1}$ and $N(\nu)$ is the population of emitted photons. Here the photons are the T.E.M. modes of a coaxial mode but other modes could be considered depending on coupling. The important thing is to note that the photon flux is a measure of the finite frequency electronic shot noise.

Now, we arrive to the main subject of this contribution: the low frequency photon noise. Practically it can be measured by recording the power fluctuations over long time corresponding to a low frequency bandwidth Δf (not to be confused with $\delta\nu$). The variance of the photon population is obtained by that of the power fluctuations

$$\left\langle(\Delta P)^2\right\rangle = \left\langle(\Delta N)^2\right\rangle (h\nu)^2\,\Delta\nu\,\Delta f$$

which is itself related to the variance of the high frequency electronic noise. The average photon population generated by the electronic shot noise at frequency ν is

$$< N >=|S_{21}|^2\frac{(1-D)}{2}\frac{eV-h\nu}{h\nu}$$

where $|S_{21}|=\frac{4RZ_C}{(R+Z_C)^2}$ represents the coupling between the quantum conductor and the transmission line (this electromagnetic scattering parameter is not to be confused with the electron scattering parameter inside the conductor).

If many photon are emitted and they can populate the same T.E.M. mode within the time $\delta\nu^{-1}$, bunching of photons results according to Bose statistics. This is the case if the frequency ν is very small compared to eV/h the rate of electron arriving on the scatter and generating current noise. This gives a super-poissonian photon noise similar to that of a black-body source and $\left\langle\Delta N^2\right\rangle \approx N(1+N)$. If however the frequency of shot noise observation is close to eV/h there is only one possible photon associated with each electron : antibunching occurs.

Although at present this photon antibunching remains a prediction, it is important to design an experimental tool to detect various photon statistics. In this contribution will be presented the results of a Hanbury-Brown Twiss photon correlation experiment performed in the GHz photon regime. The frequency range is suitable for observation of the phenomenon with low temperature controllable quantum point contacts (this will not preclude observation at higher frequency and the phenomenon could be observed also in atomic point contact and nanotubes at THz frequencies). To test the sensitivity, the few photon regime has been investigated (from N=0 to few tens).

When the conductor is at thermal equilibrium and the temperature is varied above $h\nu / k_B$, positive correlation and super-poissonian noise is measured as expected for a black body photon source. When the temperature reaches values below $h\nu / k_B$ the quantum cross-over from a linear variation with temperature to a constant is found. This shows that the quantum regime is observable.

In order to show the sensitivity to another statistics, a coherent microwave source has been used in place of the conductor but with a power corresponding to similar average photon number. Here the photon noise cross-correlation is found equal to zero and the variance of the photon population is found linear with the average population. This is what is quantitatively expected for poissonian noise. A quantum description of the amplifier used prior to detection has been performed which predicts a Fano factor of the order of $k_B T_N / h\nu$ where T_N is the noise temperature of the amplifier. This Fano factor is quantitatively observed in the experiment. But the important experimental result is that the photon cross-correlations remain unaffected by the detection details. The GHz Hanbury-Brown Twiss experiment shows enough sensitivity to detect antibunched photons. The non-classical photons should give negative cross-correlations.

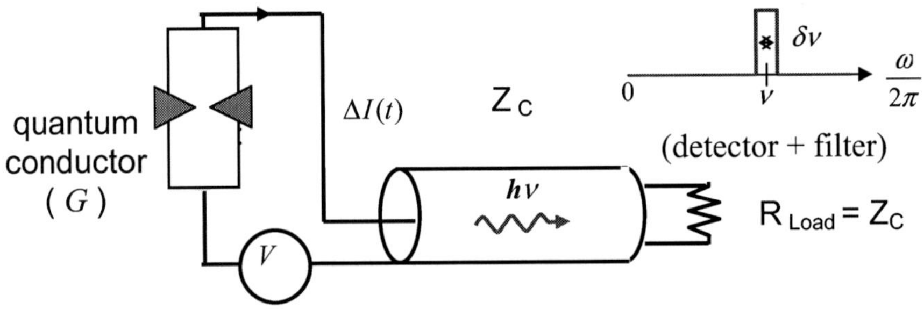

FIGURE 1. Schematic of the electromagnetic power emitted by a quantum conductor and detected by a transmission line.

REFERENCES

1. G. B. Lesovik, JETP Lett. **49**, 592 (1989) ; Y.M. Blanter and M. Buttiker, Phys. Rep. **336**, 2 (2000).
2. A. Kumar, L. Saminadayar, D. C. Glattli, Y. Jin, and B. Etienne, Phys. Rev. Lett. **76**, 2778 (1996).
3. J. Gabelli, L.-H. Reydellet, G. Fève, J.-M. Berroir, B. Plaçais, P. Roche, and D. C. Glattli, Phys. Rev. Lett. 93, 056801 (2004)
4. C. W. J. Beenakker and H. Schomerus, Phys. Rev. Lett. **93**, 096801 (2004)

Shot Noise Suppression As Indicator Of Coherent Tunneling In Resonant Diodes

V.Ya. Aleshkin*, L .Reggiani[+], M. Rosini[+]

*Institute for Physics of Microstructures of RAS, Nizhny Novgorod Russia
[+]National Nanostructure Laboratory of INFM, Dipartimento di Ingegneria dell' Innovazione, Università di Lecce, Via Arnesano s/n, 73100 Lecce, Italy

Abstract. The low frequency current spectral density and the Fano factor of a double barrier resonant diode are investigated in the case of coherent tunneling as a function of voltage just below the current peak. At for voltages near the current peak the Fano factor is predicted to become suppressed below the value of 0.5. This drop below 0.5 is interpreted as an indication of coherent against sequential tunneling transport.

Keywords: Resonant tunneling diode, shot noise.
PACS: 72.70+m

INTRODUCTION

The microscopic interpretation of electron transport in double barrier resonant diodes (DBRDs) is found to admit a coherent or a sequential tunneling approach [1]. The coherent approach [2] consists in considering the whole device as a single quantum system characterized by a transmission coefficient describing carrier transport from one contact to the other. This approach is justified when the probability of electron scattering inside the quantum well is negligible in comparison with the probability of electron tunneling to the contacts. In the opposite case, the sequential approach is used. The sequential approach [3] consists in considering tunneling through the diode as a two step process where carriers first transit from one contact into the well, then from the well to the other contact. The intriguing feature of these two approaches is that both of them are capable to explain the current voltage (I-V) experimental data [4] as well as most of the shot noise characteristics. As a consequence, there is no way to distinguish between these two transport regimes and the natural question whether tunneling transport is coherent or sequential remains an unsolved one. The coherent approach to shot noise in DBRDs has received wide attention since the first experimental evidence by Li et al [5] of shot noise suppression with a minimum value of the Fano factor $\gamma = S_I(0)/2qI = 0.5$, here $S_I(0)$ is the low frequency spectral density of current fluctuations and q the absolute value of the unit charge responsible of current. Remarkably, most of the coherent approaches developed so far (see bibliography in [1]) predict a maximum suppression $\gamma = 0.5$ even if there is clear

CP800, *Unsolved Problems of Noise and Fluctuations: UPoN 2005*
edited by L. Reggiani, C. Pennetta, V. Akimov, E. Alfinito, and M. Rosini

experimental evidence of suppression below this value [6, 7, 8] down to values of $\gamma = 0.25$. To this purpose, some authors obtained theoretical values of the Fano factor just below the value of 0.5, [9], [10]. However, the physical interpretation of these results remains mostly qualitative. We remark that the theory of shot noise in DBRDs under the sequential approach [11] provides a Fano factor of 0.5 as the minimum value of shot noise suppression. The aim of this work concerns with the prediction that suppression of shot noise with a Fano factor below 0.5 can be a signature of coherent tunneling against sequential tunneling.

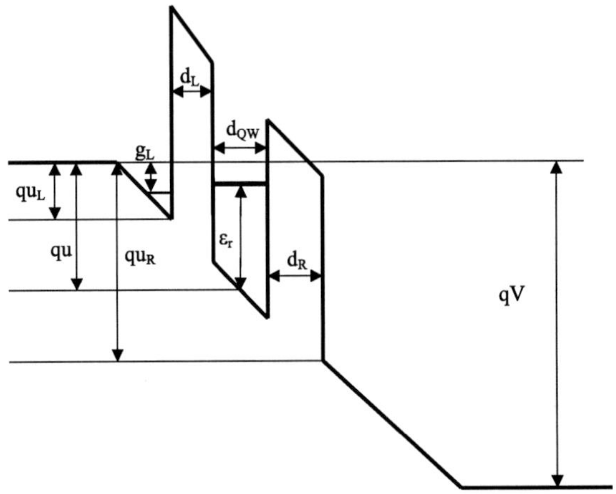

FIGURE 1. Sketch of the band profile of the double barrier structure.

MODEL OF ELECTRON TRANSPORT IN RTD UNDER COHERENT TUNNELING

The typical diode investigated here is the standard double well structure depicted in Fig. 1. We denote by $\Gamma = \Gamma_L + \Gamma_R$ the resonant states width, by ε_r the energy of the resonant level as measured from the center of the potential well and by $\Gamma_{L,R}$ the partial widths due to the tunneling through the left and the right barrier, respectively. We consider the case of coherent tunneling in the presence of only one resonant state and we assume that the diode has contacts with unit square surfaces. For convenience, calculations are carried out using the CGS system.

The kinetic model is developed by assuming that the electron distribution functions in the emitter and in the collector $f_{L,R}$ are equilibrium-like, but with different electro-chemical potentials $F_{L,R}$:

$$f_\alpha(\varepsilon, F_\alpha) = \frac{1}{1 + \exp\left(\dfrac{\varepsilon - F_\alpha}{k_B T}\right)} \tag{1}$$

here $\alpha = L$ stands for the left contact (the emitter), $\alpha = R$ for the right contact (the collector), k_B is the Boltzmann constant, T the bath temperature and ε the carrier energy. By using a second quantization formalism [12,13], we find the usual expression for the average current

$$< I >=< I_L >= -< I_R >= \frac{qm}{2\pi^3\hbar^3} \int_{-g_L}^{\infty} d\varepsilon_z D(\varepsilon_z) \int_0^{\infty} d\varepsilon_\perp [f_L(\varepsilon, F_L) - f_R(\varepsilon, F_R)] \quad (2)$$

where $\varepsilon_\perp = p_\perp^2 / 2m$ is the kinetic energy of the transverse motion, $\varepsilon = \varepsilon_z + \varepsilon_\perp$, g_L the energy gap between the bottom of the conduction band and the first quantized level in the well before the left barrier (see Fig. 1), $D(\varepsilon_z)$ the transparency of the double barrier given by:

$$D(\varepsilon_z) = \frac{\Gamma_L \Gamma_R}{(\varepsilon_z - \varepsilon_r + qu) + \Gamma^2 / 4} \quad (3)$$

To calculate the spectral density of current fluctuations at zero frequency under fixed voltage we use the expression:

$$S_I(0) = \int_{-\infty}^{\infty} dt < \delta I(0)\delta I(t) + \delta I(t)\delta I(0) > \quad (4)$$

We recall that the total current fluctuation consists of two sources. The first comes from the fluctuation of population states in the contacts and the second from the fluctuation of the electron charge in the quantum well δQ_{QW} [12] which changes the transparency of the structure. Accordingly, $S_I(0)$ is found to be given by the sum of three terms as:

$$S_I(0) = S_1 + S_2 + S_3 \quad (5)$$

The expressions for S_1, S_2, S_3 are detailed in [13] where it is proven that Eq. (5) satisfies the Nyquist theorem and $S_I(0) \Rightarrow \infty$ on the border of the instability region where $\partial < I > / \partial V \Rightarrow -\infty$.

RESULTS AND DISCUSSION

We investigate the condition of high applied voltages, when $qV > F_R$, because more close to experiments. (We recall that typical magnitudes for the relevant parameters of DBRDs are: for Γ less than a few meV, for $F_{L,R}$ less than 100 meV, and for the voltage at the peak current around 0.5 V.) Then, at zero temperature the expressions for the current and Fano factor become:

$$< I >= \frac{qm\Gamma_L\Gamma_R}{4\pi^2\hbar^2} B(f, \xi) \quad (6)$$

$$\gamma = 1 - \frac{4\Gamma_L\Gamma_R}{\Gamma^2} \left\{ 1 - \lambda \frac{\Gamma_L - \Gamma_R}{\Gamma} - \lambda^2 \frac{\Gamma_L\Gamma_R}{\Gamma^2} \right\} \frac{A(f, \xi)}{B(f, \xi)} \quad (7)$$

where

$$A=(f+\xi)\left[\frac{f+\xi}{1+(f+\xi)^2}+\arctan(f+\xi)-\frac{\xi}{1+\xi^2}-\arctan(\xi)\right]+\frac{1}{1+(f+\xi)^2}-\frac{1}{1+\xi^2},$$

$$B(f,\xi)=2(f+\xi)[\arctan(f+\xi)-\arctan(\xi)]-\ln\left(\frac{1+(f+\xi)^2}{1+\xi^2}\right)$$

and for convenience we define dimensionless chemical potential f and voltage drop ξ as

$$f=\frac{2(F_L+g_L)}{\Gamma},\qquad \xi=\frac{2(qu-\varepsilon_r-g_L)}{\Gamma}$$

The Coulomb interaction is described by the dimensionless parameter λ

$$\lambda=\frac{\hbar\Gamma}{\Gamma_L\Gamma_R}\frac{1}{(C_L+C_R+C_{QW})}\frac{\partial<I>}{\partial u} \tag{8}$$

where C_L, C_R, C_{QW} are the capacitances of the emitter, the collector and the quantum well, respectively. If $f>>1$ and the resonant level is located below the Fermi level of the emitter far from its borders F_L and $-g_L$, i.e. $\xi<<-1$, $f+\xi>>1$, then $A(f,\xi)/B(f,\xi)\approx 1/2$ and Eq. (9) recovers the expression given in [12]

$$\gamma=\frac{\Gamma_L^2+\Gamma_R^2+2\Lambda(\Gamma_R-\Gamma_L)+2\Lambda^2}{\Gamma^2} \tag{9}$$

with $\Lambda=-\lambda\Gamma_L\Gamma_R/\Gamma$. As it will be shown below, when $f>>1$ the values taken by γ in Eq. (10) are practically constant and correspond to the plateau exhibited by $\gamma(f,\xi)$. On this plateau, $1\geq\gamma\geq 1/2$ and shot noise is suppressed.

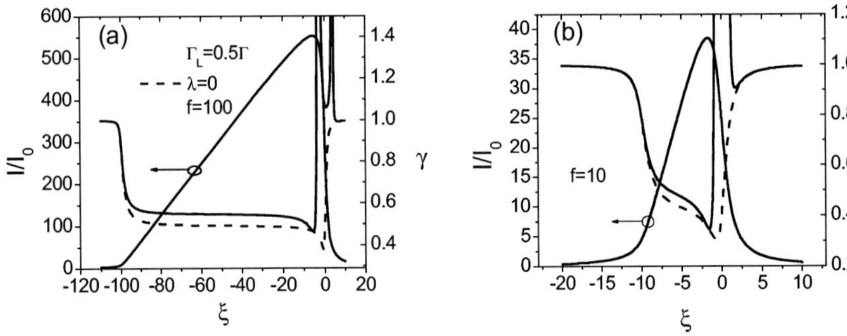

FIGURE 2. Dependence of the Fano factor and of the dimensionless current on the dimensionless electrical potential $\xi=2(qu-\varepsilon_r-g_L)/\Gamma$ at $T=0$ K. Here $f=2(F_L+g_L)/\Gamma$ and $I_0=qm\Gamma_L\Gamma_R/2\pi^2\hbar^3$.

Now we demonstrate that at voltage values for which the resonant level is close to the band edge of the emitter the Fano factor can drop below the value 1/2. To this purpose, we first consider the simplified model by taking dielectric constant $\kappa=12.9$ (GaAs), $C_L=C_R=\kappa/4\pi d$, $d=5$ nm. Figure 2 reports the dependencies of current and Fano factor on ξ for the two values $f=100$ (Fig. 2 (a)), $f=10$ (Fig. 2(b)) when

$\Gamma_L = \Gamma_R$ and in the presence (continuous curves) or absence (dotted curves) of Coulomb interaction. The figure shows that for large values of the ratio between the Fermi energy and the resonant width ($f = 100$), $\gamma(\xi)$ exhibits a wide plateau region where $\gamma \approx 0.55$ followed by a minimum with $\gamma_{\min} \approx 0.46$. By further increasing the value of f ($f = 1000$) the plateau region is found to widen and $\gamma_{\min} \approx 0.49$. Note that, as it follows from Eq. (10), Coulomb interaction always increases the noise if $\Gamma_L = \Gamma_R$. With the decrease of *f* (see Fig. 2(b) where *f*=10), the plateau region narrows and the value of $\gamma_{\min}$ decreases. Such a decrease of $\gamma_{\min}$ is due to two complementary reasons. The first is associated with the decreasing of the strength of Coulomb interaction. The second is associated with the increase of the effective barrier transparency near to the current peak and in turn with the further suppression of partition noise. For the ideal case $f << 1$, at the peak current the transparency $D \Rightarrow 1$ and in turn $\gamma_{\min} \Rightarrow 0$.

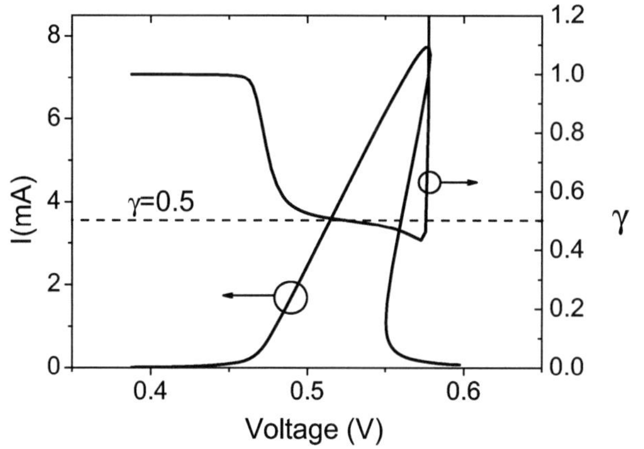

FIGURE 3. Dependence of the current and Fano factor on applied voltage for the structure of [6] at *T=4.2* K. Values of $\Gamma_{L,R}$ and ε_r are chosen by fitting the current voltage characteristic at 77 K.

To confirm the possibility of evidencing the giant suppression $\gamma_{\min} < 0.5$ in a real structure, Fig. 3 presents the calculations performed with a set of parameters related to the experiments in [6] at *T=4.2* K. The structure parameters are $n = 2 \cdot 10^{16}\,\text{cm}^{-3}$, the barrier and quantum well widths $d_L = d_R = d_{QW} = 5$ nm, squared area of contacts 50 μm^2 and $m=0.067m_0$, with m_0 the free electron mass. The only two fitting parameters are $\Gamma_L = \Gamma_R = 0.48$ meV and $\varepsilon_r = 104$ meV. Their values control the location and the amplitude of the current peak, respectively, and are chosen by optimizing the agreement between the experimental and calculated I-V characteristics at 77 K. Calculations give $\gamma_{\min} = 0.43$ in reasonable agreement with the value of 0.35 found in experiments [6]. From Fig. 3 we see that both the I-V and the noise characteristics are sensitive to the asymmetry of the structure, as expected. In any case, the main features of both characteristics are preserved.

As final remark, we notice that there is the point of view that sequential tunneling represents the semiclassical limit of coherent tunneling [14]. Accordingly, the difference in shot noise exhibited by these two limits would signal a macroscopic quantum effect [15]. We have demonstrated the possibility of shot noise suppression below one half for coherent tunneling, what is impossible for sequential tunneling. A remaining open problem is whether or not suppressed shot noise below ½ due to coherent tunneling evidences a macroscopic quantum effect?

ACKNOWLEDGMENTS

This research was performed within the project "Noise models and measurements in nanostructures" supported by the Italian Ministery of Education, University and Research (MIUR). Partial support from the NATO Science Programme (NATO LINKAGE grant PST. EAP. CLG 980629) is gratefully acknowledged.

REFERENCES

1. Y. M. Blanter and M. Buttiker, *Phys. Rep.* **336**, 1- (2000).
2. L. L. Chang et al, *Appl. Phys. Lett.* **24**, 593-595 (1974).
3. S. Luryi, *Appl. Phys. Lett.* **47**, 490-492 (1985).
4. S. Datta, *Electronic transport in mesoscopic systems*. Cambridge University Press 1995.
5. Y. P. Li, A. Zaslavsky, D. C. Tsui, M. Santos, M. Shayegan, *Phys. Rev. B* **41**, 8388-8391 (1990).
6. E. R. Brown, *IEEE Trans. Electron. Dev.* **39**, 2686-2693 (1992).
7. A. Przadka, K. J. Webb, D. B. Janes, H. C. Liu, Z. R. Wasilewski *Appl. Phys. Lett.* **71**, 530-532 (1997).
8. V. Kuznetsov, E. E. Mendez, J. D. Bruno and J. T. Pham, *Phys. Rev. B* **58**, R10159-R10162 (1998).
9. J. C. Egues, S. Hershfield and J. W. Wilkins, *Phys. Rev. B* **49**, 13517-13527 (1994).
10. M. Jahan and A. Anwar *Solid-State Electron.* **38**, 429-432 (1995).
11. G. Iannaccone, M. Macucci, and B. Pellegrini, *Phys. Rev. B* **55**, 4539-4550 (1997).
12. Y.M. Blanter and M. Buttiker, *Phys. Rev. B* **59**, 10217-10226 (1999).
13. V.Ya. Aleshkin, L. Reggiani, N. V. Alkeev and V. E. Lyubchenko, C. N. Ironside, J. M. L. Figueiredo, and C. R. Stanley, *Phys.Rev.B* **70**,115321 (2004)
14. Y.M. Blanter and M. Buttiker, *Semicond. Sci. Technol.* **19**, 663-664 (2004)
15. A.J. Leggett and A. Garg, *Phys. Rev. Lett.* **54**, 857-860 (1985)

Enhanced Shot Noise In Diffusive S/N/S Junctions

F. Lefloch*, M. Sanquer*, C. Hoffmann† and B. Pannetier†

*Département de Recherche Fondamentale sur la Matière Condensée/SPSMS/LCP
CEA/Grenoble, 17 rue des Martyrs, 38054 Grenoble cedex 09, FRANCE
†Centre de Recherches sur les Très Basses Températures-C.N.R.S. associated to Université Joseph Fourier, 25 rue des Martyrs, 38042 Grenoble, FRANCE

Abstract. Current noise measurements enligthen the transport mechanisms in mesoscopic samples in a complementary way to conductance measurements. The noise gives directly access to the charge of the current carriers and is modified by electronic interactions. This is very instructive in the context of hybrid superconductor - normal metal (S-N) structures where charge pairs generated by Andreev reflections compete with quasiparticles in the current transport. In S-N-S junctions, the shot noise is very much enhanced compared to the normal case because of the confinement of normal quasiparticles between the superconducting electrodes. The relevant time depends on the number of Andreev reflections (N_{AR}) necessary for quasiparticles to reach the superconducting gap Δ and escape the normal metal. It can therefore be tuned by the bias voltage because N_{AR} goes like $1/eV$. Diffusive junctions of intermediate size ($\xi_\Delta < L < L_\Phi$) are particularly interesting because various transport regimes can then be studied from coherent (when $eV < E_{Th} = \hbar D/L^2$) to incoherent ($eV > E_{Th}$). We present shot noise measurements in Al/Cu/Al junctions down to $100mK$, that reveal a clear mesoscopic transition when the voltage equals the Thouless energy. A good agreement with semi-classical description is found at large energy when Incoherent Multiple Andreev Reflections prevail. At low voltage, the shot noise diverges which may indicates the coherent transfer of mutiple charges. No theoretical predictions exist for such a regime in diffusive junctions of intermediate size and this point remains unsolved.

Keywords: Mesoscopic Physics, Shot noise, Superconductor - Normal metal Contacts
PACS: 74.45.+c, 72.70.+m, 74.40.+k

INTRODUCTION

Studies of the quantum properties of complex solid state devices is a continuous growing research field that emerged from mesoscopic physics together with the progress of the nanotechnologies. One can mentionned well known quantum bits [1, 2] as quantum integrated circuits but also more fundamental physics oriented realizations like Aharonov-Bohm rings for persistent currents measurements [3], Y-shape quantum conductors for Hanbury-Brown Twiss noise correlations [4, 5, 6], networks of normal [7] or superconduting wires [8], (coupled) quantum dots [9]

One of the experimental tools that can be used to investigate such quantum effects is shot noise which refers to the fluctuations of the electrical current though a device and is the consequence of the discrete nature of the charge. In a quantum conductor, the shot noise follows the Landauer-Büttiker formula [10] : $S = F2eI$ with F the Fano factor given by $F = \Sigma_n T_n(1-T_n)/\Sigma_n T_n$ where $\{T_n\}$ gives the transmission coefficients. The shot noise is therefore a sensitive measurement of the details of the transmission of the elctronic waves through a conducting sample. It is also proportional to the charge of the current carriers. Some experiments have exploited this property to establish fractional

CP800, *Unsolved Problems of Noise and Fluctuations: UPoN 2005*
edited by L. Reggiani, C. Pennetta, V. Akimov, E. Alfinito, and M. Rosini

quantization of quasiparticles in the fractional quantum Hall effect [11, 12]. In an other experiment [13, 14, 15, 16], the doubling of the shot noise in S-N contacts has been seen due to the emission of Cooper pairs from the superconducting reservoir to a normal conductor.

The mechanism which is responsible for the transfert of current in such S-N contacts is the Andreev reflection. An electron with energy $|\varepsilon| < \Delta$ with respect to the Fermi level can not escape in the superconducter due to the absence of electronic states in the gap Δ. Instead, it enters the superconductor together with a second electron to create a Cooper pair and a hole is retroreflected in the normal metal. The electron and the retroreflected hole form an Andreev pair which remains coherent, in the diffusive limit, over a distance $L_c = min(L_\phi, \xi_\varepsilon = \sqrt{\hbar D/\varepsilon})$ where D is the diffusion constant of the normal metal and L_ϕ the single particle phase coherence length.

In S/N/S junctions with a normal metal length $L > L_c$, the Andreev pair is split off and the electron and the hole behave, far from the interface, as independent quasiparticles. In this *incoherent regime* the quasiparticles produce shot noise which originates from the diffusion through the normal metal. The noise is then enhanced compared to a N/N/N system because each quasiparticle entering the normal region is successively retroreflected at the two SN interfaces (incoherent multiple Andreev reflections - IMAR) and implies many passages of quasiparticles through the junction, instead of only one in the normal case. These IMAR persist as long as the quasiparticle energy stays in the interval $-\Delta < \varepsilon < \Delta$ and when no inelastic collisions occur. The effect of the inelastic interactions on shot noise in S/N/S junctions have been studied by various groups [17, 18, 19]. It is shown that electron-electron interaction reduces the energy window of accessible states for the quasiparticles participating to the IMAR leading to a decrease of the current noise density.

In short S/N/S junctions, the situation is somewhat more complicated. Indeed, as long as $L < L_c$, and when the Thouless energy $E_{Th} = \hbar D/L^2$ exceeds the bias voltage, successive Andreev reflections at the two interfaces are coherent and the interference between quasiparticles leads to the formation of Andreev bound states. In this *coherent regime*, the bound states can carry a supercurrent and one observes dc and ac Josephson effects. In very short junctions with $E_{Th} > \Delta$ only two bound states exist and the transport via these states, can be considered as the transfer of effective charges $2\Delta/eV$. Then, the noise at low voltage can be interpreted as the shot noise of these effective charges due to Landau-Zener transitions between the bound states [20]. If now, the voltage is larger than the Thouless energy, the quasiparticles behave indepedently and one recovers the incoherent regime.

COLLISIONLESS REGIME

We recently measured shot noise in diffusive S/N/S junctions with a normal metal length between the superconducting coherence length $\xi_\Delta = \sqrt{\hbar D/\Delta}$ and the phase coherence length L_ϕ ($\xi_\Delta < L < L_\Phi$). This situation is particularly interesting because it provides a unique way to study different case of electrical transport in diffusive conductors. Indeed, because of the sensitivity of the size of the Andreev pairs to the voltage at low temperature, $L_c \simeq \sqrt{\hbar D/eV}$, one can tune the transition between the regime of coherent

pair transport and single quasiparticle transport on a single experimental run. This transition is easy to achieve experimentally but stays a difficult task on the theoretical side. The existing theories concern either diffusive junctions with negligible proximity effect, which are accessible with semi-classical models [21, 22] or the fully coherent situation [20]. The latter applies to the noise properties of coherent superconducting atomic point contacts with a small number of conducting channels [23]. In this situation, experimental results are well understood [24]. In contrast, the interpretation of the few experimental investigations of noise in multichannel S/N/S junctions available up to now [17, 18, 25, 19] stays a puzzling business.
We studied junctions made of aluminum ($480nm$) for the superconducting reservoirs and copper ($50nm$) as a normal metal with a length of $0.85\mu m$ and a width of $0.4\mu m$ for the main results and a second junctions half wide and long for comparison (see left panel on figure 1). Both samples show a zero resistance state at low temperature (below $800mK$ for the longest). On the right panel of figure 1 is plotted the resistance V/I as a function of the bias voltage for various temperature. The resistance shows a clear non-linear behavior which is very strong at low voltage when approaching the zero resistance state at low current. The overall shape of the current noise density times the resistance, as a function of bias voltage is shown in Fig. 2 together with a SEM pictre of a typical sample. We observe a pronounced minimum at $V \approx 10\,\mu V$ corresponding to the Thouless energy $E_{Th} \simeq 7\,\mu eV$ of sample 1. This minimum indicates the transition from the regime of coherent pair transport to the regime where the Andreev pairs are split off into independent quasiparticles before reaching the opposite interface.

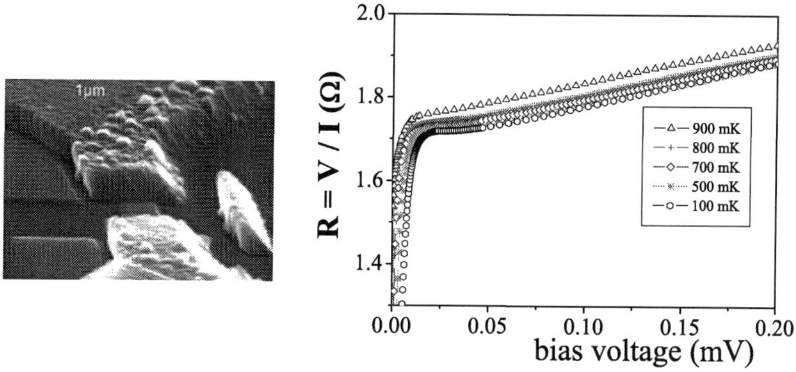

FIGURE 1. SEM picture of a typical sample and resitance of the junction $R = V/I$ as a function of the bias voltage

The noise behavior at high voltage can be understood within a simplified model. Consider an electron entering the normal metal at the energy $\varepsilon \approx -\Delta$. At the first SN-interface it is Andreev-reflected into a hole which travels the normal region a second time. At the other SN-interface the hole is again retroreflected as an electron and so on. In the incoherent case the phase information between two subsequent Andreev reflections is lost and no interference is possible. The quasiparticle energy is increased by eV when it travels from one interface to the other. Therefore, the quasiparticle can escape to the superconducting electrodes only after N passages, with $N = 2\Delta/(eV) + 1$, reaching an energy $\varepsilon \sim \Delta$.

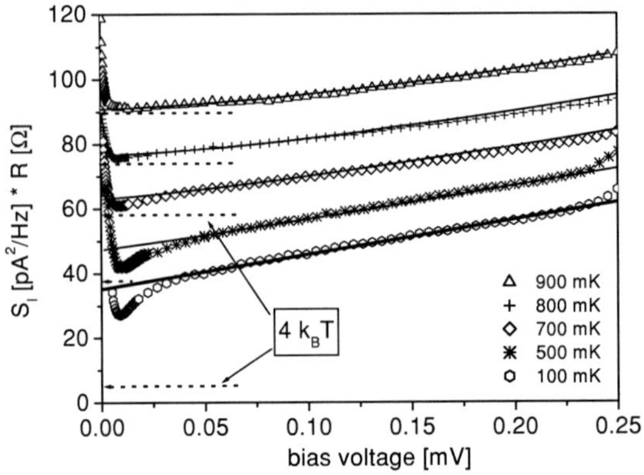

FIGURE 2. Current noise density times the resistance $R = V/I$ versus bias voltage at various temperatures (the data curves are shifted successively by $10\,pA^2/Hz$). Thick solid line: Theoretical prediction in the zero temperature limit (Eq.(1)) with $\Delta = 165\mu eV$. Thin solid lines: Theoretical predictions following Ref. [22] including noise of quasiparticles outside the gap. The arrows indicate the thermal noise level corresponding to each data curve.

Within this description, each quasiparticle entering the normal part causes a series of incoherent Andreev reflections which leads to the diffusion of N quasiparticles through the normal part. The total current noise is therefore the shot noise of a diffusive metal $\frac{1}{3}2eI$ times N:

$$S_I(V) = \frac{1}{3}2eI \times N = \frac{2}{3R}(eV + 2\Delta). \tag{1}$$

This is exactly the prediction of semiclassical theory in the zero temperature limit and when no inelastic processes take place [21, 22]. In this model, the proximity corrections are neglected. In our junctions however, such corrections persist over the whole voltage range. A noise expression taking into account the proximity effect has been derived recently in SIN junctions with a tunnel barrier at the interface [26] but is still lacking in the S/N/S case. In order to include the observed voltage dependence of the resistance, we used $R(V) = V/I$ in equation (1) rather than the normal resistance R_N and analyzed the product $S_I(V)R(V)$ (Fig. 2).
At $T = 100\,mK$ we obtain very good agreement between experiment and Eq.(1) with $\Delta = 165\,\mu eV$, shown in Fig. 2 by a thick solid line, in the range from $50\,\mu V$ up to the current induced transition at about $260\,\mu V$. Up to now this linear regime, reflecting electronic transport through IMAR, was only approximately achieved with a large scatter in the data [25] or in a very small voltage range [19].
At temperatures above $300\,mK$, the thermal noise of the quasiparticles outside the gap have to be taken into account. Along the lines of Ref. [22] we can write the total noise as a sum of this thermal noise and the subgap noise. The fits obtained using the BCS

temperature dependence of the superconducting gap, show excellent agreement with the experimental data between $T = 500\,mK$ and $900\,mK$ (thin solid lines).
So far, we considered only the linear part of the noise at high voltage. However, for decreasing voltage ($V < 50\,\mu V$) the experimental data show a nonlinear regime which extends down to its minimum at the Thouless energy. The simple model used above to derive Eq.(1) supposes that the quasiparticles reach the gap without inelastic interactions and the corresponding voltage range is therefore called "collisionless regime". However at low voltage and finite temperature, the effective length of the junction for the multiple retroreflected particles $L_{eff} = NL \sim L\Delta/V$ exceeds the inelastic length L_{in}. In this "interacting regime" e-e-collisions interrupt the successive incoherent multiple Andreev reflections before the quasiparticles reach the gap. In the case of strong interaction a Fermi distribution with an effective temperature T_e is restored. [17, 18, 19].

MESOSCOPIC TRANSITION

Decreasing further the voltage below E_{Th}, the noise increases again. The preliminary investigations of this behavior on two different samples (the one presented here and a second one half wider and longer), indicates a linear divergence of the Fano factor $F = S_I/(2eI)$ as a function of the inverse voltage. The slopes are $200\,\mu V$ and $340\,\mu V$ for sample 1 and sample 2 and the Thouless energies $7\,\mu V$ and $28\,\mu V$ respectively. In the two junctions with different Thouless energies, the cross over from coherent to incoherent or inelastic regimes appears clearly at the Thouless energy. The apparent linear divergence could be the signature of a coherent transfer of multiple charges [20, 23, 24, 27]. However some questions remain concerning how inelastic processes affect the charge transfer in the coherent regime and what is the time required to transfer a coherent charge Ze from one superconducting reservoir to the other.
In conclusion, we presented noise investigations on diffusive S/N/S junctions ($\xi_\Delta < L < L_\Phi$), which show a clear transition from a coherent pair current to an independent quasiparticle current. In these junctions the "collisionless regime" at high voltage is very pronounced and the current noise grows linearly with the bias voltage due to incoherent multiple Andreev reflections. The results are in quantitative agreement with semiclassical theory over a large temperature range. Further investigations are now required to interpret the noise divergence in the coherent regime at low voltage.

ACKNOWLEDGMENTS

We would like to thank M. Houzet, H. Courtois, V. Shumeiko, E. V. Bezuglyi and C. Strunk for fruitful discussions. The samples have been elaborated using the CNRS/NANOFAB clean room facility.

REFERENCES

1. Y. Nakamura, Y. A. Pashkin, and J. S. Tsai, *Nature*, **398**, 786 (1999).

2. T. Fujisawa, T. Hayashi, and Y. Hirayama, *Journal of Vac. Sci. Tech. B Microelec. and nm Struc.*, **22**, 2035 (2004).
3. W. Rabaud, L. Saminadayar, D. Mailly, K. Hasselbach, A. Benoit, and B. Etienne, *Phys. Rev. Lett.*, **86**, 3124 (2001).
4. M. Henny, S. Oberholzer, C. Strunk, T. Heinzel, M. Holland, and C. Schönenberger, *Science*, **284**, 296 (1999).
5. W. D. Oliver, K. Kim, R. C. Liu, and Y. Yamamoto, *Science*, **284**, 299 (1999).
6. J. Gabelli, L.-H. Reydellet, G. Fève, J. M. Berroir, B. Plaçais, P. Roche, and D. C. Glattli, *Phys. Rev. Lett.*, **93**, 056801 (2004).
7. C. Naud, G. Faini, and D. Mailly, *Phys. Rev. Lett.*, **86**, 5104 (2001).
8. E. Serret, P. Butaud, and B. Pannetier, *Europhys. Lett.*, **59**, 225 (2002).
9. F. R. Waugh, M. J. Berry, D. J. Mar, R. M. Westervelt, K. L. Campman, and A. C. Gossard, *Phys. Rev. Lett.*, **75**, 705 (1995).
10. Y. M. Blanter, and M. Büttiker, *Phys. Rep.*, **336**, 1 (2000).
11. L. Saminadayar, D. C. Clattli, Y. Jin, and B. Etienne, *Phys. Rev. Lett.*, **79**, 2526 (1997).
12. R. D. Picciotto, M. Reznikov, M. Heiblum, V. Umansky, G. Bunin, and D. Mahalu, *Nature*, **389**, 162 (1997).
13. X. Jehl, M. Sanquer, R. Calemczuk, and D. Mailly, *Nature*, **405**, 50 (2000).
14. A. A. Kozhevnikov, R. J. Schoelkopf, and D. E. Prober, *Phys. Rev. Lett.*, **84**, 3398 (2000).
15. F. Lefloch, C. Hoffmann, M. Sanquer, and D. Quirion, *Phys. Rev. Lett.*, **90**, 067002 (2003).
16. B. Reulet, A. A. Kozhevnikov, D. E. Prober, W. Belzig, and Y. Nazarov, *Phys. Rev. Lett.*, **90**, 066601 (2003).
17. T. Hoss, C. Strunk, T. Nussbaumer, R. Huber, U. Staufer, and C. Schonenberger, *Phys. Rev. B*, **62**, 4079 (2000).
18. P. Roche, H. Perrin, D. C. Glattli, H. Takayanagi, and T. Akazaki, *Physica C*, **352**, 73 (2001).
19. C. Hoffmann, F. Lefloch, and M. Sanquer, *Eur. Phys. J. B*, **29**, 629–633 (2002).
20. Y. Naveh, and D. V. Averin, *Phys. Rev. Lett.*, **82**, 4090 (1999).
21. K. E. Nagaev, *Phys. Rev. Lett.*, **86**, 3112 (2001).
22. E. V. Bezuglyi, E. N. Bratus, V. S. Shumeiko, and G. Wendin, *Phys. Rev. B*, **63**, 100501 (2001).
23. J. C. Cuevas, A. Martin-Rodero, and A. L. Yeyati, *Phys. Rev. Lett.*, **82**, 4086 (1999).
24. R. Cron, M. F. Goffmann, D. Esteve, and C. Urbina, *Phys. Rev. Lett.*, **86**, 4104 (2001).
25. C. Strunk, and C. Schönenberger, "Shot noise in diffusive superconductor/normal metal heterostructures," in *Proceedings of the NATO ARW on Quantum Noise in Mesoscopic Physics*, 2002.
26. F. Pistolesi, G. Bignon, and F. W. J. Hekking, *Phys. Rev. B*, **69**, 214518 (2004).
27. P. Dieleman, H. G. Bukkems, T. M. Klapwijk, M. Schicke, and K. H. Gundlach, *Phys. Rev. Lett.*, **79**, 3486 (1997).

Discussion Of Shot Noise Suppression In Cascaded Mesoscopic Cavities

M. Macucci, P. Marconcini

Dipartimento di Ingegneria dell'Informazione, Università di Pisa, Via Caruso 16 I-56122 Pisa, Italy

Abstract. In this contribution we discuss the issue of shot noise suppression in cascaded mesoscopic cavities, which presents some still puzzling aspects that require further investigation. While a semiclassical model would predict, assuming a few idealized hypotheses, that the Fano factor for a series of cavities should tend to 1/3 as the number of cavities increases, a numerical quantum calculation yields analogous results for unequal cavities (although with a tendency to overcome the 1/3 value) and sharply different results for identical cavities, for which the Fano factor is found to be the same as that of a single cavity or even lower. We present and discuss a series of numerical results obtained with a recursive Green's function approach, trying to find clues to a general understanding of the problem.

Keywords: Shot noise suppression, mesoscopic cavity, cascaded cavities
PACS: 72.70.+m, 73.23.-b, 75.75.+a, 75.47.Jn

INTRODUCTION

Shot noise suppression in mesoscopic cavities has raised significant interest in the recent literature; the theoretical result by Jalabert *et al.* [1], predicting a Fano factor of 1/4 (ratio of the shot noise power spectral density to that which would be expected from Schottky's theorem [2] for the same value of the average current) for a symmetric cavity, has received experimental confirmation with the work of Oberholzer *et al.* [3]. Detailed measurement have been performed also for the analysis of the behavior of the Fano factor as a function of the width of the apertures (if their width becomes non negligible compared to that of the cavity, the Fano factor decreases below 1/4) and of a perpendicular magnetic field [4].

The issue becomes more complicated when considering cascaded cavities: based on a simple semiclassical model [5], one would obtain, for a number of symmetric cascaded cavities, a suppression factor that tends to 1/3 as the number of cavities is increased, analogously to what happens in diffusive conductors [6]. More precisely, if we model each constriction with a resistor and a current noise source, and enforce Kirchhoff's laws in the resulting equivalent circuit, the Fano factor γ is given, for N identical cavities, by

$$\gamma = \frac{1}{3}\left(1 - \frac{1}{(N+1)^2}\right) . \tag{1}$$

The authors of Ref. [5] also provide an experimental result that appears to support such a theory, but they actually have available one single significant data point, that for

CP800, *Unsolved Problems of Noise and Fluctuations: UPoN 2005*
edited by L. Reggiani, C. Pennetta, V. Akimov, E. Alfinito, and M. Rosini

two cavities, due to the problems encountered because of malfunctioning gates, which prevented them from adding a third cavity.

However, a numerical calculation of shot noise suppression in cascaded cavities with a coherent quantum mechanical model [7] yields, for identical cavities, a substantially constant, or even decreasing, value [8], which is obtained also as a limit for a rectangular antidot lattice [9].

In the case of even slightly different cavities, numerical simulations [8] yield a Fano factor that does instead increase with the number of cascaded elements, reaching soon values well above the analytically predicted asymptotic result of 1/3.

This discrepancy cannot be attributed to the absence of coherence in the analytic model (Ref. [5]) vs. its presence in the numerical one (Ref. [7]), since dephasing effects have been shown not to have a significant action on noise properties, according to the recent literature [13], in particular if the number of propagating modes is large, as in our case.

MODEL

We solve the transport problem through the device with the application, to the properly discretized structure, of an optimized version of the recursive Green's function method [10, 11], which consists in computing the analytical Green's function matrix for each longitudinal section with constant potential (in the representation over a basis of transverse modes) assuming it detached from the other sections and with Dirichlet boundary conditions at the ends. The separate sections are then joined with a recursive procedure based on the application of Dyson's equation that includes the perturbation due to coupling. From the obtained overall Green's function it is straightforward[10] to derive the transmission matrix t, with elements $t_{i,j}$, representing transmission from mode j to mode i. Conductance is computed with the Landauer-Büttiker formula

$$G = \frac{2e^2}{h} \sum_{n,m} |t_{nm}|^2 , \tag{2}$$

where e is the electron charge and h is Planck's constant. The noise current power spectral density is given by[12]

$$S_I = 4\frac{e^2}{h} |eV| \mathrm{Tr} \left[t^\dagger t (I - t^\dagger t) \right] , \tag{3}$$

where V is the voltage applied across the device, As a result of the invariance properties of the trace of a matrix for rotations, we can also write

$$S_I = 4\frac{e^2}{h} |eV| \sum_i w_i (1 - w_i), \tag{4}$$

where w_i is the i-th eigenvalue of $t^\dagger t$. The modulus of the average current can be evaluated as:

$$|I| = G|V| = 2\frac{e^2}{h} |V| \sum_{nm} |t_{nm}|^2 = 2\frac{e^2}{h} |V| \sum_i w_i, \tag{5}$$

The shot noise power spectral density expected for a current I is $2eI$ [2]. therefore we can write the Fano factor γ as

$$\gamma = \frac{\sum_i w_i(1-w_i)}{\sum_i w_i}. \tag{6}$$

There are two important considerations to be made: due to the large size of the cavity compared to the Fermi wavelength (which is of the order of 50 nm) and the coherence of our model, quantum fluctuations will be present in the computed quantities, the numerator and the denominator of Eq. (6), which will need to be separately averaged (before taking the ratio) over an appropriate energy range. We point out that such averaging must not be weighed with the derivative of the Fermi function, as in the case of a quasi-equilibrium condition. In the present case, instead, a voltage $V \gg kT$ (where k is the Boltzmann constant and T is the temperature) must be applied to make shot noise observable over thermal noise. This implies that the energy interval over which the Fermi function varies from 0 to 1 is small compared to the potential difference existing between the leads of the device. Therefore constant averaging over an energy interval corresponding to the applied potential difference is the most sensible approach for the derivation of a realistic result.

RESULTS AND DISCUSSION

Most of our results are obtained for rectangular cavities, which require much less computational effort, compared to more realistic shapes, since in such a case each single cavity can be represented with just three constant potential sections. One may argue that rectangular cavities are not the best possible description of real cavities, but, according to our tests and to our explanation of the phenomenon of shot noise suppression in these structures, all of the main noise features are well described.

Each cavity acts as a quasi-reservoir, with all the states at the same energy having the same occupancy. This is possible only if there is strong mode-mixing that allows electrons entering the cavity to explore all the possible transverse modes, as a result of reflections off the entrance and exit walls. Notice that this is simply due to the presence of variations in the transverse potential where the constrictions interrupt the cavity wall. Therefore a classically regular shape, such as a rectangle, produces the very same effect, from the point of view of interest for us, as a classically chaotic shape, such as a stadium.

All of our calculations have been performed for the AlGaAs/GaAs material system, in the effective mass approximation (with $m^* = 0.067m_0$), for cavities 5 μm long and 8 μm wide.

We start with the presentation, in Fig. 1, of the results for the Fano factor as a function of the number of identical cascaded cavities, for two different values of the constriction width: 110 nm and 210 nm. It is apparent that cascading of identical cavities leads to an initially somewhat constant and then decreasing Fano factor, in any case well below the 1/3 limit predicted by the analytic theory.

If the cascaded cavities are instead slightly different in length, for the level of the bottom potential, or for confinement details, the value of the Fano factor increases with their number, even exceeding 1/3, as shown in Fig. 2 for cavities that differ in length

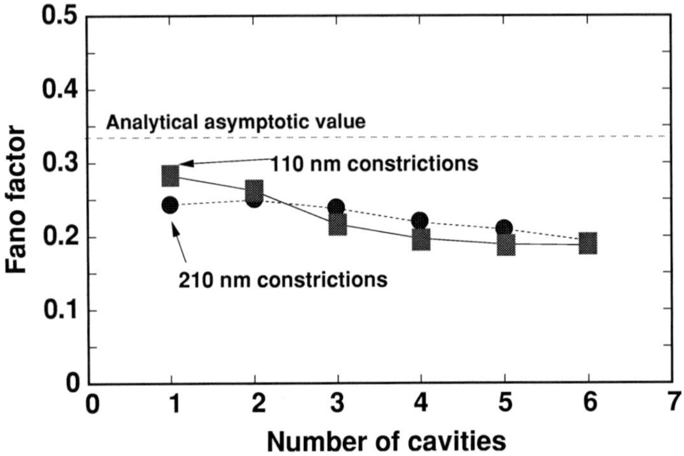

FIGURE 1. Fano factor as a function of the number of identical cascaded cavities.

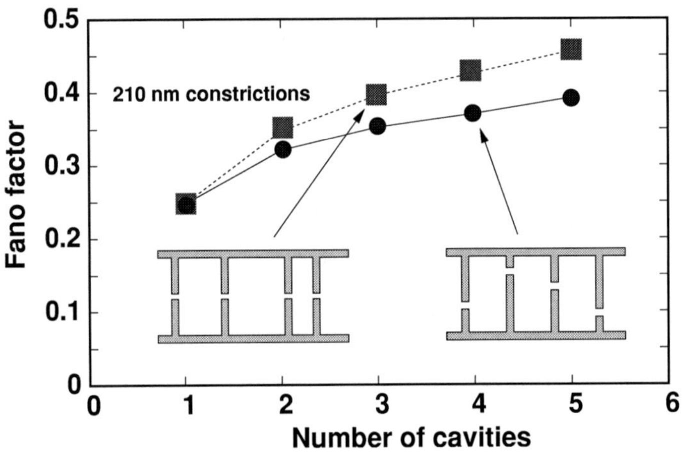

FIGURE 2. Fano factor as a function of the number of unequal cascaded cavities.

or for the vertical position of the constrictions. A variation in the length of one of the cavities by a few nanometers (less than 1% of the total) is sufficient to switch to this sharply different regime.

Some of these calculations have been repeated for stadium-shaped cavities (sketched in Fig. 3) with a radius $R = 4\ \mu$m, a central constant width region 1 μm long, and 150 nm constrictions. The results are $\gamma = 0.2511$ for a single cavity, $\gamma = 0.2550$ for two identical cavities, and $\gamma = 0.3116$ for two cavities differing because one is 100 nm longer in the middle region. Thus, these further calculations confirm the results already obtained with the rectangular cavities.

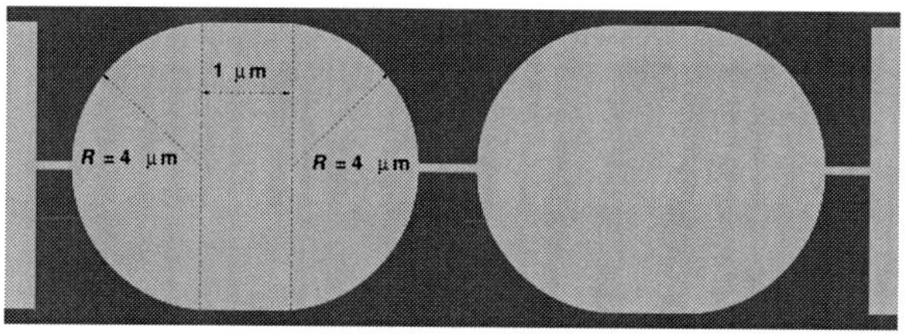

FIGURE 3. Stadium-shaped cavities

An analogous difference is found for the resistance of the overall structure: while for slightly different cascaded cavities it is equal to the sum of the resistances of the constrictions and thereby proportional to the number of cavities, for identical cavities such a sum rule does not hold any more, as shown in Fig. 4, where we report the computed resistance (in units of the reciprocal of the conductance quantum $G_0 = 77.53\ \mu S$) as a function of the number of cascaded cavities, for identical cavities, for cavities with constrictions shifted up and down in the vertical direction and for identical cavities. The differences are even more striking: when cavities are identical, the resulting resistance is only slightly increased with respect to the value of a single cavity.

We have observed that any small variation in one of the cavities suffices to determine the transition between the two regimes. Therefore one possible reason for the observed behavior of identical cavities could consist in the presence of resonances: in such a case, it should be possible to restore, for two cavities of different length, the equivalent of a specific resonance by slowly tuning, for example, the bottom potential. If more than one energy levels need to be resonant this might instead be impossible.

We have performed several test calculations for two cascaded cavities of slightly different length, introducing a gradual variation of the bottom potential of one of the cavities, without, however, ever observing a Fano factor of 0.25 or less. As already mentioned, we have also noticed a discrepancy between the result 1/3 that we theoretically expect for the case of an infinite number of cavities and the asymptotic value of the Fano factor obtained from our numerical simulations as the number of slightly different cavities is increased: the value that we obtain from calculations is larger than 1/3. This phenomenon, which has an analogous in the case of diffusive mesoscopic conductors, could be explained with the low number of propagating modes inside the constrictions and thus with the decreasing significance of the semiclassical argument on which the theoretical explanation is based.

Overall, this remains an unsolved problem, and further work is needed to explain the very peculiar noise properties that we have discussed so far. In particular, it would be important to have more experimental results available, although this may be extremely difficult or outright impossible for the case of identical cavities, which cannot be fabricated in practice.

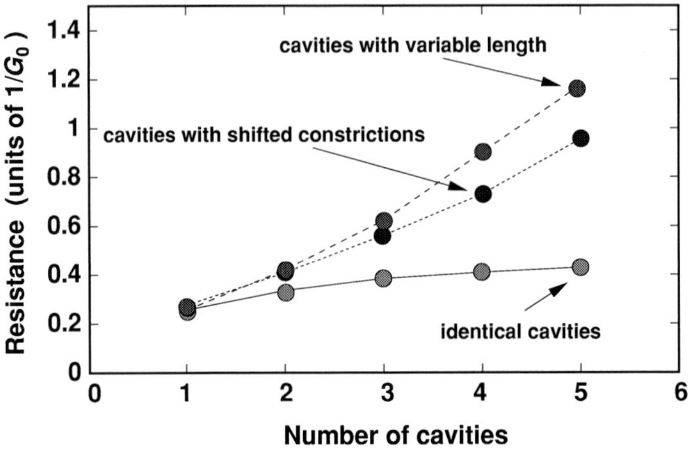

FIGURE 4. Resistance of cascaded cavities as a function of their number.

ACKNOWLEDGMENTS

Financial support from the Italian Ministry of Education, University and Research (MIUR) through the FIRB Project "Nanotechnologies and nanodevices for the information society" and through the PRIN "Excess noise in nanoscale devices" is gratefully acknowledged.

REFERENCES

1. R. A. Jalabert, J.-L. Pichard, and C. W. J. Beenakker, *Europhys. Lett.* **27**, 255 (1994).
2. W. Schottky, *Annalen der Physik* **Ann. Phys. (Leipz.) 57**, 541 (1918).
3. S. Oberholzer, E. V. Sukhorukov, C. Strunk, C. Schönenberger, T. Heinzel, and M. Holland, *Phys. Rev. Lett.* **86**, 2114 (2001).
4. S. Oberholzer, E. V. Sukhorukov, and C. Schönenberger, *Nature* **415**, 765 (2002).
5. S. Oberholzer, E. V. Sukhorukov, C. Strunk, and C. Schönenberger, *Phys. Rev. B* **66**, 233304 (2002).
6. M. J. M. de Jong, C. W. J. Beenakker, *Phys. Rev. B* **51**, 16867 (1995).
7. M. Macucci, G. Iannaccone, and B. Pellegrini, *Shot noise suppression in single and multiple ballistic and diffusive cavities*, Proceedings of the 16th International Conference on Noise in Physical Systems and 1/f Fluctuations (22-25 October 2001), Gainesville, Florida (USA), published by World Scientific (ed. G. Bosman).
8. M. Macucci, G. Iannaccone, and P. Marconcini, *Shot noise behaviour of cascaded mesoscopic structures*, Proceedings of the 27th International Conference on the Physics of Semiconductors (26-30 July 2004), Flagstaff, Arizona (USA), to be published in the AIP Conference Proceedings (ed. J. Menéndez and C. G. Van de Walle).
9. M. Macucci, *Physica B* **314**, 494 (2002).
10. F. Sols, M. Macucci, U. Ravaioli and K. Hess, *J. Appl. Phys.* **66**, 3892 (1989).
11. M. Macucci, A. T. Galick, and U. Ravaioli, *Phys. Rev. B* **52**, 5210 (1995).
12. M. Büttiker, *Phis. Rev. Lett.* **65**, 2901 (1990).
13. M. G. Pala, G. Iannaccone, *Phys. Rev. Lett.* **93**, 256803 (2004).

Shot Noise In Mesoscopic Systems Within The First Quantization Formalism: A De Broglie-Bohm Wave-Particle Description

X. Oriols

Departament d'Enginyeria Electrònica, Universitat Autònoma de Barcelona, SPAIN
E.mail: xavier.oriols@uab.es

Abstract. The experimental current measured in mesoscopic systems fluctuates around average values as a consequence of the wave-particle nature of electrons. An alternative quantum noise approach, based on Bohm trajectories, is presented. This formalism directly deals with the wave and particle nature of electrons and provides an intuitive framework to study quantum noise. The approach allows the direct consideration of coulomb correlation among electrons. In fact, most of the numerical techniques used in classical Monte Carlo simulations of electron devices can be directly adapted to our approach. Simulated noise results for interacting electrons tunnelling through double barriers are presented in agreement with experimental results. The present approach opens a new path for studying electron transport and quantum noise in mesoscopic systems.

Keywords: Noise, Bohm trajectories, Quantum transport, Monte Carlo simulation.
PACS: 05.40.Ca, 72.70.+m, 73.63.-b, 05.60.Gg,

INTRODUCTION

In general, quantum devices simulators (with few exceptions) emphasize the wave nature of electrons, but neglect its complementary particle-like behavior. In order to correctly model the current fluctuations, it is mandatory to consider both aspects of the electron. For example, in two-terminal device with tunneling barriers, the electron probability of being reflected or transmitted is determined by the wave nature of electrons (the transmission coefficient) and, on the other side, the fact that the electron is "measured" either at the left or the right contact (but not both!) shows its particle-like behavior. The Buttiker formalism [1,2] has become the standard procedure for noise predictions of quantum-based devices. It contains the essential wave and particle description of the electron. Buttiker generalized the Landauer scattering approach via the second quantization formalism. The basic elements of his approach are creation and annihilation operators. Due to the commuting algebra of these operators, the simultaneous presence of one electron at different contacts is not possible [2]. The Buttiker formalism has been successfully applied in many experimental situations [1,3,4] and it has the additional advantage that it directly includes the fermionic/bosonic symmetries when describing many particle systems [1,2].

In addition, although experimental results show that current fluctuations can be dramatically sensible to Coulomb correlations, most quantum device simulators are

CP800, *Unsolved Problems of Noise and Fluctuations: UPoN 2005*
edited by L. Reggiani, C. Pennetta, V. Akimov, E. Alfinito, and M. Rosini

based on a non-interacting picture to describe electron dynamics. In these simulators, the electron-electron interaction is averaged over all particles and a mean electrostatic potential (mean field approximation) is used. An elementary (Thomas-Fermi) density functional is the standard procedure to obtain such averaged potential. Therefore, much of the present understanding of the electron transport properties in mesoscopic devices is based on this non-interacting (Fermi liquid) picture. In spite of its unquestionable success, such single-particle approximation cannot describe coulomb correlation.

Here, we present an alternative quantum transport approach using de Broglie-Bohm (dBB) quantum trajectories [3-7]. In our approach, each electron is described by a well-defined quantum trajectory guided by a wave function. Since the dBB formalism directly deals with the wave and particle nature of electrons, it provides an excellent framework to study quantum noise. Our approach allows the introduction of the Coulomb interaction among electrons with a procedure quite similar to that of the classical Monte Carlo device technique. In fact, most of the numerical techniques used in classical Monte Carlo simulations, which are related with the microscopic description of electron trajectories [8], can be directly adapted to our dBB formalism [9-12].

OUR APPROACH

The fundamental difference between the classical Monte Carlo technique for electron devices and our approach is the computation of the electron velocity. In the former, it is directly related to the electric field (solution of the Poisson equation), while in our approach it is determined from the wave-nature of the electron. For a system of N interacting electrons, the wave nature of electrons is obtained from the many-particle Schrödinger equation, where the potential energy $U(\vec{r}_1,...,\vec{r}_N,t)$ takes into account the electron-electron Coulomb interaction:

$$i\hbar\frac{\partial\Psi(\vec{r}_1,...,\vec{r}_N,t)}{\partial t}=\left\{\sum_{k=1}^{N}\left(-\frac{\hbar^2}{2m}\nabla_k^2\right)+U(\vec{r}_1,...,\vec{r}_N,t)\right\}\Psi(\vec{r}_1,...,\vec{r}_N,t) \quad (1)$$

the vector $\vec{r}_k=(x_k,y_k,z_k)$ determines each particle position and $\Psi(\vec{r}_1,...,\vec{r}_N,t)$ is the many-particle wave-function that can be rewritten, for convenience, in a polar form $\Psi(\vec{r}_1,...,\vec{r}_N,t)=R(\vec{r}_1,...,\vec{r}_N,t)\cdot\exp(iS(\vec{r}_1,...,\vec{r}_N,t)/\hbar)$. Following Bohm development [4], the previous Schrödinger equation can be reconstructed into a system of N Newton-like equations:

$$\frac{d^2\vec{r}_k}{dt^2}=-\nabla_k\left(U(\vec{r}_1,....,\vec{r}_N,t)+Q(\vec{r}_1,....,\vec{r}_N,t)\right)\Big|_{r_i=r_i(t),i\neq k} \quad ; \quad k=1,....,N \quad (2)$$

where, in addition to the classical term, a quantum potential appears which is responsible for all non-local quantum features:

$$Q(\vec{r}_1,...,\vec{r}_N,t) = -\left\{\sum_{k=1}^{N} \frac{1}{R(\vec{r}_1,...,\vec{r}_N,t)} \left(\frac{\hbar^2 \nabla_k^2 R(\vec{r}_1,...,\vec{r}_N,t)}{2m} \right)\right\} \tag{3}$$

In general, since the computation of Bohm trajectories needs the knowledge of the quantum potential (that depends on the wave function), the extra effort needed for the trajectories seems unnecessary. However, the re-formulation of a system of (non-relativistic) interacting electrons in terms of Bohm trajectories can provide an advantaged numerical framework to study Coulomb correlations. For fermions in a small device active region, the quantum potential in (3) can be reasonably approximated by a single-particle quantum potential [13]:

$$Q(\vec{r},t) = \frac{1}{R(\vec{r},t)} \left(-\frac{\hbar^2 \nabla^2 R(\vec{r},t)}{2m} \right) \tag{4}$$

The quantum potential for each k-electron (4) can be obtained by solving the single particle Schrödinger equation:

$$i\hbar \frac{\partial \Psi_k(\vec{r}_k,t)}{\partial t} = \left\{ -\frac{\hbar^2 \nabla_k^2}{2m} + U(\vec{r}_1(t),...,\vec{r}_k,...,\vec{r}_N(t),t) \right\} \Psi_k(\vec{r}_k,t) \tag{5}$$

where the classical potential $U(\vec{r}_1(t),...,\vec{r}_k,...,\vec{r}_N(t),t)$ is time-dependent and it takes into account all electron-electron correlations because it depends on all electron trajectories. For each particle k, the quantum potential (4) is computed by solving the Poisson equation according to other electrons dBB trajectories $\vec{r}_1(t),...,\vec{r}_N(t)$. The approach allows the direct computation of the Coulomb correlation effects among a large ($N \approx 100$) number of electrons. The details will be explained somewhere else [13]. As a numerical example, we study current noise in resonant tunneling diodes (RTD). Experimental results [14,15] shows noise enhancement at the region of negative conductance due to Coulomb correlation.

NUMERICAL RESULTS

We study the role of the electron-electron Coulomb interaction in the transport properties of a typical RTD with a quantum well of GaAs with a length of 6 nm between two barrier of AlGaAs of 2nm. Numerical details of the simulation can be found in reference 12. The role of the Coulomb interaction as a source of electron correlation is shown on the noise properties of the RTD depicted in figures 1 and 2 in terms of the Fano factor. The power of the current fluctuations at low frequency, $S(0)$, is normalized to the well-known Poissonian power spectral density, $S(0) = 2 \cdot q \cdot I$ with I the average current value and q the electron charge. Therefore, the Fano factor is defined as $F = S(0)/2 \cdot q \cdot I$. We have used room temperature (300 K) for all numerical simulations.

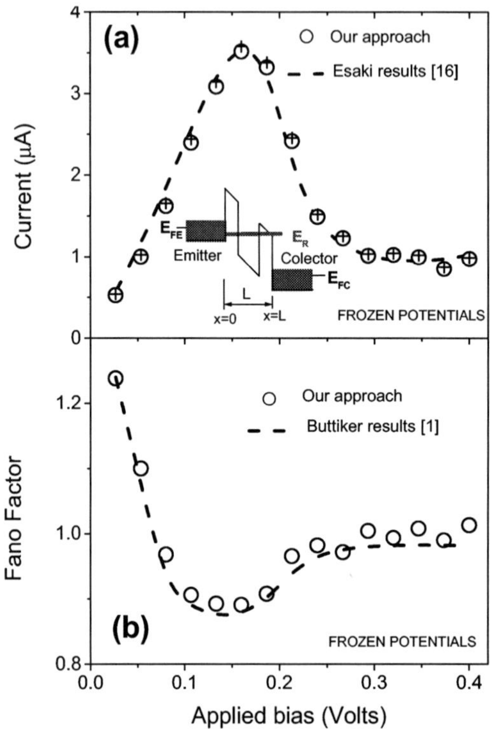

FIGURE 1. (a) Current voltage characteristic for the standard RTD diode of the text. The inset shows the static (frozen) potential profile. In dashed line, current results obtained from the Esaki expression [16]. (b) In circles, numerical values of F for each bias points computed using our Quantum Monte Carlo approach. In dashed line, Buttiker formalism [1].

In order to be able to distinguish the effects originated by the electron-electron interaction, first, in figure 1, we consider frozen potentials (i.e without using self-consistence between the electron charge and the potential energy). As we can see in the inset of figure 1, a hand-made linear approximation is used for defining the potential profile for different bias. For this simple system, with independent electrons, the excellent agreements between our numerical values and the theoretical predictions for the current (obtained from reference [16]) and the noise (obtained from reference [1]), provides a satisfactory test of the numerical implementation of our approach.

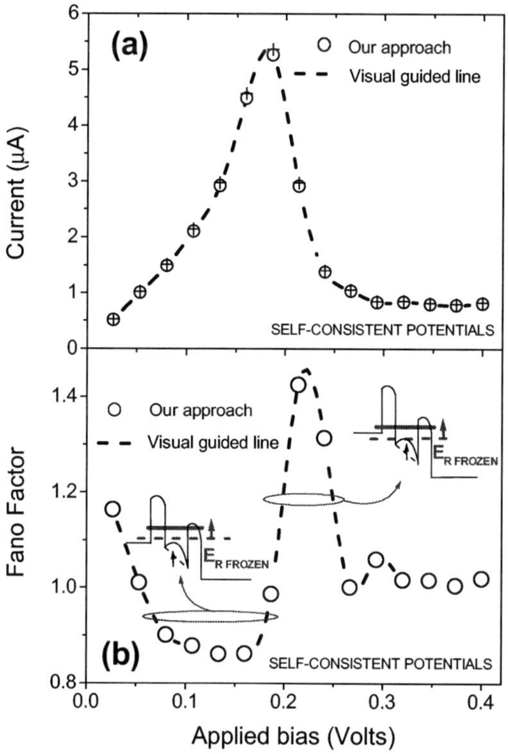

FIGURE 2. (a) Current-voltage characteristic for the RTD diode of fig. 1 when the Poisson equation is solved self-consistently with the time-dependent Schrödinger equation. (b) Fano factor for each bias point computed with our approach. The insets show the effect of time-dependent variations of the potential energy profile (due to the presence of an electron at the quantum well) on the transmission of other electrons, before and after the resonant voltage.

The results depicted in figure 2 are obtained for Coulomb interacting electrons. The Poisson and the Schrödinger equations for the same RTD considered in figure 1 are solved self-consistently. The instantaneous charge is computed from the Bohm trajectories. The effect of the Coulomb correlation between electrons is clearly manifested in the Fano factor depicted in figure 2b. Just after the resonant voltage, in the negative differential conductance region, (when the resonant energy, E_R, is below the conduction band at the emitter contact) the presence of one electron inside the quantum well raises the value E_R (see inset in fig. 2b). Thus, the transmission for the next electron is highly enhanced at this voltage. Roughly speaking, the Coulomb interaction affects the electron dynamics by trying to regroup the electrons and providing a Fano factor higher than one. The noise shows a process of bunching of electrons (i.e. super-Poissonian noise). On the contrary, just before the resonant voltage, the presence of an electron inside the quantum well obstructs the transport of the second one. Therefore, the noise presents a process of anti-bunching of electrons (i.e. sub-Poissonian noise). These results are in good agreement with experimental data [14-15].

CONCLUSIONS

In conclusion, we show that quantum noise can be naturally understood within the dBB quantum trajectories. Most of the numerical techniques used in classical MC simulations can be directly adapted to our quantum MC formalism. Simulated noise results for non-interacting and interacting electrons through double tunnelling barriers are presented, in agreement with experimental results. The present approach opens a new path for studying average values and current fluctuations in either static [9-11] or time-dependent [17] quantum scenarios.

ACKNOWLEDGMENTS

The author is really grateful to Javier Mateos, Tomás González, Emilio Méndez and Giusseppe Iannaccone for fruitful discussions. This work has been partially supported by the Ministerio de Ciencia y Tecnología and FEDER through project TIC2003-08213-C02-02.

REFERENCES

1. Blanter, Y.M. and Büttiker, M., *Phys. Rep.*, 336, 1, (2000) ; De Jong, M.J.M. and Bennakker, C.W. J., NATO ASI Series E, 345 Kluwer Academic Press (1997)
2. Büttiker, M., *Phys. Rev. Lett.* 65, 2901-2904 (1990); *Phys. Rev. B* 46(19), 12485-12507 (1992).
3. Bell, J.S., *Speakable and Unspeakable in Quantum Mechanics*, Cambridge U.P., England (1997).
4. Bohm, D., *Phys. Rev.* 85, 166-179, (1952).
5. Suñé, J. and Oriols, X., *Phys. Rev. Lett.*85, 894 (2000).
6. Oriols, X., Martín, F. and Suñé, J., *Phys. Rev. A* 54, 2594-2604, (1996).
7. Oriols, X. *Physical Review A* 71, 017801 (2005).
8. Jacoboni, C., and Reggiani, L., *Rev. Mod. Phys.*55, 645-700, (1983).
9. Oriols, X., Martín, F., and Suñé, J., *Appl. Phys. Lett.*79, 1703-1705, (2001).
10. Oriols, X., Martín, F., and Suñé, J., *Appl. Phys. Lett.*80, 4048-4050 (2002).
11. Oriols, X., *IEEE Trans. on Electron Devices*50(9), 1830 (2003).
12. Oriols, X., Trois, A., and Blouin., G., *Appl. Phys. Lett.* 85(16), 3596-3598 (2004).
13. Oriols, X., Submitted.
14. V.Ya Aleshkin *et al.*, Semicond. Sci and Technol, 18, L35 (2003); and references therein.
15. Iannaccone, G., Macucci, M., and Pellegrini, B., *Phys. Rev. B*55(7), 4539-4550 (1997); Iannaccone, G., Lombardi, G., Macucci M., and Pellegrini, B., *Phys. Rev. Lett* 80(5), 1054-1057(1998); Song, W., Méndez, E.E., Kuznetsov V., and Nielsen,B., *Appl. Phys. Lett.* 82(10), 1568 (2003).
16. Esaki, L., and Chang, L.L., *Phys. Rev. Lett.* 33, 495-498, (1974).
17. Oriols, X., Alarcón, A., and Fernández-Díaz, E., *Physical Review B* in press (2005)

Shot Noise Suppression and Enhancement at 2D Hopping and in Single-Electron Arrays

Viktor Sverdlov*, Yusuf A. Kinkhabwala†, Daniel M. Kaplan**, Alexander N. Korotkov‡, Hans Kosina* and Siegfried Selberherr*

*Institute for Microelectronics, TU Wien, Gusshausstrasse 27-29/E360, 1040 Wien, Austria
†Department of Physics and Astronomy, Stony Brook University, NY 11794-3800, USA
**Intel Corporation, 2501 NW 229th, Portland, OR 97124, USA
‡Department of Electrical Engineering, University of California Riverside, CA 92521-0204, USA

Abstract. We have carried out numerical simulations of shot noise at two-dimensional (2D) hopping and in 2D arrays of single-electron islands with and without random background charges. Such key transport characteristics as the average (dc) current $\langle I \rangle$, the single-particle density of states and the current fluctuation spectrum, have been calculated within a broad range of the applied electric field E and temperature T. Substantial Coulomb interaction effects are shown to not only suppress the average value of hopping current, but also affect its fluctuations rather substantially. In particular, at sufficiently low frequencies $(f \to 0)$ the spectral density $S_I(f)$ of current fluctuations exhibits a $1/f$-like increase which approximately follows the Hooge scaling. As f increases, there is a crossover to a broad range of frequencies in which $S_I(f)$ is nearly constant, hence allowing characterization of the current noise by the Fano factor $F \equiv S_I(f)/2e\langle I \rangle$. For sufficiently large samples and low temperature, the Fano factor is suppressed below the Schottky value ($F = 1$), scaling with the sample length L as $F = (L_c/L)^{\alpha}$. The exponent α is significantly affected by the inclusion of Coulomb interaction effects, changing from $\alpha = 0.76 \pm 0.08$ when such effects are negligible to virtually unity when they are substantial. The scaling parameter L_c, interpreted as the average percolation cluster length along the electric field direction, scales as $L_c \propto E^{-(0.98\pm0.08)}$ when Coulomb interaction effects are negligible and $L_c \propto E^{-(1.26\pm0.15)}$ when such effects are substantial, in good agreement with results of directed percolation theory.

In arrays of single-electron islands with completely random background charges, the current noise is strongly colored at low currents, and its spectral density levels off at very low frequencies. The Fano factor may be much larger than unity, due to the remnants of single-electron/hole avalanches. However, even very small thermal fluctuations reduced the Fano factor below unity for almost any bias.

Keywords: Shot noise suppression, Coulomb blockade, single-electron arrays, Coulomb gap, hopping
PACS: 72.20.Ee, 72.20.Ht, 72.70.+m, 73.23.Hk, 73.40.Rw, 85.35.Gv

INTRODUCTION

Due to the fast approach of conventional field effect transistors (FET's) to their scaling limits, single-electron transistors (SET's) and arrays of tunnel junctions have begun attracting significant attention as possible FET alternatives. In fact, the arrays are of particular interest, since they may not only be used to replace the double junctions in SET's, hence raising the single-electron energy and operating temperature, but due to their high resistance may potentially be used as a device to leak parasitic random background charge, usually a fraction of the electron charge e, away from SET single-

CP800, *Unsolved Problems of Noise and Fluctuations: UPoN 2005*
edited by L. Reggiani, C. Pennetta, V. Akimov, E. Alfinito, and M. Rosini

electron islands. It is the latter capability which is the motivating factor behind this work.

For a device to dissolve such fractional random background charge, the transport through it must be quasi-continuous. One requirement of such quasi-continuous charge transfer was formulated in Ref. [1] in the perhaps most interesting limit of negligible thermal and quantum fluctuations and uses the fact that the dynamical fluctuations of the current flowing through a mesoscopic system are more sensitive to the charge transport mechanism peculiarities than the average transport characteristics, therefore providing additional information about the conduction physics [2, 3]. In particular, the low temperature broadband current fluctuations ("shot noise") are caused by the discreteness of the charge carriers and may be characterized using the spectral density $S_I(f)$ of current I fluctuations at low frequency f. In order to transport charge quasi-continuously [1] the shot noise must be suppressed. The Fano factor definition $F \equiv S_I(0)/2eI$ is found to be convenient, since it must be considerably smaller than the discrete electron transfer (Schottky) value $F = 1$.

An alternative to a 2D array as a prospective device candidate is a 2D hopping conductor, which is potentially more promising in applications than the array due to its very high resistance. Although the variable range hopping mechanism had been described a few decades ago [4, 5, 6], little was known about its current noise properties due to the complexity of charge flow in a completely disordered system. Even now due to the needs of large-scale statistical averaging with respect to a broad range of random parameters, such as site positions, on-site energies and random background charge, supercomputers are required to simulate the charge transfer process and calculate the current noise characteristics of interest.

We present our results of extensive numerical simulations of shot noise in 2D arrays of single-electron islands and at 2D hopping within a broad range of electric fields E and temperatures T.

SHOT NOISE ENHANCEMENT IN SINGLE-ELECTRON ARRAYS

Arrays of small conducting islands separated by tunnel junctions are among the simplest systems exhibiting Coulomb blockade and the related effects of correlated single-electron tunneling.

Figure 1 shows the Fano factor F as a function of dc current I per channel for several arrays of length $N = 20$ and width $M = 20$ islands, with completely random distributions of background charges. As dc current is reduced and/or the array size is increased, the frequency at which $S_I(\omega)$ levels off decreases rather dramatically, while the saturation level $S_I(0)$ increases well beyond the Schottky value $2eI$. In 1D arrays (see inset of Fig. 1), $S_I(0)$ stays always below $2eI$, so that the Fano factor F is always below unity ($F < 1$) [1]. For $T = 0$ and $I \to 0$, the Fano factor saturates for any particular system as $f \to 0$, but at a very high level, with exponentially broad statistical distribution. At large currents the results are virtually independent of the system dimensionality; in fact, general arguments [1, 7] show that in all cases $S_I(\omega)$ tends to $2eI/N$.

The difference between results for 2D and 1D arrays may be attributed to the different topology of charge carrier motion. In the 1D case, current is carried along the array by a single stream of individual electrons (holes). The array provides a bottleneck for this

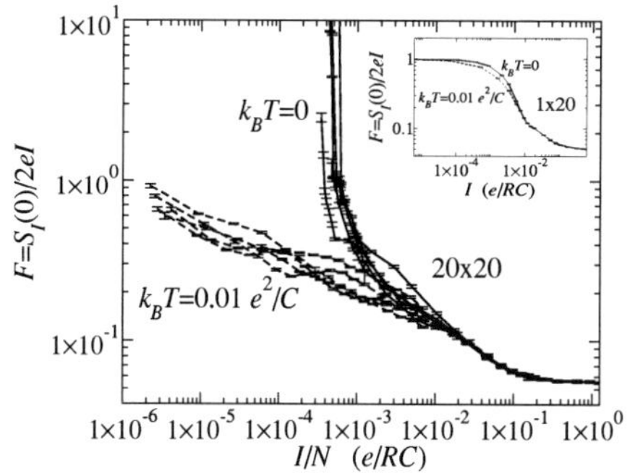

FIGURE 1. Fano factor F as a function of dc current I per channel for several 20×20 arrays, at zero and finite temperatures. The inset shows the corresponding dependences in a 1D array.

motion. At $I \to 0$, overcoming this bottleneck becomes a dominating source of current fluctuations, so that the current fluctuations are essentially the same as in a single tunnel junction, i.e. of shot noise character with $F \to 1$ [1].

On the contrary, broad 2D arrays may feature multiple streams of electrons and holes moving in opposite directions, that do not necessarily annihilate even in the absence of disorder, giving rise to long "avalanches" producing very large noise ($F \gg 1$) [8]. The disorder suppresses this effect (Fig. 1), but only partly.

It is important for applications that the Fano factor at low currents is substantially suppressed by even very small thermal fluctuations. For example, Fig. 1 shows that the temperature as low as $k_B T = 10^{-2} e^2/C$ reduces the Fano factor at low currents by several orders of magnitude, bringing it below unity, except for a very small region where $eV < 2k_B T$, where the current fluctuations are essentially thermal, rather than due to the shot noise. This behavior is very much different from that of 1D arrays (inset) where a small increase of temperature has a negligible effect on the shot noise.

NOISE AT HOPPING

Our simulations of hopping transport through 2D conductors of different geometries for a moderately strong Coulomb interaction strength $\chi = (\nu_0 a e^2/\kappa) = 0.5$ (with a the localization radius, ν_0 the density of states and κ the dielectric constant) has allowed us to calculate the frequency dependence of the current spectral density $S_I(\omega)$ (Fig. 2).

Two frequency ranges must be distinguished. At very low frequencies, the sharp $1/f$-type increase in noise is seen. This $1/f$-type contribution has been observed experimentally in a wide variety of conductors [3] and is observed to be proportional to the square of the current and in many cases found to scale approximately in accordance with the phenomenological Hooge formula [3, 9], which in 2D can be presented as

$$S_I(f) / \langle I \rangle^2 = \left(a^2/LW\right) \left(C(f)/f\right), \tag{1}$$

where L is the length of the conductor along the current flow, W is the width, and $C(f)$ is a weak function of the observation frequency f: $C(f) \propto 1/f^{(p-1)}$, where p is typically

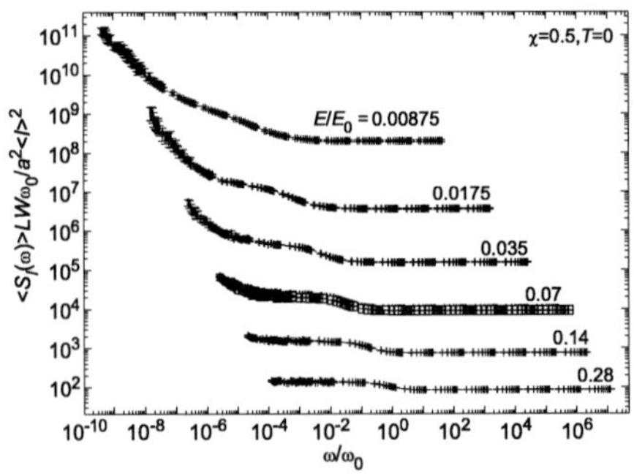

FIGURE 2. Spectral density $S_I(\omega)$ of current fluctuations at $T = 0$ and $\chi = 0.5$, normalized to the Hooge scaling factor $a^2 I^2/LW\omega_0$, as a function of observation frequency ω (measured in units of $\omega_0 \equiv g/\hbar\nu_0 a^2$) for several values of electric field. Each point represents data averaged over 48 samples.

between 1 and 2. Current spectral densities scaled in accordance with (1) are shown in Fig. 2 and seem to follow the $\langle I\rangle^2$ dependence nicely for different electric fields E and geometries. The power p is found to depend slightly on the electric field. For the smallest field accessible for numerical study the exponent p is close to 1.5, the value recently observed experimentally at 2D hopping in mesoscopic samples [10].

At moderately low frequencies, the noise spectral density as a function of f develops a plateau. This region of "broadband" noise may be crudely categorized according to which fundamental source of noise dominates, either thermal fluctuations or electron charge discreteness (shot noise). In the most interesting case of sufficiently low temperature, the thermal fluctuations are negligible, hence the broadband fluctuations are due almost entirely to the shot noise. In the case when the frequency dependence of the current spectral density $S_I(f)$ is flat, the strength of current fluctuations may be characterized by its ratio to the Schottky value $2eI$, i.e. using Fano factor $F \equiv S_I(f)/2eI$.

Recent experimental results [10, 11, 12] have shown that the shot noise may be indeed suppressed ($F < 1$). Theoretical and computational studies of shot noise at hopping in artificial space-ordered 1D [7, 13] and space-ordered and random 2D [14, 15, 16] systems have predicted that the shot noise suppression obeys

$$F = (L_c/L)^\alpha, \qquad L \gg L_c, \tag{2}$$

where L_c is a scaling constant which is interpreted as the average distance separating critical hops, which in percolation theory corresponds to a critical cluster size [5], and α is a positive exponent. Two examples of such scalings, for substantial ($\chi = 0.5$) and negligible ($\chi = 0$) Coulomb interactions, for different values E of the electric field, are shown in Fig. 3. The exponent α is found to be $\alpha = 1$ and $\alpha = 0.76 \pm 0.08$ for these two cases, respectively. Electric field dependences of L_c obtained for $\chi = 0.5$ and $\chi = 0$ using (2) are shown in Fig. 4. From Fig. 4, the shot noise suppression length $L_c(E)$

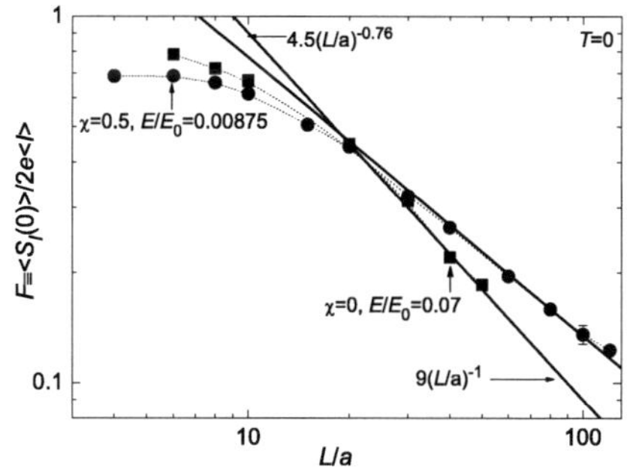

FIGURE 3. Average Fano factor F as functions of conductor length L for two values of applied field at $\chi = 0.5, T = 0$ and $W \gg L_c$.

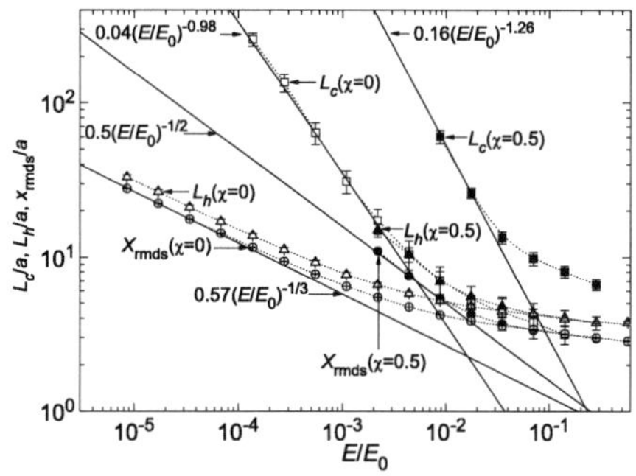

FIGURE 4. Cluster size L_c dependences on electric field E with ($\chi = 0.5$) and without ($\chi = 0$) the long-range Coulomb interaction included. Hop length x_{rmds} and high frequency scaling length L_h are also shown for comparison.

approaches the asymptote

$$L_c(E) \propto (E_0/E)^{\mu} a, \tag{3}$$

for suffi ciently small E. The power μ of shot noise suppression is quite different from that determining the variable-range hopping length $x_{\text{rmds}} \propto E^{-\beta}$, which is also shown in Fig. 4 for $\chi = 0$ ($\beta = 1/3$) and $\chi = 0.5$ ($\beta = 1/2$). Numerical values of the exponent μ are found to be $\mu = 0.98 \pm 0.08$ for zero Coulomb interaction strength $\chi = 0$ and $\mu = 1.26 \pm 0.15$ for $\chi = 0.5$, which are consistent with the estimates $\mu = \frac{1}{3}\left(1 + \nu_{\parallel}\right) \approx 0.91$ or $\mu = \frac{1}{2}\left(1 + \nu_{\parallel}\right) \approx 1.37$ obtained from the expression for the critical cluster size:

$$L_c \propto x_{\text{rmds}} \left(x_{\text{rmds}}/a\right)^{\nu_{\parallel}}, \tag{4}$$

where the critical cluster size index in the percolation direction $\nu_{\parallel}$ is 1.73 confi rming the interpretation of L_c as the average directed percolation cluster size [17, 18, 19].

CONCLUSION

In summary, results of our numerical simulations demonstrate that the shot noise in single-electron arrays may substantially exceed its Schottky value $2eI$ due to remnants of electron-hole avalanches, which are only partly suppressed by the disorder.

On the other hand, shot noise at 2D hopping is suppressed below the Schottky value, in agreement with recent experiments. The characteristic length L_c determining this suppression is found to be in agreement with its interpretation as the average percolation cluster length.

ACKNOWLEDGMENTS

A major contribution from K. K. Likharev to the discussed results is greatfully acknowledged. This work was supported by the Department of Energy, by the Semiconductor Research Corporation, and by the Austrian Science Fund FWF, project P17285-N02.

REFERENCES

1. K. A. Matsuoka and K. K. Likharev, *Phys. Rev. B* **57**, 15613 (1998).
2. Ya. M. Blanter and M. Buttiker, *Phys. Reports* **336**, 2 (2000).
3. Sh. Kogan, *Electronic Noise and Fluctuations in Solids*, (Cambridge University Press, Cambridge, 1996).
4. N. F. Mott and J. H. Davies, *Electronic Properties of Non-Crystalline Materials, 2nd Ed.*, (Oxford Univ. Press, Oxford, 1979); N. F. Mott, *Conduction in Non-Crystalline Materials, 2nd Ed.* (Clarendon Press, Oxford, 1993).
5. B. I. Shklovskii and A. L. Efros, *Electronic Properties of Doped Semiconductors* (Springer, Berlin, 1984).
6. *Hopping Transport in Solids*, edited by A. L. Efros and M. Pollak (Elsevier, Amsterdam, 1991).
7. A. N. Korotkov and K. K. Likharev, *Phys. Rev. B* **61**, 15975 (2000).
8. V. A. Sverdlov, D. M. Kaplan, A. N. Korotkov, and K. K. Likharev, *Phys. Rev. B* **64**, R041302 (2001).
9. F. N. Hooge, *Phys. Lett. A* **29**, 139 (1969).
10. A. K. Savchenko, S. S. Safonov, S. H. Roshko, D. A. Bagrets, O. N. Jouravlev, Y. V. Nazarov, E. H. Linfield and D. A. Ritchie, Phys. Stat. Sol. (B) **241**, No. 1, 26-32 (2004).
11. V. V. Kuznetsov, E. E. Mendez, X. Zuo, G. L. Snider and E. T. Croke, *Phys. Rev. Lett.* **85**, 397 (2000).
12. S. H. Roshko, S. S. Safonov, A. K. Savchenko, W. R. Tribe, and E. H. Linfield, Physica E **12**, 861 (2002).
13. B. Derrida, *Phys. Reports* **301**, 65 (1998).
14. V. A. Sverdlov, A. N. Korotkov and K. K. Likharev, *Phys. Rev. B* **63**, 081302(R) (2001).
15. Y. A. Kinkhabwala, V. A. Sverdlov, A. N. Korotkov and K. K. Likharev, *Preprint* cond-mat/0302445 (Submitted to Journal of Physics: Condensed Matter).
16. Y. A. Kinkhabwala, V. A. Sverdlov and K. K. Likharev, *Preprint* cond-mat/0412209 (Submitted to Journal of Physics: Condensed Matter).
17. D. Stauffer and A. Aharony, *Introduction to Percolation Theory, Rev. 2nd Ed.*, (Taylor and Francis Inc, Philadelphia, 1994).
18. S. P. Obukhov, Physica A **101**, 145 (1980).
19. J. W. Essam, K. De'Bell, J. Adler, and F. M. Bhatti. Phys. Rev. B **33**, 1982 (1986).

The Puzzle Of Shot Noise Suppression In A Series Of N Tunnel Barriers

M. Rosini*, V. Ya. Aleshkin†, A. Reklaitis** and L. Reggiani*,‡

*Dipartimento di Ingegneria dell'Innvazione, Università di Lecce, via Arnesano, 73100 Lecce Italy
†Institute for Physics of Microstructures, Nizhny Novgorod GSP-105 603600, Russia
**Semiconductor Physics Institute A. Goštauto 11, LT-01108 Vilnius, Lithuania
‡National Nanotechnology Laboratory NNL-INFM

Abstract. In the case of a series of N identical diodes it is usually argued that the Fano factor should be suppressed by a factor 1/N with respect to the full shot value ascribed to the single diode. To discuss the limits of validity of this conjecture, we consider a simple two terminal device of N tunnel barriers, generating shot noise. In the absence of correlations among carriers, the application of Carson theorem gives full shot noise in the limit of vanishing frequency, contrary to the above conjecture. The results of Monte Carlo simulations, on a series of two and more tunnel barriers, confirm that for independent current pulses full shot noise is obtained in the limit of vanishing frequencies, as dictated by Carson theorem. When repulsive correlations among carriers are introduced, the simulations evidence the onset of shot noise suppression. We notice that the presence of different time scales in the system, as well as the unavoidable presence of 1/f noise can make the resolution of the vanishing frequency limit difficult to be achieved experimentally.

Keywords: hopping, shot noise, tunnel junction
PACS: 72.70.+m 72.20.Ee

INTRODUCTION

The determination of shot noise in a simple circuit consisting of N identical diodes in series remains an open problem since 1996, when Landauer [1] concluded that the total shot noise arising from this circuit would be suppressed by a factor 1/N with respect to the full shot noise ascribed to the single diode. This conclusion has been used to argue that in a 1D array of N tunnel junctions the Fano factor can be reduced to 1/N of its full Poissonian value, owing to the low ratio between the junction resistance and the total system resistance [2]. In the same way shot noise reduction was shown for hopping systems of different dimensions, length and geometry [2, 3], by proposing that, in such cases, the length and geometric properties of the system are the reason of the shot noise suppression. We stress, that apparently the shot noise suppression mentioned above is not ascribed to any correlation among carriers, but mostly based on the conjecture that "as the sample length grows, the sequence of electron hops from one electrode to another looks more as continuous transport (for which shot noise vanishes)". In essence, the fact that macroscopic conductors do not exhibit in general shot noise is taken as an heuristic justification of the above conjecture. However, the recent demonstration that also linear macroscopic conductors can exhibit shot noise [4] makes such a justification not physically plausible and the problem of shot noise suppression in a series of N tunnel barriers a puzzling one.

CP800, *Unsolved Problems of Noise and Fluctuations: UPoN 2005*
edited by L. Reggiani, C. Pennetta, V. Akimov, E. Alfinito, and M. Rosini

For a more rigorous discussion of this puzzle, we consider a simple two terminal device with independent carriers entering from a left contact, crossing N tunnel barriers of transparency $D \ll 1$ (generating shot noise), and exiting from a right contact. In this case, the application of Carson theorem [5] is justified and the result gives full shot noise, in the limit of vanishing frequency, contrary to the above conjecture.

To shed some light on this controversy, we have performed a set of simulations on a series of two and more tunnel barriers and calculated the corresponding spectral density of current fluctuations. The results confirm that for independent current pulses full shot noise is obtained in the limit of vanishing frequencies, as dictated by Carson theorem. On the other hand, it is possible to obtain shot noise suppression by introducing some correlations among carriers, such as by imposing a maximum occupation number in each site between two barriers. In the following sections we consider two structures of significance for the present purposes, namely: the series of N tunnel junction and the nonresonant double barrier diode.

SERIES OF *N* TUNNEL JUNCTIONS

In this section, we discuss our approach to the problem. We consider a series of N tunnel junctions, where electrons can only jump one step forward from left to right, with a constant transition rate Γ. All junctions are equal and separated by the same distance Δ; the contacs and the junctions are supposed to behave as Poissonian injectors.

Theory

By considering the transit of one electron through the system: the current observed in the external circuit can be written as

$$I(t) = \frac{q}{N}\left[\delta(t-\tau_1) + \delta(t-\tau_1-\tau_2) + \ldots + \delta(t-\tau_1-\tau_2-\ldots-\tau_N)\right] \tag{1}$$

where $t = \tau_1$ is the time associated with an electron transition through the first junction, $t = \tau_1 + \tau_2$ the same through the second junction, and so on. For the process in hand, the probability that $t < \tau_i < t + \mathrm{d}t$ is given by

$$P(t)\mathrm{d}t = \Gamma e^{-\Gamma t}\mathrm{d}t \tag{2}$$

The Fourier image of the current is

$$I(\omega, \tau_1 \cdots \tau_N) = \int_0^\infty I(t)e^{i\omega t}\mathrm{d}t = \frac{e}{N}\left[e^{i\omega(\tau_1)} + e^{i\omega(\tau_1+\tau_2)} + \ldots + e^{i\omega(\tau_1+\tau_2+\ldots+\tau_N)}\right], \tag{3}$$

with e the unit charge. According to Carson theorem, the spectral density of current fluctuations is given by:

$$S(\omega) = \frac{2\langle I\rangle}{e}\int_0^\infty \mathrm{d}\tau_1 \cdots \int_0^\infty \mathrm{d}\tau_N \left|I(\omega,\tau_1\cdots\tau_N)\right|^2 P(\tau_1)\cdots P(\tau_N) \tag{4}$$

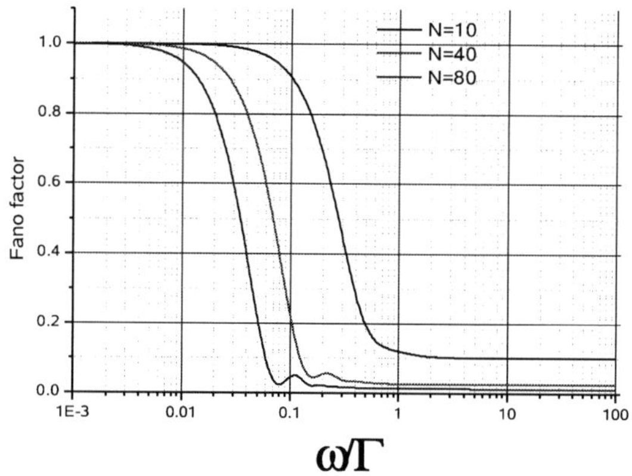

FIGURE 1. Fano factor calculated for a series of N tunnel junctions, with different values of N.

and from Eqs. (3) and (4) the frequency dependent Fano factor becomes

$$\begin{aligned} F(\omega) &= \frac{1}{q^2}\int_0^\infty d\tau_1 \int_0^\infty d\tau_2 \ldots \int_0^\infty d\tau_N P(\tau_1)P(\tau_2)\ldots P(\tau_N)\,|I(\omega)|^2 \\ &= \frac{1}{N} + \frac{2}{N^2}\mathrm{Re}\left[\sum_{m=1}^{N-1} \frac{N-m}{(1+i\omega/\Gamma)^m}\right]. \end{aligned} \tag{5}$$

Equation (5) implies that the value of the Fano factor at zero frequency be

$$F(0) = \frac{1}{N} + \frac{2}{N^2}\sum_{m=1}^{N-1}(N-m) = \frac{1}{N} + \frac{2}{N^2}\frac{N(N-1)}{2} = 1 \tag{6}$$

In the case of a great N we can approximate

$$F(\omega) = \frac{[\sin(N\omega/2\Gamma)]^2}{(N\omega/2\Gamma)^2} \tag{7}$$

with $\frac{N}{\Gamma}$ playing the role of the transit time. Figure 1 reports the plot of the Fano factor in Eq.(5) for different values of N. For the different lengths of the device considered here, we obtain full shot noise at zero frequency. Furthermore, owing to the different transit time in the three devices, we observe different cutoff frequencies, and at the highest frequencies the Fano factor saturates at the value of exactly $\frac{1}{N}$ since at these frequencies the junctions behave as if they were completely uncorrelated.

Simulation

In order to better exploit the problem, we have performed a set of Monte Carlo simulations using an ensemble of independent electrons, moving in the system described

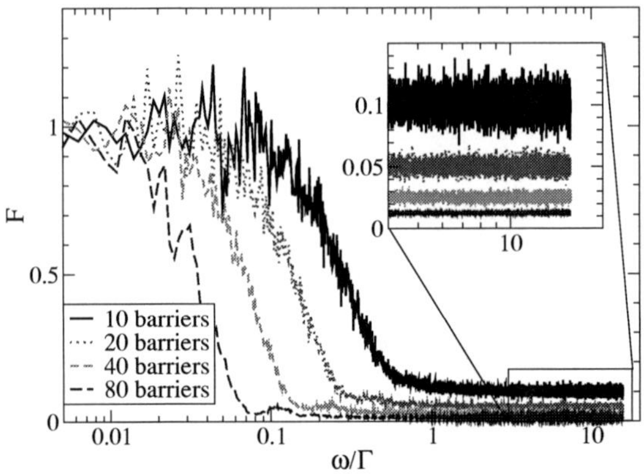

FIGURE 2. Fano factor simulated for a series of N tunnel junctions, with different values of N. These results reproduce the theoretical calculation of Fig. 1 within the numerical uncertainty

before, with different lengths corresponding to respectively, 10, 20, 40, 80 hopping junctions.

The current as measured in the external circuit, has been calculated from the Ramo-Shockley theorem [6] that, in its general form, can be written as:

$$I(t) = \frac{e}{L} \sum_{m=1}^{M(t)} \frac{\mathrm{d}x_m(t)}{\mathrm{d}t} \tag{8}$$

where $L = N\Delta$ is the length of the device, $M(t)$ is the number of electrons in the device at time t and

$$x_m(t) = \sum_i \Delta\Theta(t - t_i) \tag{9}$$

is the time law of the m-th electron in the system when the only possible hop is of one Δ step in the forward direction.

Since in the simulator the time is discretized, it is convenient to write the current in a discrete form; in this way, the measured current in the time step j is found to be

$$I_j = \frac{e}{N\mathrm{d}t} M_j \tag{10}$$

where now M_j is the number of particles that have performed a hop between the time steps $j\mathrm{d}t$ and $(j+1)\mathrm{d}t$. We notice that Eq. (10) does not express an istantaneous current, rather it represents the mean current over a time $\mathrm{d}t$ under the condition that the observation frequency should be lower than $\frac{1}{\mathrm{d}t}$.

The results of the simulations are reported in Fig. 2 and, within the numerical uncertainty, they are found to reproduce the theoretical findings of Fig. 1).

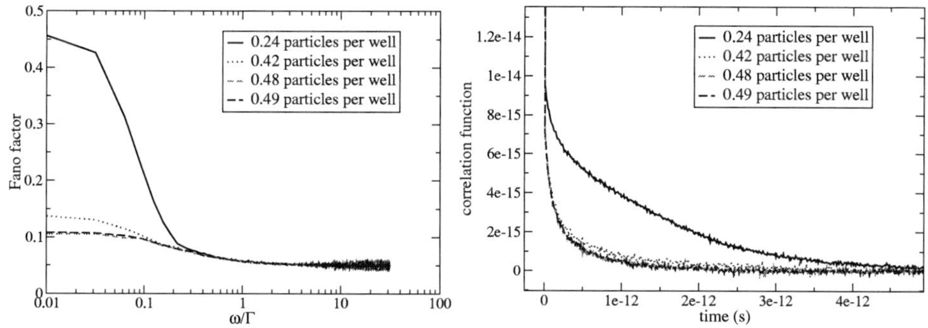

FIGURE 3. Fano factor (left) and correlation function (right) for different carrier concentration and maximum occupation number $\alpha = 1$. The shape of the correlation function tends to become triangular at lowering the correlation.

Occupation number and correlation

In the previous calculations, electrons were completely uncorrelated, no interacting potentials or physical constraint were imposed on the particle dynamics. In the present section we investigate the noise characteristics when as correlation among particles a maximum number of particles per hopping site is imposed. To this purpose we consider a device with 20 tunnel junctions and impose a maximum occupation of one particle per site, in order to reproduce the main features of a single-electron device. Similarly to the Pauli blockade, this constraint implies that a particle can jump to the next site only if it is empty. Figure 3 reports the Fano factor for such a device. Here, the different curves are calculated for different values of the carrier concentration inside the device. We have found that the greater the concentration, the lower the Fano factor. This dependence upon concentration is also evidenced by the shape of the correlation functions which are reported in Fig.3. Here, we can see that, by lowering the carrier concentration, the correlation function tends to assume the triangular shape typical of uncorrelated current pulses. We argue that as the concentration increases, the possibility for an electron to find the next site occupied becomes higher, so the interaction between electrons becomes stronger and as a consequence shot noise is suppressed. This conclusion is confirmed by a further set of simulations, where we have considered a constant concentration of carriers in the device, and changed the value of the maximum occupation number as reported in Fig. 4. In this case the full shot noise value decreases as the allowed occupation number is increased and in turn the interaction between electrons is weakened.

DOUBLE BARRIER

The second system (see Fig. 5) studied is a nonresonant double barrier structure, where electrons are injected from the left contact, with constant velocity v, and absorbed from the right one; they are only allowed to transfer from left to right. In particular, we

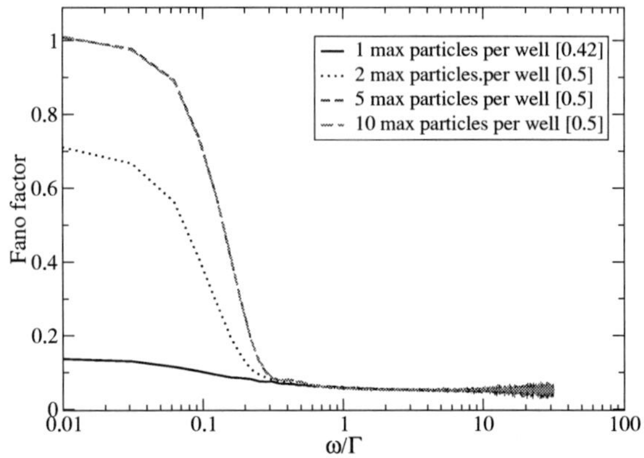

FIGURE 4. Fano factor for different occupation number α and same carrier concentration (in square brackets). The first curve refers to a slightly lower value of concentration since the simulator adjusts the concentration in a self-consistent manner.

consider three different lengths and times: L_1, τ_1 the length between the left contact and the double barrier and the time needed to run this distance; L_2, τ_2 the same for the double barrier and the right contact and, finally, L_B, τ_B the distance between the two barriers and the time spent by the carrier in this region. For simplicity we consider $L_B \ll L_1 \simeq L_2$ and $\tau_B \gg \tau_1 \simeq \tau_2$. The transparency D of the two barrier is taken to be much smaller than unity so that the number of scattering suffered by a carrier inside the barriers is significantly high.

The shot noise formula can be obtained, again, by applying the Carson theorem (4), where the current can be written in term of the Heavyside step function as

$$I(t) = \frac{q}{2\tau_B}\left[\Theta(t)\Theta(\tau_1 - t) + \Theta(t - \tau_1 - \tau_B)\Theta(\tau_1 + \tau_2 + \tau_B - t)\right] \tag{11}$$

and the probability for an electron to go from the emitter to the collector is

$$P(t)\mathrm{d}t = e^{-\frac{t}{\tau_B}}\mathrm{d}t \tag{12}$$

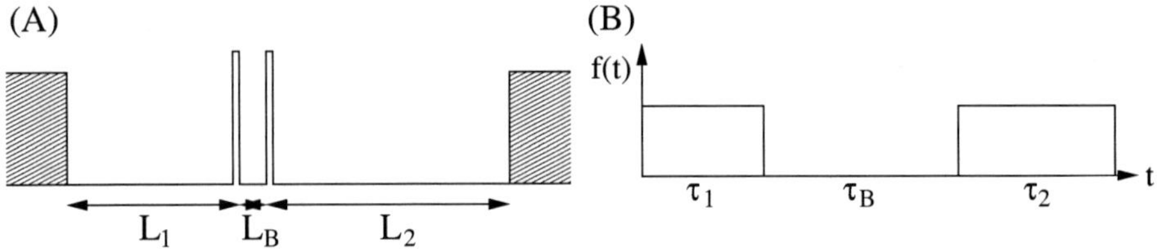

FIGURE 5. Sketch (A) of the simulated double barrier structure and (B) of the current signal.

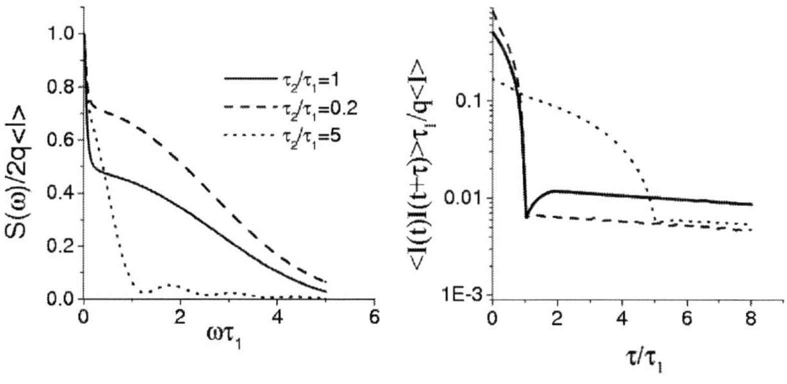

FIGURE 6. Fano factor and correlation function calculated for a double barrier described in section . Here $\frac{\tau_L}{\tau_1} = 60\,D$, for different values of τ_1 and τ_2.

where $\tau_B = \frac{2L_B}{vD}$. In this case the spectral density for current fluctuations is

$$\begin{aligned} S(\omega) &= \frac{4q\langle |I\rangle}{\omega^2(\tau_1+\tau_2)^2}\Bigg[3D\,(2-\cos\omega\tau_1-\cos\omega\tau_2)+ \\ &+\frac{-1+\cos\omega\tau_1-\omega\tau_L\sin\omega\tau_1+\cos\omega\tau_2-\omega\tau_L\sin\omega\tau_2}{1+\omega^2\tau_L^2}+ \\ &+\frac{-\cos\omega(\tau_1+\tau_2)+\omega\tau_L\sin\omega(\tau_1+\tau_2)}{1+\omega^2\tau_L^2}\Bigg] \end{aligned} \tag{13}$$

A plot of this result can be seen in Fig. 6. It is evident the spike at vanishing frequency, rising from the very long tail of the correletion function.

A Monte Carlo simulation has been implemented with the above device geometry and a very long time interval has been taken into account. The results of these simulations are shown in Fig.7: here, the theoretical predictions are confirmed, and the sharp spike at vanishing frequency is found, as expected.

CONCLUSIONS

For a series of N tunnel barriers we have demonstrated that, in case of independent particles, the Fano factor must be one at vanishing frequencies and reaches the value 1/N at frequencies higher than the inverse of the transit time. The introduction of correlations among carriers, by means of a maximum occupation number, causes strong shot noise suppression. This last case is representative of hopping transport, when the energy states are described by Pauli statistics or in the presence of coulomb interaction. In the presence of correlation the Fano factor is found to depend on the strength of the correlation and/or on the carrier concentration. Indeed, the higher the concentration, the lower the value of the Fano factor, since it is higher the probability to find the state after scattering filled up

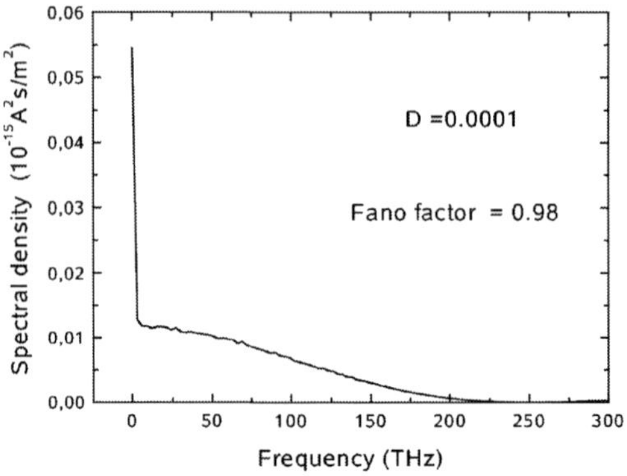

FIGURE 7. Simulation of the Fano factor for the system described in section . The low frequency spike is clear.

to the maximum occupation value. According to the Carson theorem, noise suppression is present only if there is correlation between carriers.

A still open problem is that an experimental observation of the shot noise in a series of tunnel junctions under the regime of uncorrelated transport seems to be difficult to realize experimentally. Indeed, when the time spent by the electron in a site is very long, the experiment should measure shot noise at very low frequencies, in order to account for the very long tail of the correlation function of current fluctuations. However, under these conditions it is difficult to avoid the presence of other excess noise sources such as $\frac{1}{f}$ noise which, by adding to shot noise, inhibit its detection.

ACKNOWLEDGMENTS

Work supported by the Italian Ministry of Education, University and Research (MIUR) under the project "Noise models and measurements in nanostructures".

REFERENCES

1. R. Landauer, *Physica B*, **227**, 156 (1996).
2. A. Koroktov, and K. Likharev, *Phys. Rev. B*, **23**, 15975 (2000).
3. V. Sverdlov, A. Koroktov, and K. Likharev, *Phys. Rev. B*, **63**, 81302 (2001).
4. M. Sampietro, G. Ferrari, G. Bertuccio, G. Gomila, C. Pennetta, and L. Reggiani, *Phys. Rev. Lett.*, **92**, 226601 (2004).
5. A. V. D. Ziel, *Noise source and characterization*, Prentice Hall, New Jersey, 1970.
6. S. Ramo, *J. Appl. Phys.*, **9**, 635 (1938).

NOISE AND COHERENCE

Non Breakable Data Encryption With Classical Information?

Laszlo B. Kish,[1,+] Swaminathan Sethuraman[2] and Peter Heszler[3]

[1] *Department of Electrical Engineering, Texas A&M University, College Station, TX 77843-3128, USA*
[3] *Department of Mathematics, Texas A&M University, College Station, TX 77843, USA*
[3] *Angstrom Laboratory, Solid State Physics, Uppsala University, P.O.B 534, Uppsala, S-75121, Sweden*

Abstract. With the Kish-Sethuraman (KS) cipher an attempt was made, by using special operators and communication, to reach absolutely secure classical communication. First the message is bounced back with additional encryption by the Receiver and then the original encryption is removed and the message is resent by the Sender. The mechanical analogy of this operation is using two padlocks; one by the Sender and one by the Receiver. Klappenecker has pointed out that finding an efficient software realization of the operators is equivalent of proving the $P \neq NP$ problem. Various open problems are discussed.

Keywords: data security, secure encryption, classical communication, quantum communication.
PACS: 89.70.+c

INTRODUCTION

Quantum encryption is considered to be absolutely secure. However, quantum communication requires handling single photons, thus it is:

- Slow,
- Fragile/Noisy,
- Expensive to install,
- Maintenance-needy,
- Bulky, not portable.

Therefore, we should ask: Is there any way to make a non-breakable encryption by classical information channel, which is known to be cheaper, faster and less noisy?

[+]Corresponding author, Laszlo.Kish@ee.tamu.edu

CP800, *Unsolved Problems of Noise and Fluctuations: UPoN 2005*
edited by L. Reggiani, C. Pennetta, V. Akimov, E. Alfinito, and M. Rosini

THE TWO-PADLOCK COMMUNICATOR

Recently, Kish proposed a new method, and later studied its aspects with Sethuraman [1], which is claimed to be an absolute secure data encryption, by using only classical information. Sethuraman [2] has identified a mechanical analogy of the method with a mail-carrier-box and two-padlocks, see Fig. 1. The Sender is using a padlock, see Figure 1, to lock the box (operation A) before the mail is sent. When the Receiver receives the box, he is using another padlock to lock the box again (operation B). Then the double-locked box is sent back to the Sender where the first padlock is removed (operation A^{-1}). Then the Sender sends the box again, which is still locked by the Receiver's padlock, back to the Receiver where the box is opened (operation B^{-1}) thus the mail is delivered. In conclusion, at the Kish-Sethuraman (KS) chiper (named by Klappenecker [4]), the following operations are carried out on the t-th bit (or word, etc.) $m(t)$ of the message:

$$\begin{aligned} m(t) \overset{\mathrm{S}}{\Rightarrow} A(t)m(t) \overset{\mathrm{R}}{\Rightarrow} B(t)A(t)m(t) \overset{\mathrm{S}}{\Rightarrow} A^{-1}(t)B(t)A(t)m(t) = B(t)m(t) \\ \overset{\mathrm{R}}{\Rightarrow} B^{-1}(t)B(t)m(t) = m(t) , \end{aligned} \tag{1}$$

where the arrows with S and R mean that the resulting signal was generated by the Sender and the Receiver, respectively.

FIGURE 1. Mechanical analogy of the Kish-Sethuraman cipher. The operations represent locking (A and B) and opening (A^{-1} and B^{-1}) and they are to be carried out by the Sender (A and A^{-1}) and Receiver (B and B^{-1}), respectively.

At the third step, the presumed mathematical condition is:

$$A^{-1}(t)B(t)A(t) = B(t) \quad , \tag{2}$$

which was identified by Bergou [2] as commutativity of the operators:

$$A(t)B(t) = B(t)A(t) \tag{3}$$

The commutativity is necessary to get through the message to the Receiver. Another mathematical condition concerns the security of the cipher:

It should be impossible to determine the message and the actual operator values by recording and comparing the publicly sent signals $A(t)m(t)$, $B(t)A(t)m(t)$ and $B(t)m(t)$. (4)

is that the comparison of a data sequent before and after the operator are used does not serve enough information to identify the data or the actual operator code.

Klappenecker [3] pointed out that operators in the RSA public key cryptosystem satisfy conditions (3) and (4). Further discussions with A. Singer [6] made it obvious that, for the absolute security, one more condition is required which has been implicitly assumed by not explicitly stated:

The Sender and Receiver share no information and submit no public information about the actual operators A(t) and B(t) (except their class which is public information). (5)

Indeed, the RSA system is not safe because of the shared public information about the operators, thus the system can be broken if the Evasdropper has extraordinary calculation power for factoring, such as a quantum computer.

In conclusion, with conditions (4,5) the KS cipher is absolutely secure, and condition (3) is needed for the functioning of this crypto system.

It is interesting to note that Klappenecker has pointed out that creating Klappenecker has pointed out that finding an efficient software realization of the operators is equivalent of proving the $P \neq NP$ problem.

UNSOLVED PROBLEMS

There are several unsolved problems of noise here.

i) The most intriguing question is: *What are the necessary scrambling operators for the method? How efficient are they?* The two-padlock argumentation intuitively suggests that, if there is a mechanical analogy then a software representation should also be possible. Thus, we believe that there is a solution.

ii) Heszler proposed to move toward statistically working encryption, where there are multiple communications to confirm that the received data were correct or not. The scheme is similar to quantum communication where two entangled pairs of information are received at the two sides. *But how to do that?*

iii) (*Finally, how to win the $1 million for proving the* $P \neq NP$ *problem ???*)

ACKNOWLEDGEMENTS

Valuable discussions with: Janos Bergou, Julio Gea-Banacloche, Andreas Klappenecker, Gabor Schmera, Suhail Zubairy, and Deepa Kundur are appreciated.

REFERENCES

1. L.B. Kish and S. Sethuraman, Non-breakable data encryption with classical information. *Fluctuation and Noise Letters* **4**(2), C1–C5 (2004).
2. Swaminathan Sethuraman, personal communication (June 2004).
3. Janos Bergou, personal communication (April 2004).
4. A. Klappenecker, Remark on a "Non-breakable data encryption" scheme by Kish and Sethuraman, *Fluctuation and Noise Letters* **4**(4), C25-C26 (2004).
5. L.B. Kish, Response to Klappenecker's Remark on a Non-Breakable Data Encryption Scheme by Kish and Sethuraman. *Fluctuation and Noise Letters* **4**(4), C27–C29 (2004).
6. Amit Singer, personal communication at the Unsolved Problems of Noise 2005 conference (June 2005).

Can Self-Sustaining Currents Be Induced In A System Of Mesoscopic Rings?

J. Łuczka*, J. Dajka*, M. Mierzejewski* and P. Hänggi†

*Institute of Physics, University of Silesia, 40-007 Katowice, Poland
†Institute of Physics, University of Augsburg, Germany

Abstract. Mesoscopic systems exhibit both quantum and classical features. A purely quantum and topological phenomenon is the occurrence of a persistent current. We analyze a collection of coaxial, mesoscopic rings which are coupled via mutual inductances. At temperatures $T > 0$, thermal fluctuations are taken into account. The system is described in terms of a set of Langevin equations. The unsolved problem is to find steady states in the limit of infinitely many rings (i.e. in the thermodynamic limit). The first problem is to evaluate the effective coupling constants which are determined by elements of the inverse matrix of mutual inductances. The second problem is to apply the mean-field type approximation: can it be justified? If yes, then the resulting steady states are determined by a nonlinear Fokker-Planck equation from which it follows that self-sustaining currents can be induced by interactions among rings. If no, then it is an open and much more difficult problem to identify the corresponding effective state equation.

Keywords: mesoscopic systems, persistent current, phase transition
PACS: 64.60.Cn, 05.10.Gg, 73.23.-b

DESCRIPTION OF MODEL

In mesoscopic systems [1] of ring, toroidal or cylindrical symmetry persistent currents can occur [2] as a result of the phase coherence among electrons, the so-called *coherent electrons*. Experiments [3] have confirmed the existence of such persistent currents. In the ground state, at temperature $T = 0$, the only electrons present in the system are coherent ones, possessing a non-dissipative flow. At non-zero temperature, $T > 0$, some of those electrons become "normal" and their behavior is dissipative. This induces a decrease of the amplitude of the persistent current.

We consider a system of N identical, coaxially composed mesoscopic metal rings (toroids) which are coupled via the mutual inductances. The rings are separated with a distance r between their centers. Magnetic fluxes and currents in the rings are coupled according to the formula [4]

$$\phi_i = \sum_{k=1}^{N} \mathscr{M}_{ik} I_k + \phi_{ext} , \tag{1}$$

where ϕ_i and I_i are flux and current in the i-th ring, respectively. The flux ϕ_{ext} is induced by an external uniform magnetic field B. The coupling coefficients $\mathscr{M}_{ik} = \mathscr{M}_{ki}$ (forming the matrix $\mathscr{M}$) denote the mutual inductances for $i \neq k$ and identical self-inductances $\mathscr{L} = \mathscr{M}_{ii}$ for $i = k$.

CP800, *Unsolved Problems of Noise and Fluctuations: UPoN 2005*
edited by L. Reggiani, C. Pennetta, V. Akimov, E. Alfinito, and M. Rosini

The current in the k-th ring equals a sum

$$I_k = I_k^{nor} + I_k^{coh} \tag{2}$$

of the Ohmic (dissipative) current I_k^{nor} plus the persistent current I_k^{coh}. The Ohmic current contribution $I_k^{nor} = I_{nor}(\phi_k)$ is determined by Ohm's law and Lenz's rule, i.e.,

$$I_{nor}(\phi_k) = -\frac{1}{R}\frac{d}{dt}\phi_k + \sqrt{\frac{2k_BT}{R}}\Gamma_k(t)\,, \tag{3}$$

where R denotes the resistance of a single ring [5], k_B is the Boltzmann constant and $\Gamma_k(t)$ describes the thermal, Johnson-Nyquist fluctuations of the Ohmic current. This thermal noise is modelled by a set of independent Gaussian white noises of zero average, i.e., $\langle\Gamma_k(t)\rangle = 0$ and δ-correlated correlations $\langle\Gamma_k(t)\Gamma_i(s)\rangle = \delta_{ki}\delta(t-s)$. The noise intensity $D_0 = \sqrt{2k_BT/R}$ is chosen in accordance with the classical fluctuation-dissipation theorem [6].

The coherent current can be either of paramagnetic nature for an even number N_e of coherent electrons, or of diamagnetic nature for an odd number of coherent electrons. The probability of finding a channel (ring) with an odd number of coherent electrons is denoted by P and the probability of finding a channel with an even number of coherent electrons is equal to $1-P$, respectively. The current of the coherent electrons $I_k^{coh} = I_{coh}(\phi_k, T)$ has been determined in Ref. [7], reading

$$I_{coh}(\phi_k,T) = I^*\left[Pg(\phi_k/\phi_0,T) + (1-P)g(\phi_k/\phi_0 + 1/2,T)\right]\,, \tag{4}$$

where the flux quantum $\phi_0 := h/e$ is the ratio of the Planck constant h and the electron charge e. The characteristic current $I^* = heN_e/(2l_x^2m_e)$, with N_e being the number of coherent electrons in a single current channel (ring), l_x is the circumference of the ring and m_e is the mass of electron. Moreover, [7]

$$g(x,T) = \sum_{n=1}^{\infty} A_n(T)\sin(2n\pi x) \tag{5}$$

denotes the current in a channel with an even number of coherent electrons. The amplitudes read

$$A_n(T) = \frac{4T}{\pi T^*}\frac{\exp(-nT/T^*)}{1-\exp(-2nT/T^*)}\cos(nk_Fl_x)\,. \tag{6}$$

The characteristic temperature T^* is determined from the relation $k_BT^* = \Delta_F/2\pi^2$, where Δ_F marks the energy gap and k_F is the momentum at the Fermi surface.

STOCHASTIC EVOLUTION EQUATIONS

Given eqs. (1)-(4), it follows that the evolution equations for the fluxes assume the form

$$\frac{1}{R}\sum_{k=1}^{N}\mathcal{M}_{ik}\frac{d\phi_k}{dt} = \phi_{ext} - \phi_i + \sum_{k=1}^{N}\mathcal{M}_{ik}I_{coh}(\phi_k,T) + \sqrt{\frac{2k_BT}{R}}\sum_{k=1}^{N}\mathcal{M}_{ik}\Gamma_k(t) \tag{7}$$

for $i = 1 \ldots N$. We like to remark that this set of equations possesses an untypical form: In the language of particles, the left hand side of this equation corresponds to an unusual coupling of 'velocity' degrees of freedom $d\phi_k/dt$. This note is essential because the known literature results for particle or spin systems with a position-position interaction cannot be directly applied to the set (7).

Let us recall that the interaction among the coaxially composed rings is characterized by the mutual inductances $\mathscr{M}_{ik} = f(r_{ik})$, where r_{ik} is the distance between the i-th and the k-th ring and [8]

$$f(r) = \frac{8\pi R}{b(r)} \left[\left(1 - \frac{b^2(r)}{2} \right) \mathrm{K}(b(r)) - \mathrm{E}(b(r)) \right]. \tag{8}$$

The function $b^2(r) = 4R^2/[4R^2 + r^2]$ and R is the radius of the ring. The functions $\mathrm{K}(z) =$ and $\mathrm{E}(z)$ are the complete elliptic integrals [4]. For short distances between the rings $\mathscr{M}_{ik} \propto -\ln r_{ik}$, while $\mathscr{M}_{ik} \propto r_{ik}^{-3}$ for large distances r_{ik}.

The system (7) can be reformulated in terms of an effective 'position-position' interaction. Indeed, multiplying this system of equations by the elements $\left(\mathscr{M}^{-1}\right)_{ni}$ of the inverse matrix $\mathscr{M}^{-1}$ and next summing up over i one obtains the following set of Langevin equations

$$\frac{1}{R}\frac{d\phi_n}{dt} = \sum_{i=1}^{N} (\mathscr{M}^{-1})_{ni} \left[\phi_{ext} - \phi_i\right] + I_{coh}(\phi_n, T) + \sqrt{\frac{2k_B T}{R}}\, \Gamma_n(t)\,. \tag{9}$$

Its dimensionless form reads [9]

$$\frac{dx_n}{ds} = -V'(x_n, T) - \sum_{i(\neq n)}^{N} \lambda_{ni} x_i + \sqrt{2D}\, \tilde{\Gamma}_n(s)\,. \tag{10}$$

The dimensionless flux is $x_n = \phi_n/\phi_0$ and the dimensionless time reads $s = t/\tau_0$, where $\tau_0 = \mathscr{L}/R$ is the relaxation time of the averaged Ohmic current. The prime denotes the derivative with respect to the first argument of the generalized potential $V(x_n, T)$, i.e. here with respect to x_n. The generalized potential itself is given by

$$V(x_n, T) = \frac{1}{2} a_n x_n^2 - b_n x_n - I_0 \int^{x_n} f(y, T)\, dy\,. \tag{11}$$

The coupling constants are $\lambda_{ni} = \mathscr{L}(\mathscr{M}^{-1})_{ni}$, and the parameter $a_n = \mathscr{L}(\mathscr{M}^{-1})_{nn}$ corresponds to the n-th diagonal element of the inverse matrix $\mathscr{M}^{-1}$. The re-scaled, externally induced fluxes read $b_n = \gamma_n \phi_{ext}/\phi_0$, where $\gamma_n = \mathscr{L} \sum_{i=1}^{N} (\mathscr{M}^{-1})_{ni}$. The re-scaled characteristic current is given by $I_0 = \mathscr{L} I^*/\phi_0$ and

$$f(y, T) = P g(y, T) + (1 - P) g(y + 1/2, T)\,. \tag{12}$$

The zero-mean re-scaled noise reads $\tilde{\Gamma}_n(s) = \sqrt{\tau_0}\, \Gamma_n(\tau_0 s)$ with the correlations $\langle \tilde{\Gamma}_n(s_1) \tilde{\Gamma}_m(s_2) \rangle = \delta_{nm} \delta(s_1 - s_2)$. Its intensity is $D = k_B T/2\varepsilon_0$, where $\varepsilon_0 = \phi_0^2/2\mathscr{L}$ [10].

The generalized potential (11) can either be multistable, specifically assume a symmetric monostable or also a symmetric bistable shape. The dynamics is thus equivalent to a dynamic model of a collection of one-dimensional anharmonic oscillators moving in a multistable on-site potential that are coupled among each other with bi-linear interactions.

STEADY-STATE EQUATION

From the corresponding Fokker-Planck equation for the joint probability density $p(\{x_n\},s)$ of the N-ring system described by (10) one can derive, – after integration over all variables except x_k, – the nonlinear steady-state equation for the one-dimensional stationary probability density $p_s(x_k)$, which has the form [9]

$$\frac{\partial}{\partial x_k}\left[V'(x_k,T)+\sum_{i\neq k}^{N}\lambda_{ik}\langle x_i|x_k\rangle\right]p_s(x_k)+D\frac{\partial^2}{\partial x_k^2}p_s(x_k)=0, \tag{13}$$

where $\langle x_i|x_k\rangle=\int x_i p_s(x_i|x_k)dx_i$ is a stationary, conditional mean value of x_i with respect to the conditional probability density $p_s(x_i|x_k)$. This equation is formally exact, however it is clearly not closed: it contains the unknown quantity $\langle x_i|x_k\rangle$ which can be determined only via an approximation scheme. The most popular mean-field approximation can be formulated as follows. Rewrite the conditional mean value as $\langle x_i|x_k\rangle=\langle x_i\rangle+c_{ik}$, wherein c_{ik} accounts for correlations between i-th and k-th ring. In the thermodynamic limit, when $N\to\infty$, the system becomes statistically homogeneous so that the stationary average $\langle x_k\rangle=\langle x\rangle$ no longer depends on the index k. If in this limit we shall neglect the correlations, i.e. if we put $c_{ik}=0$ then (13) renders a closed but non-linear equation. The crucial problem is whether this constitutes an approximation or whether in fact it presents an exact, limiting result. Let us recall that the coupling constants λ_{ik} are expressed by elements of the inverse matrix $\left(\mathcal{M}^{-1}\right)_{ik}$. In the thermodynamic limit, it is an infinite-dimensional matrix. How does its elements behave with respect to the dependence on the distance r_{ik} between the rings? Presently, we are not able to answer to this question. As a consequence we lack a rigorous proof that $c_{ik}\to 0$ as $N\to\infty$. If the mean-field approximation can indeed be applied, then the stationary probability density for $x=x_k$ would satisfy the non-linear Fokker-Planck equation

$$\frac{d}{dx}[V'(x,T)-\lambda\mu]p_s(x)+D\frac{d^2}{dx^2}p_s(x)=0, \tag{14}$$

where

$$\mu\equiv\langle x\rangle=\int_{-\infty}^{\infty}x\,p_s(x)\,dx \tag{15}$$

is the order parameter of the system and $\lambda=-\sum_{i\neq k}\mathcal{L}(\mathcal{M}^{-1})_{ik}$ denotes an effective coupling constant. In this sum, the index k is fixed and $i\in(-\infty,\infty)$. However, for the system composed of infinitely many rings, the result does not depend on k. The next problem involves the sign of λ. If $\lambda>0$, then we can expect a "ferromagnetic" state of the system, characterized by the parallel alignment of the magnetic moments induced by the currents flowing in the neighboring rings. The *flux state* is characterized by the non-vanishing mean flux $\mu = \langle x \rangle \neq 0$. If an external magnetic field is applied, then trivially $\mu \neq 0$. The non-trivial case emerges when the external flux ϕ_{ext} is zero but $\mu \neq 0$. Then, via Eq. (1), currents may flow without any external magnetic field and are solely induced by the interaction among the various rings. This phenomenon is what we call term self-sustaining currents. Can this situation be realized experimentally?

INTERACTION MATRICES

In Eq. (7), the coupling among the $(N+1)$-rings is described by the matrix

$$\mathscr{M} = \begin{pmatrix} \mathscr{L} & f(r) & f(2r) & f(3r) & \dots & \dots & f(Nr) \\ f(r) & \mathscr{L} & f(r) & f(2r) & \dots & \dots & f((N-1)r) \\ f(2r) & f(r) & \mathscr{L} & f(r) & \dots & \dots & f((N-2)r) \\ \vdots & & & & & & \\ f(Nr/2) & \dots & f(r) & \mathscr{L} & f(r) & \dots & f(Nr/2) \\ \vdots & & & & & & \\ f(Nr) & \dots & \dots & f(3r) & f(2r) & f(r) & \mathscr{L} \end{pmatrix} \tag{16}$$

where the function $f(r)$ is defined in Eq. (8). In Eq. (9), the 'position-position' interaction is determined by the inverse matrix $\mathscr{M}^{-1}$. As mentioned above, we are interested in the thermodynamic limit $N \to \infty$. Then the inverse matrix $\mathscr{M}^{-1}$ is infinite-dimensional. In this limit, each ring has infinitely many symmetric neighbors from above and from below. The crucial quantities are the elements $\left(\mathscr{M}^{-1}\right)_{ik}$ of the inverse matrix and their dependence on the distance r_{ik} between rings.

Mean-field theory has played a seminal role for understanding the behavior of complex and cooperative systems, in particular phase transitions. A number of rigorous results have been obtained answering the question: under what conditions is a mean-field procedure yielding a qualitatively correct prediction. Much is known for lattice models and spin systems [11, 12, 13]. There exists a widespread opinion that, for example, there are (at finite temperatures) no phase transitions occurring in one-dimensional (1D) systems possessing *short range* interactions. Van Hove's result [14], and the extension by Ruelle [15], as well as the collection of well-known, exactly solvable models (Ising or Potts models) seem to support this view. However, there are examples of 1D models with short range interactions, and very important – in presence of on-site potentials – that indeed do exhibit a true thermodynamic phase transition [11, 16]. Our reduced description of the real 3D system of coupled mesoscopic rings is based on the classical Langevin stochastic equations (10), which equivalently describes a 1D system of anharmonic, multistable oscillators. For the case of a global interaction, such systems can undergo a well-defined phase transition [17]. The interaction in (10) is not global but local and we do not know of any rigorous results for systems of the type studied in this paper. So, it is left a open question whether we have a phase transition in a one-dimensional system of coupled mesoscopic rings.

ACKNOWLEDGMENTS

Financial support by the Foundation for Polish Science via the fellowship SN-302-934/2005 (J. D), the KBN grant PBZ-MIN-008/P03/2003 (J. D.and J. Ł), the ESF (Stochastic Dynamics: fundamentals and applications) (J. Ł and P. H.) is gratefully acknowledged.

REFERENCES

1. Y. Imry, *Introduction to Mesoscopic Physics* (Oxford University Press, New York, 1997).
2. F. Hund, Ann. Phys. (Leipzig) **32**, 102 (1938); U. Eckern, P. Schwab, J. Low Temp. Phys. **126**, 1291 (2002).
3. L. P. Levy, G. Dolan, J. Dunsmuir and H. Bouchiat, Phys. Rev. Lett. **64**, 2074 (1990);
   V. Chandrasekhar *et al.*, Phys. Rev. Lett. **67**, 3578 (1991);
   D. Mailly *et al.*, Phys. Rev.Lett. **70**, 2020 (1993);
   E.M.Q. Jariwala *et al.*, Phys. Rev. Lett. **86**, 1594 (2001).
4. J. Wang and Z.-S. Ma, Phys. Rev. B **52**, 14829 (1995).
5. L. Landauer and M. Buttiker, Phys. Rev. Lett. **54**, 2049 (1985).
6. P. Hänggi and H. Thomas, Phys. Rep. **88**, 207 (1998).
7. H.F. Cheung, Y. Gefen, E.K. Riedel, W.H. Shih, Phys. Rev. B **37**, 6050 (1988).
8. J. D. Jackson, *Classical Electrodynamics* (Wiley, New York, 1975).
9. J. Dajka, J. Łuczka, M. Szopa and P. Hänggi, phys. stat. sol.(b) **242**, 196 (2005).
10. J. Dajka, J. Łuczka, M. Szopa, E. Zipper, Phys. Rev. **B67**, 073305 (2003).
11. J. A. Cuesta and A. Sanchez, J. Stat. Phys. 115, 869 (2004).
12. M. Biskup, L. Chayes and N. Crawfors, J. Stat. Phys. (2005).
13. K.H. Khoo and H.K. Sy, J. Phys.: Condens. Matter 13 (2001) 101.
14. L. van Hove, Physica **16**, 137 (1950).
15. D. Ruelle, Statistical Mechanics: Rigorous Results
16. M. Peyrard and T. Dauxois, Mathematics and Computers in Simulation **40**, 305 (1996); N. Theodorakopoulos. T. Dauxois and M. Peyrard, Phys. Rev. Lett. **85**, 6 (2000).
17. K. Kometai and H. Shimizu, J. Stat. Phys. **13**, 473 (1975);
    R.C. Desai and R. Zwanzig, J. Stat. Phys. **19**, 1 (1978);
    D.A. Dawson, J. Stat. Phys. **31**, 29 (1983);
    M. Shiino, Phys. Rev. **A 36**, 2393 (1987).

Berry phase in the presence of external noise

Robert S. Whitney*, Yuriy Makhlin†, Alexander Shnirman** and Yuval Gefen‡

*Departement de Physique Théorique, Université de Genève, 24 quai Ernest-Ansermet, 1211 Genève 4,Switzerland.
†Landau Institute for Theoretical Physics, Kosygin st. 2, 117940 Moscow, Russia
**Institut für Theoretische Festkörperphysik, Universität Karlsruhe, D-76128 Karlsruhe, Germany
‡Department of Condensed Matter Physics, The Weizmann Institute of Science, Rehovot 76100, Israel

Abstract. Here we summarise and place in context two of our earlier works. There we investigate the geometric phase or Berry phase (BP) acquired by a spin-half which is subject to external noise in addition to a slowly varying magnetic field (which generates the Berry phase). The noise may be due to the fluctuations of either quantum or classical degrees-of-freedom in the vicinity of the spin. We find that the noise *modifies* the Berry phase and that this modification is of a geometric nature. While the original BP (for an isolated system) is the flux of a *monopole*-field through the loop traversed by the magnetic field, the noise-induced modification of the BP is the flux of a *quadrupole-like* field. The noise-induced phase is *complex* and its imaginary part giving a geometric contribution to dephasing; its sign depends on the direction of the loop. Unlike the BP, this *geometric dephasing* is gauge invariant for open paths of the magnetic field. We discuss the consequences of this for proposals to use Berry phases to control qubits, in particular solid state (superconducting) qubits.

Keywords: geometric phases, dissipative quantum mechanics, decoherence, qubits
PACS: 03.65.Vf, 03.65.Yz, 85.25.Cp

Introduction: Rather than present new results in this conference proceedings, we take the opportunity to summarise and place in context two of our earlier works [1, 2].

There has recently been signifi cant progress in building quantum devices over which one can have complete *phase-coherence* control [3]; much of the interest in such devices is in their potential as qubits [4]. This work has renewed interest in the effect of external noise on quantum systems. It has been known for a long time that such noise — induced by the coupling of the system to environmental degrees-of-freedom beyond experimental control — leads to non-Hermitian evolution of the state of the system [5, 6, 7]. For a two-level system, weak external noise causes energy-relaxation on a timescale T_1, loss of coherence on a timescale T_2, and a Lamb-like shift of the energy-levels, $\delta B_{\rm Lamb}$. Even relatively weak external noise can induce a T_2 which is too short to allow coherent control of the quantum system. The challenge is to understand how the noise affects quantum systems, with a view to fi nding methods for minimising the loss of coherence.

Here we study the effect of external noise on the Berry phase [8, 9]. Berry showed that the phase acquired by an eigenstate of a slowly varying Hamiltonian $H(t)$ is related to the geometric properties of the loop traversed by $H(t)$. It was suggested [10] that the *Berry phase would be insensitive to external noise*. The argument was that the noise-induced fluctuations of geometric properties of the loop traversed by $H(t)$ would average away on the long times required for an adiabatic Berry phase experiment. It had been shown

CP800, *Unsolved Problems of Noise and Fluctuations: UPoN 2005*
edited by L. Reggiani, C. Pennetta, V. Akimov, E. Alfinito, and M. Rosini

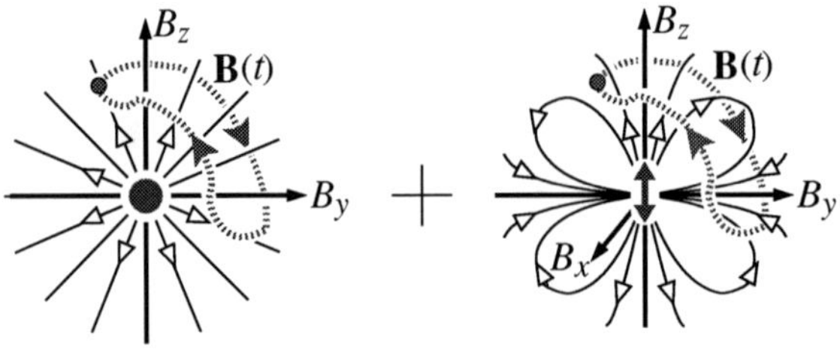

FIGURE 1. Without external noise, the Berry phase is the flux enclosed by the path of $\mathbf{B}(t)$ (shown as the discontinuous line) due to a monopole field (the continuous lines in the left-hand plot). For a system experiencing anisotropic external noise, the Berry phase is the flux enclosed by the path of $\mathbf{B}(t)$ due to the sum of the monopole field *and a quadrupole-like field* (the latter is shown on the right). The quadrupole-like field is the *noise-induced* Berry phase; its magnitude goes like the strength of the coupling to the environment squared. The prefactor on this term is *complex*, leading to geometric dephasing.

that Berry phases could be used for complete control of the state of a two-level system (qubit) [11, 12]. Thus it seemed that even in the presence of external noise coherent control of a qubit could be achieved using Berry phases.

Independently we approached the problem from exactly the opposite direction. It is well known that to manipulate a quantum system adiabatically one must change the Hamiltonian at a rate much less than the spectral gap. However a quantum system in a noisy environment has broadened levels and hence no true gap. Thus one would naively assume that the condition for adiabaticity can never be satisfied, and hence even *infinitesimal external noise would destroy the Berry phase*. This is clearly too naive, all real systems experience some external noise and yet Berry phases have been observed.

The contradiction between these two naive arguments led us to investigate the exact conditions for observing a Berry phase in a system which experiences external noise [1], these are summarised below. We then proceeded to investigate the nature of the Berry phase, and have shown that it is *modified by the external noise* [1], that this modification is *geometric and complex*[2].

Berry phase without noise: Berry [8] considers a two-level spin-half system in a magnetic field, which is varied slowly along a closed path: $H_{\rm spin} = -\frac{1}{2}\mathbf{B}(t)\sigma$ [13]. The rate of the field's change is characterized by the time to complete the loop, $t_{\rm P}$. In the adiabatic limit, $Bt_{\rm P} \gg 1$, the relative phase acquired by the eigenstates is a sum $\Phi = \oint|\mathbf{B}(t)|dt + \Phi_{\rm BP}$ of the dynamical and Berry phases. The latter is geometric, it depends on the geometry of the loop but not on the details of its traversal (for an isolated spin-half it is given by the solid angle subtended by the loop $\mathbf{B}(t)$). In the spin language, the evolution is a rotation of the spin by an angle Φ about $\mathbf{B}$. Note there are non-adiabatic correction to the phase (and to the amplitude to return to spin-up at $t_{\rm P}$) which go like $(Bt_{\rm P})^{-1}$, these can only be ignored in the adiabatic limit, $Bt_{\rm P} \gg 1$.

Observing the Berry phase in the presence of external noise: We consider a spin-half coupled to a time-dependent field that we slowly rotate along a closed loop, $\mathbf{B}(t)$ and a noisy-field along the z-axis, $\hat{\mathbf{X}}(t) \equiv \hat{X}(t)\mathbf{z}$. The Hamiltonian reads [13]

$$\hat{H} = -\tfrac{1}{2}\mathbf{B}(t)\hat{\sigma} - \tfrac{1}{2}\hat{X}\hat{\sigma}_z + \hat{H}_{\rm env}(\hat{X})\ . \tag{1}$$

Below we discuss the generalisation of this to multiple noisy-fields coupled to multiple axes, for now we consider a single noise-field as in the above Hamiltonian.

We showed in [1] that the condition for observing the Berry phase of a system is that the external noise is weak enough that $B \gg T_2^{-1}$. Then one can carry out a BP experiment slowly enough for non-adiabatic effects to be ignored ($t_P \gg B^{-1}$) while fast enough that the noise has not completely destroyed the coherence, i.e. it has not destroyed the phase information ($t_P \lesssim T_2$). If the noise is stronger than this there is no t_P for which one can observe the BP of the quantum system.

Why anisotropic noise is highly-relevant to experiments: In our works we consider anisotropic noise; indeed in [2] we noted that for isotropic external noise the modification of the BP is *zero*, see below. Thus one should ask whether anisotropic noise is relevant to experiments. However we argue that situations in which the noise is *isotropic* will be extremely rare, in most real solid-state devices the noise is strongly anisotropic. Take for example the super-conducting charge qubit of Nakamura *et al* [3] (often called a Cooper-pair box); it is an effective two-level system which has either n or $(n+1)$ Cooper pairs on a superconducting island which forms one arm of a SQUID. One can treat it as a pseudo-spin-half with the z-axis effective field given by a gate voltage and the x-axis effective field given by a flux through the SQUID. It is then clear that there is no reason that the magnitude of the noise in the x- and z-directions should be related to each other; the former being caused by charge fluctuations and the latter being caused by noise in the applied magnetic-field. More generally, the strongest source of noise is often fluctuations of the field used to control the quantum system, since these fields are not isotropic, it seem unlikely that the noise in them will be.

Difference between external noise of quantum and classical origin: We define two noise correlators $S_\pm(t;\tau) = \langle [\hat{X}(t+\tau)\hat{X}(t) \pm \hat{X}(t+\tau)\hat{X}(t)] \rangle$, where for noise of classical origin $\langle \cdots \rangle$ indicates averaging, while for noise of quantum origin $\langle \cdots \rangle$ indicates an expectation value. For noise generated by an environment containing many non-identical degrees-of-freedom, one often finds that these correlators have a finite correlation time; i.e. they decays on a timescale $t_{\rm mem}$, often called the memory time. If in such cases $\langle \hat{X}^2 \rangle t_{\rm mem} \ll B$ then the effect of the noise is weak enough that it is given by golden-rule type calculations in which all effect of the noise on the system's dynamics are given by these two noise correlators [14]. All other details of the environment (oscillators/two-level-systems/chaotic-system, fermionic/bosonic, etc) are irrelevant.

The only difference between noise of classical origin and noise of quantum origin, is that in the former case $S_-(t;\tau) = 0$ while in the latter case this may be non-zero. We showed in [2] (see also [15]) that the noise-induced modification of the Berry phase is independent of $S_-(t;\tau)$. Thus the Berry phase is insensitive to whether the noise is of quantum or classical origin, it depends only on the functional form of $S_+(t;\tau)$.

Modification of Berry phase: In the absence of noise, the BP [8] is given by the surface integral $\Phi_{\rm BP} = \int {\rm d}\mathbf{S}\, \mathbf{b}(\mathbf{B})$, where the field $\mathbf{b}(\mathbf{B})$ is that of a charge-one monopole, $b_B = 1/B^2$. We find that the noise induces an additional BP of the form $\delta\Phi_{\rm BP} = \int {\rm d}\mathbf{S}\, \delta\mathbf{b}(\mathbf{B})$ where the field $\delta\mathbf{b}(\mathbf{B})$ has two non-zero components;

$$\delta b_B = B^{-2} F(B)(3\cos^2\theta - 1) \qquad \delta b_\theta = -B^{-1} F'(B) \sin\theta \cos\theta \tag{2}$$

The function $F(B)$ is a rather ugly function of the Fourier transform of $S_+(\tau)$, see [2].

The angular dependence of $\delta\mathbf{b}$ resembles that of a *quadrupole*. For slow environment modes ($\Omega \ll B$), $\mathrm{Re}[F(B)] \propto B^{-2}$ and hence $\delta\mathbf{b}(\mathbf{B})$ is a quadrupole field. For other environment modes, $\delta\mathbf{b}$ has non-zero curl, thus it is not a sum multipoles but rather the field generated by a pseudo-current [2].

Here we briefly summarise the steps of the derivation of the above result, and refer those interested in details to [2]. We go to a rotating frame, as used in [16], which is chosen in such a way that $\mathbf{B}(t)$ remains static in the new frame. The price we pay is that this new frame is non-inertial, rotating with angular velocity vector, ω; so in this frame the spin experiences an additional pseudo-magnetic-field, ω. However since we are in the adiabatic limit of slow rotation, $\omega \ll B$ and so the total (noiseless) field in this frame, $(\mathbf{B}+\omega)$ is approximately time-independent. Thus we can use the usual techniques for introducing external noise into an otherwise static Hamiltonian [5], which lead to a dissipative master equation for the spin-half. The time-dependence of ω and the axis that X couples to only slightly complicate the calculation. The total (complex) phase acquired by the off-diagonal elements of the density matrix is given by the Lamb-shifted (complex) energy gap (in the rotating frame) integrated over time. The imaginary part of this energy is T_2^{-1} and gives the width of the levels. However both the real part of the noise-induced shift, δB_{Lamb}, and the imaginary part, T_2^{-1}, are a function of the initial gap, which in the rotating frame is $|\mathbf{B}+\omega|$; thus expanding $(\delta B_{\mathrm{Lamb}}+\mathrm{i}T_2^{-1})$ to first order in ω leads to an order $X^2\omega$-term which when integrated over time gives a *complex* geometric phase [17]. Thus one can write the total (complex) BP for a closed path as

$$\Phi_{\mathrm{BP}} = \oint \mathrm{d}\varphi \, \frac{\mathrm{d}}{\mathrm{d}B_z}(B+\delta B_{\mathrm{Lamb}}+\mathrm{i}T_2^{-1}) \, . \tag{3}$$

where B_z is the component of $\mathbf{B}$ along the axis that the noise couples to (the z-axis).

Geometric dephasing: The imaginary part of a phase causes the decay of off-diagonal elements of the density matrix, and hence is dephasing. In eq. (3) we clearly see that the noise introduces an imaginary part to the BP, this imaginary geometric term is *geometric dephasing*.

The first intriguing feature of this geometric dephasing is that it may be of *either sign* (it changes sign if $\mathbf{B}(t)$ traverses the same loop in the opposite direction). If this were the only source of dephasing, it would be unphysical because it could lead to the purity of the density matrix becoming greater than one. However it is not the only source of dephasing; it is a small modification of the total dephasing. Our analysis is in the adiabatic limit $Bt_{\mathrm{P}} \gg 1$, and the total dephasing goes like $\langle \hat{X}^2 \rangle B t_{\mathrm{P}}$ while the geometric dephasing only goes like $\langle \hat{X}^2 \rangle$. None-the-less this asymmetry in the dephasing (between going one way or the other way around a loop) is unexpected and intriguing.

The second intriguing feature of the geometric dephasing is that it is gauge-*independent* for *open* paths of $\mathbf{B}(t)$. This is absolutely different from the real part of the BP, its gauge-*dependence* is a consequence of ambiguity in the choice of instantaneous basis of cartesian coordinates for a given $\mathbf{B}(t)$; $\mathbf{B}(t)$ defines the instantaneous z-axis but the x-axis may lie anywhere in the plane perpendicular to $\mathbf{B}(t)$. Gauge transformations rotate this x-axis in this plane. The ambiguity is absent for closed loops, where the final basis must coincide with the initial one. However one can monitor the magnitude of the transverse spin component (dephasing); this magnitude is independent of the choice of the instantaneous x-axis, thus *all dephasing is gauge-invariant for open paths*.

Note that the BP above, in eqs. (2,3), is *only* for *closed* paths of $\mathbf{B}(t)$. This is because it was derived using Stokes' theorem to write the line-integral along a *closed* path as a surface integral. The result for $\mathrm{Im}[\Phi_{\mathrm{BP}}]$ for *open* paths is given in [2] and is rather uglier than those above.

Multiple sources of external noise: Above we considered a single noise field coupled to the z-component of the spin. However since $\mathbf{B}(t)$ was left arbitrary the analysis is equally applicable if the noise field couples to any axis, i.e. to $\sigma_r \equiv \mathbf{r}.\sigma$ where $\mathbf{r}$ could be any unit vector. In this case the axis of the quadrupole-like field would simply be along the $\mathbf{r}$-axis instead of the z-axis.

It is then straightforward to generalise our analysis to multiple noisy-fields coupling to multiple arbitrary components of the spin. In this case we see that each noise field's effect on the BP is independent (at least to the order in coupling that we do the calculations) and so their effects on the BP just sum up. Thus the resulting noise-induced BP is the sum of the quadrupole-like fields each with a different magnitude and each with their axis pointing in a different direction.

There is a interesting special case of this, and that is an *isotropic* environment. This is modelled by an environment coupling of the form $\mathbf{X}.\sigma$ where $\mathbf{X}$ has three components and the environment Hamiltonian is of the form $\hat{H}_{\mathrm{env}}(|\mathbf{X}|)$. It is clear to see that this Hamiltonian has no preferred direction in space, so there can be no quadrupole-like field. Indeed the only field which can have no preferred direction in space is the monopole-field. Thus one might expect the magnitude of the monopole field to be modified by the presence of the environment. However the BP must be insensitive to whether one defines it via the interior or exterior solid angle (up to factors of 2π); hence the monopole field must have a magnitude of one and so it cannot be modified by the environment. A more pedestrian way to see this is to assume the isotropic environment has a Hamiltonian of the form $\hat{H}_{\mathrm{env}}(X_x)+\hat{H}_{\mathrm{env}}(X_y)+\hat{H}_{\mathrm{env}}(X_z)$. Then the x-component of the environment generates a quadrupole-like field with its axis in the x-direction, the y-component generates one with its axis in the y-direction, etc. The three quadrupoles all have the same magnitude, and they cancel each other when we sum them up.

Acknowledgments: This work is part of the CFN (DFG) and was supported by the EC RTN Spintronics, and by the ISF of the IAS. RW was supported by the EPSRC and the NSF. YM was supported by the AvH Foundation (S. Kovalevskaya award). and YG is supported by the AvH Foundation (Max-Planck Award). RW, YM, AS acknowledge partial support from the Minerva Einstein center (DFG) during their visits to the Weizmann Institute, while YM and AS also thank the Transnational Access program for the support during a visit to the WIS.

REFERENCES

1. R.S. Whitney and Y. Gefen, *Phys. Rev. Lett.* **90** 190402 (2003).
2. R.S. Whitney Y. Makhlin, A. Shnirman and Y. Gefen, *Phys. Rev. Lett.* **94** 070407 (2005).
3. For experimental work on superconducting qubits see: Y. Nakamura, Yu. A. Pashkin and J.S. Tsai *Nature* **398** 786 (1999): D. Vion *et al.*, *Science* **296** 886 (2002); I. Chiorescu *et al.*, *Science* **299** 1869 (2003).

4. See e.g. *Introduction to quantum computation and information* ed. H.-K. Lo, S. Popescu and T. Spiller (World Scientific, Singapore,1998)
5. F. Bloch, Phys. Rev. **105**, 1206 (1957). A. G. Redfield, IBM J. Res. Dev. **1**, 19 (1957).
6. A.J. Leggett *et al*, *Rev. Mod. Phys.* **59** 1 (1987), and references therein.
7. see e.g. C. Cohen-Tannoudji, J. Dupont-Roc and G. Grynberg, *Atom-photon interactions* (Wiley, New York, 1992); U. Weiss, *Quantum dissipative systems* (World Scientific, Singapore, 1999); H.-P. Breuer and F. Petruccione, *The theory of open quantum systems.* (Oxford University Press, Oxford, 2002).
8. M.V. Berry, *Proc. R. Soc. Lond.* **392** 45 (1984).
9. For reviews of the Berry phase and its applications see: J. Anandan, J. Christian and K. Wanelik, *Am. J. Phys.* **65** 180 (1997); A. Shapere and F. Wilczek (Eds.), *Geometric Phases in Physics* (World Scientific, Singapore, 1989); J. Zwanziger, *The Geometric Phase in Quantum Systems: Foundations, Mathematical Concepts, and Applications in Molecular and Condensed Matter Physics* (Springer-Verlag, New York, 2003).
10. S. Lloyd, *Science* **292** 1669 (2001)
11. J.A. Jones *et al*, *Nature* **403**, 869 (2000). A. Ekert *et al*, *J. Mod. Opt.* **47**,2501, (2000).
12. G. Falci *et al*, *Nature* **407** 355 (2000).
13. All energies and magnetic fields are in units of inverse time ($g\mu_B = \hbar = 1$).
14. Another way to say this is that in this limit Wicks theorem works, so we can write higher-order correlators in terms of $S_{\pm}(t;\tau)$.
15. R.S. Whitney Y. Makhlin, A. Shnirman and Y. Gefen, *Proc. NATO ARW St. Petersburg* (2003), eprint: *cond-mat/0401376.*
16. M.V. Berry, Proc. R. Soc. Lond. A, **414**, 31 (1987).
17. The possiblility for a Berry phase to be complex was first discussed, in the context of non-Hermitian Hamiltonians, by J.C. Garrison and E.M. Wright, *Phys. Lett. A* **128**, 177 (1988). However here we observe it in evolution which preserves the trace of the density matrix (which non-Hermitian Hamiltonians do not).

NOISE AND CHAOS

Coupling-Induced Oscillations In Overdamped Bistable Systems In The Presence Of A Noise Floor

A. R. Bulsara*, V. In*, A. Kho*, J. Lindner†, W. -J. Rappel** and S. Baglio, B. Ando, V. Sacco‡

*Space and Naval Warfare Systems Center San Diego, Code 2363, 53560 Hull Street, San Diego, CA 92152-5001, USA.

†Physics Dept., The College of Wooster, Wooster, OH 44691, USA.

**Physics Dept., Univ. of california at San Diego, La Jolla, CA 92037, USA.

‡Dipartimento di Ingegneria Elettrica Elettronica e dei Sistemi, Univ. degli Studi di Catania, Viale A. Doria 6, 95125 Catania, ITALY

Abstract. Recently, we have shown the emergence of oscillations in overdamped undriven nonlinear dynamic systems subject to carefully crafted coupling schemes and operating conditions. Here, we outline the existing theoretical results obtained on a system of $N = 3$ coupled fluxgate magnetometers, taken as the representative system of interest; the oscillatory behavior is triggered when the coupling constant exceeds a threshold value (bifurcation point), and the oscillation frequency exhibits a characteristic scaling behavior with the "separation" of the coupling constant from its threshold value, as well as with an external target dc magnetic flux signal. We also present the first (numerically obtained) results on the effect of a noise-floor on the oscillations.

Keywords: overdamped nonlinear system, coupling, noise, oscillations
PACS: 05.10.-a, 05.40.-a, 05.45.-a

INTRODUCTION

Overdamped bistable dynamics, of the generic form $\dot{x} = -\nabla U(x)$, underpin the behavior of numerous systems in the physical world. The most-studied example is the overdamped Duffing system, the dynamics of a particle in a bistable potential $U(x) = -ax^2 + bx^4$. Absent an external forcing term, the state-point $x(t)$ will rapidly relax to one of two stable attractors, for any choice of initial condition. It has been shown [1], however, that coupling similar elements via a linear uni-directional coupling with cyclic boundary conditions, can lead to oscillatory behavior past a critical value of the coupling coefficient. Typically, this behavior is dictated by symmetry conditions [2], and is generated via Hopf bifurcations; it appears to occur in any coupled system of overdamped bistable elements, none of which would oscillate when isolated and undriven, subject to the appropriate choice of parameters and operating conditions (albeit through different bifurcation mechanisms).

To better understand this behavior, we focus on a specific nonlinear dynamic system in which the state point $x(t)$ represents the (suitably normalized) magnetic induction in a ferromagnetic sample; the dynamical model is obtained via the continuum limit of a discrete spin model of individual domain dynamics and has been recently used [3] to characterize the response of a specific magnetic measurement system, the fluxgate magne-

CP800, *Unsolved Problems of Noise and Fluctuations: UPoN 2005*
edited by L. Reggiani, C. Pennetta, V. Akimov, E. Alfinito, and M. Rosini

tometer. The single fluxgate magnetometer can be treated [3] as a nonlinear dynamic system by assuming the core as approximately single-domain, and writing down an equation for the evolution of the macroscopic magnetization parameter $x(t)$: $\dot{x}(t) = -\nabla_x U(x)$ in terms of the potential energy function $U(x,t) = \frac{x^2(t)}{2} - \frac{1}{c}\ln\cosh c[(x(t)+h(t)]$. c is a temperature-dependent nonlinearity parameter which controls the topology of the potential function: the system becomes monostable, or paramagnetic, for $c < 1$ corresponding to an increase in the core temperature past the Curie point. The overdot denotes the time-derivative, and $h(t)$ is an external signal. In single fluxgate magnetometers, $h(t) = h_1(t) + \varepsilon$. Here, ε is the target signal (taken to be dc and quite small compared to the energy barrier height U_0 of the potential) to be detected, and $h_1(t) = A\sin\omega t$, a *known* bias signal that is applied to make the device switch readily between its saturation states (the target signal is too small to accomplish this alone), so that the asymmetry introduced by the dc target signal manifests itself as a (measureable) difference in the residence times spent in the up and down saturation states of the core hysteresis loop.

In two recent papers [4], we have demonstrated that coupling an *odd* number of overdamped bistable elements in a ring, with unidirectional coupling, can lead to oscillatory behavior when the coupling strength exceeds a critical value. Clearly, the practical importance of this effect lies in the potential sensitivity enhancement when the system is "tuned" very close to the oscillation threshold (i.e. in the regime of very low frequency). We now provide an overview of the (deterministic) results [4] of our investigations to date, followed by preliminary results on the role of background noise. Laboratory experiments [5] have confirmed the occurrence of the oscillations, as well as all the other predictions of the deterministic theory; however, we will not discuss these results in this work.

COUPLED FLUXGATE MAGNETOMETERS: THE DETERMINISTIC DYNAMICS

We start by writing down the dynamics for three unidirectionally (cyclically) coupled fluxgate magnetometers ($i = 1..3, x_4 \equiv x_1$):

$$\tau\dot{x}_i = -x_i + \tanh(c(x_i + \lambda x_{i+1} + \varepsilon)), \qquad (1)$$

where $x_i(t)$ represents the (suitably normalized) magnetic flux at the output (i.e. in the secondary coil) of unit i, and $\varepsilon \ll U_0$ is an externally applied dc "target" magnetic flux, U_0 being the energy barrier height (absent the coupling) for each of the elements (assumed identical for theoretical purposes). τ represents the device time constant (inverse bandwidth); it is a measure of how rapidly the switching events occur between the two steady states of each core. In well-fabricated "single domain" cores, this switching rate is quite large ($\tau \approx 10^{-6}$) with the hysteresis loop being narrow and sharp. It is important to note [4] that the oscillatory behavior occurs even for $\varepsilon = 0$, however when $\varepsilon \neq 0$, the oscillation characteristics change (since the underlying potentials of each device are now asymmetric); these changes can be exploited for signal quantification purposes, the motivation for this work. We reiterate that the oscillations do *not* occur for $\lambda = 0$ due to the overdamped nature of the dynamics (1).

It is clear that the oscillatory behavior actually corresponds to the periodic switching back and forth of the state point of each element, between the stable (or saturation) states of the input-output (hysteretic) transfer characteristic. The following features are observed, and are readily derived through an analysis [4] of the deterministic system (1): (1) The oscillations commence when the coupling coefficient exceeds a threshold value $\lambda_c = -\varepsilon - x_{inf} + \frac{1}{c}\tanh^{-1}x_{inf}$, with $x_{inf} = \sqrt{(c-1)/c}$; note that in our convention, $\lambda < 0$ so that oscillations occur for $|\lambda| > |\lambda_c|$. (2) The individual oscillations (in each elemental response) are separated in phase by $2\pi/N$, and have period $T_i = \frac{N\pi}{\sqrt{cx_{inf}}}\left[\frac{1}{\sqrt{\lambda_c-\lambda}} + \frac{1}{\sqrt{\lambda_c-\lambda+2\varepsilon}}\right]$, which shows a characteristic dependence on the inverse square root of the bifurcation "distance" $\lambda_c - \lambda$, as well as the target signal ε; these oscillations can be experimentally produced at frequencies ranging from a few Hz to high kHz. (3) The *summed* output oscillates at period $T_+ = T_i/N$ and its amplitude (as well as that of each elemental oscillation) is *always supra*threshold i.e., the emergent oscillations are strong enough to drive each core between its saturation states. Increasing N changes the frequency of the individual elemental oscillations, but the frequency of the summed response is seen to be independent of N. (4) The residence times difference, i.e. the difference in the times spent by the system in its two stable magnetization states, can be computed as $\Delta t \approx \frac{\pi}{\sqrt{cx_{inf}}}\left[\frac{1}{\sqrt{\lambda_c-\lambda}} - \frac{1}{\sqrt{\lambda_c-\lambda+2\varepsilon}}\right]$, which vanishes (as expected) for $\varepsilon = 0$, and can be used as a quantifier of the target signal, analogous to the time-domain operation of the single fluxgate [3]. (5) The system responsivity, defined via the derivative $\partial\Delta t/\partial\varepsilon$, is found to increase dramatically as one approaches the critical point in the oscillatory regime; this suggests that careful tuning of the coupling parameter so that the oscillations have very low frequency, could offer significant benefits for the detection of very small target signals. Coupling, therefore, allows one to exploit the target-signal dependence of the emergent oscillations for detection and quantification purposes. Note also, that the oscillations also occur for N even and large; we do not consider this case in this paper. Forthcoming papers will include the cases of (1) no noise but random initial conditions, and (2) low noise, random initial conditions, and large N.

THE EFFECTS OF A NOISE FLOOR

In any practical application, one always has to contend with noise that arises from two sources: the (magnetic) noise floor of each sensor, and noise that contaminates the target signal. Here we will consider the effects, solely, of the noise floor, assuming it to be gaussian band-limited noise having zero mean, correlation time τ_c, and variance σ^2. The noise appears as an additive term on the rhs of each of the equations (1).

We start with a rapid recapitulation of some theoretical [3] and experimental [6] observations stemming from our work with single RTD fluxgates. The voltage signal from the fluxgate has been shown [6] to have a noise component that can be well-approximated by a gaussian distribution. However, the individual residence times have noise components which are, in general, non-gaussian; they have noise-dependent tails and, with increasing noise intensity the tails get longer. This is a feature that is quite common to two-state devices; in particular, the inverse gaussian distributions that describe the dynamics of "integrate-fire" neurons have this structure. Our earlier experimental work [3] showed these features, and also showed that decreasing the noise intensity (alternatively, increasing the bias amplitude A) reduced the tail and made the distributions more gaussian-like. Of course, this comes at the price of reduced sensitivity (since the sensitivity in this case is inversely proportional to the bias amplitude). A careful optimization of geometrical and other core parameters can also lower the noise in the voltage signal (see e.g. [7]).

To better understand the ramifications of background noise, we have introduced [3] the (critical, to a practical system) observation time T_{ob}, and defined a response signal-to-noise ratio (SNR) which is directly proportional to $\sqrt{T_{ob}}$. A longer observation time leads to an enhanced response (to very weak target signals), however, practical constraints may limit T_{ob}. One can increase the bias frequency ω (when dealing with only a single device), thereby increasing the number of crossing events and improving the statistics of the measurement process, however this implies a larger power constraint, once again. Hence, in a practical application, one must strike a balance between the physical constraints (e.g. onboard power, noise from the bias signal generator) and the need to carry out a reliable measurement of the mean RTD. The practical configuration is also, of course, heavily dependent on the amplitude (relative to the energy barrier height) of the target signal ε to be quantified.

In the presence of a small noise floor, one observes that the noise floor manifests itself in fluctuations of the "rest" states of the magnetometer, about the deterministic mean values $\pm b$ which can readily be calculated [4] from the single fluxgate potential energy function. Numerical simulations of the coupled system show that the threshold crossings are quite sharp and unambiguous (as long as the noise is not too strong). This is a direct consequence of the very low time constant τ (usually $\ll \tau_c$) of the fluxgate; to all intents and purposes, the device behaves like a static nonlinearity, with near-instantaneous switching events.

We now present the first results from the numerical simulations of the coupled system (1) in the presence of additive gaussian noise, taken to have the same variance σ^2 and bandwidth (τ_c^{-1}) (but uncorrelated) in all 3 elements. For convenience, the nonlinearity parameter c is also taken to be the same throughout. Further, we consider only the case of zero target signal (i.e. $\varepsilon = 0$) so that the system is symmetric.

The power spectral density (psd) of any of the solutions $x_i(t)$ shows interesting features, as seen in figure 1. Specifically, it appears to share much in common with what is observed in renewal processes [8]. Given the near-instantaneous transition between the upper and lower thresholds, preceded by a relatively slow transition to the threshold (this is readily observable in the deterministic time series [4]), and coupled with the (not-unreasonable) assumption of independent crossing events, the renewal description seems to be a good one. The features of the psd are better explained with reference to figure 2 which shows the modal switching frequency (the location, f_M, of the fundamental peak in the corresponding psd) of a *single* element in the $N = 3$ ring, as a function of the noise intensity for two values of the coupling coefficient bracketing its critical value. On this figure, we also plot (for comparison purposes) the characteristic (Kramers) noise-induced hopping rate, which is numerically computed for a single uncoupled element in the array. For the subthreshold case, the psd for any choice of parameters shows a "knee" at the fundamental (noise-induced) oscillation frequency f_M; with increasing noise, the spectrum becomes flatter, approaching the limit where the maximum spectral amplitude occurs at zero frequency for large noise. For the suprathreshold case we have,

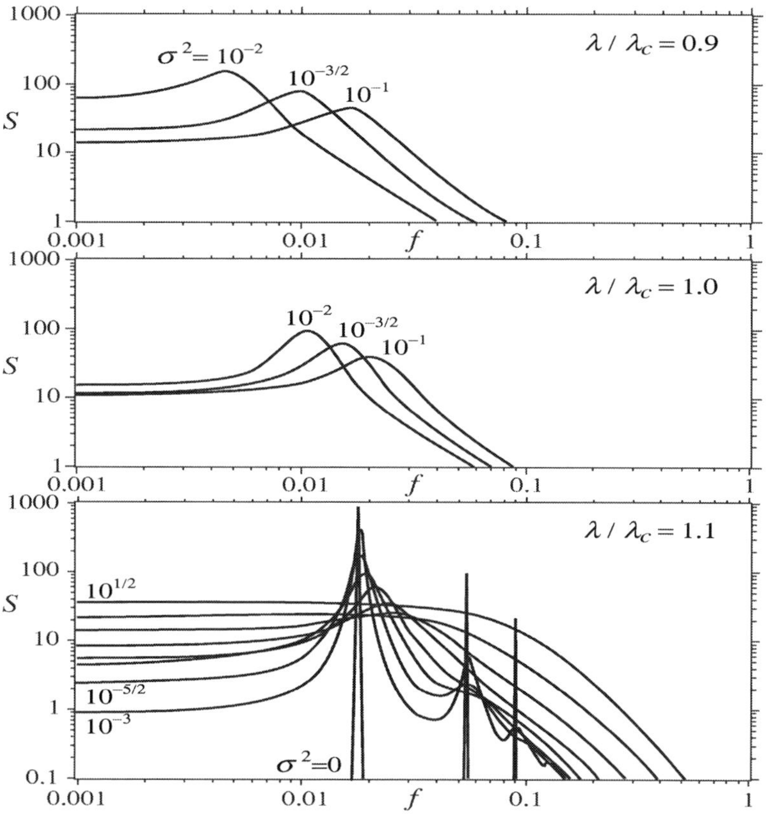

FIGURE 1. Power spectral densities (psds) for three values of the coupling coefficient, with λ_c computed as given in section 1. For each panel, the noise intensity corresponding to each psd is indicated. $c = 3, \tau_c = 1, \lambda_c = -0.4345$. Sharp peaks (bottom panel) correspond to the deterministic oscillation frequency and its (odd) higher harmonics.

already, a deterministic oscillation frequency (shown as a spike, together with its 3rd. and 5th. harmonics in figure 1), and the effect of the noise is to replace this spike with a broad (and shifted) peak and its odd (because the system is symmetric) harmonics; note that the higher harmonics are not visible unless the noise is very small. For very small noise intensities, the modal oscillation frequency (i.e. the inverse mean of the period distribution function) does not deviate appreciably from the deterministic oscillation frequency; this is also evident in the top curve of figure 2 wherein one observes a finite oscillation frequency even for zero noise, as expected. With increasing noise intensity, however, the mean oscillation frequency becomes a function of the noise and separates itself from the deterministic frequency. Simultaneously (with increasing noise), the peak amplitude (in the psd) decreases until, past the Kramers rate f_K, the psd has its maximum amplitude at zero frequency (for any value of λ), as would be expected; this is clearly evident in the bottom panel of figure 1. The above features are the subject of ongoing

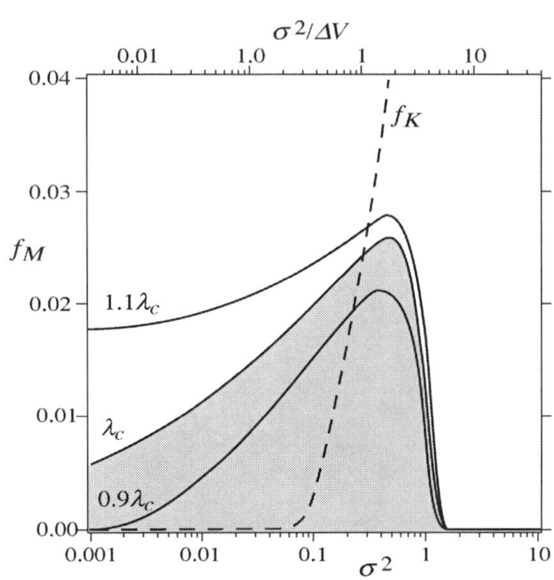

FIGURE 2. Modal oscillation (or switching) frequency f_M of a *single* fluxgate in the $N = 3$ configuration as a function of noise intensity σ^2 (normalized to potential barrier height ΔV of a single uncoupled fluxgate, on top scale). f_M denotes location of peak in the corresponding psd (see e.g. figure 1). Dashed curve is the (numerically computed) Kramers rate f_K for a single *uncoupled* element. Other parameters as figure 1.

investigations.

The occurrence of hopping events in the sub-critical regime, where the deterministic system is quiescent is, clearly, a case of purely noise-induced switching. In figure 3 we plot the relative spectral amplitude $\Delta S = [S(f_M) - S(f=0)]/S(f=0)$ of the frequency feature at f_M in the psd as a function of noise; for the subthreshold case, this amplitude is taken at the "knee" of the psd. Note the Stochastic Resonance (SR)-like feature in the subthreshold case; this behavior is not seen in the suprathreshold case, quite analogous to observations on driven bistable systems [9].

Finally, we address the issue of increasing the number of elements in the coupled array. As already mentioned (section 1), changing N leads to a proportionate scaling of the individual frequencies of the component elements, however, the summed output oscillates at a frequency independent of N. Figure 4 shows the normalized spectral amplitude at the fundamental frequency feature in the output psd of a single element for 5 different N values, in the subthreshold regime. Clearly, analogously to the Array Enhanced SR effect [10], increasing N leads to an enhancement in the spectral response of *individual* elements in the array but not for the summed output. These features are, currently, being investigated.

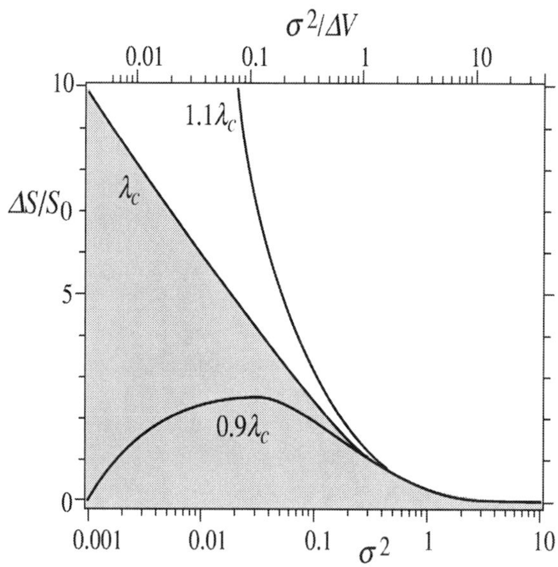

FIGURE 3. Spectral amplitude at frequency feature normalized to the spectral amplitude at zero frequency (see text), vs. noise intensity σ^2 (normalized by the energy barrier height on top scale), for same values of coupling λ as previous figure. $c = 3, \lambda_c = -0.4345$.

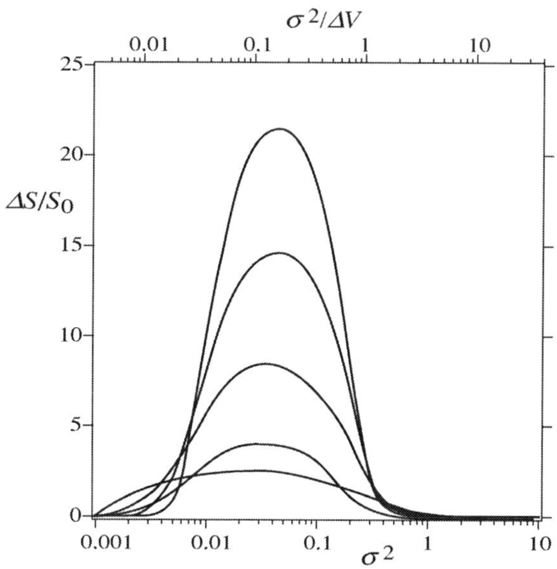

FIGURE 4. Same as figure 3 but showing the effects of increasing N; (bottom to top) $N = 3,5,7,9,11$. $c = 3, \lambda = 0.9\lambda_c = -0.39105$.

CONCLUDING REMARKS

We have presented preliminary results on the effects of Langevin noise in a unique system: N (odd) overdamped bistable elements coupled unidirectionally, with the coupling parameter controlling the onset of oscillatory behavior in the deterministic case. With noise, the system oscillates even in the regime where the deterministic system is quiescent; in this case, the "oscillations" are, more correctly, the noise-induced hopping events between the stable states of the potential energy function characterizing each element. The fact that the noise-induced hopping events lead to a "resonance" (as a function of the noise variance σ^2) in the spectral amplitude at the frequency feature, is reminiscent of the stochastic resonance effect. The modal hopping frequency also displays (for any value of the control parameter λ) a peak at, roughly, the Kramers rate for each individual element.

Clearly, the results shown here beg numerous questions, all of which cannot be answered in the framework of this paper. We will address this behavior in much greater detail in forthcoming publications.

ACKNOWLEDGMENTS

Support from the U.S. Office of Naval Research (Code 331) is gratefully acknowledged.

REFERENCES

1. V. In, A. Kho, J. Neff, A. Palacios, P. Longhini, B. Meadows; *Phys. Rev. Lett.* **91**, 244101 (2003).
2. M. Golubitsky, I.N. Stewart, and D.G. Schaeffer; "Singularities and Groups in Bifurcation Theory: Vol. II." in *Appl. Math. Sci.* **69**,(Springer-Verlag, New York, 1988). D.G. Aronson, M. Golubitsky, and M. Krupa; *Nonlinearity* **4**, 861 (1991). M. Golubitsky, I. Stewart; *Symmetry and Pattern Formation in Coupled Cell Networks*, (Springer-Verlag, New York 1999).
3. A.R. Bulsara, C. Seberino, L. Gammaitoni, M. Karlsson, B. Lundqvist, J. Robinson; *Phys. Rev.* **E67**, 016120 (2003).
4. V. In, A. Bulsara, A. Palacios, P. Longhini, A. Kho, J. Neff; *Phys. Rev.* **E68**, 045102(R) (2003). A. Bulsara, V. In, A. Kho, P. Longhini, A. Palacios, W-J. Rappel, J. Acebron, S. Baglio, B. Ando; *Phys. Rev.* **E70**, 036103 (2004).
5. V. In, V. Sacco, A. Kho, S. Baglio, B. Ando, A. Bulsara, A. Palacios; *IEEE Sensors*, submitted (2005).
6. B. Ando, S. Baglio, A. Bulsara, V. Sacco; in Proceedings of the 24th IEEE Instrumentation and Measurement Conference, Como, Italy Pg. 1419 (2004). B. Ando, S. Baglio, A. Bulsara, V. Sacco; *IEEE Sensors* (2005), in press. B. Ando, S. Baglio, A. Bulsara, V. Sacco; *Transaction on Instrumentation and Measurement* (2005) in press.
7. R. Koch, and J. Rozen, *Appl.Phys. Lett*, **78**, 1897 (2001).
8. T. Lukes, *Proc. Phys. Soc. London* **78**, 153 (1961); S. Lowen and M. Teich, *Phys. Rev.* **E47**, 992 (1993).
9. See e.g. L. Gammaitoni, P. Hänggi, P. Jung, and F. Marchesoni, *Rev. Mod. Phys.* **70**, 223 (1998) for a good review.
10. J. Lindner, B. Meadows, W. Ditto, M. Inchiosa, A. Bulsara; *Phys. Rev. Lett.* **75**, 3 (1995), *Phys. Rev.* **E53**, 2081 (1996).

Quantum Irreversibility And Noise In Mesoscopic Devices

F. Borondo*, D. A. Wisniacki† and R. M. Benito**

*Dep. de Química C–IX, Universidad Autónoma de Madrid, Cantoblanco–28049 Madrid, Spain.
†Departamento de Física "J. J. Giambiagi", FCEN, UBA, 1428 Buenos Aires, Argentina.
**Dep. de Física, ETSI Agrónomos, Universidad Politécnica de Madrid, 28040 Madrid, Spain.

Abstract. Irreversibility in quantum processes in the Bunimovich stadium and rectangular billiard in the presence of noise is studied. For this purpose, a novel method based on Loschmidt echo and quantum trajectories, as defined in the de Broglie–Bohm formulation, is used. Our results indicate that the dynamics along the diagonal of the billiard is most sensitive to noise when the wave packet (or alternatively the quantum trajectories) collide with the corners of the billiard.

Keywords: Quantum irreversibility, quantum chaos, mesoscopic systems, electron transport
PACS: 05.45.Mt, 73.21.-b, 73.23.Ad

INTRODUCTION

In the last years there has been a growing interest in the study of chaotic phenomena in mesoscopic systems [1]. Technical developments have made possible the manufacturing of micro and nanostructures that allow charge transport without loss of electron phase coherence [2]. For example, the ballistic transport of electrons in semiconductor heterostructures has been recently studied experimentally [3, 4]. Moreover, when the system is confined in all dimensions and the sample is sufficiently clean, the correlations in the energy spectrum against variations of parameters, such as the sample geometry or the intensity of an external field, are very similar to those found in quantum billiards [5], thus making these simple systems excellent models for this kind of systems.

The classical dynamics in billiards is often chaotic [6], and "quantum chaos" is an active field of research [7]. Most problems in quantum chaos are of theoretical nature, falling within the category of semiclassical theories [8]. One important topic in quantum chaos that has received recently a great deal of attention is irreversibility [9, 10, 11]. This concept has been be related to the theory of chaos, which is classically interpreted as the result of exponential separation of trajectories. However, in quantum mechanics, this sensitivity is meaningless due to unitarity. For this reason, Peres proposed to consider the sensitivity to perturbations in quantum systems, as a mean to investigate the instability of quantum motion [12]. Such quantity, called Loschmidt echo (LE), in reference to the famous Loschmidt–Boltzmann correspondence, is defined as

$$M(t) = |\langle\psi|\exp(\mathrm{i}\hat{H}t)\exp(-\mathrm{i}\hat{H}_0 t)|\psi\rangle|^2, \tag{1}$$

($\hbar$ is set equal to unity throughout this paper). It measures the ability of a system to return to an initial state, $|\psi\rangle$, after a forward evolution with a Hamiltonian, $\hat{H}_0$, followed by a (imperfect) reverse evolution with a perturbed Hamiltonian $\hat{H} = \hat{H}_0 + \Sigma$. Alternatively,

CP800, *Unsolved Problems of Noise and Fluctuations: UPoN 2005*
edited by L. Reggiani, C. Pennetta, V. Akimov, E. Alfinito, and M. Rosini

LE can be thought of as comparing the evolution of an initial state under different Hamiltonians (sensitivity to perturbations). For a given range of perturbation strengths, the LE decays exponentially at a rate given by the smallest quantity between the mean Lyapunov exponent and the level broadening following from the golden rule [9, 10].

Summarizing, two basic ingredients in the idea of LE are irreversibility and sensitivity to perturbations. However, the definition of LE's relies only on magnitudes evaluated at the end of the propagation process, an then is not able to provide any information about the involved history and the associated underlying physical mechanisms. (Actually, this can be a serious drawback as pointed out in [11].)

To avoid this problem, we propose here a new method to study irreversibility. It keeps the same basic philosophy of the original LE, but it is based in the use of quantum trajectories, as defined in the causal de Broglie–Bohm (BB) formulation of quantum mechanics [13]. This complementary quantum theory of motion [14] combines both the accuracy of the standard quantum description with intuitive explanations derived by a trajectory formalism, thus providing a powerful tool to understand the physics underlying microscopic phenomena (see for example Ref. [15]). This method has also an important additional advantage, since within its framework it is very easy to consider realistic perturbations, such as noise. This is particularly interesting in connection with billiards since, as stated before, they constitute ideal model for the transport of electrons in mesoscopic systems [7].

MODEL AND CALCULATIONS

The fundamental equations in the BB theory are derived by introducing the wave function in polar form, $\psi(\mathbf{r},t) = R(\mathbf{r},t)\ e^{iS(\mathbf{r},t)}$ into the time–dependent Schrödinger equation, obtaining the continuity and "quantum" Hamilton–Jacobi equations. The last one contains the so–called quantum potential which, together with V, determines the total forces acting on the system. Also, from it a quantum equation of motion can be defined: $m\dot{\mathbf{r}} = \nabla S$, from which quantum trajectories are obtained by numerical integration.

These orbits are used to define our measure of irreversibility. Starting from an ensemble of initial conditions reproducing the initial probability density distribution [16], we propagate them forward in time until a final value, t_f. We then propagate them backwards, introducing in the process a perturbation consisting of a kick given after every integration step. To avoid confusions, we will denote this "new" reversed time by τ. The effect of the kick consists of a displacement of the particle to a new position, randomly chosen within a given circle around the landing point. In this process we assume that the pilot wave function do not changes during the kick. Finally, the distance in configuration space, d, between both orbits is monitored as a function of τ. In this way, we compute a comparison, followed in time, of the unperturbed and perturbed dynamics of the system. From this, information about the mechanisms of irreversibility can be obtained. The idea behind this procedure is to mimic noise.

The systems that we have chosen to study are two degrees of freedom models consisting of a particle of mass 1/2 enclosed in a desymmetrized stadium billiard of radius $r = 1$ and area $1 + \pi/4$ with Dirichlet boundary conditions, and a rectangular billiard of side length $l = 1$ and the same area. The dynamics of the first system is known to be

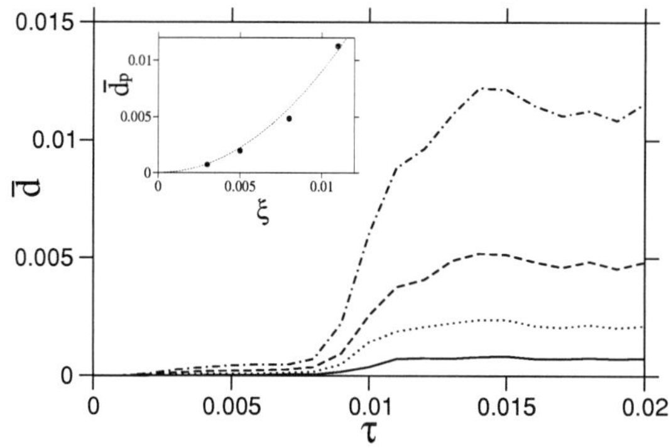

FIGURE 1. Averaged distance between forward and reversed quantum orbits as a function of the reverse time, τ, for different values of the perturbation strength parameter, ξ: 0.003 (full line), 0.005 (dotted line), 0.008 (dashed line), and 0.011 (dashed–dotted line). A value of $t_f = 0.02$ was used in the calculations. A fitting to a quadratic expression of the averaged distance at the plateaus (observed for $t > 0.015$) is shown in the inset.

classically ergodic and the second integrable.

RESULTS AND DISCUSSION

We first examine the dynamics of a wave packet running along the diagonal orbit going from the upper left (square) corner to the lower right (round) corner of the stadium at an energy value of E=2304, for which the period is $T = 0.0466$. As described in Ref. [17], the packet initially moves following the classical path with a slight dispersion. After the first rebound at the lower right corner ($t \simeq 0.009$), the packet spreads ($t \sim 0.019$), experiencing the well known defocalization effect described in [18]. Afterwards, the dispersed wave collides with the upper left corner, giving rise to a noticeable series of horizontal fringes ($t \simeq 0.028 - 0.047$). For subsequent times other rebounds take place, originating at $t \geq 0.066$ a complicated structure in the distribution of the quantum probability density.

In Fig. 1 we show the distance, $\overline{d}$, averaged over 20 quantum trajectories propagated, using a Gear stiff method with tolerance control, up to a final time of $t_f = 0.02$, for different values of the perturbation strength (parameterized by the kicking radius ξ). Notice that the origin of the reversed time, $\tau = 0$, corresponds to the final point of the forward propagation, t_f. As can be seen, the behavior of the four curves is the same. For the lowest times, $t \leq 0.008$, they grow very slowly and linearly. After that, the averaged distance increases dramatically in a cubic fashion for times up to the order of $t \sim 0.01$. And finally, for $0.015 < t < 0.020$, the values of $\overline{d}$ stabilize oscillating slightly around some sort of plateaus.

More interesting is the relationship existing between these results and the dynamics of the original packet [17]. The first part of the $\overline{d}$ vs. τ plots (showing a linear behavior) corresponds to a period of time in which the particle moves from the center of the stadium to just before the lower right corner. The packet here is in a semiclassical

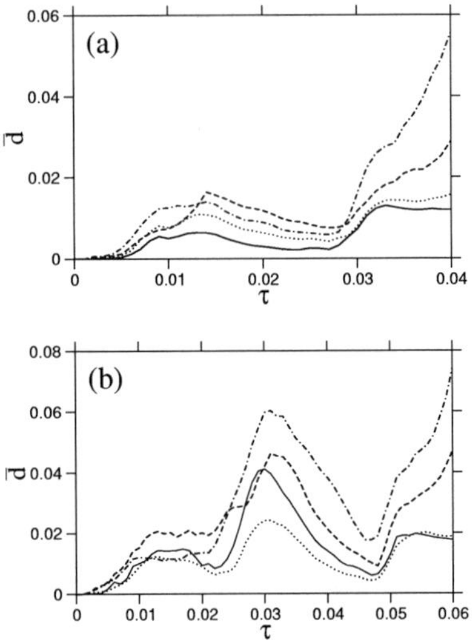

FIGURE 2. Averaged distance between forward and reversed quantum orbits as a function of the reverse time, τ, for the same values of the perturbation strength considered in Fig. 1 and values of $t_f = 0.04$ (a), and 0.06 (b).

regime, in which no much dispersion (irreversibility) is expected, in perfect agreement with our numerical findings. On the contrary, in the interval $\tau = 0.008 - 0.012$ the, now, perturbed dynamics include the bounce with the corner. Here a lot of interference of the packet with itself happens, and a great dispersion due to the perturbation takes place. This corresponds to the big, cubic growth observed in the computed values of $\overline{d}$. Also, for $\tau \sim 0.02$ we are at the echo, and then our results can be compared with those that would be obtained from the usual LE theory [9, 10]. This is done in the inset to the figure, where the functional form of $\overline{d}$ at the plateaus, $\overline{d}_p$, as a function of the magnitude of the perturbation, ξ, is shown. As it is seen, this dependence is quadratic with a very good accuracy, thus indicating that we are in a regime controlled by the Fermi golden rule [10]. Accordingly, we can conclude that noise–type perturbations like ours should be considered as generic from the point of view of the LE.

Let us discuss now what happens when longer final times, t_f, are considered, thus allowing a more complicated dynamics to enter into play. The results are shown in Fig. 2 for two values of this parameter. As can be seen similar results are obtained [17], i.e. big increases in $\overline{d}$ and then irreversibility, takes place only for those values of the time for which trajectories collide with the corners of the stadium, being there where the effect of the perturbation is stronger. Also, the plateaus in Fig. 2 are not completely flat, but rather they show a conspicuous decreasing behavior that, for example, in the case of Fig. 2(b) is quite important in the range $0.03 < \tau < 0.047$. The reason for this behavior can be understood if one considers that in these ranges of the reverse time, the packet is travelling from the upper left to the lower right corners, where it takes place a dynamics influenced by the self–focal point. This creates a quantum potential that forces the packet to return close to the original unperturbed path, which makes the separation

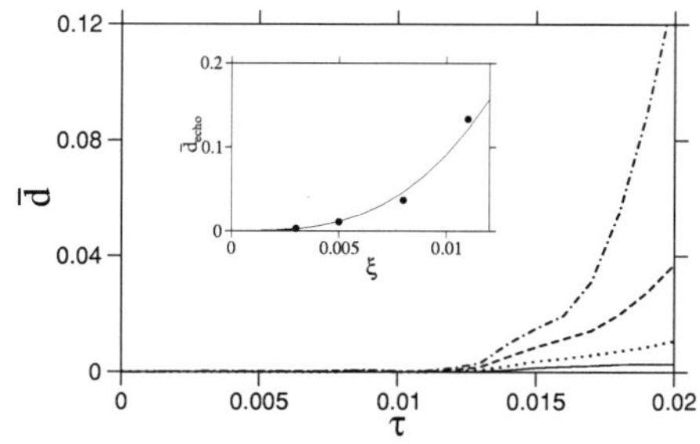

FIGURE 3. Same as Fig. 1 for a rectangular billiard of side length unity and area $1+\pi/4$. The inset shows a cubic fit of the values of $\bar{d}$ at the echo time, $\tau = 0.02$, with the perturbation parameter, ξ.

$\bar{d}$ to go down.

To conclude this section, let us compare the results obtained so far with those corresponding to the rectangular stadium, for which the dynamics is completely integrable, and then regular and non–chaotic. The results are shown in Figs. 3 and 4. As can be seen the values for the trajectories separation is always greater than in the case of the (chaotic) stadium billiard considered before. This is in agreement with the results obtained by Prosen [19] for the standard LE. Moreover, the calculations shown in Figs. 3 and 4 do not show the presence of any plateau, as opposed to what happens in the case of the stadium. Finally, the values of $\bar{d}$ at the echo time increase cubically (see inset to Fig. 3) withe the parameter ξ controlling the perturbation strength, again behaving differently that in the case of the stadium billiard.

CONCLUDING REMARKS

In this paper we have presented a novel method to study irreversibility in quantum processes. This method is similar in spirit to the LE introduced by Peres [12], but recast in terms of quantum trajectories, as defined in BB theory. In this way, useful information about the history and mechanisms involved in the perturbation process are obtained.

This method has been applied to the stadium billiard with noise, a model which is adequate for electrons transport in mesoscopic cavities. Our principal results are

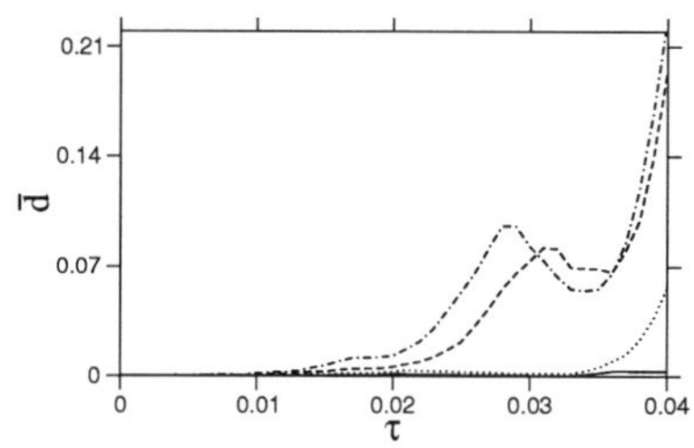

FIGURE 4. Same as Fig. 2(a) for a rectangular billiard of side length unity and area $1+\pi/4$.

summarized as following. First, the dynamics is sensitive to the perturbation mainly when the particle is bouncing at the corners, points in which the trajectories separate from each other cubically in time, on average. Second, the noise–type perturbation that has been used in the present work behaves in a totally generic way, as it is indicated by the fact that the Fermi golden rule regime is found. Third, comparison with the results obtained for a rectangular billiard, in which the dynamics is regular, shows that in this later case the growth of $\overline{d}$ is greater than for the classically chaotic stadium billiard, in agreement with previous results obtained for the usual LE.

Finally, we should remark that in our calculation times beyond the Ehrenfest time have not been considered. For these longer times, large interference effects in the pilot wave guiding the quantum trajectories, and then the complexity of the associated quantum potential, is much higher and widespread over all configuration space. This point is very interesting and deserves further investigation, which will imply an enormous computational effort if a reasonable average over initial conditions is to be maintained.

ACKNOWLEDGMENTS

We acknowledge support from MCyT and MECD (Spain) under contracts no. BQU2003–8212 and SAB2000–0340, and Fundación Antorchas (Argentina).

REFERENCES

1. K. Richter, *Semiclassical Theory of Mesoscopic Quantum Systems* (Springer–Verlag, Berlin, 2000); R. A. Jalabert, in *New Directions in Quantum Chaos*, edited by G. Casati, I. Guarneri, and U. Smylanski (IOS Press, Amsterdam, 2000).
2. C. M. Marcus, A. J. Rimberg, R. M. Westervelt, P. F. Hopkins, and A. C. Gossard, Phys. Rev. Lett. **69**, 506 (1992).
3. T. M. Fromhold, P. B. Wilkinson, F. W. Sheard, L. Eaves, J. Miao, and G. Edwards, Phys. Rev. Lett. **75**, 1142 (1995); P. B. Wilkinson, T. M. Fromhold, L. Eaves, F. W. Sheard, N. Miura, T. Takamasu, Nature **380**, 608 (1996).
4. Y. Takagaki and K. H. Ploog, Phys. Rev. E **62**, 4804 (2000).
5. B. D. Simons and B. L. Altshuler, Phys. Rev. Lett. **70**, 4063 (1993); E. J. Heller, M. F. Crommie, C. P. Lutz, and D. M. Eigler, Nature **369**, 464 (1994).
6. L. E. Reichl, *The Transition to Chaos* (Springer Verlag, New York, 2004).
7. H.–J. Stöckmann, *Quantum Chaos: An Introduction* (Cambridge U. Press, Cambridge, 1999).
8. M. Brack, R. K. Bhaduri, *Semiclassical Physics* (Perseus Books, Reading MA, 1997)
9. R. A. Jalabert and H. M. Pastawski, Phys. Rev. Lett. **86**, 2490 (2001).
10. Ph. Jacquod, P. G. Silvestrov, and C. W. J. Beenakker, Phys. Rev. E **64**, 055203(R) (2001).
11. M. Hiller, T. Kottos, D. Cohen, and T. Geisel, *Phys. Rev. Lett.* **92**, 010402 (2004).
12. A. Peres, Phys. Rev. A **30**, 1610 (1984).
13. D. Bohm, Phys. Rev. **85**, 166 and 194 (1952).
14. P. R. Holland, *The Quantum Theory of Motion* (Cambridge University Press, Cambridge, 1993).
15. C. L. Lopreore and R. E. Wyatt, Phys. Rev. Lett. **82**, 5190 (1999);
16. D. A. Wisniacki, F. Borondo, and R. M. Benito, Europhys. Lett. **64**, 441 (2003).
17. D. A. Wisniacki, F. Borondo, and R. M. Benito, Phys. Rev. A (submitted).
18. E. J. Heller, in *Chaos and Quantum Physics*, edited by M. J. Giannoni, A. Voros, and J. Zinn–Justin (Elsevier, Amsterdam, 1991).
19. M. Prosen, J. Phys. A **35**, 1455 (2002).

Suppression Of Shot-Noise In Quantum Cavities: Chaos Vs. Disorder

Ph. Jacquod* and Robert S. Whitney*

*Département de Physique Théorique, Université de Genève, CH-1211 Genève 4, Switzerland

Abstract. We investigate the behavior of the shot-noise power through quantum mechanical cavities in the semiclassical limit of small electronic wavelength. In the absence of impurity scattering, the Fano factor F, giving the noise to current ratio, was previously found to disappear as more and more classical, hence deterministic and noiseless transmission channels open up. We investigate the behavior of F as diffractive impurities are added inside the cavity. We find that F recovers its universal value provided (i) impurities cover the full cavity so that only a set of zero measure of classical trajectories may avoid them, and (ii) the impurity scattering rate exceeds the inverse dwell time through the cavity. If condition (i) is not satisfied, F saturates below its universal value, even in the limit of strong scattering. Our results corroborate the validity of the two-phase fluid model according to which the electronic flow splits into two well separated components, a classical deterministic fluid and a stochastic quantum-mechanical fluid. Only the latter carries shot-noise.

Keywords: mesoscopic transport, noise, quantum chaos, random matrix theory
PACS: 73.23.-b, 74.40.+k, 05.45.Mt

Time-resolved transport measurements through quantum mechanical systems invariably observe current fluctuations, even in the (experimentally unrealistic) situation of a noiseless measurement apparatus and at zero temperature. This intrinsically quantal noise is usually referred to as shot-noise. It results from the quantization of charge together with the statistical nature of quantum mechanical transport [1]. As but one of the consequences of the quantum-classical correspondence at large quantum numbers, it has been predicted that, shot-noise through a chaotic ballistic cavity disappears as the system becomes more and more classical, i.e. when the ratio of the electronic Fermi wavelength to the linear cavity size vanishes $\lambda_F/L \to 0$ [2]. The purpose of this article is to discuss when and how shot-noise starts to be reduced by an emergent classical, deterministic behavior.

Recent technological advances have made it possible to make electronic systems small and clean enough that the resulting electronic mean free path is larger than the size of the confining potential defining the device [3]. The electronic motion in these quantum dots is thus ballistic, and provided that their wavelength is short enough, the electrons have a dynamics strongly related to the dynamics that a classical particle would have. When this classical dynamics is chaotic, that is, when the shape of the dot differs significantly from a circle or an ellipse, the transport properties are usually universal and well-captured by the Random Matrix Theory (RMT) of transport [4]. The starting point of RMT is the scattering approach [5], which relates transport properties to the system's scattering matrix

$$\mathscr{S} = \begin{pmatrix} \mathbf{r} & \mathbf{t}' \\ \mathbf{t} & \mathbf{r}' \end{pmatrix}. \tag{1}$$

CP800, *Unsolved Problems of Noise and Fluctuations: UPoN 2005*
edited by L. Reggiani, C. Pennetta, V. Akimov, E. Alfinito, and M. Rosini

Here we consider a symmetric two terminal geometry (the cavity is connected to two external leads with equal number N of propagating channels) for which $\mathcal{S}$ is a 2-block by 2-block matrix, written in terms of $N \times N$ transmission ($\mathbf{t}$ and $\mathbf{t}'$) and reflection ($\mathbf{r}$ and $\mathbf{r}'$) matrices. From $\mathcal{S}$, the system's conductance is given by $g = \mathrm{Tr}(\mathbf{t}^\dagger \mathbf{t}) = \sum_n T_n$ (g is expressed in units of e^2/h and the T_n's are the N eigenvalues of $\mathbf{T} = \mathbf{t}^\dagger \mathbf{t}$). RMT provides a statistical theory of transport where $\mathcal{S}$ is assumed to be uniformly distributed over one of Dyson's circular ensemble of random matrices [6]. Transport properties can be calculated from this sole assumption. For instance, within RMT, and in the limit $N \gg 1$ the transmission eigenvalues have a probability distribution [4]

$$P_{\mathrm{RMT}}(T) = \frac{1}{\pi} \frac{1}{\sqrt{T(1-T)}} \tag{2}$$

for any $T \in [0,1]$. Note that classical particles would be either deterministically transmitted, $T = 1$ or reflected $T = 0$.

The distribution of transmission eigenvalues is all one needs to get fluctuations and higher moments of the current at low frequency. The zero-frequency shot-noise power in particular is given by $S = 2eV \sum_n T_n(1-T_n)$ [1]. According to (2), S is suppressed below its Poissonian value of $S_{\mathrm{p}} = 2e\langle I \rangle$ (V is the applied voltage and $\langle I \rangle$ the time-averaged current) by the Fano factor which reads

$$F = \frac{\sum_n T_n(1-T_n)}{\sum_n T_n} \quad ; \quad F_{\mathrm{RMT}} = \frac{1}{4}. \tag{3}$$

The RMT predictions (2) and (3) have been confirmed in various transport experiments and numerical simulations on open chaotic cavities [1, 4].

In closed chaotic systems, the semiclassical limit $\lambda_{\mathrm{F}}/L \equiv \hbar_{\mathrm{eff}} \to 0$ usually results in a better and better agreement with the Hamiltonian RMT of spectral fluctuations [7]. One may thus expect that the same applies to transport in open systems. That is not so, as illustrated in Fig. 1. The numerical data presented there (see also Ref. [8]) show that instead of the RMT prediction of Eq. (2), the transmission eigenvalues appear to be distributed according to

$$P_\alpha(T) = \alpha P_{\mathrm{RMT}}(T) + \frac{1-\alpha}{2}\left[\delta(T) + \delta(1-T)\right], \tag{4}$$

with an increasingly deterministic behavior $\alpha \to 0$ as $\hbar_{\mathrm{eff}} \to 0$ (all classical parameters being fixed). The presence of δ-peaks at $T = 0$ and $T = 1$ in $P_\alpha(T)$ becomes evident once the integrated distribution $I(T) = \int_0^T P(T')dT'$ is plotted. One has

$$I_\alpha(T) = \frac{2\alpha}{\pi} \sin^{-1}\sqrt{T} + \frac{1-\alpha}{2}(1+\delta_{1,T}), \tag{5}$$

so that $I_\alpha(0) = (1-\alpha)/2$ vanishes only for $\alpha = 1$. It turns out that the parameter α is well approximated by $\alpha \approx \exp(-\tau_e/\tau_d)$, in term of the new time scale $\tau_e = -\lambda^{-1}\ln[\hbar_{\mathrm{eff}}\tau_d^2]$ and the average dwell time τ_d through the cavity [8]. In short, for a classically fixed configuration (i.e. considering an ensemble of systems with fixed

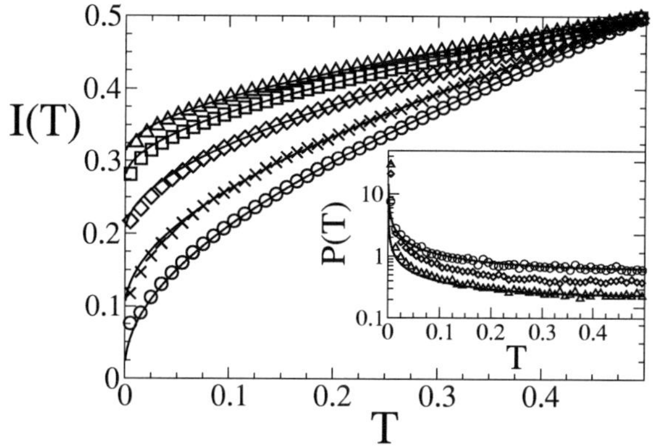

FIGURE 1. Integrated probability distribution $I(T)$ of transmission eigenvalues for $\tau_D = 25$, $\hbar_{\rm eff}^{-1} = 2048$ and $\tau_E \simeq 0$ (circles; distribution calculated over 729 different samples); $\tau_D = 5$, $\hbar_{\rm eff}^{-1} = 128$ and $\tau_E = 0.16$ ($\times$; 1681 samples), $\hbar_{\rm eff}^{-1} = 1024$ and $\tau_E = 1.5$ (diamonds; 729 samples), $\hbar_{\rm eff}^{-1} = 8192$ $\tau_E = 2.8$ (squares; 16 samples), $\hbar_{\rm eff}^{-1} = 65536$ and $\tau_E = 4.1$ (triangles; 2 samples). The solid curves give the distribution I_α of Eq. (5), with $\alpha \approx 0.98, 0.81, 0.6, 0.45$, and 0.385 (from bottom to top). Inset: Probability distribution $P(T)$ of transmission eigenvalues for the same set of parameters as in the main panel (squares and $\times$ have been removed for better visibility). The solid curves give the universal distribution $P_{\rm RMT}$ of Eq. (2) and that of Eq. (4), with $\alpha = 0.39$ (from top to bottom). Note that $P(T)$ is symmetric around $T = 0.5$.

λ and τ_d), the fraction $(1-\alpha)/2$ of deterministic transmission eigenvalues $T = 0, 1$ increases as one goes deeper and deeper into the semiclassical limit, $\hbar_{\rm eff} \to 0$. The rate of the crossover is set by a partially quantum, partially classical time scale, the *Ehrenfest time* τ_e [9]. Compared to closed systems, the emergence of a finite τ_e has more profound an impact on transport properties once it becomes comparable to the dwell time $\tau_d \propto (N\hbar_{\rm eff})^{-1}$.

Inserting (4) into (3) with the numerically extracted value $\alpha \approx \exp(-\tau_e/\tau_d)$, one directly recovers $F \propto \exp[-\tau_e/\tau_d]$, in agreement with the analytical prediction of Refs. [10, 11], the experimental results of Ref. [12] and the numerical data of Ref. [13]. They can be qualitatively understood by first realizing that complex quantum systems split into the two classes of quantum chaotic and quantum disordered systems [14]. Ballistic cavities in particular belong to the first class, for which electronic wavepackets are carried along very few classical paths until the time τ_e after which they have a finite probability to be found on trajectories that their center of mass would not follow classically. This dynamical diffraction process restores quantum mechanical stochasticity for larger times, however it does not affect short trajectories with $\tau < \tau_e$. Thus, in quantum chaotic systems, two classes of classical trajectories emerge, depending on their dwell time through the cavity. Short trajectories with $\tau < \tau_e$ are able to carry an electronic wavepacket deterministically through the cavity (i.e. with transmission probability $T = 0$ or 1). If the electronic wavepacket sits on longer trajectories with $\tau > \tau_e$ on the other hand, diffraction splits it into pieces before its exit, and quantum mechanical stochasticity (with $T \in]0, 1[$) prevails. The fraction of scattering trajectories in the

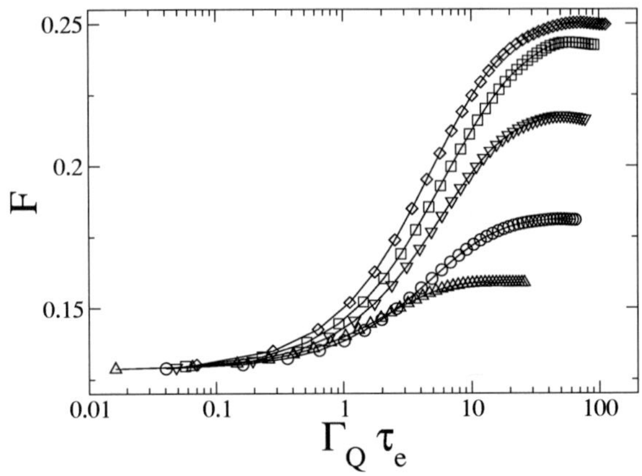

FIGURE 2. Crossover from quantum chaotic behavior (left) to quantum disordered behavior (right) of the Fano factor F as a function of $\Gamma_Q \tau_e$. Each set of data corresponds to an average over four different samples with $\hbar_{\rm eff}^{-1} = 8192$, $K = 7.65$, $\tau_D = 5$ and $\xi = 1250$ (upward triangles), 2500 (circles), 3750 (downward triangles), 5000 (squares) and 7500 (diamonds).

stochastic subset is obtained via the dwell time distribution $\rho(\tau)$ by

$$\alpha \equiv \int_{\tau_e}^{\infty} \rho(\tau) \mathrm{d}\tau. \tag{6}$$

In a chaotic system one has $\rho(\tau) = \tau_d^{-1} \exp[-\tau/\tau_d]$, hence the fraction of stochastically transmitted channels gives $\alpha \approx \exp[-\tau_e/\tau_d]$ for the weight α of Eq. (4), in agreement with the numerics shown in Fig. 1. This is the essence of the *two-phase-fluid model* [8], originally proposed in Ref. [11] and given a microscopic foundation in Ref. [16]

Fig. 1 provides us with a direct evidence for the validity of the two-phase fluid. Other evidences of this kind have been found in investigations of the excitation spectrum of Andreev billiards, i.e. ballistic cavities in contact with a superconductor [15]. In a previous work [16], we used an approach based on a semiclassical expansion for the Green's function in term of a sum over classical trajectories. Together with the construction of a quantum-mechanical phase-space basis, this allowed us to import classical concepts such as Liouville conservation and determinism into quantum mechanics, thereby providing with a microscopic foundation for the two-phase fluid model [17]. Our purpose in the reminder of this article is to check numerically the model by investigating its quantum chaos – quantum disorder crossover.

To this end, we use the open kicked rotator model and follow the procedure described e.g. in Ref. [13]. So far, the model has been implemented only in its quantum chaotic version. Here we add a diffraction term to it so that its Floquet (time-evolution) operator has matrix elements

$$\begin{aligned} U_{m,m'} &= \hbar_{\rm eff}^{1/2} \, e^{-(iK/4\pi\hbar_{\rm eff})[\cos(2\pi m\hbar_{\rm eff})+\cos(2\pi m'\hbar_{\rm eff})]} \, e^{-(i/4\pi\hbar_{\rm eff})[\eta(m)+\eta(m')]} \\ &\quad \times \sum_l e^{2\pi i l(m-m')\hbar_{\rm eff}} e^{-(\pi i\hbar_{\rm eff}/2)l^2}, \end{aligned} \tag{7}$$

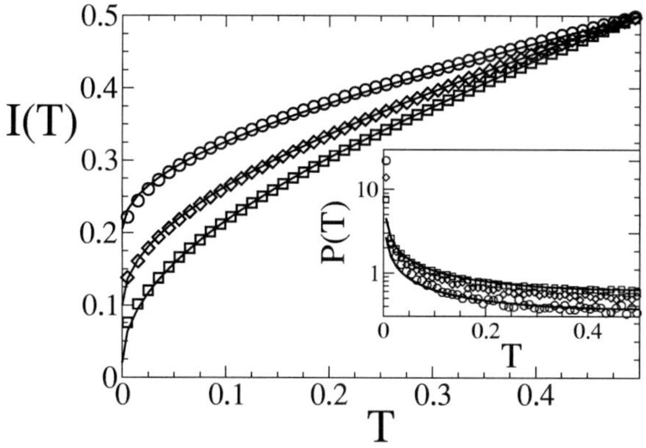

FIGURE 3. Integrated probability distribution $I(T)$ of transmission eigenvalues for the open kicked rotator model with diffractive disorder of Eq. (7). The data have been obtained from 16 different samples with parameters $\hbar_{\rm eff}^{-1} = 8192$, $K = 7.65$ $\tau_D = 5$ and $u = 0.4$, with $\xi = 0$ (circles), 2500 (diamonds), and 7500 (squares). The solid curves give the distribution I_α of Eq. (5), with $\alpha \approx 0.99$, 0.76, and 0.52 from bottom to top. Inset: Probability distribution $P(T)$ of transmission eigenvalues for the same set of parameters as in the main panel. The solid curves give the universal distribution $P_{\rm RMT}$ of Eq. (2) and that of Eq. (4), with $\alpha = 0.59$ respectively.

with a randomly distributed function $\langle \eta(m)\eta(m')\rangle = u^2\, \delta_{m,m'}\, \Theta(|m-m_0|-\xi/2)$. Such a point-like diffractive impurity potential corresponds to the extreme quantum disorder limit. The length scale ξ allows to cover all or only a part of the system, and u sets the strength of the impurity scattering. We extracted the scattering rate Γ_Q from the width of the Local Spectral Density of States (LDOS) induced by $\eta(m)$ for the closed kicked rotator. As is commonly the case [18], we found that the LDOS has a Lorentzian shape with a width $\Gamma_Q \approx 0.016\, u^2 \sqrt{\xi}/\hbar_{\rm eff}$ in a wide range of parameter. One expects that, as Γ_Q becomes comparable to the Ehrenfest time, quantum diffraction effects start to eat away the determinism of short trajectories.

We first show in Fig. 2 the behavior of the Fano factor as the diffractive scattering rate is cranked up, all other parameter being fixed. For $\Gamma_Q = 0$, $F < 0.25$ lies below its universal value, indicating a finite τ_e, thus a finite fraction of non-diffractive scattering orbits. This fraction gets reduced once Γ_Q increases, and accordingly $\partial F/\partial \Gamma_Q > 0$. The most striking feature of Fig. 2, however, is that as long as the diffractive disorder does not cover the whole system volume, i.e. for $\xi < \hbar_{\rm eff}^{-1}(1-\tau_d^{-1})$, F saturates below its universal value, even in the limit $\Gamma_Q \tau_e \to \infty$. This reflects the fact that some short trajectories are able to avoid the diffractive potential and thus remain deterministic. The existence of two separated phase-space fluids allows only that part of the deterministic fluid which directly scatters off the impurities to become stochastic. It is not clear from our numerics if, for $\xi > \hbar_{\rm eff}^{-1}(1-\tau_d^{-1})$, one has an exponential, or a power-law behavior of the Fano factor as predicted in Ref. [12, 19]. More detailed investigations are necessary to draw definite conclusions.

We finally show on Fig. 3 the behavior of the distribution of transmission eigenvalues in the saturated regime $\Gamma_Q \tau_e \gg 1$. Clearly, the distribution and integrated distribution follow Eqs. (4) and (5). In contrast to the quantum chaotic case, we qualitatively found a dependence $\alpha \propto \exp[-\tau_e/\tau_d]\xi\hbar_{\rm eff}/(1-\tau_d^{-1})$, again reflecting a reduction of that part of the deterministic component which directly touches the diffractive potential.

All these findings support the two-phase fluid hypothesis, that is, the splitting of the cavity into a stochastic and a deterministic cavity. The latter being noiseless, shot-noise is suppressed by a factor reflecting its phase-space measure relative to the total phase-space. This measure is reduced by the presence of diffractive disorder, however, only homogeneously spread impurities are able to diffract all trajectories, thus only in this case does one recover universality at large diffractive scattering rate. This complement recent investigations of shot-noise with homogeneously spread diffractive disorder in regular cavities [20].

ACKNOWLEDGMENTS

This work has been supported by the Swiss National Science Foundation.

REFERENCES

1. For a review on shot-noise, see : Ya.M. Blanter and M. Büttiker, Phys. Rep. **336**, 1 (2000).
2. C.W.J. Beenakker and H. van Houten, Phys. Rev. B **43**, R12066 (1991).
3. L.P. Kouwenhoven, C.M. Marcus, P.L. McEuen, S. Tarucha, R.M. Westervelt, and N.S. Wingreen, *Electron Transport in Quantum Dots*, Nato ASI conference proceedings, L.P. Kouwenhoven, G. Schön, and L.L. Sohn Eds. (Kluwer, Dordrecht, 1997); Y. Alhassid, Rev. Mod. Phys. **72**, 895 (2000).
4. C.W.J. Beenakker, Rev. Mod. Phys. **69**, 731 (1997).
5. R. Landauer, Phil. Mag. **21**, 863 (1970); M. Büttiker, Phys. Rev. Lett. **57**, 1761 (1986).
6. M. L. Mehta, *Random Matrices* (acadamic, New York, 1991).
7. F. Haake, *Quantum Signatures of Chaos*,(Springer, Berlin, 2000).
8. Ph. Jacquod, and E.V. Sukhorukov, Phys. Rev Lett. **92** 116801 (2004).
9. G.M. Zaslavsky, Phys. Rep. **80**, 157 (1981); M.G. Vavilov and A.I. Larkin, Phys. Rev. B **67**, 115335 (2003).
10. O. Agam, I. Aleiner and A. Larkin, Phys. Rev. Lett. **85**, 3153 (2000).
11. P.G. Silvestrov, M.C. Goorden, and C.W.J. Beenakker, Phys. Rev. B **67**, 241301(R) (2003).
12. S. Oberholzer, E.V. Sukhorukov, and C. Schönenberger, Nature **415**, 765 (2002).
13. J. Tworzydlo, A. Tajic, H. Schomerus, and C.W.J. Beenakker, Phys. Rev. B **68**, 115313 (2003).
14. I.L Aleiner and A.I. Larkin, Phys. Rev. B **54**, 14423 (1996).
15. M.C. Goorden, Ph. Jacquod, and C.W.J. Beenakker, cond-mat/0505206.
16. R.S. Whitney, and Ph. Jacquod, Phys. Rev. Lett. **94**, 116801 (2005).
17. We note that the prediction of the two-phase fluid model that coherent effects such as weak localization do not depend on τ_e have recently been criticized; see S. Rahav and P.W. Brouwer, cond-mat/0505250. It is not our purpose here to discuss this point in detail. Instead we will present elsewhere a semiclassical calculation of weak localization contradicting the results of Rahav and Brouwer and supporting the survival of the weak localization corrections to conductance at their universal value independently of τ_e; Ph. Jacquod and R.S. Whitney, in preparation (2005).
18. Ph. Jacquod and D.L. Shepelyansky, Phys. Rev. Lett. **75**, 3501 (1995).
19. E.V. Sukhorukhov and O.M. Bulashenko, Phys. Rev. Lett. **94**, 116803 (2005).
20. F. Aignier, S. Rotter, and J. Burgdörfer, cond-mat/0502417.

Chaos And Noise-driven Emergence Of Order From Disorder

Maide Bucolo[1], Riccardo Caponetto[1], Luigi Fortuna[1], Mattia Frasca[1], Alessandro Rizzo[2]

[1]*Dipartimento di Ingegneria Elettrica Elettronica e dei Sistemi*
Università degli Studi di Catania, viale A. Doria 6, 95125, Catania, Italy
E-mail: mfrasca@diees.unict.it

[2]*Dipartimento di Elettrotecnica ed Elettronica*
Politecnico di Bari, Via Re David, 200 - 70125 Bari, Italy

Abstract. This paper investigates several case studies in which noise signals or random sequences play an important role in the emergence of a self-organized behaviour or a paradoxal behaviour. We considered the effects induced by using either a chaotic signal instead of noise or a chaotic sequence instead of a random one. We show that in many cases noise and chaos have similar effects and in some cases chaos work better than noise.

Keywords: Chaos, Noise, Nonlinear Dynamics, Game Theory, Coupled nonlinear systems.
PACS: 05.45.Ac; 43.50.+y; 05.45.-a

INTRODUCTION

Deterministic chaos and random signals share the property of long term unpredictable irregular behaviour and broad band spectrum. The question if chaos and noise can be distinguished in real data time series is widely investigated and still does not have a definitive answer. Another important question is related to those phenomena and algorithms in which the presence of noise helps the emergence of order from disorder: are the effects of a chaotic signal comparable to that of a noise source? For instance, let us consider the well-known example of order from disorder arising in array of locally coupled pendulums [1]. In this case a moderate increase in the spatial disorder of the array can lead to significant improvements in the synchronization process. This spatial disorder can be driven by either a chaotic or stochastic signal. Another important open question is if in these cases does chaos work better than noise. In this work we investigate several case studies, in which the presence of noise is fundamental, with the aim of answering this question. In particular, different applications, self-organization in arrays of locally coupled systems in which a chaotic dissymmetry is present, chaos-driven optimisation strategies in game theory and noise stabilization in nonlinear circuits are treated. This means to investigate the effect of the introduction of either deterministic chaotic or random sequences in different types of phenomena. In many of our experiments we observed

CP800, *Unsolved Problems of Noise and Fluctuations: UPoN 2005*
edited by L. Reggiani, C. Pennetta, V. Akimov, E. Alfinito, and M. Rosini

either similar effects of introducing chaos instead of noise or several benefits derived by the use of chaos instead of noise.

SELF-ORGANISATION IN ARRAYS OF DYNAMICAL SYSTEMS

In this section the topics related to the role of chaotic induced spatial diversity versus random one are introduced in relation to the self-organising behaviour of arrays of dynamical systems. In particular, this section deals with an example concerning arrays of Josephson junctions. It has already been shown [6] that spatial diversity helps the attitude to self-synchronisation of the junctions: if an array of identical junctions is considered, in-phase periodic orbits in fact exist for a certain range of the parameter set, but they are not asymptotically stable. On the other hand, Braiman et al. [1-2] also showed that a moderate increase in the spatial disorder of the array can lead to significant improvements in the synchronization process. The results obtained parallel those obtained by the same authors for mechanical systems [2], such as arrays of damped, forced pendula, in which a moderate, random, spatial diversity introduced in the length of the pendula enhances the synchronisation capability. In particular, the same array of pendula considered in [2] has been synchronised by introducing a moderate amount of diversity generated by chaotic systems.

An array of non identical, locally connected chaotic Josephson junctions is taken into account. It is known that, under particular conditions, a single junction can exhibit chaotic behaviour. In this work the chaotic behaviour of the single junction is achieved by feeding the junction with a periodic driving bias. The global behaviour of the array is investigated in the case of identical junctions, where spatio-temporal chaos emerges, and in presence of spatial diversity, generated by either random or chaotic law. In the latter case, it has been confirmed that a certain amount of disorder enhances the self-organization capability. In particular, in agreement with the results presented in [3-4], the introduction of diversity generated by a chaotic law leads to a further improvement in synchronization.

From a mathematical point of view the array of Josephson junctions [6] can be described by the following equations:

$$\ddot{\phi}_j + a_j\dot{\phi}_j + \sin\phi_j = I_0 + k(\phi_{j+1} - 2\phi_j + \phi_{j-1}) + I \cdot \sin(\omega t) \quad j = 1..N \tag{1}$$

where ϕ is the phase difference between the two quantum mechanical wave functions of the two layers of the junction (see Appendix), a_j is a parameter related to the physical properties of the junction, the term $k(\phi_{j+1} - 2\phi_j + \phi_{j-1})$ represents the coupling with the neighbouring junctions and $I_0 + I \cdot \sin(\omega t)$ is the current bias.

Under these assumptions, when the array consists of identical junctions ($a_j = a \ \forall j$), as time elapses, more and more disordered spatio-temporal patterns emerge, denoting a chaotic behaviour. This can be noticed in Fig. 1(a), where a colour map codes the variable characterising the behaviour of each junction in terms of the variable ϕ versus time. The prevalence of blue in the colour map is due to the d.c. term in the forcing torque. The evolution appears disordered, and is in particular chaotic. By introducing a random, symmetrical disorder in the a_j (in a range of 10-20% of the nominal value), periodic spatio-temporal patterns can be observed (Fig. 1(b)). Our analysis deals now with the effects induced by the introduction of deterministic, non-organised dissymmetry, like the ones generated by a chaotic attractor. In this experiment, a portion of a Chua's attractor is sampled and adequately scaled in order to superimpose a deterministic disorder in the junction parameters. The result of simulation shows (Fig. 1(c)) that a chaotic variation on junction parameters leads the array towards a collective organisation. Junctions in the central region of the array are synchronised both in space and time, oscillating with the same frequency of the forcing current, while in the external bands a regular spatial wave propagates. Therefore, spatio-temporal chaos disappeared, leading to a periodic behaviour by simply perturbing the symmetry of the system.

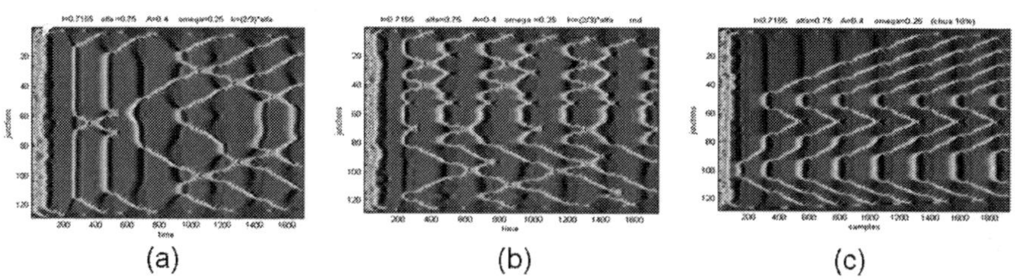

FIGURE 1. (a) Chaotic behaviour of identical Josephson junctions; (b) Self-synchronisation by random dissymmetry; (c) Self-synchronisation by non-organised deterministic dissymmetry.

CHAOTIC SEQUENCES IN GAME THEORY

Should chaos play a central role also in game theory? This question is investigated by considering the Parrondo's Paradox. This paradox has been introduced by Parrondo as a pedagogical illustration of the Brownian ratchet. It states that a resulting winning game can be obtained by playing in a random or periodic fashion two games which are separately losing [7,8].

The two losing games are called Game A and Game B. Game A consists of a coin having p as winning probability ($p < \frac{1}{2}$ is chosen to obtain a losing game). Game B consists of two alternatively played coins: if the present played capital is multiple of a given integer M then the coin to be played will have a winning probability $p1$, otherwise (the present capital is not multiple of M) the coin to be played has a winning probability $p2$. The two games, when played independently from each other, are losing games if the following conditions hold:

$$\frac{1-p}{p} > 1 \tag{3}$$

for Game A, and

$$\frac{(1-p_1)(1-p_2)^{M-1}}{p_1 p_2^{M-1}} > 1 \qquad (4)$$

for Game B. The paradox consists in constructing a game in which Game A and Game B are alternatively played with a probability γ: this composite game is winning.

Parrondo showed that the paradox occurs if the following condition holds:

$$\frac{(1-q_1)(1-q_2)^{M-1}}{q_1 q_2^{M-1}} > 1 \qquad (5)$$

where $q_1 = \gamma p + (1-\gamma)p_1$ and $q_2 = \gamma p + (1-\gamma)p_2$.

While in the original Parrondo's Paradox the strategy is based on a random choice of one of the two games with the probability γ, in our work this choice is based on the value given by a chaotic sequence. Many chaotic sequence generators have been considered. Fig. 2 illustrates the trend of the capital gained by playing the two games separately, either alternated by a random law or by chaotic laws. It is evident as performance is much increased when a chaotic law is adopted.

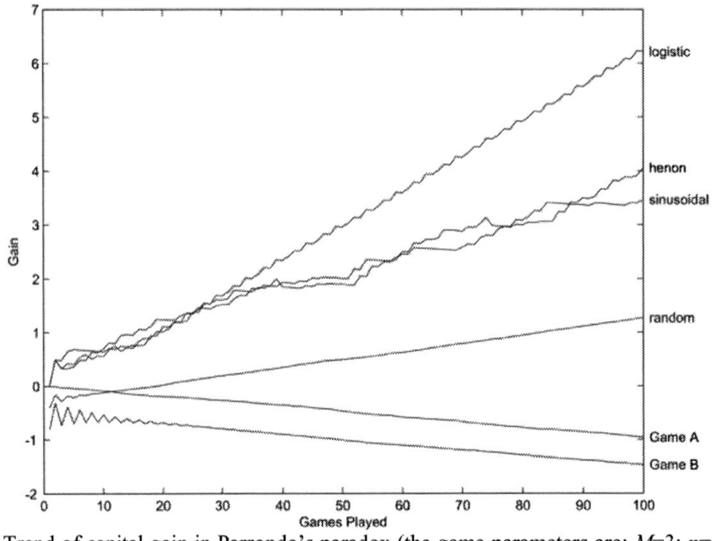

FIGURE 2. Trend of capital gain in Parrondo's paradox (the game parameters are: M=3; p=1/2-0.005; p_1=1/10-0.005; p_2=3/4-0.005).

NOISE AND CHAOS-DRIVEN STABILIZATION IN NONLINEAR CIRCUITS

The third case study discussed in this work deals with an experiment illustrating the role played by noise and chaos in nonlinear circuits. In a recent paper [5], Cooke et al. have shown in an electronic circuit an interesting phenomenon, called noise stabilization. They showed a nonlinear dynamical system that is unstable when it is noise-free. This circuit becomes stable when noise is applied.

In particular, the equations implemented by the circuit are the following:

$$\dot{x} = (y-1)x + \sigma\xi(t)$$

$$\dot{y} = 0.24y^{\frac{3}{2}} - x^2 y$$

$$y > 0$$

This system, when noise-free (i.e. $\sigma=0$), has two unstable spirals associated to exploding oscillatory motions, which are separated by a line of symmetry (the y-axis). When a suitable amount of noise is added, the system switches from one half-plane to the other and the system exhibits bounded behavior which resembles (but it very different to) low order chaotic behavior.

The question investigated in our work is whether chaos and noise have the same effects on this system or not. In particular, the experiment we carried out was to compare the effects of adding noise generated by the PC audio output and adding a chaotic signal generated by the well-known Chua's circuit. Before performing the experiment, the Fast Fourier Transform of the two signals was evaluated and the chaotic circuit parameters were adjusted so that the bandwidths of the two signals are comparable. Moreover, the output of the Chua's circuit (one of its state variables) was preprocessed so that the amplitude of this signal can be manually adjusted.

It was found that the chaotic signal stabilizes the circuit and the "attractor" shown in the phase plane x-y is very similar to that obtained by using noise (as shown in Fig. 3). The experiment allows us to conclude that the two signals (chaos and noise) have similar effects on this circuit.

- Noise (generated by the PC audio output)
- Chaos (generated by a Chua's circuit)

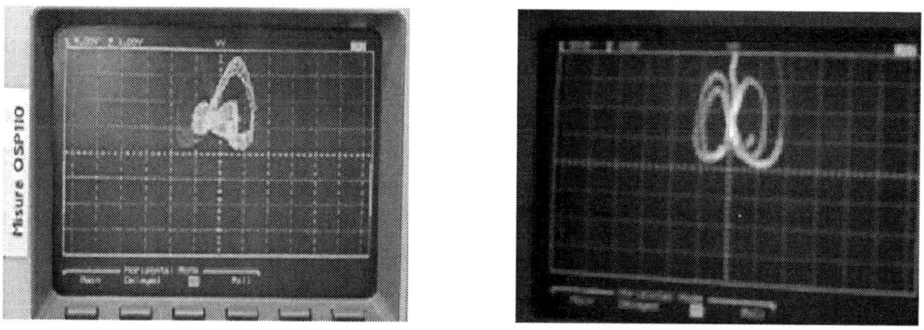

FIGURE 3. Stabilization of a nonlinear circuit by using either chaos or noise.

CONCLUSIONS

In this paper we investigated several case studies in which deterministic chaos is used instead of noise with the aim of answering the question if chaos and noise have the same effects and if there are cases in which chaos work better than noise. The examples reported in this work belong to very different fields. The first example deals with self-organization in arrays of coupled Josephson junctions, showing that chaotic dissymmetries lead to a self-organized global behavior. The second example deals with game theory and shows that chaotic sequences are more efficient than random ones in the generation of a winning game from two losing games (the so-called Parrondo's paradox). Finally, the third example deals with noise stabilization in nonlinear circuit and shows that chaos may be used to stabilize an unstable nonlinear circuit and has similar effects to the application of additive noise.

More than a real conclusion, these experiments leads to many open questions and unsolved problems which may be investigated in the future. When chaos work better than noise and what are the reasons of this difference are some of the interesting questions to be answered.

ACKNOWLEDGMENTS

This work was partially supported by the Italian "Ministero dell'Istruzione, dell'Università e della Ricerca" (MIUR) under the project FIRB RBNE01CW3M.

REFERENCES

1. Y. Braiman, J. F. Lidner, and W. L. Ditto, "Taming spatiotemporal chaos with disorder", *Nature*, Vol. 378, 30 November, pp. 465-467, 1995.
2. Y. Braiman, W. L. Ditto, K. Wiesenfeld, M. L. Spano, "Disorder-enhanced synchronization", Physics Letters A, vol. 206, pp. 54-60, 1995.
3. P. Arena, R. Caponetto, L. Fortuna, A. Rizzo, "Nonorganized deterministic dissymmetries induce regularity in spatiotemporal dynamics", *International Journal of Bifucation and Chaos*, vol. 10, No. 1, pp. 73-85, 2000.
4. M. Bucolo, R. Caponetto, L. Fortuna, M. Frasca, A. Rizzo, "Does chaos works better than noise?", *CAS Magazine*, vol. 2 no. 3, pp. 4-19, 2002.
5. W.E.Cooke, A.S.Richardson, E.R.Tracy, W.Yang, J.M.Finn, "Noise stabilization in nonlinear circuits", 8th Experimental Chaos Conference, AIP Conference Proceedings 742, pp.63-68.
6. K. Y. Tsang, R. R. Mirollo, S. H. Strogatz, K. Wiesenfeld, "Dynamics of a Globally Coupled Oscillator Array", *Physica D*, Vol. 48, pp.102-112, 1991.
7. G. P. Harmer and D. Abbott, "Parrondo's Paradox", *Statistical Science* Vol. 14, pp. 206-213, 1999.
8. G. P. Harmer and D. Abbott, "Losing strategies can win by Parrondo's paradox", *Nature*, Vol. 402, p. 864, 1999.

Divergence Of The Chaotic Layer Width And Acceleration Of The Chaotic Transport

Stanislav M. Soskin*,†, Oleg M. Yevtushenko† and Riccardo Mannella**

*Institute of Semiconductor Physics, 03028 Kiev, Ukraine, and Department of Physics, Lancaster University, Lancaster LA1 4YB, UK
†Abdus Salam ICTP, 34100 Trieste, Italy
**Dipartimento di Fisica, Università di Pisa and INFM UdR Pisa, 56100 Pisa, Italy

Abstract. We show that, despite conventional belief, a *weak* perturbation in a Hamiltonian system may lead to a *wide* chaotic layer and *fast* chaotic transport. This occurs in a spacially periodic Hamiltonian system subject to a dipole-type time-periodic perturbation with a *small* frequency. We explain this and develop an explicit theory for the layer width, in nice agreement with simulations.

The challenging unsolved problems relate to the chaotic transport in the system and to numerous applications, in particular to the noise-induced diffusion of particles on crystalline surfaces.

Keywords: Chaos, diffusion, transport
PACS: 05.45.Ac, 05.45.Pq, 75.70.cn

INTRODUCTION

The behaviour of chaotic systems is similar in some respects to the behaviour of noise-driven systems, in particular transport in chaotic systems is diffusion-like on coarse time scales. An important class of chaotic systems is the class of integrable Hamiltonian systems subject to a weak perturbation. The magnitude of the perturbation is analogous to a noise intensity in noise-driven systems. If the noise intensity is small, then the diffusion in the noise-driven systems is slow. It is natural to assume that, by analogy, if the perturbation magnitude in Hamiltonian systems is small, then the diffusion rate should necessarily be small too. Such natural assumption seemed to be confirmed by numerous examples: to the best of our knowledge, in *all* former works considering cases with a weak perturbation, the chaotic transport was *slow* (see e.g. the textbooks [1, 2, 3]).

However, as we show in [4], the situation may be drastically different in a rather general and important class of systems, namely in periodic Hamiltonian systems subject to a dipole-type time-periodic perturbation (such systems are wide-spread in nature): if the perturbation frequency goes to zero, then both the width of the chaotic layer and the speed of chaotic transport within the layer *diverge* (cf. Figs. 1-2).

RESULTS

To demonstrate the divergence of the chaotic layer width, we choose an archetypal example [2]: a pendulum driven by a coordinate-independent (often called dipole [2, 5])

CP800, *Unsolved Problems of Noise and Fluctuations: UPoN 2005*
edited by L. Reggiani, C. Pennetta, V. Akimov, E. Alfinito, and M. Rosini

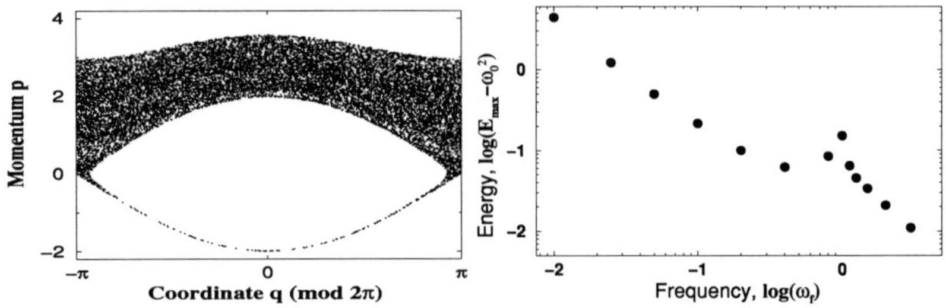

FIGURE 1. (left) The trajectory in the stroboscopic (for $t = n2\pi/\omega_f$ with $n = 0, 1, 2, ...$) Poincaré section for the system (1) with $\omega_0 = 1, h = 0.01, \omega_f = 0.01$ and the initial conditions $(p(0) = 0, q(0) = \pi)$. The number of points is 20000. (right) Spectral dependence of the maximum E_{max} of energy $E \equiv p^2/2 + U_0(q)$ in the chaotic layer, $\omega_0 = 1, h = 0.01$.

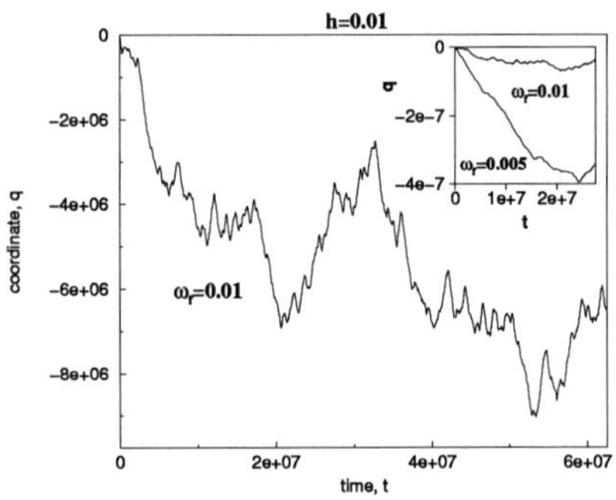

FIGURE 2. The dependence $q(t)$ on long time-scales, for the same system as in Fig. 1, left. The inset compares its initial part with that for $\omega_f = 0.005$ (other parameters are the same).

time-periodic weak force:

$$\dot{q} = p, \qquad \dot{p} = -dU_0/dq - h\omega_0^2 \sin(\omega_f t) \equiv -dU/dq, \tag{1}$$
$$U_0 \equiv U_0(q) = -\omega_0^2 \cos(q), \quad U \equiv U(q,t) = U_0(q) + qh\omega_0^2 \sin(\omega_f t), \quad h \ll 1.$$

The mechanism of chaos onset in our system is similar to that in the case of a conventional (narrow) chaotic layer in the adiabatic case [6] and is related to the motion near the unperturbed separatrix. Its rigorous consideration is very complicated (cf. [6]) but it is not essential for the quantity of the main interest in our work i.e. for the *width* of the chaotic layer. Indeed, as shown below, a chaotic trajectory in such a system spreads far beyond the vicinity of the separatrix. Obviously, the width is determined just by those (regular-like) parts of the chaotic trajectory that lie far from the separatrix.

To get an insight into the physical origin of the phenomenon, and to obtain a qualitative explanation of Figs. 1-2, we first consider the case

$$\omega_f/\omega_0 \ll h. \tag{2}$$

Since ω_f is small, the system may be considered at each instant t as moving in the quasi-stationary "wash-board" potential U with a slope $h\omega_0^2 \sin(\omega_f t)$. The absolute value of the slope is $\sim h\omega_0^2$ for most time while its sign is positive during odd half-periods of the perturbation, changing to the opposite one in the even half-periods. If the system is initially at the top of the potential barrier (i.e. $p = 0, q = \pi$), then, owing to the condition (2), even a small slope $\sim h\omega_0^2$ is sufficient to accelerate it during the first half-period up to a (negative) velocity of a *large* absolute value. In the second half-period, the slope changes its sign and therefore the acceleration changes into a braking effect, and at the end of the period the kinetic energy returns to zero or close to zero. Assuming that the maximum value $\mathscr{K}_{max}$ of the kinetic energy $\mathscr{K} \equiv p^2/2$ greatly exceeds the spatial modulation of $U(q,t)$ (the result will confirm this: see Eq. (6)),

$$\mathscr{K}_{max} \gg \omega_0^2, \tag{3}$$

one may neglect the term $U_0(q)$ in $U(q,t)$ while describing the main part of the trajectory (namely when $\mathscr{K} \gg \omega_0^2$) during the *first* period of perturbation. Then the equations of motion reduce to those for the free particle driven by a time-periodic force, which are solved exactly:

$$\begin{aligned} q(t) &= A + \left(B - \frac{h\omega_0^2}{\omega_f}\right)t + \frac{h\omega_0^2}{\omega_f^2}\sin(\omega_f t), \\ p(t) \equiv \dot{q} &= B - \frac{h\omega_0^2}{\omega_f}(1 - \cos(\omega_f t)), \end{aligned} \tag{4}$$

where A and B are constants of integration which may be easily written as:

$$A = q(t = n2\pi/\omega_f), \qquad B = p(t = n2\pi/\omega_f), \qquad n = 0, 1, 2, ... \tag{5}$$

If the initial state is $(p(0) = 0, q(0) = \pi)$, then $B = 0$ so that the velocity p in Eq. (4) oscillates from 0 to $-2h\omega_0^2/\omega_f$ while the kinetic energy $\mathscr{K} \equiv p^2/2$ oscillates from 0 to

$$\mathscr{K}_{max} = 2h^2\omega_0^4/\omega_f^2 \tag{6}$$

(note that Eq. (6) confirms the assumption (3), owing to condition (2)).

During next periods of perturbation, the major part of the trajectory is still described by Eq. (4) but the value of B in each subsequent period slightly shifts from that of the previous period. The shift occurs when the kinetic energy becomes small. Moreover, the sequence of shifts of B from period to period is irregular and, on long time-scales, B is densely distributed in the interval $[B_{min}, B_{max}]$, where $B_{min} = -2\omega_0$ and $B_{max} = 2h\omega_0^2/\omega_f \gg |B_{min}|$. Note that B_{max} diverges as $\omega_f \to 0$. Given the second equality in (5), we immediately conclude that the upper boundary of the chaotic layer in the Poincare section diverges (cf. Fig. 1). Because of the irregularity of the shifts in B, the trajectory is chaotic on long time-scales (cf. Fig. 2) while the maximum energy on it is described by Eq. (6) i.e. it diverges as $\omega_f \to 0$.

In a more general case when the condition (2) may not necessarily be satisfied (e.g. like for parameters related to Figs. 1 (left), 2), the neglection of the term U_0 in U in the analysis of the regular-like parts of the chaotic trajectory is not valid anymore. We develop for this case a more sophisticated approach which allows to derive a result that is in excellent agreement with results of simulations (Fig. 3): the upper energy of the chaotic layer E_{max} is the solution of the following transcendental equation,

$$\frac{\mathrm{E}(x)}{x} = 1 + \frac{\pi h}{2}\frac{\omega_0}{\omega_f}, \qquad \mathrm{E}(x) \equiv \int_0^{\frac{\pi}{2}} \mathrm{d}\varphi \sqrt{1 - x^2 \sin^2(\varphi)}, \qquad x \equiv \frac{1}{\sqrt{1+\frac{E_{max}-\omega_0^2}{2\omega_0^2}}} \tag{7}$$

where $\mathrm{E}(x)$ is the full elliptic integral of the 2nd order. The quantity E_{max} is a monotonously decreasing function of the argument $h\omega_0/\omega_f$ (it decreases from infinity to ω_0^2 as the argument increases from zero to infinity), and it can be easily found numerically (see Fig. 3). We recall that $\mathscr{K}_{max} \equiv \max(p^2/2) = E_{max} - \omega_0^2$.

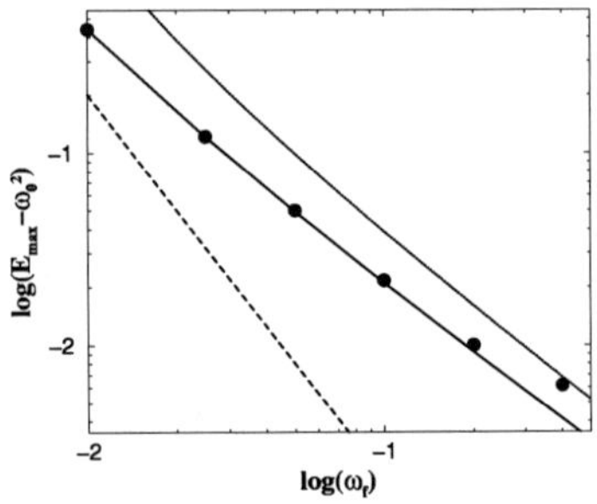

FIGURE 3. Simulations (circles) and theory for the upper energy boundary of the chaotic layer: solid, dashed and dotted lines correspond to the adiabatic formula (7) and asymptotes (6) for $\omega_f/\omega_0 \ll h$ and (9) for $\omega_f/\omega_0 \gg h$ respectively.

The solution (7) is valid in the frequency range

$$\omega_f \ll \frac{\omega_0}{\ln(1/h)}. \tag{8}$$

In the asymptotic limit $\omega_f \to 0$, i.e. when the condition (2) holds, the solution of Eq. (7) reduces to the explicit asymptote (6). In the asymptotic limit $h \to 0$, i.e. when $\omega_f/\omega_0 \gg h$, the solution of Eq. (7) reduces to the following explicit asymptote:

$$\frac{E_{max} - \omega_0^2}{\omega_0^2} \approx \frac{32a}{\ln(1/a)}, \qquad a \equiv \frac{\pi h \omega_0/\omega_f}{8} \ll 1. \tag{9}$$

UNSOLVED PROBLEMS

1. One part of the unsolved problems relates to chaotic *transport* in the system.

On long time-scales, the chaotic trajectory provides transport in *both* directions, unlike regular trajectories. For chaotic transport, the mean-square displacement depends on time generally as [7] $\langle (q(t) - q(0))^2 \rangle = Dt^b$ with $0 < b < 2$: brackets $\langle ... \rangle$ mean averaging of the initial conditions over the chaotic layer. The *larger* the coefficients D and b are, the *faster* the transport is. It is suggested in [8], basing on numerical results, that $b \to 1$ as $\omega_f \to 0$, i.e. the diffusion on coarse time scales seems to be *normal*. But this should be proved/disproved *rigorously*. The situation with the coefficient D is much more interesting: we suggest that it should *diverge* (in contrast to conventional cases, where it vanishes as $\omega_f \to 0$: cf. [9]). The intuitive arguments are the following. Chaotic trajectories generally spend most of time close to boundaries with regions of a regular motion [7]. In our case, it means that the trajectory moves most of time close to either the lower border of the chaotic layer, where the average velocity is negative while its absolute value is $\sim \sqrt{\mathscr{K}_{max}} \propto \omega_f^{-1}$, or the upper border, where the average velocity is positive with the same absolute value. Such regimes may be called *acceleration-braking flights*, in analogy with [7]. Intervals between the flights take a *small* portion of time which tends to zero as $\omega_f \to 0$. So the duration of the flight grows unbounded as $\omega_f \to 0$. The coefficient D should diverge yet faster than the absolute value of the average velocity during a single flight or the duration of the flight: cf. the so called classical diffusion of electrons in solids [10] where $D \propto v_F^2 \tau$: here v_F is a ballistic velocity (corresponding to the Fermi energy) and τ is the mean free path time. The inset in Fig. 2 demonstrates how faster the chaotic transport becomes as ω_f decreases. But the *rigorous* calculation of D is an unsolved challenging problem.

It is also an intriguing problem to explain why some memory about the initial state may keep for very long times (cf. Fig. 2), much longer than the time during which the system is situated close to one of the boundaries of the layer.

2. Another part of the unsolved problems relates to applications.

(a). It would be interesting to apply the theoretical results obtained in our work to real systems as well as to test on them our conjecture about the drastic acceleration of the chaotic transport. To name just a few, we may mention mechanical (pendulum-like) systems [11], Josephson junctions [12], 2D electron gas in a 1D magnetic superlattice [13], and particles moving on crystalline surfaces [14].

(b). Transient chaos decreases the activation energy for noise-induced escapes/transitions [15, 16]. In the limit of small dissipation, the region of transient chaos in phase space reduces to the chaotic layer for the non-dissipative case. Therefore, there should be a drastic decrease of the activation energy for noise-induced transitions to a given high energy if the driving frequency is decreased. It would be important to test this conjecture since it may be relevant to *threshold* devices.

(c). Apart from the effect on the activation energies, transient chaos strongly affects also the *prefactor* in the noise-induced escape/transition rate. In our case, it should dramatically affect the transition rate to far wells: the rate should drastically increase if the driving frequency is decreased. Our theory may be immediately generalized for spatially periodic systems of a higher dimension and, therefore, it shows that the adiabatic driving may be an easy tool for the drastic acceleration of the noise-induced diffusion on crystalline surfaces [14].

SUMMARY

1. We show that, for any spatially periodic potential system in a homogeneous time-periodic field, the width of the chaotic layer diverges as the driving frequency goes to zero. The effect is easily generalized to any spatially periodic Hamiltonian system of any dimension subject to a time-periodic perturbation, if its magnitude in the Hamiltonian diverges as the coordinate along some direction goes to infinity.
2. We provide intuitive arguments, supported by simulations, in favour of the conjecture that the speed of the chaotic transport in our system diverges in the adiabatic limit. It is an important and challenging problem to prove this rigorously.
3. It is an intriguing problem to explain why, if the initial state is close to a boundary of the layer, a distinctly non-zero drift on a coarse time-scale keeps for a very long time.
4. We suggest numerous applications of the above results:

(a) acceleration of the chaotic transport in real systems, e.g. pendulum, Josehson junctions, 2D electron gas in a magnetic superlattice, particles on crystalline surfaces;

(b) drastic decrease of the activation energy for the noise-induced transition to a high energy, relevant e.g. to threshold devices;

(c) drastic increase of the transition rates for noise-induced transitions to far wells, relevant e.g. to the diffusion on 2D surfaces.

ACKNOWLEDGEMENTS

S.M.S. acknowledges the hospitality of the Abdus Salam ICTP and Pisa University where some part of this work was done.

REFERENCES

1. A.J. Lichtenberg and M.A. Liebermann, *Regular and Stochastic Motion*, Springer, New York, 1992.
2. G.M. Zaslavsky, R.D. Sagdeev, D.A. Usikov and A.A. Chernikov, *Weak Chaos and Quasi-Regular Patterns*, Cambridge University Press, Cambridge, 1991.
3. G.M. Zaslavsky, *Physics of Chaos in Hamiltonian systems*, Imperial Colledge Press, London, 1998.
4. S.M. Soskin, O.M. Yevtushenko, and R. Mannella, "Divergence of the Chaotic Layer Width in the Adiabatic Limit and Acceleration of the Chaotic Transport", submitted to *Phys. Rev. Lett.*
5. L.D. Landau and E.M. Lifshitz, *The Classical Theory of Fields*, Pergamon, Oxford, 1975.
6. Y. Elskens and D.F. Escande, *Nonlinearity* **4**, 615 (1991).
7. G.M. Zaslavsky, *Physics Reports* **371**, 461 (2002).
8. R. Harrish, S. Rajasekar, K.P.N. Murthy, *Phys. Rev. E* **65**, 046214 (2002).
9. D.L. Bruhwiler and J.R. Cary, *Physica D* **40**, 265 (1989).
10. S. Datta, *Electronic transport in mesoscopic systems*, Cambridge, University Press, 1995.
11. L.D. Landau and E.M. Lifshitz, *Mechanics* , Pergamon, London, 1976.
12. *Dynamics of Josephson Junctions and Circuits*, Gordon and Breach, Philadelphia, 1986.
13. S.M. Soskin, O.M. Yevtushenko, and R. Mannella, *Phys. Rev. Lett.* **90**, 174101 (2003).
14. J.M. Sancho, A.M. Lacasta, K. Lindenberg, I.M. Sokolov, A.H. Romero, *Phys. Rev. Lett.* **92**, 250601 (2004); *Phys. Rev. Lett.* **94**, 188902 (2005); A.M. Lacasta, J.M. Sancho, A.H. Romero, I.M. Sokolov, K. Lindenberg, *Phys. Rev. E* **70**, 051104 (2004); J.D. Bao, Y. Zhou, *Phys. Rev. Lett.* **94**, 188901 (2005); J.M. Sancho, A.M. Lacasta, A.H. Romero, K. Lindenberg, *Phys. Rev. Lett.* **94**, 160601 (2005).
15. S.M. Soskin, R. Mannella, M. Arrayás and A.N. Silchenko, *Phys. Rev. E* **65**, 051111 (2001).
16. S.M. Soskin, R. Mannella, and P.V.E. McClintock, *Physics Reports* **373**, 247 (2003).

CONSTRUCTIVE ROLE OF NOISE

Attention Excludes Noise. Does It Exclude Stochastic Resonance?

Lawrence M. Ward*† and Keiichi Kitajo*§

*Psychophysics and Cognitive Neuroscience Laboratory, Department of Psychology, The University of British Columbia, 2136 West Mall, Vancouver, BC V6T1Z4, Canada
University of British Columbia
†The Brain Research Centre, University of British Columbia
§Educational Physiology Laboratory, Graduate School of Education, The University of Tokyo, 7-3-1 Hongo, Bunkyo-ku, Tokyo 113-0033, Japan

Abstract. The most highly confirmed models of attentional psychophysics imply that focused attention, as compared to unfocused attention, always excludes, or filters out, both external and internal noise. This filtering action of attention would appear to exclude the possibility of stochastic resonance (SR) in the detection and discrimination of attended signals. In contrast, recent experiments have definitively demonstrated the presence of SR in detection and discrimination of (presumably attended) sensory signals by humans. Here we discuss this apparent conflict between theory and experiment, propose several alternative views of attention that do allow SR, and present the results of preliminary experiments intended to test our ideas.

Keywords: noise, stochastic resonance, attention, psychophysics, vision, threshold
PACS: 87.10.+e, 87.19.Dd, 87.19.La, 05.40.Ca , 42.66.-p

STOCHASTIC RESONANCE AND PSYCHOPHYSICS

Stochastic resonance (SR), the improvement of weak signal transmission through a nonlinear system by the addition of noise, has been demonstrated in both animal and human sensory systems (e.g., [1]). These data are typically modeled by some form of threshold SR, in which the relevant nonlinearity is a Heaviside function and the weak signal is, in the absence of noise, not able to cross the nonlinearity. More recently, soft threshold SR, in which the threshold nonlinearity is more gradual, as in a logistic function, has been thoroughly studied [2]. Moreover, in line with earlier speculations about the role of internal noise in sensory SR (e.g., [3]), a model combining influences of both internal noise, arising in the system itself, and external noise, arising from the environment, on visual signal detection in human sensory systems has been proposed [4], based on a model of Simonotto for the perception of visual SR ([5]; cf. also [6]). In this model internal and external noise add, and detection of subthreshold signals is improved under very low external noise conditions because of the presence of the internal noise. The model expresses the minimum contrast, A, at which a sine-wave grating of luminance can just be discriminated from a uniform luminance field. (Contrast expresses the ratio of the peak-to-trough luminance difference of the grating to the sum of the peak and trough luminances.) The model is

CP800, *Unsolved Problems of Noise and Fluctuations: UPoN 2005*
edited by L. Reggiani, C. Pennetta, V. Akimov, E. Alfinito, and M. Rosini

$$A = k(\sigma_E^2 + \sigma_I^2)^{1/2} \exp\left[\frac{d^2}{2(\sigma_E^2 + \sigma_I^2)}\right] \quad (1)$$

where k and d are constants and σ_E and σ_I are external and internal noise, respectively. Notice that the noise amplitudes occur in two places in the equation; it is the interplay of the multiplicative noise term and the exponential noise term that generates the typical SR curve. Moreover, even when $\sigma_E = 0$ internal noise will affect performance. This model provides a very good fit to sensory SR data [4], particularly the small deviations in signal detection performance at very low external noise levels from predictions by models (e.g., [3], [5]) that do not include internal noise. It has, however, no provision whatsoever for the action of attention.

VISUAL ATTENTION AND NOISE

In contrast to the SR picture of noise amplifying weak sensory signals, pretty much the entire literature of psychophysics considers noise to have only negative effects on processing of sensory signals. For example, processing of sine-wave gratings of luminance is degraded by the addition of pixel noise, and more degraded the higher the contrast of the noise (Figure 1b). Of special interest in the context of unsolved problems of noise is the fact that attention appears to weaken the usual masking effects of noise on detection of such gratings. That is, when pixel noise is added to sine-wave gratings, performance is better at high noise levels when the gratings appear in attended locations in the visual field than when they appear in unattended locations (i.e., when attention is focused elsewhere than the location of the grating; e.g., [7], [8]). These results confirm a model of visual attention in which attention acts at various places in the sensory processing system to exclude, or attenuate, noise, both internal and external. The model is presented pictorially in Figure 1a and the equation for d', an unbiased measure of detection performance [9], looks like this:

$$d' = \frac{(\beta c)^\gamma}{\sqrt{(A_{ext} N_{ext})^{2\gamma} + N_{mul}^2[(\beta c)^{2\gamma} + (A_{ext} N_{ext})^{2\gamma}] + (A_{\text{int}} N_{\text{int}})^2}} \quad (2)$$

where d' is the measure of detection performance, c is the contrast of the grating, β and γ are constants reflecting sensory processing, N_{ext}, N_{mul}, and N_{int} are external, multiplicative and internal noise, respectively, and $0 < A_{ext}, A_{int} < 1$ are attention attenuation factors.

Without explaining exactly how the model of Equation 2 was derived, or how it functions specifically in this context, it is clear from its form that SR cannot occur as a function of N_{ext}, N_{mul}, or N_{int}, because as any one of these increases while A_{ext} and A_{int} remain constant, d' decreases monotonically. This is similar to previous work with signal detection theory, in which the conventional definition of d' excludes SR [10]. Moreover, in this model paying attention excludes noise, because both attention

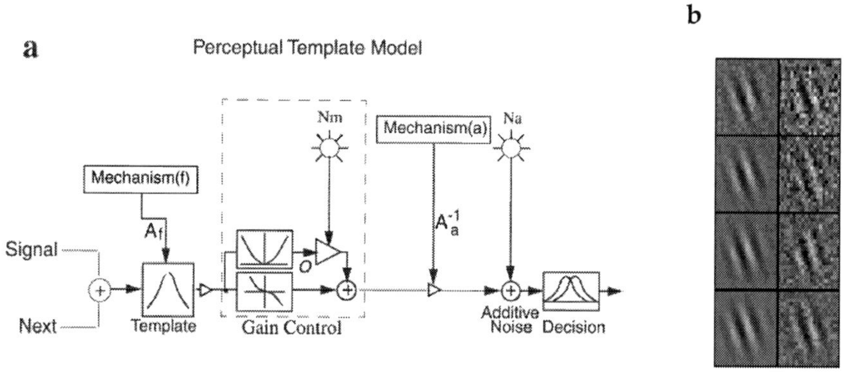

FIGURE 1. (a) Model of Dosher and Lu for exclusion of noise by attention. "Mechanism(f)" and "Mechanism(a)" indicate where noise exclusion mechanism acts (from [8] with permission). (b) Examples of sine-wave gratings with different amounts of pixel noise added. Noise is minimal in the lower left panel and increases bottom to top and left to right, being maximal in the upper right panel (from [8] with permission).

Multipliers are larger for unfocused than for focused attention. The first unsolved problem is: *What is the mechanism by which the attention multipliers operate?* The developers of this model [7] are vague as to how the noise exclusion might work, and also silent as to where the control signals might come from.

Assuming there are control signals from somewhere, it is possible that attention could vary as a function of noise, for example attention could be more focused in moderate noise than in low or high noise conditions, making the denominator smaller and thus d' larger in moderate noise. This, however, is the opposite of SR, since performance is being enhanced by a *decrease* in noise. Some experiments make this adjustment difficult, moreover, since trials under the different noise conditions are presented at random, removing the opportunity to adjust the level of attention in advance. Very brief stimulus presentation, e.g. 85 ms in [8], precludes adjustment during the stimulus presentation itself because it takes about 300 ms to 500 ms to adjust the attention state. Moreover, in such experiments attention is required to be focused on a particular spatial location, meaning that it is not focused anywhere else. That such is the case is demonstrated by the excellent fit of the model of Equation 2 to the data: r^2 for such fits is typically near or above 0.95. It would seem that not only does attention exclude noise, *this model of how attention excludes noise also excludes SR.*

IS STOCHASTIC RESONANCE EXCLUDED BY ATTENTION?

As far as we know, all previous experiments demonstrating SR in human sensory psychophysics have assumed that the subjects were attending to the stimuli they were attempting to detect. Certainly this has been the case in our own experiments (e.g., [1], [4]. How is it then that we obtained SR? *This is the primary unsolved problem to which this contribution is addressed.* According to the model of Equation 2 paying attention should *exclude* the external noise, or at least attenuate it. Would this effect simply translate the optimal level of external noise to a higher level? Or is the assumption that attention only excludes noise too simple? Perhaps attention modulates the level of internal noise to an optimal level for stimulus detection (e.g., [6]). Neither of these possibilities is allowed by Equation 2. A third possibility is a model that includes both an attentional noise-exclusion mechanism and a SR mechanism. One such possible model is as follows:

$$d' = \frac{(\beta c_{tot})^{\gamma}}{\sqrt{(A_{ext}N_{ext})^{2\gamma} + N_{mul}^{2}[(\beta c_{tot})^{2\gamma} + (A_{ext}N_{ext})^{2\gamma}] + (A_{\text{int}}N_{\text{int}})^{2}}} \qquad (3)$$

where all symbols are the same as in Equation 2 except for c_{tot}, which stands for the total effective amplitude of the signal,

$$c_{tot} = b_c c + b_{SR}\left[\int_{-\infty}^{a/N_{ext}} f(z)\,dz - \int_{-\infty}^{(a-c)/N_{ext}} f(z)\,dz\right] \qquad (4)$$

where a is a Heaviside threshold, b_c and b_{SR} are constants, $b_c + b_{SR} = 1.0$, and $f(z)$ is the density function of the external noise. Equation 4 for c_{tot} represents the additional (to signal contrast, c) contribution of SR to the effective signal amplitude. The difference-of-integrals SR term in brackets is based on the statistical model of stochastic resonance created by Greenwood, Ward and Wefelmeyer [11] and is the numerator of an equation for d' in a SR model from unpublished work by Greenwood, Müller and Ward. In this form, the SR mechanism is assumed to function prior to the "gain control" stage of the standard model, in parallel to the "template" mechanism upon which external noise attenuation operates (Figure 1a). The weighting constants b_c and b_{SR} regulate the contribution of the SR mechanism to the effective signal strength. Generally, $b_{SR} > b_c$ because the contribution of the SR mechanism must be greater than the ordinary signal contribution at low signal and noise intensities. The difference-of-integrals is positive when $c < a$ and negative when $c > a$ and its absolute value is always less than 0.5 for $f(z)$ a Gaussian. It is usually close to zero and is dominated by the other part of Equation 4 except where the mutual tuning of $(a - c)$ and external noise creates a maximum, generating the SR curve. It approaches zero both as external noise approaches zero, when both upper limits of integration approach $+\infty$ and both integrals approach +1, and at high noise intensities when both upper limits of integration approach 0 and both integrals approach the same number (0.5 for a Gaussian noise distribution). Thus, the difference of integrals of Equation 4 shows SR when $c < a$, in that it is maximal for some non-zero noise level. Just how the difference in weighting of this part of Equation 4 from that of the ordinary signal contribution could be accomplished, however, is unknown, another unsolved problem.

A representative computation showing how d' according to Equation 3 varies with N_{ext} and c is shown in Figure 2a for $f(z)$ the standard Gaussian. Here $b_c = 0.17$, $b_{SR} = 0.83$, $a = 0.3$, $N_{int} = 0.1$, $N_{mul} = 0.5$, $\beta = 3.0$, $\gamma = 1.0$, $A_{ext} = 1.0$ and $A_{int} = 1.0$, where the values for parameters common to Equations 2 and 3 are similar to those fitted from experimental data. This computation confirms that the hybrid model of Equation

3 can show SR (for $c < a$) and also that it shows the usual monotonic decrease of d' with noise amplitude (for $c > a$). Figure 2b shows the same computation but with external noise exclusion by attention, $A_{ext} = 0.7$, and Figure 2c shows the computation with internal noise exclusion by attention, $A_{int} = 0.7$. There are only subtle differences between the three figures. To better appreciate these differences, Figure 2d shows juxtaposed curves from each of the figures for the same below-threshold signal. Noise exclusion by the attention mechanisms does affect the curves, even for a below threshold signal, but does not eliminate SR. Clearly, in this model at least, SR can exist along with attentional noise exclusion mechanisms.

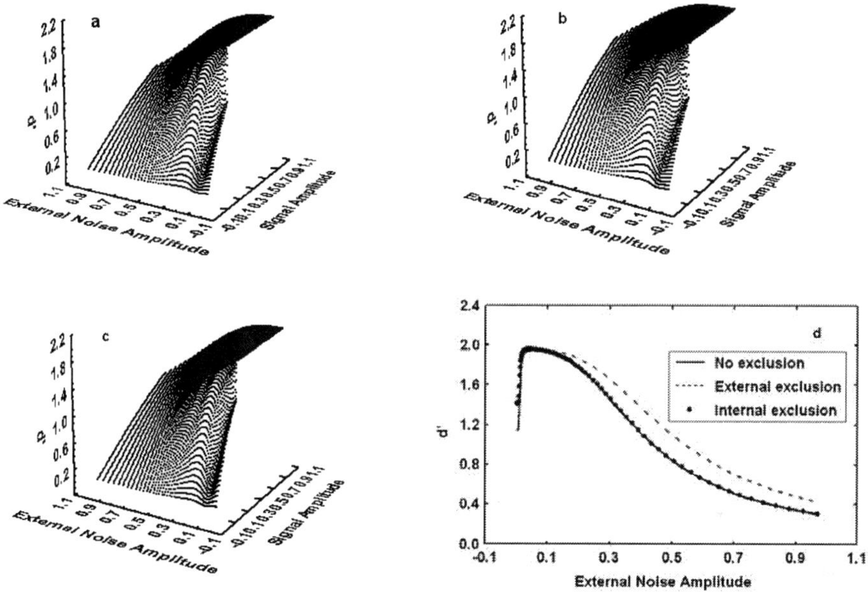

FIGURE 2. (a) d' as a function of external noise and signal amplitude with no noise exclusion ($A_{ext} = A_{int} = 1.0$). The non-monotonic change in d' at small values of c as external noise increases indicates that SR has occurred below the threshold $a = 0.3$, whereas only masking occurs for $c > a$, indicated by the monotonic decrease of d' with increasing noise amplitude in this region. (b) Same as (a) but with external noise exclusion, $A_{ext} = 0.7$. (c) Same as (a) but with internal noise exclusion, $A_{int} = 0.7$. (d) Comparison of SR curves for various attentional noise exclusion cases. Curves were extracted from curves in (a), (b), and (c) for $c = 0.28$.

STOCHASTIC RESONANCE IN ATTENDED AND UNATTENDED LOCATIONS: A FIRST EXPERIMENT

We ran an experiment to see whether stochastic resonance would occur for weak, near-threshold stimuli at an attended location, as predicted by models like Equation 3, or whether stochastic resonance would be excluded by attention and would only occur at an unattended location. To create both attended and unattended locations within the same display we presented human subjects with the display shown in Figure 3a, in which an arrow pointed to the marker box that was most likely (67%) to contain a target on a particular trial. Experimental subjects were told to switch their attention from the fixation cross to the marker box indicated by the arrow on each trial. Targets were 500 ms-duration luminance increments or decrements of equal size occurring inside one of the two boxes. Subjects were required to press one key if the target was an increment and a different key if it was a decrement. Subjects performed 800 discrimination trials in attended locations and 400 discrimination trials in unattended locations, in each noise condition over five 1-hour sessions (6000 trials in all). Under these conditions, attended targets typically are detected and discriminated with greater accuracy than are unattended targets, and responses to attended detected targets are made with a shorter reaction time [12].

We also presented dynamic noise inside the marker boxes by adding a random sample from a Gaussian distribution of gray levels with zero mean and standard deviation ranging from 0 to 4 to the background gray level (64). The gray level of the marker box was refreshed continuously at a frequency of 30 Hz, that is, every 33.3 ms. Thus, during a target duration the gray level would be changed 15 times around the mean target gray level. We attempted to fix the d' in the zero-noise condition at 1.0 by determining target gray level using an adaptive staircase procedure. We found, however, that performance in the zero-noise condition in the experiment proper was usually significantly greater than 1.0, often over 2.5, making it difficult to observe any significant noise-induced increase in performance in the low-noise conditions. We thus ran additional sessions with some of the subjects in which we decreased the target luminance difference, obtaining decreased performance in the zero noise condition. We report here only data from sessions (minimum of 4) in which $d' < 2.549$ in the attended location in the zero-noise condition. We also discarded trials on which eye movements occurred (<10%). Resulting data points were based on approximately 580 to 860 (attended) and 290 to 430 (unattended) trials per noise condition.

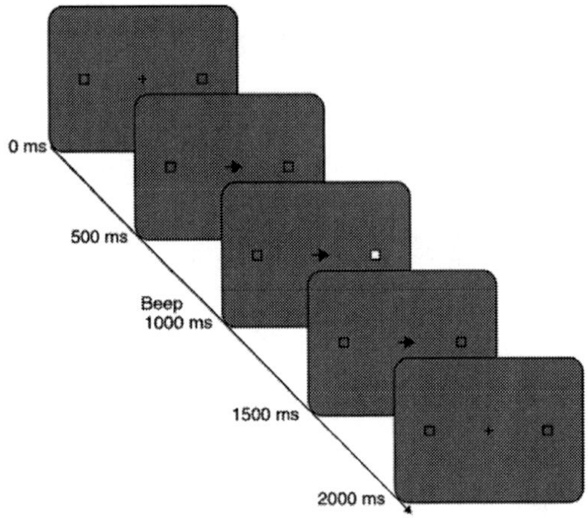

FIGURE 3. The sequence of visual displays used in the experiment. Both dynamic (flicker) noise and targets appeared in the marker boxes. Subjects fixated the central cross or arrow at all times.

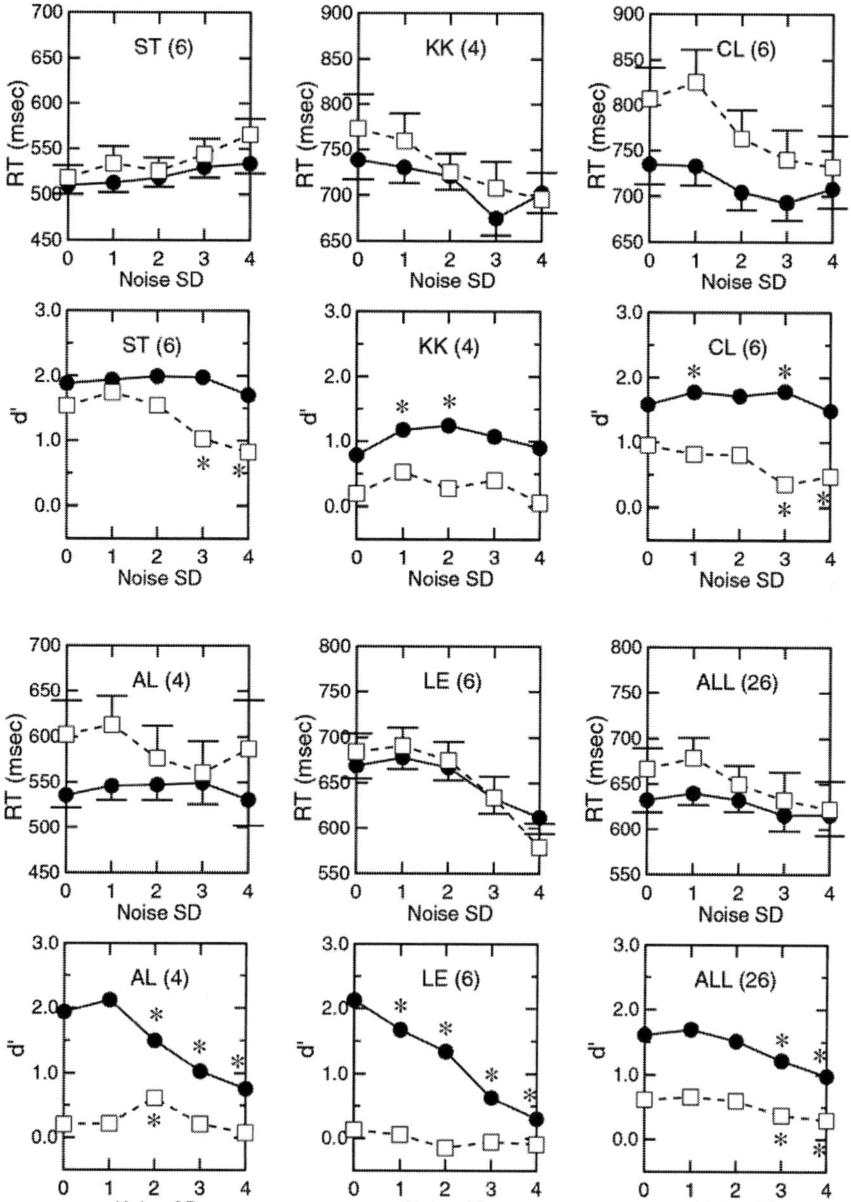

FIGURE 4. Response time (RT) for correct discrimination responses and *d'* to discriminate target luminance, both as a function of external noise level (Noise SD), for attended (solid line, filled dots) and unattended (dashed line, open squares) conditions for individual subjects (initials, e.g., ST) and the average of all subjects (ALL). Number of sessions is in parentheses beside initials. Error bars (RT) represent one standard deviation. * (*d'*) means the data point is statistically reliably different from that of the zero noise condition at $p < 0.05$ [9].

Figure 4 summarizes our findings. The average data (ALL) clearly replicate the usual finding that attended targets are discriminated with greater accuracy and in a shorter time than are unattended targets. They do not replicate the earlier finding [8] that, at higher levels, external noise is excluded by attention; just the opposite is found as indicated by the slightly shallower slope of the *d'* versus noise curve for the unattended than for the attended condition. Moreover, the average data show scant evidence of SR in either attended or unattended conditions. The individual data, however, tell a different story. Looking only at the *d'* results, one subject (LE) showed classic masking results in the attended condition and failed to reliably discriminate unattended targets in any noise condition, and one other subject (ST) showed no SR in either condition. Two subjects (KK and CL), however, showed statistically reliable SR in the attended condition only, whereas one other (AL) showed SR in the unattended condition only. In 7 of the 10 *d'* curves, moreover, a noise condition other than the zero noise condition showed the largest *d'*. Thus, there is some, albeit weak, evidence that SR occurred in this experiment. Variability across individuals for both RT and *d'* results is striking. For example, for only one subject (KK) was the attended-unattended difference in *d'* stable across noise conditions; for AL and LE the difference decreased with increasing noise whereas for ST and CL the difference increased with increasing noise. RT patterns were uncorrelated with *d'* patterns, indicating that speed-accuracy tradeoffs cannot explain the differences. Thus, strategic adjustments of attention as a function of noise condition, presumably via the attention multipliers in Equation 3, must have interacted with the SR mechanism to yield the observed performances. This is the final unsolved problem of this paper: *How can we gain control over the various mechanisms implied by Equation 3 to unambiguously display their action and interaction in determining performance in such experiments?*

ACKNOWLEDGMENTS

This research was supported by grants from the Natural Sciences and Engineering Research Council (NSERC) of Canada and the Toyota Corporation.

REFERENCES

1. Moss, F., Ward, L.M., and Sannita, W.G., *Clin. Neurophysiol.* **115**, 267-281 (2004).
2. Greenwood, P.E., Müller, U.U., and Ward, L.M., *Phys. Rev. E* **70**, 065411(1-11) (2005).
3. Wiesenfeld, K., and Moss, F., *Nature* **373**, 33-36 (1995).
4. Ward, L.M., *Fluct. Noise Letters* **4**, L11-L21 (2004).
5. Simonotto, E., et al., *Phys. Rev. Letters* **78**, 1186-1189 (1997).
6. Linkenkaer-Hansen, K., et al., *J. Neurosci.* **24**, 10186-10190 (2004).
7. Lu, Z-L., and Dosher, B.A., *Vis. Res.* **38**, 1183-1198 (1998).
8. Dosher, B.A., and Lu, Z-L., *Psychol. Sci.* **11**, 139-146 (2000).
9. Macmillan, N.A., and Creelman, C.D., *Detection Theory: A User's Guide,* New York: Cambridge University Press, 1991.
10. Ward, L.M., Neiman, A., and Moss, F., *Biol. Cybernet.* **87**, 91-101 (2002).
11. Greenwood, P.E., Ward, L.M., and Wefelmeyer, W., *Phys. Rev. E* **60**, 4687-4695 (1999).
12. Wright, R.D., and Ward, L.M., "The control of visual attention" in *Visual Attention* edited by R.D. Wright, Oxford: Oxford University Press, 1998, pp. 132-186.

Multiphoton Antiresonance And Quantum Activation In Driven Systems

M. I. Dykman

Department of Physics and Astronomy, Michigan State University, East Lansing, MI 48824, USA

Abstract. We show that nonlinear response of a quantum oscillator displays antiresonant dips and resonant peaks with varying frequency of the driving field. The effect is a consequence of special symmetry and is related to resonant multiphoton mixing of several pairs of oscillator states at a time. We also discuss escape from a metastable state of forced vibrations. Two important examples show that the probability of escape via diffusion over quasienergy is larger than via dynamical tunneling provided the relaxation rate exceeds both of them. Diffusion dominates even for zero temperature, so that escape occurs via quantum rather than thermal activation. The effects can be studied using Josephson junctions and Josephson-junction based systems.

Keywords: Rabi oscillations, quasienergy, antiresonance, activated escape
PACS: 05.45.-a, 05.60.Gg, 74.50.+r, 33.80.Wz

INTRODUCTION

Much progress has been made recently in experimental studies of periodically modulated vibrational systems. Examples include optically bistable systems, electrons in Penning traps, Josephson junctions, and various nano- and micro-mechanical resonators [1]-[8]. All these systems display bistability of forced vibrations. Because of thermal or externally applied noise, there occurs switching between coexisting stable vibrational states. The measured switching probabilities of noise-induced transitions are in a good agreement with the theoretical predictions.

In the present paper we are interested in the dynamics of a quantum oscillator. The generality of the oscillator as a model system and the current interest in quantum computing and coherent phenomena lead to two major questions: (i) does a resonantly driven oscillator display coherent quantum effects that would qualitatively differ from those in two-level systems, and (ii) in the presence of relaxation, what is the probability of switching between coexisting stable states due to quantum fluctuations? These two questions are addressed in the present paper, which is based on the results [9]-[11].

A weakly nonlinear oscillator is a multi-level quantum system with nearly equidistant energy levels E_n. Therefore a periodic force of frequency ω_F can be nearly resonant for many transitions at a time, i.e., $\hbar\omega_F$ can be close to the interlevel distance $E_{n+1} - E_n$ for many n. This leads to strong nonlinearity of the response even to comparatively weak resonant fields. A well-known quantum effect of the oscillator nonlinearity is the onset of Rabi oscillations due to resonant multiphoton transitions. Multiphoton Rabi oscillations occur when the spacing between remote energy levels n and m coincides with the energy of $n-m$ photons, $E_n - E_m = (n-m)\hbar\omega_F$ [12]. The multiphoton transition amplitude is resonantly enhanced, because an $m \to n$ transition occurs via a sequence of virtual field-

CP800, *Unsolved Problems of Noise and Fluctuations: UPoN 2005*
edited by L. Reggiani, C. Pennetta, V. Akimov, E. Alfinito, and M. Rosini

induced transitions $k \to k+1$ (with $m \leq k \leq n-1$), all of which are almost resonant. An associated classical effect, in the presence of dissipation, is hysteresis of the amplitude of forced vibrations as function of the field amplitude A and ω_F.

In this paper (see also [10]) we show that multiphoton transitions in the oscillator are accompanied by a new effect, *antiresonance* of the response. When the frequency of the driving field adiabatically passes through a resonant value, the vibration amplitude displays a sharp minimum or maximum, depending on the initially occupied state. We argue that the antiresonance and the multiphoton Rabi oscillations can be observed in Josephson junctions.

If the frequency ω_F is close to twice the oscillator frequency, then $\hbar\omega_F$ is close to $E_{n+2} - E_n$ for many n at a time. This leads to parametric resonance in the oscillator, in which it oscillates at frequency $\omega_F/2$ in response to the driving. Such oscillations are intrinsically bistable, because their phase can take on two values that differ by π.

We will be most interested in the semiclassical behavior of the oscillator, which, on the one hand, stretches all the way to the classical region, and on the other hand, works well for oscillators even deep in the quantum domain. In the semiclassical picture, resonant multiphoton transitions correspond to tunneling between Floquet states of the oscillator with equal quasienergies. [The quasienergy ε gives the change of the wave function $\psi(t)$ when time is incremented by the modulation period τ_F, $\psi(t+\tau_F) = \exp(-i\varepsilon\tau_F/\hbar)\psi(t)$]. The occurrence of equal-quasienergy states is related to the bistability of forced vibrations of a classical oscillator.

Tunneling of a driven oscillator is a carefully studied[13] example of dynamical tunneling [14]. The WKB analysis gives an important insight into the origin of the antiresonance, which goes beyond the perturbation theory in the driving field.

Dynamical tunneling also leads to transitions between coexisting metastable states of forced vibrations, which emerge in the presence of dissipation due to coupling to a thermal reservoir. In terms of quantum mechanics, dissipation is due to interlevel oscillator transitions with energy being transferred to (or from, for nonzero temperature) the reservoir. It turns out that dissipation may also lead to transitions between metastable states of forced vibrations, even for zero temperature [9].

For $T = 0$ there occur only interlevel transitions where the oscillator energy goes to the reservoir (but the energy loss is compensated by the driving field, in the stationary regime). However, the quasi-energy may increase or decrease as a result of a coupling-to-reservoir induced transition, although with different probabilities. Therefore along with drift over quasienergy towards a metastable state, which results from more probable transition, there emerges diffusion away from this state as a sequence of less probable transitions. The diffusion may lead to activated-like escape. Activation in this case has purely quantum nature, and therefore we call it quantum activation.

THE MODELS

The Hamiltonian of a nonlinear oscillator with mass $M = 1$ has the form

$$H(t) = \frac{1}{2}p^2 + \frac{1}{2}\omega_0^2 q^2 + \frac{1}{4}\gamma q^4 + H_F(t). \tag{1}$$

We will consider two types of periodic modulation, $H_F = H_F^{(r,p)}$, which correspond to resonant and parametric driving,

$$H_F^{(r)}(t) = -qA\cos(\omega_F t), \qquad \delta\omega^{(r)} \equiv \omega_F - \omega_0 \ll \omega_0, \tag{2}$$
$$H_F^{(p)} = \frac{1}{2}q^2 F\cos(\omega_F t), \qquad \delta\omega^{(p)} \equiv \frac{1}{2}\omega_F - \omega_0 \ll \omega_0$$

(in what follows we set $\gamma > 0$).

It is convenient to analyze the dynamics in the rotating wave approximation by switching from the fast oscillating operators q, p to slowly varying operators Q, P using transformations

$$q = \alpha^{(r)}(Q\cos\omega^{(r)}t + P\sin\omega^{(r)}t), \qquad p = -\alpha^{(r)}\omega^{(r)}(Q\sin\omega^{(r)}t - P\cos\omega^{(r)}t)$$

for resonant driving and

$$q = \alpha^{(p)}(P\cos\omega^{(p)}t - Q\sin\omega^{(p)}t), \qquad p = -\alpha^{(p)}\omega^{(p)}(P\sin\omega^{(p)}t + Q\cos\omega^{(p)}t)$$

for parametric driving, with

$$\omega^{(r)} = \omega_F, \qquad \alpha^{(r)} = (8\omega_F\delta\omega^{(r)}/3\gamma)^{1/2},$$

and

$$\omega^{(p)} = \omega_F/2, \qquad \alpha^{(p)} = (2F/3\gamma)^{1/2}.$$

The variables Q, P are the appropriately scaled coordinate and momentum. The commutation relation for them has a simple form

$$[P, Q] = -i\lambda, \qquad \lambda = \hbar\left(\omega^{(r,p)}\right)^{-1}\left(\alpha^{(r,p)}\right)^{-2}. \tag{3}$$

The parameter λ plays the role of the effective Planck constant. We note that it is proportional to the oscillator nonlinearity γ scaled by the comparatively small detuning of the field frequency, in the case of nearly resonant driving, or the comparatively small field amplitude, in the case of parametric driving.

The dynamics of Q, P in the two cases is described by effective Hamiltonians

$$H^{(r)} = \omega_F\delta\omega^{(r)}\left(\alpha^{(r)}\right)^2 g^{(r)},$$
$$H^{(p)} = (F/4)\left(\alpha^{(p)}\right)^2 g^{(p)}. \tag{4}$$

Their eigenvalues are equal to the quasienergies ε_n of the oscillator. The functions $g^{(r,p)}$ have the forms

$$g^{(r)}(P, Q) = \frac{1}{4}(Q^2 + P^2 - 1)^2 - \beta^{1/2}Q, \quad \beta = \frac{3\gamma A^2}{32\omega_F^3\left(\delta\omega^{(r)}\right)^3}, \tag{5}$$

$$g^{(p)} = \frac{1}{4}(Q^2 + P^2)^2 + \frac{1}{2}(1-\mu)P^2 - \frac{1}{2}(1+\mu)Q^2, \quad \mu = 2\omega_F\delta\omega^{(p)}/F. \tag{6}$$

They are shown in Fig. 1. Each of them depends on one parameter. In the region of bistability of period one vibrations, $0 < \beta < 4/27$, the function $g^{(r)}$ has a shape of a tilted Mexican hat, with a maximum at the top of the central dome and a minimum at the lowest point of the rim. For a parametrically excited oscillator in the region $-1 < \mu < 1$ the function $g^{(p)}$ has two symmetrical minima. These extrema of $g^{(r,p)}$ correspond to metastable states of the driven oscillator in the presence of weak dissipation. The saddle points of $g^{(r,p)}$ correspond to unstable states of forced vibrations.

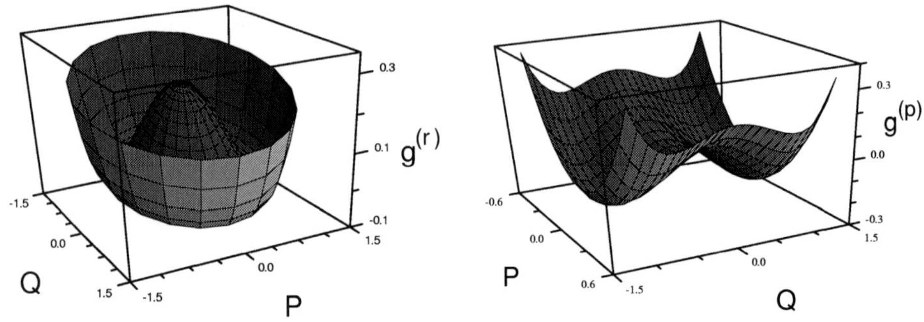

FIGURE 1. The scaled classical quasienergy of the oscillator for resonant and parametric driving (left and right panels, respectively) as function of the slowly varying coordinate and momentum Q,P. The plot of $g^{(r)}$ refers to the reduced field intensity $\beta = 3\gamma A^2/32\omega_F^3(\delta\omega^{(r)})^3 = 1/27$. The plot of $g^{(p)}$ refers to $\mu = 2\omega_F\delta\omega^{(p)}/F = -0.1$.

MULTIPHOTON ANTIRESONANCE

We will start with the studies of the coherent response of the oscillator to a nearly resonant field. When the field amplitude $A \to 0$, the eigenstates $|n\rangle$ of the Hamiltonian $H^{(r)}$ coincide with the Fock states of the oscillator, and the quasienergies are

$$\varepsilon_n = -\delta\omega^{(r)} n + \frac{1}{2}Vn(n+1), \quad V = \frac{3\hbar\gamma}{4\omega_0^2}. \tag{7}$$

We keep only the lowest-order term in V, which corresponds to the weak nonlinearity approximation. In this approximation the energy of an Nth oscillator state for $A = 0$ is $E_N = \hbar\omega_0 N + VN(N+1)/2$. The N-photon resonance $N\hbar\omega_F = E_N - E_0$ occurs, in terms of $\delta\omega^{(r)}$, for

$$\delta\omega^{(r)} = \delta\omega_N^{(r)} = V(N+1)/2.$$

For the corresponding field frequency ω_F the quasienergies ε_0 and ε_N are equal.

The field leads to mixing of the wave functions of resonating states and to level anticrossing. This anticrossing is clearly seen in the upper right panel of Fig. 2. The minimal splitting of the levels ε_0 and ε_N is given by the multiphoton Rabi frequency Ω_R. For weak field it can be obtained by perturbation theory [12]. To the lowest order in the field amplitude A in the limit of large N

$$\Omega_R = V\,(A/A_N)^N N^{5/4}(2\pi)^{-3/4}, \qquad A_N = (2\hbar\omega_0)^{1/2}|V|N^{3/2}\exp(-3/2)/2. \tag{8}$$

The Rabi frequency depends on N exponentially, $\Omega_R \propto A^N$. This dependence works well numerically in the whole region $A < A_N$ [10].

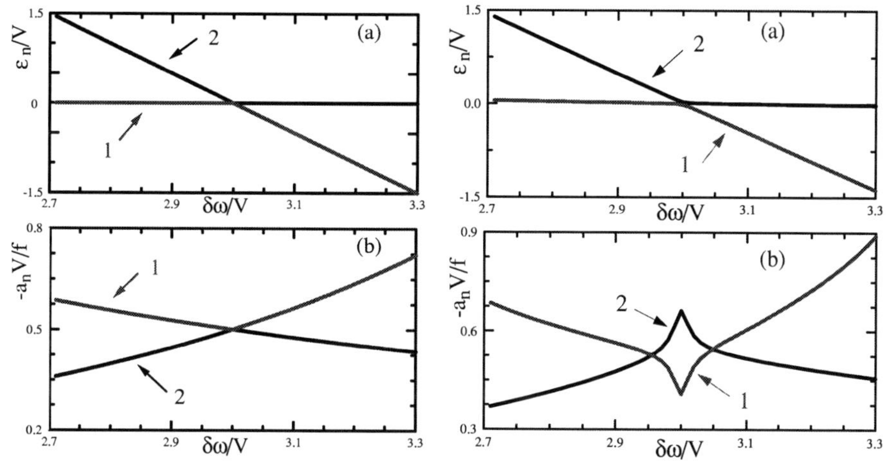

FIGURE 2. Quasienergy levels ε_n and reduced susceptibilities in the resonating states of a driven nonlinear oscillator. The plots refer to a 5-photon resonance, $N = 5$. The labels 1 and 2 correspond to the ground ($n = 0$) and 5th ($n = 5$) oscillator states in the absence of driving, for $\delta\omega \equiv \delta\omega^{(r)} < \delta\omega_5^{(r)}$. Left panel: the limit of zero driving amplitude, $A \to 0$. The quasienergy levels cross for $\delta\omega_N^{(r)}/V = 3$. The susceptibilities in the resonating states also cross at that same frequency. Right panel: level repulsion and antiresonance of susceptibilities for comparatively weak modulation. The data refer to $A/A_N = 1/2$. It is seen from the figure that the susceptibility antiresonance is much more pronounced than the level repulsion. The scaling factor for the amplitude a_n of oscillator vibrations in the states $n = 0,5$ is $f = (8\hbar\omega_0)^{-1/2}A$.

Coherent response of the oscillator to the driving field is characterized by the expectation value of the coordinate q. If the oscillator is in an eigenstate $|n\rangle$ of the Hamiltonian $H^{(r)}$, this value has the form

$$q_n = (\hbar/2\omega_0)^{1/2} a_n e^{-i\omega_F t} + \text{c.c.}, \qquad q_n = \langle n|q|n\rangle. \tag{9}$$

To first order in the field, the reduced amplitude of forced vibrations in an nth state a_n is

$$a_n = -f\delta\omega^{(r)}/\left((\delta\omega^{(r)} - Vn)[\delta\omega^{(r)} - V(n+1)]\right), \qquad f = (8\hbar\omega_0)^{-1/2}A. \tag{10}$$

Remarkably, for $\delta\omega^{(r)} = \delta\omega_N^{(r)}$ the vibration amplitudes in the resonating states coincide with each other, $a_{N-n} = a_n$ for $0 \le n < N/2$, see the left lower panel in Fig. 2.

Field-induced multiphoton mixing leads not only to splitting of the quasienergy levels, but also to repulsion of the vibration amplitudes. It can be calculated by diagonalizing the Hamiltonian $H^{(r)}$ and is shown in the right lower panel of Fig. 2 as a function of frequency detuning $\delta\omega \equiv \delta\omega^{(r)}$. One of the involved resonating states is the ground state of the oscillator $n = 0$ in the limit $A \to 0$. The quantities plotted in Figs. 2(b) are susceptibilities, they are proportional to the ratio of the vibration amplitude to the modulation amplitude a_n/A.

The antiresonant splitting of the expectation values of the vibration amplitudes is by far the most interesting feature of Fig. 2. It occurs at the adiabatic passage of $\delta\omega^{(r)}$ through resonance, where the system switches between the ground and excited states. In particular, the amplitude displays an antiresonant dip if the oscillator is mostly in the ground state for $(\delta\omega^{(r)} - \delta\omega_N^{(r)})/V < 1$ or in the state N for $(\delta\omega^{(r)} - \delta\omega_N^{(r)})/V > 1$. The magnitude and sharpness of the dip are determined by Ω_R/V and depend very strongly on the field and N. With decreasing Ω_R/V the dip (and peak) start looking like cusps located at resonant frequency. We note that, in contrast to the case of energy levels, there is no reason for repulsion (anticrossing) of susceptibilities. In fact, as seen from Fig. 2 the susceptibilities do cross, although away from the resonant frequency. The effect of antiresonance is due purely to specific quantum interference [10].

The dip in the oscillator response has no analogue in two-level systems. There, for nearly resonant driving, the coherent response in the two adiabatic states differs only in sign. It displays a peak when the radiation frequency adiabatically passes through the transition frequency.

The WKB picture of the antiresonance

In the WKB approximation, Rabi oscillations correspond to tunneling between the states with nearly equal quasienergies. Such semiclassical states can be found from the Hamiltonian $H^{(r)}$ (4) by using the Bohr-Sommerfeld quantization condition applied to the mechanical action $\oint P\,dQ$ for trajectories $g(Q,P) = \text{const}$, with $\hbar$ replaced by λ, Eq. (3). It is seen from the left panel of Fig. 1 that, in a certain range of g, there are two types of trajectories with the same g, those on the internal dome and those on the external part of the Mexican hat $g(Q,P)$. If, as a result of the Bohr-Sommerfeld quantization, the quantized values of g on the two parts of the surface $g(Q,P)$ coincide with each other, then there may occur resonant tunneling between the corresponding quantum states. The resulting tunneling splitting [13] is the Rabi frequency.

Interestingly, one can show that the average value of the coordinate

$$Q(g) = \tau^{-1}(g) \int_0^{\tau(g)} Q(t)\,dt$$

[$\tau(g)$ is the period of oscillations for a given g] is the same for the internal and external trajectories with the same g. This corresponds to the susceptibilities of the resonating quasienergy states being the same, in the neglect of tunneling-induced mixing of the states. We emphasize that the fact that the susceptibilities are equal is not a result of the perturbation theory in the field amplitude, as in the case of Eq. (10), they are equal in all orders of the perturbation theory in A as long as tunneling is disregarded. Tunneling-induced state mixing leads to the antiresonance of the response [10].

ESCAPE OF A DRIVEN SYSTEM: TUNNELING OR QUANTUM ACTIVATION?

We will now briefly outline the new effects and unanswered questions that emerge when dissipation is taken into account [9, 11]. We will assume that, even though dissipation is weak, the dissipation rate exceeds the tunneling rate. The problem of fairly general interest that will be addressed is switching between metastable states of forced vibrations of a quantum oscillator. We will consider the most interesting situation where there are many quasienergy states between the extrema of the quasienergy surface. In the case of escape of a particle from a potential well it corresponds to a well with many energy levels.

In systems in thermal equilibrium, the rate of tunneling decay of a metastable state for low temperatures is given by the probability of a tunneling transition from the ground state in a metastable potential well. In the case of a resonantly driven oscillator this corresponds to dynamical tunneling from the top of the dome of the quasienergy surface $g^{(r)}$ to the state on the external orbit with the same quasienergy, see the left panel of Fig. 1. The tunneling is shown schematically in the central panel of Fig. 3. For a parametrically driven oscillator the corresponding tunneling occurs between the minima of the surface $g^{(p)}$, as shown in the right panel of Fig. 3.

For higher temperatures, again in the case of equilibrium systems, one has to take into account tunneling from excited intrawell states. Escape may occur also via thermal activation over the potential barrier. One of these escape mechanisms dominates, depending on temperature [15]. In the case of a driven oscillator tunneling from excited states corresponds to tunneling with quasienergies that differ from those at the extrema of $g(Q,P)$. In addition, there is a probability of activation over the quasienergy barrier. However, since the distribution over quasienergy is not of the Boltzmann form, it is not clear which of the escape mechanisms dominates at a given temperature of the bath.

For weak dissipation the distribution can be described by the balance equation for the occupations ρ_n of quasienergy (Floquet) states $|n\rangle$,

$$\dot{\rho}_n = -\sum\nolimits_m W_{nm}\rho_n + \sum\nolimits_m W_{mn}\rho_m. \tag{11}$$

The transitions probabilities W_{nm} can be calculated as matrix elements of the operator that describes relaxation of the oscillator. The wave functions $|n\rangle$ can be found from the Bohr-Sommerfeld approximation disregarding tunneling. A standard WKB calculation allows one to express W_{nm} in terms of the Fourier components of the coordinate and momentum $Q_{m-n}(g_n), P_{m-n}(g_n)$ for a given quasienergy g_n [the functions $Q(t), P(t)$ are periodic functions of time for a given g, with period $\tau(g)$]. In the simple case of linear friction, which corresponds to relaxation transitions between nearest energy (not quasienergy) levels of the oscillator shown in the left panel of Fig. 3, W_{nm} are simply quadratic in $Q_{m-n}(g_n), P_{m-n}(g_n)$. They exponentially decay with $|n-m|$.

The probabilities W_{nm} are organized so that it is more likely for a system to make a transition toward the value of g in the metastable state rather than away from it. This is why the state is metastable. However, in contrast to systems in equilibrium, the probabilities W_{nm} do not satisfy the condition $W_{nm} = W_{mn}\exp[(g_n - g_m)/kT]$. Even for $T \to 0$ there is a nonzero probability to make a transition in the direction opposite to the metastable state. This is a consequence of the fact that the Floquet states $|n\rangle$ are linear combinations of the Fock states of the oscillator. Therefore, even where all transitions between the Fock states go in one direction in energy, as in the left panel of Fig. 3, transitions between the Floquet states go in different directions in quasienergy, although with different probabilities.

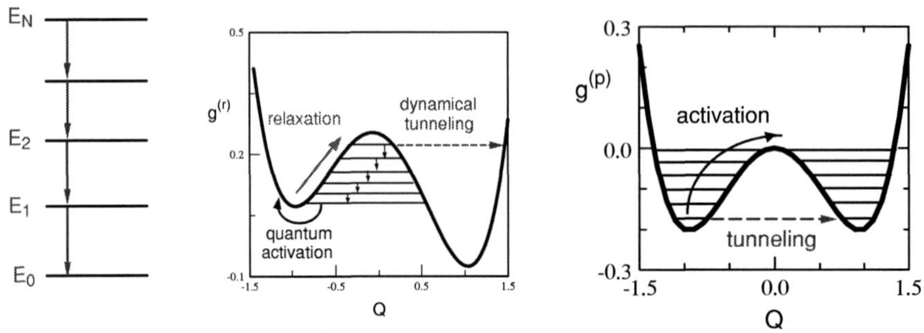

FIGURE 3. Left panel: transitions between the energy levels of the oscillator E_n due to coupling to a thermal reservoir. For $T = 0$ there occur only transitions to lower levels with emission of excitations of the reservoir. In the simplest case relaxation is due to transitions between neighboring oscillator levels. Central panel: the cross-section of the surface of the scaled quasienergy $g^{(r)}$ shown in Fig. 1 by the plane $P = 0$. Quasienergy levels for orbits that lie on the dome in Fig. 1 are shown schematically. Relaxation to the metastable state at the top of the dome corresponds to transitions to higher quasienergy, whereas less probable transitions to lower quasienergy lead to an activation-type escape from the metastable state. Right panel: the cross-section of the quasienergy surface of the parametrically driven oscillator $g^{(p)}$, Fig. 1, by the plane $P = 0$, with schematically shown quasienergy levels. Relaxation drives the oscillator down to the lowest quasienergy levels at the bottom of the wells of $g^{(p)}(Q,P)$, but there is a possibility of activation even for $T = 0$.

Transitions in the "wrong" direction lead to diffusion over quasienergy away from the metastable state. Their immediate consequence is a finite width of the stationary distribution over quasienergy even for $T \to 0$. Another closely related consequence is a nonzero probability to reach quasienergy of the saddle of the functions $g^{(r,p)}$ starting from a metastable state. The saddle of $g^{(r,p)}$ is similar to the top of a potential barrier for a particle in a metastable potential. The logarithm of the probability of reaching the saddle gives the activation energy of escape as a result of diffusion over quasienergy. We call this process quantum activation, since it occurs even for $T = 0$.

For resonantly and parametrically driven oscillators the transition rates W_{nm} can be calculated explicitly using the fact that the classical trajectories in (Q,P)-variables can be expressed in terms of the Jacobi elliptic functions [9, 11]. We have compared the activation energies of escape with the tunneling exponents. When the activation energy is smaller than the tunneling exponent in the absolute value, escape occurs via quasienergy diffusion. We found that, both for a resonantly and a parametrically excited oscillator, escape from metastable states occurs via activation, not tunneling. This holds true for all parameter values where a nonlinear oscillator has coexisting stable states. We found an unusual behavior of the distribution for an underdamped parametrically driven oscillator for $T = 0$. For some quasienergy the distribution displays a sharp decrease, and at the same time the tunneling rate goes to zero.

The physical origin of the fact that escape occurs via activation, not dynamical tunneling, remains not fully understood, the existing arguments are formal [9]. Apparently, there must be a crossover from activation to tunneling when the system goes to equilibrium, but the models that we have discussed are strongly nonequilibrium, the very presence of coexisting metastable states is due to periodic driving.

In conclusion, we have shown that a simple system, a driven nonlinear oscillator, displays unusual quantum coherent phenomena and unusual switching behavior. The considered effects have no analogue in two-level systems and are qualitatively different from what has been known about switching in thermal equilibrium systems. They are not only of fundamental interest, but are also important for many applications, in particular in sensing and quantum measurements.

ACKNOWLEDGMENTS

This research was supported in part by the Institute for Quantum Sciences at Michigan State University and by the NSF through grant ITR-0085922.

REFERENCES

1. P. D. Drummond and D. F. Walls, *J. Phys. A* **13**, 725 (1980).
2. G. Gabrielse, H. Dehmelt, and W. Kells, *Phys. Rev. Lett.* **54**, 537 (1985).
3. A.I. Larkin and Yu.N. Ovchinnikov, *J. Low Temp. Phys.* **63**, 317 (1986); B.I. Ivlev and V.I. Mel'nikov, *Phys. Lett. A* **116**, 427 (1986); S. Linkwitz and H. Grabert, *Phys. Rev. B* **44** 11888, 11901 (1991); M.H. Devoret, D. Esteve, J.M. Martinis, A. Cleland, and J. Clarke, *Phys. Rev. B* **36**, 58-73 (1987).
4. A. Wallraff, T. Duty, A. Lukashenko, and A. V. Ustinov, *Phys. Rev. Lett.* **90**, 037003 (2003); M. V. Fistul, A. Wallraff, and A. V. Ustinov, *Phys. Rev. B*, **68**, 060504(R) (2003).
5. I. Siddiqi, R. Vijay, F. Pierre, C. M. Wilson, M. Metcalfe, C. Rigetti, L. Frunzio, and M. H. Devoret, *Phys. Rev. Lett.* **93**, 207002 (2004).
6. J. S. Aldridge and A. N. Cleland, *Phys. Rev. Lett.* **94**, 156403 (2005); cond-mat/0406528 (2004).
7. R. L. Badzey, G. Zolfagharkhani, A. Gaidarzhy, and P. Mohanty, *Appl. Phys. Lett.* **86**, 023106 (2005).
8. C. Stambaugh and H. B. Chan, cond-mat/0504791 (2005).
9. M.I. Dykman and V.N. Smelyanskiy, *Zh. Eksp. Teor. Fiz.* **94**, 61 (1988) [*Sov. Phys. JETP* **67**, 1769 (1988)].
10. M.I. Dykman and M.V. Fistul, *Phys. Rev. B* **71**, 140508(R) (2005).
11. M.I. Dykman and M. Marthaler, unpublished.
12. D.M. Larsen and N. Bloembergen, *Opt. Communications* **17**, 254 (1976).
13. A.P. Dmitriev and M.I. D'yakonov, *Zh. Eks. Teor. Fiz.* **90**, 1430 (1986) [*Sov. Phys. JETP* **63**, 838 (1986)].
14. M. J. Davis and E. J. Heller, *J. Chem Phys.* **75**, 246 (1981).
15. A. I. Larkin and Y. N. Ovchinnikov, J. Stat. Phys. **41**, 425 (1985)

Escapes And Transitions In Overdamped Systems On Short Times: General Solution

Stanislav M. Soskin*,†, Valentin I. Sheka *, Tatiana L. Linnik * and Riccardo Mannella**

*Institute of Semiconductor Physics, 03028 Kiev, Ukraine
†Abdus Salam ICTP, 34100 Trieste, Italy; and Department of Physics, Lancaster University, Lancaster LA1 4YB, UK
**Dipartimento di Fisica, Università di Pisa and INFM UdR Pisa, 56100 Pisa, Italy

Abstract. We find the complete general solution of the noise-induced escape/transition problem on time scales less than or order of the relaxation time. The theory is checked in simulations. The most interesting unsolved problems are: (i) calculation of the first-order correction and matching of our results with those by Shneidman obtained for larger time-scales by a different method; (ii) generalization to transitions in a bistable potential driven by a moderately slow and strong periodic force, which may be relevant for bilogical applications.

Keywords: Noise-induced escape/transition, overdamped, optimal fluctuation, path-integral
PACS: 05.40.-a, 02.50.-r, 05.20.-y, 82.40.Bj

INTRODUCTION

The overdamped stochastic equation is commonly defined as [1]

$$\frac{\mathrm{d}x}{\mathrm{d}t} = -\frac{\mathrm{d}U(x)}{\mathrm{d}x} + f(t), \qquad \langle f(t)\rangle = 0, \quad \langle f(t)f(0)\rangle = 2D\delta(t). \tag{1}$$

The beginning of theoretical studies on such a stochastic process dates back to the celebrated works by Einstein [2] who studied the Brownian motion of a free particle which may formally be considered as an overdamped motion in a parabolic potential. A more general study of overdamped stochastic processes was started by Smoluchowski [3] who formulated the equation of motion for the probability density in an arbitrary overdamped system: this equation bears his name nowadays. The next milestone was the work by Kramers [4] who, in particular, formulated the problem of the noise-induced *escape* for an overdamped system from a metastable potential well and derived its quasi-stationary solution: the value of the quasi-stationary escape flux was obtained in [4] using a stationary solution of the Smoluchowski equation:

$$J_{qs} = A_{qs}\mathrm{e}^{-\frac{\Delta U}{D}}, \qquad D \ll \Delta U, \tag{2}$$

where ΔU is the potential barrier (assumed to be much less than the noise intensity D) and A_{qs} is certain prefactor which depends on D weakly in comparison with the exponential (activation) factor.

As follows from [4], the escape flux becomes quasi-stationary when time greatly exceeds a characteristic value $t_{qe} \sim t_r \ln(\Delta U/D)$ where t_r is a characteristic relaxation

CP800, *Unsolved Problems of Noise and Fluctuations: UPoN 2005*
edited by L. Reggiani, C. Pennetta, V. Akimov, E. Alfinito, and M. Rosini

time. But what is the escape flux $J(t)$ on time scales $t \lesssim t_{qe}$? Such a question becomes relevant e.g. in the case when noise acts on the system only during a short time. Also, the relevance of short time scales for real experiments and applications grows as modern technologies develop.

There were only a few theoretical works on the escape in overdamped systems on short time scales. The most general of such works was the work by Shneidman [5] who solved the non-stationary Smoluchowski equation for an arbitrary potential using the method of the Laplace integral transformation while assuming that the quasi-equilibrium in the vicinity of the bottom of the well has been formed. The latter assumption is valid only for times significantly exceeding the relaxation time t_r while, for shorter times, results of [5] are invalid. Our *present* work does cover the latter range. We note also that, unlike [4, 5], our theory may be valid for large D too.

In parallel to the development of the escape problem in the 90th, there was an interesting discussion [6, 7, 8] on the *transition* problem on short times. Unlike the case of the escape, both the initial and final points of the transition differ from the stationary points of the noise-free system and, if they lie on one and the same slope of the potential, the transition may possess features distinctly different from those of the escape. Thus, basing on the method of *optimal fluctuation* [9] (for a recent review see [10]), the authors of [6] suggested that the short-time transition uphill the slope of the potential barrier may first relax nearly to the bottom of the well and only then go to the final point. They supported their suggestion by analytic calculations for the parabolic approximation of the potential and, seemingly, by the numerical calculations for the exact potential. However it was shown in [7] (also by means of numerical calculations within the optimal fluctuation method) that the transition path which first climbes up to the barrier top and only then relaxes to the final point may provide an exponentially larger activation factor. Is it possible to resolve this problem rigorously? And, if yes, may the most probable transition path (MPTP) possess more than one turning point? And may the MPTP switch its form jump-wise as the transition time changes? Our work answers all these questions too. Though the transition problem is more an academic problem, it may have interesting applications as discussed in the end of the paper.

BASIC EQUATIONS AND GENERAL RESULTS

To solve the problems formulated above, we use the method of optimal fluctuation [9, 10] i.e. we seek the flux in the form

$$J(t) = P(D,t)\mathrm{e}^{-\frac{S_a(t)}{D}}, \qquad D \ll S_a, \tag{3}$$

where the prefactor depends on D much more weakly than the activation (exponential) factor while the activation energy S_a is a minimum of the functional S, called action,

$$S_a = \min_{[x(\tau)]} S, \qquad S \equiv S[x(\tau)] = \int_0^t d\tau\, L, \qquad L = \frac{1}{4}\left(\dot{x} + \frac{dU}{dx}\right)^2, \tag{4}$$
$$x(0) = x_0, \qquad x(t) = x_f.$$

The necessary condition for a minimum of the functional is the equality of the variation ΔS to zero. The latter is equivalent to the Euler equation (EE):

$$\frac{\partial L}{\partial x} - \frac{\mathrm{d}}{\mathrm{d}t}\left(\frac{\partial L}{\partial \dot{x}}\right) = 0, \tag{5}$$

which reads, for a Lagrangian of the form (4)

$$\ddot{x} + \frac{\mathrm{d}\tilde{U}}{\mathrm{d}x} = 0, \qquad \tilde{U} = -\frac{1}{2}\left(\frac{\mathrm{d}U}{\mathrm{d}x}\right)^2. \tag{6}$$

So, the solutions of the EE are the trajectories of the auxiliary mechanical system (6).

The *quasi-energy*

$$E \equiv -\partial S/\partial t = \dot{x}(\partial L/\partial \dot{x}) - L = \left(\dot{x}^2 - (dU/dx)^2\right)/4 \tag{7}$$

is conserved along a solution of the EE, so that one easily derives from (7):

$$\dot{x} = \pm\sqrt{4E + (dU/dx)^2}. \tag{8}$$

It also follows from (7) that the range of allowed energies is:

$$E \geq E_{min} \equiv -\min_{[x_0,x_f]}\left[(dU/dx)^2/4\right]. \tag{9}$$

Eq. (8) can be integrated in quadratures. The activation energy S_a can be expressed in quadratures too.

Prefactor

To calculate the prefactor, we use the path-integral approach [9]. By means of the standard scheme [11] (its more general form may be found e.g. in [12]; cf. also [13, 14]), the numerical procedure is reduced to the numerical solution of some linear 2nd-order ordinary differential equation for the auxiliary function $Q_k(\tau)$ with certain initial conditions:

$$\frac{\mathrm{d}^2 Q_k}{\mathrm{d}\tau^2} + 2\frac{\mathrm{d}Q_k}{\mathrm{d}\tau}\left.\frac{\mathrm{d}^2 U(x)}{\mathrm{d}x^2}\right|_{x=x_k} + Q_k \left.\frac{\mathrm{d}^3 U(x)}{\mathrm{d}x^3}\right|_{x=x_k}\left(\frac{\mathrm{d}x_k}{\mathrm{d}\tau} - \left.\frac{\mathrm{d}U(x)}{\mathrm{d}x}\right|_{x=x_k}\right) = 0, \tag{10}$$

$$Q_k(\tau=0) = 0, \qquad \frac{\mathrm{d}Q_k(\tau=0)}{\mathrm{d}\tau} = 1,$$

where $x_k \equiv x_k(\tau)$ is the kth solution of the EE.

At the lowest order in D/S_a, the flux is expressed as:

$$J_0(t) = \sum_k P_k(D,t)\exp(-S_a^{(k)}/D), \qquad P_k = \frac{1}{2}\frac{\mathrm{d}x_k(t)}{\mathrm{d}t}\frac{1}{\sqrt{4\pi D Q_k}} \tag{11}$$

(typically, only the term with the lowest $S_a^{(k)}$ dominates in the sum).

Setup of simulations

To measure the flux in computer simulations, we use the following scheme. We put the system in $x = x_0$ and integrate the stochastic equation until some given time limit t_l is reached. Then the system is reset to the bottom and everything is repeated. After a sufficient statistics is achieved, the flux is calculated as

$$J(t) = (N_{res})^{-1}\frac{\Delta N_+(t) - \Delta N_-(t)}{\Delta t}, \tag{12}$$

where N_{res} is the overall number of resets while $\Delta N_+(t)$ and $\Delta N_-(t)$ are numbers of crossings of x_f during a short interval $[t, t+\Delta t]$ in the positive and negative direction respectively. The interval duration Δt should be chosen small in comparison with any relevant time-scale but large enough for $\Delta N_\pm(t)$ to greatly exceed 1.

DETAILED RESULTS

Escape problem

The escape problem may be defined as that when the initial state is at the bottom of the well i.e. $x_0 = x_w$ (which is the most natural initial state since the system stays in the absence of noise just in this state), while the final coordinate x_f is at the top of the barrier, x_b, or beyond it, for example at the bottom of the neighboring well (cf. Fig. 1(a)).

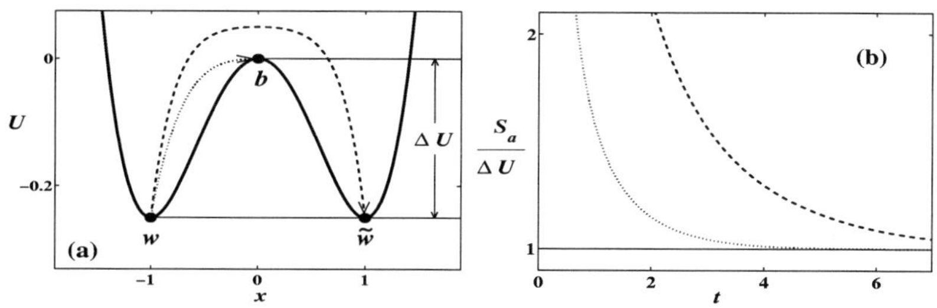

FIGURE 1. (a) The Duffing potential $U(x) = -x^2/2 + x^4/4$ (thick solid line) and schematic graphs for the escape from the left well (dotted line) and for the inter-well transition (dashed line); dots mark the minima of the potential wells and the top of the barrier. (b) The time dynamics of the normalized activation energy for the escape (dotted line) and for the inter-well transition (dashed line), calculated using Eq. (15); the asymptotic large-time value is shown by the solid line.

It follows from Eq. (9) that, if the interval $[x_o, x_f]$ contains at least one stationary point (i.e. the point where $dU/dq = 0$), then $E_{min} = 0$. Therefore the quasi-energy E for the solution of the EE for the escape problem is necessarily non-negative. This means in turn that, for each given escape time t: there is only one solution of the EE, $x(\tau)$, it is monotonous (i.e. without turning points), and it is given by the equation

$$\tau(x) = \int_{x_w}^{x} dq\, z(q,E), \qquad z(q,E) \equiv \text{sign}[x_f - x_0] \frac{1}{\sqrt{4E + (dU(q)/dq)^2}}, \qquad x \in [x_w, x_b], \tag{13}$$

where the quasi-energy E is determined by the final condition

$$t = \int_{x_w}^{x_f} dq\, z(q,E). \tag{14}$$

The activation energy can be expressed in quadratures too:

$$S_a(t) = \int_{x_0}^{x_f} dq\, \eta(q,E(t)), \qquad \eta(q,E) = \frac{1}{2}\left(\frac{\text{sign}[x_f - x_0](2E + (dU/dq)^2)}{\sqrt{4E + (dU(q)/dq)^2}} + \frac{dU}{dq}\right), \tag{15}$$

where $E \equiv E(t)$ is determined by Eq. (14).

The activation energy monotonously decreases with time since $dS_a/dt = -E$. Examples of $S_a(t)$ (for the Duffing oscillator) are shown in Fig. 1(b).

Fig. 2 shows the zero-order flux calculated by Eq. (11): there is only one term in the sum over k where the path $x(\tau)$ is given by Eqs. (13)-(14) while the corresponding activation energy is qiven by Eqs. (14)-(15). In the range of short times $t \lesssim t_r \equiv 1$, it is clearly in good agreement with the flux derived from the simulations, in contrast with the Shneidman theory [5] which drastically differs from the simulation results.

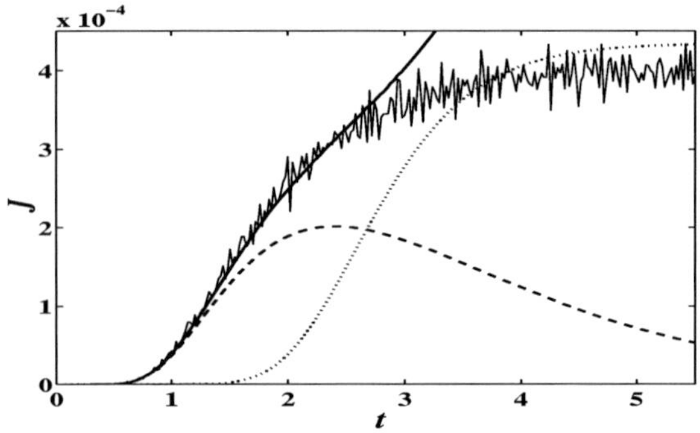

FIGURE 2. Dynamics of the escape flux from the bottom of the Duffing potential (as in Fig. 1) to the top of the barrier, for the Langevin equation (1) with $D = 0.04$. The flux was measured in computer simulations (thin jagged line), and calculated: (i) by Eqs. (11) and (16) (thick solid line), (ii) by Eq. (11) (dashed line), and (iii) by the Shneidman theory i.e. by Eq. (17) of [5] (dotted line).

Assuming that the first-order (in D/S_a) correction in the probability density is equal to zero, the first-order correction to the flux $J_0(t)$ reduces to

$$\Delta J^{(1)}(t) = \frac{1}{2} D \frac{\partial \ln(Q)}{\partial x_b} \frac{\exp(-S_a/D)}{\sqrt{4\pi D Q}}. \tag{16}$$

This formula nicely describes the simulations up to $t \sim 3 \equiv t_{qe}$. However the assumption on which it is based is not obvious, and it is a challenging problem either to rigorously prove it or, otherwise, find the true first-order correction to the probability density.

Transition problem

Consider the transition within a potential $U(x)$ monotonous in the interval $[x_0, x_f]$ (apart from academic interest [6, 7, 8], such a case may have an interesting application discussed at the end of the paper). Then E_{min} (9) is negative and, therefore, the solution of the EE may possess a *negative* quasi-energy E. This means, in turn, that a solution of the EE may possess turning points and, generally speaking, there may be more than one solution of the EE for a given transition time t. Detailed results are given in [15]. Fig. 3 demonstrates this for the transition within the left slope of the barrier in the Duffing potential (as in [6, 7, 8]). Note that, for the time $t = 8$ considered in [6, 7, 8], there is a third solution (corresponding to two turning points in the transition path) which was missed in [6, 7, 8]. At the same time, the action along this solution is not minimal and therefore its statistical weight in real transitions is negligible. Moreover, we have rigorously proved a general theorem stating that, for arbitrary potential and transition time and initial and final points, the action along a solution of the EE with two or more turning points cannot be minimal. Thus, the *MPTP can either be monotonous or possess one turning point*. As time varies, the MPTP may either vary continuously (cf. Fig.3) or switch jump-wise at the instant t_j when the actions along MPTPs of different topologies coincide. In the latter case, the function $S_a(t)$ possesses a bend at $t = t_j$; this scenario may occur even in the Duffing potential [15] provided the initial and final points of the transition are properly chosen, but its most pronounced manifestation occurs for the potential where the second derivative d^2U/dq^2 is non-monotonous: cf. Fig. 4.

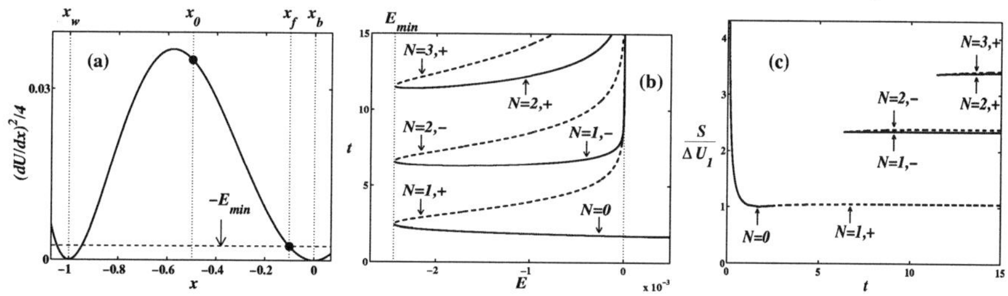

FIGURE 3. (a) The function $(dU(x)/dx)^2/4$ (thick solid line) for the Duffing potential as in Fig. 1; the dots show the points on the curve which correspond to $x_0 = -0.5$ and $x_f = -0.1$; $-E_{min}$ (9) is indicated by the dashed line. (b) Different branches of $t(E)$, corresponding to different topologies of the extremal path, are shown by solid/dashed lines with labels indicating a number of turning points and a sign of an initial velocity $\dot{x}(0)$ multiplied by the sign of $x_f - x_0$. (c) Different branches of action $S(t)$ (normalized by $\Delta U_1 \equiv U(x_f) - U(x_0) = 0.1044$) are marked similarly to the corresponding branches of $t(E)$ in (b).

Fig. 5 demonstrates the switching of the MPTP in simulations: 5 of 6 sequentially simulated paths with $t = 14.1$ concentrate near the corresponding theoretical path *without* turning points while all 6 simulated paths with $t = 15.2$ concentrate near the theoretical path which *turns back* close to $x = x_w = -1$.

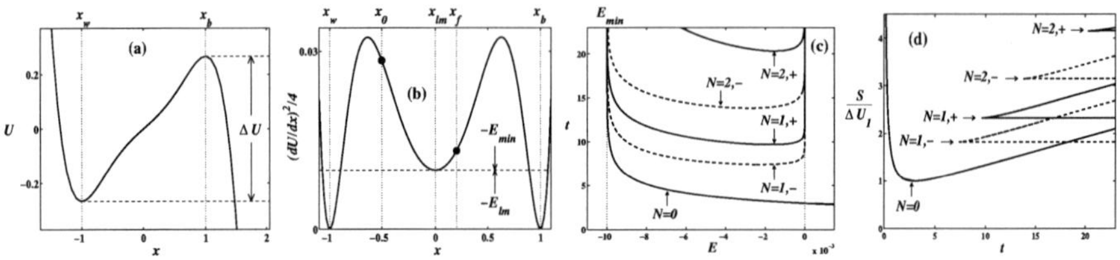

FIGURE 4. (a) The potential $U(x) = -x^5/5 + 0.8x^3/3 + 0.2x$; and (b) the corresponding function $(dU(x)/dx)^2/4$ (dots indicate its values at $x_0 = -0.5$ and $x_f = 0.2$); E_{min} coincides with the singularity energy E_{lm} related to the local minimum of $(dU(x)/dx)^2/4$. Figures (c) and (d) are analogous to Figs. 3(b) and 3(c) respectively. The normalization in (d) is: $\Delta U_1 \equiv U(x_f) - U(x_0) \approx 0.16915$.

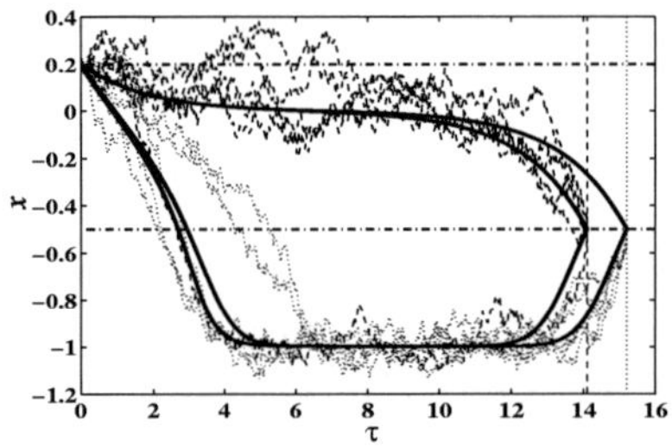

FIGURE 5. Two series of sequentially measured transition paths $[x(\tau)]$ in the potential as in Fig. 4, with $D = 0.01$, $x_0 = 0.2$, $x_f = -0.5$, for the final instant $t = 14.1$ (jagged dashed lines) and $t = 15.2$ (jagged dotted lines) respectively. The coordinates 0.2 and -0.5 are marked by the dash-dotted lines; the instants 14.1 and 15.2 are marked by the thick dashed and dotted lines respectively; the theoretical extremal paths corresponding to the branch "$N = 0$" and to the lower part of the branch "$N = 1, -$" (cf. Fig. 4(d)) are drawn by the thick solid lines.

UNSOLVED PROBLEMS

We propose to solve problems of two kinds.

1. Theory. It would be important to match our results in the asymptotic limit $D \to 0$ with those by Shneidman [5] and Kramers [4] (see also [13] for the re-derivation of the Kramers result within the path-integral approach). In particular, it would be important to find rigorously the 1st-order correction (see the discussion in the end of the subsection "Escape problem").

2. Generalizations and Applications. It would be interesting to generalize our theory for the following situation. Let the bistable potential be driven by a weak noise and a moderately slow periodic force of an amplitude such that the instantaneous potential has no barrier for a short interval in the middle of the driving period. The probability during this interval to transit to the opposite side of the potential (and then to turn out in the opposite well at the end of the period) is rather high but, at the same time, it may strongly and non-trivially depend on the noise intensity as simulations demonstrate [16]. A generalization of our theory for short-time transitions would be very helpful for predicting the ultimate transition probability as a function of noise intensity and driving parameters. This problem may have a relation with the hearing problem [16] (cf. also the discussion of the relevance of a similar model to neurons [17]).

SUMMARY

We have presented the complete general theoretical description of the escape/transition problem in overdamped systems on time scales less than or order of the relaxation time. The theory is successfully verified in simulations. In case of the escape, i.e. when the starting state is the stable state of the noise-free system, the most probable escape path $q(\tau)$ does not possess turning points while the flux J monotonously grows as function of the escape time t. In case of a transition within one and the same slope of the potential, (i) the most probable transition path (MPTP) can be either monotonous or possess just one turning point, (ii) as the transition time changes, the MPTP may either change continuously at any time or switch jump-wise at some instant, (iii) the function $J(t)$ is non-monotonous. Unsolved problems include: (i) matching of our theory with the theory [5], which is seemingly valid in the complementary range of escape times in the asymptotic limit of weak noise, and with the quasi-stationary limit within the path-integral approach [13]; (ii) generalization of our theory for the case when the bistable potential is driven by a moderately slow and strong periodic force.

ACKNOWLEDGEMENTS

S.M.S. and T.L.L. acknowledge the hospitality of Pisa University where some part of this work was done.

REFERENCES

1. H. Risken, *The Fokker-Planck Equation*, 2nd ed., Springer-Verlag, Berlin, 1992.
2. A. Einstein, Ann. Physik **17**, 549 (1905) and **19**, 371 (1906).
3. M. v. Smoluchowski, Ann. Physik **48**, 1103 (1915).
4. H. A. Kramers, Physica **7**, 284 (1940).
5. V.A. Shneidman, Phys. Rev. E **56**, 5257 (1997).
6. B. E. Vugmeister, J. Botina, and H. Rabitz, Phys. Rev. E **55**, 5338 (1997).
7. R. Mannella, Phys. Rev. E **59**, 2479 (1999).
8. B. E. Vugmeister, J. Botina, and H. Rabitz, Phys. Rev. E **59**, 2481 (1999).
9. R.P. Feinman, A.R. Hibbs, *Quantum Mechanics and Path Integrals* (McGraw-Hill, New York, 1965).
10. D.G. Luchinsky, P.V.E. McClintock, M.I. Dykman, Rep. Prog. Phys. **61**, 889 (1998).
11. I.M. Gel'fand, A.M. Yaglom, J. Math. Phys. **1**, 48 (1960).
12. J. Lehmann, P. Reimann, P. Hanggi, Phys. Rev. E **62**, 6282 (2000).
13. K. M. Rattray and A. J. McKane, J. Phys A: Math. Gen. **24**, 4375 (1991).
14. K. Kirsten and A. J. McKane, J. Phys. A: Math. Gen. **37**, 4649 (2004).
15. S.M. Soskin, V.I. Sheka, T.L. Linnik, R. Mannella, in *Noise in Complex Systems and Stochastic Dynamics*, eds. L. Schimansky-Geier, D. Abbott, A. Neiman and C. Van den Broeck, Proceedings Series volume 5114, SPIE, Washington, 2003, pp. 289-300.
16. N.G. Stocks, private communication.
17. B. Lindner, contribution to this volume.

NOISE IN BIOLOGICAL SYSTEMS

"Fatal Scream" Of Bacteria Infected By Phages: Nanoscale Detection Of Bacteriophage Triggered Ion Cascade

Maria D. King,[1] Sungkyu Seo,[2] Jong Kim,[2] Mosong Cheng,[2] Ryland Young,[1] Robert J. Biard,[2] Sergey M. Bezrukov,[3] Claes-Goran Granqvist[4] and Laszlo B. Kish[2,+]

[1] *Department of Biochemistry and Biophysics, Texas A&M University, 2128 TAMU, College Station TX 77843-2128*
[2] *Department of Electrical Engineering, Texas A&M University, College Station, TX 77843-3128, USA*
[3] *Laboratory of Physical and Structural Biology, NICHD, National Institutes of Health, Bethesda, MD 20892, USA*
[4] *Angstrom Laboratory, Solid State Physics, Uppsala University, P.O.B 534, Uppsala, S-75121, Sweden*

Abstract. A rapid, inexpensive and specific identification of arbitrary bacteria under field conditions is urgently needed. To this end, we have introduced and tested a new technology, called SEPTIC, SEnsing of Phage-Triggered Ion Cascade. In its prototype form based on a nanowell chip, SEPTIC has already been shown to be capable of unambiguous identification of live bacteria on a time scale of seconds to minutes, many times faster than any other system. The technology is based on using noise analysis to detect the massive ionic fluxes associated with the initial step of bacteriophage infection, the injection of the phage DNA into the cell. Here we show the results and pose a number of unsolved problems of noise. Ultimately, sensors based on this new technology would be able to save many lifes.

Keywords: Bacterium identification; Bacteriophage, Biochip; Nanowell; Stochastic signals.
PACS: 87.19.St, 07.50.Hp

INTRODUCTION

Recently, a method for the rapid detection and identification of bacteria was proposed [1]. The new method, SEnsing of Phage-Triggered Ion Cascade (SEPTIC), is detecting and analyzing the electrical field caused by the stochastic emission of ions during phage infection by measuring the microscopic voltage fluctuations in a nanowell device [1-4], in less than 10 minutes. Virtually, a single bacterium can be detected and identified [5,6]. Because the phage infection is a very selective process, where only bacteria of a specific strain are infected, a SEPTIC-biochip containing an array of sensors where each sensor is

[+] Corresponding author, Laszlo.Kish@ee.tamu.edu

CP800, *Unsolved Problems of Noise and Fluctuations: UPoN 2005*
edited by L. Reggiani, C. Pennetta, V. Akimov, E. Alfinito, and M. Rosini

sensitized with a different phage, can detect and identify all relevant bacteria with extraordinary speed and selectivity. Though the method works with high reproducibility, presently there are more questions about its biology and physics than answers.

ABOUT THE BIOLOGICAL AND PHYSICAL BACKGROUND

In this talk, we will discuss the phage biology behind the phenomenon and also consider some practical physical conditions which may play an important role in determining the observed effects. Unfortunately, the exact details of the ion channels and the dynamics of ion emission, which are important to understand the possible charging effects, are unknown.

Bacteriophages are the most numerous biological entities, estimated at 10^{31} in the biosphere, and are unimaginably diverse [7]. Attempts to exploit the specificity of phages in detection and identification of pathogenic bacteria have been burdened by the requirement of culturing the target bacteria, growing the infected culture, and assaying the production of progeny virions, processes which at minimum require hours and also knowledge of the culture conditions is required. However, when we considered the fundamental pathway of the phage infection process, a potential way to avoid these limitations was suggested. The committed step in bacteriophage infection is irreversible adsorption. For double-stranded DNA (dsDNA) phages, this results from interactions between the specific adsorption apparatus, usually tail fibers, with specific receptors on the surface of the host cell [8,9,10]. For two of the three main morphotypes of dsDNA phages, the myophages with contractile tails and the siphophages with flexible tails, the injection of DNA into the host cell follows rapidly and involves the transitory formation of a channel through which the phage DNA passes into the target cytoplasm [12, 13, 14]. Concomitant with injection is a short-lived membrane depolarization and an efflux of ions, including a substantial fraction of the ~0.2M potassium salts present in the cytoplasm, at a rate of $\sim10^6$ /sec per infected cell [11, 14]. This phenomenon represents an ideal opportunity for bacterial diagnostics, because it not only takes advantage of the well-known specificity available in bacteriophage but it can occur, given sufficient phage concentration, within seconds after admixture of the virions and cells [15]. Moreover, it requires no culturing of the analyte sample and detects only live bacteria.

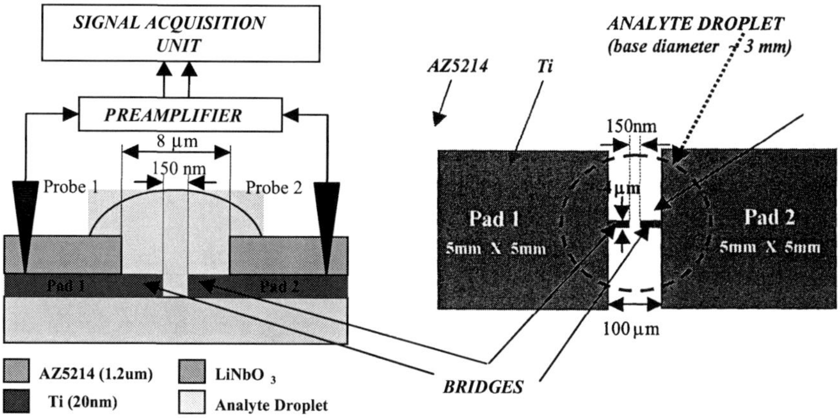

FIGURE 1. Experimental setup and the nanowell. The nanowell with the analyte droplet. Side view (left) and upper view (right).

The nanowell is a lateral capacitor with a 150nm gap between the two 4x4 micron size Ti plates. The power density spectrum of the electrical field fluctuations when there is reaction (phage infection) has $1/f^2$ shape, see Figure 2, which implies correlation times beyond the time window of observation.

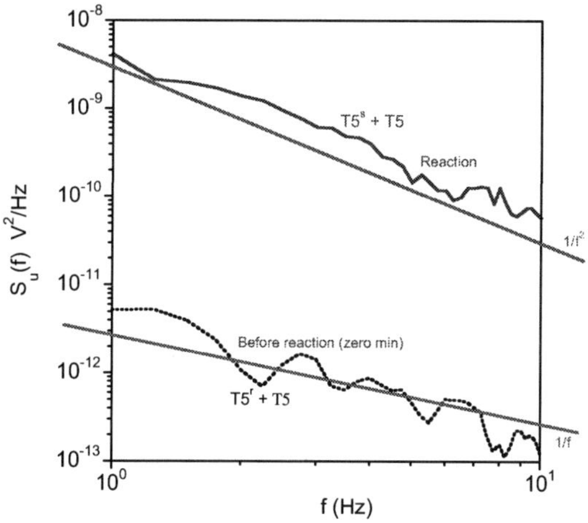

FIGURE 2. Power density spectrum before the DNA passage and during the injection. T5 phages and the sensitive *E. coli* bacteria. The reaction takes place in two steps. During the First Step Transfer (FST) 8% of the phage DNA enters the cytoplasm, followed by a 4min pause, during which proteins A1 and A2 are synthesized to facilitate the Second Step Transfer (SST) of the rest (92%) of the phage DNA in a 1-5 minute time period.

Taking the above-mentioned facts into the account, the following aspects are to be considered during the talk:
i) The phage attachment to the specific receptors on the bacterial surface, triggering the opening of ion channels.
ii) The difference between the DNA injection mechanisms of the two siphophages (lambda versus T5).
iii) Currents and ion numbers: the net ion charge vs. the current observed during the measurements.
iiii) Mean distance between bacteria: about 10 micron during the experiments published so far.
v) Charging energy and mirror force: if the ion emission is not neutral, the bacteria may get charged. Considerations of the minimal effect (based on the equipartition energy, kT/2) indicate that the charging is of the order of 1000*e. That would yield a considerable mirror force within the Debye screening length. Assuming an electrochemical process with 1 eV characteristic energy behind the ion transport would allow an ionization number up to 40,000.
vi) Debye length < 1 micron (the deionized water has about 1 micron Debye length); therefore mirror forces seem to be screened and can not serve as a possible explanation of bacteria attraction to the surface.
vii) Diffusion coefficients/times of ions. D is in the order of 10^{-9} m^2/s, thus to diffuse out of the nanowell's vicinity (8 micron), they would need less than 0.1 second.
viii) Diffusion coefficients/times of bacteria. D is in the order of 10^{-12} m^2/s, thus bacteria need about 40 seconds to diffuse out of the nanowell's vicinity.
ix) Debye relaxation time and high-pass filter response. Charge changes within the Debye relaxation time can get through the Debye screening; however, this phenomenon is characterized by strong high-pass filtering (time-derivative) characteristics.

EARLIER MODEL CONSIDERATIONS AND UNSOLVED PROBLEMS

The physical basis of the fluctuations detected using the nanowell device has not been unambiguously determined, however, a reasonable model can be proposed. During the process of DNA injection of a siphophage or a myophage, each irreversibly-adsorbed virion triggers the opening of a single channel in the bacterial envelope, by as yet unknown components, through which the phage DNA molecule passes. After the DNA passes through into the cytoplasm, the channel is sealed, usually within 1 minute and, again, by unknown processes. Bulk solution measurements have shown that injection is coupled to transient cellular depolarization requiring ion flows on the order of 10^8 ions per infected cell. The emitted ions will execute a rapid Brownian motion and many will be able to escape from the vicinity of the bacterium. Together with the randomness of both the timing and the spatial orientation of the ion emission, these ion leakage events would generate stochastic spatiotemporal electrical field fluctuations at the micron or submicron scale, as detected in the nanowell device. In this part, we list a few earlier ideas about the noise generation mechanisms. Most of them do not explain the experimental results and, for the valid picture, see the next section. A valid picture has to explain both the magnitude and the slow dynamics of the noise.

a) Thermal noise? No, and neither it is the Brownian motion of ions. Reason: that would decrease by increasing ion concentration because of the corresponding lower resistance.

b) Excess motion of bacteria? No, they are not motile.

c) Spatiotemporal charge density fluctuations caused by the emitted ion clouds? No, that dissipates too fast, see point vii above.

d) Displacement currents? Due to the high-pass filter effect, to produce a $1/f^2$ noise by fast charging spikes is possible only if there are strong inter-spike correlations (e.g., spike frequency modulation), however the efficiency would be very low. So far, no indication exists that the bacteria would produce a sequence of charging spikes with the proper modulation. Still, because this effect may cause mirror forces beyond the Debye length, it should be further investigated.

e) Concentration cell? A bacterium asymmetrically located in the nanowell and slowly emitting ions may yield a voltage due to the concentration cell effect. Because of the slow diffusion dynamics of bacteria (see point viii), this effect is a possible candidate. Open questions are if, during phage infection, the bacteria leak ions for a prolonged time, which would be necessary for this effect to be a proper explanation.

UNDERSTANDING WHY THE METHOD WORKS WITH PASSIVE ELEMENTS

Recently, a breakthrough took place in the understanding of the underlying working mechanism of SEPTIC [5,6]. A DC current flowing through the fluid between the two sensor electrodes acts as a very efficient collection mechanism of the ionized bacteria via drift in the electric field. This mechanism makes it possible that the SEPTIC method works without active (amplifying) sensing elements. It is easy to generate such an electrical field which is able to collect all ionized bacteria at the electrode within the measurement time window [5,6]. Thus, after optimization and proper design, the detection of a few bacteria in 10 microliter fluid will be possible [5,6].

The proposed enhancement by the DC current [5,6], is consistent with subsequent experimental observations. It turned out that the only reason, why the SEPTIC method was working at the experiments reported earlier, was a parasitic small DC current through the input shunt resistance of the preamplifier due to some electrochemical asymmetry of the nanowell, which was producing a DC voltage in the mV range. Applying the preamplifier in AC mode should result in the termination of the input DC current, and consequently, in the disappearance of the SEPTIC signal.

UNSOLVED PROBLEMS FOR MEDICAL APPLICATIONS

There are many important unsolved problems about the SEPTIC method. Some examples follow here:

1. What is the optimal electrical field and geometry [5,6]?
2. Effect of the salt concentration?
3. Response of other phages and bacteria?
4. The timing, dynamics and ion composition of ion emission?
5. How to construct a pen-size biolab with a disposable biochip for the instantaneous detection and identification of a library of bacteria?

REFERENCES

1. M. D. King, S. Seo, J.U. Kim, R.F. Young, M. Cheng and L.B. Kish, *Journal of Biological Physics and Chemistry* **5**, 3-7 (2005).
2. L.B. Kish, M. Cheng, J.U. Kim, S. Seo, M.D. King, R. Young, A. Der and G. Schmera, *Fluctuation and Noise Lett.*ers **5** (1): L105-L108 (2005).
3. M.D. King, M. Cheng, S. Seo, J.U. Kong, L.B. Kish and R.Young, *Proc. 3rd International SPIE Symposium on Fluctuations and Noise, Conference on Noise in Biological Systems*, Austin, TX, May 24-26, (2005).
4. L.B. Kish, S.M. Bezrukov, A. Der, M. Cheng, M.D. King, R. Young, S. Seo and J. Kim. *Proc. 3rd International SPIE Symposium on Fluctuations and Noise, Conference on Noise and Information in Nanoelectronics, Sensors and Standards*, Austin, TX, May 24-26, (2005).
5. R.J. Biard and L.B. Kish, *Fluctuations and Noise Letters* **5** (June 2005), in press.
6. R.J. Biard and L.B. Kish, Enhancing the sensitivity of the SEPTIC bacterium detection method by concentrating the phage-infected bacteria via dc electrical current, TAMU-TEES Patent Disclosure, May 18, 2005.
7. R.W. Hendrix, M.C. Smith, R.N. Burns, M.E. Ford and G.F. Hatfull, *Proc Natl Acad Sci.USA*. **96**, 2192-2197 (1999).
8. E.B. Goldberg, L. Grinius and L. Letellier, *Molecular biology of bacteriophage T4. In* Karam, J. D. et al. (eds.), American Society for Microbiology, Washington D.C., 1994, pp. 347-357.
9. A.A. Lindberg, *Annu Rev Microbiology* 27, 205-241 (1973).
10. U. Henning and S. Hashemolhosseini, *Molecular biology of bacteriophage T4. In* Karam, J. D. et al. (eds.), American Society for Microbiology, Washington D.C., 1994.
11. L. Letellier and P. Boulanger, *Biochimie* **71**, 167-174 (1989).
12. S. Silver, E. Levine and P.M. Spielman, *J Virology* **2**, 763-781, (1968).
13. D.J. MacKay and V.C. Bode, *Virology* **72**, 154-166 (1976).
14. P. Boulanger and L. Letellier, *J Biol Chemistry* **263**, 9767-9775 (1988).
15. M. Schwartz, *J Mol Biology* **103**, 521-536 (1976).
16. E.V. Kalasauskaite, D.L. Kadisaite, R.J. Daugelavicius, L.L. Grinius and A.A. Jesaitis, *Eur J Biochemistry* **130**, 123-130 (1983).

Can Electrical Vestibular Noise Be Used for the Treatment of Brain Diseases?

Yoshiharu Yamamoto*, Rika Soma*, Zbigniew R. Struzik* and Shin Kwak†

*Educational Physiology Laboratory, Graduate School of Education, The University of Tokyo, 7–3–1 Hongo, Bunkyo-ku, Tokyo 113–0033, Japan
†Department of Neurology, Graduate School of Medicine, The University of Tokyo, 7–3–1 Hongo, Bunkyo-ku, Tokyo 113–0033, Japan

Abstract. The therapy currently available for the treatment of degenerative neurological diseases is far from satisfactory, and a novel therapeutic strategy, especially for pharmacologically unresponsive patients, would be welcomed. The vestibular nerves are known to influence neuronal circuits in the medullary cardiovascular areas and, through the cerebellar vermis, the basal ganglia and the limbic system. By means of noisy galvanic vestibular stimulation (GVS), it may now be possible to ameliorate blunted responsiveness of degenerated neuronal circuits in the brains of multiple system atrophy (MSA) and/or Parkinson's disease (PD) patients, through a mechanism known as stochastic resonance. We evaluate the effect of 24-hour noisy GVS on long-term heart rate dynamics in seven MSA patients, and on daytime locomotor activity dynamics in twelve patients with either PD or levodopa unresponsive parkinsonism. Short-range heart rate variability and long-range anti-correlation of trunk activity are significantly increased by the noisy GVS compared with sham stimulation, suggestive of improved autonomic and motor responsiveness. The noisy GVS is effective in boosting the neuro-degenerative brains of MSA and/or PD patients, including those unresponsive to standard levodopa therapy.

INTRODUCTION

The beneficial role of noise and fluctuations in biological systems has been extensively studied over the past decade. One of the fundamental phenomena considered in this respect is that of stochastic resonance (SR) [1, 2], which is generally defined as the optimization of responses of a non-linear system to a weak input signal caused by external noise. For instance, in neural systems SR has been shown to optimize the sensitivity of sensory neurons [3, 4, 5, 6], which may further lead to an enhancement of perception and/or behavior in animals [7] and humans [8, 9, 10]. In addition, we have recently demonstrated that externally added noise can also enhance input-output coherence within the human brain [11, 12, 13].

A natural extension of this line of research is an inquiry into whether a similar mechanism would facilitate impaired neural transmission associated with intractable brain diseases, for which a novel therapeutic strategy, especially for pharmacologically unresponsive patients, is still needed.

In this paper, we first introduce our experimental observations on SR at the level of the brain stem, where noise added mechanically to the brain can help trigger changes in heart rate in response to small changes in blood pressure [11, 12], i.e., the baroreflex response. This is followed by an introduction to our clinical findings [14] showing that similar

CP800, *Unsolved Problems of Noise and Fluctuations: UPoN 2005*
edited by L. Reggiani, C. Pennetta, V. Akimov, E. Alfinito, and M. Rosini

mechanically noise-enhanced responses of the brain can also be used in the treatment of neurological patients with primary autonomic failure (PAF) [15].

Then, we describe our recent study [16] showing that, by means of noisy electrical or galvanic vestibular stimulation (GVS) applied by a portable device, it may be possible to ameliorate the blunted autonomic responsiveness of degenerated neuronal circuits of multiple system atrophy (MSA) [17] patients and further motor responsiveness of Parkinson's disease (PD) [18] patients. Finally, we discuss some practically and theoretically "unsolved problems" for the application of the SR concept in neurological sciences.

SR IN HUMAN BAROREFLEX

In the first study of functional SR in the brain, in 2000, we demonstrated experimentally that noise can enhance the homeostatic function in the human blood pressure regulatory system [11]. We studied the human baroreflex system, in which an increase (or a decrease) in blood pressure is automatically compensated for by decreases (or increases) in heart rate and vascular resistance via the autonomic nervous system.

The baroreflex system has two kinds of pressure-sensitive nerve cell receptors. By tilting a subject back and forth on a horizontal table, as described in Ref. [11], we periodically moved blood to the lower part of the body. The draining of blood from the chest area stimulated the cardiopulmonary baroreceptors to fire a weak repeating signal, which the brain interpreted as a drop in (central) venous blood pressure. To create neural noise, we randomly applied and removed mechanical pressure to and from the neck. This caused the arterial baroreceptors to fire randomly, as the artery wall pressure, which normally indicates blood pressure in the arteries, was increasing and decreasing randomly (Fig. 1).

The results showed that the compensatory heart rate response to a weak periodic signal introduced at the venous blood pressure receptor was optimized by applying noise to the arterial blood pressure receptor. We thus concluded that this *functional* SR most likely resulted from the interaction between the noise and the signal in the brain stem, where the neuronal inputs from these two different receptors first join together.

Recently, we further conducted a similar experiment, but replaced the mechanical noise added to the arterial baroreceptor with an electrical one applied to the vestibular afferent nerves [12]. With this GVS, electrical current is delivered transcutaneously to the vestibular afferents through electrodes placed over the mastoid bones, modulating their continuous firing levels [19]. Through the vestibular nuclei in the brain stem, these afferents influence neuronal circuits in the medullary cardiovascular areas [20, 21] and thus the heart rate [22], just like when we stimulate the arterial baroreceptors (Fig. 1).

There are two advantages to using GVS. The first is a practical one. As shown in the later sections, SR invoked by externally added sensory noise can be used to treat patients with central nervous system dysfunction. While the mechanical device used in our previous studies [11, 14] cannot be used for this purpose, given its large size and immobility, the vestibular afferents can easily be stimulated by using a portable GVS apparatus. Secondly, the electrical stimulator is capable of applying more precisely

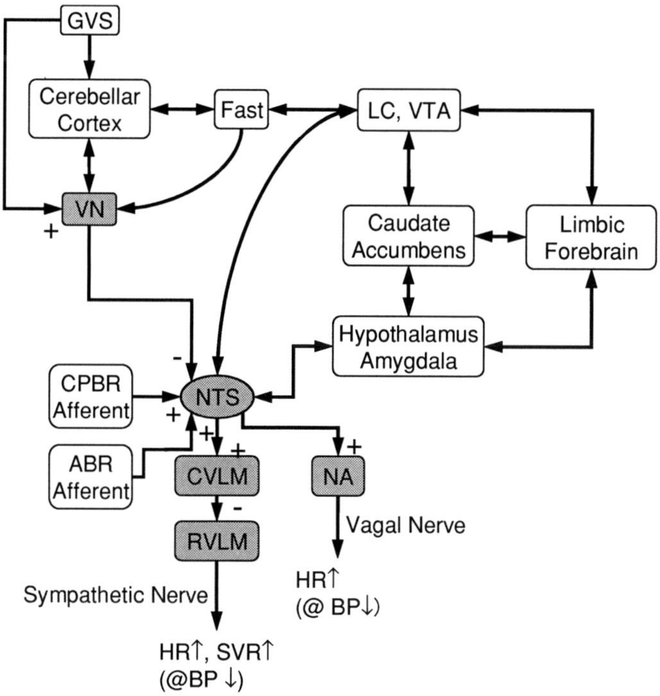

FIGURE 1. Schematic diagram showing connectivity within the brain responsible for autonomic and motor functions. GVS: galvanic vestibular stimulation; VN: vestibular nuclei; Fast: fastigial nuclei; LC: locus coeruleus; VTA: ventral tegmental area; CPBR: cardiopulmonary baroreceptors; ABR: arterial baroreceptors; NTS: nucleus tractus solitarius; NA: nucleus ambiguus; CVLM: caudal ventrolateral medulla; RVLM: rostal ventrolateral medulla; HR: heart rate; BP: blood pressure; SVR: systemic vascular resistance. The brain stem neuronal structures involved in autonomic functions, with their connections (+, facilitatory; −, inhibitory), are shaded in gray. Note that most of the other higher brain structures responsible for motor/affective, as well as autonomic, functions have a network-type connectivity.

controlled input noise than the mechanical actuator. In Ref. [12], we tested whether externally added Gaussian $1/f$ noise, in the frequency range of 0.01–2.0 Hz, where spike trains in single medullary neurons exhibiting spontaneous activity show $1/f$-type behavior [23], could more effectively sensitize the medullary cardiovascular areas than Gaussian white noise. In other words, we tested whether the noise mimicking *innate* fluctuations of the brain would be superior to *artificial* white noise.

When we examined the compensatory heart rate response to a weak periodic signal introduced via venous blood pressure receptors, while adding $1/f$ or white noise with the same variance to the brain stem through GVS, in both cases, this noisy GVS optimized the covariance between the weak input signals and the heart rate responses. However, the optimal level with $1/f$ noise was significantly lower than with white noise, suggesting a functional benefit of $1/f$ noise for neuronal information transfer in the brain.

NEUROLOGICAL APPLICATION FOR PAF

The above-mentioned baroreflex system plays a crucial role in maintaining adequate blood perfusion to the brain, and failure can easily lead to fainting, an annoying and potentially dangerous symptom of patients with autonomic failure. While most previous studies on sensory SR have dealt with the noise-induced sensitization in the responses of *normal* sensory neurons to weak input signals [3, 4, 5, 6, 7, 8, 9], we hypothesized that, in neurological diseases with blunted neural responses even to non-weak signals, noise might ameliorate the impaired neural transmission, just like the noise-enhanced detection of small signals in normal sensory systems. Thus, we first studied [14] whether adding noise to the arterial baroreceptors might facilitate cardiovascular responses to even strongly hypotensive stimuli in neurological patients with PAF [15]. PAF is an intractable disorder in which symptoms of autonomic dysfunction, including orthostatic hypotension, are primarily the result of progressive neuronal degeneration of unknown cause.

The patients were tested for their transient responses of heart rate and systolic and diastolic blood pressures to head-up tilt, with and without the continuous application of beat-to-beat mechanical noise to the arterial baroreceptors. The PAF patients exhibited marked drops in blood pressure and a blunted increase in heart rate upon transition from a supine to a head-up position. The addition of noise to the arterial baroreceptors significantly increased the magnitude of the heart rate increment and diminished that of drops in blood pressure. We thus concluded that the addition of external noise to baroreceptor signaling ameliorated the marked postural hypotension seen in patients with PAF.

NEUROLOGICAL APPLICATIONS USING GVS

Recently, we further examined [16], by means of noisy GVS applied by a portable device, whether it might be possible to ameliorate the blunted responsiveness of degenerated neuronal circuits of MSA [17] patients, the majority of our pool of PAF patients. In this study, we evaluated the effect of 24-hour noisy GVS, with the $1/f$ spectrum as used in Ref. [12], on long-term heart rate dynamics in MSA patients.

The magnitude of fluctuation in heart rate, or *heart rate variability*, has been widely used to evaluate the cardiac autonomic responsiveness in laboratory as well as ambulatory settings [24]. As the heart rate variability depends significantly on the time scale for which the magnitude of fluctuation is calculated [25], we first obtained the *scale-dependent* variability—the root-mean-square fluctuations (ms)—from whole day heart rate time series by using detrended fluctuation analysis (DFA) [26, 27]. DFA eliminates *non-stationarity* in the input data in the form of changes in the baseline and trends within the data windows at different scales, which could affect the calculation of the magnitudes of fluctuation, thus making this approach suitable for the analysis of long-term data such as those collected in that study.

With DFA, *scale-independent* characterization of the input data is also possible with the *(power-law) scaling exponent*, obtained as the slope of a straight line fit in the

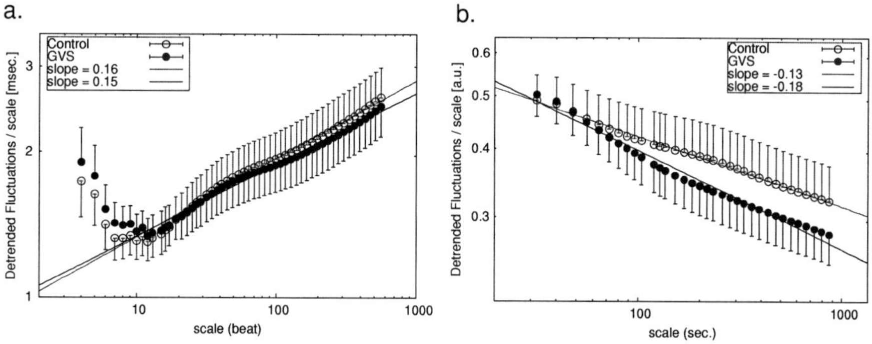

FIGURE 2. **a**, Scale dependency of the mean detrended fluctuation (normalized by the scale) of heart rate, measured by the heart interbeat intervals, with and without noisy GVS. The detrended fluctuations were calculated after linear trend removal for sliding windows at each scale [26, 27]. The vertical bars represent the standard deviations from the group means. **b**, The same as **a**, but for the trunk activity time series. Slopes of these plots are the (power-law) scaling exponents for heart rate increments (a) and trunk activity dynamics (b). Our exponent range coincides with the Hurst exponent [26, 27] range $[0-1]$ for "walks" such as integrated fluctuations, e.g., Brownian motion yields exponent 0.5. Therefore, our exponent becomes negative for "noise" or raw fluctuations, e.g., -0.5 for white noise. One can convert our exponent values to those as used in Peng et al [26] by adding 1.0.

double-logarithmic plot of time scales vs. magnitudes of fluctuation (Fig. 2) [26, 27]. This index reflects the probability of a simultaneous increase or decrease of the variability at two distant points in time in the time series, applied for all the distances up to *long-range* time scales, hence probing the nature of "switching" patterns between high and low values in a statistical sense. Larger scaling exponents indicate positive temporal auto-correlation or *persistency* in the increase or decrease, and the lower values correspond with negative auto-correlation or *anti-persistency*. The scaling exponent of heart rate variability is indeed reported to be greater in patients with neuro-degenerative autonomic failure than in healthy subjects [27], suggestive of an impaired autonomic control of heart rate. We calculated the scaling exponent of heart rate variability for time scales > 17 beats, where acceptable linearity was observed (Fig. 2a).

The application of noisy GVS resulted in increased heart rate variability at short time scales (< 10 beats) in MSA patients (Fig. 2a), with a significant increase from the value of 1.77 ms to 1.91 ms at a 4 beat resolution. We observed a slight decrease in the scaling exponent by noisy GVS. On an individual basis, four of the seven patients showed a significantly smaller scaling exponent for GVS than for the control. The increased short-range heart rate variability, especially at around 4 beats, typically corresponding to respiratory periods, is known to be a robust indicator of increased responsiveness of the parasympathetic nervous system [24]. Further, the decrease in the scaling exponent by noisy GVS, or an increase in long-range anti-persistency of the time series, suggests an alteration of heart rate dynamics in the direction of that of healthy individuals [27] with more "complex" variability [28]. Thus, we concluded that the noisy GVS increases the autonomic, especially parasympathetic, responsiveness in MSA patients.

The vestibular nerves are also known to influence neuronal circuits in the basal ganglia and the limbic system [29, 30] through the cerebellar vermis [20, 31] (Fig. 1). This possible vestibular effect on the basal ganglia and the limbic system is intriguing when considering clinical observations in PD [18] consistent with the lack of a limbic-to-motor link within the basal ganglia, i.e., the disinclination to move by dissociating motivation from executive (spontaneous) motor movement, of which the expressionless face of a PD patient is a common clinical example [32, 33].

Thus, in the same study [16], we further hypothesized that the noisy GVS could alleviate the disinclination to move, or the *bradykinesic* symptom, in PD and MSA patients. We evaluated the effect of 24-hour noisy GVS on daytime trunk activity dynamics in patients with either PD or pharmacologically unresponsive parkinsonism. Long-range anti-persistency of trunk activity patterns calculated by DFA [26, 27] for time scales > 32 s was significally increased by the noisy GVS (Fig. 2b), suggestive of a quickening of bradykinesic rest-to-active transitions [34]. Taken together, we concluded that the noisy GVS is effective in boosting the neuro-degenerative brains of MSA and/or PD patients, including those unresponsive to standard pharmacological therapy, and in improving their autonomic and motor responsiveness.

DISCUSSION—UNSOLVED PROBLEMS

In the above study, we hypothesize that, in central neuro-degenerative disorders with a blunted neural responsiveness, input noise might ameliorate the impaired neural transmissions, and confirm a scenario analogous to that of the noise-enhanced detection of weak signals in normal sensory systems in most neurobiological SR studies. Thus, the most plausible mechanism for the increased autonomic and motor responsiveness in MSA and/or PD patients is that of SR. However, further studies are needed to determine the exact mechanism of the improved neural transmissions caused by the noisy GVS. SR theory predicts that there will be an optimal level of noise, with higher noise levels possibly being deleterious [35]. Thus, to prove the SR scenario, further studies with a higher level of noisy GVS would be needed, although such an intensive application of GVS may also lead to significant postural responses [36] as a side effect.

The theory of SR has been rigorously studied [2] in dynamical systems, such as overdamped bistable systems or monostable systems with a response threshold—the latter being used as a model for sensory neurons. However, neuronal networks within the brain may not have a definite threshold or bistability, and they rather can be regarded as mutually coupled excitable and/or oscillatory units [37], as shown in Fig. 1, capable of showing multistable dynamics. Thus, we believe extended theories of SR in complex networks are needed to understand how the noisy GVS works.

We find that non-nociceptive applications of noisy GVS by a portable device increase short-range heart rate variability and long-range anti-persistency of trunk activity in MSA and PD patients, suggestive of an increased responsiveness in the relevant neuronal circuits. The patients examined in study [16] were hospitalized and medicated, and this possibly made them less symptomatic, and to some extent masked the effect of noisy GVS. However, even in these challenging conditions, we found consistent effects of noisy GVS; note in addition that half of the patients showing parkinsonism were unresponsive to standard levodopa therapy [38]. Further evaluation is needed of the effect of the portable noisy GVS on the symptoms and quality of life in ambulatory patients with central neuro-degenerative disorders.

CONCLUSION

We have shown here that noise added to the human brain is effective in boosting the neuro-degenerative brains of MSA and/or PD patients, including those unresponsive to standard levodopa therapy. Although the exact mechanism of this therapeutic effect is currently unknown, it may be possible to use external sensory noise in treating intractable and prevalent neurological diseases such as PD.

ACKNOWLEDGMENTS

This study was in part supported by the Japan Science and Technology Agency and the Toyota Motor Corporation.

REFERENCES

1. Wiesenfeld, K., and Moss, F., *Nature*, **373**, 33–36 (1995).
2. Gammaitoni, L., Hänggi, P., Jung, P., and Marchesoni, F., *Rev. Mod. Phys.*, **70**, 223–287 (1998).
3. Douglass, J. K., Wilkens, L., Pantazelou, E., and Moss, F., *Nature*, **365**, 337–340 (1993).
4. Levin, J. E., and Miller, J. P., *Nature*, **380**, 165–168 (1996).
5. Collins, J. J., Imhoff, T. T., and Grigg, P., *J. Neurophysiol.*, **76**, 642–645 (1996).
6. Nozaki, D., Mar, D. J., Grigg, P., and Collins, J. J., *Phys. Rev. Lett.*, **82**, 2402–2405 (1999).
7. Russell, D. F., Wilkens, L. A., and Moss, F., *Nature*, **402**, 291–294 (1999).
8. Collins, J. J., Imhoff, T. T., and Grigg, P., *Nature*, **383**, 770 (1996).
9. Richardson, K. A., Imhoff, T. T., Grigg, P., and Collins, J. J., *Chaos*, **8**, 599–603 (1998).
10. Simonotto, E., Riani, M., Seife, C., Roberts, M., Twitty, J., and Moss, F., *Phys. Rev. Lett.*, **78**, 1186–1189 (1997).
11. Hidaka, I., Nozaki, D., and Yamamoto, Y., *Phys. Rev. Lett.*, **85**, 3740–3743 (2000).
12. Soma, R., Nozaki, D., Kwak, S., and Yamamoto, Y., *Phys. Rev. Lett.*, **91**, 078101–1–4 (2003).
13. Kitajo, K., Nozaki, D., Ward, L. M., and Yamamoto, Y., *Phys. Rev. Lett.*, **90**, 218104–1–4 (2003).
14. Yamamoto, Y., Hidaka, I., Iso-o, N., Komai, A., Soma, R., and Kwak, S., *Brain Res.*, **945**, 71–78 (2002).
15. Bannister, R., and Mathias, C. J., "Clinical features and evaluation of the primary chronic autonomic failure syndromes," in *Autonomic Failure, A Textbook of Clinical Disorders of the Autonomic Nervous System*, edited by C. J. Mathias and R. Bannister, Oxford University Press, Oxford, 1999, pp. 307–316, 4 edn.
16. Yamamoto, Y., Struzik, Z. R., Soma, R., Ohashi, K., and Kwak, S., *Ann. Neurol.*, **to appear** (2005).
17. Gilman, S., Low, P. A., Quinn, N., Albanese, A., Ben-Shlomo, Y., Fowler, C. J., Kaufmann, H., Klockgether, T., Lang, A. E., Lantos, P. L., Litvan, I., and Mathias, C. J., *J. Neurol. Sci.*, **163**, 4–5 (1999).
18. Barbeau, A., "Parkinson's disease: clinical features and etiopathology," in *Handbook of Clinical Neurology*, edited by P. Vinken, G. W. Bruyn, and H. L. Klawans, Elsevier Publishers, Amsterdam, 1986, vol. 49 (Revised Series 5) Extrapyramidal disorders, pp. 87–152, 4th edn.
19. Fitzpatrick, R. C., and Day, B. L., *J. Appl. Physiol.*, **96**, 2301–2316 (2004).
20. Barmack, H. H., *Brain Res. Bull.*, **60**, 511–541 (2003).

21. Balaban, C. D., and Beryozkin, G., *Exp. Brain Res.*, **98**, 200–212 (1994).
22. Radtke, A., Popov, K., Bronstein, A. M., and Gresty, M. A., *Lancet*, **356**, 736–737 (2000).
23. Lewis, C. D., Gebber, G. L., Larsen, P. D., and Barman, S. M., *J. Neurophysiol.*, **85**, 1614–1622 (2001).
24. Task Force of the European Society of Cardiology and the North American Society of Pacing and Electrophysiology, *Circulation*, **93**, 1043–1065 (1996).
25. Peng, C. K., Mietus, J., Hausdorff, J. M., Havlin, S., Stanley, H. E., and Goldberger, A. L., *Phys. Rev. Lett.*, **70**, 1343–1346 (1993).
26. Peng, C. K., Havlin, S., Stanley, H. E., and Goldberger, A. L., *Chaos*, **5**, 82–87 (1995).
27. Struzik, Z. R., Hayano, J., Sakata, S., Kwak, S., and Yamamoto, Y., *Phys. Rev. E.*, **70**, 050901(R)–1–4 (2004).
28. Goldberger, A. L., *Lancet*, **347**, 1312–1314 (1996).
29. Albert, T. J., Dempesy, C. W., and Sorenson, C. A., *Biol. Psychiatry*, **20**, 1267–1276 (1985).
30. Anderson, C. M., Polcari, A., Lowen, S. B., Renshaw, P. F., and Teicher, M. H., *Am. J. Psychiatry*, **159**, 1322–1328 (2002).
31. Newlands, S. D., and Perachio, A. A., *Brain Res. Bull.*, **60**, 475–495 (2003).
32. Chaudhuri, A., and Behan, P. O., *J. Neurol. Sci.*, **179**, 34–42 (2000).
33. Joseph, R., *Neuropsychiatry, Neuropsychology, and Clinical Neuroscience*, Williams & Wilkins, Baltimore, 1996, 2nd edn.
34. Ohashi, K., Amaral, L. A. N., Natelson, B. H., and Yamamoto, Y., *Phys. Rev. E*, **68**, 065204(R)–1–4 (2003).
35. Moss, F., Ward, L. M., and Sannita, W. G., *Clin. Neurophysiol.*, **115**, 267–281 (2004).
36. Pavlik, A. E., Inglis, J. T., Lauk, M., Oddsson, L., and Collins, J. J., *Exp. Brain Res.*, **124**, 273–280 (1999).
37. Varela, F., Lachaux, J. P., Rodriguez, E., and Martinerie, J., *Nat. Rev. Neurosci.*, **2**, 229–239 (2001).
38. Teychenne, P. F., Bergsrud, D., Racy, A., Elton, R. L., and Vern, B., *Neurology*, **32**, 577–583 (1982).

Background processes and the generation of the flicker noise in nanochannel transport

I. D. Kosińska and A. Fuliński

Jagiellonian University, M. Smoluchowski Institute of Physics Reymonta 4, PL 30-059 Kraków, Poland

Abstract. Abundance of $1/f$ (flicker) noise is now a well-established and well-known empirical fact. Still, in spite of about 40 years of investigations, the generating mechanism (or mechanisms) leading to the flicker noise remains largely unknown. It was shown recently (Z. Siwy, and A. Fuliński, Phys. Rev. Lett. 89, 158101 (2002)) by comparison of the synthetic and biological channels that the power spectrum of the nanochannel's ionic currents $S(f) \sim 1/f^{\alpha}$ with $\alpha \approx 1$ originates from the channel's opening-closing process and is related to the motion of charged constituents of the walls of the pore. Here we present the model based on that idea, treating these motions as the restricted Brownian process in constant external electric field (applied voltage), and influenced by time-dependent electric fields exerted by the charges (ions) moving through the channel. Our aim is to find which of these conditions and processes are sufficient, and which are necessary to reproduce the appearance of the flicker noise.

Keywords: 1/f noise; nanochannels; Brownian Dynamics
PACS: 05.40.Ca, 05.40.Tp, 87.16.Ac, 05.40.-k

INTRODUCTION

Recent results [1] seem to support the conjecture [2] that the flicker noise in biological ionic channels is caused by "the equilibrium conductance fluctuations related to the conformational flexibility of the channel pore structural constituents" and that it is not an "inherent property of ion transport". Here we attempt a more detailed theoretical examination of this conjecture by the Brownian Dynamics simulations of the single-file motion of K^+ ions through the fluctuating gate in a charged nanochannel. We limit ourselves to the modelling of processes in the synthetic nanochannel described in [1], where the motions of subunits constituting the channel wall, and resulting in the changes of the effective diameter of the pore and of its local charge, are relatively simple in comparison with the motions and charge shifts in the peptides forming the biological channels. The process of preparation of the synthetic pore reported in [1] w results in the formation of negatively charged carboxylate groups being the end groups of – sticking from the pore walls – loose ends of cleaved polymer strands, so-called "dangling ends" [3], which move continuously in a more or less random fashion [4]. The pore consisted of two cones (lengths about 20 nm and 10 μm) linked at their narrow apertures ("gate", "selectivity filter", less than 2 nm diameter in its maximally open state). The flicker noise related to the process of random (nonmarkovian) process of the opening/closing of the channel was observed when the ionic flow through the channel was induced by voltages about 180 mV.

The basic assumptions of the model considered in this work are as follows: (i) The

CP800, *Unsolved Problems of Noise and Fluctuations: UPoN 2005*
edited by L. Reggiani, C. Pennetta, V. Akimov, E. Alfinito, and M. Rosini

process of the opening/closing of the channel is understood simply as the the changes of the diameter of the narrow part (gate) of the pore; (ii) only the processes in the narrow part of the pore are taken into account; (iii) the pore is cation-selective, therefore we consider solely the passage of K^+ ions; (iv) the cations pass through the gate in a single file (v) the changes of pore's effective diameter result from the changes in the inclination of the "dangling ends" with respect to the pore's walls; (vi) we do not postulate any specific form of the correlation or distribution functions of the considered processes, couplings, etc.

MODEL – BROWNIAN DYNAMICS

The Langevin-type equations of motion for the cations moving in a single file along the z-axis of the channel and for the "dangling ends" in the gate read, respectively:

$$m_i \ddot{z}_i = -\gamma_i \dot{z}_i + R_i + F_i, \tag{1}$$

$$I\dot{\vec{\omega}}_k = -\tilde{\gamma}_k l_k^2 \vec{\omega}_k + \vec{l}_k \times \vec{R}_k + \vec{l}_k \times F_k, \tag{2}$$

where m_i is the mass of i-th ion, $\vec{\omega}_k$, I_k and l_k – angular velocity, inertia and length of the k-th dangling end, γ_i, $\tilde{\gamma}_k$ – the friction coefficients, F_i, $\vec{F}_k$ – sums of deterministic forces, and R_i, $\vec{R}_k$ – the random forces assumed here to be the thermal noise represented by the Gaussian white noise.

Friction coefficients are assumed to result from the water viscosity η via the Stokes law: $\gamma_i = 6\pi r_i \eta$, $\tilde{\gamma}_k = 6\pi f_0 r_k \eta$, where r_i, r_k are the radii of the cations and of the negative charges (carboxylate groups) at the end of the dangling ends, and f_0 is the correction resulting from the non-spherical shape of the dangling end (we assumed $f_0 = 2$).

The deterministic forces experienced by the cations and by the carboxylate groups consist of (i) the applied external force (from voltage U extended ove the length L):

$$\vec{F}_{ext,j} = q_j U/L, \quad j = i,k \tag{3}$$

(ii) the internal Coulomb force between all the charges j (cations and dangling ends):

$$\vec{F}_{int,j} = \sum_{m \neq j} \vec{F}_{C,jm} = \sum_{m \neq j} \frac{q_i q_j}{4\pi\varepsilon} \frac{\vec{r}_{jm}}{r_{jm}^3}, \quad j,m = i,k \tag{4}$$

and (iii) the short-range repulsive forces [5]:

$$\vec{F}_{SR,j} = \sum_{m \neq j} \vec{F}_{SR,jm} = F_{SR} \sum_{m \neq j} \frac{\vec{\tilde{r}}_{jm}}{\tilde{r}_{jm}^{11}}, \quad j,m = i,k \tag{5}$$

with $F_{SR} = 2 \times 10^{-10}$ N [5], r_{jm} – the distance between charges q_j and q_m, and $\tilde{r}_{jm}$ – the contact distance corrected by 0.16 nm (i. e., $\tilde{r}_{jm} = r_{jm} + 0.16$ nm) for like ions [5].

Besides, we add the "harmonic" force acting on the dangling ends, which prevents the dangling ends to remain close to the channel walls, i. e., simulates repulsion from other negative charges on the channel's walls, not taken into account directly in our model:

$$\vec{F}_{harm,k} = -\,|\vec{F}_{ext,k}|\,\hat{\mathbf{s}}, \tag{6}$$

with $\hat{\mathbf{s}}$ being the radial direction.

RESULTS

Values of the parameters used in the calculations are: masses and radii of the (bare, i. e. non-hydrated) potassium ions i and of the carboxylate groups k $m_i = 39$ a.u., $m_k = 40$ a.u. (a.u. – atomic units), $r_i = 0.133$ nm., $r_k = 0.14$ nm, length of the “dangling end" -CO-(C_6H_4)-COO^- 0.584 nm, radius of the channel in the most narrow part (the gate) $r_0 = 0.85$ nm, double-conical channel opening angle $\alpha = 0.05$ [1], length of the modelled fragment of the channel = 20 nm, with the gate located in the middle, the external voltage $U = 0.18$ V, extended over length $L = 10\mu$m [1], and the force constant B in Eq.(6) chosen to be equal and of opposite sign to the force from the external voltage. Other physical parameters: temperature $T = 298$ K, $D = k_BT/\gamma_{K^+} = 48 \times 10^{10}$ s kg^{-1}, $\varepsilon = 80.1\varepsilon_0$, ε_0 being the vacuum dielectric susceptibility, water viscosity $\eta \approx 9 \times 10^{-4}$ N s m^{-2}.

In the numerical integration of equations of motion (Brownian Dynamics) the random forces $R_j(t)$ were generated with the Gaussian probability distribution with the width

$$\langle R_j^2(t)\rangle = 2\gamma_j k_B T/\delta t\,, \quad j = i,k. \tag{7}$$

δt being the time-step used in the integration of the Langevin equations, chosen to be equal to m_j/γ_j [5]. For potassium ions K^+ this gives $\delta t \approx 30$ fs.

We assume the following boundary conditions: cations can both enter and leave the simulated channel's fragment depending on the temporary conditions inside the channel, i. e., on the presence/absence of other ions in the neighbourhood the channel's ends. Besides, the cations cannot jump over each other (single-file motion) and cannot pass through closed gate. The gate is closed when the distance between the negative charge and the channel's axis is smaller than some prescribed value Δ^*. In the calculations the value $\Delta^* = 1.1(r_i + r_k)$ was used.

Simulations were performed for the series consisting of 5 000 000 time-steps $\delta t = 30$ fs. It was found that (i) the number of cations $N_c(t)$ present inside the 20 nm channel varies from 1 to 41 (maximal registered number in the whole series), (ii) the cations leave the right end of the channel in bursts consisting of $N_n \in [1,8]$ ions, (iii) bursts are separated by time lags $\Delta t_n = M_n\delta t$ with observed $M_n \in [1,3675]$. The distribution of M_n has the long heavy tail – the histogram of the initial part (up to $M = 600$) of this distribution is shown in Fig. 1.

The current $I(t)$ through the channel was calculated as the number of cations (elementary charges e) in the given burst per unit time: $I(t) = eN_n/\Delta t_n$.

The momentary values of all the computed quantities (angles, positions of cations, diameter $\Delta R(t)$ of the channel at the gate, $I(t)$, $\Delta(t)_n$, $N_c(t)$, etc.) form various stochastic series. Fig. 2 shows teh power spectra $S(f)$ for the stochastic variables $I(t)$ and $\Delta R(t)$ defined above. The current fluctuations show distinct flicker-noise behavior $S(f) \sim f^{-\alpha}$ with $\alpha \approx 1.05 \pm 0.1$, which is in a very good agreement with the results found in [1]

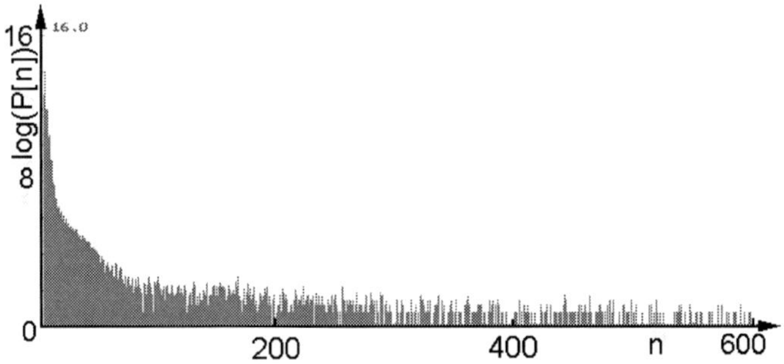

FIGURE 1. Fig. 1. Histogram of $\Delta t_n = m_n \times 30$ fs between subsequent bursts of cations through the right end of the channel, for the first 600 values of n.

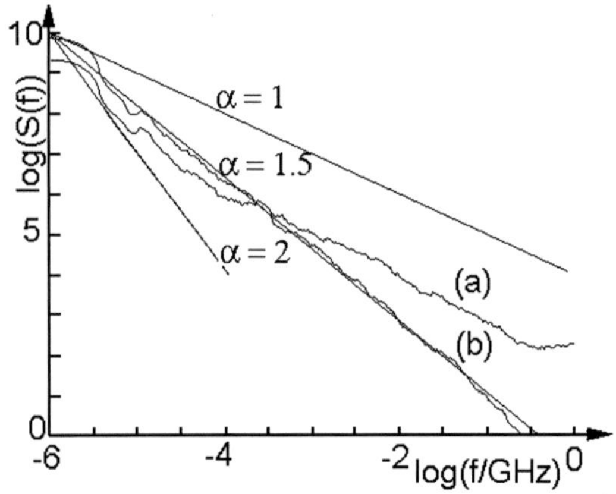

FIGURE 2. Fig. 2. Power spectra of the calculated current $I(t)$ and the width of the channel $\Delta R(t)$.

for the experimental data both for the biological and the synthetic channels. The gate diameter, as well as some other quantities fluctuate in time with $\alpha \approx 1.5$.

It was found in [1, 6] that the experimental data mentioned above are nonmarkovian. We calculated the nonmarkovian index (measure) G for the series $I(t)$ in the same way as in [1, 6] and found $G \approx 0.85$ in good agreement with the values found for biological channel ($G \approx 0.85$ and synthetic channel ($G \approx 0.9$). Such values of the index G imply strongly nonmarkovian character of these series (for purely random dichotomous series of the comparable lenght $G \approx 0.0014$ [6]).

Another important stochastic characteristics is the so-called Hurst exponent H: if the stochastic process $X(t)$ has the property $X(at) = a^H X(t)$, the process $X(t)$ is self-similar with index H. We found $H = 0.93 \pm 0.03$ for the series $I(t)$, whereas for the biological channel reported in [6] the Hurst index is $H = 0.84 \pm 0.08$ [7]. Such values imply the existence of the long memory in the examined stochastic series [7], which seems to correspond with the strong nonmarkovianity of these series reported above.

The self-similar character of the series $I(t)$ was confirmed by the calculation of the power spectrum for the cumulated currents, i. e. by summing the currents over longer time intervals $\tilde{t}_m = m\delta t$, and found that the power spectra for $m = 10$ and $m = 100$ do not differ significantly from the original $S(f)$.

COMMENTS AND CONCLUSIONS

The various stochastic characteristics calculated from the Brownian Dynamics model simulation of the motion of cations through very narrow nanochannel show remarkable similarity to such characteristics found earlier both for the ionic currents flowing through biological [1, 6, 7] and synthetic [1] nanochannels. The most notable is the $1/f$-noise characteristics of the power spectra of these currents. Our calculations imply that most crucial for the appearance of the flicker noise are the single-file condition and the stopping of the ions by the closed gate. Other important factor is the presence (though not the intensity) of the repulsive forces between ions. On the other hand, reasonable changes in the numerical values of various parameters, e. g. the use of the values of radii and masses of hydrated (instead of bare) ions, of the exact values of the external voltage gradient and of the force constant B or shape of the functionf_{harm} in Eq.(6), and even of the exact value of the gate's radius Δ^* or of the noise intensity, influence little the stochastic characteristics described above.

These observations seem to confirm earlier conjectures that (i) this type of noise is produced hy fluctuator dynamics [2], (ii) it is not important whether the channel is made of synthetic polymer or of biological proteins [1], and (iii) the crucial factor is the motion of the "dangling ends" [1] or of structural elements of the channel-forming proteins [2].

The values of the nonmarkovianity index and of the Hurst exponent, together with the observed heavy long tails in the distributions of calculated stochatic variables (Fig. 1) seem to suggest that the underlying microscopic processes are a superposition of stochastic and chaotic dynamics, the first resulting from the presence of thermal noise, the second – from the nonlinear deterministic forces. This is an open question which deserves further analysis.

REFERENCES

1. Z. Siwy, and A. Fuliński, Phys. Rev. Lett. 89, 158101 (2002); "1/f Noise in Ion Transport through Nanopores: Origins and Mechanism", in: *Unsolved Problems of Noise and Fluctuations (Proc. Third Int. Conf. on Unsolved Problems of Noise in Physics, Biology, and High Technology, Washington 2002)*, edited by S. M. Bezrukov, Melville, New York, 2003, pp. 273-282.
2. S. M. Bezrukov, and M. Winterhalter, Phys. Rev. Lett. 85, 202-205 (2000); S. M. Bezrukov, "The status of 1/f noise research in biological systems: empirical picture and theories", in *Proc. First Int. Conf. on Unsolved Problems of Noise, Szeged 1996*, edited by C. R. Doering, L. B. Kiss, and M. F. Schlesinger, World Scientific, Singapore, 1997, pp. 263-274.
3. A. Wolf, N. Reber, P. Yu. Apel, B. E. Fischer, and R. Spohr, *Nucl. Instrum. Meth. B* **105**, 291-293 (1995).
4. M. Doi, *Introduction to Polymer Physics*, Oxford University Press (1997); K. Schmidt-Rohr, W. Hu, and N. Zumdulyadis, *Science* **280**, 714-717 (1998).
5. G. Moy, B. Corry, S. Kuyucak, and S.-H. Chung, *Biophys. J.* **78**, 2349 (2002); S. Kuyucak, O. S. Andersen, and S.-H. Chung, *Rep. Prog. Phys.* **64**, 1427 (2001).

6. A. Fuliński, Z. Grzywna, I. Mellor, Z. Siwy and P. N. R. Usherwood, *Phys. Rev. E* **58**, 919 (1998).
7. S. Mercik, and K. Weron, *Phys. Rev. E* **63**, 051910 (2001).

Signal processing problems of neurocardiological fluctuations

Zoltán Gingl*, László Rudas† and Péter Makra*

*Department of Experimental Physics, University of Szeged, Dóm tér 9, Szeged,H-6720 Hungary
†Medical Intensive Care Unit, Cardiology Centre, Faculty of Medicine, University of Szeged, Korányi fasor 7, Szeged, H-6720 Hungary

Abstract. The spectral analysis of electrocardiogram (ECG) and blood pressure fluctuations is an important tool both in medical diagnostics and in theories that endeavour to account for the complex, feedbacked patterns of control mechanisms that regulate human circulation. In the case of ECG signals, the information is in the rate of heartbeats, which means that it is not the whole signal but only a so-called RR signal, containing the distances of beats, that serves as the basis of spectral processing. This RR signal is inherently discrete and unevenly sampled, which introduces a number of methodological problems: fast Fourier transformation (FFT) needs even sampling, and this means that some re-sampling (inserting new samples at equal distances) is necessary. In the case of blood pressure signals, the information is both in the amplitude itself and in the rate, consequently both an RR-like and a full recording-based treatment are usable here. In the present paper we review the different interpolation methods used in re-sampling, along with the Lomb periodogram that does not require re-sampling at all, and contrast the results they yield both for a model sine and for real signals. We also argue for a full recording-based treatment of blood pressure fluctuations, as they require no peak detection and thus avoid the artefacts that peak detection may cause.

Keywords: Heart rate variability, power spectral analysis
PACS: 87.19.Hh

INTRODUCTION

The analysis of neurocardiological signals and their fluctuations has long been established as an important tool of medical diagnostics. Apart from heart diseases, this tool may provide information on the functioning of many other organs. The signals most frequently measured and studied include the electrocardiogram (ECG), blood pressure, respiration rate and blood flow. These signals can now be treated using sampling techniques, which has opened up great possibilities for an efficient analysis of their fluctuations, making it feasible to use these as sources of information. Now we know that the analysis of heart rate fluctuations may warn of heart failure and other cardiac diseases, and may also help to cast light on the intricacies of the regulatory system that controls human circulation and respiration [1, 2, 3].

Processing neurocardiological signals involves a number of methodological difficulties. At the heart of these problems lies the fact that we should like to map a net of very complex sensory and regulatory mechanisms, which are supposed to be continuous in time, solely on the basis of a signal which is inherently discrete and unevenly sampled, since it is based on heartbeats. This raises a number of difficult questions as one needs to clarify whether the long-standing methods which are of great use in other areas (such as FFT and re-sampling) are fully adequate here, and if not, what should replace them.

CP800, *Unsolved Problems of Noise and Fluctuations: UPoN 2005*
edited by L. Reggiani, C. Pennetta, V. Akimov, E. Alfinito, and M. Rosini

In what follows, we set out to explore the problems associated with the analysis of sampled ECG and blood pressuresignals and weigh the pros and cons of different re-sampling methods used here, along with their potential alternatives. We also recommend a method for the spectral analysis of blood pressure signals that is easier to use and less sensitive to artefacts than the techniques most widely used now.

RE-SAMPLING METHODS AND THEIR ALTERNATIVES IN THE SPECTRAL ANALYSIS OF ECG FLUCTUATIONS

When analysing ECG signals, one rarely takes the whole ECG voltage time series as the basis of examinations. It is more informative rather to concentrate on the distance between heartbeats; as the most prominent component of a heart cycle is the so-called *R wave*, the separation between successive R waves, the *RR intervals*, lends itself as an appropriate measure of heart rhythm (see Figure 1). An RR interval time series is a series of ordered pairs of time instances where R waves occur and the corresponding time intervals that separate them from previous R waves. A *heart rate* series is very similar, only it contains the reciprocals of RR intervals at the time instances corresponding to R waves.

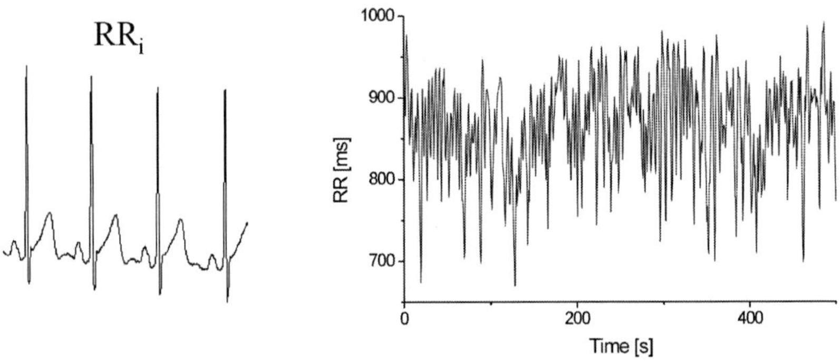

FIGURE 1. R waves and an RR interval time series.

As heartbeats are somewhat irregular (in fact, the information they contain lies in their fluctuations, their very irregularity), RR intervals are inherently unevenly distanced (unevenly 'sampled,' so to speak, if we conceive of them as resulting from the sampling of a supposed continuous regulatory signal in the background). This uneven sampling poses no problem to time-domain analysis, which uses mainly statistical indices of the fluctuations of RR intervals, but if one wants to carry out spectral investigations to explore the rhythmic components of the control mechanisms behind heart rate, it introduces a number of difficulties. The most obvious option for spectral assessment would be FFT, but it is fully dependent on even sampling (equal distances between samples).

There have been several approaches established in neurocardiological practice to circumvent this problem. One way is to enable the use of FFT through *re-sampling*:

inserting new samples at equal distances. Here it is still variable how to render RR values to the newly inserted sampling instances: the options are interpolation (linear or spline) or zero padding (simply inserting zeros as new samples).

This approach has met considerable resistance: opponents have argued that re-sampling means the insertion of new heartbeats that have nothing to do with reality. In our view, this argument fails to take into account that after re-sampling, the data points are no longer connected to the literal meaning off RR intervals (the distance between successive R waves)—otherwise an equally sampled RR sequence would suggest perfectly regular heartbeats, which is, of course, not the case. It is more fruitful to conceive of this re-sampling as a simple calculation tool, as part of the assessment of the regulatory mechanisms controlling human circulation (see Figure 2), wherein the lengthening of RR intervals indicates the dominance of inhibitory trends, and *vice versa*, and when we insert a new sample we simply evaluate the regulatory signal at an intermediate point on the basis of neighbouring points.

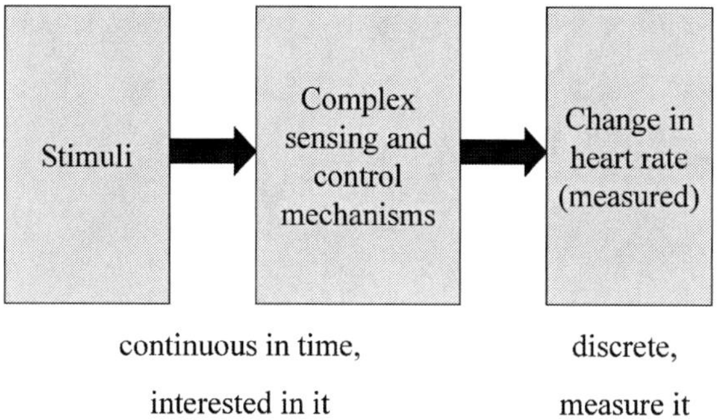

FIGURE 2. Heart rate modulation: assessing a continuous control mechanism through the analysis of a discrete signal.

The other major approach drops the idea of FFT completely and applies other spectral techniques that do not require even sampling. The most widespread of these is the *Lomb periodogram*, which is basically a Discrete Fourier Transform (DFT) made invariant to time rescaling [4, 5]. This makes re-sampling unnecessary, of course.

The major problem of assessing the adequacy and general applicability of these techniques lies in the fact that we do not know what the correct spectrum looks like (it is exactly this spectrum that we should like to obtain). So we applied a model signal whose spectrum we exactly know (a sine) and compared the results the different methods yield. We also examined whether these methods diverge significantly in the case of real RR signals. What we have found reinforced the view that there is no significant difference between the general performance of these methods, although they may differ in the degree of sensitivity to various artefacts (eg, small number of samples).

THE SPECTRAL ANALYSIS OF BLOOD PRESSURE FLUCTUATIONS

Apart from the ECG recordings, the continuous monitoring of blood pressure is also an important source of information both in diagnosis and in research aiming at exploring the intricate patterns of control mechanism that links heart rate to blood pressure and to respiration rate. Often ECG and blood pressure signals are recorded simultaneously.

To conform to the treatment of ECG recordings, many researchers substitute the series of systolic blood pressure peaks (SBPs) for the whole blood pressure signal (see Figure 3). This, of course, creates a discrete and unevenly sampled signal like RR, with the same methodological problems. Here we discuss the practical and theoretical concerns associated with this practice.

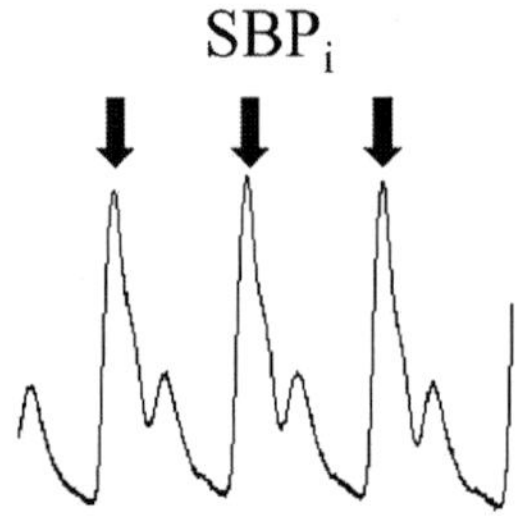

FIGURE 3. Systolic blood pressure peaks.

First, one may argue that, in contrast to ECG signals, the information in blood pressure recordings lies not only in the rate of peaks, but in the totality of the signal with each instantaneous value taken into account. This suggests that the SBP-based treatment essentially means losing information.

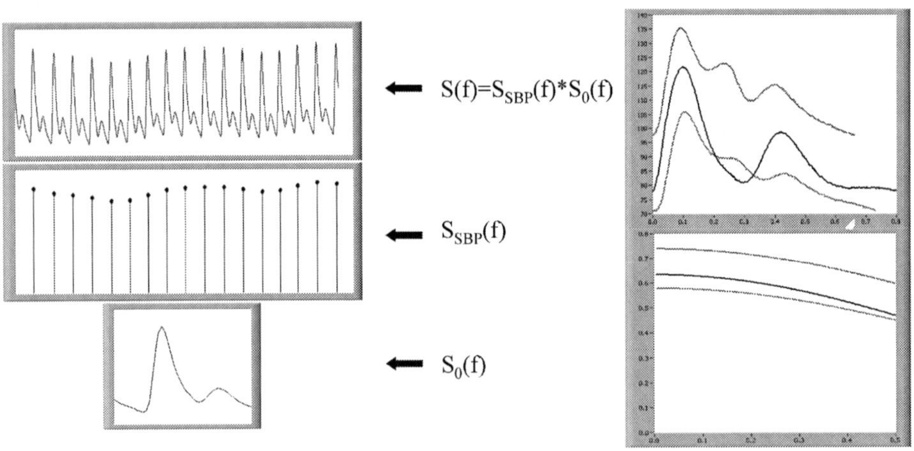

FIGURE 4. The link between the full blood pressure spectrum and the SBP-based spectrum.

The problem seems less severe if one considers that a continuous blood pressure signal can be approximated as the convolution of the series of SBP peaks and a one-period blood pressure waveform. In the spectral domain, it means that the spectrum of the full blood pressure recording is approximately the product of the SBP series spectrum and the spectrum of a one-period waveform. The latter is approximately constant in the frequency range of interest (below 0.5 Hz), therefore the difference between the SBP-based and the full recording-based treatment is basically a constant multiplier here (see Figure 4). So one may argue that it is more practical to take the whole blood pressure signal and Fourier transform it, regardless of theoretical considerations, as it requires no pre-processing (the recordings are evenly sampled) and consequently it avoids the artefacts that peak detection may introduce (missing or spurious peaks), while the spectrum it yields is essentially proportional to the SBP-based spectrum at the relevant frequencies (see Figure 5).

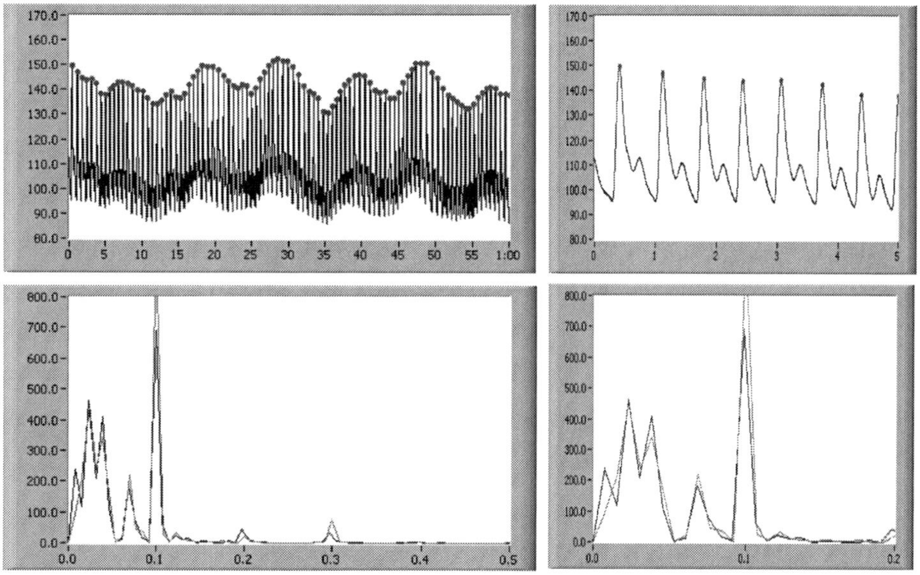

FIGURE 5. Full blood pressure spectra versus SBP spectra. The upper panels show the blood pressure in time domain, with solid dots representing SBP values. The lower panels depict the full signal based and the SBP-based spectra of these blood pressure signals together.

UNSOLVED PROBLEMS

The area of the interdependent regulation systems that controls heart rate, blood pressure and respiration rate is still riddled with open questions. There has not been a comprehensive model established yet which would adequately account for the complex feedbacked interactions that link these bodily functions. Consequently, it is problematic to judge which of the established methods offers the most faithful description of the fluctuations that inform us about these regulatory mechanisms. It is also an open question whether the non-linear descriptors stemming from chaos theory, which are gaining hold in noise and fluctuation research, may be better measures, and if they are, whether their use can be introduced to practical diagnostics.

ACKNOWLEDGMENTS

Our research has been supported by OTKA (Hungary), under grant T037664.

REFERENCES

1. F. Lombardi, G. Sandrone, S. Pernpruner, R. Sala, M. Garimoldi, S. Cerutti, G. Baselli, M. Pagani, and A. Malliani, *Americal Journal of Cardiology*, **60**, 1239–1245 (1987).
2. J. P. Saul, Y. Arai, R. D. Berger, L. S. Lilly, W. S. Colucci, and R. J. Cohen, *Americal Journal of Cardiology*, **61**, 1292–1299 (1988).
3. G. Parati, J. P. Saul, M. Dirienzo, and G. Mancia, *Hypertension*, **25**, 1276–1286 (1995).
4. N. R. Lomb, *Astrophysics and Space Science*, **39**, 447–462 (1976).
5. P. Laguna, G. B. Moody, and R. G. Mark, *IEEE Transactions on Biomedical Engineering*, **45**, 698–715 (1998).

Dynamics Of α-Helix Proteins: Highly Structured Noise Generated By Soliton Modes

Marcelo Silva and Per-Anker Lindgård

The QuP Centre ,Institute of Physics, The Danish Technical University, 2800 Lyngby, Denmark

Abstract. The great advantages of using the new pump-probe techniques and free electron tunable lasers for studying dynamics in proteins, has renewed the interest in more realistic modeling and theories for the non-linear phenomena in bio-molecules. In particular effects originating from solitonic behavior and/or trapped modes may give distinct signatures in the measured spectra (pioneered by Austin *et al* [1] and Hamm *et al* [2]). We have constructed a fairly realistic 3D α-helix model and studied the dynamics, including the effect of thermal noise by means of computer simulations. We use various perturbations to create excitations; in particular a kick at one residue – simulating a local lattice deformation caused by the chemical stimulus from a molecule changing quantum states. This generates a multitude of modes, including at least two kinds of solitons: one moving on the Hydrogen-chains along the extremities of the helix, and one Davidov-like soliton moving along the back bone and carrying along a packet of the peptide-bond excitations (Amino-I C-O stretch mode), as well as phononic modes. The unsolved problem with respect to noise is to understand and describe the highly structured propagation of energy, which is found to occur in the model □-helix – and which is probably also present in real α-helix proteins.

Keywords: Dynamics, proteins, structured noise.
PACS: 05.45.-a, 05.40.Ca, 36.20.-r, 82.53.Hn

INTRODUCTION

Noise plays undoubtedly a large role in the physics of biological systems: both in a constructive way and the 'absence of a destructive' way. We shall consider an example of the latter here with respect to the dynamics in proteins.

Proteins are amino-acid chains that fold naturally into functional shapes, first in secondary – then in the native tertiary structures. Among the most common secondary structures that emerge from this folding are the α-helices and the β-sheets. Such folded structures, interconnected by seemingly unstructured parts (the loops), are vital for the protein functionality[3], thus featuring proteins like nano-machines. The α-helix is a special conformation with the peptide-bonds forming the backbone of the helix (called the principal helix), which is stabilized by angular limitations (or effectively, second neighbor interactions) and the H-bonds that connect third neighbors. These H-bonds form 3 slightly winding lines along the perimeter of the α-helix. One does not fully understand how a protein is capable of transforming chemical and radiation energy into mechanical motion and also the capability of transporting energy, without loss, over large distances[4]. The proteins reside in a noisy surrounding of water, protein

CP800, *Unsolved Problems of Noise and Fluctuations: UPoN 2005*
edited by L. Reggiani, C. Pennetta, V. Akimov, E. Alfinito, and M. Rosini

or membrane, and must be functional at room temperatures. It is therefore a challenge to understand why noise does not destroy the functionality to a larger degree than is observed. It is not solved here, but we shed some light on the problem by doing rather realistic 3D simulations of the α-helix dynamics. In the past there has been numerous more idealized and mathematical studies, which, however, are not sufficiently close to the experimental realities. Thus, we have here found several new effects in our simple, direct simulation.

New pump-probe experiments on the prototypical membrane protein Bacteriorhodopsin reveals information which may be of relevance for this question. In particular the measurements allow determining the lifetime of certain excitations. Two frequencies seem to play an important role in such transport. The high frequency amide-I mode (C=O stretch at $\nu \sim 1600$ cm^{-1}), which was proposed by Davydov[5] as a candidate for soliton transport, and the low frequency ($\nu \sim 115$ cm^{-1}) long-lived excitations measured in pump/probe experiments on Bacteriorhodopsine[1,2], which consists mainly (80%) of seven long α-helices.

More realistic 3D models of an α-helix have been investigated before[6,7,8,9] with respect to the Davydov soliton mechanism. Hennig[8] used a confining cage and initiated soliton solutions by non-linear map methods. Christensen *et al*[7] studied the existence of solitons on an unsuspended 3D helix with Morse interactions. In a previous work[9,11], some of the authors modeled an α-helix in three dimensions by three coupled Toda[10] lattices with stable soliton solutions.

MODEL GEOMETRY AND INTERACION POTENTIALS

α-helix geometry

The geometry of our models is shown in Figure 1. In previous works[9,11] the α-helix was simplified and considered all H-bonds parallel like Figure 1, left, with the number of residues per turn, n, exactly 3. The α-helix found in proteins is slightly more refined. In this paper we investigate a setup around more realistic angles and distances. Now the principal helix is built with a number of residues per turn n=3.61 and a pitch of 5.41 Å. The equilibrium distance along the peptide-bond is set as 3.8 Å (our standard model is shown in Figure 1, right, but we have also studied for example the 3-10 helix etc.). In our constructions all residues are considered as point masses with the average weight m=1.7 10^{-22} g.

Helix interaction potentials

It is important to consider anharmonic, realistic potentials like for example the Lennard Jones potential. One can show the Toda potential fits exactly the Lennard-Jones potential in a region around the equilibrium[11]. The reason for using the Toda potential is that a lot of exact, analytic information (in 1D) is available – which can be shown to apply to a good approximation also for the Lennard Jones potential and the

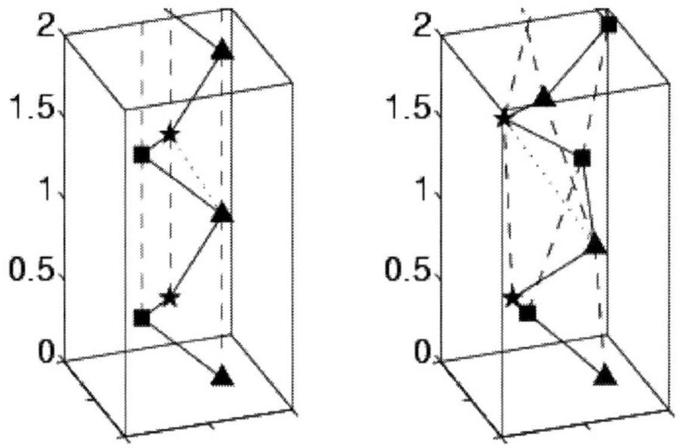

FIGURE 1. Left, simplified α-helix model used previously[9,11], right the more realistic one used here.

3D case. Hence we use the Toda-lattice potential, for an equilibrium position r_0 and parameters, a,b

$$V(r_j - r_0) = \frac{a}{b}\sum_{j=1}^{N} e^{-b(r_j - r_0)} + b(r_j - r_0)$$

In our helix they are different for each of the three interactions: peptide-bond interaction (P, nearest neighbor), a H-bond interaction (H, third neighbor) and an angular-bond interaction (T, second neighbor). Expanding the potential we can compare the product ab to the harmonic spring constant k and use r_0 to obtain $a=kr_0/21$, $b=21/r_0$. That means the potential sharpens and becomes more harmonic with the increase of r_0. We have used $k_H =$ 1.41 10^4 dyn/cm and $k_T=k_H$ and two values for the much stronger peptide bond to see possible effects: $k_P= 16\ k_H$ as in ref. 9,11 and $k_P = 512\ k_H$ giving the amide-I frequency near 1600 cm^{-1}.

Because of the long length of the effective H-bond including the peptide platelet: r_0 (=6.7 Å) in the proper helix structure compared to a real O-H-bond potential distance $r_0 =$ 1.9 Å we have introduced a rigid stick of length 4.8 Å to attain both the correct distance and the correct potential properties for the H-bond. With this, our model still considers the real length of the H-bond at the equilibrium position. The equations of motion that come from the Hamiltonian were solved numerically with the use of MATLAB and a variable order Adams-Bashforth-Moulton PECE solver. We have used both open boundary conditions of the 120 residue long α-helix model, and fixed boundary conditions for a 126 long α-helix model with the first and last three having very large masses. There is essentially no difference in the results in the two cases, so we show here only the first case.

COMPUTER SIMULATIONS OF EXCITATIONS IN THE HELIX MODEL: RESULTS

Results of a kick perturbation

A possible way to perturb an α-helix model is to add a momentum different from zero at some site of the helix in equilibrium at time t=0 and check the evolution in time of displacements,

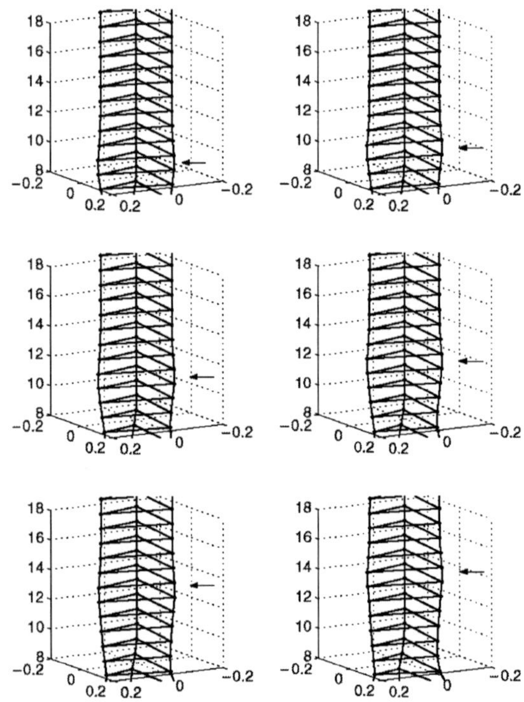

FIGURE 2. Evolution with time of the H-bond soliton in the simplified model. It can be perfectly described by the analytical expressions for the Toda model.

velocities and energies. In the previous simulations[9,11](simplified α-helix Figure 1, left) the perturbation was made in the H-bond-direction, which for that straight geometry was also the z-direction. Our perturbations on (Figure 1, right) are also along the H-bond-direction, now deviating from the z-direction. The reason of our choice is that a H-bond-direction perturbation gives us a clearer signal among several tries (for instance, peptide-bond-direction, angular-bond-direction, z-direction and so on). The result of a perturbation in the H-bond direction at one site in the middle of a 120 residue long helix with initial energy of 100 cm^{-1} can be seen on the space-time energy density plot on Figure 3. The x-coordinate is the residue label - or `site' - and the y-

coordinate is the time in the simulation (here up to 10 ps). Observe the formation, almost at the same time, of pairs of traveling signals, with opposite directions. Both are compression pulses. So although the kick is introduced in the right direction, the recoil introduces a similar pulse going in the opposite direction. Such pair formation behavior was absent in the previous model[11] where the kick was done at one end of a 60 site long helix. In that study we found the ultrasonic moving 'soliton' moving along the H-bonds. Even if the kick is done at one site, the pulse is transmitted to all three H-bond lines and a synchronised pulse is moving as a bulge a long the helix, see Figure 2. The pulse is reflected at the boundaries.

Here, we have additionally found two or more slower moving 'solitons'. A frequency analyses reveals they involve the excitations of the peptide bonds, which are then moving along the α-helix backbone as a pulse. (The frequency spectrum has a strong $1/\nu^2$ component, to been discussed elsewhere). It is obvious that the energy transported by these are much larger than by the faster H-solitons. We call these 'solitons' modes although they may not be so in the strict mathematical sense. However, they have the same characteristics: move undisturbed as a pulse, which can be reflected at either an open or fixed boundary – and further they can pass each other without being disintegrated. The slow 'solitons' have velocities below that of sound, and correspond probably to what has been termed 'envelope' solitons – or Davidov solitons (although a much richer behavior is found here, than previously discussed).

What is interesting in the noise context is that the 'solitons' slowly emit phonons and create a weak noise (heat) after passing. However, they are still able to travel through the noisy background. This means the 'solitons' are long-lived and can feed energy slowly into the low frequency modes. That is what probably has been observed by Austin *et al.*[1] The described modes do essentially not move the α-helix as a whole, but are internal excitations. Hence, the modes will be minimally effected by the surrounding water, protein or membrane (noisy) conditions.

Results of a noise perturbation

It is then interesting to simulate if the system was excited by noise. This is done by assuming at time zero all residues are behaving as an ideal gas with velocities distributed in all directions, statistically, according to the chosen temperature. The result using the same total energy as before is shown on figure 3. It shows the noise at t=0 has different size at all sites. However the propagation in time is far from random, but clearly the 'soliton-like' modes found from the single kick can be followed at least for 10 ps in this noisy environment. Such an unexpected behavior was already found by Fermi-Pasta-Ulam[12] for a perfect 1D line, but clearly the phenomena exists also in the more realistic 3D α-helix. In the real α-helix there are of course many more modes to which the system can disperse the energy.

CONCLUSION

We have briefly described the results of a computer simulation of a rather realistic α-helix model. We have found a multitude of new 'solitonic' modes which are robust against mutual collisions and against noise in the background. As a result we have found the noise spectrum for the biologically very relevant α-helix structure is very unusual and highly structured. Describing the details of this is a remaining and challenging problem for the 'noise-community'.

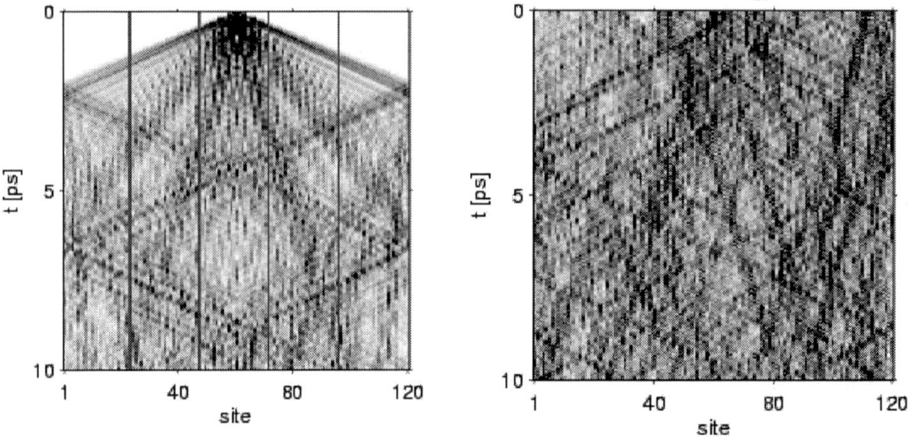

FIGURE 3. Left, energy distribution (dark means high total energy) after a kick at site 60 in the right direction along the H-bond. Notice the same fast H-bond solitons as shown in figure 2 – but also the appearance of two slower, higher energy (darker) peptide-bond solitons. Right, the system excited by random kicks at all sites in random directions. Notice the energy is again distributed by the various 'soliton-like' modes. Notice the noise is highly structured.

ACKNOWLEDGMENTS

We would like to thank Francesco d'Ovido for useful discussions.

REFERENCES

1. A. Xie, A.F.G. van der Meer, and R.H. Austin, Phys. Rev. Lett. **88**, 018102 (2002).
2. J. Edler, R. Pfister, V. Pouthier, C. Falvo, and P. Hamm, Phys. Rev. Lett. **93**, 106405 (2004)
3. C. Branden and J. Tooze, Introduction to Protein structure (Garland, NY) (1991).
4. S. Yomosa. ``Toda-Lattice Solitons in α-Helical Proteins". J. Phys. Soc. Jpn. **53**, No. 10, 1984, pp. 3692-3698.
5. A. S. Davydov, J. Theor. Biol. **38**, 559 (1973).
6. O.H. Olsen, M.R. Samuelsen, S.B. Petersen and L. Nørskov, Phys. Rev. A **38**, 5856 (1988), A. La Magna, R. Pucci, G. Piccitto, and F. Siringo, Phys. Rev. B **52**, 15273 (1995). A. V. Zolotaryuk, P. L. Christensen, and A. V. Savin, Phys. Rev. E **54**, 3881 (1996). M. P. Brown and K. Austin, *The New Physique,* pp. 25-30.
7. P. L. Christiansen, A. V. Zolotaryuk, A. V. Savin. ``Solitons in an isolated helix chain". Phys. Rev. E **56**, No. 1, 1997, pp. 877-889.
8. D. Hennig. ``Energy transport in α-helical protein models: One-strand versus three-strand systems". Phys. Rev. B **65**, 174302 (2002).
9. F. d'Ovidio, H. G. Bohr, P.-A. Lindgård. ``Solitons on H-bonds in proteins". J. Phys.: Condens. Matter **15** No 18 (14 May 2003)
10. M. Toda. ``Theory of Nonlinear Lattices". Springer-Verlag, Berlin, 1981.
11. F. d'Ovidio, H. G. Bohr, P.-A. Lindgård, Phys. Rev. E, **71**, 026606-1 (2005).
12. E. Fermi, J. Pasta, S. Ulam, and M Tsingou, in The Many Body Problems, edited by D. C. Martin (World Scientific, Singapore, 1993) (reprinted).

Why Going Zig-Zag?

Udo Erdmann*, Sebastian Göller* and Lutz Schimansky-Geier*

**Humboldt-Universität zu Berlin, Institut für Physik, Newtonstr. 15, 12489 Berlin*

Abstract. We report a theory to describe the motion of zooplankton. Apart from moving just randomly like a classical Brownian particle, zooplankton like *Daphnia* or *Copepods* pick their turning angle from a distribution which is far from being Gaussian or equally distributed. This leads to different behavior in the motion compared to normal diffusion. The question which can be asked here is: Is there an evolutionary reason to forage for food in the aforementioned manner? First attempts to answer the question are made throughout this manuscript although it still remains an unsolved problem.

Keywords: active motion, persistent random walk, distribution of turning angles, optimal food consumption, zooplankton
PACS: 87.19.St, 87.23.Kg

INTRODUCTION

Daphnia, a freshwater zooplankton, are creatures which swim in a sequence of discrete hops while foraging for food. In contrast to a usual random walker these species do not turn equally weighted with an turning angle between 0 and 360 degrees. As has been discovered in experiments, *Copepods*, the saltwater analog of *Daphnia*, hop with a preferred turning angle [1] and *Daphnia* have an exponential turning angle distribution (TAD) [2]. Due to the fact that this has been observed not just for one type of species, the question which occurs is about the existence of an evolutionary advantage of going zig-zag as described.

In [3] we applied a random walk model to find the effective diffusion coefficient of the random motion with preferred turning angles. Starting at the space correlation function of such an random walker, using the the approach of KAREIVA and SHIGESADA [4], one could derive the diffusion coefficient depending on the angular correlation function. This angular correlation depends on the turning angle which the zooplankton chooses after every hop. Knowing the TAD the diffusion coefficient can be easily calculated. It was shown that for a preferred turning angle around 30 degrees a particle could cover a larger area. This is why the diffusion coefficient for a persistent random walk as described above is larger than the diffusion coefficient for the classical Brownian motion.

First, we shortly summarize the connection between a random walk and the motion of an active particle. Out of a continuous model which is based on the stochastic equations of motion in two dimensions, we can derive a formula for the spatial diffusion depending on the mean turning angle and the strength of the fluctuations.

Than, we derive a simple theory of food consumption which is compared with random walk simulations and experimental data from [2]. Finally we will discuss that the exponential distribution could be of evolutionary advantage while foraging for food.

CP800, *Unsolved Problems of Noise and Fluctuations: UPoN 2005*
edited by L. Reggiani, C. Pennetta, V. Akimov, E. Alfinito, and M. Rosini

Having in mind that a water flea moves with a straight swimming stroke of approximately constant length within a approximately fixed time, we could assume that this particle has a constant velocity. If the stationary velocity is a non-vanishing one the system has to be far from equilibrium. Hence energy has to be up-taken. If this is not the case, a permanent motion could not be obtained. Living species, of course, take up energy out of the environment and convert this after some dissipative losses into energy of motion. This is taken into account with the theory of Active Brownian particles [5, 6]. The equations of motion of such a particle are

$$\dot{\vec{r}} = \vec{v}\,, \qquad \dot{\vec{v}} = \left(\gamma_1 - \gamma_2 v^2\right)\vec{v} + \vec{\omega}\times\vec{v} + \sqrt{2D_v}\,\vec{\xi}(t) \tag{1}$$

Here, the simplest assumption of gaining energy has been made [7]. The up-taken energy leads to acceleration with strength γ_1 while all dissipations are assumed to enter quadratically depending on the velocity with rate γ_2. This leads exactly to the stationary situation of permanently moving particles as has been observed in the Zooplankton experiments. The stationary velocity in terms of these two parameters is $v_0 = \sqrt{\gamma_1/\gamma_2}$. The randomness of the fleas in the Active Brownian particle approach enters via a Gaussian white noise term. Although the motion of Zooplankton looks more like a random walk the description with the help of Langevin equations is equivalent and has its advantages as well. While the constant velocity of the random walker is sort of God given, in the Active Brownian particle approach the permanent motion of a particle can be motivated biologically out of the internal energy household. Furthermore, the stochastic differential equation gives us the opportunity to apply additional interactions very convenient with just adding an additional force term. This is exactly what has been done to obey the preferred TAD at the Zooplankton motion. The term in Eq. (1) which reflects the a preferred turning angle is the Lorentz-like force $\vec{\omega}\times\vec{v}$. We assume that $\vec{\omega}$ is directed perpendicularly to the plane of motion, i.e. $\vec{\omega} = \{0,0,\omega\}$ and a Larmor-like rotation is generated. The sign of ω determines clockwise or counter clockwise motion. Later we will assume that half of a population is running clockwise and the second half counter clockwise.

In [8] it has been shown that out of a system of particles obeying Eq. (1) the mean square displacement of these particles can be calculated. At this, Eq. (1) was transformed into polar coordinates and then with the help of the stationary solution of the corresponding Fokker-Planck equation

$$\left\langle(\vec{r}(t)-\vec{r}(0))^2\right\rangle = 2\tilde{v}^2\frac{D_\phi}{D_\phi^2+\omega^2}\,t + O\left(\text{const}, \cos(\omega t), \exp(-D_\phi t)\right) \tag{2}$$

was obtained. Here, we use $\tilde{v}$ as the most probable value of the velocity in the stationary regime, which in the noise free case this value is equal to v_0. With the well-known connection of mean square displacement and diffusion, the diffusion coefficient becomes

$$D_r = \frac{\tilde{v}^4}{2D_v}\left(\frac{1}{1+\omega^2\,\tilde{v}^4/D_v^2}\right), \tag{3}$$

which is presented in Fig. 1. This result is very similar to the behavior of the spatial diffusion coefficient obtained with a random walk model in [3]. To understand the full derivation of Eq. (3) we refer to [8]. Depending on the convenience of the handling

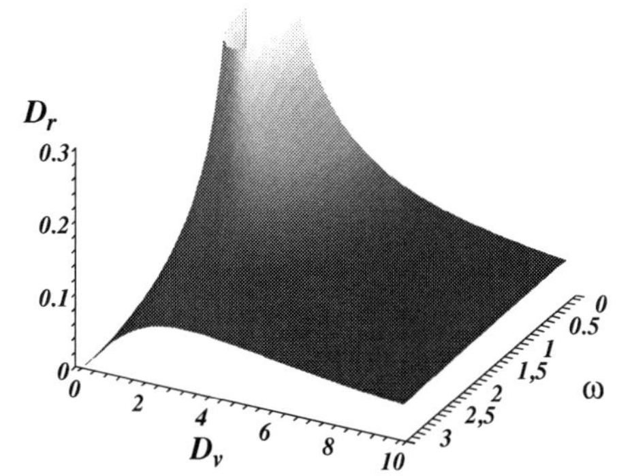

FIGURE 1. Spatial diffusion D_r depending on the external noise D_v and the preferred turning angle ω.

the random walk or the Active Brownian particle approach could be used due to their equivalence.

FOOD CONSUMPTION IN FINITE FOOD PATCHES

Theory

The result of the previous section shows that a TAD different from a distribution with equally distributed turning angles shows an increase of the diffusion coefficient due to a steeper slope of the mean square displacement over time. Having the hypothesis in mind that the experimentally obtained TAD's [1, 2] are due to evolution, the ballistic flight would be optimal if the maximum of the diffusion would the optimum. This means maximal diffusion coefficient could not be a motion with a non vanishing mean turning angle. But, if we assume that food is not equally distributed all over the ocean or the lakes, another confining property has to be taken into account, the size of a food patch. It is quite reasonable that food is distributed in patches. Now, we want to present a simple theory of food consumption. First, we assume that a *Daphnia* has a probability $\rho(\vec{r},t)$ to be at time t at position $\vec{r}$. If the animal is a random walker one could additionally assume the its probability distribution $\rho(\vec{r},t)$ obeys the diffusion equation

$$\frac{\partial \rho}{\partial t} = D_r \Delta \rho \,. \tag{4}$$

Here D_r is the diffusion coefficient which has been calculated in the previous sections. In the two-dimensional case the solution of Eq. (4) is:

$$\rho(\vec{r},t) = \frac{1}{4\pi D_r t} \exp\left[-\frac{\vec{r}^2}{4D_r t}\right] . \tag{5}$$

Let us assume that the particle during its random motion consumes food C with constant rate k. The food is given by its density $c(\vec{r},t)$. If we additionally assume that the food does not move within the time scale a particle is inside the patch, the consumption of food can be described by the dynamics

$$\frac{\partial}{\partial t}c(\vec{r},t) = -kc(\vec{r},t)\rho(\vec{r},t) \tag{6}$$

If the solution Eq. (5) is inserted, this equation can be solved exactly:

$$c(\vec{r},t) = c_0(\vec{r}) \exp\left[-\int_{t_0}^{t} k\rho(\vec{r},t')\,\mathrm{d}t'\right]. \tag{7}$$

With the definition of the exponential integral, $\mathrm{E}_1(a) = \int_a^\infty \frac{e^{-t}}{t}\mathrm{d}t$ [9], we obtain in compact shape

$$c(\vec{r},t) = c_0(\vec{r}) \exp\left[-\frac{k}{4\pi D_r}\mathrm{E}_1\left(\frac{\vec{r}^2}{4D_r t}\right)\right]. \tag{8}$$

Here we have put $t_0 = 0$.

Let us look at bounded distributed food patches now. As a prototypical distribution we take spheres with radius R where C is present. Outside of the circle no food could be found. For simplicity we assume the circle positioned at the origin, i.e. $c_0(\vec{r}) = c_0$ if $|\vec{r}| < R$ and vanishing, elsewhere. Then after a fixed time T we determine the overall food which is still left. It results from integration of Eq. (8):

$$C(T) = 2\pi c_0 \int_0^R \exp\left[-\frac{k}{4\pi D_r}\mathrm{E}_1\left(\frac{r^2}{4D_r T}\right)\right] r\,\mathrm{d}r. \tag{9}$$

The consumed food as a function of the spatial diffusion coefficient D_r is shown in Fig.2. It shows that there exists a minimum with respect to the diffusion coefficient D_r.

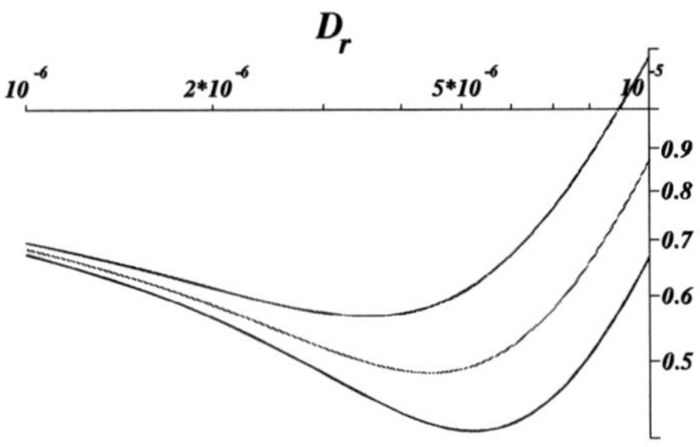

FIGURE 2. Amount of left food in a bounded patch depending on the spatial diffusion coefficient within a fixed eating time. It was assumed that a flea is foraging for food for 8 hours. The radius of the patch is about 1.3m in diameter. The figure shows the amount of left-over food for three different eating rates k (red: 0.0003, green: 0.0004, blue: 0.0005)

Since D_r depends monotonously on the mean turning angle ω of the zig-zag hopping it implies a minimum with respect to the mean angle, too. Thus, the variation of the mean angle with change of the effective diffusion coefficient gives rise to a change of the food consumption. A simple explanation can be given as follows. If particles hop straight forwardly they will leave the food region, going forwardly and backwardly is also of disadvantage (see also Fig. 3 (left) of the next section). Hence there is a optimal angle which the particle spends most time in the food patch with. It should not not matter whether they start at the center of the food or not.

Random Walk Simulations

In contrast to the *Copepod* experiments [1] *Daphnia* choose their turning angle out of an exponential distribution $P = \exp(q|\alpha|)$. The experiments come up with a slope of this distribution around $q = 1.54$ [2]. Using the approach of [3] one could show that the spatial diffusion coefficient depends monotonically on q. Putting this result into Eq. (9) one should find a minimum in the left-over food as well. Of course, one could talk about a maximum in the food eaten, too. This extremum is what we looking for in random walk simulations described in this section.

We assume a motion in two dimensions with a constant hop length L and no pause time between the hops. The turning angles between two hops are equally distributed to the left and right and are given by an exponential distribution. The simulated motion starts at the origin of a cycle (radius R) with a constant velocity and a random starting angle. The cycle is considered a food patch which is divided into boxes of size one. If one of these boxes is hit by the trajectory the whole box is marked. After a given time ($T = 500$) the marked area versus the area of the whole cycle is calculated in percent and considered the consumed food.

With regard to the TAD's with a large q the flea moves quiet straight on and leaves the food patch after a short time. Considering small q the flea becomes a random walker. Within the simulation time it stays close to the origin where after a short time all food is already eaten (see left picture of Fig. 3).

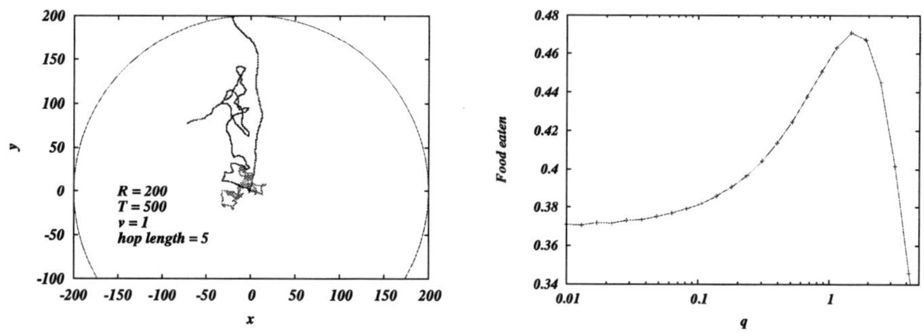

FIGURE 3. *left:* sample trajectories for different values of q, representing the sharpness of the TDA (green: $q = 0.1$, blue: $q = 1.5$, magenta: $q = 10$) *right:* food eaten over the width of the exponential distribution of turning angles

Random walk simulations of the food consumption of a single animal varying the width q of the show a maximum of the food eaten with q.

CONCLUSIONS

In this article, we summarize recent results describing the zig-zag motion of zooplankton like *Copepods* or *Daphnia*. First, the theory of Active Brownian particles [5, 6] was extended with terms which obey the motion of individuals with preferred turning angle [8]. The preferred TAD comes into play as a Lorentz-like force within the Langevin equation. This translation of the preferred TAD into the concept of stochastic differential equations makes the treatment of such biological systems more handy in terms of physical interpretations of the influencing forces. With the help of the, now more generalized, theory of Active Brownian particles a broader variety of swarming systems can be described via physical models. The calculated spatial diffusion coefficient shows a maximum with the noise strength for preferred turning angles between -90 and +90 degrees (see Fig. 1). This result can be concluded as follows: With a preferred turning angle a individual can cover a larger area within a given time.

Furthermore we developed a theory of food consumption within a bounded food patch. Assuming a Gaussian distributed density for a single individual, a minimum in the left over food with the spatial diffusion coefficient has been found. Hence, to cover the maximum of a given area going zig-zag is many times more beneficial than just normal diffusion (see Figs. 2 and 3). The analytical results could have been proven qualitatively with random walk simulations. It could have been shown that the parameter of the measured exponential distribution of turning angles coincides with the maximum of the consumed food. As has been pointed out in [2] the hypothesis of evolutionary advantage which is still an unsolved problem. If the parameter q for the exponential TAD remains around 1.5 for more than just two species, one might speculate that nature has found by this way the reason to optimize the zig-zag hopping.

REFERENCES

1. F. G. Schmitt, and L. Seuront, *Physica A*, **301**, 375–396 (2000).
2. R. Garcia, F. Moss, A. Nihongi, and J. R. Strickler, Are mean turning angles selected for survival of zooplankton?, paper within these proceedings (2005).
3. N. Komin, U. Erdmann, and L. Schimansky-Geier, *Fluctuation and Noise Letters*, **4**, L151–L159 (2004).
4. A. Okubo, and S. A. Levin, *Diffusion and Ecological Problems: Modern Perspectives*, vol. 14 of *Interdisciplinary Applied Mathematics*, Springer, New York, 2001, 2 edn.
5. W. Ebeling, F. Schweitzer, and B. Tilch, *BioSystems*, **49**, 17–29 (1999).
6. U. Erdmann, W. Ebeling, L. Schimansky-Geier, and F. Schweitzer, *European Physical Journal B*, **15**, 105–113 (2000).
7. J. W. S. Rayleigh, *The Theory of Sound*, vol. I, MacMillan, London, 1894, 2 edn.
8. L. Schimansky-Geier, U. Erdmann, and N. Komin, *Physica A*, **351**, 51–59 (2005).
9. M. Abramowitz, and I. A. Stegun, *Pocketbook of Mathematical Functions*, Harri Deutsch, Frankfurt/Main, 1984.

Are Mean Turning Angles Selected For Survival Of Zooplankton?

Ricardo Garcia[1], Frank Moss[1], Ai Nihongi[2] and J. Rudi Strickler[2]

[1] *Center for Neurodynamics*
University of Missouri at St. Louis
St. Louis, MO 63121, USA

[2] *Great Lakes WATER Institute*
University of Wisconsin at Milwaukee
Milwaukee, WI 53204, USA

Abstract. Zooplankters are aquatic creatures that swim in a sequence of discrete hops, each followed by a directional change represented by a turn through some angle. We observe that the mean turning angles of foraging *Daphnia* are similar across species suggesting that the mean turning angles may have some adaptive value.

Keywords: Zooplankton, *Daphnia,* Turning Angles, Foraging
PACS: 87.19.St, 87.23.Kg

INTRODUCTION

In this study, we focus on the zooplankter *Daphnia,* and investigate two species: *D. magna* and *D. pulex*. These creatures have a short generation time (weeks), are parthenogenetic (easy to produce clones), and represent food for fish in ponds and lakes. *Daphnia* move in a sequence of hops interspersed with pauses and directional changes made by turns through angles. The latter we call the turning angles. The hop lengths, pause times and turning angles seem random, and so the swimming and feeding motions of *Daphnia* resemble random walks [1, 2]. We focus here on the distributions of the turning angles. We find the mean turning angles to be quite similar in spite of the fact that these two species are different in size and display somewhat different swimming characteristics. These observations have led us to speculate that the mean turning angles have been selected over evolutionary time scales for survival. A specific example – the focus of a theoretical paper presented also in these proceedings [3] – suggests that random patterns of foraging with non-zero turning angles optimize food gathering in a finite sized food patch [4, 5]. We note that our combined work that includes systematic modeling of observed animal behavior has a long and fruitful history [6, 7].

Zooplankters are aquatic creatures that mostly drift with the current, but within their surrounding water volume wander about a confined location while feeding. They are attracted to visible light, flee from ultra violet, and are largely blind to infra-red. They swim in a sequence of discrete hops, and while feeding mainly confine themselves to movements within horizontal planes. *Daphnia* normally feed near the

CP800, *Unsolved Problems of Noise and Fluctuations: UPoN 2005*
edited by L. Reggiani, C. Pennetta, V. Akimov, E. Alfinito, and M. Rosini

surface of the water at night when visual predators are a lesser risk than in day time. The hops are motivated by beats of a pair of swimming appendages, or locomotory antennae, located near the head as indicated by the picture of a *D. magna* individual shown in Fig. 1 (a). The length of the hops, as well as the turning angle, α, between two hops, is defined by the schematic in Fig. 1 (b). An example of the other species examined here, *D. pulex,* is shown in (c). Note that *D. pulex* is typically about 1/4th to 1/3rd the size of *D. magna*, and this difference is also reflected in the swimming characteristics other than the mean turning angle, for example swimming velocity in units of body lengths per second.

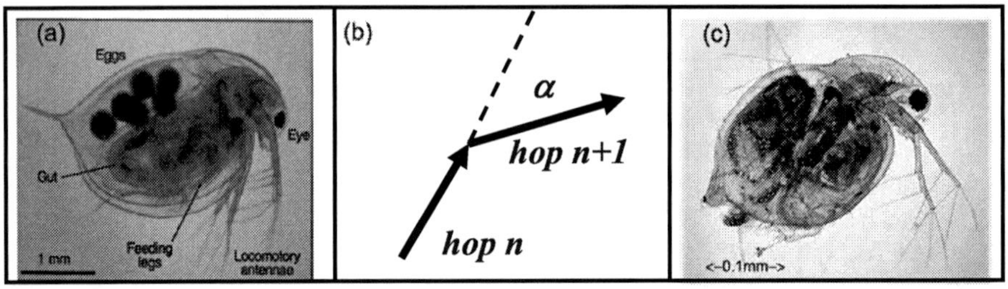

FIGURE 1. (a) The freshwater crustacean zooplankton, *Daphnia magna*, showing a typical body size, feeding and swimming (locomotory) appendages, an eye, eggs and the gut. Most zooplankters are sensitive to light. They swim in sequences of hops that result from beats of the swimming appendages. (b) A schematic representation of two hops indicating the *turning angle,* α*,* which is the focus of this research. Each hop is represented here by a vector of length l_n, and takes place during a time, t_n. (c) An example of *Daphnia pulex.* It is similar to *D. magna* but much smaller in size.

THE EXPERIMENT

In this work, we measure the statistical distributions of turning angle and hop length as obtained from video records made in the apparatus shown in Fig. 2. A shallow aquarium measuring 26 by 28 cm length and width by 4.5 cm deep (or a Petri dish 12 cm diam. by 2 cm deep) was provided to which a standard feeding solution 2 cm deep was added. Typically 30 to 60 *Daphnia* were added to this solution. They were illuminated from the side with 30 W infra red, long wavelength pass illuminators with a cutoff wavelength of 800 nm [8]. Because *Daphnia* in the wild feed at night, it was necessary to condition our experimental subjects in the dark (with the IR switched on) before recording their motions. The recording camera was a Sony DCR-TRV80 operating at 30 frames per second. In the "Night Shot" mode this camera can record in the IR at the wavelengths available from the illuminators. Typically, recordings of 1 to 2 minutes in the bottom view, lower camera in Fig. 2, provided sufficient data for the analyses presented here. An example frame from one video is shown in Fig. 3. Motions from multiple animals were analyzed. Side views were sometimes recorded in order to insure that the motions were approximately confined to the two horizontal dimensions.

FIGURE 2. About 40 *Daphnia* are placed in a shallow aquarium in well water with a standard feeding solution, 13.3 mg/l freeze dried *Spirolina* algae. The *Daphnia* are conditioned in the dark for ½ hour then their swimming motions are recorded with infra-red illumination, to which they are blind. Side views (upper camera position) served to indicate that *Daphnia* swim mostly in horizontal planes while feeding. The data reported here were obtained from the bottom view (lower camera) recordings. The camera is a Sony camcorder with IR capability operating at 30 frames per sec. Recordings were analyzed using *Track-It* software.

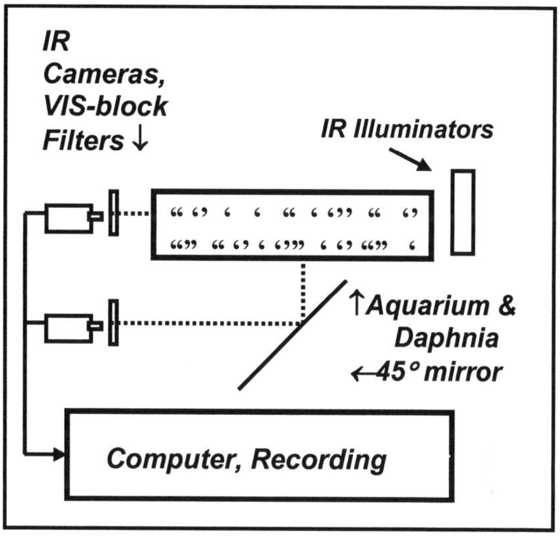

Here we are interested in the distributions and mean values of the turning angle, α, for different species of zooplankton having different swimming and physical characteristics. Little quantitative data exist, though turning angle distributions (TADs) have been previously reported for *D. magna* [9, 10] and for an oceanic copepod [11].

Figure 3. An example single frame as obtained with the apparatus shown in Fig. 2. The larger animals are adult *D. magna*. The small individuals are juveniles less than 24 hours from birth. The juveniles are genetic clones as *Daphnia* reproduce asexually under ideal environmental conditions. Under poor conditions sexual reproduction commences. The object in the center field is a centimeter square length calibrator. Only adults were used in the analyses reported here.

THE RESULTS

Figure 4 shows distribution functions (histograms) of turning angles and hop length for *D. magna*.

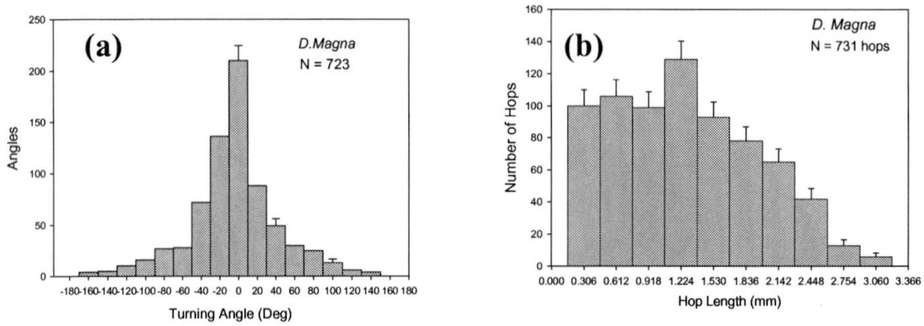

Figure 4. Turning angle (a) and hop length (b) distributions for *D. magna*. The size of this animal is typically 1.5 to 2.5 mm in length. From the turning angle distribution shown, $\langle \alpha \rangle = -0.31 \pm 49.6°$. One-sided distributions of $|\alpha|$ (not shown) were analyzed, from which we obtained the mean value, $\langle |\alpha| \rangle = 34.9 \pm 35.2°$. The mean hop length is $\langle l \rangle = 1.1 \pm 0.7$ mm.

Figure 5 shows the histograms of turning angles and hop length for *D. pulex*.

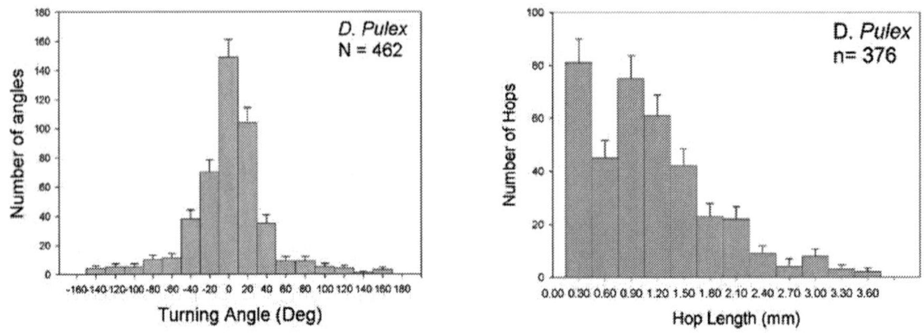

Figure 5. (a) Turning angle, α, and (b) hop length, l, distributions for *D. pulex*. The size of this animal is typically 0.3 to 1.0 mm in length. From the turning angle distribution shown, $\langle \alpha \rangle = -0.95 \pm 39.4°$. The one-sided distribution of $|\alpha|$ (not shown) was analyzed, from which we obtained the mean value, $\langle |\alpha| \rangle = 26.0 \pm 29.6°$. The mean hop length is $\langle l \rangle = 0.91 \pm 0.7$ mm.

The swimming motions, and hence the turning angles and lengths, were obtained from video recordings of projections on a 2-dimensional horizontal plane. Frames at the beginning and end of each hop were exported and the tracking software, *Track-It,* was used to extract the vectors shown in Fig. 1 (b). In addition to the turning angles and hop lengths, we also obtained (from the video frame rate) the hop times and pause times, data that will be reported in a future publication. (At the end of each hop, *Daphnia* tend to pause briefly before commencing the next hop.)

Though *D. pulex* is comparatively gracile, being only about 1/4th to 1/3rd the size of *D. magna*, we note from the above that the hop lengths are comparable. Our hop time data indicate that *D. pulex* swims with approximately the same velocity as *D. magna,* though in units of body lengths per second they swim about 4 to 5 times faster. Therefore, in a given time period *D. pulex* will cover approximately the same area as *D. magna.* We note also from the TADs that both the mean and standard deviation of the turning angles are somewhat larger for *D. magna* though probably not significantly so.

These results represent some preliminary data for the two species studied. Experiments on TADs and swimming characteristics are currently in progress for three additional species of *Daphnia* including two different forms of one of them [12] and for juvenile *D. magna* as shown in Fig. 3.

DISCUSSION AND SUMMARY

Our motivation is to provide experimental data upon which one can base analytical models for feeding and swimming zooplankton as, for example, has been done in references [3-6]. In particular we wish to test the hypothesis, put forward by the Humboldt University group, that a non-zero mean turning angle optimizes foraging success in a finite size food patch [3-5]. The idea is simply stated as follows: If the TADs as shown in Figs. 3 (a) and 4 (a) collapse into delta functions at the origin, then the animal does not turn at all but instead follows a straight line thus soon escaping the food patch. In contrast if the TADs are constant on the complete support interval [−180°,+180°] then the animal executes an uncorrelated, or "pure", random walk as described by Einstein's diffusion law [13]. In the latter case the animal covers too much area near the origin (eating and depleting much of the food there) and too little time exploring regions of the food patch at greater distances. If food is gathered in proportion to the distance covered or time elapsed before escaping the patch, then it seems clear that a TAD of some optimal width, $0 < \sigma_{opt} < 180°$, similar to those shown in Figs. 3 (a) and 4 (a) exists and maximizes food gathering.

This conjecture leads us to a statement of our unsolved problem of noise: Are the TADs observed here the result of Darwinian natural selection for survival? [15]. If the answer to this question should turn out to be yes, then the TADs of feeding zooplankton with their optimal mean turning angles represent yet another instance of the beneficial effects of noise.

ACKNOWLEDGEMENTS

RG and FM extend warm thanks to our coauthors, Prof. J. Rudi Strickler and Ai Nihongi for their kind hospitality and for making their laboratory facilities freely available during which time we collected data on additional species of *Daphnia* that will be reported later. JRS acknowledges support by the National Science Foundation, grant OCE-0352264.

We are indebted to the Humboldt group, Profs. Lutz Schimansky-Geier, Werner Ebeling and Dr. Udo Erdmann, for many stimulating discussions, for their kind hospitality on numerous occasions and for their continuing patience in explaining their ideas to us, particularly to FM.

REFERENCES

1. Berg, H. C., *Random Walks in Biology*, (Princeton University Press, Princeton, New Jersey, 1993).
2. Okubo, A. and Levin, S. A. *Diffusion and Ecological Problems: Modern Perspectives,* 2nd ed. (Springer, New York, 2001).
3. Erdmann, U., and Schimansky-Geier, L., Why going zig-zag?, paper this conference proceedings..
4. Schimansky-Geier, L., Erdmann, U. and Komin, N., Advantages of hopping on a zig-zag course. *Physica A,* **351**, 51-59, 2005.
5. Komin, N., Erdmann, U. and Schimansky-Geier, L., Random walk theory applied to *Daphnia* motion. *Fluct. & Noise Lett.* **4,** L151-L160 (2004).
6. Keiyu, A. Y., Yamazaki H. and Strickler, J. R., A new modeling approach for zooplankton behavior. Deep-Sea Research II 41, 171-184 (1994).
7. Alt, W., Elements of a systematic search in animal behavior and model simulations. *BioSystems* **34,** 11-26 (1995).
8. The animals used in this experiment, *D. magna and D. pulex*, were observed to be still slightly phototactic at this wavelength, that is, about ½ the population would respond in some way to the IR illumination usually over a time scale of typically 15 min. The American Dynamics illuminators are described at: http://www.americandynamics.net/products/cameras_IR.aspx.
9. Ordemann, A., Balazsi, G., and Moss, F., Motions of Daphnia in a light field: Random walks with a zooplankton, *Nova Acta Leopoldina* NF *88,* Nr. **332**, 87 - 103 (2003) (Proc. German Academy of Science).
10. Ordemann, A., Balazsi, G., and Moss, F., Pattern formation and stochastic motion of the zooplankton *Daphnia* in a light field, *Physica A*, **325**, 260-266 (2003).
11. Schmitt, F. G., and Seurout, L., Multifractal random walk in copepod behavior. *Physica A,* **301,** 375-396 (2000)
12. All experimental data reported here were obtained in the laboratories of the Center for Neurodynamics in St. Louis. Data on the additional species were obtained at the Great Lakes WATER Institute in Milwaukee and will be analyzed and reported in a future publication.
13. A. Einstein, *Investigations on the Theory of the Brownian Movement,* (Dover, New York, 1956)
14. Experiments such as ours and the theory advanced in Refs [3-5] cannot answer this question. But our results offer support for the conjecture.

How Can An Olfactory System Deal With Fluctuations At Different Scales?

Hans Liljenström[ab] and Yuqiao Gu[cd]

[a]*Dept. of Biometry and Engineering, SLU, P.O. Box 7032, S-750 07 Uppsala, Sweden, and*
[b]*Agora for Biosystems, P.O. Box 57, SE-193 22 Sigtuna, Sweden*
[c]*School of Health Information Sciences, University of Texas Health Science Center at Houston, USA*
[d]*School of Automation and Energy Engineering, Tianjin University of Technology, P.R.China*

Abstract. Neural systems have to function efficiently in the presence of noise, which occurs at all levels of neural information processing. We use a computational model of the insect olfactory system in order to investigate how this system can deal with external and internal noise. We compare our results with previous models of the mammalian olfactory system. We address questions, such as, how can the system suppress noise, or maybe even make use of it? Could noise-induced oscillations and stochastic resonance also exist in the insect olfactory system? Our simulation results indicate a positive role for noise also in the insect olfactory system, but many questions are still open.

Keywords: Fluctuations, olfaction, antennal lobe, neural networks, stochastic resonance.
PACS: 87.18.Sn, 87.18.Bb, 87.19.Bb, 87.19.La

INTRODUCTION

Biological systems are generally characterized by a great degree of order, but there are also fluctuations in processes at different organizational levels. The nervous system is not an exception, despite its function in accurate and fast information processing. A certain degree of randomness in the neural activity seems inevitable at every level of organization. The question is if it is also of an advantage to the system, and in that case, what could this advantage be?

The olfactory system provides an ideal model to consider the issues of robust sensory signal transmission and efficient information processing within neural systems. At the periphery, at the olfactory receptors, there is a constantly changing and fluctuating flux of mixed odours, which the animal should discriminate among and base its behavioural decisions on. The degree of success is often a matter of survival, for the individual and for the population, since olfaction for many animals is essential for feeding, as well as for mating. The essential computation made by the olfactory system is an analysis of the fluctuations of the odour intensity with time, and the relation of these to fluctuations in the local wind direction. In addition to these fluctuations at the peripheral, there are fluctuations at every level of the information processing, in the activity patterns of the neural network structures involved, in the neural spike trains transmitted between neurons, as well as in the synaptic discharges. A major task is to analyze and separate the fluctuations of multi-source environments

CP800, *Unsolved Problems of Noise and Fluctuations: UPoN 2005*
edited by L. Reggiani, C. Pennetta, V. Akimov, E. Alfinito, and M. Rosini

into individual odours, but fluctuations also seem to be essential for odour learning, as both experimental and computational evidence suggest.

We have previously developed a computational model of the mammalian olfactory cortex [1], and recently of the olfactory system of insects, in order to complement experimental studies of animal olfaction with applications to agriculture, biological control, and artificial odour sensory systems. In particular, we address the general problem of stability and sensitivity: how can a system maintain sufficient stability, and not over-react, while still be sensitive to weak signals? Another, but related issue concerns the mechanisms for amplification and control of input stimuli. With simulated cortical neurodynamics we have shown how internal noise can result in state transitions [2,3]. An increased noise level in all network units can result in a transition from a stationary to an oscillatory state, or vice versa, or alternatively, in a shift between two oscillatory states. Under certain circumstances, a few noisy network nodes can induce global oscillations in the entire network [3]. Internal noise can even increase the efficiency in the olfactory information processing, where the rate of odour recognition is maximized for some optimal, non-zero noise level [4].

We now carry this computational paradigm into models of insect olfactory system, in particular of the antennal lobe, where we model how the spatio-temporal representation patterns of sex pheromone components and host plant odours emerge in the glomeruli. These patterns appear to depend on the interplay of synchronous oscillations in local and global circuits of the antennal lobe. We further investigate the dose-response characteristics of glomerular activity in relation to the network structure levels and the complex neurodynamics, and study how weak signals can be amplified (sensitivity to individual odour molecules) together with how a specific odour mixture can occur, based on neuronal and network properties and stochastic resonance (SR) dynamics [5]. Here, we primarily focus on the role of noise and how the insect olfactory system deals with noise. How can the system detect weak signals in the presence of environmental, as well as neural fluctuations? Can the system suppress the noise, or maybe even make use of the noise? Could noise-induced oscillations and SR also exist in the insect olfactory system?

There are many similarities in the (gross) structure of the mammalian and the insect olfactory systems, e.g. the antennal lobe and mushroom body of the insect correspond to the mammalian olfactory bulb and olfactory cortex, respectively. It is less clear to what extent the dynamics and functions are similar. However, the simpler structure of the insect olfactory system allows more realistic computational models, with relevance also to higher organisms and artificial systems. The increased understanding, as well as technological implementations, can lead to improved biocontrol of insects, important for agricultural production and ecological protection. The results may also lead to improved general electronic nose systems.

GLOMERULAR MODEL

Our previous model of the mammalian olfactory cortex used network nodes with a continuous output, representing the mean firing rates of populations of neurons [1]. In this work, we use a model of the insect glomerular network, with spiking neurons of Hodgkin-Huxley type. Fig. 1 shows the synaptic connection topology within the glomerular network model. Insects rely mainly on their olfactory system to memorize, identify and classify sex pheromonal odours and plant odours for mating and finding host plants. In the insect olfactory system, the odour information is processed through three known stages in the olfactory pathway: the antenna, the antennal lobe (AL), and the mushroom body (MB). The olfactory receptor neurons (RNs) in the antenna perform the basic function of recognizing the odour molecules [6].

The AL, which consists of local inhibitory interneurons (LNs) and output excitatory projection neurons (PNs), is the primary odour information processing centre in the insect olfactory system. Amplification, suppression, discrimination and integration of the odour stimulus is apparently carried out by the complex circuit dynamics of the AL network.

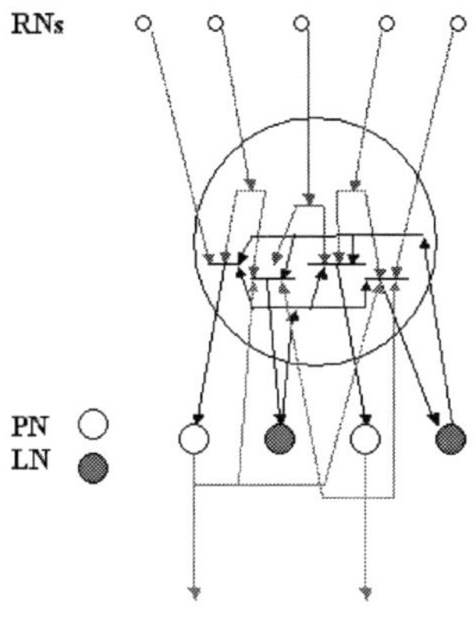

FIGURE 1. Synaptic connection topology within a glomerular network, where each PN and LN neuron receives input from RN neurons (purple arrow lines), the LNs send inhibitory input to each other and to PNs (blue arrows), the PNs send excitatory input to LNs and to MB and LH (yellow arrows).

The membrane potentials, V_{PN} and V_{LN}, of the PNs and LNs, respectively, are described by the following dynamical equations [7]:

$$\begin{aligned} C_m \frac{dV_{PN}}{dt} &= -g_L(V_{PN} - E_L) - I_{Na} - I_K - I_A \\ &\quad - g_{KL}(V_{PN} - E_{KL}) - I_{GABA_A} - I_{slow-inh} - I_{nACh} - I_{stim} \\ C_m \frac{dV_{LN}}{dt} &= -g_L(V_{LN} - E_L) - I_{Ca} - I_{K(Ca)} - I_K \\ &\quad - g_{KL}(V_{PN} - E_{KL}) - I_{GABA_A} - I_{nACh} - I_{stim} \end{aligned} \tag{1}$$

The parameters, C_m, g_L, g_{KL}, E_L and E_{KL} represent membrane capacitance, leak conductance, leak potasium conductance, and reversal potentials for the leak and leak potassium currents, respectively. The values of these parameters for PN and LN are taken from physiological data [7]. The intrinsic currents I_{Na}, I_K, I_A, I_{Ca} and $I_{K(Ca)}$

are fast sodium current, fast potassium current, transient potassium A current, transient Ca^{2+} current, and calcium-dependent potassium current, respectively.

SIMULATION RESULTS

We use the glomerular network model described above as the basis for our studies of the relations between structure, dynamics and function of insect olfaction. By connecting several such glomerular networks, we can get a more realistic picture of the function of the insect antennal lobe. We compare the results with our previous results of the mammalian system, and discuss the similarities and differences.

To study the effect of noise in detecting weak signals in our model, we add a white noise current with zero mean to a subthreshold constant stimulus signal, which alone cannot be detected by the network. The white noise could either represent the external background stimulus, or the intrinsic network noise. The PNs respond with random spiking activity, as shown in Fig. 2a. However, the local field potential (LFP), calculated as the mean membrane potential of all PNs, oscillates below zero, as seen in Fig. 2b. Such random spiking activity of PNs disappeared with a lower signal amplitude. With an increased noise variance, random spiking activity re-appears.

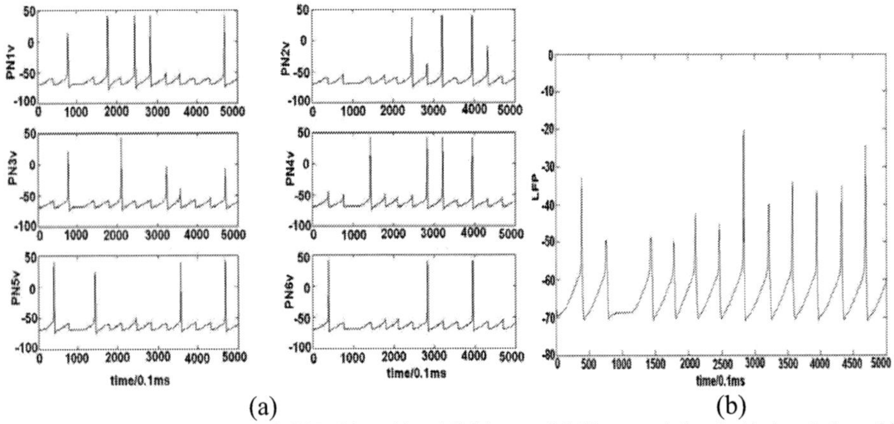

FIGURE 2. The response of six PNs (a) and local field potential (b) to a subthreshold signal plus white noise with zero mean.

When there is no odour signal and the mean of the background noise is subthreshold, with a small variance, the PNs show subthreshold and low amplitude oscillations. When two subthreshold odour pulses are added to the background noise, the PNs show superthreshold rhythmic oscillation. The simulation result of the LFP is shown in Fig. 3.

These results show that the glomerular model network can enhance weak signals by the added noise, thus demonstrating a stochastic resonance phenomenon, similar to what is found in many other sensory systems [5].

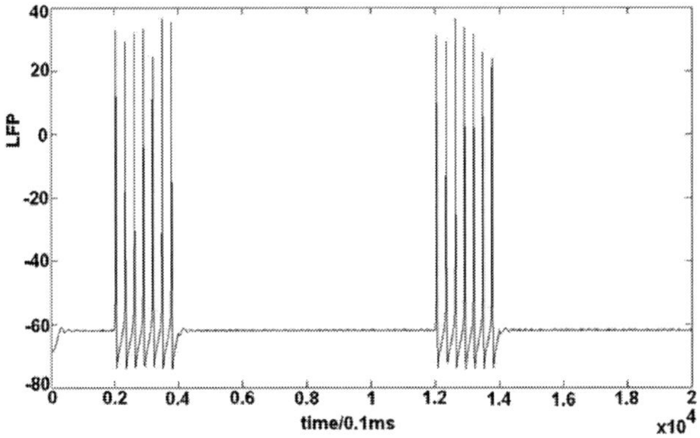

FIGURE 3. Noise-facilitated oscillations. Two subthreshold odour pulses with constant amplitude, which were presented during periods 200ms to 400ms and 1200ms to 1400ms, induce high frequency oscillations under background noise (with subthreshold mean and small variance).

When the white noise level is increased, computer simulations show that the PNs and LNs oscillate slowly in a random way under the noise background input. During the odour pulses are presented to the network, the neurons oscillate in a rhythmic way with higher frequency. The result is shown in figure 4.

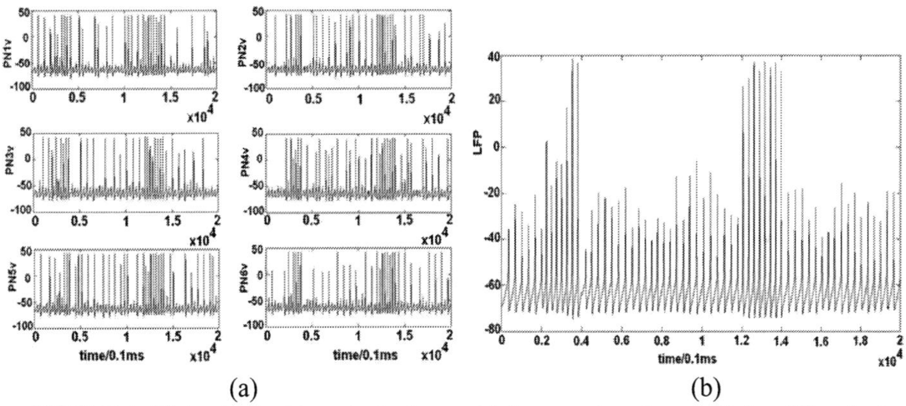

FIGURE 4. Higher level white noise background induces random slow oscillation. When the two odour pulses are added to the noise background, the PNs show higher frequency rhythmic oscillation.

With our glomerular model, we were unable to get synchronized noise-induced oscillations, unless we introduced PN-PN connections (akin to the pyramidal-pyramidal interconnections in the mammalian system). This may have functional significance that will be discussed below.

We currently combine several model glomeruli into larger networks, in order to study discrimination of odours in mixtures. With slow synaptic inhibition included to each LN-PN connection, the oscillation frequency decreased. The combination of slow inhibition with noise can result in larger differences between the timing patterns of PNs in the different glomeruli. In this sense noise and slow inhibition enhance the odour discrimination ability of the network.

CONCLUSIONS AND DISCUSSION

In this work, using a neural network model with an anatomically realistic circuit and a biophysically realistic neuron model of the insect olfactory system, we have investigated relationships between structure, dynamics, and function. In particular, we have investigated the role of noise in insect olfaction, and how the olfactory system deals with it, as compared to the mammalian counterpart. We demonstrate that a model network mimicking the general circuitry of the insect glomeruli can effectively suppress noise and enhance weak signals. We have also shown that weak, subthreshold signals can be detected when certain levels of noise is added, as in typical stochastic resonance. However, if we artificially include connections between the PNs in the model circuit, low level noise can induce synchronous oscillations. This is probably not desired, and hence such connections are avoided.

Oscillatory neurodynamics seems to be paramount for the mammalian cortical structures, presumable linked to cognitive functions. Our previous model of the olfactory cortex shows that intrinsic noise can induce oscillations in the absence of external input, suggesting that mental activity can be sustained in mammals even without any external stimuli. The fact that we could not induce synchronized oscillatory activity in our insect model, unless we artificially introduce connections between the excitatory PN cells, might indicate that (synchronized) neural oscillations are not important, and may even be undesired to the insect in the absence of external stimuli. This is in accordance with the richer mental capacity of the mammalian brain.

In a model with four glomeruli, and with no PN-PN connections, the system can easily identify a specific odour stimulus to one of the glomeruli, while it suppresses another weaker stimulus to another glomerulus. A network with (strong) PN-PN connections is not able to identify specific odour. A larger network with eight glomeruli, each with six PNs and two LNs is able to discriminate the dose difference of the different components in an odour mixture.

These results indicate that the interplay of the nonlinear feature of the micro-membrane ionic circuits of PNs and LNs and the macro-glomerular circuits plays an important role in dealing with noise signals.

ACKNOWLEDGMENTS

This work was made possible through a grant from Vinnova, the Swedish Agency for Innovation Systems.

REFERENCES

1. H. Liljenström, *Int. J. Neural Systems* **2**,1-15 (1991).
2. H. Liljenström, *Int. J. Neural Systems* **7**,497-505 (1996).
3. H. Liljenström and P. Århem, "Investigating Amplifying and Controlling Mechanisms for Random Events in Neural Systems". in *Computational Neuroscience*, (Ed. J.M. Bower), New York: Plenum Press (1997).
4. H. Liljenström and X. Wu, *Int. J. Neural Systems*. **6**,19–29 (1995).
5. F. Moss, L. M. Ward, W. G. Sannita, *Clinical Neurophysiology* **115**,267–281 (2004).
6. B. Hansson, Bill, *Insect Olfaction*, Springer (1999).
7. M. Bazhenov, M. Stopfer, M. Rabinovich, R. Huerta, H. D.I. Abarbanel, T. J. Sejnowski,and G. Laurent, *Neuron*, **30**, 553–567 (2001).

Correlations In Sequences Of First-Passage Times

Benjamin Lindner

Max-Planck Institute for the Physics of Complex Systems
Nöthnitzer Str. 38, 01187 Dresden, Germany

Abstract. The correlations in the sequence of first-passage times (FPTs) is discussed for two noise-driven systems. First, results for a perfect integrate-and-fire neuron driven by dichotomous process (DP) are reviewed. The FPT sequence in this case corresponds to the interspike intervals separating neural voltage discharges (action potentials). Furthermore, simulation results for the correlations of residence times in a bistable potential driven by a DP are presented. In both systems, the amplitude of correlations in the FPTs sequence decays exponentially with a rate inversely proportional to the correlation time of the driving process. While the neuron model shows a strictly positive correlation between all FPTs, the bistable system displays correlations that oscillate with the lag between intervals.

Keywords: Noise, First-Passage times, Dichotomous Process
PACS: 05.40.Ca, 87.19.La

INTRODUCTION

In neural modeling, researchers face the problem of high variability shown by many neurons *in vivo*. Neurons transmit information by means of stereotypical action potentials or spikes, the timing of which can exhibit a considerable jitter; spikes are even generated in the absence of any stimulus (spontaneous activity). In theoretical models it is thus important to incorporate the different kinds of noise (see, e.g. [1] for a review on noise in neural systems) and to study the resulting variability in the neural output, i.e. in the spike train.

In studies of stochastic neurons, most researchers have focussed on the variability of a *single* interspike-interval (ISI) which is the period separating two adjacent spikes. The ISI variability can be discussed in terms of its probability density, which shows in the spontaneous case often a unimodal shape; over the last decades, many different models have been investigated to properly describe this kind of statistics (cf., for instance, Ref. [2] and the more recent Ref. [3] for some experimental data and theoretical models). In contrast to this, *correlations among ISIs* have attracted attention only recently. ISI correlations have been reported, for instance, from experiments on auditory neurons and sensory neurons in different kinds of weakly electric fish [4, 5, 6] and have been studied in simple stochastic neuron models [5, 7, 8, 9, 10, 11, 12, 13, 14]. It was shown that correlations in the ISI sequence of the spontaneous activity of a neuron can have serious implications for neural signal transmission. In particular, negative ISI correlations can lead to significant enhancement of the signal transmission as shown in theory [5, 8, 13] and experiment [15].

Correlations among ISIs can be characterized by the serial correlation coefficient (SCC)

CP800, *Unsolved Problems of Noise and Fluctuations: UPoN 2005*
edited by L. Reggiani, C. Pennetta, V. Akimov, E. Alfinito, and M. Rosini

[16]

$$\rho_k = \frac{\langle I_{j+k} I_j \rangle - \langle I_{j+k} \rangle \langle I_j \rangle}{\langle I_j^2 \rangle - \langle I_j \rangle^2}. \tag{1}$$

Here, the numerator gives the covariance between two ISIs I_{j+k} and I_j that are lagged by an integer k; in the SCC this covariance is normalized by the variance of the single interval.

In most neuron models, the ISI can be mathematically described as a first-passage time (FPT). Hence, in neurobiology the problem of finding the correlations between two ISIs is equivalent to the problem of finding a correlation between FPTs. As shown in this paper, however, FPTs correlations can be observed *outside the context of neurobiology*. Such correlations can be expected for FPT processes driven by external signals or involving internal slow feedback variables. Examples include the residence times of a Brownian particle in a bistable potential driven by a a periodic force or a colored noise (for reviews on driven stochastic bistable systems, see e.g. [17, 18], Brownian motion in periodic potentials with time-dependent driving or the dynamics of stochastic excitable systems under external stimulation.

First the concept of the correlations of FPTs will be introduced for a simple neuron model. In the next section, numerical results for the overdamped Brownian motion in a bistable potential driven by a dichotomous process will be discussed. In this case, the FPTs correspond to resident times of the metastable states. It will be shown that the SCC of these residence times is an oscillating function of the lag k with an amplitude decay determined by the time-scale of the dichotomous driving. Such information might be absent in the residence time density (see Ref. [18]) commonly studied. Put differently, looking at the correlation of the residence times may enable us to estimate the time scale of the driving if we do not know it.

POSITIVE FIRST-PASSAGE TIME CORRELATIONS IN A PERFECT INTEGRATE-AND-FIRE MODEL

Consider a perfect integrate-and-fire neuron driven by a noisy current

$$\dot{v} = \mu + \eta(t) \tag{2}$$

with $\mu > 0$ being a constant input. The model is supplemented by the spike generator rule: whenever the voltage reaches a certain threshold v_T a δ spike is generated and the voltage is reset to zero. The periods between subsequent spikes define the ordered sequence of interspike intervals (ISIs) $I_i, I_{i+1}, \ldots$, these ISIs are given by the FPTs from 0 to v_T.

If $\eta(t)$ is a white noise then all ISIs will be uncorrelated implying that

$$\rho_k = 0, \quad \text{for} \quad k > 0. \tag{3}$$

The reason for this is simple: the reset of the voltage erases any memory carried by the voltage variable and the input noise does not have any correlations (it is white and thus uncorrelated).

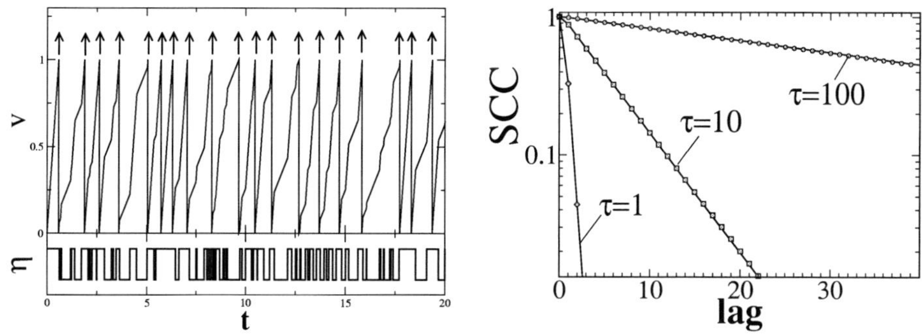

FIGURE 1. Left: Trajectory of the perfect integrate-and-fire neuron with spikes (arrows) and the dichotomous noise $\eta(t)$ (bottom, $\tau = 1$); ISIs are the periods between adjacent spikes. Right: Serial correlation coefficient of the perfect integrate-and-fire model driven by a dichotomous process with different correlation times as indicated. The analytical expression eq. (5) (solid lines) is compared to results of numerical simulations of the model (symbols). Parameters: $v_T = 1, \mu = 1, \sigma^2 = 0.5$.

Recently, the model eq. (2) was studied in the case where $\eta(t)$ is a dichotomous noise [14] (for an illustration of the dynamics in this case, see Fig. 1 (l.h.s.)). Such noise is exponentially correlated

$$\langle \eta(t)\eta(t') \rangle = \sigma^2 \exp[-\frac{|t-t'|}{\tau}], \tag{4}$$

where σ^2 and τ are the variance and correlation time of the noise, respectively. For $\sigma^2/\mu^2 \leq 1$, one can derive an exact expression for the SCC of the ISI sequence [14]

$$\rho_k^{DP} = 2e^{-k\beta} \sinh^2(\beta/2)/(\beta - 1 + e^{-\beta}), \;\; k > 0 \tag{5}$$

with the abbreviation

$$\beta = \frac{v_T \mu/\tau}{\mu^2 - \sigma^2}. \tag{6}$$

The theoretical expression is plotted and compared to results of numerical simulations in Fig. 1. It can be easily seen in the plot as well as inferred from eq. (5) that the SCC is strictly positive and decays exponentially with increasing lag. The rate of decrease is proportional to β and the latter in turn is inversely proportional to the correlation time of the noise. Hence, in this simple example the exponential correlation in the driving dichotomous noise results in a simple exponential correlation in the output ISI sequence. The longer the correlation time in the input noise, the longer will be the effective "correlation lag" $1/\beta$ in the ISI sequence. Similar findings were made for an Ornstein-Uhlenbeck process instead of a dichotomous noise [14]. The SCC seems to be solely determined by the temporal correlations of the driving process but not by the stationary probability density of the latter.

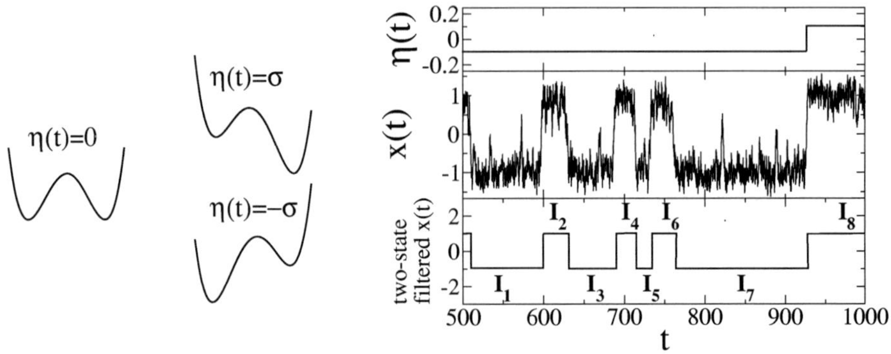

FIGURE 2. Left: The bistable potential $U(x,t)$ is shown in the absence of driving (l.h.s.) and for the two possible values of a finite dichotomous driving ($\sigma = 0.1$, r.h.s.). Right:Trajectory and driving force of the bistable system. From top to bottom: the dichotomous process with correlation time $\tau = 1000$ (only one switching of the driving occurs in the shown time window); the bistable dynamics undergoing several transitions from left to right and vice versa; the two-state filtered bistable dynamics, the residence times in the metastable states are indicated. Parameters: $D = 0.1, \sigma = 0.1, \tau = 1000$.

OSCILLATING CORRELATIONS IN THE TRANSITION TIMES OF A DRIVEN BISTABLE PROCESS

Here we consider the overdamped Brownian motion in a bistable potential driven by dichotomous noise. The dynamics is given by the equation of motion

$$\dot{x} = x - x^3 + \eta(t) + \sqrt{2D}\xi(t) \tag{7}$$

The process $\eta(t)$ is again a dichotomous process obeying the correlation function eq. (4) and assumed to be weak ($\sigma \ll 1$) such that at any time two metastable states exist. The last term in the above equation denotes white Gaussian noise; D is the noise intensity. The time-dependent potentials $U(x,t) = x^4/4 - x^2/2 - x\eta(t)$ corresponding to the driven motion is illustrated in Fig. 2 together with a typical trajectory of the system showing several transitions from one stable state to the other.

The transitions in $x(t)$ are due to the white noise $\xi(t)$ because although the dichotomous driving modifies the potential, it does not lead to a monostable potential (cf. Fig. 2, l.h.s.). The switching process is characterized by the sequence of first-passage times from -1 to 1 (residence time in state $x \approx -1$) and from 1 to -1 (residence time in state $x \approx 1$, cf. Fig. 2, r.h.s.). In the following we discuss the correlations in the ordered sequence of residence times; here we *do not* distinguish to which of the two states the residence time belongs.

It can be seen, that for a long-correlated dichotomous driving, at a given time transitions from one state to the other are favored compared to the inverse transition. Consequently, the residence times will be different: for $\eta(t) = \sigma > 0$, for instance, a transition from left to right will take a shorter period than the mean residence time $\langle I \rangle$ of both states. On the other hand, the residence time in the right state will be longer than $\langle I \rangle$. This implies a negative correlation for adjacent residence times. The first and the third residence times,

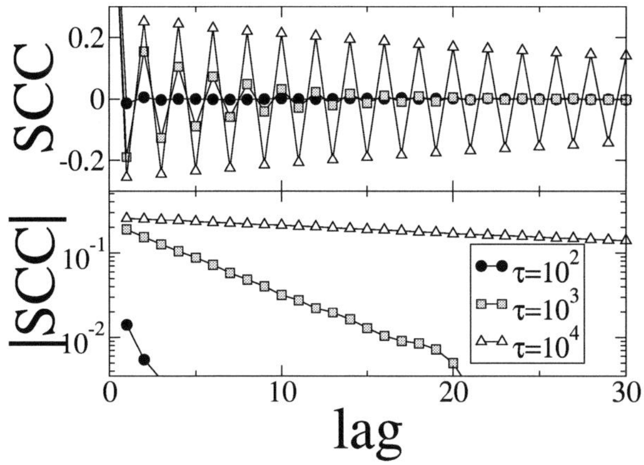

FIGURE 3. Serial correlation coefficient (top) and absolute value of this function (bottom) versus lag. Simulation result for the transition times in a bistable potential with white noise and dichotomous driving.

however, are associated with the same state and are therefore positively correlated.

So far we have assumed that we are in a given state of the dichotomous driving ($\eta(t) = \sigma > 0$). A jump of the dichotomous process will reverse the ratio between the mean values of the two different residence times associated with the two states and thus destroy the correlations discussed above. Consequently, pronounced correlations of the residence times will be observed only if the dichotomous process remains at the same value for a period covering many transitions, i.e. for a correlation time much larger than the mean residence time[1]. We can also expect that for small σ the correlations extend roughly over $\tau/\langle I\rangle$ residence times.

These conjectures are confirmed by simulation results of the model shown in Fig. 3. The serial correlation coefficient of the transition times shows a damped oscillation with negative correlation at lag one (corresponding to the correlation between adjacent residence times associated with the left and the right states), a positive correlation at lag two (corresponding to the correlation between closest residence times in the same state) and so on. An increase in the correlation time τ of the driving process results in a stronger correlation of the residence times. This is true with respect to the amplitude of correlations at a given lag, but also with respect to the "correlation lag". The logarithmic plot in Fig. 3 (lower panel) reveals an exponential decay with respect to the lag. In fact all our data can be described by the formula

$$\rho_k \approx A(\tau)\exp\left[-\frac{\alpha k}{\tau}\right](-1)^k, \quad k > 0. \tag{8}$$

Apart from the change in the sign indicated by the last term, this has the same structure as the SCC of the ISI sequence of the perfect integrate-and-fire neuron eq. (5). The

[1] More exact, given the two mean residence times associated with the states favored/not favored due to the dichotomous driving, τ has to be much larger than the larger one of these times.

amplitude $A(\tau)$ grows with τ and saturates for large values of τ. Both the amplitude $A(\tau)$ and the parameter α depend on the noise intensity D and the variance σ^2 of the driving process.

The exponential decay in the correlations reflects the randomness introduced by the dichotomous driving. For a static ($\eta(t) \equiv 0$) but *asymmetric* potential, we would obtain an infinite "correlation lag" because in this case the alternation between long and short intervals is never interrupted. There are certainly more interesting limits than this static limit. In the case of a weak white noise ($D \ll 1$), for instance, the difference between the mean residence periods of the favored and the unfavored state grows exponentially. In this case, one could detect a very weak dichotomous driving by looking at the correlations among the residence times. In real systems in which one does not know the time scale of a weak and slow underlying driving process, this high sensitivity of interval correlations could be exploited to identify the unknown time scale.

SUMMARY AND CONCLUSIONS

We have discussed correlations in the sequence of first-passage times for a neuron model and for a noisy bistable dynamics. We have found significant correlations in the case that these systems are driven by a slow dichotomous noise. The decay rate for these first-passage time correlations is inversely proportional to the correlation time of the driving process. Analytical work on the dependence of correlations in driven first-passage time processes is an exciting problem for future studies.

REFERENCES

1. H. C. Tuckwell, *Stochastic Processes in the Neuroscience*, Society for industrial and applied mathematics, Philadelphia, Pennsylvania, 1989.
2. H. C. Tuckwell, *Introduction to Theoretical Neurobiology*, Cambridge University Press, Cambridge, 1988.
3. N. Brenner, O. Agam, W. Bialek, and R. de Ruyter van Steveninck, *Phys. Rev. E*, **66**, 031907 (2002).
4. S. B. Lowen, and M. C. Teich, *J. Acoust. Soc. Am.*, **92**, 803 (1992).
5. R. Ratnam, and M. E. Nelson, *J. Neurosci.*, **20**, 6672 (2000).
6. S. Bahar, J. Kantelhardt, A. Neiman, H. Rego, D. Russell, L. Wilkens, A. Bunde, and F. Moss., *Europhys. Lett.*, **56**, 454 (2001).
7. M. J. Chacron, A. Longtin, M. St-Hilaire, and L. Maler, *Phys. Rev. Lett.*, **85**, 1576 (2000).
8. M. J. Chacron, A. Longtin, and L. Maler, *J. Neurosci.*, **21**, 5328 (2001).
9. Y.-H. Liu, and X.-J. Wang, *J. Comp. Neurosci.*, **10**, 25 (2001).
10. M. J. Chacron, K. Pakdaman, and A. Longtin, *Neural Comp.*, **15**, 253 (2003).
11. B. Lindner, and A. Longtin, "Nonrenewal spike trains generated by stochastic neuron models," in *Noise in Complex Systems and Stochastic Dynamics*, edited by L. Schimansky-Geier, D. Abbott, A. Neiman, and C. V. den Broeck, SPIE, Bellingham, Washington, 2003, vol. 5114, p. 209.
12. J. W. Middleton, M. J. Chacron, B. Lindner, and A. Longtin, *Phys. Rev. E*, **68**, 021920 (2003).
13. M. J. Chacron, B. Lindner, and A. Longtin, *Phys. Rev. Lett.*, **92**, 080601 (2004).
14. B. Lindner, *Phys. Rev. E*, **69**, 022901 (2004).
15. M. J. Chacron, L. Maler, and J. Bastian, *Nature Neuroscience*, **8**, 673 (2005).
16. D. R. Cox, and P. A. W. Lewis, *The Statistical Analysis of Series of Events*, Chapman and Hall, London, 1966.
17. P. Hänggi, and P. Jung, *Adv. Chem. Phys.*, **89**, 239 (1995).
18. L. Gammaitoni, P. Hänggi, P. Jung, and F. Marchesoni, *Rev. Mod. Phys.*, **70**, 223 (1998).

Rhythm Analyses Of Melodies Used To Obtain Women Marathon Gold Medal

Munecazu Tacano[1], Saburo Yokokura[1], Yoko Kajiwara[2], Jan Pavelka[1], Nobuhisa Tanuma[1], Tatsuhisa Uemura[1], Sumihisa Hashiguchi[3], Josef Sikula[4]

[1] *Meisei University, Hino, Tokyo, 191-8506*
[2] *Bunkyo University, Koshigaya, Saitama, 343-8511 Japan*
[3] *Yamanashi University, kofu, yamanashi, japan*
[4] *Brno University of Technology , Bruno Czech REP.*

Abstract. In Athena Olympics in 2004 a Japanese girl got the gold medal in Women Marathon games. Just before the beginning, she was listening to some domestic melodies in order to concentrate on the race. The rhythm or power of that music is found to have the typical $1/f$ noise characteristics. The $1/f$ music is found effective to concentrate as well as to relax themselves for a fairly long time range, while some short time trial runner uses a kind of white noise like music.

Keywords: $1/f$ music, rhythm analyses, power of music, concentration and relaxation.
PACS: 87.54 Hk, 43.50 Cb, 43.50 Op

INTRODUCTION

Classic music is known to exhibit typical $1/f$ noise characteristics in its audio power, that is the instantaneous loudness of the music, while modern music is expected to exhibit rather white noise characteristics [1]. Whereas a quantity with a white spectral density is uncorrelated with its past, and a quantity with a $1/f^2$ spectral density depends very strongly on its past, a quantity with a $1/f$ spectral density has an intermediate behavior and presents some correlation over all times. Studies of interaction between music and human beings will give some improved use of the music as an advanced healing media. Woman marathon runners are commonly listening to their favorite melodies in order to concentrate their mind. We are then interested in the nature of these melodies in terms of their rhythm properties.

The gold medalist in Athena Olympics women marathon games was listening to some domestic melody, this melody has been analyzed, in terms of the rhythm frequency, and the results obtained are compared with those from other melodies.

We have three ways of analyses using personal computer. The most direct method is to Fast Fourier Transform (FFT) analyze the melodies and their squared power by computer. The 2nd one is to get the squared signals by an audio detecting device, and to analyze the electrical output by FFT. The third one is to make the zero crossing analyses of the melodies. In this last case, several wave files, from 1 to 3 minutes long, are sampled out from the melodies, and the frequency of the zero crossing intervals

CP800, *Unsolved Problems of Noise and Fluctuations: UPoN 2005*
edited by L. Reggiani, C. Pennetta, V. Akimov, E. Alfinito, and M. Rosini

were FFT analyzed for a certain period (40 ms). Squared frequency of the inverse of zero crossing intervals are plotted and their average gradients were numerically calculated to obtain the exponent of the frequency. To this purpose we used the program prepared by Musha [2].

ANALYSES

Part of the electronic output and of the music pertaining to the song favored by the gold medalist is given in Fig. 1. The entire song is analyzed by PC as reported in Fig. 2. This music is a kind of losing love song composed anew by a famous singer, known by every young people. Figure 2(a) shows the direct FFT analyses of the melody. The melody, electronically obtained from the CD, contains frequency components only within the audio frequency range while the other frequencies are filtered out electronically, as it is shown remarkably in the figure. The electrical output of the melody was then squared at each time and FFT analyzed by a commercial computer software. The power spectrum density is shown in Fig. 2(b), where the 1/*f* dependence clearly appears at frequency values between 0.1 and 10 Hz. The low frequency part below 10 Hz corresponds to the rhythm of a melody and somehow resonates with the brain waves. Some kind of resonance of the rhythm with the brain waves will make us to feel easy and composed.

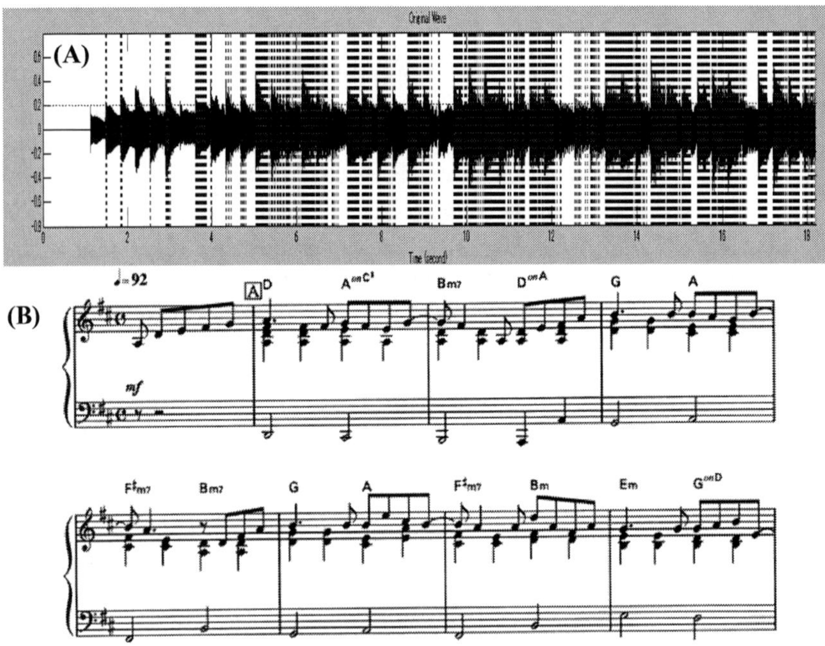

FIGURE 1. (A) CD output wave file of the starting part, and (B) corresponding music of the melody listened by the gold medalist of Athena Olympic Women Marathon game.

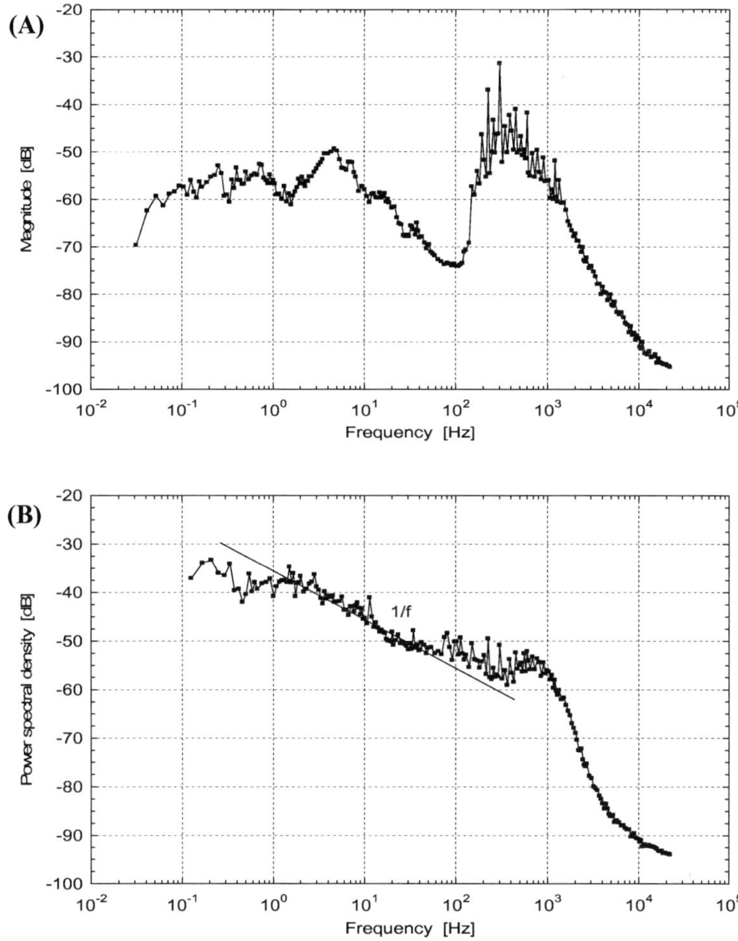

FIGURE 2. (A) Spectrum analysis of original CD recording, and (B) Power spectral density of squared signal listened by the gold medalist (Tsunami)

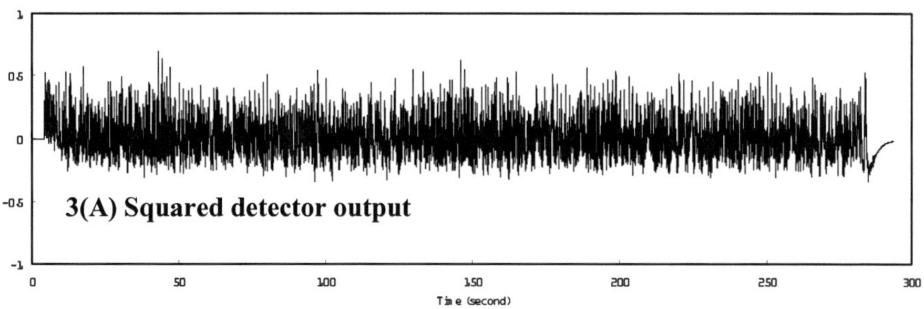

(B)

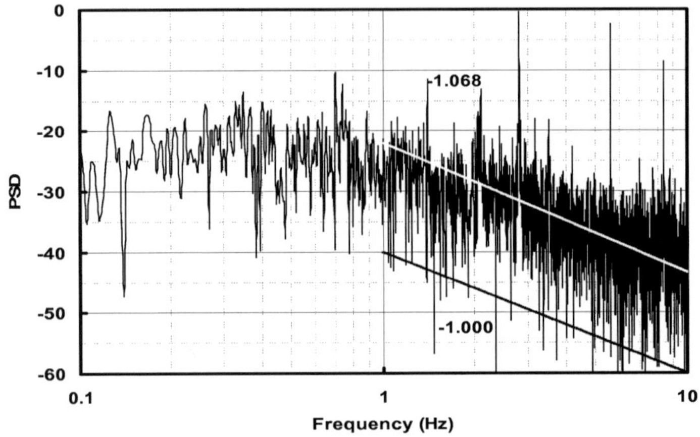

FIGURE 3. (A) Square detector output and (B) its FFT analyses. Least square approximation analyses give a frequency exponent of –1.07.

Next we obtained the analog output through a squaring detector diode, and the whole melody was FFT analyzed. Figure 3(A) shows the squared detector output and (B) its FFT analyzed data between 0.1 and 10 Hz. The least square of the data between 1 and 10 Hz gives a frequency exponent of -1.07.

(A)

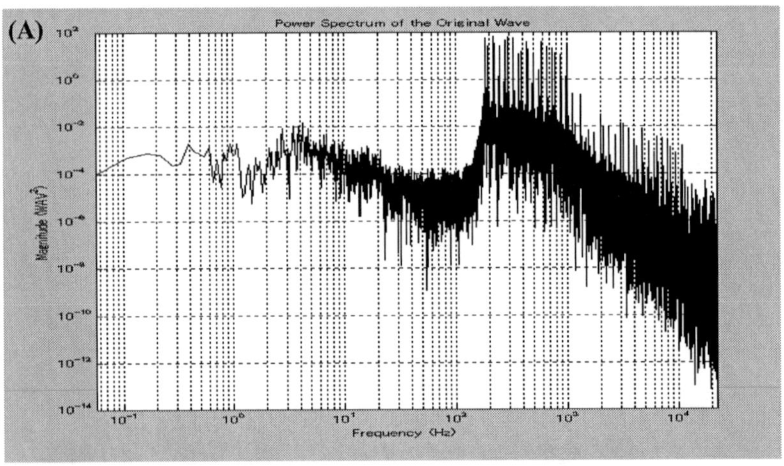

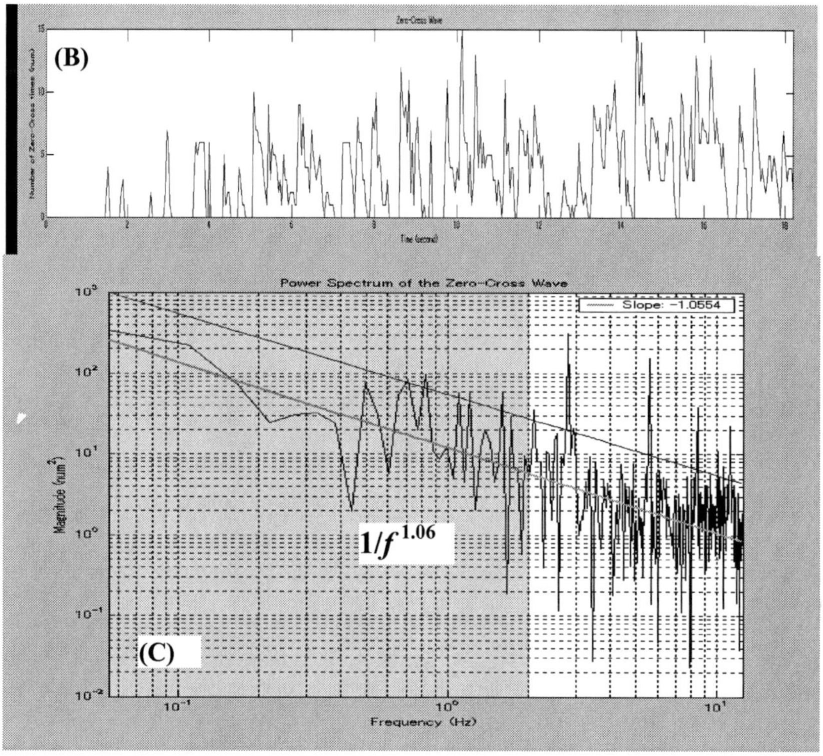

FIGURE 4. (A) Direct FFT analyses of original CD recording, essentially the same as Fig. 2(A). (B) Zero cross frequency of the signal during 40 ms versus time. (C) Its square versus frequency. The exponent of f is -1.06.

TABLE 1. Summary of Analyses

Music kind	Number of songs	Number of data	*f* exponent
RAP R&B	30	318	-0.85
J-POP	99	939	-0.90
Piano	58	534	-0.94
Disney	16	154	-0.95
Orchestra	40	480	-1.02
European	26	208	-1.03
Opera	50	412	-1.04
Lieder	38	268	-1.32
Overall Average	357	3313	-1.005

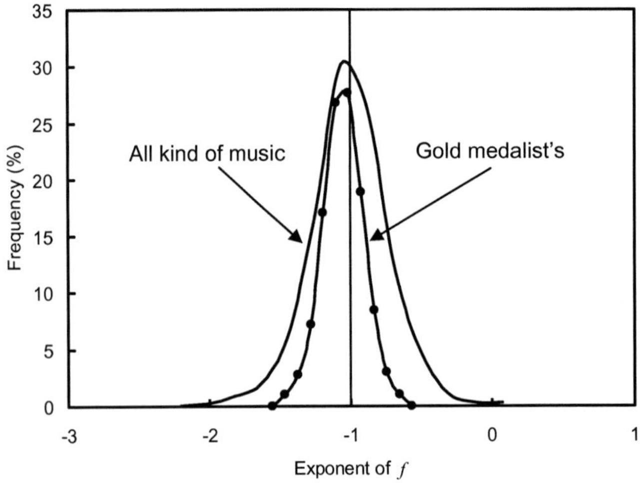

FIGURE 5. Distribution of f exponents of Gold medalist's favorite music in comparison with those of all kind of music. Gold medalist liked typical 1/f fluctuating music to concentrate her mind just before the start of the competition.

We have analyzed other various kinds of music: more than 350 songs corresponding to more than 3300 sampling points. As summarized in Table 1, the average of all kind of music is centered exactly at $1/f^{1.00}$. Here, the melody favored by the medalist is the typical $1/f$ music because of the smaller variation of 0.28. From the data in Table 1 we find that the so called modern music, like Pop and Rap, surely exhibits f exponents below –1.00, while old or folk music exhibits f exponents above –1.00.

Figure 5 shows the distribution of the exponent of this melody in comparison with those of various music and melodies, for a total amount of more than 3300 samples. It is apparent from the analyses that the gold medalist was listening to a typical $1/f^{1.07}$ melody. All kinds of music, at least the kind of easy listening music, show a Gaussian distribution centered at $f^{-1.00}$.

For the sake of completeness, we would like to show one more analyses of the most famous Italian song, O sole mio.

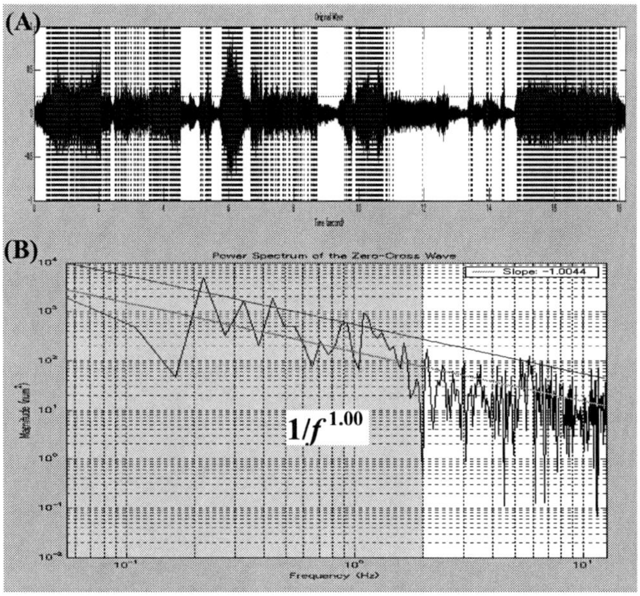

FIGURE 6. (A) CD output wave file of O Sole Mio, *"Ma natu sole cchiu bello, chine o sole o sole mio"*. (B) Square of zero cross frequency versus frequency. Exponent of *f* is found to be of –1.004 for *O sole mio*

The second paragraph was analyzed both by the zero cross analysis software and computer. As shown in Fig. 6(B), the exponent is –1.00 for about 20 minutes, and the whole melody has –1.2 between 30 and 0.1 Hz as shown in Fig. 7(b).

CONCLUSION

The rhythm analyses of the melody used to win the gold medal in Athena Olympic women marathon showed the typical $1/f^{\,1.07}$ fluctuation with very small standard deviation. The average of various kind of music over 3300 samples gave just exactly $1/f^{\,1.00}$ characteristics. The so called modern music, like Pop and Rap, has small f exponents below the value of –1.00, while old and folk music exhibit large f exponents above the value of –1.00. Independently of the kind of music, easy listening melodies are characterized by an f exponent value of –1.00.

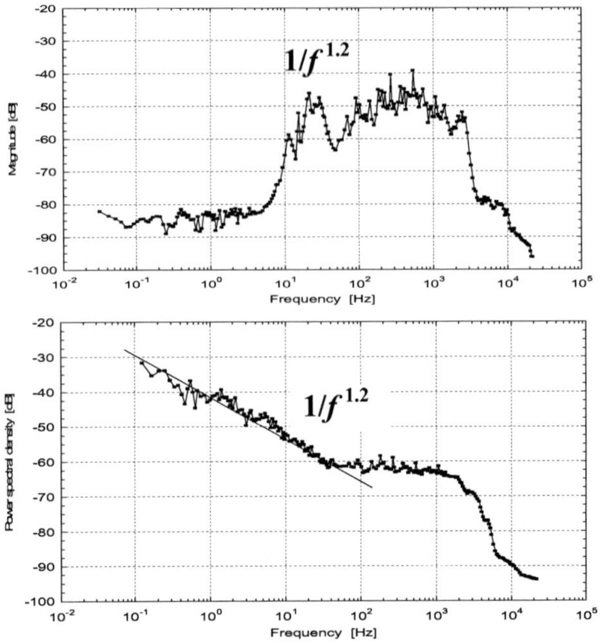

FIGURE 7. Frequency spectrum of the original CD recording and power spectral density of the squared signal (O Sole Mio)

ACKNOWLEDGMENTS

Authors are greatly indebted to Prof. T. Musha for allowing his zero cross analyses program to be used and for many fruitful discussions on the results.

REFERENCES

1. R. F. Voss, *Proceedings of the Symposium on 1/f fluctuations*, pp. 199-205 (1977)
2. T. Musha (private communication)

Single-Molecule Biochemical Analysis Using Channel Current Cheminformatics

Stephen Winters-Hilt

Computer Science Department, University of New Orleans, 2000 Lakeshore Drive, New Orleans, LA 70148;
and
Research Institute for Children, Children's Hospital 200 Henry Clay Avenue, New Orleans, LA 70118 USA

Abstract. A single nanometer-scale protein channel, residing in a bilayer, is used as a single-molecule measurement device. Single molecule kinetic information can be directly obtained with this approach via observation of single-molecule channel current blockades. A nanopore-based detector can also measure molecular characteristics indirectly, by changes in the blockades resulting from a changing bound-molecule complex. In essence, the heart of chemistry – the nature of the chemical bond – is now accessible via a new, computationally intensive, single-molecule observation method. In this work: (i) analysis of blockade signals is done using a variety of bioinformatics and machine learning tools; (ii) antibody blockade signals are examined and preliminary data on the characterization of antibody-antigen binding is briefly explored; and (iii) aptamer-based drug-discovery screening prospects are explored. The initial feature identification and extraction of blockade signals involves HMMs for level identification, HMM-EM for level projection, and time-domain FSAs for processing of the level-projected waveform. HMMs are then used for feature extraction and an SVM decision tree for multiclass discrimination. A new family of SVM variants is used, based on regularized-divergence kernels, and restriction is also made to feature vectors that can be interpreted as probability vectors. A web interface to the Channel Current Cheminformatics tools (unoCCC) and the Support Vector Machine classifier (unoSVM) will also be described.

Keywords: Single-molecule biophysics and biochemistry, nanopore detection, cheminformatics; biomolecules – structure and function; biomolecules – dynamics and conformational changes; biomedical engineering; biological signal processing.
PACS: 82.37.-j, 82.39.-k, 02.50.Ga, 07.05.Mh, 87.15.By, 87.15.He, 87.80.-y, 87.80.Tq

INTRODUCTION

Nanopore-based single-molecule detection has recently become established as a new tool in single molecule biophysics [1-7]. This has been made possible by two recent advances: (**i**) low-noise measurement of ionic currents through single, stable, nanoscopic channels (Fig. 1), and (**ii**) the development of modern cheminformatics methods for channel current feature extraction and pattern recognition (Fig. 2).

Using a nanopore detector it is possible to obtain at least Angstrom-level resolution at the termini of duplex DNA molecules, as shown in Fig. 1 [7]. Preliminary results show similar "telegraph" blockade signals occur with single antibody molecule blockades (Fig. 2). Furthermore, a clear transition in channel current blockade signal is observed in the nanopore detector upon binding and dissociation of antigen to the

CP800, *Unsolved Problems of Noise and Fluctuations: UPoN 2005*
edited by L. Reggiani, C. Pennetta, V. Akimov, E. Alfinito, and M. Rosini

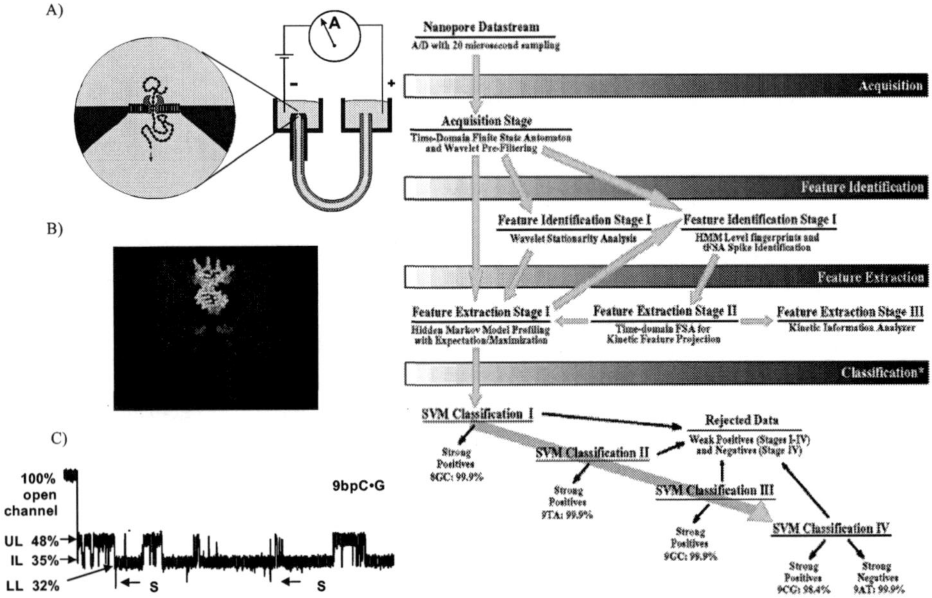

FIGURE 1. [left side] (A) shows a nanopore device based on the α–hemolysin channel. It has been used for analysis of single DNA molecules, such as ssDNA, shown, and dsDNA. A nine base-pair DNA hairpin is shown in (B) superimposed on the channel geometry. The channel current blockade trace for the nine base-pair DNA hairpin blockade from (B) is shown in (C). **[right side]** Shows the signal processing architecture. Classification was done by Support Vector Machine on five DNA molecules: four DNA hairpin molecules with nine base-pair stem lengths that only differed in their blunt-ended DNA termini, and an eight base-pair DNA hairpin. The accuracy shown is obtained upon completing the 15th single molecule sampling/classification (in approx. 6 seconds).

captured antibody (Fig. 3). It is hypothesized that nanopore-based detection can be used to study single molecule dynamics of the antibody-antigen interaction and analyze conformational changes that occur in the antibody upon binding to antigen. If proven, this hypothesis presents numerous ramifications for understanding antibody interactions. Because antibody flexibility appears to play a key role in the initiation of effector function, the ability to measure flexibility may also allow prediction of the abilities of different antibodies to initiate effector functions, thus enhancing the utility of the antibody-nanopore detector in multiple ways.

BACKGROUND AND METHODS

The notion of using channels as detection devices dates back to the Coulter counter [8], where pulses in channel flow were measured in order to count bacterial cells. Since its original formulation, channel sizes have reduced from millimeter scale to nanometer scale, and the detection mechanism has shifted from measurements of hydrostatically driven fluid flow to measurements of electrophoretically driven ion flow.

Angstrom precision structures for numerous DNA, RNA, and protein molecules have been revealed by X-ray diffraction analysis and NMR spectroscopy. These approaches, however, rely upon average properties of very large numbers of molecules and are often biased towards crystallization and NMR conformer structures different from those present in solution under physiological conditions. With the introduction of single molecule techniques in the early 1990's, such as atomic force microscopy and laser tweezers, new explorations into polymer structure and dynamics have begun.

Channel Current Cheminformatics. Nanopore-based detection offers a new approach to single molecule analysis that can directly, or indirectly, report on single molecule kinetics. The nanopore-based detector can *directly* measure molecular characteristics in terms of the blockade properties of individual molecules – this is possible due to the kinetic information that is embedded in the blockade measurements, where the adsorption-desorption history of the molecule (to the surrounding channel), and the configurational changes in the molecule itself, imprint on the ionic flow through the channel [4-7]. This offers prospects for DNA sequencing and single nucleotide polymorphism (SNP) analysis. Nanopore methods may even have potential for *single-molecule* sequencing at some point in the future. Sequencing and SNP analysis, however, represent a small fraction of the immense potential of such a device. This is because a nanopore-based detector can also measure molecular characteristics *indirectly*, by using a reporter molecule that binds to certain molecules, with subsequent distinctive blockade by the bound-molecule complex. In essence, the heart of chemistry – the nature of the chemical bond – is now accessible to a new, single-molecule, observation method that can reveal the kinetics of binding events (and conformational changes) by observation of state transition events as seen via changes of channel blockade.

The prototype signal processing architecture developed in [7], and shown in Fig. 1b, is designed to rapidly extract useful information from noisy blockade signals. Hidden Markov Models (HMMs) are used for the channel current feature extraction. Classification of feature vectors obtained by the HMM (for each individual blockade event) is then done using SVMs, an approach which automatically provides a confidence measure on each classification. SVMs are fast, easily trained, discriminators [9], for which strong discrimination is possible without the over-fitting complications common to neural net discriminators [9]. The SVM approach encapsulates a significant amount of model-fitting information in the choice of kernel in the SVM. In references [4,7] novel, information-theoretic, kernels were successfully employed for notably better performance over standard kernels. Further details on the signal processing are in [6,7].

Bioinformatics and Machine Learning Tools. New, web-accessible, channel current analysis tools, that make use of bioinformatics and machine learning methods, are described for extracting kinetic information. The computational tools are for analysis of digitized channel current data and make use of finite-state automata (FSAs), wavelets, and Hidden Markov Models (HMMs). For very short time-scales, a time-domain FSA has been devised. At intermediate scale in the signal structure, time-domain FSAs, wavelet-domain FSAs, and HMM-EM-projection approaches are applicable. At long timescales, HMM methods are applicable. The HMM and HMM-EM-projection approaches require the least tuning and are preferable for this reason. At the intermediate structure level this motivates use of HMMs for level identification, HMM-EMs for level projection, and time-domain FSAs for processing of the level-projected waveform. This provides a robust kinetic feature extraction formalism with a minimal amount of tuning.

SVM-based channel blockade classification has been mentioned, use is also made of SVMs in an entirely new way for clustering. Unlike the DNA hairpins that were designed with a unique capture orientation (the stem-end), protein molecules and non-designed molecules in general, can have multiple capture orientations. This greatly increases the complexity of the bookkeeping, and clustering, on channel current blockades, but is easily managed by computer.

Antibody Conformations And Antibody-Antigen Binding. As a test case for protein interaction studies, antibody conformation and antigen-binding was chosen as the focus of study. The basic domain structure of antibody is an elongated globular domain consisting of beta-pleated sheets. Antigen binding occurs in the amino-terminal domain, while the effector functions are localized to the C-terminal domains. A flexible hinge joins the amino-terminal domains of the molecule (Fab: two domains of the heavy chain and the entire light chain) with the carboxy-terminal portion (Fc).

Antibody effector function. Immunoglobulin (Ig) is a bifunctional molecule. Antigen is bound by the Fab portion. Upon binding to antigen, a series of events are initiated by the interaction of the antibody Fc region with serum proteins and cellular receptors. Biological effects resulting from Fc interactions include activation of the complement cascade, binding of immune complexes by Fc receptors on various cells, and the induction of inflammation. One goal of these experiments is to demonstrate whether allosteric changes are transmitted through the molecule upon antigen binding.

Aptamer-based Drug Discovery. Aptamers are nucleic acids selected for their binding properties to target molecules of interest. There is rapidly growing interest in aptamers due to their likely future use as a whole new class of medicines -- and the nanopore detector holds a novel position from which single aptamer binding efficacy studies might be pursued. In many respects, the DNA hairpins used in previous studies [7], and now used as controls, can be viewed as "dumb" aptamers in that they are nucleic acids with no binding properties. In the case of the DNA hairpins, and lengthy dsDNA molecule variants, it is known that the large cumulative charge along the backbone of the molecule leads to a channel blockade at a fixed, usually deep, level. Thus, much of the highly informative "telegraph" signal that tracks changes between binding states, is lost. The same appears to be the case for DNA aptamers recently examined. This is known as the state "sticking" problem, which appears to be solved by simply driving the system with laser pulsing (see Results for details).

RESULTS

New Channel Current Cheminformatics Tools. Web-accessible tools for HMM-based feature extraction and SVM classification [7] are now accessible via http://logos.cs.uno.edu/~nano/. The channel current feature extraction can help in identification of blockade levels, the level transition and lifetime characteristics, and the fast blockade "spike" characteristics. The SVM classification is of general use for any kind of classification problem, and a number of novel kernels and novel implementation are employed. SVM-based clustering is also implemented in a novel way to yield a non-parametric clustering approach, which is used to cluster signals into multiple classes (particularly important for complex multi-orientation data-sampling situations such as with an antibody).

Preliminary Antibody Blockade Characterization. The blockade signals for a murine IgG1 monoclonal antibody (labeled T6), directed against the synthetic polypeptide antigen (Y,E)-A--K have been examined most extensively thus far. The antibody preparation was grown in hybridoma supernatant and purified by antigen affinity chromatography, and thus represents a highly purified population of antibody. Upon introduction into the chamber, the antibody exhibits more than one blockade signal, suggesting that the different parts of the molecule can be drawn in. Since the immunoglobulin domain fold is an elongated barrel, it is possible that capture events could occur at the narrow ends of the domain, as individual chains, or portions of both chains, within the Fab and Fc, and possibly, the carbohydrate attached to the heavy chain. Figure 2 shows recurring blockade signals that have been observed. Similar findings, have been observed with other purified antibodies.

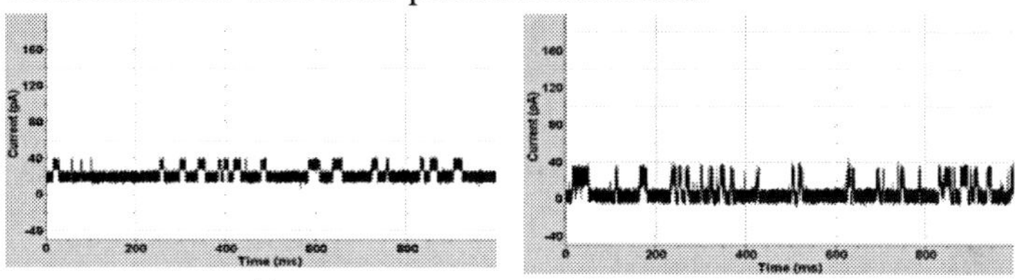

(a) (b)

Figure 2. (a) and (b) are recurring blockade signals for T6 antibody.

Antibody-Antigen Binding/Biosensing. It is found that the antibody blockade signal alters shortly after introduction of antigen, as Fig. 3 shows upon addition of a moderately high concentration (100 μg/ml) of 200kD multivalent synthetic polypeptide (Y,E)-A—K.

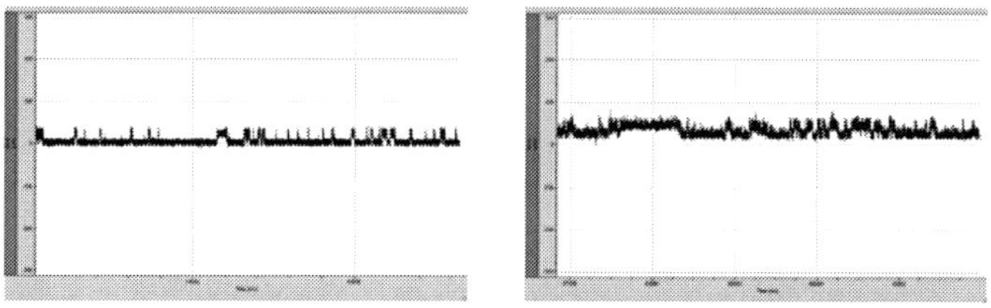

(a) (b)

Figure 3. (a) Blockade due to T6 antibody. (b) T6 antibody blockade one minute after exposure to antigen.

Presumably, these changes are the result of antibody binding to antigen. The time before the blockade signals are altered is also interesting; it ranges from seconds to minutes (not shown). This presumably is a reflection of antibody affinity.

Aptamer. Binding analysis with aptamers is much easier than with antibody-antigen due to the strong charge distributed along a DNA molecule, leading to a much stronger electrophoretic force interaction. The problem is that this force can be too strong when the molecule is actually captured, stifling sensitive blockade transitions. A possible solution is indicated by the results shown in Fig. 4, where laser pulsing is used to drive the system and keep it out of a "stuck" blockade mode:

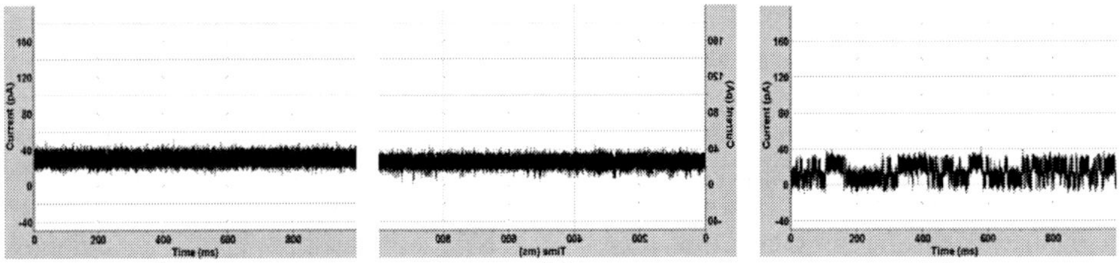

(a) (b) (c)

Figure 4. (a) Blockade with 20 base-pair DNA hairpin. (b) 20 base-pair hairpin with magnetic bead attached (biotin-streptavidin). (c) 20 base-pair hairpin with magnetic bead pulsed by laser.

CONCLUSION

Single-molecule biochemical analysis using channel current cheminformatics appears to be a viable technology. To maintain the existing biologically-based channel, however, there are salt and pH restrictions on the analyte solutions. More versatility is clearly needed here, but the significant strengths of this approach, with informatics methods leveraging the useful sensitivity of the detector, suggest great potential utility in this approach.

ACKNOWLEDGMENTS

Thanks to Dr. Seth Pincus for providing the antibodies used in the experiment. Thanks to the student nanopore technicians that helped gather data: Philip Romero, Eric Morales, and Iftekhar Amin. Thanks to the cheminformatics programming team: Srikanth Sendamangalam, Charlie McChesney, John Millet, and Anand Prabhakaran. Funding provided by LA Board of Regents grants LEQSF (2004-07)-RD-A-39; LEQSF (2004-05)-ENH-TR-108; and LEQSF (2004-05)-LASPACE-REA-14.

REFERENCES

1. M. Akeson, D. Branton, J.J. Kasianowicz, E. Brandin, D.W. Deamer. 1999. Microsecond Time-Scale Discrimination Among Polycytidylic Acid, Polyadenylic Acid, and Polyuridylic Acid as Homopolymers or as Segments Within Single RNA Molecules. Biophys. J. 77(6):3227-3233.
2. S. M. Bezrukov, I. Vodyanoy, V.A. Parsegian. 1994. Counting polymers moving through a single ion channel. Nature 370 (6457), pgs 279-281.
3. J. J. Kasianowicz,, E. Brandin, D. Branton, and D.W. Deamer. 1996. Characterization of Individual Polynucleotide Molecules Using a Membrane Channel. Proc. Natl. Acad. Sci. USA 93(24), 13770-73.
4. W. Vercoutere, S. Winters-Hilt, H. Olsen, D.W. Deamer, D. Haussler, and M. Akeson. 2001. Rapid discrimination among individual DNA hairpin molecules at single-nucleotide resolution using an ion channel. Nat. Biotechnol. 19(3):248-252
5. W. Vercoutere, S. Winters-Hilt, V. S. DeGuzman, D. Deamer, S. Ridino, J. T. Rogers, H. E. Olsen, A. Marziali, and M. Akeson. 2003. Discrimination Among Individual Watson-Crick Base-Pairs at the Termini of Single DNA Hairpin Molecules. Nucl. Acids Res. 31, pg 1311-1318.
6. S. Winters-Hilt and M. Akeson, "Nanopore cheminformatics," DNA and Cell Biology, Oct. 2004.
7. S. Winters-Hilt, W. Vercoutere, V.S. DeGuzman, D.W. Deamer, M. Akeson, and D. Haussler. 2003. Highly
8. W. H. Coulter, 1957, High speed automatic blood cell counter and cell size analyzer: proceedings of the National Electronics Conference, p. 1034-1042.
9. V. N. Vapnik. 1998. The Nature of Statistical Learning Theory (2nd ed.). Springer-Verlag, New York.

Functional Sub-Conformations in Protein Folding: Evidence from Single-Channel Experiments

Lisen Kullman[a,b], Mathias Winterhalter[c], and Sergey M. Bezrukov[a]

[a] *Laboratory of Physical and Structural Biology, NICHD, NIH, Bethesda, MD 20892-0924, USA*
[b] *Department of Medical Cell Biology, Uppsala University, Box 571, 751 23 Uppsala, Sweden*
[c] *International University of Bremen, PO Box 750 561, 28725 Bremen, Germany*

Abstract. We study fluctuations in ion conductance and enzymatic activity of a channel-forming protein, Maltoporin, at the single-molecular level. Specifically, we analyze time-persistent variability of these parameters from channel to channel and discuss our findings in the context of static disorder in protein folding. The natural function of Maltoporin is passive sugar translocation. Reconstituted into planar lipid bilayers this channel shows large fluctuations in both its ion conductance and sugar binding strength. The dynamic features of conductance fluctuations, representing the dynamic disorder, were studied previously and showed a complex power-law behavior. In the present paper, trying to elucidate the impact of channel's static disorder on its enzymatic function, we have analyzed more than 130 single channels. We conclude that fluctuations of the binding strength in time (same single channel) are much smaller than its fluctuations in ensemble (different single channels). Importantly, the disorder in ion conductance shows no significant correlation with the disorder in the binding strength. Variations in protein folding that are responsible for variations in ion conductance do not seem to affect sugar binding. This leads us to a conclusion that Maltoporin and, generally, other proteins can exist in a wide set of fully functional (but distinctly different) sub-conformations.

Keywords: static disorder; fluctuations; noise analysis.
PACS: 82.37.Np; 87.14.Ee; 87.15.Cc; 87.15.He; 87.15.Kg; 87.15.Ya

INTRODUCTION

Protein folding and misfolding is an exiting subject of theoretical and practical importance [1]. With the upcoming new technologies to observe the functioning of single proteins, it is now possible to monitor variables that are usually hidden in the ensemble averages. Among new questions to address is that of the static and dynamic disorder in protein folding. In experiments with single protein molecules, static disorder is seen as a time-persistent deviation of reaction rates from their average values. Dynamic disorder is manifested by the rate variations in time.

Reconstitution of ion channels into planar lipid bilayers is arguably the oldest single-molecule technique of them all [2,3]. In a typical modern experiment spontaneous insertion of a single channel is seen as a sharp step in the ion current through the membrane with a typical time resolution of about 10 μs [4]. From the very first reconstitution experiments which were performed with the "excitability

CP800, *Unsolved Problems of Noise and Fluctuations: UPoN 2005*
edited by L. Reggiani, C. Pennetta, V. Akimov, E. Alfinito, and M. Rosini

inducing material" [2,3] it became clear that the "discrete conductance steps" show certain variation around their average value and that this variation is significantly larger than the accuracy of the measurement. Nevertheless, with one notable exception [5], for more than 30 years of experimentation this persistent variability of conductance from channel to channel has never managed to attract any in-depth discussion of its origin. Among the possible reasons of such neglect could be the following two: (i) most of the researcher's attention was concentrated on the average conductance, and (ii) a proper language to discuss this functional variability was absent.

In this paper we focus on the time-persistent fluctuations in the functional properties within the ensemble of 150 Maltoporin trimers reconstituted into lipid bilayers and interpret our findings using the language of static disorder in protein folding [6,7]. Channel conductance fluctuations in time, related to dynamic disorder [8], were analyzed previously and demonstrated the stretched-exponential or 1/f type of behavior [9]. Maltoporin (also called LamB) is a member of a family of solute-specific channels that facilitate exchange of maltosugars across bacterial membranes [10,11]. The crystal structure of Maltoporin reveals its homotrimeric organization [12]. Each monomer is formed by an 18-stranded β-barrel with short turns at the periplasmic side and large irregular loops at the outside of the cell. The third loop, L3, folds inside the barrel and forms a constriction at the middle of the channel, giving the pore an hourglass shape. The pore has a slight helical twist following the secondary structure of larger maltodextrins. The structure of sugar-soaked crystals of Maltoporin shows a specific sugar translocation pathway [13].

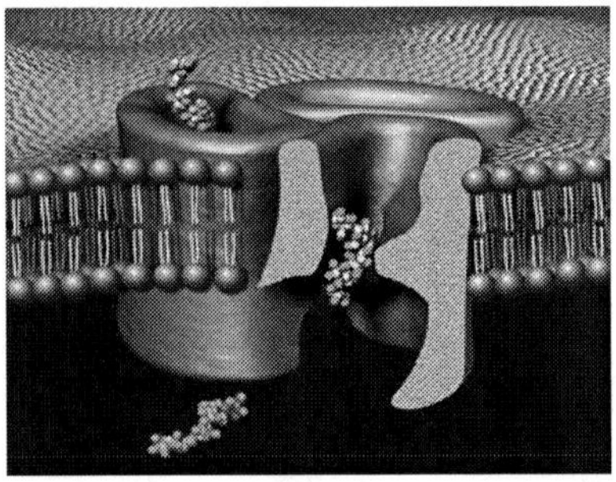

FIGURE 1. A sketch of a single trimeric Maltoporin channel reconstituted into a lipid bilayer. Each subunit is about 80Å long.

RESULTS AND DISCUSSION

Maltoporin channels have been extensively studied using planar lipid bilayers, a technique which offers the advantages of a well-controlled environment. When reconstituted into a lipid membrane, Maltoporin forms ion-permeable channels [14,15]. Penetrating sugars bind to Maltoporin and block the small-ion permeation [16,17]. Recently we investigated the "enzymatic activity" of Maltoporin – that is, its ability to facilitate maltodextrin transfer – at the single-molecule level [18,19]. We have shown that: (i) single molecular events of maltodextrin translocation can be observed as time-resolved transient interruptions in the ion current through the Maltoporin channel; (ii) one sugar molecule completely blocks ion current through one of the monomers of the trimer; (iii) monomers bind sugars independently of each other, which results in a binomial distribution of the probabilities to find the channel in a particular blocked state. Thus, we can study channel enzymatic activity simultaneously with its ability to conduct small ions, e.g., potassium and chloride.

Figure 1 gives a simplified schematic presentation of the Maltoporin trimer in the presence of permeating sugars. Sugar molecules randomly diffuse through the channel pore interacting with its interior via van der Waals forces and hydrogen bonding. The constructive role of this attractive interaction was recently elucidated in a theoretical model for single particle diffusion in the potential of mean force [20,21].

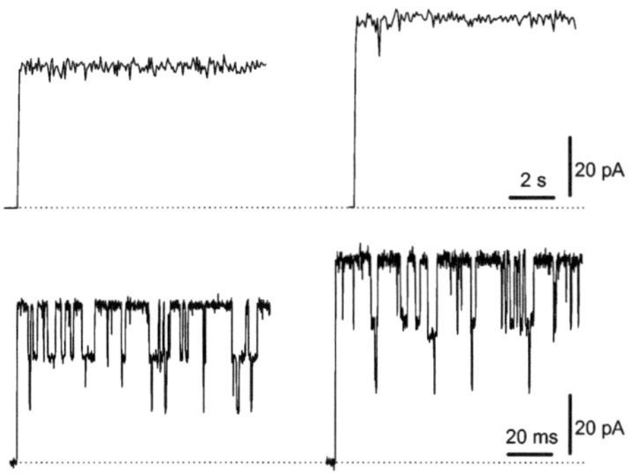

FIGURE 2. Currents through two different Maltoporin channels in the presence of sugar in 1M KCl solution show large difference in conductance. Permeating sugar molecules induce partial blockage of channel small-ion currents. The higher resolution recordings at the bottom show that conductances of the fully open channels differ by about 20%.

While in the pore, the permeating sugars block it for the small-ion conduction. The upper tracks in Fig. 2 show currents through two single Maltoporin channels at 200 mV of applied voltage and 50 ms time resolution in the presence of 10 μM maltohexaose. The channels were reconstituted at identical experimental conditions; however, the channels differ by their conductance. This could be related to the difference in the initial channel conductance (before sugar addition), the difference in sugar binding strength, or both. Higher resolution recordings shown at the bottom of Fig. 2 allow us to discriminate between these possibilities. The sugar binding events are represented by discrete reversible transitions to 2/3 or 1/3 of the initial current, corresponding to one or two pores blocked simultaneously, so that the conductance of a fully open channel can be accurately measured in the presence of sugar [18,19].

Figure 2 (bottom) also demonstrates that the dwell time of a sugar molecule in the channel is quite large, of the order of $10^{-3} - 10^{-4}$ s. Spectral analysis of the single-channel current noise (Fig. 3) demonstrates that the dwell time depends on the applied transmembrane voltage and varies from 0.2 ms at -200 mV to about 1 ms at the positive potentials. It is interesting to compare this time with the characteristic time of diffusional relaxation expected for a molecule of this size in a channel of 5 nm length.

For a cylindrical channel of length L and radius a, the average lifetime of a particle that is initially uniformly distributed in the channel is given by [22]:

$$\tau_{DR} = \frac{L^2}{12D}\left(1 + \frac{3\pi a}{2L}\right) \quad (1)$$

where it is assumed that particle diffusion in the channel and in the bulk is described by the same diffusion coefficient D. This expression accounts for the finite probability of particle return to the channel at finite L/a ratios. For the case of a long and narrow channel, $L >> a$, it gives a well-known estimate $L^2/12D$ [23]. Substituting $L \approx 5$ nm and $D \approx 5\text{x}10^{-10}$ m^2s^{-1}, we obtain $\tau_{DR} \approx 5\text{x}10^{-9}$ s. In the absence of specific interactions between the particle and the pore, diffusion within the pore can be slowed down by an order of magnitude [24], but even then the calculated time is about 4 orders of magnitude smaller that the time obtained in the experiment.

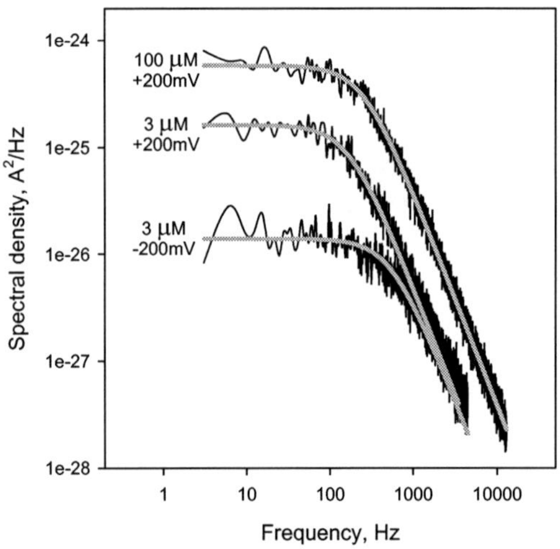

FIGURE 3. Examples of spectra for two different sugar concentrations and two applied voltages obtained from the same Maltoporin channel. Gray solid lines are Lorentzians fitted to the data.

This strong particle/pore interaction allows us to resolve single-molecular events of sugar translocation and search for possible correlations between the small-ion conductance of the channel and its enzymatic activity. Main parameters of the enzymatic reaction are the on- and off-rates which can be found as:

$$k_{off} = 2\pi f_c(1-p), \quad (2)$$

$$k_{on} = \frac{2\pi f_c p}{[c]} \quad (3)$$

where f_c is the corner frequency of a corresponding noise spectrum (Fig. 3), $[c]$ is sugar concentration, and p is the probability to find a monomer of the channel in the blocked state, measured as the relative average current decrease due to sugar addition (Fig. 2). The binding constant is the ratio k_{on} / k_{off}.

In Fig. 4 we plot the sugar binding constant versus the trimer conductance. The data were collected in 130 independent experiments. The vertical error bars represent the accuracy of the binding constant measurements; the error of the conductance measurement is close to the symbol size. It is seen that deviations in the sugar binding constant and small-ion conductance from their average values are poorly correlated.

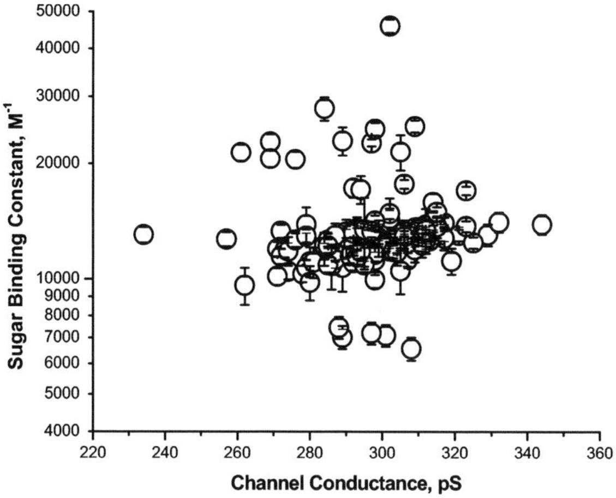

FIGURE 4. The static disorder in sugar binding and channel conductance show no correlation.

Applying linear correlation analysis, we find that the correlation coefficient between the small-ion conductance and logarithm of the sugar binding constant is smaller than 0.05. Thus ion conductance, which presumably reflects deviations in the pore aperture, fluctuates independently from the sugar binding strength. Similar results were obtained for the rate constants (data not shown).

CONCLUSIONS

To interpret these findings we use the language of protein folding disorder [6,7]. The absence of any significant correlations between the two parameters leads us to conclude that the static folding disorder that is responsible for the time-persistent deviations of the channel conductance from its average value does not perturb the channel ability to bind sugars. We find that the channel remains enzymatically functional in many sub-conformations as witnessed by their different conductance. This finding is analogous to static disorder found in enzymatic turnover rates of single enzyme molecules [25-27].

One of the unsolved problems is to understand what particular sub-conformations would result in such features. The 18-stranded Maltoporin β-barrel itself is recognized to be very rigid and, therefore, is an unlikely candidate. It seems that the short turns at the periplasmic side do not have enough freedom either. However, we think that the large irregular loops at the channel entrance facing outside of the cell could introduce a significant degree of static disorder, contributing to persistent fluctuations in both the small-ion conduction and the sugar translocation studied here.

REFERENCES

1. Dobson, C.M., *Seminars Cell Develop. Biol.*, **15**, 3-16 (2004).
2. Bean, R.C., W.C. Shepard, H. Chen, and J. Eichner, *J. Gen. Physiol.* **53**, 741-757 (1969).
3. Ehrenstein, G., H. Lecar, and R. Nossal, *J. Gen. Physiol.* **55**, 119-133 (1970).
4. Nestorovich, E.M., C. Danelon, M. Winterhalter, and S.M. Bezrukov, *Proc. Natl. Acad. Sci. U.S.A.* **99**, 9789–9794 (2002).
5. Sawyer, D.B., R.E. Koeppe II, and O.S. Andersen, *Biochemistry* **28**, 6571-6583 (1989).
6. Frauenfelder, H., S.G. Sligar, and P.G. Wolynes, *Science* **254**, 1598-1603 (1991).
7. Frauenfelder, H., and D.T. Leeson, *Nat. Struct. Biol.* **5**, 757-759 (1998).
8. Zwanzig, R., *Acc. Chem. Res.* **23**, 148-152 (1990).
9. Bezrukov, S.M., and M. Winterhalter, *Phys. Rev. Lett.*, **85**, 202-205 (2000).
10. Nikaido, H., and M. Vaara, *Microbiol. Rev.* **49**, 1-32 (1985).
11. Benz, R., *Ann. Rev. Microbiol.* **42**, 359-393 (1988).
12. Schirmer, T., T.A. Keller, Y.-F. Wang, and J.P. Rosenbusch, *Science* **267**, 512-514 (1995).
13. Dutzler, R., Y.-F. Wang, P.J. Rizkallah, J.P. Rosenbusch, and T. Schirmer, *Structure* **4**, 127-134 (1995).
14. Schindler G., and J.P. Rosenbusch, *Proc. Natl. Acad. Sci. USA* **75**, 3751-3755 (1978).
15. Benz, R., A. Schmid, T. Nakae, and G.H. Vos-Scheperkeuter, *J. Bact.* **165**, 978-986 (1986).
16. Benz, R., A. Schmid and G.H. Vos-Scheperkeuter, *J. Membrane Biol.* **100**, 21-29 (1987).
17. Nekolla, S., C. Andersen, and R. Benz, *Biophys. J.* **66**, 1388-1397 (1994).
18. Bezrukov, S.M., L. Kullman, and M. Winterhalter, *FEBS Letters,* **476**, 224-228 (2000).
19. Kullman, L., M. Winterhalter, and S.M. Bezrukov, *Biophys. J.,* **82**, 803-812 (2002).
20. Berezhkovskii, A.M., M.A. Pustovoit, and S.M. Bezrukov, *J. Chem. Phys.* **116**, 9952-9956 (2002).
21. Berezhkovskii, A.M., and S.M. Bezrukov, Biophys. J. **88**, L17-L19 (2005).
22. Berezhkovskii, A.M., M.A. Pustovoit, and S.M. Bezrukov, *J. Chem. Phys.* **119**, 3943-3951 (2003).
23. Feher, G., and M. Weissman, *Proc. Nat. Acad. Sci. USA* **70**, 870-875 (1973).
24. Bezrukov, S.M., *J. Membrane Biol.* **174**, 1-13 (2000).
25. Xue, Q.F., and E.S. Yeung, *Nature* **373**, 681-683 (1995).
26. Craig, D.B., E.A. Arriaga, J.C.Y. Wong, H. Lu, and N.J. Dovichi, *Anal. Chem.* **70**, 39A-43A (1998).
27. Dyck, A.C., and D.B. Craig, *Luminescence* **17**, 15-18 (2002).

“1/*f* Spectra”: Noise, Chaotic Dynamics – Or Phase Coupled Oscillators?

Aneta Stefanovska[*,†] and Igor Khovanov[†]

[*] *Group of Nonlinear Dynamics and Synergetics, Faculty of Electrical Engineering, University of Ljubljana, Tržaška 25, 1000 Ljubljana, Slovenia*
[†] *Department of Physics, Lancaster University, Lancaster LA1 4YB, UK*

Abstract. Coupled nonlinear oscillators operating in the presence of noise are widespread in nature. Examples include brain waves and cardiovascular oscillations. They can be amplitude or frequency-coupled, and are characterised by time-varying frequencies of oscillation. The time series related to cardiovascular and brain oscillations have been extensively analysed in the past decades and often associated with $1/f$ spectra. Their origin has been characterised variously as noise, chaotic dynamics and phase coupled oscillations. Here we present and discuss an unsolved problem of how to define and calculate instantaneous phases and frequencies of noisy coupled oscillators.

Keywords: noise, coupled oscillators, phase coherence, wavelet transform
PACS: 05.40.Ca, 05.45.Xt

MOTIVATION

The underlying motivation for this work is the quest to reach an understanding of cardiovascular and brain dynamics. Basically, we study oscillatory components that contribute to the circulation of blood through the cardiovascular system. The latter can be perceived as a closed system of tubes within which an amount of blood equivalent to the total volume in the human body ($\sim$5 ℓ) is circulated by the heart every minute [1]. We evaluate the contributions of noisy oscillators involved in the blood circulation and their interactions [2, 3, 4].

We present in this paper a discussion of an unsolved problem motivated by these studies of cardiovascular dynamics: *How to define and calculate the instantaneous phases and frequencies of noisy coupled oscillators?* We first describe briefly the existing methods. Then we choose a model of three coupled van der Pol oscillators to illustrate the problem. We present synchronization analyses and decomposition techniques. Then we introduce the wavelet-based phase coherence function as a possible tool and compare it to the coherence function based on the Fourier transform. Finally, we provide a summary and identify more unsolved problems.

Existing approaches. One useful approach is that of phase interaction [5]. It is based on the analytical signal concept and it is widely applied for the analysis of random processes [6]. It uses a representation of the signal (time-series) $y(t)$ in the form $y(t) = a(t)\cos(\varphi(t))$, i.e. in the form of a harmonic function with time-dependent amplitude $a(t)$ and phase $\varphi(t)$ (the phase derivative defines an instantaneous frequency). The amplitude $a(t)$ characterizes the energetic property of the signal $y(t)$, whereas the phase $\varphi(t)$ describes its time scale. In some cases, for example, if the amplitude is a slow

CP800, *Unsolved Problems of Noise and Fluctuations: UPoN 2005*
edited by L. Reggiani, C. Pennetta, V. Akimov, E. Alfinito, and M. Rosini

variable (see [5, 7] for details), the phase $\varphi(t)$ alone is sufficient to describe oscillations of $y(t)$ and, consequently, the analysis of the initial time series is simplified [5]. Moreover, the phase interaction approach provides a very sensitive tool for characterizing the interaction between coupled systems, including non-stationary and noisy processes. The key point in this approach lies in finding a method to determine the phase $\varphi(t)$.

Cardiovascular signals have frequency content spanning a wide interval. It is obvious that, in this case, we cannot apply the phase interaction approach directly, since we have several (not just one) time scales and we therefore have to introduce several phases. Thus we must either decompose the signals into narrow-band intervals, or take explicit account of the wide frequency band and the time-variable nature of basic frequencies, i.e. go far beyond the conventional phase interaction approach. We present below an example that demonstrates some of the problems that typically arise in the analysis of processes occurring on several time scales.

THE MODEL IN TIME AND FREQUENCY DOMAIN

As a simple representative example, we consider two Van der Pol oscillators coupled via a third Van der Pol oscillator:

$$
\begin{aligned}
\ddot{x} - \varepsilon_x \dot{x}(1-x^2) + \omega_x x + \gamma_{xy}(y-x) &= \sqrt{D}\xi_x(t),\\
\ddot{y} - \varepsilon_y \dot{y}(1-y^2) + \omega_y y + \gamma_{xy}(x-y) + \gamma_{zy}(z-y) &= \sqrt{D}\xi_x(t),\\
\ddot{z} - \varepsilon_z \dot{z}(1-z^2) + \omega_z z + \gamma_{zy}(y-z) &= \sqrt{D}\xi_z(t).
\end{aligned} \tag{1}
$$

Here ε_i and ω_i are oscillator parameters, γ_{ij} specifies their coupling and $\xi_i(t)$ is white noise of intensity D. For our numerical simulations, we use parameter values $\varepsilon_x = 2$, $\varepsilon_y = 0.5$, $\varepsilon_z = 1$, $\omega_x = 1$, $\omega_y = 0.3$, $\omega_z = 0.11$, $D = 0.001$. Note that there is no direct connection between the x-oscillator and the z-oscillator and that the only communication between them takes place via the y-oscillator, Eq. (1). Our choice of parameters places the oscillators in a region where noise affects their phase dynamics rather than their amplitude behaviour.

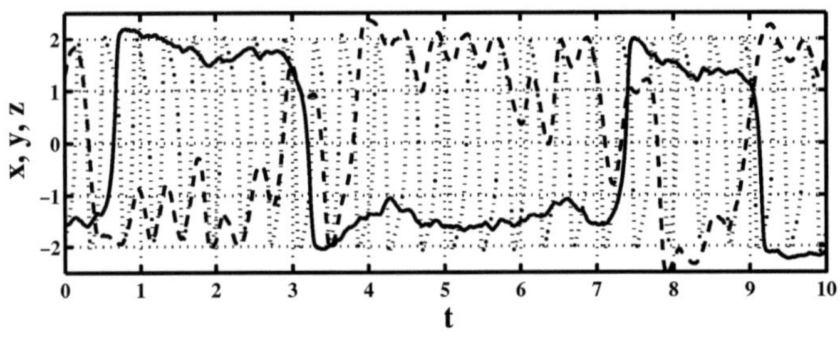

FIGURE 1. The time series of oscillators. The solid line corresponds to z-oscillator, the dashed line to y-oscillator and the dotted line to x-oscillator.

The system of coupled oscillators (1) demonstrates a variety of regimes, and we choose the regime of relatively strong coupling: the values of coupling coefficients are $\gamma_{xy} = 0.15$ and $\gamma_{zy} = 0.05$. If we consider noise-free dynamics of neighbouring oscillators with either ($\gamma_{xy} = 0.15$, $\gamma_{zy} = 0$) or ($\gamma_{xy} = 0$, $\gamma_{zy} = 0.05$) then, for these values of coupling, we will see synchronization between them. The study of strong interactions of this kind is motivated by experimentally observed interactions between respiration, cardiac and brain oscillations and waves.

Realizations of the corresponding signals are presented in Fig. 1. Note, that we do not introduce non-stationarity in this example, so that the standard Fourier approach is still valid.

Let us characterize the interaction between oscillators by using the phase approach. It can be seen (Fig. 1) that the oscillations of the y-oscillator have two components, and therefore the representation of the signal in the form: $y(t) = a(t)\cos(\varphi(t))$ can be problematic, although the Hilbert transform is still formally valid. To avoid the problem, we therefore decompose the signal $y(t)$ into two parts: $y(t) = a_1(t)\cos(\varphi_1(t)) + a_2(t)\cos(\varphi_2(t))$. We effect low- and high-pass filtration using the wavelet transform (though the precise method of filtration is not of crucial importance at this point). During high-pass filtration we remove from $y(t)$ the low-frequency component corresponded to the z-oscillator; low-pass filtration removes the high-frequency component corresponded to the x-oscillator. As a result we have after filtration two new signals $y_H(t)$ and $y_L(t)$ (see Fig. 2).

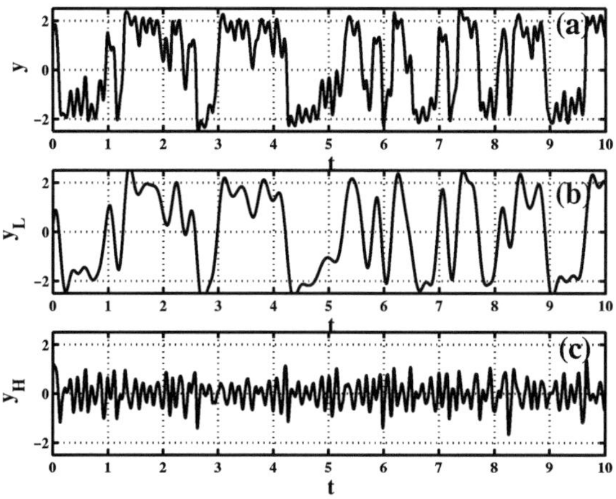

FIGURE 2. (a) The realization of y-oscillator, (b) the high-pass filtered realization, (c) the low-pass filtered realization.

Now let us compare the phase behaviour of all three signals: without filtration $y(t)$, and with filtration $y_L(t)$ and $y_H(t)$. We calculate the phases by Hilbert transform. The results are presented in Fig. 3. We see dramatic differences in the ratios between the phases. Both filtered components of the y-oscillator are synchronized (there are episodes of synchronization and noise-induced phase slips), whereas the phase of the initial signal $y(t)$ demonstrates non-synchronized and synchronized behaviours with different ratios between phases. Additional detailed inspection has proved the correctness of the results sahown for the filtered signals. So, the decomposition allows us to use the phase approach based on the Hilbert transform, and to describe double simultaneous synchronization.

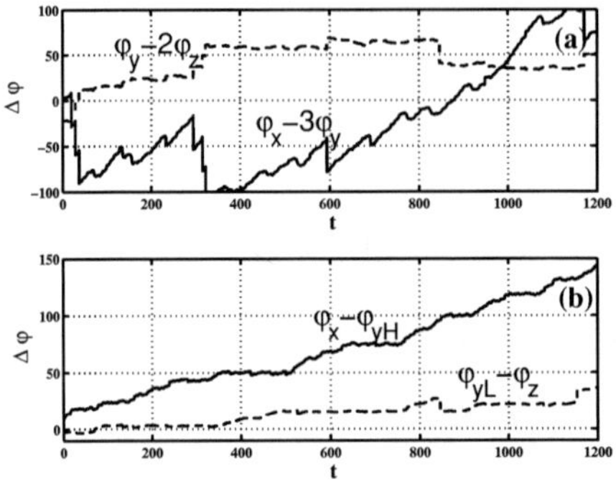

FIGURE 3. The phase difference for (a) the initial signal, (b) the filtered signals.

However for the cardiovascular system we have more than two time scales and, moreover, these scales vary in time. To illustrate multiscale time variability we present the power spectrum of the heart rate calculated by the FFT and the Morlet wavelet (Fig. 4). It can be seen that the power spectrum has several peaks superimposed on a high background. As a result, the spectrum approximates the so-called $1/f$ spectrum [8]. It is clear, that for such kinds of process, it is difficult to decompose signals into different components, although decomposition can be useful if we are interested in an interaction between only a few subsystems of the cardiovascular system, e.g. the heart and respiration rates. However in the general case we need an approach that will enable us to avoid the problem of decomposition. As a possible candidate we suggest the so-called phase coherence function [4] based on wavelet transform.

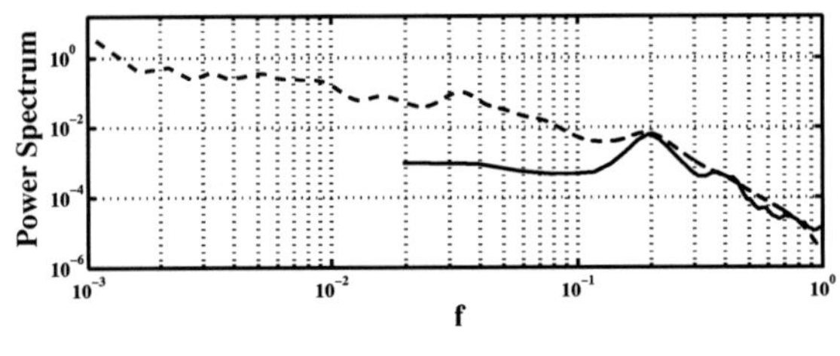

FIGURE 4. A typical power spectrum of the heart rate variability signal; solid line corresponds to FFT transform, dashed line to Morlet wavelet transform.

WAVELET-BASED PHASE COHERENCE FUNCTION

The wavelet-based phase coherence function has the following features –

- The splitting of the time interval T should be different for different basic frequencies ω_k.
- The scaling of ω_k can be logarithmic or linear.
- The continuous Morlet wavelet is used for the window function.

- For each time t_n and frequency ω_k, the phase $\varphi_{k,n} = \arctan\left(b_{k,n}/a_{k,n}\right)$ are evaluated, where $b_{k,n}$ and $a_{k,n}$ are respectively the imaginary and real parts of the complex wavelet transform.
- The time-averaged wavelet phase coherence function is then defined as

$$C_{\phi}(\omega_k) = \sqrt{\langle \cos\Delta\varphi_{k,n}\rangle^2 + \langle \sin\Delta\varphi_{k,n}\rangle^2},$$

  where $\Delta\varphi_{k,n}$ is the phase difference between the wavelet phases of two signals under consideration.

We note that in [9] an approach was suggested for calculation of the phases by use of the wavelet transform. Here, however we do not use bandpass filtration (decomposition) of the signal before application of wavelet transform. In Fig. 5 the phase coherence function $C_{\varphi}(\omega)$ is shown, reflecting interactions between the oscillators.

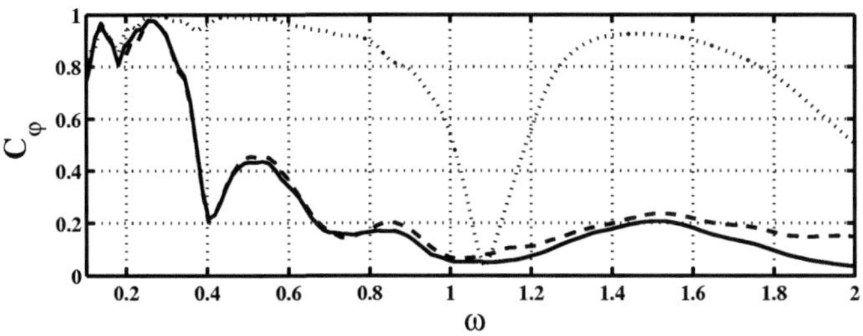

FIGURE 5. The phase coherence function. The solid line corresponds to the function $C_{\varphi,XZ}(\omega)$, the dashed line to $C_{\varphi,YZ}(\omega)$ and the dotted line to $C_{\varphi,XY}(\omega)$.

Relationship to the coherence function based on the Fourier transform. The usual FFT-based coherence function (or the coherence) is defined as:

$$C^2_{FFT}(\omega) = \frac{|S_{xy}|^2}{|S_{xx}||S_{yy}|}, \tag{2}$$

where S_{xy} is the cross-spectrum, and S_{xx} and S_{yy} are power spectra of two realizations of x and y. The coherence describes a linear interaction between two realizations. The phase coherence function $C_{\varphi,XZ}(\omega)$ considered is similar to the FFT-based coherence as demonstrated in Fig. 6.

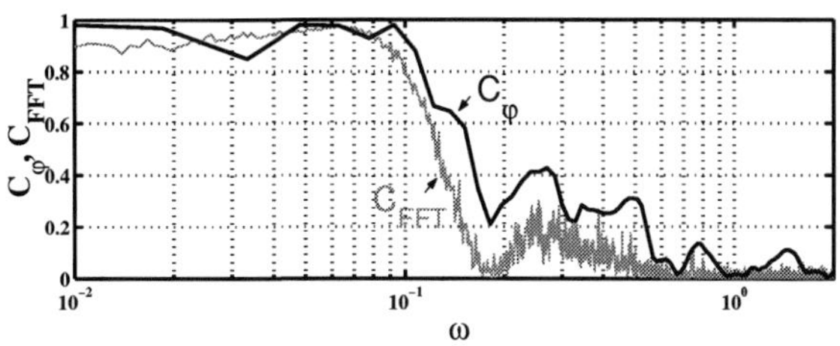

FIGURE 6. The phase coherence function $C_{\varphi,XZ}(\omega)$ and FFT-based coherence $C^{XZ}_{FFT}(\omega)$.

The main difference between them is that, for calculation of the wavelet-based function $C_{\varphi,XZ}(\omega)$, we used a single realization, whereas calculation of the FFT-based function $C_{FFT}(\omega)$ involved averaging over more than 400 realizations. Another advantage of the wavelet-based phase coherence function is that it is significantly more stable, statistically, so that it is therefore more suitable for non-stationary time-series.

SUMMARY AND MORE QUESTIONS

The results show that the suggested phase coherence function describes well the general interaction between the systems, and that the function is not limited by the number of time scales in a signal. The measure is applicable to any system, including systems where it is difficult (maybe impossible) to introduce a unique instantaneous phase or set of phases. Moreover, it is possible to extend the function to the time domain. In the latter case we can select the phase by using the energy spectrum of the signal (since the wavelet transform gives a time-frequency spectral representation) and, consequently, apply the phase approach [5] for analysis.

However the following three questions still cannot be clearly answered: (i) Are typical "$1/f$ spectra" (similar to those presented in Fig. 4) in fact the spectra of noisy coupled oscillatory processes with time-varying frequencies spanning a wide frequency interval, e.g. $\gg$ 100:1 for the cardiovascular system (see [1] for review)? (ii) Are phase-coupled oscillators a more plausible choice for the description of cardiovascular dynamics? and (iii) Might the phase description alone be sufficient?

ACKNOWLEDGMENTS

This study was supported by the Wellcome Trust and the EPSRC (UK) and the ARRS (Slovenia). The authors are grateful to Alan Bernjak and Peter McClintock for useful discussions.

REFERENCES

1. A. Stefanovska, and M. Bračič, *Contemporary Phys.*, **40**, 31–55 (1999).
2. M. Bračič Lotrič, and A. Stefanovska, *Physica A*, **283**, 451–461 (2000).
3. A. Stefanovska, H. Haken, P. V. E. McClintock, M. Hožič, F. Bajrović, and S. Ribarič, *Phys. Rev. Lett.*, **85**, 4831–4834 (2000).
4. A. Bandrivskyy, A. Bernjak, P. McClintock, and A. Stefanovska, *Cardiovascular Engineering*, **4**, 89–93 (2004).
5. A. Pikovsky, M. Rosenblum, and J. Kurths, *Synchronization – A Universal Concept in Nonlinear Sciences*, Cambridge University Press, Cambridge, 2001.
6. L. Vainshtein, and D. Vakman, *The Separation of Frequencies in the Theory of Oscillations and Waves (in Russian)*, Nauka, Moscow, 1983.
7. R. Stratonovich, *Topics in the Theory of Random Noise (Vol. 1)*, Gordon and Breach, 1967.
8. P. C. Ivanov, L. A. N. Amaral, A. L. Goldberger, S. Havlin, H. E. S. M. G. Rosenblum, and Z. R. Struzik, *Chaos*, **11**, 641–652 (2001).
9. M. Le Van Quyen, J. Foucher, J. P. Lachaux, E. Rodriguez, A. Lutz, J. Martinerie, and F. J. Varela, *J. Neurosci. Methods*, **111**, 83–98 (2001).

Role of Noise in Complex Networks of FitzHugh-Nagumo Neurons

Luigi Fortuna*, Mattia Frasca* and Manuela La Rosa*,†

*Dipartimento Elettrico Elettronico e dei Sistemi, Università degli Studi di Catania, viale A. Doria 6, 95125 Catania, Italy
†SST Group, STMicroelectronics, Stradale Primosole 50, 95121 Catania,Italy

Abstract. This paper deals with the open question related to the role of noise in complex networks of interconnected FitzHugh-Nagumo neurons. In this paper this problem is faced with extensive simulations of different network topologies. The results show that several topologies behave in an optimal way with respect to the range of noise level leading to an improvement in the stimulus-response coherence, while other with respect to the maximum values of the performance index. The best results in terms of both the suitable noise level and high stimulus response coherence have been obtained when a diversity in neuron characteristic parameters has been introduced and the neurons have been connected in a small-world topology.

Keywords: Noise, FitzHugh-Nagumo neurons, Networks
PACS: 84.35.+i; 05.45.-a; 05.40.Ca

INTRODUCTION

The role of noise in complex networks of interconnected neuron units is a topic currently under investigations, since the seminal work of Collins et al. [7] appeared. Collins et al. showed that in a summing network of many uncoupled noisy FitzHugh-Nagumo (FHN) units the so-called phenomenon of stochastic resonance without tuning occurs. However, an open question in this research field is the role of noise in complex networks of interconnected FHN neurons.

In this paper this problem is faced with extensive simulations of different network topologies. This work thus extends previous studies which only consider idealized connectivity [7] and [1] to several connection topologies. Different topologies of populations of FHN neurons have been taken into account in order to investigate the role played by the noise in the network, where each neuron is subjected to an independent source of noise. In these conditions the role of noise and the global behavior of the population depend on the connection among the elements.

Precedent studies on stochastic resonance [8, 2] in summing networks of neurons modelled by FHN equations have demonstrated that the sensitivity of neuron populations to under-threshold inputs enhances with the number of neurons involved in the computation [7]. It should be pointed out that a scientific debate is currently open on the effects of the presence of noise in systems driven by subthreshold signals slower than the typical time scales of the system itself [10] and on the most appropriate index to study these phenomena. The focus of this paper is on the investigation of the role of topology and parameter dissymmetries in neuron networks in presence of noise. We will refer to

CP800, *Unsolved Problems of Noise and Fluctuations: UPoN 2005*
edited by L. Reggiani, C. Pennetta, V. Akimov, E. Alfinito, and M. Rosini

the enhancement in the stimulus response, due to the presence of noise, as a generalized stochastic resonance effect. We retain that the correlation-like index introduced in the following which takes into account information coded in spike timing is an appropriate measure of the noise-induced effects since the system under investigation is driven by a signal slower than the typical time scales of the system itself. The phenomenon studied in this paper is also related to noise-enhanced phase synchronization [14].

NEURONS AND STOCHASTIC RESONANCE

Real neurons are characterized by a complex dynamics regulated by the effects of several ion channel flows through the neuron membrane. The behavior of the membrane potential is high-dimensional and its description requires accurate mathematical models. A simple model capturing the salient features of the neuron dynamics is the well-known FitzHugh-Nagumo neuron. The neuron populations investigated in this work are therefore modelled by FHN equations. The equations include two inputs, a noise term $\xi(t)$ and a "signal" input $s(t)$, as follows:

$$\begin{cases} \varepsilon \frac{dv_i}{dt} = v_i(v_i - a)(1 - v_i) - w_i + A + s(t) + D\xi_i(t) \\ \frac{dw}{dt} = v_i - w_i - b \end{cases} \quad (1)$$

In eq. (1) the variable $v_i(t)$ is a fast variable mimicking the membrane potential, $w_i(t)$ is a slow variable modelling the recovery dynamics and A is a constant current bias. Moreover, $s(t)$ represents the input signal and $\xi(t)$ is a Gaussian white random signal. In absence of noise ($D = 0$) the system (1) is excitable, when the input signal is chosen over the threshold characteristic of the system a spike is emitted. The FHN neuron shows characteristic signatures of stochastic resonant behavior due to the presence of this threshold-dependent dynamics. It can be clearly identified a noise value optimal in the sense that the system responds to a sub-threshold stimulus with a coherent emission of spikes. This phenomenon allows to detect signals under the threshold and is therefore very important in neurophysiological sensory systems. In the following, the parameters are chosen according to: $A = 0$, $\varepsilon = 0.005$, $a = 0.5$ and $b = 0.15$. A periodic, underthreshold signal is considered as input.

Coherence between input and related firing has been quantified by an index taking into account that the information in neural systems is coded in spikes. Therefore, ideally at each input peak should correspond a spike or a spike train. By counting the number of true spikes P_{true} (i.e. spikes in correspondence of an input peak) and false spikes P_{false} (i.e. spikes occurring without any input excitation) the following index is defined:

$$C = \left| c_1 \frac{P_{true}}{P_{tot}} - c_2 \frac{P_{false}}{P_{tot}} \right| \quad (2)$$

where P_{tot} is the total number of input peaks. In general the system may respond to a given stimulus peak with a burst of spikes. The weights c_1 and c_2 take into account the density of true spikes elicited by the same input peak and the density of false spikes, respectively. More precisely, let us indicate with N_{burst} the number of spikes in a burst of spike elicited by the same input peak, with N_{true} and N_{false} the number of true and false spikes in a burst, respectively. The coefficients c_1 and c_2 are given by $c_1 = \frac{N_{true}}{N_{burst}}$ and $c_2 = \frac{N_{false}}{N_{burst}}$. It should be noticed that $C = 1$ when the spikes and input peaks are coherent or perfectly in anti-phase. In Fig. 1 the value of the index C against the noise level D for a single FHN neuron is reported. This values is maximized for a given noise level, but the maximum value $C_{max} \simeq 0.28$ is quite low.

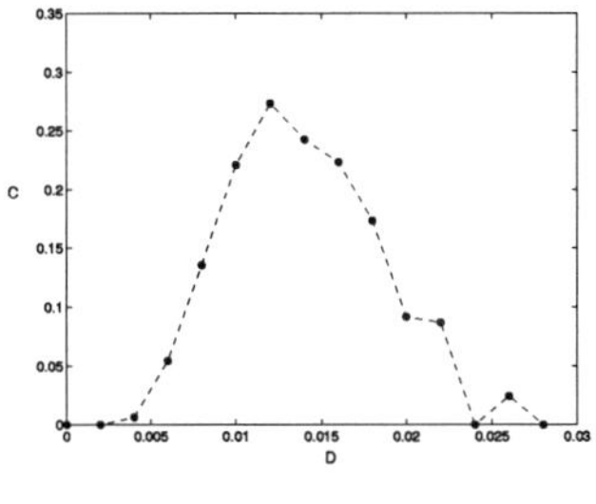

FIGURE 1. Values of the index C for a single FHN neuron.

BEHAVIOR OF REGULARLY COUPLED NEURONS

Regular network architectures have been first investigated. The equations of the FHN neuron have been therefore extended to include the coupling terms as follows:

$$\begin{cases} \varepsilon \frac{dv_i}{dt} = v_i(v_i - a)(1 - v_i) - w_i + A + s(t) + \xi_i(t) + \frac{J}{N} \sum_{j=i-r}^{j=i+r} (v_j - v_i) \\ \frac{dw}{dt} = v_i - w_i - b \end{cases} \tag{3}$$

where r is the radius of the neighborhood. By varying r the network architecture changes from local connections to all-to-all global coupling. We also considered the case of uncoupled neurons (represented by $r = 0$). In all the cases the same input has been applied to each element of the population, whereas independent noise source for each neuron has been taken into account. The response to the stimulus of the entire network has been monitored by considering the average of each membrane variable v_i, namely $V(t) = \sum_i v_i(t)$. The calculation of the index C has been performed by evaluating the coherence between the common input $s(t)$ and $V(t)$.

First of all, the case in which the neurons are not coupled has been investigated. Fig. 2 reports the value of the index C with respect to increasing values of the noise. As it can be noticed when the number of neurons increases the curve shows a plateau. The results agree with [7] where this phenomenon is referred as stochastic resonance without tuning in order to emphasize the low sensitivity of the large summing networks with respect to the level of the noise. However, the hypothesis of uncoupled neurons is far from the biological case of neuron assemblies, where coupling is the natural solution to enhance information exchange and robustness within the network. For this reason we investigate the emergence of stochastic resonance without tuning in the case of regular networks.

Architectures of globally coupled neurons show different behaviors according to the value of the coupling strength defined by the parameter J in equations (3). When low coupling values ($J = 0.01$) are considered, performance similar to the case of uncoupled

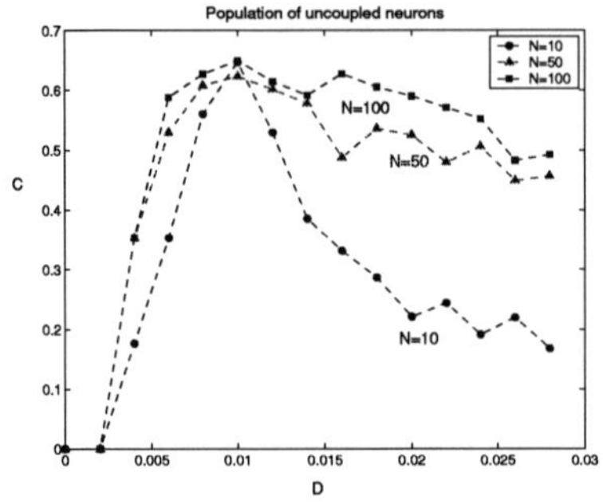

FIGURE 2. Values of the index C for a population of uncoupled FHN neurons.

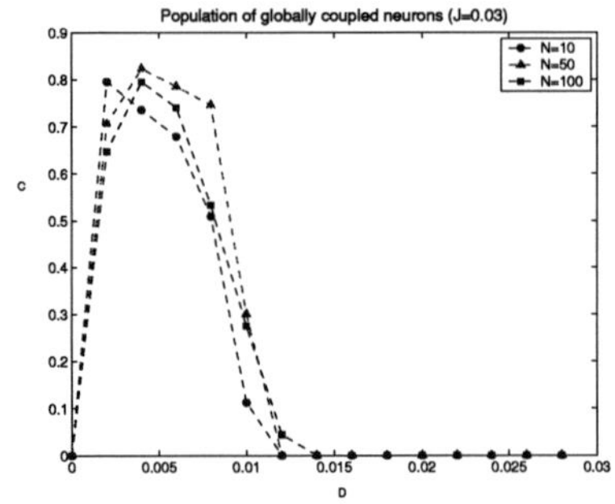

FIGURE 3. Values of the index C for a population of globally coupled FHN neurons: (a) $J = 0.01$; (b) $J = 0.03$.

neurons has been observed. By increasing the value of the coupling to $J = 0.03$ as shown in Fig. 3, the index assumes higher values in spite of a reduction of the range of noise leading to stochastic resonance. This phenomenon is due to the fact that the coupling strength reflects the amount of information exchanged among neurons that leads to collective behavior. As more information is exchanged, the coherence of the response to the stimulus is enhanced, thus leading to higher values of the index. On the other hand, stochastic resonance without tuning is a prerogative of the presence of many sources of noise acting on independent neurons. Thus, the attitude to synchronization [11, 4, 5], that is higher for increasing values of the coupling, has the effect of enhancing the performance of the system and its sensitivity to the noise. In fact, the minimum noise level detectable is higher in the case of $J = 0.03$.

RANDOM NETWORKS AND ARCHITECTURES WITH DETERMINISTIC DISSYMMETRY

Regular architectures, so far investigated, represent an abstraction of real neuron assemblies, but allow to focus on the dynamical features of the population in a simple connection pattern. On the opposite side, random networks are characterized by asymmetric structures and capture in an idealized way the features of many real systems. These networks have a short average path length which derives from the fact that, starting from each node, any other node of the network can be reached in a small number of links [12]. An example of the behavior of a random network is shown in Fig. 4, where it can be noticed that these networks have an high maximum value of the index C and a larger interval of suitable noise values. Small-world networks have been also taken into account. In this case the stimulus response coherence is lower than in the case of random networks and higher than regular networks.

In recent works [3, 6] it has been demonstrated that parametric and structural dissymmetries may enhance the synchronization properties of spatially extended networks. Moreover, dissymmetries obtained by using deterministic processes have been compared with random ones. These deterministic dissymmetries are obtained starting from a chaotic process and exploiting the noise-like spreadband spectrum of a chaotic signal. A chaotic state variable is sampled to generate a sequence of unpredictable values. Such a sequence can be used to introduce dissymmetry in a system in two main ways: parameter and structural dissymmetries. In a large number of cases the introduction of a deterministic fluctuation in the parameters of the units constituting the complex system has lead to an improvement of the self-organizing properties. On the other side deterministic sequences may also be used to generate the structure of connections in a small-world topology.

Both cases have been investigated in FHN populations. The results are summarized in Fig. 4, where are also reported the cases of uncoupled neurons (i.e. the best case as regards the range of suitable noise levels) and random connections (i.e. the best case as regards the maximum level of stimulus response coherence). The introduction of structural dissymmetries has several advantages: with respect to the case of random connections, it increases the range of suitable noise and mantains an high value of the maximum level of C. But the improvement given by the introduction of an unhomogeneity in the parameters of each single neuron is even greater. When non identical FHN neurons arranged in a small-world network (regulated by a deterministic structural dissymmetry) are taken into account, the network behaves optimally both with respect to the maximum level of C and to the range of suitable noise. In particular, the results have been obtained by considering A as the parameter accounting for the unhomogeneity of the neuron population and letting it variable in the range of [-0.01, 0.01], while the chaotic sequence has been generated by using the peak-to-peak dynamics [13] of the variable x of a Double Scroll Chua's Attractor.

CONCLUSIONS

In this paper networks of noisy neurons have been analyzed from the viewpoint of stochastic resonance. Several topologies from regularly connected to random connected networks have been studied. These systems behave in a very different way. It is still an open problem the role played by the topology of connections in these networks.

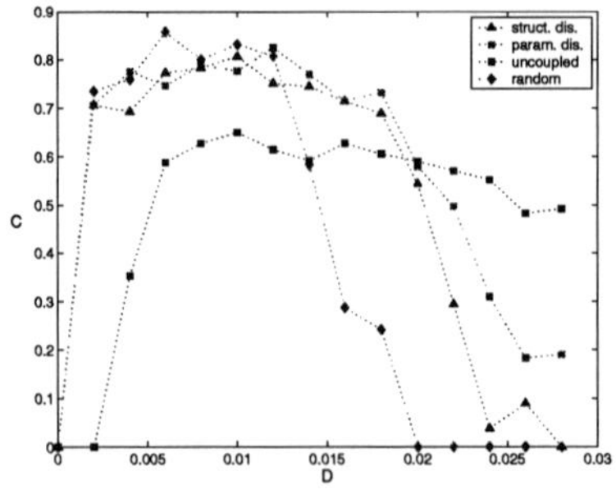

FIGURE 4. Comparison between different architectures of FHN populations: with structural dissymmetry, with parametric dissymmetry, random and uncoupled.

ACKNOWLEDGMENTS

This work was partially supported by the Italian "Ministero dell'Istruzione, dell'Università e della Ricerca" (MIUR) under the Firb project RBNE01CW3M.

REFERENCES

1. J. A. Acebrón, A. R. Bulsara, W.-J. Rappel, "Noisy FitzHugh-Nagumo model: From single elements to globally coupled networks", *Physical Review E* **69**, (2004) 0262202.
2. V.S. Anishchenko, V. Astakhov, A. Neiman, T. Vadivasova, L. Schimansky-Geier, *Nonlinear Dynamics of Chaotic and Stochastic Systems*, Springer Series in Synergetics, 2003.
3. Y. Braiman, J. F. Lidner, W. L. Ditto. "Taming Spatiotemporal Chaos with Disorder", *Nature* **378**, (1995) 465–467.
4. S. Boccaletti, J. Kurths, G. Osipov, D.L. Valladares, C.S. Zhou "The synchronization of chaotic systems", *Physics Reports* **366**, (2002) 1–101.
5. M. Bucolo, L. Fortuna, M. La Rosa. "Network self-organization through "small-worlds" topologies", *Chaos, Solitons and Fractals*, **14** (2002) 1059–1064.
6. M. Bucolo, R. Caponetto, L. Fortuna, M. Frasca, A. Rizzo. "Does chaos works better than noise?", *CAS Magazine* **2**, 3 (2002) 4–19.
7. J.J. Collins, C. C. Chow, T. T. Imhoff. "Stochastic resonance without tuning", *Nature*, **376** (1995) 236–238.
8. L. Gammaitoni, P. Hänggi, P. Jung, F. Marchesoni. "Stochastic resonance", *Mod. Phys. Rew.*, 70, 23, (1998).
9. M. A. L. Hodgkin, A. F. Huxley. "A quantitative description of membrane current and its application to conduction and excitation in nerve", *J. Physiol. Lond.*, 117, pp. 500–544, 1952.
10. A. Longtin, D. R. Chialvo. "Stochastic and Deterministic Resonances for Excitable Systems", *Physical Review Letters*, **81**, 18 (1998) 4012–238.
11. A. Pikosky, M. Rosenblum, J. Kurths. "Synchronization: A universal concept in nonlinear sciences", Cambridge University Press, (2001).
12. S. H. Strogatz. "Exploring complex networks", *Nature* **410** (2001).
13. S. H. Strogatz, *Nonlinear Dynamics and Chaos*, Perseus Publishing, 2000.
14. C. S. Zhou, J. Kurths, E. Allaria, S. Boccaletti, R. Meucci and F. T. Arecchi, "Noise-Enhanced Phase Synchronization of Chaotic Oscillators", *Physical Review Letters*, **89**, 1, (2002), 014101.

Robustness Vs Redundancy In Biological Systems

Enrico Capobianco

Biomedical Engineering Department, Boston University, 44 Cummington St., Boston, MA 02215 USA

Abstract. Gene networks offer a wealth of data; this is mainly due to the genomic dimensionality rather than the samples, as the latter usually come from measurements obtained under a few experimental conditions or time points. It is therefore a challenging task to design suitable statistical models and to develop effective reverse engineering algorithms. The signature of noise is pervasive in genetic networks. For instance, in perturbation experiments only a few genes change expression value, while most genes are either noisy or constant. Consequently, a genetic regulatory network is a redundant system, due to the high-dimensionality and the dependence between genes, and also a sparse system through the gene-gene interaction matrix only partially active. In order to explore these two aspects, redundancy and sparsity, independent component analysis (ICA) is proposed as a flexible approximation model targeted to dimensionality reduction and gene feature selection.

Keywords: Genetic Networks; Independent Component Analysis; Dimensionality Reduction; Feature Selection; Redundancy and Robustness
PACS: 89.75-K; 02.60. Pn; 02.70. Rr.

INTRODUCTION

Network inference and reverse engineering deal with an inverse problem that is stimulating a great deal of biomedical research in systems biology: reconstructing gene-gene interactions from measurements obtained under specific experimental conditions.

One peculiar aspect of gene networks is their huge dimensionality combined with very limited samples, mainly due to the cost of carrying out experiments.Typically, a gene network is monitored through the changes in gene expression values after injecting some type of drug.

The measurements are usually taken at an initial time and then regularly until the genes reach a steady state. The gene temporal patterns are subject to several kinds of fluctuations, and only some of them are induced by the drug, directly or indirectly as a cascade effect, while most genes show just random patterns.

This work shows how to decompose and approximate a gene network in a sparse way and through a few independent components. In order to explore the complex gene network dynamics, two key aspects are investigated: sparsity and intrinsic dimensionality. Exploiting the level of sparsity inherent to the gene-gene interaction matrix is a key goal since most of the experimental evidence suggests the presence of redundancy and the need to target just subsets of genes when measuring the effects of the experiments.

Sparsity links to dimensionality too, through the possibility of approximating the truly intrinsic dimension of the input space. This may bring some valuable experimental insight too, for instance how to calibrate the perturbations or which genes may become

CP800, *Unsolved Problems of Noise and Fluctuations: UPoN 2005*
edited by L. Reggiani, C. Pennetta, V. Akimov, E. Alfinito, and M. Rosini

targets.

ICA ROLE IN GENOMICS

Standard data pre-processing methods from classical matrix analysis techniques [1, 8] like singular value decomposition (SVD), or from statistics, like principal component analysis (PCA), represent a base for many other methods; one of them is ICA [3, 4, 6], which represents a novel direction of research in genomics (see recently some published work on microarray data [5, 12, 13], gene feature detection [2], and classification [9]).

ICA is a powerful exploratory method that induces dimensionality reduction through flexible models and non-linear search in some optimization space; it is *ad hoc* for exploring high-dimensional data endowed with non-Gaussian distributional features. This applies to gene data, for instance; when looking at the dimensionality of genomic profiles, these can be seen as mixtures of statistically independent components most likely linked to various biological processes that regulate their observed network dynamics.

ICA overcomes the limitations of PCA, whose role is valuable under Gaussianity. While PCA exploits the second-order statistics related to the data covariance function, and thus addresses the linear independence (or in other terms decorrelates the data), ICA exploits higher moments of the data distribution, thereby stronger dependence.

The goal of ICA is to extract from the data a set of independent sources of information which are observed as mixed together. The two unknown variables are thus the independent components and the matrix mixing. Consider at time t an $\varepsilon-$ noise based system such as $X = AS + \varepsilon$. The mixture signals X represent $k-$dimensional vectors $x_i, i = 1, \ldots, n$, and the independent non-Gaussian sources (or components) S represent $k-$dimensional vectors $s_j, j = 1, \ldots, m$. The latter are mixed up linearly through the mixing matrix A; the system may be thought to be an approximation of a non-linear system such as $Y = f(X)$. Both S and A must be estimated; the dimensions of the model (for $k = 1$) are $S \in R^m, X \in R^n$, and $A \in R^{n \times m}$.

The assumption underlying all ICA models is that the sources are statistically independent[1]; the sources S are thus assumed to generate a joint probability density function (*pdf*) $P(S)$ factorized as $P(S) = \Pi_{i=1}^m p(s_i(t))$. ICA thus finds a separating transform, or matrix under linear mixing, then the sources can be readily obtained by the inverse matrix (if $n = m$) or by the pseudo-inverse (when $n > m$). In the under-determined case when $m > n$, there is no unique inverse; this case requires constraining the sources or transforming the model so as to get an approximate solution.

In order to estimate both the mixing matrix and the sources, several computational algorithms can be efficiently applied, such as the Joint Approximate Diagonalization of Eigenmatrices (for Real signals) or *JadeR* [4] and the *fastICA* [10, 11]. Once computed the de-mixing matrix B, through its pseudoinverse one finds an estimate of A, and through $Y = BX$ approximates S. Under conditions of perfect separation we would have

[1] Theoretical studies have indicated that if the number of observed mixtures (n) is greater than or equal to the number of sources (m), then it is possible to separate statistically independent sources provided that at most one is Gaussian [3, 10].

$Y = BAS = S$, but in real cases such solution holds only up to permutation P and scaling D matrices, i.e., $Y = DPS$.

INFORMATION RETRIEVAL FROM NOISY GENOMIC SIGNALS

Noise is a strong conditioning factor for the gene temporal patterns; the smoother ones are usually of interest compared to those where a random signature is dominant. While the correlation can be more or less distinct, sharp time-localized fluctuations lead to discard many patterns and focus just on the genes changing expression values smoothly with conditions or time.

A gene expression matrix X has rows representing the different genes involved in the regulatory network under exam, while the columns list the measurements (six in our study) at successive time points. For each available sample point, a mixture of genes of the genome dimensionality is considered, and feature sets are identified by linear combinations of genes which are delivered by each estimated component.

An important tool is thresholding, which deals with the separation of signal from noise. It identifies homogeneous gene groups by selecting for each estimated components the genes that are differentially expressed with regard to noise-dependent pattern shapes, and with regard to the average of the pivotal estimated component. Thus, a gene is considered outlying if its expression value computed at the estimated component j falls outside a certain interval range R calculated from the entire genome.

Due to the inherent biological characterization of the estimated components, thresholding validates or contradicts for each gene the hypothesis that the observed expression values are significantly different from certain test statistics. Therefore, each gene can be distinguished on average because of an expression value resulting inside or outside a certain interval computed from a mean (median) statistic through the standard (absolute) deviation and the cutoff values determined by standard confidence levels, as in usual significance tests.

The gene selection is done over the estimated components via fixed intervals, by taking a scalar q times the standard deviation from the mean of the pivotal estimated component (approximately the 95% or 99% confidence levels) and thus identifying as outliers the genes whose values are outside the interval R. The steps that are repeated for each estimated component require to: (1) Fix a pivot component j; (2) Calculate the interval R for level fixed by q equal to 2 or 3; (3) Determine the mean and the standard deviation over the pivotal $k-$dimensional genomic profile. The corresponding formulation is through the following equation, where mq represents the gene expression mean value calculated as log-ratio:

$$mq_k^j = mean[s_k^j] \pm q * std[s_k^j]. \tag{1}$$

Given $j = 1, \ldots, m$ sources, the number of thresholds computed is $2m$, due to q. Thresholding represents a suitable instrument for gene selection in the context of ICA, as it works by exploiting the discriminatory power of each estimated component. The selected genes that are significantly deviating over the gene profile in each estimated component, support the rejection of the null hypothesis of no presence of outlying effects within the IC-based genomic profile. These genes are outliers, i.e., significantly and differentially expressed in that specific component.

DATA ANALYSIS

The data set refers to the *Escherischia coli* genome (reduced to the analysis of the most significantly changing 2330 genes) after perturbation experiments[2]. The gene data have been log-transformed after previous normalization of the measurements at the initial time. The gene selection results that are reported in the tables refer to both the estimated independent components and the z-scores computed for each gene across the sources.

The z-scores are used to standardize the data; in particular, they are applied to the four possible expression values assigned to each gene by the estimated components. Thus, the z-scores are computed over the k−dimensional genome of the $j = 1,\ldots,m$ sources for each gene, whose value is now transformed to s_{j_k}:

$$s_{j_k} = \frac{(s_{j_k} - mean[s_{j_k}])}{std[s_{j_k}]}. \tag{2}$$

With regard to gene pre-processing, one typical practice is to turn to a whitening (sphering) transform applied to the data (via SVD) or to their covariance (through PCA), which delivers unit variance uncorrelated variables. The method is also used for de-noising, based on excluding the smallest eigenvalues. SVD and PCA are thus aimed to obtain the smallest possible signal subspace (S) from the originally noisy space (Y) where the data lie, i.e., $Y = S + \varepsilon$.

However, due to the limited information in the sample, one may just be able to approximate the optimal signal space, and reach a dimension of this linear space smaller than the real number of sources characterizing the system dynamics. In order to explore this possibility, we have embedded the system with some redundancy through the inclusion of an higher number of independent components compared to that suggested by the previously executed PCA step.

The application of the ICA algorithms yields four independent components. which represent a minimal system able to capture the gene network dynamics. The estimated mixing matrix is reported in Table 1[3]; the third and fourth columns are quite sparse, and this suggests a minor role or contribution of these sources in explaining the dynamics.

Table 2 reports the size of the gene groups selected from each component. The systems with four and five estimated components are compared. The dimensionality problem is suitably managed by ICA and by the thresholding step; the global number of selected genes is orders of magnitude smaller than the initial sample size.

The genes involved in the same biological pathway are expected to show relatively strong connectivity, correlation and homogeneity within the groups, and some genes

---

[2] The perturbation was carried out by gene-calibrated drug injection at the initial time. In particular, the cells were treated with $10\mu g/ml$ *Norfloxacin* and then six measurements were taken every twelve minutes. The microarray data were subject to the *Affy MAS 5.0* normalization algorithm.

[3] The first row of the estimated mixing matrix lists zeros due to data normalization and log-transformation.

TABLE 1. Estimated Mixing Matrix with 4 IC.

Estimates	IC-1	IC-2	IC-3	IC-4
t=1	0	0	0	0
t=2	0.8914	0.065	0.4032	0.0437
t=3	1.0206	0.1678	−0.0577	0.0006
t=4	1.0554	0.4532	−0.0834	0.0882
t=5	1.006	0.6381	0.0029	−0.0654
t=6	0.908	0.6125	0.0171	−0.1867

TABLE 2. Genes from significance intervals (m2-m3). The numbers indicate the size of the groups.

4 Components	IC-1	IC-2	IC-3	IC-4	
m2	106	79	124	104	
m3	16	26	40	34	
Z-scores for 4 Components	ICz-1	ICz-2	ICz-3	ICz-4	
m2	−−	−−	−−	−−	
m3	−−	−−	−−	−−	
5 Components	IC-1	IC-2	IC-3	IC-4	IC-5
m2	112	79	120	119	100
m3	17	24	41	49	35
Z-scores for 5 Components	ICz-1	ICz-2	ICz-3	ICz-4	ICz-5
m2	46	66	19	29	55
m3	−−	−−	−−	−−	−−

appear to belong to different groups, due to the simultaneous presence of different underlying biological influences. By looking at the groups selected under four and five components, we also account for those genes differentially expressed in more than one group (i.e., IC), and thus most likely co-regulated by different biological factors. They are reported in Table 3, and identified by the frequency with which an overlap of some genes between 2, 3, or 4 components occurs.

TABLE 3. Score by frequency: differentially expressed genes belonging to 4,3,2 groups, from m2 - m3 intervals. The size of the groups are indicated.

Score by frequency	4	3	2
4 IC - m2	1	13	85
4 IC - m3	−−	1	9
5 IC - m2	3	20	142
5 IC - m3	−−	3	21

TABLE 4. Differentially expressed genes in m2 - m3 intervals. Cross-links (c-l) indicate gene overlaps; scores are assigned according to the frequency. Aligned genes (a-g) are reported for std = 1 (0.1).

Gaussian std.1 (std. 0.1)	m2	m3	c-l score 3	c-l score 2
IC-1	108 (114)	16 (5)	0(2)	14(22)
IC-2	102 (77)	11 (22)	0(2)	15(23)
IC-3	113 (113)	10 (26)	0(3)	22(22)
IC-4	118 (107)	9 (17)	0(2)	11(21)
Alignment across intervals	m3	m3 pert.	a-g	
IC-1	16 (16)	16 (5)	1 (3)	
IC-2	26 (26)	11 (22)	1 (19)	
IC-3	40 (40)	10 (26)	2 (17)	
IC-4	34 (34)	9 (17)	0 (11)	

The presence of genes selected in intervals from z-scores across components deserves some consideration in terms of redundancy and robustness to noise in the system. Figure 1 illustrates the difference between a system with 4 (top) and 5 IC, for both the estimated sources (left) and the z-scores. The z-scores project the component values onto a sphere and compact them in an area of limited energy dispersion; the spherical shape is typically obtained by random projection techniques. The smoother sphere's surface marks a difference in the dispersion of energy, and indicates a different degree of randomness in the two systems.

The redundancy is considered in terms of system stability and signal-to-noise ratio (see [7] for reliability aspects). Exploratory work is carried out through ad-hoc perturbation analysis, and the results are reported by Table 4.

Figure 2 confirms that the system with four IC is more noise resistant compared to that with 5 IC; we have injected zero mean Gaussian noise of different magnitude (std 1, 0.1 and 0.01), both directly to the estimated components before taking the z-scores, and to the gene data, from which the sources are then estimated and the z-scores computed. The level of noise affects the system to a different degree; a small amount of noise enables a consistent reconstruction and a good alignment with the gene groups previously selected by thresholding, but higher noise magnitudes only gene reshuffling effects are derived. Thus, with regard to system robustness to noise, more than minimal components deteriorate the discriminatory power of thresholding when forming gene groups. Then, the z-scores reveal that a noisy system boosts the gene expression values.

REFERENCES

1. O. Alter, P. Brown and D. Botstein, *PNAS*, **97(18)**, 10101-10106 (2000).
2. J.A. Berger, S. Hautaniemi, H. Edgren, O. Monni, S.K.Mitra, O. Yli-Harja and J. Astola, *Proc. IEEE NNSP*, Toulose, FR, 2003, pp. 81-90.
3. J. Cardoso, *Proc. ASSP*, 1989, pp. 2109-2112.
4. J. Cardoso and A. Souloumiac, *IEE Proc. F.*, **140(6)**, 1993, pp. 771-774.
5. P.Chiappetta, M.C.Roubaud and B. Torresani, *J. Comput. Biol.*, **11(6)**, 1090-1109 (2004).

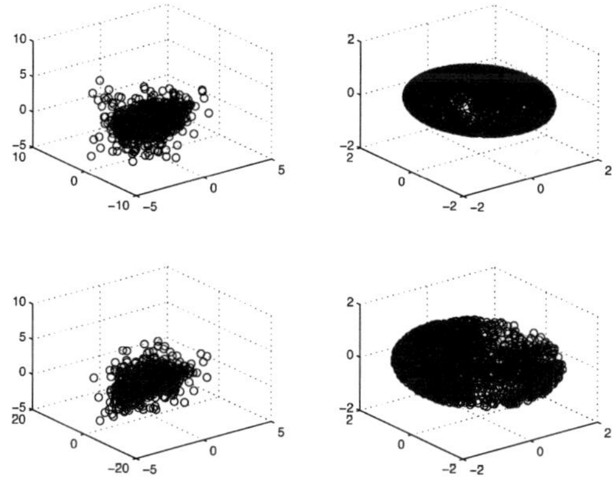

FIGURE 1. Scatter plot of $IC-4$ (top-left) and ICz_4 (top-right) vs corresponding plots (bottom) of $IC-5$ and $ICz-5$. The axes report the first three components. Note the smoother and denser sphere with 4 IC, due to the presence of less randomness.

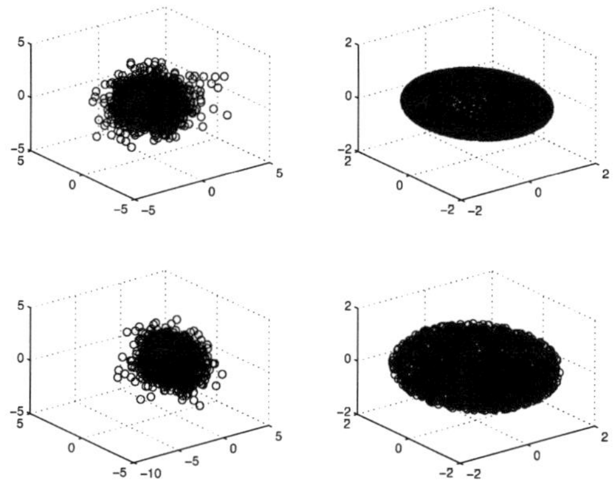

FIGURE 2. Scatter of three IC under system perturbation: $IC-4$ (top-left), $ICz-4$ (top-right), and corresponding plots (bottom) for $IC-5$ and $ICz-5$. The axes report the first three components. Note how the system with 4IC results more noise resistant.

6. P. Comon, *Sign. Proc.*, **36(3)**, 287-314 (1994).
7. S. Harmeling, F. Meinecke, and K.R. Muller, *Sign. Proc.*, **84**, 255-266 (2004).
8. N. Holter, A. Maritan, M. Cieplak, N. Fedoroff and J. Banavar, *PNAS*, **98 (4)**, 1693-1698 (2001).
9. G.Hori, M. Inoue, S. Nishimura, and H. Nakahara, *Gen. Inform.*, **12**, 255-256 (2001).
10. A. Hyvarinen and E. Oja, *Neural Comput.*, **9(7)**, 1483-1492 (1997).
11. A. Hyvarinen, *IEEE Trans. Neural Netw.* **10(3)**, 626-634 (1999).
12. S. Lee and S. Batzoglou, *Gen. Biol.*, **4**:R76 (2003).
13. W. Liebermeister, *Bioinform.*, **18**, 51-60 (2002).

Informative Essence Of Chaos

S.F. Timashev,* G.V. Vstovsky,* A.B. Solovieva

L.Ya. Karpov Institute of Physical Chemistry, Moscow, Russia
**N.N. Semenov Institute of Chemical Physics, Russian Academy of Science, Moscow*

Abstract. Flicker-Noise Spectroscopy (FNS) as a new phenomenological approach for extracting information hidden in chaotic signals is presented. According to FNS, the information is stored in sequences of distinct types of irregularities – spikes, jumps, and discontinuities of derivatives of different orders of chaotic signals. FNS application to analysis of Parkinsonian tremor dynamics studied by using laser systems is demonstrated. FNS could be considered as a new instrument to early diagnostics of patient states.

Keywords: Flicker-Noise Spectroscopy, information, chaotic signals, Parkinson disease.
PACS: 07.50.Hp, 72.70.+m

FLICKER-NOISE SPECTROSCOPY: BASIC PRINCIPLES

The paper contains the answers to the general questions: what type of general phenomenological information is in fact contained in chaotic time series $V(t)$ (t is a time) which is given at an interval T as well as what is the physical sense of this information and what are the tools and methods allowing everyone to extract this information in a desirable quantity. We assume that a new image of information hidden in complex temporal signals has to be introduced for extracting the information. Such image is introduced in Flicker-Noise Spectroscopy (FNS) [1, 2]. The FNS uses some artificial constructions similar to those introduced by Schuster [3]. A set of space-temporal hierarchy levels of complex open dynamic dissipative systems is introduced. It is supposed also the dependencies $V(t)$ of dynamic variables are essentially irregular for the every ith hierarchical level. We postulat that the most general description of the $V(t)$ for every ith space-temporal level can be represented in the form of intermittency with irregularities of different types (spikes, jumps, discontinuities of derivatives). Within the characteristic time intervals of the ith spatio-temporal level of the system evolution, there are small intervals (δ_i^k-intervals) that contain the main information about the sequence (k = 1, 2, 3, ...) of structural-energetic states of the system at the ith hierarchical level. The assumption that the behavior of the system obeys the hypothesis of self-similarity implies that the shape of the irregular temporal evolution is similar in every space-time scales of the process.

We consider the irregularities of the studied signals as informative carriers. This assumption suggests that not all points in the time axis are informatively equivalent. The transition of the system from one informative state (one δ_i^k-interval) during $T_0^{\,i}$ time interval to another (adjacent δ_i^{k+1}-interval), which differs from the previous in structural-energetic characteristics, is characterized by the onset of irreversibility for

CP800, *Unsolved Problems of Noise and Fluctuations: UPoN 2005*
edited by L. Reggiani, C. Pennetta, V. Akimov, E. Alfinito, and M. Rosini

this evolution "step" ("arrow of time"), as suggested by the Trieste-theory of von Weizsäcker [4]. To use these physical ideas in developing an approach for time series analysis, it is necessary to idealize the system evolution by considering one hierarchical space-temporal level only (the ith level) and condensing all the informative δ_i^k-intervals to points. Irregularities at each "time point" should be considered as carriers of information about the ith hierarchical level of the considered evolution at the moment of time it marks, i.e., the irregularities should act as markers of various types for the ith hierarchical level of the signal $V(t)$. The zero duration of each "time point" suggests that the function under study should contain an actual (the Dirac δ-functions and their derivatives) or potential singularity (the Heaviside θ-functions and functions with discontinuities of the first-, second-, and higher-order derivatives) at each of these points. We assume that the main information about the system evolution is contained in the postulated singularities and these are considered the only markers of the evolution process.

The ability to distinguish the irregularities means that the parameters or patterns characterizing the totality of properties of the irregularity sequences, are extracted from the power spectra $S(f)$ (f – frequency):

$$S(f) = \left| \int_{-T/2}^{T/2} \langle V(t)V(t+t_1)\rangle \cdot \exp(2\pi i f t_1) dt_1 \right|, \quad \langle(\ldots)\rangle = \frac{1}{T}\int_{-T/2}^{T/2}(\ldots)dt, \tag{1}$$

and the difference moments $\Phi^{(p)}(\tau)$ (structural functions) of the pth order ($p = 1, 2, \ldots$):

$$\Phi^{(p)} = \left\langle \left[\int_{t}^{t+\tau} \frac{dV(x)}{dx} dx \right]^p \right\rangle \tag{2}$$

where τ is a time delay parameter.

In this case, $\Phi^{(p)}(\tau)$ is formed exclusively by jumps of the dynamic variable at different space-time hierarchical levels of the system under consideration, and $S(f)$ is formed by spikes and jumps. In other words, the power spectra and difference moments of the 2nd order carry out different information, which complement each other. (It is different from the standard point of view, which could be adequate only for smooth signals). The introduced representation of the function $V(t)$ was used to obtain a rather simple analytical expressions for the dependences $S(f)$ and $\Phi^{(p)}(\tau)$ for stationary processes in the case $T \to \infty$ [1, 2]. In the low-frequency limit $2\pi f T_0^i \ll 1$, the interpolation equation for $S_S(f)$, which expresses the contribution of only the spike-type irregularities to the power spectrum $S(f)$ is given by:

$$S_S(f) = 2\int_0^\infty \Phi_0^i(\eta) \cdot \cos(2T_0^i f \eta \pi) d\eta \ . \tag{3}$$

Here $\Phi_0^i(z)$ is an effective probability density functions for the spikes in the temporal interval T. Owing to the dimension $\{[V]^2 \cdot \text{sec}]\}$, one can write:

$$\Phi_0^i(\eta) = \frac{g_0}{4\pi} T_0^i \chi_0(b_0^i \eta), \tag{4}$$

where $\chi_0(b_0^i\eta)$ is a dimensionless function, b_0^i is a dimensionless scaling factor for the ith spatio-temporal level, and g is a constant with the dimensionality $[V]^2$. Then,

$$S_S(f)=\frac{g_0}{2K_0\pi}\int_0^\infty \chi_0(x)CosZxdx\ ;\quad K_0\equiv\frac{b_0^i}{T_0^i}\ ,\quad Z\equiv\frac{2\pi f}{K_0}\ . \tag{5}$$

We postulate that the introduced parameter K_0 with the dimensionality [sec] is independent of the number of the spatio-temporal level in the case of a stationary process. In this case Eq. (5) could be considered as a general $S_S(f)$ dependence for a real process (not only for the ith hierarchical level of the considered evolution), and K_0 can be regarded as an invariant frequency for the process under investigation [2]. For analyzing a contribution of $S_S(f)$ to the power spectrum $S(f)$ of a real process, it is necessary to choose approximations for the $\chi_0(Z)$ function, which adequately represent the possible types of the prolonged non-exponential nonlinear relaxation in the considered systems. The interpolation equation for $S_S(f)$, which expresses the contribution of only the spike-type irregularities to the power spectrum $S(f)$ in the case of stationary processes, is presented in [2]. The contribution $S_R(f)$ of only the jump-type irregularities to $S(f)$ can be derived by using the expression for $\Phi^{(2)}(\tau)$:

$$S_R(f)=\int_0^\infty Cos(2\pi f\tau)\left[\Phi^{(2)}(\infty)-\Phi^{(2)}(\tau)\right]d\tau. \tag{6}$$

It is easy to see from Eq. (2), the structural functions $\Phi^{(2)}(\tau)$ for the dynamic variable $V(t)$ describing the evolution process at the ith spatio-temporal level are defined by potential singularities which are the Heaviside θ-functions ("jumps") of the dynamic variable $V(t)$. The corresponding equations for $\Phi^{(2)}(\tau)$ and $S_R(f)$ in the case of a stationary evolution process are presented in [2]. The characteristic information extracted from the $S(f)$ and $\Phi^{(p)}(\tau)$ dependencies are the "passport parameters" which are the correlation times, parameters characterizing the loss of "memory" for these correlation times, characterizing the sequences of "spikes", "jumps" and discontinuities of derivatives of different orders (in the latter case, time series for "quasi-derivatives" are formed). Note, in the most real cases, $S(f)$ and $\Phi^{(p)}(\tau)$ dependences manifest their complexity and non-stationary behavior. In these cases, the introduction of "passport parameters" to describe the temporal evolution is useless. At the same time, the behavior of the real $S(f)$ and $\Phi^{(p)}(\tau)$ dependences is very specific and individual for each studied case. These dependences can be considered as "characteristic passport patterns" of the state of systems under study, especially, as these dependencies have a very clear physical sense. Below we demonstrate in what way FNS with getting information by analyzing the $S(f)$ and $\Phi^{(2)}(\tau)$ dependences can be used to determine the "passport patterns" characterizing dynamics of complex non-stationary biological signals.

Frequently there is a problem of smoothing of initial signals for getting information contained in the studied time series. Usually for the filtration of a signal with allocation of the low-frequency component, the smoothing polynoms and wavelets are used. We proposed a new method of splitting the signal into low frequency $V_R(t)$ and high frequency $V_F(t)$ components [2]. Derivation of low frequency components is based on a "relaxation procedure" by analogy with the solution of diffusion (heat conductivity) equation presented in the form of a finite difference equation [2]. Taking the original signal as the initial condition we obtain the low frequency component V_R. Taking it off from the initial signal, we obtain the high frequency component $V_F\equiv V(t)-V_R(t)$. The main advantage of the procedure is that it does not require to set any *a priori* form of the trend line to calculate the component V_R. Now, we can calculate $S(f)$ and $\Phi^{(p)}(\tau)$ for each of the $V_J(t)$ ($J = R$ or F), where the subscripts R or F refer to $V_R(t)$ and $V_F(t)$, respectively. It gives a possibility to find new characteristic "low-frequency" and "high frequency" patterns of the studied signals.

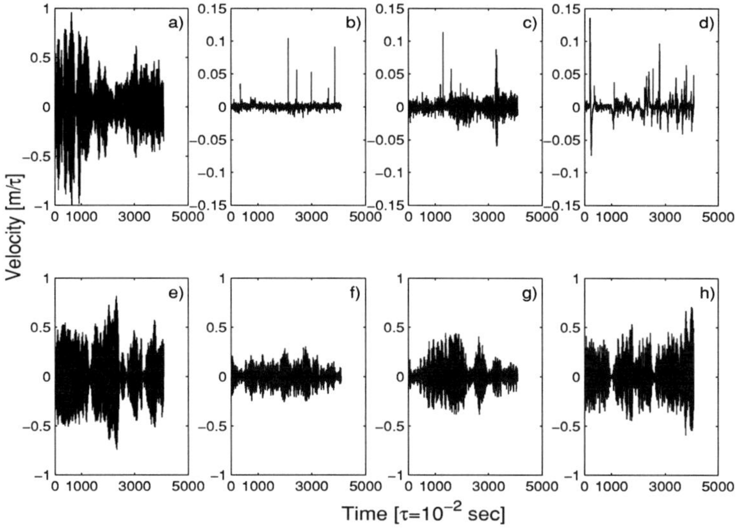

FIGURE 1. The pathological tremor velocity of the left hand index finger of a patient with Parkinson's disease. The registration of the Parkinsonian tremor velocity was carried out for the following conditions: (*a*) "OFF-OFF" (no treatment), (*b*) "ON-ON", (*c*) "OFF-ON", (*d*) "ON-OFF", (*e*) - (*h*) "15 OFF", "30 OFF", "45 OFF", "60 OFF".

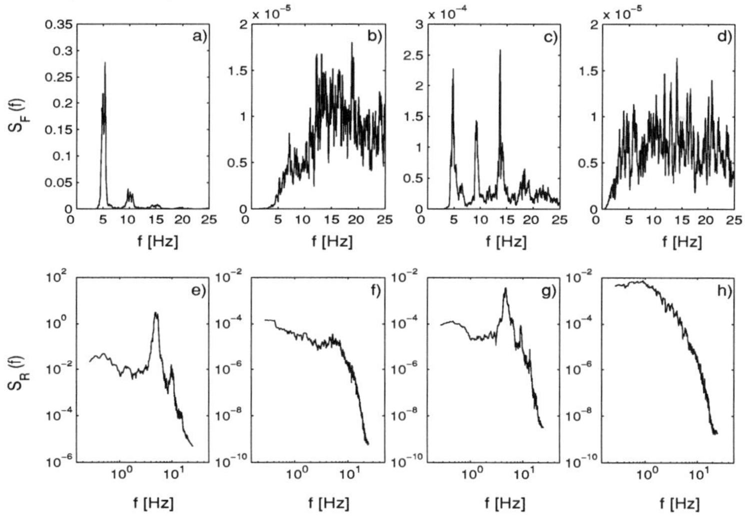

FIGURE 2. High-frequency (*a-d*) and low-frequency (*e-h*) power spectrum components of the tremor time series (Fig. 1*a* - *d*) for "OFF-OFF" condition (*a*, *e*) and for various conditions of treatment: "ON-ON" (*b*, *f*), "ON-OFF" (*c*, *g*), "OFF-ON" (*d*, *h*). The power spectra of high frequency components are submitted in a linear scale (*a-d*), power spectra of low frequency components are submitted in log-log scale (*e-h*). The set of eigenfrequencies in power spectra is connected to the periodic nature of the Parkinsonian tremor.

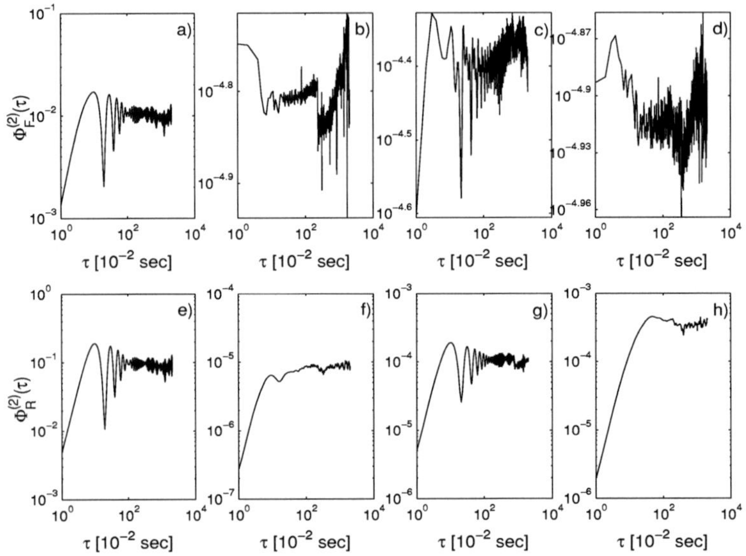

FIGURE 3. The $\Phi_F^{(2)}(\tau)$ and $\Phi_R^{(2)}(\tau)$ dependences for the tremor time series (Fig. 1*a* - *d*): (*a*, *e*) - no treatment ("OFF-OFF"); (*b*, *f*)- complex treatment ("ON-ON"); (*c*, *g*) – DBS only ("OFF-ON"); (*d*, *h*) - medicaments only ("ON-OFF").

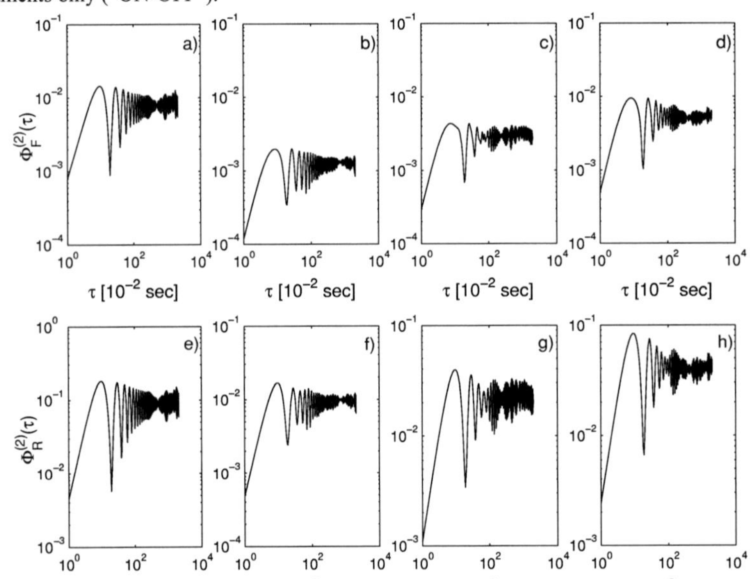

FIGURE 4. The $\Phi_F^{(2)}(\tau)$ and $\Phi_R^{(2)}(\tau)$ dependences for the tremor time series (Fig. 1*e* - *h*) after the DBS switching-off: (*a*, *e*) 15 minutes; (*b*, *f*) 30 minutes; (*c*, *g*) 45 minutes; (*d*, *h*) 60 minutes. Medicaments are not used.

FNS APPLICATION TO ANALYSIS OF THE PARKINSONIAN TREMOR

Here we analyze (see also [5]) time series of pathological finger tremor velocity [6] for a patient with Parkinson's disease to demonstrate FNS possibilities in determining the informative essence of complex chaotic signals. The patient took a dopaminergic therapy (L-Dopa), and deep brain stimulation (DBS). Tremor was recorded [7] with a velocity-transducing laser under two conditions of DBS (on-off), two conditions of medication (L-Dopa, on-off) and under four conditions of stopping DBS after 15, 30, 45, 60 min: "OFF-OFF" regime (no medication, no stimulation), "ON-ON" regime (on medication, on stimulation); "ON-OFF" regime (on L-Dopa medication, no stimulation); "OFF-ON" regime (DBS stimulation only); "15 OFF", "30 OFF", "45 OFF", "60 OFF" regimes – the patient's states after 15, 30, 45, 60 minuets correspondingly in case that the DBS is switching-off, no medication. The time series of the Parkinsonian tremor velocity of the patient index finger are presented in Fig. 1. The dada demonstrate the sharp differences in tremor velocity for patient's state, when the treatment is not used (Fig. 1*a*) and in cases of any ways of treatment (Figs. 1*b*-*h*).

The power spectra of $S_F(f)$ and $S_R(f)$ which could be considered as "informative patterns" are presented in Fig. 2. The Figs. 2*a* and 2*e* demonstrate that a set of several own frequencies is the main feature of the patient tremor. In accordance with Figs. 2*c*, *g* the maximal power spectrum values are reduced at the DBS treatment ("OFF-ON" regime). The situation is qualitatively changed at the use of medicaments ("ON-OFF" regime, Fig. 2*d*, *h*), and also at the combined use of DBS and medicaments ("ON-ON" regime, Fig. 2*b*, *f*). As follows from Fig. 3-4, the $\Phi_F^{(2)}(\tau)$ and $\Phi_R^{(2)}(\tau)$ dependences also contain the valuable information about the tremor dynamics. We see the combined use of DBS and medicaments (ON-ON, Figs. 3*b*, *f*) results in the strongest changes in tremor velocity. Fig. 4 characterizes dynamic changes of the pathological tremor after the DBS switching-off. The submitted results demonstrate the FNS analysis of power spectra and difference moments could be considered as a new effective tool for obtaining operative information about functional states of complex biological systems.

ACKNOWLEDGMENTS

Authors thank Prof. R.Yulmetyev and Dr. S.Demin for collaboration in analysis of the Parkinsonian tremor signals. The work was supported by the ISTC grant (2280) and the Russian Foundation for Basic Researches (Grants No. 02-05-64340, 04-02-16850).

REFERENCES

1. Timashev S.F. in: *Annals of the New York Academy of Science*, **879** 129-143 (1999).
2. Timashev S.F., and Vstovsky G.V. *Russian J. Electrochemistry*, **39**, 156-169 (2003).
3. Schuster H.G. *Deterministic chaos. An introduction*, Physik-Verlag, Weinheim (1984).
4. C.F. von Weizsäcker C.F. von. In: *Time, Temporality, Now.* Ed. H. Atmanspacher and E. Ruhnau, Springer-Verlag, Berlin, Heidelberg, New York (1997), 91-104.

5. Yulmetyev R.M., Demin S.A., Panischev O.Yu., P. Hänggi, Timashev S.F., Vstovsky G.V. *Physica A*, in press.
6. Titcombe M.S., Glass L., Guehl D., Beuter A. *Chaos*, **11(4)**, 766-773 (2001).
7. Norman K.E., Edwards R., Beuter A. J. Neurosci. Meth., **92**, 41-54 (1999).

Randomness Of Amoeba Movements

S. Hashiguchi, Siti Khadijah, T. Kuwajima, M. Ohki,
M. Tacano*, J. Sikula**

University of Yamanashi, Japan
**Meisei University, Japan*
*** Brno University of Technology, Czech*

Abstract. Movements of amoebas were automatically traced using the difference between two successive frames of the microscopic movie. It was observed that the movements were almost random in that the directions and the magnitudes of the successive two steps are not correlated, and that the distance from the origin was proportional to the square root of the step number.

Keywords: Amoeba movements, Random walk
PACS: 05.04.Fb

INTRODUCTION

The purpose of the present study is to see if there are some differences between movements of living things and the others. The movements of living things sometimes look like "random". We tried to determine whether such movements are statistically random or not by tracing the movements of amoeba (*Dictyostelium discoideum*)[1].

The simple two-dimensional model of "random walk" is characterized by the randomness of the direction of the unit step [2]. We developed a program to trace the movements of such an object as amoeba whose outline is not sharp and whose shape varies according to its movements. We checked their movements if the direction and the size of each step are random, and if the distance from the origin is proportional to the number of steps.

TRACING

The microscopic movie of the amoeba was taken a frame every 20 s. Fig.1 shows the first frame of a movie consists of 600 frames. The tracing program was organized to locate each amoeba's central position (x_i, y_i) automatically.

CP800, *Unsolved Problems of Noise and Fluctuations: UPoN 2005*
edited by L. Reggiani, C. Pennetta, V. Akimov, E. Alfinito, and M. Rosini

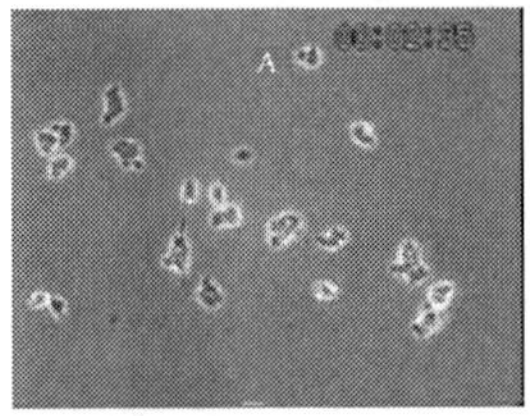

FIGURE 1. Amoebas in a frame of movie [1]. The mark "A" shows the amoeba described here.

RESULTS AND DISCUSSIONS

Figure 2 shows the trace of amoeba "A". Each segment shows the displacement in a step. Fig.3 shows the variation of the step size. There are 400 zero-size steps, where no displacement was observed between two consecutive frames of the movie, within total 560 steps. Fig.3(a) shows the variation of step size with zero-size steps, while Fig.3(b) shows the variation of step size without zero-size steps. There are almost always several zero-size steps between two non-zero steps.

Figure 4 shows the variation of the direction of the displacement. A point on the circular periphery in Fig.4(b) or (c) shows the direction of each step, and a segment of line shows the connection between two successive directions. Fig.4(b) includes the zero-size steps which are placed at the origin. It is seen that most of the segments are radial. This means that there is almost always a zero-size step between two non-zero steps. Fig.4(c) does not include zero-size steps. Segments show the variation of direction between two consecutive non-zero steps. As is seen in the figure the variation is almost random and does not depend on the past.

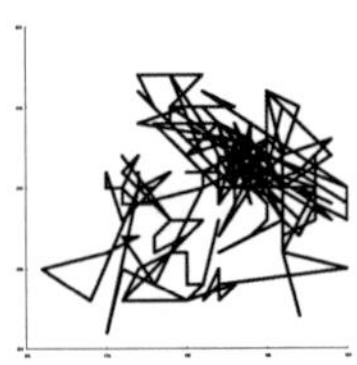

FIGURE 2. Trace of Amoeba "A"

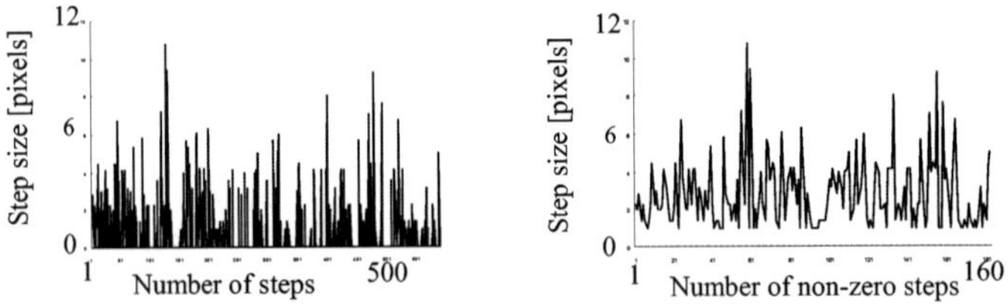

(a) Including zero-size steps

(b) Excluding zero-size steps

FIGURE 3 Variation of Step Size

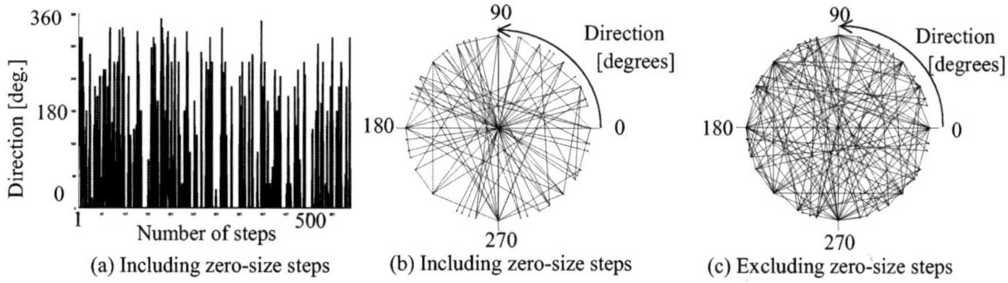

FIGURE 4. Variation of Direction

Figure 5 shows the distribution of the step size and the direction. The are about 400 zero-size steps which are not shown in the figure. The most of the steps have the step size less than 5 pixels. The average size is 0.885 pixels for all steps including zero-size steps, while it is 2.85 for non-zero steps.

Figure 6 shows the correlation of the step size between consecutive steps. Fig.6(a) includes zero-size steps, while Fig.6(b) does not. It is seen that the correlation is not high.

Figure 7 shows the correlation of the direction between consecutive steps. The correlation is not high.

From these figures it is concluded that the steps are independent from each other in any respect. This is the result of the intermissions (the zero-size steps). The amoeba may loose its memory of the direction and step size when it comes to an intermission.

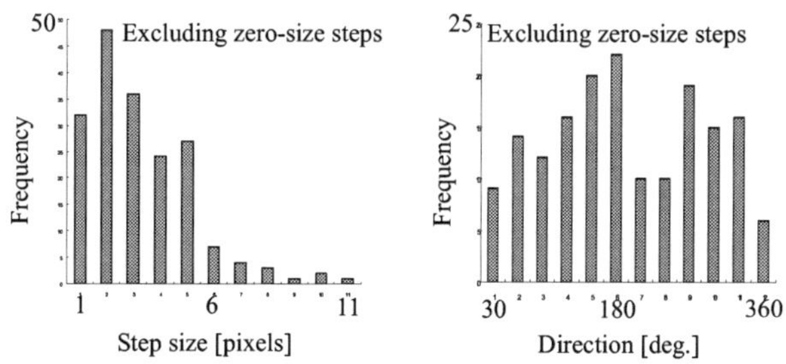

FIGURE 5. Distributions of Step Size and Direction

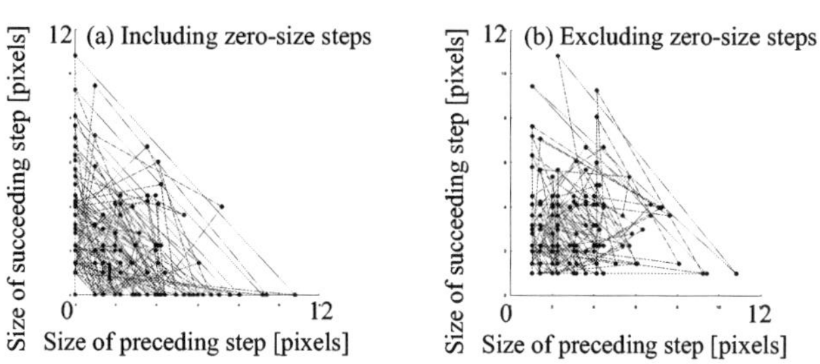

FIGURE 6. Correlation of Step Size between Two Consecutive Steps

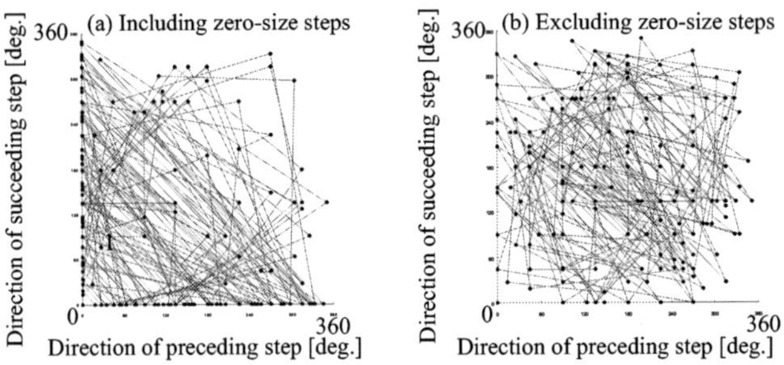

FIGURE 7. Correlation of Direction between Two Consecutive Steps

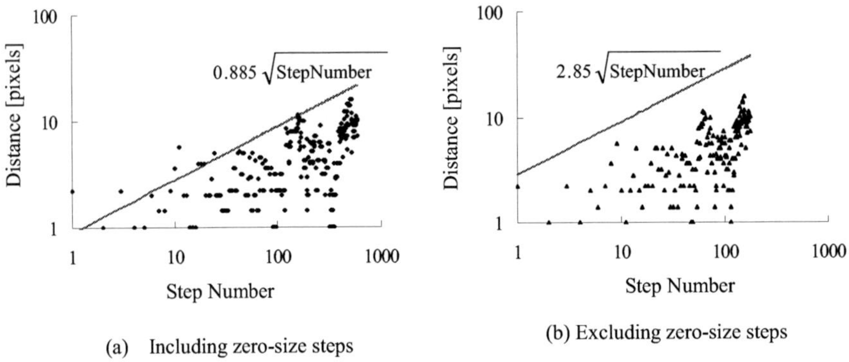

FIGURE 8. Distance from the Start Point

Figure 8 shows the distance from the start point. Fig.8(a) includes zero-size steps, and the line indicates the proportionality to square root of the step number. The proportionality factor 0.885 is the average step size. Fig.8(b) does not include zero-size steps. The proportionality factor 2.85 is the average over non-zero steps. It can be said that the distance from the start point is proportional to the square root of the step number. This is one of the characteristics of random walk. The other 16 amoebas also showed resembling behaviors as shown in Fig.A1 in Appendix.

From these results the movements of amoebas are a kind of random walk, characterized in that the size and the direction of a step do not depend on those of the preceding step, and that the distance from the start point is proportional to the square root of the step number. The independence of each step may originate from the zero-size steps, where the amoeba loses its memory of the preceding step.

We expected to find some correlation of the direction between steps of the movements of a living amoeba. The longer steps might consist of several correlated shorter steps if the sampling period is made shorter.

CONCLUSION

The purpose of the present study is to see if there are some differences between movements of living things and the others. It was found that the amoeba movements were just "Brownian" with respect to the correlation between steps. The correlation might be found by taking movies at higher frame rates.

OPEN QUESTIONS

Are the movements of living things different from those of thermally agitated non-living things?

What is the crucial parameter characterizing the living things in respect of the movements?

ACKNOWLEDGMENT

The authors are thankful to Prof. K. Hatakeyama of Hirosaki University for his permission of use of the amoeba movies.

REFERENCES

1. http://nature.cc.hirosaki-u.ac.jp/lab/2/colltech/nenkin/
2. Papoulis, A., *Probability, Random Variables, and Stochastic Processes*, McGraw-Hill, Inc., 1965, pp.290-296, 515-553

APPENDIX

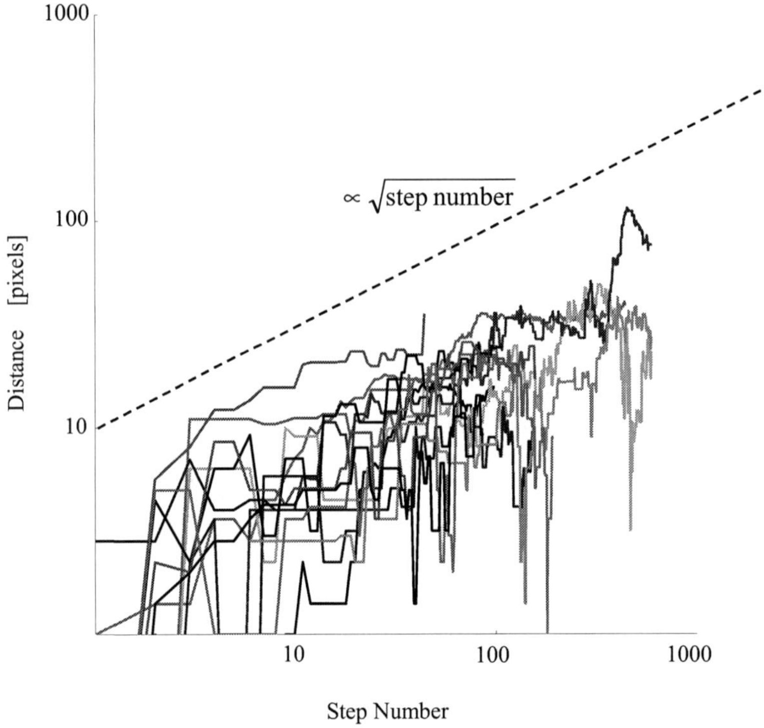

FIGURE A1. Distance from Start Point

Thermal Fluctuations Of A GPCR: A Two Force Constant Model

E. Alfinito*, V. Akimov*, C. Pennetta*, L. Reggiani* and G. Gomila†

*Dipartimento Ingegneria dell'Innovazione and National Nanotechnology Laboratory - INFM, Università di Lecce, Via Arnesano, 73100 Lecce, Italy
†Department d'Electronica and Research Centre for Bioelectronics and Nanobioscience, Universitat de Barcelona, C/ Josep Samitier 1-5, 08028 Barcelona, Spain

Abstract. Many efforts in nanobiotechnologies are devoted to the realization of nanodevices based on single receptors for the detection of specific ligands. An interesting tool to monitor the detection process, is provided by the study of the electrical properties of such a device. To estimate the sensitivity of this nanotool, we simulate the current response to an applied AC voltage of a nanodevice realized by a single G-protein-coupled receptor (GPCR) in contact with two metallic electrodes. To this purpose, recently, we have proposed a model based on a coarse grained approach that describes the protein as a static network of elementary impedances. The nodes of the network correspond to the positions of the C_α atoms of each protein amino acid and an elementary impedance is associated with each link. Here, we present an extension of this model which accounts for the impedance noise due to thermal fluctuations of the atomic positions within a rhodopsin molecule. In our model, the C_α atoms are treated as a set of independent harmonic oscillators, characterized by two different values of the force constant for α-helix and for loops. The impedance noise is then calculated for different temperatures and compared with the impedance variation due to the conformational change from the ground to the activated state of rhodopsin.

Keywords: Thermal fluctuations, proteins, G-protein coupled receptors, complex networks
PACS: 87.15Ya, 87.15.Aa, 87.14Ee, 89.75.-k

INTRODUCTION

Recently, many scientists, from different disciplinary areas, are engaged into the realization of nanodevices based on single receptors for the detection of specific ligands. Many of these studies focus on the optical response to the ligand capture [1]. However, another possibility to monitor the detection process, is provided by the study of the electrical properties of such a nanodevice [2]. Among receptors, the so-called G-protein-coupled receptors (GPCRs) play a role of primary interest, being the first cellular tool for the detection and trasduction of external signals [3]. For this reasons, a huge quantity of work has concerned this subject [3, 4, 5, 6]. Actually, GPCRs can catch specific molecules or particles (ligands) like smell molecules, photons, hormones, but also drugs and viruses [3, 4, 5]. In the cell, the detection and trasduction process begins with the conformational change of the GPCR associated with the capture of the specific ligand. Then, this conformational transition activates the so called G proteins, giving rise to a complex sequence of biological mechanisms which ends in the reception of an electric pulse by neurons [3]. In nanodevices, the detection process bypasses this sequence of biological events and it aims to achieve a sensitive and reliable monitoring of the conformational change by the direct analysis of the optical or the electrical response of the

CP800, *Unsolved Problems of Noise and Fluctuations: UPoN 2005*
edited by L. Reggiani, C. Pennetta, V. Akimov, E. Alfinito, and M. Rosini

nanodevice itself. We are interested to follow this direct way of proceeding by investigating the electrical properties of such a device [7]. More precisely, in order to estimate the sensitivity of the detection, we simulate the current response to an applied AC voltage of a device realized by a rhodopsin molecule, embedded in its membrane and in contact with two metallic electrodes [7, 8]. Rhodopsin, a photonic receptor present in the mammalian retina, is one of the most studied and best known GPCRs [3, 4, 5, 9]. In particular, in the case of rhodopsin, the atomic positions are known for both, the basic state (in the dark) [9, 10] and the most stable excited state (in light) metarhodopsin II [10, 11] (these two states in the following will also referred as Rho and Meta II). It is important to remark that all GPCRs share the same secondary structure consisting of seven transmembrane α-helices connected by six loops and two terminals: one extracellular (N-terminal) one cytoplasmic terminal (C-terminal) [3, 4, 5, 6, 10]. Due to this common structure, rhodopsin can be used as a template [3] for all the other GPCR, for which the atomic positions are generally unknown. In this framework, a coarse grained-complex network approach [8, 12], which ignores the detailed atomic structure of the protein and looks only at the protein topology, seems an appropriate tool for providing predictions for all the other GPCR.

MODEL

In a previous paper [8] we have introduced a model, based on a coarse grained approach [12, 13] and which describes the protein as a static network of elementary impedances. The network is built starting from the positions of the C_α atoms of the amino acids in the Rho and in the Meta II states, as available on the protein data bank [9, 10, 11]. Precisely, these atoms are taken as the nodes of the network [8, 14]. We assume that the amino acids interact electrically among them and that charge transfers between neighboring residues [1] and/or changes of their electronic polarization [15] affect these interactions. Accordingly, a link is drawn between any pair of nodes neighboring in space within a given a distance, d, determined by a cut-off radius, R_a, with $d = 2R_a$, and an elementary impedance is associated with each link [8]. Moreover, two extra nodes can be added to mimic the electric contacts, each of these two nodes can be linked with one or several of the C_α nodes, depending on the geometry of the electrical contacts in the actual device. The elementary impedance, $Z_{i,j}$, associated with the link between the i-th and j-th nodes at a distance $l_{i,j} \leq d$, is taken as the impedance of a RC parallel circuit (the most usual equivalent passive AC circuit) [8]:

$$Z_{i,j} = \frac{l_{i,j}}{\pi(R_a^2 - l_{i,j}^2/4)} \frac{1}{(\rho^{-1} + i\varepsilon_{i,j}\varepsilon_0\omega)} \tag{1}$$

where ω is the frequency of the external voltage, ρ the resistivity of the resistor, ε_0 the vacuum permittivity and $\varepsilon_{i,j}$ the relative dielectric constant of the capacitor [8] expressed in terms of the intrinsic polarizabilities α_i and α_j of the corresponding amino acids [15]. By taking the values $R_a = 12.5$ Å and $\rho = 10^9$ Ωm, we have found [8] that the conformational change of the receptor induced by the capture of the ligand (i.e. the transition Rho → Meta II) implies a significant variation of the impedance at all frequencies, and in particular we have found a variation of about 28 % in the value of $Re[Z]$ at $\omega = 0$ [8]. This prediction concerns a "frozen" network (static network) in which the atomic positions are fixed at the equilibrium values. On the other hand, fluctuations of the atomic positions, due to thermal motion [12, 16, 17] imply an impedance noise, whose level, in comparison with the impedance variation due to the conformational transition and with the electrode/amplifier noise, is crucial to the actual detection of the ligand capture by the GPCR. Therefore, here we extend the previous model to include the effect of the thermal motion on the electrical response to an AC field. The nodes of the network are then allowed to fluctuate around their equilibrium positions with an amplitude dependent on the temperature. For the sake of simplicity and to get a first qualitative estimation, we describe the system of fluctuating nodes as a set of independent, isotropic, harmonic oscillators, with common values of the force constant γ and of the proper frequency, ω_0, and in contact with a thermal bath at temperature T. By denoting with $\vec{r}_{n,eq}$ and $\delta\vec{r}_n = \vec{r}_n - \vec{r}_{n,eq}$, respectively the equilibrium position and the displacement from the equilibrium of the n-th oscillator, the mean square displacement, $<(\delta\vec{r})^2>$, is given by:

$$<(\delta\vec{r})^2> = \frac{3}{2}\frac{k_B\theta}{\gamma} + \frac{k_B\theta}{\gamma}\frac{1}{\exp[\theta/T]-1} \tag{2}$$

where $\theta = \hbar\omega_0/k_B$. When $T \gg \theta$, Eq. (2) simplifies in: $<(\delta\vec{r})^2> \approx k_BT/\gamma$. For an arbitrary temperature the wave function of the oscillator is a superposition of several excited states and its probability density cannot be expressed in a simple form. Again for simplicity, we take the following expression for the probability density of presence of the n-th oscillator around its equilibrium position:

$$P(\delta\vec{r}_n) = \frac{1}{[2\pi <(\delta x_n)^2>]^{3/2}} \exp\left[-\frac{(\delta\vec{r}_n)^2}{2<(\delta x_n)^2>}\right] \tag{3}$$

where: $<(\delta x_n)^2> = <(\delta y_n)^2> = <(\delta z_n)^2> = <(\delta\vec{r})^2>/3$. According to the values commonly used in literature [12, 17, 18], we have chosen $\gamma = 2.5$ KJ mole^{-1} Å^{-2}. This choice provides $\theta = 12$ K. Thus, the condition $T \gg \theta$ is satisfied at room temperature. The choice of an unique force constant for all the amino acids of the protein is a common assumption in the literature, as in the well known Gaussian Network Models [12, 17]. On the other hand, it is also well known that α helices are much more stable under thermal fluctuations than loops and terminals [3, 4, 5]. For this reason, we have analyzed the effect on the impedance fluctuations of relaxing the above assumption, by introducing two different force constants: γ_1 for α helices and $\gamma_2 < \gamma_1$ for loops and terminals. The values of γ_1 and γ_2 have been chosen by fixing the mean value of the force constant to the previous value and then by considering different values of the ratio γ_1/γ_2. Thus, the greater the value of this ratio, the more flexible the loops and terminals will be with respect to the helices. The fluctuation of the impedance network is then calculated by a Monte Carlo simulation which at each step generates, starting by the static one, a random network with the position of nodes chosen stochastically according to a mean square displacement satisfying Eq. (2).

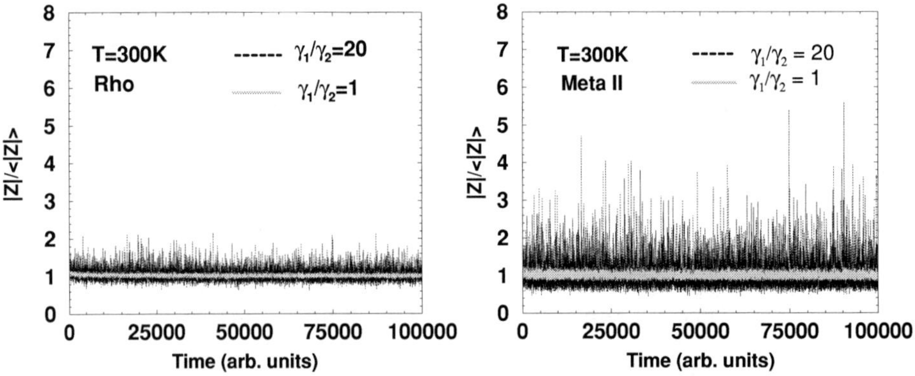

FIGURE 1. Simulation of the impedance modulus versus time for rhodopsin (left) and metarhodopsin II (right). The impedance modulus has been normalized to its average value and the time is expressed in simulation steps. The grey and the black curves are calculated by taking the ratio γ_1/γ_2 equal to 1 and 20, respectively.

RESULTS AND CONCLUSIONS

The main results of this study are illustrated by the following figures. Figure 1 displays the results of calculations of the impedance modulus versus time for rhodopsin (left) and metarhodopsin II (right) at room temperature. The impedance modulus has been normalized to its average value. The curves corresponding to two different values of the ratio γ_1/γ_2 are reported in grey (for $\gamma_1/\gamma_2 = 1$) and in black (for $\gamma_1/\gamma_2 = 20$). Figure 1 shows that the Meta II state is characterized by impedance fluctuations significantly wider than those in the Rho state. This is particularly true when the force constant of the α helices is twenty times bigger than the force constant of loop/terminals. This result reflects the minor thermal stability of the Meta II state on respect to the Rho state, probably due to an expansion of the entire structure, consequent to the conformational change [19]. This conclusion is also confirmed by Fig. 2 which reports the probability distribution function, ϕ, of the impedance fluctuations at room temperature calculated for rhodopsin (left) and metarhodopsin II (right), by taking the ratio γ_1/γ_2 equal to 1 (circles) and 20 (triangles). The probability densities have been calculated by analyzing signals $Z(t)$ made of 5×10^5 data. Here σ is the root mean square deviation from the average value of the impedance modulus $< |Z| >$. This normalized representation has been adopted for convenience because it makes the distribution independent of its first and second moments [20]. Figure 2 shows that in the Meta II state impedance fluctuations are non-Gaussian even when $\gamma_1/\gamma_2 = 1$ (for contrast, the fluctuations are Gaussian in the Rho state). Moreover, strong non-Gaussian tails are present when $\gamma_1/\gamma_2 = 20$ and in particular in the Meta II state. This strong non-Gaussianity at room temperature for $\gamma_1 \gg \gamma_2$ (and in particular in the Meta II state) is due to the presence of big spikes corresponding to the loss of links crucial to ensure the connectivity of the network. Finally, Fig. 3 reports the calculated values of the relative root mean square fluctuation of the impedance modulus, $\sigma/ < |Z| >$, as a function of the temperature. We can see that once

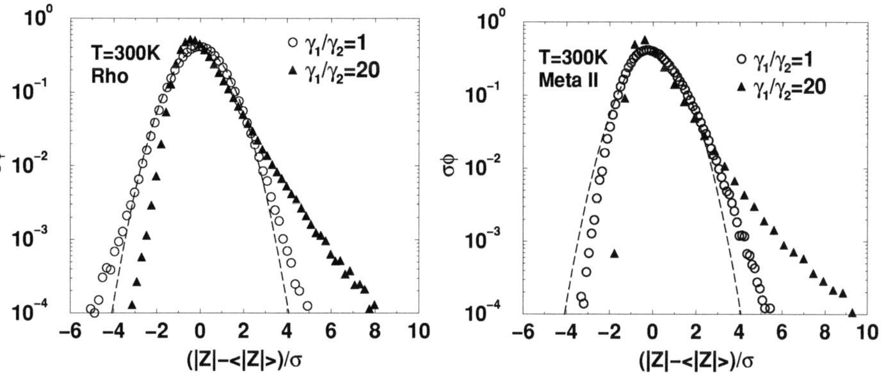

FIGURE 2. Normalized PDF of impedance modulus fluctuations at T=300K for rhodopsin (left) and metarhodopsin II (right). The curves shown by circles and triangles are calculated by taking the ratio γ_1/γ_2 equal to 1 and 20, respectively. The dashed curve represents the Gaussian distribution.

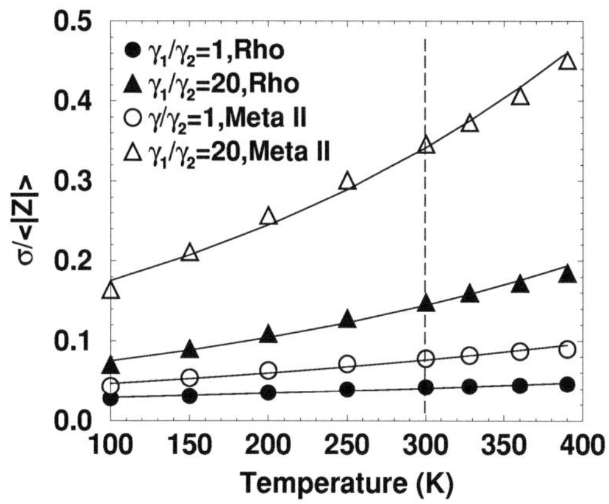

FIGURE 3. Relative root mean square deviation of the impedance modulus as a function of the temperature for rhodopsin (full symbols) and metarhodopsin II (open symbols). The data shown by circles and triangles are calculated by taking the ratio γ_1/γ_2 equal to 1 and 20, respectively. The dashed vertical line evidences the values of $\sigma/<|Z|>$ at room temperature and the continuous lines represent the best-fit of the data with an exponential law.

the higher flexibility of loops and terminals is accounted for by taking two different values for the force constants γ_1 and γ_2, the relative fluctuation of the impedance becomes strongly sensitive to the temperature. In particular, for $\gamma_1/\gamma_2 = 20$, the relative root mean square fluctuation of the impedance modulus in the Rho state is about 15 %, while in the Meta II state is about 35 %. Such high levels of noise would make problematic to detect the conformational change in terms of variation of the impedance of the nanodevice, being this variation roughly of the same order [8]. However, the increase of the impedance noise itself in the Rho $\rightarrow$ Meta II transition (more than a factor two at room temperature) could in principle provide a way to detect this transition.

A first conclusion from this study is that a careful modeling of thermal atomic motion [12, 16] is necessary to provide reliable estimates of the impedance noise whose level is crucial to the purpose of an electrical detection of the ligand capture by the GPCR. However, our simple model shows clearly that the transition Rho $\rightarrow$ Meta II is followed by a significant increase of the impedance noise level, which, in principle, would offers an interesting possibility to detect the transition itself using the logic: the noisier is the signal.

ACKNOWLEDGMENTS

This work has been performed within the SPOT NOSED project IST-2001-38899 of E.C. Partial support from the cofin-03 project "Modelli e misure in nanostrutture" financed by Italian MIUR is also acknowledged. Authors thank M. Sampietro, G. Ferrari and L. Fumagalli (Politecnico di Milano, Italy) and E. Pajot-Augy, R. Salesse and J. Minic (INRA- Jouy en Josas, France) for helpful discussions.

REFERENCES

1. H. Yang, G. Luo, P. Karnchanaphanurach, T.M. Louie, I. Rech, S. Cova, L. Xun, and X. S. Xie. Protein conformational dynamics probed by single-molecule electron transfer. *Sciences*, 302:262–266, 2003.
2. F. Patolsky, G. Zheng, O. Hayden, M. Lakadamyali, X. Zhuang, and C. M. Lieber. Electrical detection of single viruses. *PNAS*, 101:14017–14022, 2004.
3. R. J. Lefkowitz. The superfamily of heptahelical receptors. *Nature Cell Biology*, 2:E133–E136, 2000.
4. J. Lameh, R. I. Cone, S. Maeda, M. Philip, M. Corbani, L. Nádasdi, J. Ramachandran, G. M. Smith, and W. Sadée. Structure and function of g protein coupled receptors. *Pharmaceutical Research*, 7:1213–1221, 1990.
5. U. Gether and B.K. Kobilka. G protein-coupled receptors. *The Journ. Bio. Chem.*, 273:17979–17982, 1998.
6. N. Vaidehi, W. B. Floriano, R. Trabanino, S. E. Hall, P. Freddolino, E. J. Choi, G. Zamanakos, and W. A. Goddard III. Prediction of structure and function of g protein coupled receptors. *PNAS*, 99:12622–12627, 2002.
7. SPOT-NOSED. *Single Protein Nanobiosensor Grid Array*. V Framework Programme of European Community, http://www.nanobiolab.pcb.ub.es/projectes/Spotnosed/, 2003-2005.
8. C. Pennetta, V. Akimov, E. Alfinito, L. Reggiani, and G. Gomila. Fluctuations of complex networks: Electrical properties of single protein nanodevices. In J. M. Smulko, Y. Blanter, M. I. Dykman, and L. B. Kish, editors, *Noise and Information in Nanoelectronics, Sensors and Standards II*, number 5472 in Proceedings of SPIE, pages 172–182, Bellingham, 2004. Int. Soc. Opt. Eng.
9. K. Palczewski, T. Hori, C. A. Behnke, H. Motoshima, B. A. Fox, I. Le Trong, D. C. Teller, T. Okada, R. E. Stenkamp, M. Yamamoto, and M. Miyano. Crystal structure of rhodopsin: A g protein-coupled receptor. *Science*, 289:739–745, 2000.
10. H.M. Berman, J. Westbrook, Z. Feng, G. Gilliland, T.N. Bhat, H. Weissing, I.N. Shindyalov, and P.E. Bourne. Protein data bank. *Nucleic Acids Research*, 28:235–242, 2000.
11. G. Choi, J. Landin, J. F. Galan, R.R. Birge, A.D. Albert, and P.L. Yeagle. Structural studies on metharhodopsin ii, the activated form of the g-protein coupled receptor,rhodopsin. *Biochemistry*, 41:7318–7324, 2002.
12. A. R. Atilgan, S. R. Durell, R. L. Jernigan, M. C. Demirel, O. Keskin, and I. Bahar. Anisotropy of fluctuation dynamics of proteins with an elestic network model. *Biophys. J.*, 80:505–515, 2001.

13. G. Lattanzi and A. Maritan. Force dependent transition rates in chemical kinetics models for motor proteins. *J. Chem. Phys.*, 117:10339–10349, 2002.
14. R. Albert and A. L. Barabasi. Statistical mechanics of complex networks. *Rev. Mod. Phys.*, 74:47–97, 2002.
15. X. Song. An inhomogeneous model of protein dielectric properties: intrinsic polarizabilities of amino acids. *J. Chem. Phys.*, 116:9359–9383, 2002.
16. F. G. Parak. Physical aspects of protein dynamics. *Rep. Prog. Phys.*, 66:103–129, 2003.
17. M. M. Tirion. Large amplitude elastic motions in proteins from a single-parameter atomic analysis. *Phys. Rev. Letl.*, 77:1905–1908, 1996.
18. M. Delarue and Y.-H. Sanejouand. Simplified normal mode analysis of conformational transitions in dna-dependent polymerase: The elastic network model. *Journal of Molecular Biology*, 320:1011–1024, 2002.
19. S. T. Menon, M. Han, and T.P. Sakmar. Rhodopsin: Structural basis of molecular physiology. *Physiological Review*, 81:1659–1688, 2001.
20. C. Pennetta, E.Alfinito, L. Reggiani, and S. Ruffo. Non-gaussian resistance noise near breakdown, in granular material. *Physica A*, 340(1-3):380–387, 2004.

Noise Properties Of Single Open Ion Channels: An Atomistic Computational Approach

F. Affinito[1], E. Piccinini[2], A. Bigiani[3], R. Brunetti[1], C. Jacoboni[1] and M. Rudan[2]

[1] *INFM-S3 Research Center and Dipartimento di Fisica, Università di Modena e Reggio Emilia, Italy*
[2] *Dipartimento di Ingegneria Elettronica, Informatica e Sistemistica, Alma Mater Studiorum Università di Bologna, Italy*
[3] *INFM and Dipartimento di Scienze Biomediche, Università di Modena e Reggio Emilia, Italy*

Abstract. This paper presents the first results of a computational analysis of the noise associated with ion current in single open ion channels. This analysis is performed by means of a coupled Molecular Dynamics-Monte Carlo approach able to simulate the conduction process on the basis of all microscopic information today available from protein structural data and atomistic simulations. The case of potassium ions permeating the KcsA channel is considered in the numerical calculations. The obtained results evidence a noise spectrum different from what is theoretically predicted for uncorrelated ion-exit events (Poisson noise), confirming the existence of correlation in ion motion within the channel.

Keywords: ion channels, KcsA, current noise, Molecular Dynamics, Monte Carlo method
PACS: 87.14.Ee, 87.15.Ya, 87.16.Ac, 87.16.Uv

INTRODUCTION

Ion channels are nanometric macromolecular pores in the cell membranes. The availability of X-ray crystallographic structures with atomic resolution [1] produced a number of Molecular Dynamics (MD) analyses [2]. MD provides atomistic information about the system and confirms that the permeation process takes place as a single-file concerted motion of ions. MD, however, is unable to analyse the electrical properties of ionic flux due to the long time scale involved in the physiological process (ms). The aim of this work is to study the conduction and noise properties of potassium ions in the KcsA channels [3] by means of a Monte Carlo (MC) simulation that yields both current and its noise for a single channel under open-gate conditions. A multi-ion model is used, where the ion binding sites inside the protein and at the protein boundaries, the allowed transitions between different occupancy states and the associated transition rates are obtained from the atomic structure and from MD simulations. Six binding sites are included in the model and the transition rates are evaluated by means of free-energy profiles for the possible ion-occupancy configurations of the channel [4]. Computational results include current-voltage characteristics at different ion concentrations, comparable with experiments [5] without fitting parameters, and noise power spectra at different bias and

CP800, *Unsolved Problems of Noise and Fluctuations: UPoN 2005*
edited by L. Reggiani, C. Pennetta, V. Akimov, E. Alfinito, and M. Rosini

concentrations. The analysis of open-channel current fluctuations yelds useful information about the degree of correlation between consecutive ion exits from the channel.

THE THEORETICAL FRAMEWORK

Conduction Model

It has been established [6] that the KcsA channel exhibits six stable binding sites, usually referred to as $S_0,...,S_4,S_{cav}$, in which alternatively potassium ions and water molecules are found. Furthermore, due to the strong electrostatic repulsion between ions inside the filter, ion hopping inside the channel takes place as sincronous concerted motion of many ions, each ion moving from one site to an adjacent one. The existence of a free-energy barrier between two given ion occupancy configurations suggests that only a restricted set of transitions actually determines the conduction properties of the channel. The identification of the relevant configurations must account for a number of constraints imposed by what we know today about the channel protein. The configurations used in the simulation of the permeations process have been selected taking into account the following constraints [2]: a) single-file ion motion is assumed; b) no more than four ions can be found in a given configuration and no more than two of them occupy the selectivity filter (sites S_1-S_4); c) transitions involving configurations where ions move inside the selectivity filter are allowed only through concerted motion. All the transition rates can in principle be obtained by MD simulations as averages over many occurrences of the same pathway in the configuration space. In practice each transition event requires MD simulation times too long to be spontaneously observed and techniques designed in order to map specific regions of the free-energy profile (typically: umbrella sampling [7] or steering MD [8]) must be applied to circumvent this difficulty. The probability per unit time $k_{A\to B}$ to undergo a transition from channel configuration A to configuration B is then calculated using Kramers' formula for the transition-state theory [9]:

$$k_{A\to B}=\frac{M\omega_1\omega_2 D}{2\pi KT}\exp(-\beta\varepsilon_b) \tag{1}$$

where D is the ion diffusion coefficient inside the channel, ε_b the free-energy barrier separating the two states, M the ion mass, ω_1,ω_2 represent the parabolic-approximation parameters around the maximum and the minimum of the energy profile as functions of the appropriate reaction coordinate, and $\beta=1/KT$, being K the Boltzmannn constant and T the temperature. All the above microscopic parameters can be estimated from MD simulations.

Noise Analysis

Our noise analysis is restricted to the contribution coming from the electric fluctuations generated by single-file diffusion through the very narrow channel. These current fluctuations are expected to arise from the discrete nature of current flow (ions move in discrete steps across the membrane) yielding "shot noise" analogous to that seen in electronic devices. In our calculations we will assume that the channel is permanently in the open configuration. Furthermore we will not consider the fluctuations in excess shot-noise coming from internal motions of the channel protein, which can produce fluctuating energy barriers to ion permeation [10]. Theoretical calculations of current-noise spectra during single-file transport have been performed in the past on the basis of the analitic solution of macroscopic time-dependent single-file trasnport equations, based on a number of limiting hypotheses concerning the number of binding sites and ions inside the channel [11]. These calculations suggest that the presence of interactions between the ions in the channel can produce damped oscillations in the autocorrelation function of the microscopic current fluctuations. The availability of protein structural information and atomistic simulators allows us to calculate these noise properties on the basis of a more sound and general physical scheme, and in the future, to compare computational results with experimental results, now achievable with modern noise equipments. The autocorrelation function $C(\tau)$ of the current fluctuations can be numerically computed from the MC simulation assuming a stationary process. The spectral density of the current is then obtained either by Fourier transformation of $C(\tau)$ or by directly averaging the squared Fourier transform of the current fluctuations $\delta I(\omega)$.

The classical theory of shot noise [12] assumes that charge transport is unidirectional, occurs in an instantaneous process, and that the charge movements are totally uncorrelated in time. The spectral density of the fluctuations is given by Schottky's formula: $S_{shot} = 2qI$, being q the charge of the moving ion. Usually the spectral density is normalised to S_{shot}, and the Fano factor [12]: $F = S_I(\omega)/S_{shot}$ is taken as a measure of the degree of correlation in the ion motion which arises because of the single-file motion and the repulsion of electrons in the channel invalidates the assumption of uncorrelated events.

COMPUTATIONAL PROCEDURE

Our simulations are based on the structure of the KcsA potassium channel from *Streptomyces lividans* solved at 2.0 Å [1] (PDB ID code 1K4C) embedded into a water-octane-water bilayer. The crystallographic structure has been obtained with the channel in the closed-state; the open-state channel structure is supposed not to modify the binding-site conformation inside the selectivity filter, but only the inner part of the protein [14]. A total of 34434 atoms have been considered, including protein, octane bilayer, water and ions. All MD simulations were carried out using the GROMACS 3.2.1 package [15,16] with its standard internal force field. Periodic boundary conditions have been applied and electrostatic interactions have been calculated with the particle mesh Ewald summation. The x-ray crystallographic structure has initially been fully equilibrated at room temperature for 1 ns, then computational free-energy profiles have been obtained with the steering-dynamics pulling code available in GROMACS.

The conduction process is simulated by generating a sequence of transitions between different channel configurations through a Monte Carlo algorithm. Transition rates, in absence of any applied electrical field, are assumed symmetrical. A linear drop of the potential inside the channel is considered in presence of an applied field, as also reported in [17].

We estimated the entrance probability at the outer and inner mouth of the selectivity filter from the classical kinetic theory for gases as:

$$k_{entry} = \frac{n\sigma}{6}\sqrt{\frac{KT}{2M}}\exp(-\beta\varepsilon_{in}) \tag{2}$$

n being the ion concentration, ε_{in} the energy barrier at the protein mouths, and σ the cross sectional area of the vestibule. We assumed an equal concentration of potassium ions in both the intracellular and extracellular reservoirs. The conduction has finally been evaluated as the net flux of incoming and outcoming ions at the channel boundaries in a defined time interval Δt.

RESULTS

As an example of application of the proposed technique, we report in Fig. 1 the current-voltage characteristics of potassium ions across the channel at different potassium concentrations, together with available experimental results from [5]. At present our model is unable to predict the experimental current saturation, due

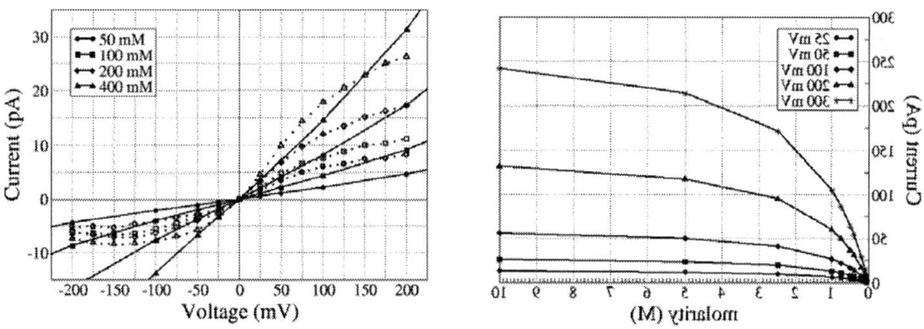

FIGURE 1. a) I(V) characteristics for K+ ions flowing across the KcsA potassium channel at different ion concentrations; b) ion current as a function of K+ concentration outside the protein at different biases. Equal concentrations are assumed in intra- and extra-cellular reservoirs, according to the experimental conditions.

probably to the simple model assumed for ion capture. Quantitative discrepances between simulated and experimental data can be related to the still partial knowledge of the microscopic parameters of each transition. More accurate MD calculations are in progress.

Fig. 2-a shows the number of ion exits as a function of the time interval between two successive ion exits, as obtained from the simulation (bias=100 mV, ion concentration: 100mM). The average time $\langle T_{exit} \rangle$ between two successive ion exits is

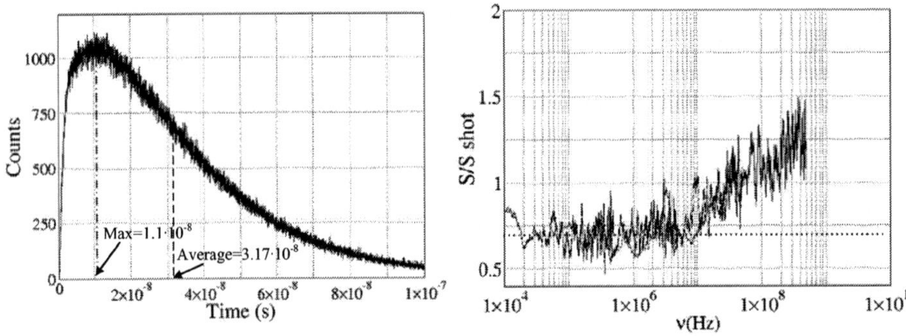

FIGURE 2. a) number of ion exits vs. time interval between two successive ion exits from the simulation; b) Power spectrum of the current fluctuations as a function of frequency . Both results refer to the case V=100 mV and c=100 mM.

approximately $3x10^{-8}$ s. An example of calculated noise power spectrum as a function of frequency is shown in Fig.2-b. White noise is found until about $1x10^{7}$ Hz with a Fano factor of about 0.7, clearly indicating that correlation in the ion motion lowers the noise spectrum with respect to the Poissonian shot-noise value. At frequencies of the order of $1/\langle T_{exit} \rangle$ an increase of the spectrum is observed.

In Fig. 3 the dependence of the power spectrum in the low-frequency limit is plotted as a function of external bias (a) and ion concentration (b). At low voltages the noise is influenced by the existence of ion motions in opposite directions. If we

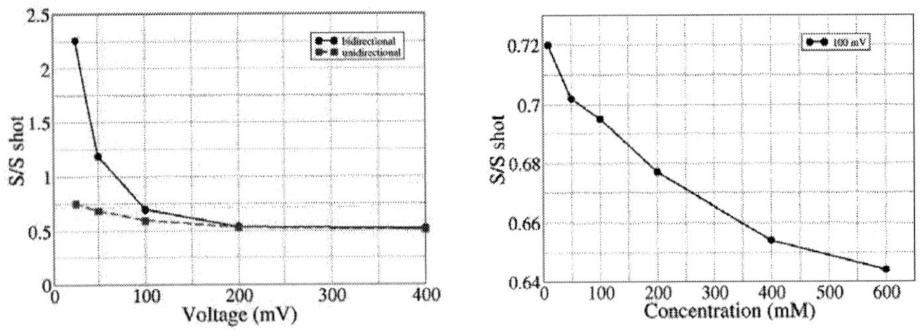

FIGURE 3: Power spectrum in the low-frequency limit normalized to the power spectrum theoretically predicted for a Poisson process a) as a function of the applied bias at fixed (100mM) concentration and, b) as a function of the ion concentration at fixed (100 mV) bias.

consider in the calculation only ions moving along a given orientation on the channel axis this enhancement is largely reduced, as it is theoretically expected.

CONCLUSIONS

A numerical procedure for the evaluation of conduction and noise associated to ion flow across membrane channels has been presented. The occupancy configurations of the channel, the transition rates between different configurations and the parameters contained in the transition-rate formula have been evaluated through MD simulations. Numerical results obtained from a MC procedure for the case of potassium ions permeating KcsA channel provide I(V) characteristics comparable with experimental data without adjustable parameters. The current noise power spectrum exibits deviations from shot noise spectrum, confirming the existence of significant correlation between successive ion exits. This computational analysis suggests that noise measurements in single open channels, today achievable with modern equipment, can contribute to the comprehension of the permeation mechanisms inside the channel.

ACKNOWLEDGMENTS

We wish to thank J. Kona, M. Rosini, and F. Sigworth for useful discussions.

REFERENCES

1. Zhou Y., Morais-Cabral J., Kaufman K., and MacKinnon R., *Nature* **414**, 43-48 (2001).
2. For a review see, e.g.: Roux B., *Curr. Opin. Struct. Biol.* **12**, 182-189 (2002).
3. Compoint M., Carloni P., Ramseyer C., Girardet C., *Biochim. Biophys. Acta*, **1661**, 26-39 (2004).
4. Affinito F., Brunetti R., Jacoboni C., Piccinini E., Rudan M., Bigiani A., and Carloni P., accepted for publication on *J. Comp. Elect*, 2005.
5. LeMasurier M., Heginbotham, and C. Miller, *J. Gen. Physiol.*, **118**, 303-313 (2001).
6. Samson M.S.P., Shrivastava I., Bright J.N., Tate J., Capener C.E., Biggin P.C., *Biochim. Biophys. Acta* **1565**, 294-307 (2002).
7. Cascella M.J., Guidoni L., Maritan A., Rothlisberger U., and P. Carloni, *Phys. Chem B* **106**, 13027-13032 (2002)
8. Roux B. *Comp. Phys. Commun.* **91**, 275-282 (1995)
9. Hanggi P., Talkner P., and Borkovec M., *Rev. Mod. Phys.* **62(2),** 251-341 (1990).
10. Heinemann S.H. and Sigworth F.J. *Biophys. J.* **57**, 499-514 (1990).
11. Frehland E. and Stephan W. *Biochimica et Biophysica Acta* **553**, 326-341 (1979).
12. Shottky W. *Ann. Phys. (Leipzig)* **57**, 541-567 (1917).
13. Fano U. *Phys. Rev.* , 26-29 (1947).
14. Allen T.W. and Chung S. H., *Biochim. Biophys. Acta*, 1515, 83-91 (2001).
15. Berendsen H.J.C., van der Spoel D., Van Drunen R. *Comp. Phys. Comm.* **91**, 43-56 (1995).
16. Lindahl E., Hess B., van der Spoel D. *J. Mol. Mod.* **7**, 306-317 (2001).
17. Roux B. , Allen T., Berneche S., and Im W. *Quart. Rev. Biophys.* **37**(1), 15-103 (2004).

Synchronization of Limit Circle Oscillators by Telegraph Noise

Denis S. Goldobin

Department of Physics, University of Potsdam, Postfach 601553, D-14415 Potsdam, Germany
Department of Theoretical Physics, Perm State University, 15 Bukireva str., 614990, Russia

Abstract. We study the influence of telegraph noise on synchrony of limit circle oscillators. Adopting the phase description for these oscillators, we derive the explicit expression for the Lyapunov exponent. We show that either for weak noise or frequent switching the Lyapunov exponent is negative, and the phase model gives adequate analytical results. In some systems moderate noise can desynchronize oscillations, and we demonstrate this for the Van der Pol–Duffing system.

Keywords: Noise; Synchronization; Lyapunov Exponent
PACS: 05.40.-a; 02.50.Ey; 05.45.Xt

INTRODUCTION

In autonomous systems exhibiting a stable periodic behavior (in other words, limit circle oscillators), deviations along the trajectory asymptotically in time neither decay nor grow, i.e. they are neutral. This neutrality is due to the time homogeneity, and may disappear as soon as this homogeneity is broken by means of a time-dependent external forcing. The phenomenon of synchronization of oscillators by periodic signal is well known and quite understood, here the oscillations follow the forcing (e.g., they attain the same frequency). When the role of this time-dependent forcing is played by a stochastic noise, the situation becomes less evident.

The first effect of noise on periodic oscillations is the phase diffusion: the oscillations are no more periodic but posses finite correlations [1, 2]. However, a nonlinear system may somehow follow the noisy force. Although it is not so evident how the synchronization between the system response and noise can be detected for one system, this synchronization can be easily detected by looking on whether the responses of a few identical systems driven by the common noise signal are identical or not. With such an approach, synchrony (asynchrony) of driven systems was early treated in the works [3, 4, 5, 6, 7, 8]. The mathematical criterion for synchronization is the negative leading Lyapunov exponent (LE; it measures the average exponential growth rate of infinitesimally small deviations from the trajectory) in the driven system.

In different fields the effect of synchronization of oscillators by common noise is known under different names. In neurophysiology the property of a single neuron to provide identical outputs for repeated noisy input is treated as "reliability" [9]. In the experiments with noise-driven ND:YAG lasers [10] this synchronization was called "consistency". When driving signal is related to not stochastic but deterministic chaos, one considers generalized synchronization [11]. In the last case the driven system is often chosen to be chaotic. In fact, in the above mentioned examples there is no limit

CP800, *Unsolved Problems of Noise and Fluctuations: UPoN 2005*
edited by L. Reggiani, C. Pennetta, V. Akimov, E. Alfinito, and M. Rosini

circle oscillators at the noiseless limit: for experiments described in [9, 10] the noiseless system is stable, i.e. LE is negative, and for generalized synchronization in chaotic systems, LE is positive. Evidently, in this cases, LE preserves its sign at sufficiently weak noise.

A more intriguing situation takes place when LE in noiseless system is zero (limit circle oscillators). Analytical and numerical treatments for different types of noise show weak noise to play an ordering role: LE shifts to negative values, and oscillators become synchronized [3, 4, 5, 12, 13]. In the work [14] the nonideal situations are considered: slightly nonidentical oscillators driven by an identical noise signal, and identical oscillators driven by slightly nonidentical noise signals; and additionally, positive LE was reported for large noise in some smooth systems similarly to how it was in the works [3, 4].

Note that in [3, 4] LE was calculated for oscillators driven by a random sequence of pulses, in [5, 12, 13, 14] the white Gaussian noise was considered. A noise of other nature is the telegraph one. By a normalized telegraph noise we mean the signal having values ± 1 and switching instantaneously between these values time to time. The distribution of time intervals between consequent switchings is exponential with the average value τ. The case of telegraph noise may be interesting not only because it completely differs from the previous two, but also because it allows to "touch" the question of relations between periodic and stochastic driving, e.g. to compare results for telegraph noise with the average switching time τ and the stepwise periodic signal of the same amplitude and the period 2τ. This is why we consider synchronization by telegraph noise.

PHASE MODEL

A limit circle oscillator with a small external force is known to be able to be well described within the phase approximation [15], where only dynamics of the system on the limit circle of the noiseless system is considered[1]. The system states on this limit circle can be parameterized by a single parameter, phase φ. With a stochastic force the equation for the phase reads

$$\dot{\varphi} = \omega + \varepsilon f(\varphi)\xi(t)\,, \tag{1}$$

where $2\pi/\omega$ is the period of the limit circle in the noiseless system, ε is the amplitude of noise, $f(\varphi)$ is the normalized sensitivity of the system to noise [$(2\pi)^{-1}\int_0^{2\pi} f^2(\varphi)d\varphi = 1$], and ξ is a normalized telegraph noise.

Master equation

Studying statistical properties of the system under consideration, one can introduce two probability density functions $W_{\pm}(\varphi, t)$ defining the probability to locate the system

[1] Noteworthy, the phase approximation is valid not only for a small external force, but also for a moderate one if only the leading Lyapunov exponent of the limit circle is negative and large enough.

in vicinity of φ with $\xi = \pm 1$, correspondingly, at the moment t. Then the Master equations of the system read

$$\frac{\partial W_+(\varphi,t)}{\partial t} + \frac{\partial}{\partial \varphi}[(\omega + \varepsilon f(\varphi))W_+(\varphi,t)] = \frac{1}{\tau}W_-(\varphi,t) - \frac{1}{\tau}W_+(\varphi,t), \quad (2)$$

$$\frac{\partial W_-(\varphi,t)}{\partial t} + \frac{\partial}{\partial \varphi}[(\omega - \varepsilon f(\varphi))W_-(\varphi,t)] = \frac{1}{\tau}W_+(\varphi,t) - \frac{1}{\tau}W_-(\varphi,t). \quad (3)$$

In the terms of $W \equiv W_+ + W_-$ and $V \equiv W_+ - W_-$ the last system takes the form of

$$\dot{W} = -\omega W_\varphi - \varepsilon (fV)_\varphi, \qquad \dot{V} = -\omega V_\varphi - \varepsilon (fW)_\varphi - \frac{2}{\tau}V. \quad (4)$$

For steady distributions the probability flux S is constant:

$$S = \omega W(\varphi) + \varepsilon f(\varphi) V(\varphi);$$

and system (4) with periodic boundary conditions has the solution

$$V(\varphi) = -\frac{\varepsilon \omega C}{\omega^2 - \varepsilon^2 f^2(\varphi)} \int_{\varphi}^{\varphi+2\pi} d\psi f'(\psi) \exp\left(\frac{2}{\tau}\int_{\varphi}^{\psi} \frac{d\theta}{\omega^2 - \varepsilon^2 f^2(\theta)}\right), \quad (5)$$

where C is defined by the normalization condition:

$$C^{-1} = 2\pi \left(\exp\left(\frac{2}{\tau}\int_0^{2\pi} \frac{d\theta}{\omega^2 - \varepsilon^2 f^2(\theta)}\right) - 1\right) + \varepsilon^2 \int_0^{2\pi} d\varphi \int_{\varphi}^{\varphi+2\pi} d\psi \frac{f(\varphi) f'(\psi)}{\omega^2 - \varepsilon^2 f^2(\varphi)} \exp\left(\frac{2}{\tau}\int_{\varphi}^{\psi} \frac{d\theta}{\omega^2 - \varepsilon^2 f^2(\theta)}\right). \quad (6)$$

The probability flux reads

$$S = \omega \left(\exp\left(\frac{2}{\tau}\int_0^{2\pi} \frac{d\theta}{\omega^2 - \varepsilon^2 f^2(\theta)}\right) - 1\right) C.$$

Lyapunov exponent

Studying stability of solutions of the stochastic equation (1), one has to consider behavior of a small perturbation α:

$$\dot{\alpha} = \varepsilon f'(\varphi) \alpha \xi(t).$$

The Lyapunov exponent (LE) measuring the average exponential growth rate of α can be obtained by averaging the corresponding velocity

$$\begin{aligned}\lambda &= \langle\frac{d}{dt}\ln\alpha\rangle = \langle\varepsilon f'(\varphi)\xi(t)\rangle = \varepsilon\int_0^{2\pi} f'(\varphi)V(\varphi)d\varphi \\ &= -\varepsilon^2\omega C\int_0^{2\pi} d\varphi \int_{\varphi}^{\varphi+2\pi} d\psi \frac{f'(\varphi)f'(\psi)}{\omega^2-\varepsilon^2 f^2(\varphi)} \exp\left(\frac{2}{\tau}\int_{\varphi}^{\psi}\frac{d\theta}{\omega^2-\varepsilon^2 f^2(\theta)}\right). \qquad (7)\end{aligned}$$

Let us remind that LE determines the asymptotic behavior of small perturbations, and describes whether close states diverge or converge with time. This process is not necessarily monotonous, i.e., close trajectories can diverge at some time intervals while demonstrating asymptotic convergence, and vice versa.

When $\tau \ll 1$ or $\varepsilon \ll \omega$, the eq. (7) can be simplified:

$$\lambda_{\text{app}} = -\frac{\varepsilon^2}{2\pi\omega}\left(\exp\left(\frac{4\pi}{\tau\omega^2}\right)-1\right)^{-1}\int_0^{2\pi} d\varphi\int_0^{2\pi} d\psi f'(\varphi)f'(\psi+\varphi)\exp\frac{2\psi}{\tau\omega^2}. \qquad (8)$$

The last expression is strictly negative. Indeed, in the Fourier space it reads

$$\lambda_{\text{app}} = -\omega\tau\varepsilon^2\sum_{k=1}^{\infty}|C_k|^2\frac{k^2}{1+(k\tau\omega^2/2)^2},$$

where $C_k = (2\pi)^{-1}\int_0^{2\pi} f(\varphi)e^{-ik\varphi}d\varphi$.

COMPARISON TO NUMERICAL SIMULATION

We found that either for weak noise or frequent switching LE is negative regardless to the properties of the smooth function $f(\varphi)$ (as for weak white Gaussian noise in similar systems [12, 13]). In the works [14], moderate white Gaussian noise was shown to be able to lead to instability even in smooth systems. In the light of above facts, it is interesting (i) *what is the region of validity of our analytical theory*, (ii) *whether there is some footprints of the synchronization by periodic forcing in the stochastic synchronization*, and (iii) *whether telegraph noise can desynchronize oscillators*.

For the two first purpose we performed simulation of a modified Van der Pol oscillator:

$$\ddot{x}-\mu(1-x^2-\dot{x}^2)\dot{x}+x = \varepsilon\sqrt{2}\xi(t), \qquad (9)$$

where $\xi(t)$ is either a telegraph noise with the average switching time τ or a periodic stepwise signal with the period 2τ, i.e. the constant switching time τ. The forcing-free modified Van der Pol oscillator has the round stable limit circle of the unit radius for all $\mu > 0$. Nevertheless, the phase equation (1) with $\omega = 1$ and the simple function

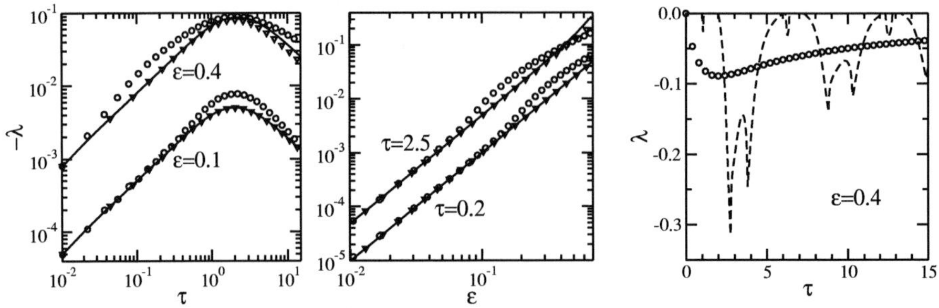

FIGURE 1. Samples of dependencies $\lambda(\varepsilon,\tau)$ for the modified Van der Pol oscillator (9) at $\mu = 0.1$. The solid lines present the analytical results of phase description, the triangles plot results of the approximation (8), the circles correspond to numerical simulation of the noisy modified Van der Pol oscillator, and the dashed line corresponds to numerical simulation of the periodically driven one.

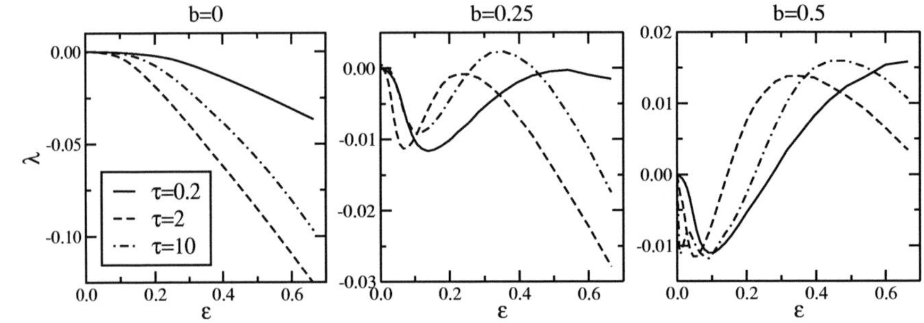

FIGURE 2. Samples of dependencies $\lambda(\varepsilon,\tau)$ for the Van der Pol–Duffing oscillator (10) at $\mu = 0.1$. The values of the parameters b and τ are indicated above the plots.

$f(\varphi) = \sqrt{2}\cos\varphi$ may be correctly adopted only if the phase speed is near-constant all over the limit circle, which is possible at small μ only.

In Fig. 1 one can see that our analytical theory is fortunately in good agreement with the results of analytical simulation not only for weak noise; and the dependence $\lambda(\varepsilon,\tau)$ for the stochastic driving has no footprints of the one for the periodic driving.

While the dynamical system (9) does not exhibit positive LEs at any noise intensity and any μ, they can be observed for a Van der Pol–Duffing model[2]:

$$\ddot{x} - \mu(1-2x^2)\dot{x} + x + 2bx^3 = \varepsilon\sqrt{2}\xi(t), \tag{10}$$

where "Duffing parameter" b describes nonisochronicity of oscillations. In Fig. 2 one can see that at large enough b positive LE appears in a certain range of parameters.

[2] A similar situation occurs for white Gaussian noise [14].

CONCLUSIONS

Having considered the phenomenon of synchronization of limit circle oscillators by common telegraph noise, we can summarize:
— *Either for weak noise or frequent switching the Lyapunov exponent is negative;*
— *For some systems, the phase model gives quite adequate results even for moderate noise levels and values of the average switching time;*
— *The dependence $\lambda(\varepsilon, \tau)$ for stochastic driving does not look to have any footprints of the one for periodic driving;*
— *In some systems, moderate telegraph noise can desynchronize oscillations.*

Here we do not present results for the nonideal situations (like in [14]): slightly nonidentical oscillators driven by an identical noise signal, and identical oscillators driven by slightly nonidentical noise signals. The reason is that for weak noise these results appear to be the same as in [14] but with λ_{app} given by Eq. (8) instead of λ.

ACKNOWLEDGMENTS

The work has been supported by DFG (SFB 555). I would like to thank Arkady Pikovsky for fruitful discussions and helpful comments.

REFERENCES

1. R. L. Stratonovich, *Topics in the Theory of Random Noise*, Gordon and Breach, New York, 1963.
2. A. N. Malakhov, *Fluctuations in Self–Oscillatory Systems*, Nauka, Moscow, 1968, (In Russian).
3. A. S. Pikovsky, *Radiophys. Quantum Electron.* **27** (5), 576 (1984).
4. A. S. Pikovsky, "Synchronization and stochastization of nonlinear oscillations by external noise", in *Nonlinear and Turbulent Processes in Physics*, edited by R. Z. Sagdeev, Harwood Acad. Publ., Singapore, 1984, pp. 1601–1604.
5. A. S. Pikovsky, M. Rosenblum, and J. Kurths, *Synchronization—A Unified Approach to Nonlinear Science*, Cambridge University Press, Cambridge, UK, 2001.
6. L. Yu, E. Ott, and Q. Chen, *Phys. Rev. Lett.* **65**, 2935 (1990).
7. L. Baroni, R. Livi, and A. Torcini, *Phys. Rev. E* **63**, 036226 (2001).
8. P. Khoury, M. A. Lieberman, and A. J. Lichtenberg, *Phys. Rev. E* **54**, 3377 (1996).
9. Z. F. Mainen and T. J. Sejnowski, *Science* **268**, 1503 (1995).
10. A. Uchida, R. McAllister, and R. Roy, *Phys. Rev. Lett.* **93**, 244102 (2004).
11. H. D. I. Abarbanel, N. F. Rulkov, and M. M. Suschik, *Phys. Rev. E* **53**, 4528 (1996).
12. J. Teramae and D. Tanaka, *Phys. Rev. Lett.* **93**, 204103 (2004)
13. D. S. Goldobin and A. S. Pikovsky, *Phisica A* **351**, 126 (2005).
14. D. S. Goldobin and A. Pikovsky, *Phys. Rev. E* **71**, 045201(R) (2005).
15. Y. Kuramoto, *Chemical Oscillations, Waves and Turbulence*, Dover, New York, 2003.

Unidirectional Flux In Brownian And Langevin Simulations Of Diffusion

A. Singer*, Z. Schuss* and B. Nadler†

*Department of Mathematics, Tel-Aviv University, Ramat-Aviv, 69978 Tel-Aviv, Israel
†Department of Mathematics, Yale University, 10 Hillhouse Avenue P.O. Box 208283, New Haven, Connecticut 06520-8283, USA

Abstract. Brownian and Langevin simulations of ions in solution require the maintenance of average fixed concentrations at the interface between the simulation volume and the surrounding continuum. This requires the injection of new trajectories into the simulation, which creates a unidirectional flux at the interface. The Wiener path integral splits the net diffusion flux into infinite unidirectional fluxes, whose difference is finite, as in classical diffusion theory. The infinite unidirectional flux is an artifact of the diffusion approximation to Langevin's equation, which fails on time scales shorter than the relaxation time $1/\gamma$. The probability of Brownian trajectories that cross a point in one direction per unit time Δt equals that of Langevin trajectories if $\gamma\Delta t = 2$. This result is relevant to Brownian dynamics simulation of particles in a finite volume inside a large bath.

Keywords: Wiener's path integral, diffusion, Brownian simulations, Langevin
PACS: 31.15.Kb, 02.50.-r, 05.40.-a.

The simulation of a test volume in an ionic solution is an important field of chemical physics, especially in computing permeation of ions in protein channels of biological membranes [1]-[4]. Such a simulation is necessarily limited to a relatively small number of particles, due to computational difficulties. Computer simulations of diffusing particles, such as ions in solution, raise the question of connecting a small discrete simulation volume to a continuum bath. There are many algorithms, procedures and protocols for particles at the interface between the simulation region and the continuum baths (see [5] for a complete list of references on simulations). However, none of them takes into account the actual physics at the interface. The failure of these attempts calls for a theory of the interface that is compatible with the physics and for the design of simulations based on such a theory.

The mathematical model of the interface is expected to reproduce the physical conditions that actually exist on the boundary of the simulated volume. These physical conditions are not merely the average electrostatic potential and local concentrations at the boundary of the volume, but also their fluctuation in time. It is important to recover the correct fluctuation, because the stochastic dynamics of ions in solution are nonlinear, due to the coupling between the electrostatic field and the motion of the mobile charges, so that averaged boundary conditions do not reproduce correctly averaged nonlinear response. However, in a system of noninteracting particles incorrect fluctuation on the boundary may still produce the correct response outside a boundary layer in the simulation region.

The boundary fluctuation consists of arrivals of new particles from the bath into, and of the recirculation of particles in and out of the simulation volume. The random

CP800, *Unsolved Problems of Noise and Fluctuations: UPoN 2005*
edited by L. Reggiani, C. Pennetta, V. Akimov, E. Alfinito, and M. Rosini

motion of the mobile charges brings about fluctuations in both the concentrations and the electrostatic field. Since the simulation is confined to the volume inside the interface, the new and the recirculated particles have to be fed into the simulation by a source that imitates the random influx across the interface. The interface does not represent any physical device that feeds trajectories back into the simulation, but is rather an imaginary wall, which the physical trajectories of the diffusing particles cross and recross any number of times. The efflux of simulated trajectories through the interface is observed in the simulation, however, the influx of new trajectories, which is the unidirectional flux (UF) of diffusion, has to be calculated so as to reproduce the random influx with the correct statistical properties of this stochastic process, as mentioned above. Thus the UF is the source strength of the influx, and also the stochastic process that counts the number of trajectories that cross the interface in one direction per unit time. We consider here simulations of particles that interact with a mean field. We note, however, that the mean field approximation to interacting particles is valid in the bulk at biological concentrations. The channel and its immediate surrounding baths are expected to be within the simulation region.

Our main results are (i) The discovery of the precise range of validity of diffusion theory (Brownian motion) as a description of ionic motion in solution. We found the correct way to use diffusion theory in simulations. Mathematically this is expressed in the determination of the range of validity of the Smoluchowski approximation to the Langevin equation; (ii) The design of Brownian and Langevin simulations that do not form spurious boundary layers, which are ubiquitous in molecular simulations in biology, chemistry, and physics. This design is based on two new mathematical insights into the theory of diffusion. One is the splitting of the probability flux of the Brownian motion into unidirectional components, which is the result of the definition of probability flux in terms of path integrals. The other is the application of methods of renewal theory to the theory of diffusion.

The dynamics of Brownian particles is described by the Langevin equation

$$\ddot{x}+\gamma\dot{x}=f(x)+\sqrt{2\varepsilon\gamma}\dot{w}. \tag{1}$$

The unidirectional flux of Langevin trajectories is expressed in terms of the space density $p(x,v,t)$ as

$$\begin{aligned} J_{LR}(x,t) &= \int_0^\infty vp(x,v,t)\,dv, \quad J_{RL}(x,t)=-\int_{-\infty}^0 vp(x,v,t)\,dv \\ J_{\text{net}}(x,t) &= J_{LR}(x,t)-J_{RL}(x,t)=\int_{-\infty}^\infty vp(x,v,t)\,dv. \end{aligned} \tag{2}$$

In the limit of large friction

$$J_{LR,RL}(x,t) = \sqrt{\frac{\varepsilon}{2\pi}}p(x,t)\mp\frac{1}{2\gamma}\left[\varepsilon\frac{\partial p(x,t)}{\partial x}-f(x)p(x,t)\right]+O\left(\frac{1}{\gamma^2}\right) \tag{3}$$

Both UFs in (3) are finite and proportional to the marginal density at x. The net flux is the difference

$$J_{\text{net}}(x,t) = J_{LR}(x,t) - J_{RL}(x,t) = -\frac{1}{\gamma}\left[\varepsilon\frac{\partial p(x,t)}{\partial x} - f(x)p(x,t)\right], \tag{4}$$

as in classical diffusion theory [6], [7].

Classical diffusion theory describes the motion of a Brownian particle in a field of force by the (overdamped) Smoluchowski equation [8]

$$\gamma\dot{x} = f(x) + \sqrt{2\varepsilon\gamma}\dot{w}. \tag{5}$$

The transition probability density of $x(t)$ is defined by the propagator

$$p(x,t+\Delta t\,|\,x_0) = \tag{6}$$

$$\int \frac{\sqrt{\gamma}}{\sqrt{4\pi\varepsilon\Delta t}}\exp\left\{-\frac{\gamma[x-y-f(y)\Delta t/\gamma]^2}{4\varepsilon\Delta t}\right\}p(y,t\,|\,x_0)\,dy\,(1+o(1)) \quad \text{for} \quad \Delta t \ll 1.$$

The unidirectional probability current (flux) density at a point x_1 is the probability per unit time of trajectories that propagate from the ray $x < x_1$ into the ray $x > x_1$. It is therefore given by

$$J_{LR}(x_1,t) = \lim_{\Delta t\to 0} J_{LR}(x_1,t,\Delta t), \tag{7}$$

where

$$J_{LR}(x_1,t,\Delta t) = \frac{1}{\Delta t}\int_x^\infty dx\int_{-\infty}^x \frac{\sqrt{\gamma}dy}{\sqrt{4\pi\varepsilon\Delta t}}\exp\left\{-\frac{\gamma[x-y-f(y)\Delta t/\gamma]^2}{4\varepsilon\Delta t}\right\}\times$$

$$p(y,t\,|\,x_0)\,dy\,(1+o(1)) \quad \text{for} \quad \Delta t \ll 1. \tag{8}$$

It was shown in [9] that

$$J_{LR,RL}(x,t,\Delta t) = \sqrt{\frac{\varepsilon}{\pi\gamma\Delta t}}\,p(x,t) \pm \frac{1}{2\gamma}\left(f(x)p(x,t) - \varepsilon\frac{\partial p(x,t)}{\partial x}\right) + O\left(\frac{\sqrt{\Delta t}}{\gamma^{3/2}}\right). \tag{9}$$

If $p(x,t) > 0$, then both $J_{LR}(x,t)$ and $J_{RL}(x,t)$ are infinite, in contradiction to the finite unidirectional fluxes (3). However, the net flux density is finite and is given by

$$\begin{aligned} J_{\text{net}}(x,t) &= \lim_{\Delta t\to 0}\{J_{LR}(x,t,\Delta t) - J_{RL}(x,t,\Delta t)\} \\ &= -\frac{1}{\gamma}\left[\varepsilon\frac{\partial}{\partial x}p(x,t) - f(x)p(x,t)\right], \end{aligned} \tag{10}$$

which is (4).

The apparent paradox is due to the idealized properties of the Brownian motion. More specifically, the trajectories of the Brownian motion, and therefore also the trajectories of

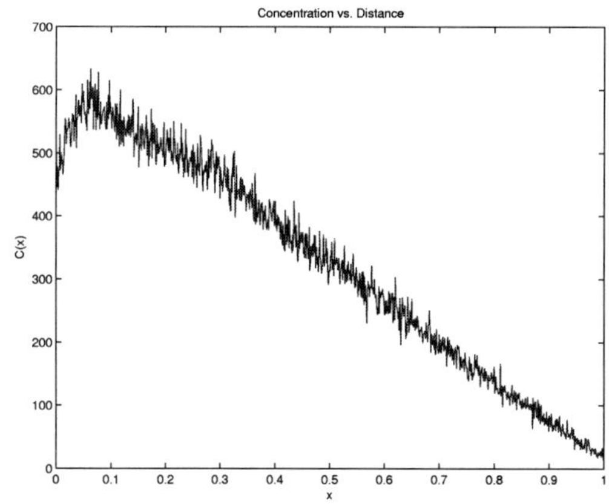

FIGURE 1. The concentration profile of Brownian trajectories that are initiated at $x = 0$ with a normal distribution, and terminated at either $x = 0$ or $x = 1$.

the solution of eq.(5), are nowhere differentiable, so that any trajectory of the Brownian motion crosses and recrosses the point x infinitely many times in any time interval $[t, t+\Delta t]$ [10]. Obviously, such a vacillation creates infinite UFs. We note that setting

$$\gamma\Delta t = 2 \tag{11}$$

in equations (9) and gives (3). This means that the diffusion approximation to the FPE on time scales shorter than $1/\gamma$ gives rise to non-physical artifacts, however, it is valid on longer time scales. The unidirectional flux in a simulation of diffusion agrees with that of the FPE only if (11) holds, otherwise, the unidirectional fluxes are missed, though the net flux is still correct.

Next, we design and analyze a BD simulation of particles diffusing between fixed concentrations. For simplicity, we consider the free Brownian motion (i.e., $f = 0$ in eq.(4)) in the interval $[0, 1]$. The trajectories were produced as follows: (I) According to the dynamics (4), new trajectories that are started at $x(-\Delta t) = 0$ move to $x(0) = \sqrt{\frac{2\varepsilon}{\gamma}}\,|\Delta w|$; (II) Inside the interval the dynamics progresses according to the Euler scheme $x(t+\Delta t) = x(t) + \sqrt{\frac{2\varepsilon}{\gamma}}\Delta w$; (III) The trajectory is terminated if $x(t) > 1$ or $x(t) < 0$. The parameters are $\varepsilon = 1$, $\gamma = 1000$, $\Delta t = 1$. We ran 10,000 random trajectories and collected their statistics with the results shown in Figure 1.

The simulated concentration profile is linear, but for a small depletion layer near the left boundary $x = 0$, where new particles are injected. This is inconsistent with the steady state diffusion equation (DE), which predicts a linear concentration profile in the entire interval $[0, 1]$. The discrepancy stems from part (a) of the numerical scheme, which assumes that particles enter the simulation interval exactly at $x = 0$. However, $x = 0$ is just an imaginary interface. Had the simulation been run on the entire line, particles would hop into the simulation across the imaginary boundary at $x = 0$ from the left,

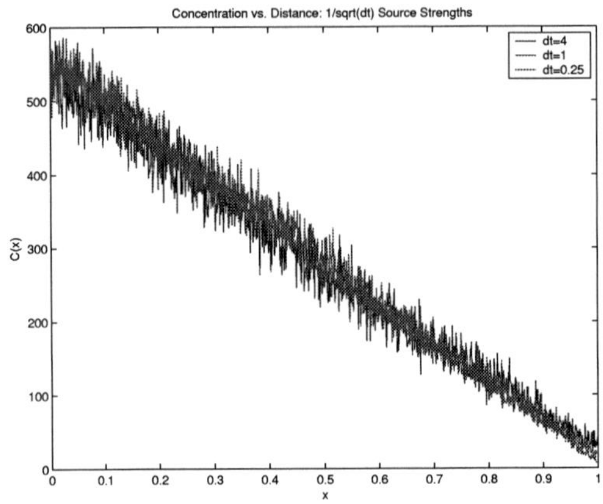

FIGURE 2. The concentration profile of Brownian trajectories that are initiated at $x = 0$ and terminated at either $x = 0$ or $x = 1$. Three different time steps ($\Delta t = 4, 1, 0.25$) are shown, and the injection rate of new particles is proportional to $1/\sqrt{\Delta t}$. Refining the time step does not alter the concentration profile.

rather than exactly at the boundary. This situation is familiar from renewal theory [11]. The probability distribution of the distance an entering particle covers, not given its previous location, is not normal, but rather it is the residual of the normal distribution, given by

$$\phi(x) = C\int_{-\infty}^{0} \exp\left\{-\frac{(x-y)^2}{2\sigma^2}\right\} dy, \tag{12}$$

where $\sigma^2 = \dfrac{2\varepsilon\Delta t}{\gamma}$ and C is determined by the normalization condition $\int_0^\infty \phi(x)\,dx = 1$. This gives

$$\phi(x) = \sqrt{\frac{\pi}{2\sigma}}\,\mathrm{erfc}\left(\frac{x}{\sqrt{2}\sigma}\right). \tag{13}$$

Rerunning the simulation with the entrance pdf $\phi(x)$, we obtained the expected linear concentration profile, without any depletion layers (see Figure 3). Injecting particles exactly at the boundary makes their first leap into the simulation too large, thus effectively decreasing the concentration profile near the boundary.

Next, we changed the time step Δt of the simulation, keeping the injection rate of new particles constant. The population of trajectories inside the interval was depleted when the time step was refined (see Figure 4).

A well behaved numerical simulation is expected to converge as the time step is refined, rather than to result in different profiles. This shortcoming of refining the time step is remedied by replacing the constant rate sources with time-step-dependent sources, as predicted by eqs.(9). Figure 2 describes the concentration profiles for three different values of Δt and source strengths that are proportional to $1/\sqrt{\Delta t}$. The concentration profiles now converge when $\Delta t \to 0$. The key to this remedy is the calculation of the UF in diffusion.

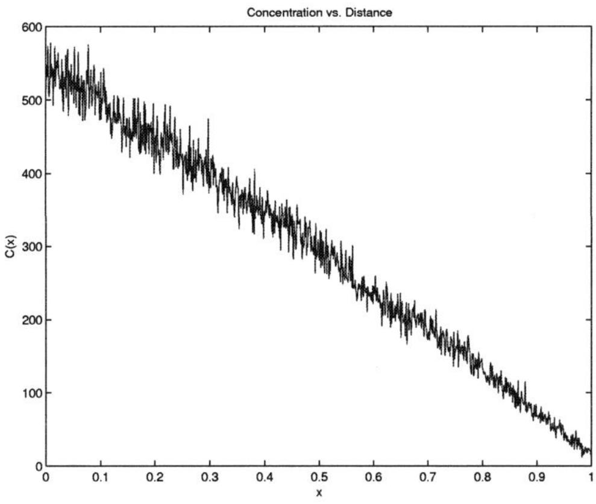

FIGURE 3. The concentration profile of Brownian trajectories that are initiated at $x = 0$ with the residual of the normal distribution, and terminated at either $x = 0$ or $x = 1$.

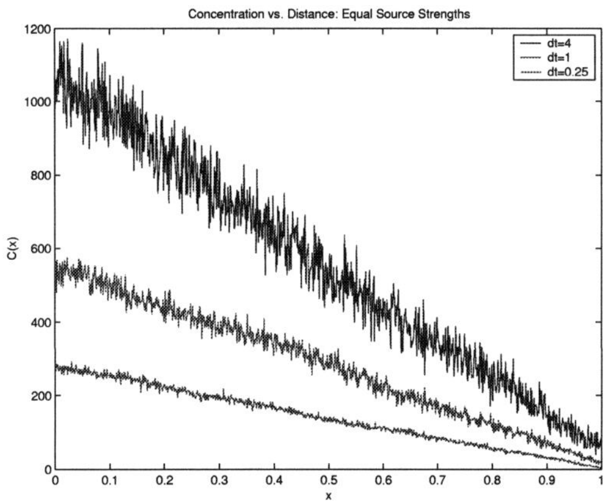

FIGURE 4. The concentration profile of Brownian trajectories that are initiated at $x = 0$ and terminated at either $x = 0$ or $x = 1$. Three different time steps $(\Delta t = 4, 1, 0.25)$ were used, but the injection rate of new particles remained constant. Refining the time step decreases the concentration profile.

A similar boundary layer arises at the interface in Langevin simulations [13]. A similar analysis of these simulations lead to a similar remedy that removes the spurious layer [14] (see figure 5).

ACKNOWLEDGMENTS

The authors are partially supported by research grants from the Israel Science Foundation, US-Israel Binational Science Foundation, and DARPA.

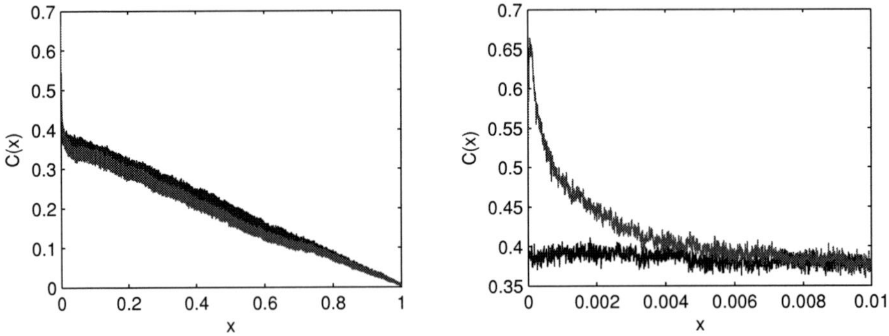

FIGURE 5. Left panel: Concentration against displacement of a LD simulation with injecting particles according to the residual distribution (blue), and according to the Maxwellian velocity distribution exactly at the boundary (red). The two graphs are almost identical, except for a small boundary layer near $x = 0$ in red. Right panel: Zoom in of the concentration profile in the boundary layer $x < 0.01 = \sqrt{\varepsilon}/\gamma$.

REFERENCES

1. B. Hille, *Ionic Channels of Excitable Membranes*, pp. 1–814, Sinauer Associates Inc., Sunderland, 3rd ed., 2001.
2. W. Im and B. Roux, "Ion permeation and selectivity of ompf porin: a theoretical study based on molecular dynamics, brownian dynamics, and continuum electrodiffusion theory," *J. Mol. Bio.* **322** **(4)**, pp. 851–869 (2002)
3. B. Corry, M. Hoyles, T. W. Allen, M. Walker, S. Kuyucak, and S. H. Chung, "Reservoir boundaries in brownian dynamics simulations of ion channels," *Biophys. J.* **82**, pp. 1975–1984 (2002).
4. S. Wigger-Aboud, M. Saraniti, and R. S. Eisenberg, "Self-consistent particle based simulations of three dimensional ionic solutions," *Nanotech* **3**, p. 443 (2003).
5. T. Naeh, *Simulation of ionic solution*, Ph.D. dissertation, Department of Mathematics, Tel-Aviv University (2001).
6. R. S. Eisenberg, M. M. Kłosek, and Z. Schuss, *J. Chem. Phys.* **102**, pp. 1767–1780, 1995.
7. C.W. Gardiner, *Handbook of Stochastic Methods*, 2-nd edition, Springer, NY 1985.
8. Z. Schuss, *Theory and Applications of Stochastic Differential Equations*, Wiley, NY, 1980.
9. A. Marchewka, Z. Schuss. *Phys. Rev. A* **61** 052107 (2000).
10. K. Itô and H.P. McKean, Jr., *Diffusion Processes and their Sample Paths*, Springer, NY 1996.
11. S. Karlin and H.M. Taylor, *A First Course in Stochastic Processes*, Academic Press, NY (1975).
12. A. Singer and Z. Schuss, *Phys. Rev. E* **71** 026115 (2005).
13. A. Singer, Z. Schuss, B. Nadler, and R. S. Eisenberg, *Phys. Rev. E* **70**, 061106 (2004).
14. B. Nadler, Z. Schuss, A. Singer, *Phys. Rev. Lett.* **94**, 218101 (2005).

Fluctuation Models Of Irregular Impedance Networks

Vladimir Akimov and Lino Reggiani

Dipartimento di Ingegneria dell'Innovazione, Universita' di Lecce and National Nanotechnology Laboratory of CNR-INFM, Via Arnesano s/n, 73100 Lecce, Italy

Abstract. An impedance irregular network is proposed to investigate electrical fluctuations of biomolecules anchored between Ohmic contacts. We take rhodopsin as prototype of a biomelecule and as source of fluctuations we consider the random oscillations of the distances between aminoacids (link oscillation model) or alternatively those of the position of the aminoacids around their static value (node oscillation model). The network impedance and its variance are then investigated as function of the amplitude of the fluctuations. Analogies and differences between the results of the two models are discussed and some open problems are identified.

Keywords: biomolecules, fluctuations, impedance, noise, proteins, rhodopsin, variance.
PACS: 87.15Ya, 87.15.Aa, 87.14Ee, 89.75.-k

INTRODUCTION AND MOTIVATION

Impedance networks are useful approaches to mimic the electrical properties of complex systems like composite materials, thin films resistors and large biomolecules. In this work we report on two complementary models able to introduce the intrinsic fluctuation of the network total impedance associated with the fluctuations of each elementary impedance pertaining to the network. To this purpose, we consider an irregular network consisting of a given number of nodes and a fluctuating value of elementary impedances, which link any pair of nodes. The source of noise of any elementary impedance is described by the oscillations of: the length of the links (link oscillation model LOM) or the position of the nodes (node oscillation model NOM) with respect to their static value. The two models are then applied to a network which mimics the impedance property of a rhodopsin molecule anchored between two Ohmic contacts. We have investigated the behavior of key output data (number of links, impedance, etc.) versus all model parameters.

IRREGULAR IMPEDANCE NETWORK AND ITS APPLICATION TO PROTEINS

We have considered the spatial structure of the protein molecule in contact with two metallic electrodes through which an external voltage can be applied as shown in

CP800, *Unsolved Problems of Noise and Fluctuations: UPoN 2005*
edited by L. Reggiani, C. Pennetta, V. Akimov, E. Alfinito, and M. Rosini

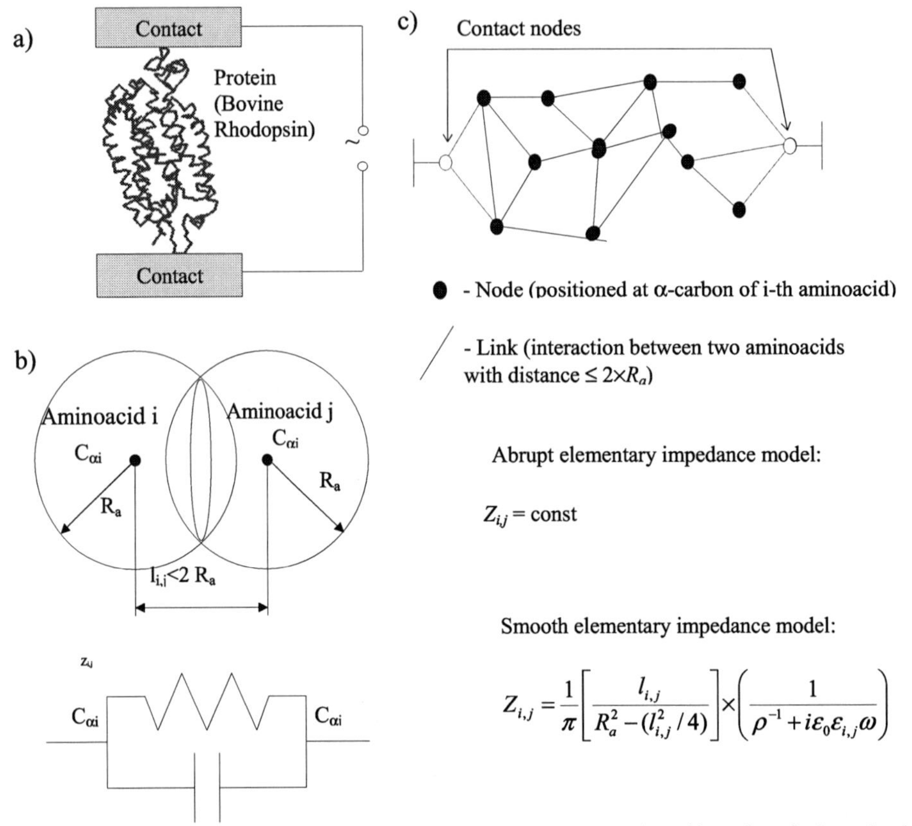

FIGURE 1. a) Protein between contacts; b) Overlap between two aminoacids and equivalent circuit element. $C_{\alpha i}$ identifies the center of the sphere corresponding to the alpha carbon atom of the *i*-th aminoacid. Parameter R_a is an interaction radius for the aminoacids. Z_{ij} is an impedance of the link between i-th and j-th nodes. c) Example Graph for the equivalent circuit network.

Fig. 1, a) [1]. By associating an elemental RC impedance to each link between two neighboring amino acids distant from an interaction radius R_a (Fig. 1, b), we have constructed the corresponding impedance network [2,3], which mimics the electrical response of the rhodopsin [4]. The network impedance depends on the radius of interaction, which is chosen as the independent variable. Then the network has been analyzed in terms of its graph properties and the equivalent network circuit solved to determine its global impedance.

FLUCTUATION MODELS

To introduce the electrical fluctuations in this network we propose two models. Both of them use the probability distribution given by the classical harmonic oscillator:

- In the case of the link oscillation model (LOM) the variation of the network characteristics (e.g. number of links, impedance) is provided by the independent change of the effective distance between each pair of nodes.

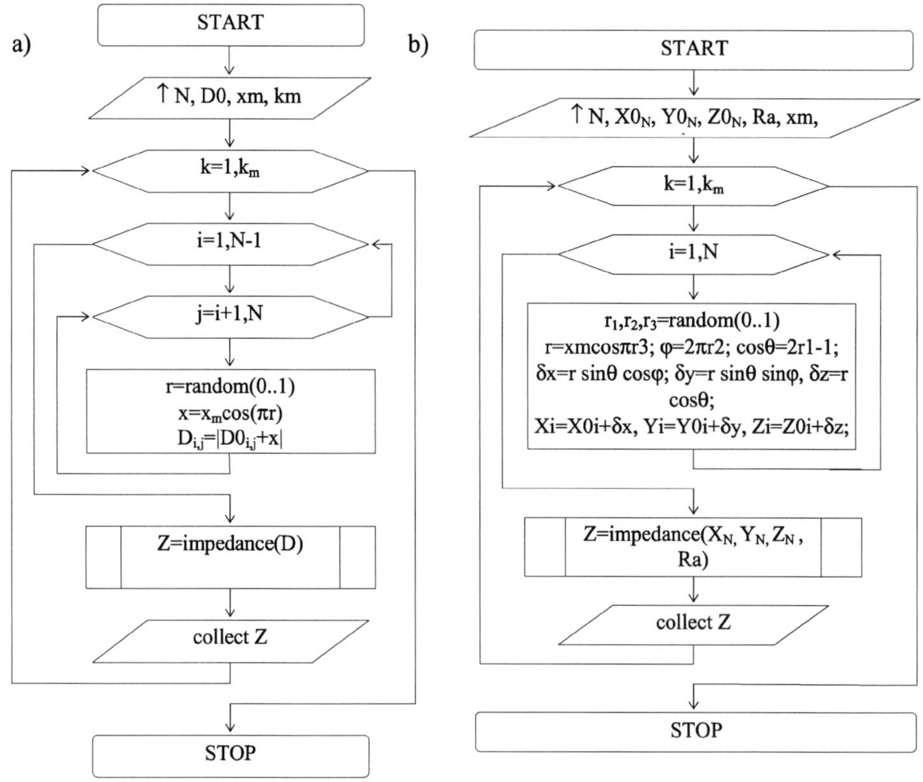

FIGURE 2. a) Flowchart of the **link oscillation model**. N – number of nodes, k_m – number of iterations, D0 – static matrix of distances, D – current matrix of distances, x_m – amplitude of oscillations.
b) Flowchart of the **node oscillation model**. $X0_N$, $Y0_N$, $Z0_N$ – static matrixes of the coordinates of nodes, X_N, Y_N, Z_N – current matrixes of the coordinates of nodes, Ra – maximum interaction radius.

- In the case of the node oscillation model (NOM) the variation of the network characteristics is provided by the isotropic variation of the position of each node in 3D space.

Thermal fluctuations: link oscillation model

In this approach the state of the molecule is fully described by the matrix D_0, which contains the distances between each pair of nodes. To calculate the impedance of the structure, we compare each distance with the interaction length and, if it is less than $2R_a$, we set the link with the impedance given by the corresponding elementary impedance model, otherwise the link is cancelled. To describe the fluctuations, we produce a number of random matrices D on the base of the static matrix D_0 by making the distances fluctuating as a harmonic oscillator.

The state of a harmonic oscillator at time t can be expressed as

$$x = x_{\max} \sin \omega t \,. \qquad (1)$$

where x_m is the maximum amplitude and ω the angular frequency of the oscillator

Its probability to take the value x_a at randomly selected times is proportional to

$$P(x_a) \sim \frac{\partial t}{\partial x}\Big|_{x=x_a} \,. \qquad (2)$$

The correspondence between a number r randomly distributed from 0 to 1 and x_a is

$$r = \frac{\int_{-x_{\max}}^{x_a} P(x)dx}{\int_{-x_{\max}}^{x_m} P(x)dx} \,. \qquad (3)$$

Accordingly, we obtain $x_a(r) = -x_{\max} \cos \pi r$, or equivalently

$$x_a(r) = x_{\max} \cos \pi r \qquad (4)$$

Equation (4) can be used to select randomly the state of the harmonic oscillator. For the details see the flowchart in Fig. 2.

From the above, it follows that the length of a link at each step is given by:

$$l=|l_0+x_a(r)|= |l_0+x_{max}\cos \pi r|. \qquad (5)$$

To consider a thermal fluctuation, according to classical statistics we can assume that the energy of the oscillator is proportional to $k_B T$: $x^2_{max} \sim T$, and thus:

$$x_{\max} = A\sqrt{T} \,, \qquad (6)$$

where A is a fitting parameter. Generally, A depends on the elastic properties of the whole structure and thus on the distance between aminoacids in this structure, but as a first approach we can assume that A is the same for each pair of aminoacids. By recalling that in our model the impedance (absolute value) is a decreasing function with increasing interaction radius, we can construct the minimum impedance structure with characteristic radius of interaction R_a+x_{max}, and, correspondingly, the maximum impedance structure with radius R_a-x_{max}. A real structure will contain all the links which satisfy the condition $l_0< R_a-x_{max}$, no links which satisfy the condition $l_0> R_a+x_{max}$ and some random links which satisfy the condition $R_a+x_{max} <l_0< R_a-x_{max}$. This approach preserves the structure geometry and at the same time includes a source of fluctuations.

Thermal fluctuations: node oscillation model

In this approach we made the nodes fluctuate around the given static position in an isotropic random direction. In this case each random state is described by the array of the absolute coordinates of nodes ***X***, ***Y***, ***Z***. For each i-th node:

$$X_i=X0_i+\delta x,\ Y_i=Y0_i+\delta y,\ Z_i=Z0_i+\delta z;$$
$$\delta x=r\sin\theta\cos\varphi;\ \delta y=r\sin\theta\sin\varphi,\ \delta z=\mathrm{r}\cos\theta;$$
$$r=r_m\cos\pi r_3;\ \varphi=2\pi r_2;\ \cos\theta=2r_1\text{-}1.$$

Here ***X0***, ***Y0***, ***Z0*** are matrixes describing the static state, r_1, r_2, r_3 are random [0..1] numbers. Then on the base of R_a and this new system of nodes the matrix of distances is built and the impedance determined. All other steps of the simulation were performed in the same way as for the link oscillation model. For the details see the flowchart in Fig. 2 b).

RESULTS OF AVERAGE QUANTITIES

In this section we report the results of the average quantities: fraction of links and modulus of the impedance as function of the maximum amplitude of the oscillations for the two models. For the LOM the number of links can be calculated analytically as following. The probability for the link to exist is

$$P_l=\frac{1}{\pi}\left(\arcsin c_2-\arcsin c_1\right), \tag{7}$$

where c_1=min(1,max($\frac{-2R_a-l_0}{x_{\max}}$,-1)), and c_2= max(-1,min($\frac{2R_a-l_0}{x_{\max}}$,1)). Accordingly, the estimated average number of links in the fluctuating structure is equal to the sum of P_l over all pairs of nodes. Fig. 3 reports the results for different values of the interacting radius. Theoretical values are in perfect agreement with Monte-Carlo simulations. We note that the number of links associated with a maximum amplitude systematically decreases at increasing the interacting radius.

For the NOM the number of links was calculated only by Monte-Carlo. Fig. 3 reports the results for different values of the interacting radius. One can see that in contrast to the LOM the number of links decreases with amplitude and increases with interacting radius.

The calculated average impedance and relative variance versus x_{max}/R_a for both models are shown in Fig. 4 . Here we detected a significant decrease of the impedance for the case of LOM in contrast to a significant increase for the case of NOM.

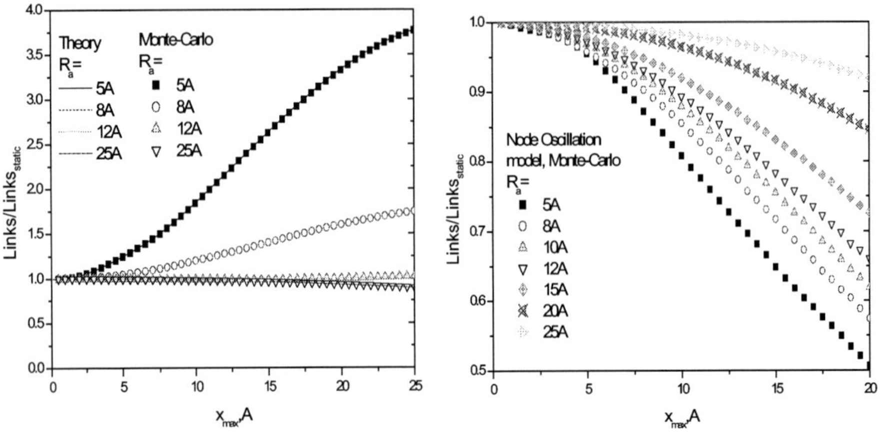

Link oscillation model. Node oscillation model.

FIGURE 3. Normalized average number of links versus the maximum amplitude for different R_a.

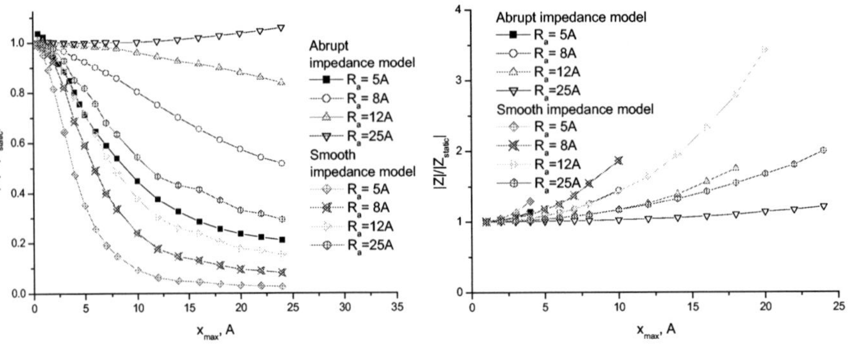

FIGURE 4. Impedance versus x_{max} for different R_a and elementary impedance models.

VARIANCE OF IMPEDANCE FLUCTUATIONS

In this section we report the results of the variance of impedance fluctuations as function of the maximum amplitude of the oscillations for the two model. The calculated relative variance versus x_{max}/R_a are shown in Fig. 5. We recall that x_{max} can be transformed into the temperature within a multiplication factor. Furthermore, it should be noted that in our algorithm iterations are independent from each other and hence these graphs do not contain any time correlation but only statistical fluctuations. For the LOM we found that the variation law for the smooth impedance model at

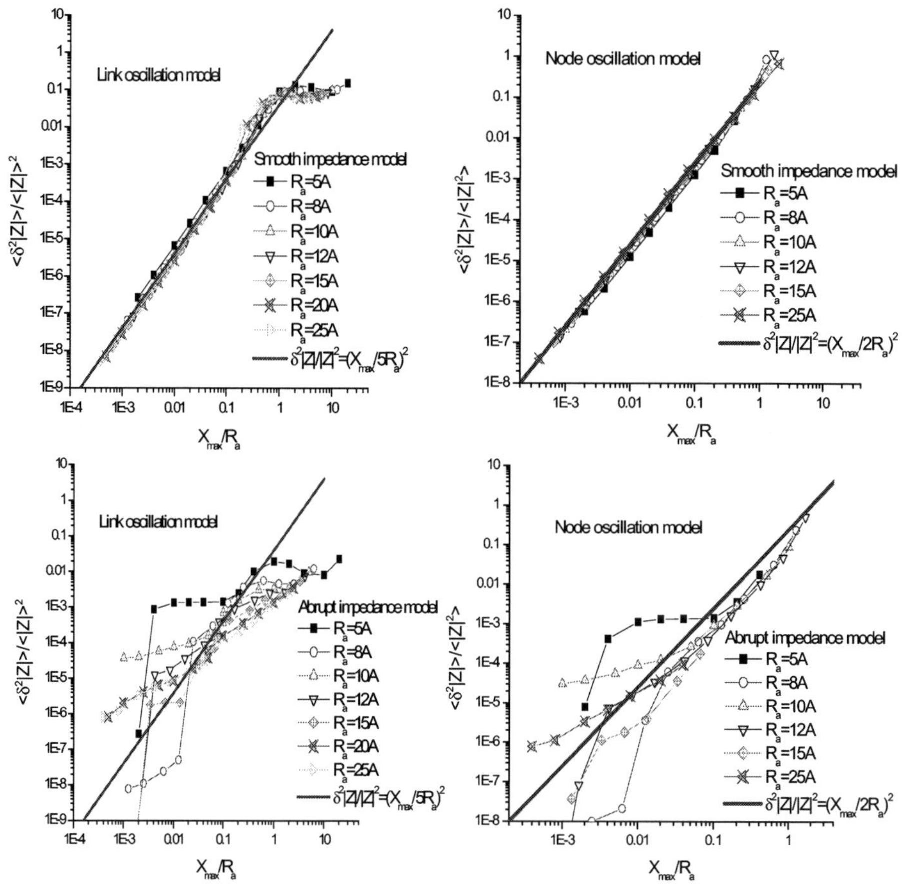

FIGURE 5. Impedance variation versus x_{max}/R_a.

small x_{max} fits reasonably well with the formula $\frac{\delta^2 |Z|}{|Z|^2} = \left(\frac{x_{max}}{5R_a}\right)^2$. By contrast the abrupt impedance model evidences significant different behaviors which are not predictable. One reason of this difference is the presence of a telegraph noise due to the existence of bottlenecks in the link distribution inside the structure which increases the variance at low x_{max} depending on the structure peculiarities. A second reason is the weaker sensitivity of the smooth impedance model to fluctuations, which dominate mainly at high x_{max}. We conclude that the abrupt impedance model provides results more sensitive to the structure peculiarities and thus it can be of interest to discriminate among different structures of the same type. The smooth impedance model shows more general and uniform properties presumably related to the whole class of these structures.

For the NOM we found that the variation law for the smooth impedance model fits reasonably well with the formula $\frac{\delta^2 |Z|}{|Z|^2} = \left(\frac{x_{max}}{2R_a}\right)^2$. It is the same law of the LOM but with a different numerical coefficient. This difference can be partly explained (within a factor of 2) by the non-equivalence of the x_{max} between the two models The remaining discrepancy is associated with a cumulative effect due to the influence of a node displacement on the displacements of all connected distances.

Again, contrast the abrupt impedance model evidences significant different behaviors which are a signature of the structure but not predictable.

CONCLUSIONS

We have developed fluctuation models of irregular impedance networks and applied them to the case of a protein (rhodopsin as prototype) sandwiched between Ohmic contacts. The results of simulations can be summarized as follows.

The average quantities show: (i) an opposite behavior when considering the link oscillation model or the node oscillation model; (ii) an analogous behavior when considering the smooth and abrupt impedance models. The microscopic reason for the opposite behavior is at present an unsolved problem.

The variance of impedance fluctuations shows: (i) analogous behaviors for the link oscillation model or the node oscillation model, (ii) a universal behavior for the smooth impedance model, (iii) peculiar behaviors for the abrupt impedance model that signify the structure of the network. The microscopic reason of the universal behavior is at present an unsolved problem.

ACKNOWLEDGMENTS

This work has been performed within the SPOT NOSED project IST-2001-38899 of EC and the project "Models and noise measurements in nanostructures" financed by Italian MIUR.

REFERENCES

1. SPOT-NOSED. *Single Protein Nanobiosensor Grid Array*. V Framework Programme of European Community, http://www.nanobiolab.pcb.ub.es/projectes/Spotnosed/, 2003-2005.
2. Pennetta C et al "Fluctuation of complex networks: Electrical properties of single nanodevice", *Spie Proc*, **5472**, p 172-184, (2004) eds. J.M. Smulko, Y. Blanter, M.I. Dikman, L.B. Kish.
3. Alfinito E. et al "Thermal Fluctuations of a GPCR: A two force constant model", this volume.
4. Palczewski K. et al "Crystal structure of rhodopsin: A G protein-coupled receptor". *Science*, 289:739–745, 2000.

Noise-Driven Switching Between Limit Cycles And Adaptability In A Small-Dimensional Excitable Network With Balanced Coupling

Leonid A. Safonov* and Yoshiharu Yamamoto*

*Educational Physiology Laboratory, Graduate School of Education, The University of Tokyo, 7-3-1 Hongo, Bunkyo-ku, Tokyo 113–0033, Japan

Abstract. We study the adaptability of coupled non-linear oscillators to external forcing by using a system of globally coupled FitzHugh-Nagumo equations. Each unit is either excitatory or inhibitory. If the numbers of units of both types are in a specific relation (balanced coupling), we observe the presence of multistable oscillatory states with different excitation or *firing* rates. In the presence of noise, there is noise-driven switching between these states. The selection between higher- and lower-frequency oscillations depends on the input, which results in increasing coherence between the periodic input and the system's output.

Keywords: stochastic resonance, chaotic itinerancy, balanced excitation/inhibition
PACS: 87.10.+e, 82.40.Bj, 05.45.-a

Biological oscillators are generally stable. This is the main reason why they are usually modeled by limit cycles [1, 2]. Limit cycles show *stability* in their "activity", when evaluated by both the amplitude and frequency, over a wide range of control parameters, including external force. For example, in neural models such as FitzHugh-Nagumo (FHN) [3, 4] and Hodgkin-Huxley [5] equations, the activity does not change much when they are forced by inputs above their "firing" threshold.

In contrast, networking limit cycles [1, 6] or general non-linear elements [7, 8, 9] are known to produce diverse spatio-temporal (synchronized) patterns with different frequencies and amplitudes, giving rise to variable activity levels over the whole network. Each pattern corresponds (in a broad sense) to a stable attractor, and the existence of multistable attractors with different activity levels accounts for *diversity* or *variability* in real-world oscillators, including biological ones.

We consider a system of coupled units whose dynamics is described by the FHN equations:

$$\begin{aligned}\varepsilon \dot{v}_i &= v_i(v_i-a)(1-v_i)-w_i+I+S(t)+\frac{1}{N}\sum_{j=1}^{N}k_{ij}(v_j-v_i)+\xi_i(t),\\ \dot{w}_i &= v_i-w_i-b,\end{aligned} \tag{1}$$

where $\varepsilon = 0.005$, $a = 0.5$, $b = 0.15$, I is a constant input, $S(t)$ is a time-dependent input, $i, j = 1,\ldots,N$, $\xi_i(t)$ is Gaussian white noise with $\langle \xi_i(t)\xi_j(s)\rangle = 2D\delta_{ij}\delta(t-s)$, where $\langle \ldots \rangle$ denotes the ensemble average, and $2D \cdot \Delta t = \langle \xi_i^2(t)\rangle = \sigma^2$ in the discrete case. $k_{ij} > 0$ if the jth unit is excitatory and $k_{ij} < 0$ if it is inhibitory.

CP800, *Unsolved Problems of Noise and Fluctuations: UPoN 2005*
edited by L. Reggiani, C. Pennetta, V. Akimov, E. Alfinito, and M. Rosini

For a single uncoupled unit, there is a stable stationary solution for small I. At $I_0 \approx 0.11$ a Hopf bifurcation produces a stable limit cycle, which corresponds to the excited (or *firing*) state of the unit.

The adaptability of excitation or *firing* rates in the suprathreshold regime to changing levels of I can best be understood by considering the system's dynamics for $N = 3$ with four possible choices of the number of excitatory units (N_E).

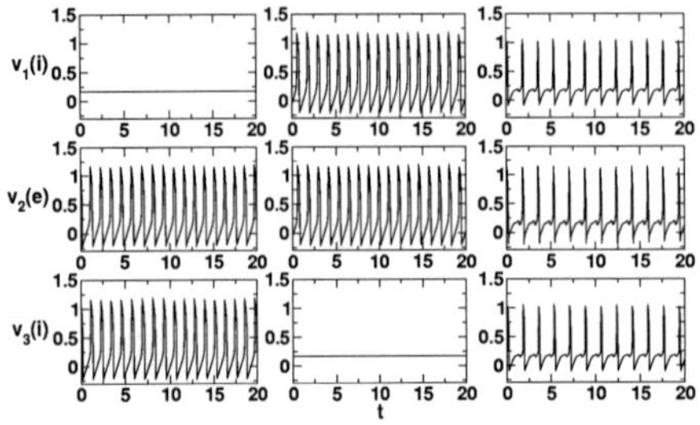

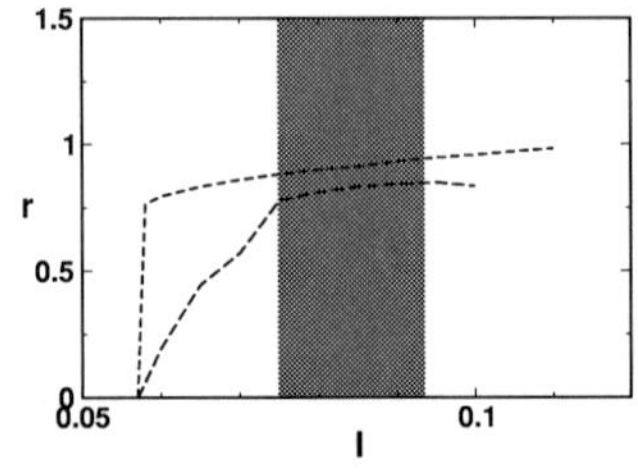

FIGURE 1. Firing regimes for attractors *1* (left), *2* (center) and *3* (right). $N = 3, N_E = 1, I = 0.07$. The second unit is excitatory (**e**), the others are inhibitory (**i**).

FIGURE 2. Dependence of the firing rate on the input I for $N = 3$ and $N_E = 1$. Upper line - high-frequency regimes (cycles *1* and *2*). Lower line - low-frequency regime (attractor *3*).

First, we study the case of three units, one of which is excitatory and the other two are inhibitory, without noise. The coupling strength of each connection is taken to be equal to $|k| = 0.3$ with a corresponding sign.

At $I_1 \approx 0.057$ there is a double resonant Hopf bifurcation, at which two stable limit cycles emerge, which have two firing units and one silent. The plots of v-variables of these cycles are shown in Fig. 1 (left and central columns). Along with the two cycles, there is a third stable attracting set (with all three units firing) emerging at or near the double Hopf bifurcation. With a growing parameter I, it has the form of a limit cycle for most of the parameter values. In the following, we refer to these three attractors as attractors *1*, *2* and *3* respectively. Attractors *1*, *2* are also referred to as limit cycles *1*, *2*, because they do not take any other form.

With increasing I, we observe that the firing rate of cycles *1* and *2* grows rapidly over a short parameter interval, after which it changes only insignificantly (Fig. 2). In contrast, the firing rate of attractor *3* grows slowly with I for a long interval, after which it stabilizes. At $I_0 \approx 0.11$ (as in the single unit case) there is an ordinary Hopf bifurcation, where the stationary solution becomes globally unstable and a new limit cycle emerges. In the present study we consider values of I between I_1 and I_0, where the firing rates of all the attractors are relatively stable (shown by a shaded area in Fig. 2). In this regime, no stable attractor solely has a sufficient rate coding ability for the external input I; see flat curves in Fig. 2.

In the presence of weak noise, the system performs random transitions between domains of attraction of the three attractors (Fig. 3). This is similar to a phenomenon known as chaotic itinerancy (CI) [7, 8, 9], although in our system we have limit cycles, instead of chaotic attractors, and the role of noise is to induce itinerant dynamics. It has been suggested in [7, 8, 9] that hopping between attractors in CI is not random but

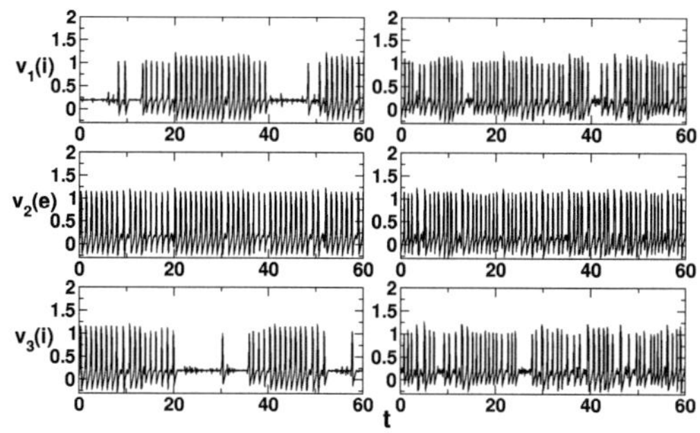

FIGURE 3. Itinerant oscillations in the presence of noise. $N = 3$, $N_E = 1$, $I = 0.09$, $\sigma^2 = 10^{-7}$ (left), 10^{-5} (right). With strong noise random firings dominate over limit cycle oscillations (right).

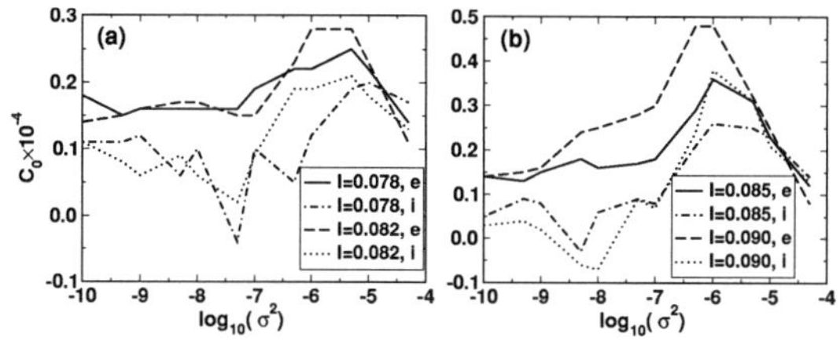

FIGURE 4. Stochastic resonance in a system with $N = 3$, $N_E = 1$ for excitatory and inhibitory units. $S = 0.003\sin(2\pi t/30)$. Results are averaged over 200 trials. Total time for each trial $T = 600$.

adheres to certain constraints. Also in our case, transitions between cycles *1* and *2* do not occur, and the only transitions possible are *1* ↔ *3* ↔ *2*.

The system also demonstrates the phenomenon of SR in rate coding. We apply periodic input $S(t)$ to all units and plot the covariance measure $C_0 = \overline{S(t)R(t)}$, (where $R(t)$ is the instantaneous firing rate (IFR) [1] and overbar denotes the time-average) against the noise intensity. Fig. 4 shows the dependence of C_0 on the noise intensity for different baseline input and amplitude of periodic input. We can observe the bell-shape curve, especially for the IFR of excitatory units, characteristic of SR. This phenomenon resembles suprathreshold SR studied previously [10, 11], but has a different dynamical nature, as explained below.

Figs. 5(a), (b) and 5(c), (d) show sample spike trains of an excitatory and an inhibitory unit for weak and moderate noise intensities respectively. Although for most of the time the system stays near attractors *1* or *2*, occasionally it jumps to attractor *3*, and the probability of such a transition is higher for lower input than for higher input.

Limit cycle oscillations on all attractors consist of spike (action potential) time and silent (resting potential) time. With the increase of firing rate, the spike time remains

[1] Hanning window filtered IFR is defined as follows. If t_i are firing times, then the spike train is $s(t) = \sum \delta(t - t_i)$, and the IFR is $r(t) = \int_{-T/2}^{T/2} \cos[4\pi(t-\tau)/T]s(\tau)d\tau$, where T is the window width, T=10.

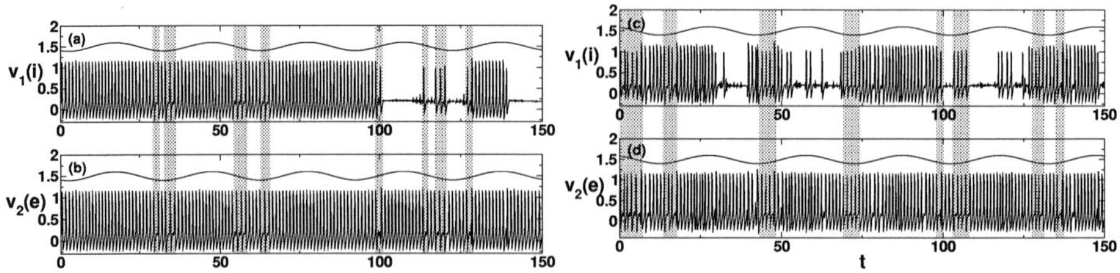

FIGURE 5. Time series of excitatory and inhibitory units with a periodic input. $\sigma^2 = 10^{-8}$: (a) inhibitory unit, (b) excitatory unit; $\sigma^2 = 5 \cdot 10^{-7}$: (c) inhibitory unit, (d) excitatory unit. $N = 3$, $N_E = 1$, $I = 0.09$, $S(t) = 0.003\sin(2\pi t/30)$. Shaded areas denote visits to attractor *3*. Sinusoid lines schematically represent the input.

almost the same, while the resting time decreases. Transitions between the attractors are most likely during the resting time when the distance to other attractors is minimal. Since this time is longer for lower input, a switch to another attractor is more likely to happen with a lower input. The average duration of a visit to attractor *3* is small compared to that of other attractors, therefore a transition to attractor *3* near the minimal input is followed by a switch to another attractor while the input is still close to the minimum.

With increasing noise the transition probability becomes higher, and the time spent by the orbit near attractor *3* is greater. Therefore the resulting IFR of an excitatory unit is more coherent with the input for moderate noise [Fig. 5(d)] than for weak noise [Fig. 5(b)]. Thus, an increase of noise from weak to moderate values leads to an increase of the ability of excitatory units to encode the incoming signal (Fig. 4). When the noise becomes too strong, the coding ability deteriorates, giving the bell-shape curve of C_0, characteristic of SR.

Unlike excitatory units, inhibitory ones have three possible firing regimes — a high-frequency and a silent one at attractors *1* and *2*, and a low-frequency one at attractor *3*. Therefore, the IFR of an excitatory unit coincides with that of an inhibitory unit only for part of the time, otherwise it is zero. This results in a worse coding ability compared to that of an excitatory unit, which is shown by the lower SR-curve in Fig. 4.

For $N_E = 0$ we put $|k| = 0.1$. With this condition fulfilled, there is a double Hopf bifurcation at the same value of I as for $N_E = 1$ and $|k| = 0.3$. There are three limit cycles emerging at this bifurcation, all of which are 2:1 synchronized, and there are no 3:0 synchronized attractors. Moreover, they are all globally stable and no noise-driven switching is observed. For $N_E = 2$ and $N_E = 3$ there are no bifurcations for $I < I_0$, hence no stable attracting set other than the stationary state exists in that interval.

Therefore the case of $N_E = 1$ has special significance. For $N_E = 1$ we observe CI-like dynamics with small noise and improved coding ability with increased levels of noise, while for lower and higher N_E the system's dynamics is much simpler and does not qualitatively depend on noise. Our simulations show that this is the case for $N > 3$ as well. In order for the switching to be observed, it must hold that

$$N_E = N_E^0 = \lceil N/2 \rceil - 1. \tag{2}$$

Relation (2) gives the definition of balanced coupling, for which the system has specific properties described above. For $N_E > N_E^0$ the firing threshold of the entire system is the same as that of a single uncoupled unit (no Hopf bifurcations take place for $I \leq I_0$), while for $N_E < N_E^0$ the new threshold $I_1 < I_0$ is present and multistability for $I_1 < I < I_0$ is observed, but there are no transitions between the cycles. It has been observed that the balancing of excitation and inhibition can bring about complex dynamics in neuronal models [12, 13], which is related to variability of cortical neuronal dynamics.

Simulations show that for any $N_E \leq N_E^0$ it is necessary to have

$$|k| \approx 0.1 \cdot N/(N - 2N_E) \tag{3}$$

in order to keep the first Hopf bifurcation (I_1)at the same parameter value (or to keep the system in the same firing range) as in the case described above for $N = 3$, $N_E = 1$ and $|k| = 0.3$ (with $I_1 \approx 0.057$). Stronger coupling leads to a greater decrease in the firing threshold.

These rules also hold for the trivial case of $N = 2$, where there are no bifurcations for $N_E = 1$ and 2, while for $N_E = 0$ there is an ordinary bifurcation at I_1 and there is no multistability.

Cases with higher odd N can be reduced to the case of $N = 3$ provided balancing conditions (2) and (3) hold. For all N there is a multiple Hopf bifurcation at I_1 with $N - 1$ pairs of equal eigenvalues crossing the imaginary axis, producing a number of different limit cycles. However, transitions take place only between N_E of N:0 synchronized attractors (analog of attractor *3*) and $N - N_E$ of $(N-1)$:1 synchronized attractors with one silent unit (analogs of attractors *1* and *2*). The dynamics of an excitatory unit is not different from that for $N = 3$ (it also iterates between high- and low-frequency oscillations). For an inhibitory unit the situation is different. Since a shorter time is spent in the silent regime with growing N (because only one of the units is silent on $(N-1)$:1 synchronized attractors), the dynamics of an inhibitory unit becomes more similar to that of an excitatory unit. Figs. 6(a) and (b) respectively show the SR-diagrams for $N = 15$ and $N = 31$ with the same input.

For even N, the dynamics is more complicated, but in the regime where stable high- and low-frequency attractors coexist, the SR-like phenomenon in the rate coding is also observed (shown in Figs. 6(c) and (d) for $N = 4$). This holds in general for higher even-dimensional cases.

It should be noted that for higher dimensions, the findings described (multistability, CI-like dynamics and adaptability to the input) are observed only if the balancing condition (2) is satisfied. Changing the number of excitatory units by just one results in a dramatic change in the dynamics, i.e. for higher N the case of $N_E = N_E^0 - 1$ is equivalent to the case of $N_E = 0$ for $N = 3$, and the case of $N_E = N_E^0 + 1$ is equivalent to the case of $N_E = 2$ for $N = 3$. This is related to the mode of coupling in Eq. (1). For each equation the coupling term is equivalent to the same sum of excitatory and inhibitory influences regardless of the system size, and the removal or addition of a single unit results in the same changes for any N. Whether different equations other than FHN with dynamical coupling would lead to a "looser" balancing condition requires further research.

Collections of non-linear oscillators encountered in biological organisms are frequently quite adaptable to changes in the external environment. The brain, for instance, is a highly adaptable organ in spite of the least *short time* plasticity embedded in sin-

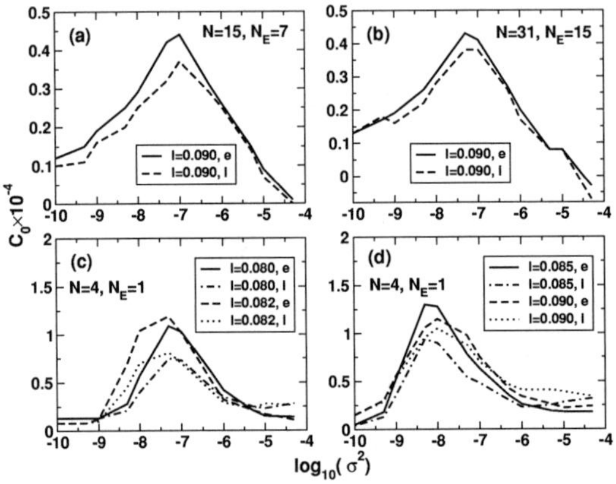

FIGURE 6. Stochastic resonance in a balanced system with (a) $N = 15$, (b) $N = 31$ and (c, d) $N = 4$ ($N_E = 1$) for excitatory and inhibitory units. $S(t) = 0.003\sin(2\pi t/30)$. Results are averaged over 200 trials. Total time for each trial $T = 600$.

gle oscillators and/or their connection strengths. Indeed, recent studies have shown that brain-wide, frequency-specific neural synchrony and asynchrony, or the dynamical attractor selection among multiple neural oscillators, seem to be involved in the generation of perceptual, cognitive and behavioral processes in response to external and/or internal stimuli [14]. We believe that ours can be a simple model of this sort of emergent adaptability by using small dimensional non-linear oscillators.

We would thank D. Nozaki, T. Nomura and A. Takamatsu for discussion. This study was supported by the Japan Society for Promotion of Science (JSPS).

REFERENCES

1. A. Winfree, *The Geometry of Biological Time*, Springer, Berlin, 1990.
2. L. Glass, *Nature*, **410**, 277–283 (2001).
3. R. FitzHugh, *Byophys. J.*, **1**, 445–466 (1961).
4. J. Nagumo, S. Arimoto, and S. Yoshizawa, *Proc. IRE*, **50**, 2061–2070 (1960).
5. A. L. Hodgkin, and A. F. Huxley, *J. Physiol.*, **117**, 500–544 (1952).
6. Y. Kuramoto, *Chemical Oscillations, Waves, and Turbulence*, Springer, Berlin, 1984.
7. K. Ikeda, K. Matsumoto, and K. Otsuka, *Prog. Theor. Phys. Suppl.*, **99**, 295–324 (1989).
8. K. Kaneko, *Physica D*, **41**, 137–172 (1990).
9. K. Kaneko, and I. Tsuda, *Chaos*, **13**, 926–936 (2003).
10. N. G. Stocks, *Phys. Rev. E*, **63**, 041114 (2001).
11. D. Rousseau, D. F., and F. Chapeau-Blondeau, *Phys. Rev. E*, **68**, 031107 (2003).
12. C. van Vreeswijk, and H. Sompolinsky, *Science*, **274**, 1724–1726 (1996).
13. C. van Vreeswijk, and H. Sompolinsky, *Neural Computation*, **10**, 1321–1371 (1998).
14. F. Varela, J.-P. Lachaux, E. Rodriguez, and J. Martinerie, *Nat. Rev. Neurosci.*, **2**, 229–239 (2001).

NOISE IN DEVICES

TeraHertz Emission and Noise Spectra in HEMTs

[1]J. Mateos, [1]S. Pérez, [1]D. Pardo, [1]T. González, J. Łusakowski[2,3], N. Dyakonova[2], W. Knap[2], S. Bollaert[4], Y. Roelens[4], A. Cappy[4], J.F. Millithaler[5], L. Varani[5]

[1]*Universidad de Salamanca, Dpt. de Física Aplicada, Plaza de la Merced s/n, 37008 Salamanca, Spain*
[2]*GES CNRS-Université Montpellier II UMR 5650, 34095 Montpellier, France*
[3]*Institute of Experimental Physics, Warsaw University, Hoża 69, 00-681 Warsaw, Poland*
[4]*IEMN-DHS UMR CNRS 8520, Avenue Poincaré, 59652 Villeneuve d'Ascq France*
[5]*CEM2 CNRS-Université Montpellier II UMR 5507, 34095 Montpellier, France*

Abstract. In this work we investigate, by means of Monte Carlo simulations, the physics of the recently observed emission of THz radiation at low temperature (4.2 K) from InGaAs/InAlAs lattice-matched high electron mobility transistors, HEMTs [1]. The spectrum of the emitted signal consists of two maxima; the one at lower frequency (around 1 THz) is sensitive to U_{DS} and U_{GS}, while the higher frequency peak (around 5 THz) is not. The gate current noise spectra obtained with room-temperature Monte Carlo simulations of similar HEMTs also show those features, although some quantitative discrepancies arise from the different temperature conditions. The signature of plasma oscillations is known to appear in the current noise spectrum [2], producing peaks appearing at very high frequencies. On the other hand, peaks at frequencies around 1 THz can be associated with the formation of Gunn domains within the transistors. Therefore, the analysis of the noise spectra of these devices can provide useful information that can help to understand the processes underlying the THz emission.

Keywords: Terahertz emission, plasma oscillations, Gunn Effect, HEMTs.
PACS: 85.30.Tv, 72.30.+q, 72.70.+m

INTRODUCTION

The part of the electromagnetic spectrum between millimeter waves and far infrared is referred to as terahertz radiation or T-rays [3]. Much of the recent interest in terahertz radiation stems from its ability to penetrate deep into many organic materials without the damage associated with ionizing radiation such as X-rays. Also, because terahertz radiation is readily absorbed by water, it can be used to distinguish between materials with varying water content. These properties can lead to very appealing applications in biomedical imaging [4].

THz radiation can penetrate poor weather, dust and smoke far better than infrared or visible systems. That is why defence and security applications are also envisaged. Imaging in this band offers the opportunity to be able to navigate and perform surveillance in these conditions of poor visibility. Furthermore the ability to penetrate

CP800, *Unsolved Problems of Noise and Fluctuations: UPoN 2005*
edited by L. Reggiani, C. Pennetta, V. Akimov, E. Alfinito, and M. Rosini

dielectrics such as windows, paper, clothing and in certain instances even walls, has opened up the opportunity to reveal concealed explosives; metallic and non-metallic weapons (such as ceramic, plastic or composite guns and knives), chemical weapons and other threats hidden in packages or on personnel. THz detection is also a very interesting field for astronomers; much of the photons generated in the Big Bang fall into the submillimeter and far-infrared region of the spectrum. Non-destructive testing of integrated circuits is also a future area of application [5].

However, the THz range is difficult to be reached both from electronic and optical approaches. The most interesting, from the application point of view, seems to be THz sources based on solid state devices that offer better possibilities of integration with other optoelectronic devices within a single chip. At present, there exist a few methods of generating THz radiation from solids: some of them are based on plasma excitations by femtosecond laser pulses in low dimensional [6] and bulk [7] semiconductors or superconductors [8], while other ones rely on mixing of laser beams with close frequencies [6]. These methods require advanced experimental equipments and their performance depends on a strong optical excitation of an appropriate semiconductor structure. It is worth noticing that all these sources are generating the THz radiation through plasma excitations in semiconductors. On the other hand, THz radiation can also be generated by semiconductor p-Ge [10] and quantum cascade [11] lasers.

An alternative mechanism of emission of THz radiation, proposed about ten years ago by Dyakonov and Shur [12], is plasma excitation in field effect transistors (FETs). The theory is based on an exact formal analogy between the equations of electron transport in a transistor channel and those of shallow water, or acoustic waves in music instruments. Instabilities in the electron flow will create waves, whose frequency depends on the concentration of the electron fluid and the length of the transistor gate. If the boundary conditions are adequate, i. e. the electron concentration is sufficiently high (at least of the order of 10^{12} cm^{-2}) and the gate length is sufficiently short (in the nanometer range) those plasma waves can reach the THz domain with sufficient amplitude to generate emission of radiation detectable from outside the device. The idea put forward by Dyakonov and Shur found recently an experimental confirmation in measurements performed in GaAs and InGaAs HEMTs [12-15]. Moreover, the resonant THz detection by Si MOSFETs at room temperature [16] represents a result particularly important in view of a possible integration of THz transistors in silicon-based circuits and systems.

The aim of this paper is to give a more complete description and analysis of the observed THz emission in HEMTs. In particular, we compare the experimental data with high-frequency noise spectra obtained by Monte Carlo simulations. A qualitative agreement of measured and simulated spectra allows a deeper interpretation and understanding of the observed emission, which we relate not only to plasma oscillations but also to the formation of Gunn domains. The paper is organized as follows. The main characteristics of the Monte Carlo simulator are provided in Section III. Section II describes the experimental system and the samples used; the experimental results and the simulations are discussed in Section IV and the main conclusions are given in Section V.

MONTE CARLO SIMULATION

For transistors with ultra short dimensions, macroscopic simulators as well as analytical models (as the Dyakonov–Shur theory) are not appropriate to describe the details of electron transport. That is why, to have a better physical insight into the microscopic carrier dynamics in the transistor investigated, we performed Monte Carlo simulations, which are the most adequate tool to describe the transport in ultra short devices. As a matter of fact, due to the nanometric gate length, very high electric fields appear locally and lead to the presence of hot carrier phenomena and ballistic transport. It is well known that such phenomena can be accurately modelled in electronic devices using an Ensemble Monte Carlo model self-consistently coupled to a Poisson solver [17]. For the special case of heterostructure devices, the physical scenario becomes particularly complex due to the appearance of quantum effects associated with energy quantization in the transversal dimension within the channel, degeneracy due to the high carrier concentration and tunnelling from the channel into the gate due to the very thin barrier layer. In our simulations degeneracy effects are locally taken into account using the rejection technique [18], while other quantum effects have been neglected in order to have reasonable CPU times, since a dynamic self-consistent solution of Poisson and Schrödinger equations represents a formidable task even for up-to-date computers. However, being the plasma oscillations a collective motion arising from Coulomb interaction among carriers, the main features of this phenomenon are taken into account correctly in this semi-classical Monte Carlo model. The microscopic simulator has been already validated through comparison with experimental results performed on similar devices and concerning static characteristics, small-signal parameters and noise performances [19, 20].

In order to detect and emphasize the presence of plasma oscillations, a special attention is devoted to the calculation of noise spectra due to their extreme sensitivity to microscopic features of carrier dynamics [21-23] and the possibility to perform a frequency analysis of electrical fluctuations [24]. For instance, the appearance of a peak in the frequency noise spectrum is a signature that the system is “ready” to develop macroscopic oscillations at that frequency if appropriate boundary conditions are fulfilled. Indeed, some features of the noise spectra presented further on are caused by macroscopic plasma oscillations in the transistor channel. To exclude from the noise spectra the contribution of numerical noise associated with the intrinsic stochastic nature of the Monte Carlo method an accurate algorithm based of the Ramo-Shockley theorem [25] and its specific formulation for the case of transistors [26] has been employed for the computation of the instantaneous currents. Such an algorithm has already provided good results for the case of MESFETs [27] and HEMTs [20].

Another interesting mechanism known to be able to produce oscillations in III-V semiconductor devices is the Gunn Effect. Under the effect of a strong electric field the electron effective mass increases as a consequence of intervalley transfer (i.e. negative differential resistance mechanism), thus originating the formation of a Gunn domain, that under certain conditions can produce current oscillations. However, up to now, the only way for Gunn-diode based circuits to reach the THz range is by means of frequency multipliers [28]. However, as we will show in this paper, Gunn oscillations in nanometer-scale devices can reach the THz range. And, which is more important, the noise spectra can be used for predicting these oscillations even when they are not clearly visible in the current (for example for biasings lower than the threshold for the Gunn Effect onset). Our Monte Carlo model, accounting for the processes at the origin of Gunn Effect and providing information in the time domain, will be a valuable tool for the analysis of the possibility to use this effect for producing a tuneable THz source.

In this context, it is important to emphasize that the capability of noise spectra to give precise indications on the onset of collective phenomena due to interparticle correlations has already been pointed out [24]. As a consequence, the calculation of gate and drain current noise spectra of the HEMTs investigated in this work will be used to complement the experiments and confirm the appearance of a component in the carrier dynamics associated with plasma or Gunn oscillations.

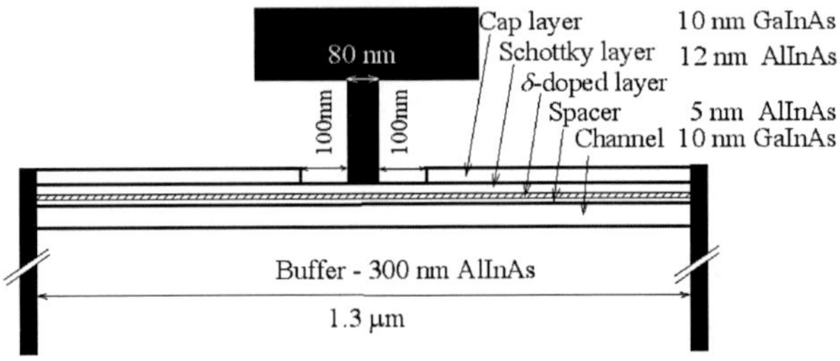

FIGURE 1. Schematic structure of the transistor investigated.

EXPERIMENTAL SYSTEM

A lattice-matched InGaAs/InAlAs High Electron Mobility Transistor (HEMT) on InP substrate grown by Molecular Beam Epitaxy is used in this study. The gate length is around $L_G = 80$ nm, the source-drain space is $L_A = 1.3$ μm, and the width of the gate is $W = 50$ μm. The investigated transistor has a recess length of 100 nm at both sides of the gate. The measurements were carried out at 4.2 K with the transistor placed in a liquid helium cryostat. Two Keithley 2400 source–meters were used for the application of the drain-to-source (U_{DS}) and gate-to-source (U_{GS}) voltages and measurement of the drain and gate currents. The transistor was placed in a cyclotron emission spectrometer [29] designed to perform a spectral analysis of a weak THz radiation. The spectrometer consisted of two superconducting coils located one above the other and independently power supplied. The photoresponse signal of the detector originates from the absorption of photons with the energy $E=hf$ equal to the cyclotron resonance energy $E=heB/2\pi m$ in the InSb bulk layer (f is the photon frequency and other symbols have their usual meaning). The absorption process causes a change of the detector resistance that is monitored by a standard lock–in technique. The THz source and the detector were placed into both ends of a 22 cm long copper waveguide placed in such a way that each device was in the centre of an appropriate coil. Both the source and the detector of the THz radiation were cooled by a helium exchange gas to 4.2 K and totally isolated from the 300 K background radiation. To satisfy the boundary conditions required for the plasma emission, the gate was short circuited with the source by the shortest possible gold wire directly on the transistor chip (that decreased the gate-source impedance) and the transistor was driven into saturation by a sufficiently large U_{DS} (that increased the gate-drain impedance).

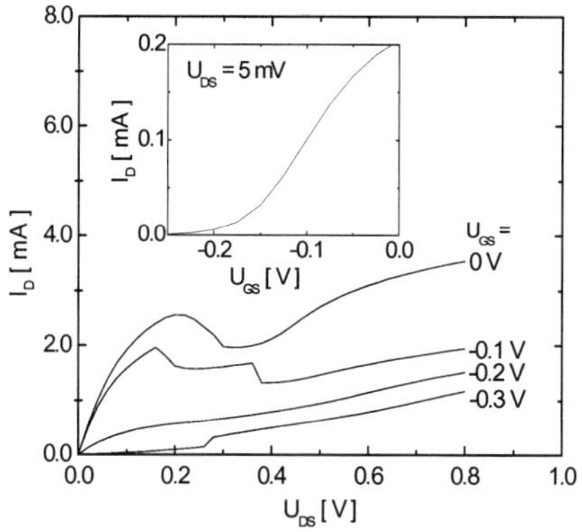

FIGURE 2. Output characteristics at $T = 4.2$ K, $B = 0$ T of the transistor investigated at the indicated U_{GS}. The inset shows the transfer characteristic at $U_{DS} = 5$ mV.

RESULTS

Figure 2 shows the output and transfer (in the inset) characteristics of the transistor before the gate-source short circuit was realized. The *I-V* curves exhibit some strange features when U_{DS} exceeds a certain threshold value U_{TH} (that increases for higher U_{GS}).

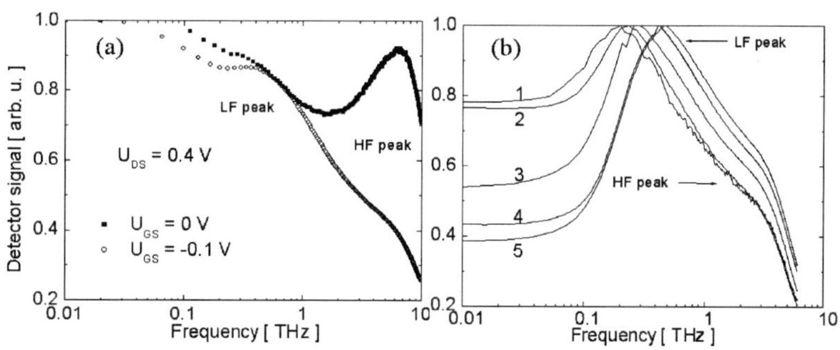

FIGURE 3. THz emission spectra of the HEMT for different gate bias at U_{DS}=0.4 V. Solid symbols are for U_{GS}=0.0 V and open ones for U_{GS} =-0.1 V. The spectra are normalized at f = 0.01 THz. (b) shows the results for the HEMT with the gate short circuited with the source by a gold wire on the chip. The spectra from 1 to 5 correspond to U_{DS} equal to 0.3 V, 0.45 V, 0.6 V, 0.7 V and 0.8 V, respectively. The spectra are normalized to the maximum of the LF peak.

Interestingly, when measuring the THz emission from these devices, it also appears in a threshold–like manner, i.e. when U_{DS} is larger than a given value, which just coincides with the appearance of the kinks in the *I-V* curves. Figs. 3(a) and (b) show the HEMT emission spectra (i) for different U_{GS} and (ii) for different U_{DS} when the gate was short circuited with the source, respectively. The obtained spectrum exhibits two peaks, the low frequency (LF) one being more evident with the gate short circuited to the source, indicating the importance of the transistor's boundary conditions for the emission spectra. The position of the low frequency (LF) peak moves to higher frequencies with the increase of U_{DS}, while the position of the high frequency peak (HF) does not change.

Fig. 4 shows the gate current noise spectra, S_{ig}, for several values of U_{DS} and U_{GS}=-0.05 V obtained with Monte Carlo simulations of similar HEMTs (those studied in [30]). There are two maxima on each spectrum: the low frequency (LF) one around 2 THz and the high frequency (HF) one at 5 THz. The position of the HF maximum does not depend on the drain voltage while the LF maximum moves to higher frequencies when the drain voltage increases. This is exactly what is observed in the experimental spectra shown in Fig. 3. On the other hand, there are also some discrepancies. The experimentally observed frequency of the LF peak is smaller than that given by simulations, while the position of the HF peaks coincides.

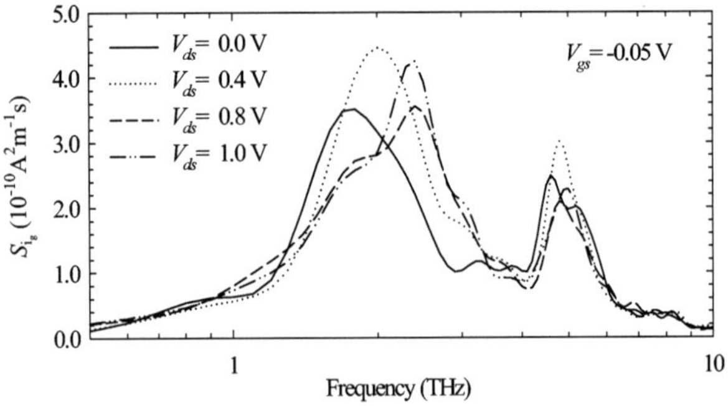

FIGURE 4. Gate current noise spectra for the indicated U_{DS} voltages at a constant U_{GS} =-0.05 V.

Looking at the *I-V* curves shown in Fig. 2, one may think that the detected emission could take place due to Gunn oscillations (their onset leads to a decrease of the drain current, effect that has already been experimentally detected in HEMTs [31]). However, Gunn oscillations at such high frequencies have never been observed and some more investigation is needed to confirm this hypothesis. In order to explain the origin of the emission we have performed Monte Carlo simulations of HEMTs with different geometries and we have observed that, under certain conditions, Gunn oscillations take place when U_{DS} is higher than a given threshold value. The drain and gate current noise spectra of a HEMT where Gunn oscillations appear are shown in Fig. 5. The Gunn oscillations appear for frequencies around 1.5 THz, which confirms that this mechanism can be responsible for the LF peaks observed in the measurements (Fig. 3). It is also interesting to note that the noise spectra for U_{DS} below the Gunn onset show a peak at frequencies around 1THz, even if the amplitude of the oscillations is not detectable yet [the onset provokes a huge increase of the peak, note the different range of axes in (a) and (c)].

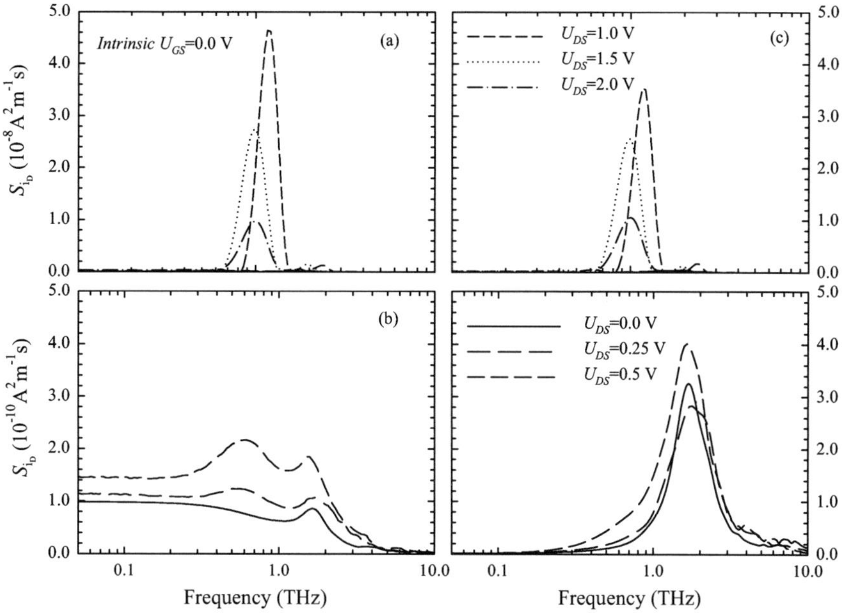

FIGURE 5. Monte Carlo results for the noise spectra of (a) and (b) drain and (c) and (d) gate currents of a HEMT biased with U_{DS} voltages (a) and (c) above, and (b) and (d) below the threshold for the onset of Gunn oscillations (intrinsic U_{GS} =0.0 V).

In contrast, the HF peak of the measured spectrum can only be interpreted as a consequence of plasma oscillations in the device, which are not much dependent on the biasing. It is so because the plasma frequency is associated with the concentration of electrons in the whole channel, which is nearly constant.

CONCLUSIONS

Terahertz emission in InGaAs/AlInAs HEMTs was observed. In order to be detected, the transistor must be biased with a sufficiently high U_{DS}, and its spectrum consists of two peaks with frequency around 6 THz and around 1 THz. Monte Carlo simulations showed also the existence of these two peaks in the noise spectra, which are interpreted as resulting from Gunn effect (whose onset also produce kinks in the *I-V* curves) and plasma oscillations. However, this agreement is unexpectedly good due to several reasons:

1. Different temperature conditions: 4.2 K for the emission measurements and 300 K for the noise simulation
2. Semiclassic nature of the Monte Carlo model: at 4.2 K quantum effects in the channel of the HEMT can be very important but they are not considered in the simulation
3. Different geometry of the simulated and measured devices.
4. Different nature of the processes: the current noise related to electronic transport and the emission dealing with electromagnetic radiation.

Therefore, Monte Carlo simulations can provide valuable information about the optimization of the HEMTs for having tuneable THz emission with high enough power, but many questions are still waiting for their answer.

AKNOWLEDGEMENTS: This work has been partially supported by the Dirección General de Investigación (Ministerio de Educación y Ciencia) and FEDER through the project TEC2004-05231.

REFERENCES

1. W. Knap, J. Łusakowski, T. Parenty, S. Bollaert, A. Cappy, V. V. Popov and M.S. Shur, *Appl. Phys. Lett.* **84**, 2331 (2004)
2. L. Varani, L. Reggiani, T. Kuhn, T. González and D. Pardo, *IEEE Trans. Electr. Dev.* **41**, 1916 (1994)
3. *Sensing with THz Radiation* edited by D. Mittleman, Springer, Berlin, 2003
4. M. C. Beard, G. M. Turner and C. A. Schmuttenmaer, *Phys. Med. Biol.* **47**, 3841 (2002)
5. See for example the webpage of Teraview, www.teraview.co.uk, a commercial company that exploits several application of Terahertz light.
6. N.Sekine, K. Hirakawa, M. Vosseburger, P. H. Bolivar, H. Kurz, *Phys. Rev. B* **64,** 201323/1-4 (2001); R. Bratschitsch, R. Kersting, T. Muller, G. Strasser, K. Unterrainer, W. Fischler, R. A. Hopfel, *Physica B* **272,** 375 (1999)
7. J. N. Heyman, P. Neocleous, D. Hebert, P. A. Crowell, T. Muller, K. Unterrainer, *Phys. Rev. B* **64,** 085202/1-7 (2001); D. Hashimshony, A. Zigler, K. Papadopoulos, *Phys. Rev. Lett.* **86,** 2806 (2001); T. Dekorsy, H. Auer, H. J. Bakker, H. G. Roskos, H. Kurz, *Phys. Rev. B* **53**, 4005 (1996)
8. I. Iguchi, *Superconductor Science Technology*. **13**, 93 (2000)
9. S.Hoffmann, M. Hofmann, E. Brundermann, M. Havenith, M. Matus, J. V. Moloney, A. S. Moskalenko, M. Kira, S. W. Koch, S. Saito, K. Sakai, *Appl. Phys. Lett.* **84,** 3585 (2004)
10. A. A. Andronov, V. A. Kozlov, S. A. Pavlov, S. G. Pavlov, *Soviet Technical Physics Letters* **14**, 891 (1988)
11. R. Kohler, A. Tredicucci, C. Mauro, F. Beltram, H. E. Beere, E. H. Linfield, A. G. Davies, D. A. Ritchie, *Appl. Phys. Lett.* **84**, 1266 (2004)
12. M. Dyakonov and M. S. Shur, Phys. Rev. Lett. **71**, 2465 (1993); M. Dyakonov and M. S. Shur in "Terahertz Sources and Systems", Kluwer Academic Publishers-Netherlands ed. by R. E. Miles, p. 187 (2001)
13. W. Knap, Y. Deng, S. Rumyantsev, J.–Q. Lü, M. S. Shur , C. A. Salor and L. C. Brunel, *Appl. Phys. Lett* **80**, 3433 (2002); W. Knap, Y. Deng, S. Rumyantsev and M. S. Shur, *Appl. Phys. Lett.* **81**, 4637 (2002)
14. X. G. Peralta, S. J. Allen, M. C. Wanke, N. E. Harff, J. A. Simmons, M. P. Lilly, J. L. Reno, P. J. Burke and J. P. Eisenstein, *Appl. Phys. Lett.* **81**, 1627 (2002)
15. J. Lusakowski, W.Knap, N. Dyakonova, L. Varani, J. Mateos, T. Gonzalez, T. Parenty, S. Bollaert, A. Cappy and K. Karpierz, *Journal of Applied Physics* **97**, 064307 [1-7] (2005)
16. W. Knap, F. Teppe, Y. M. Meziani, N. Dyakonova, J. Łusakowski, F. Boeuf, T. Skotnicki, D. Maude, S. Rumyantsev and M.S. Shur, Appl. Phys. Lett. **85**, 675 (2004)
17. C. Jacoboni and P. Lugli, *The Monte Carlo Method for Semiconductor Device Simulation*, (Springer, Vienna, 1989)
18. P. Lugli and D.K. Ferry, *IEEE Trans. Electron. Dev.* **ED-32**, 2431 (1985)
19. J. Mateos, T. González, D. Pardo, V. Hoel and A. Cappy, *Semicond. Sci. Technol.* **14**, 864 (1999)
20. J. Mateos, T. González, D. Pardo, V. Hoel and A. Cappy, *IEEE Trans. Electron. Dev.* **47**, 1950 (2000)
21. Sh. Kogan, *Electronic Noise and Fluctuations in Solids*, Cambridge University Press, Cambridge, 1996.
22. *Noise and Fluctuations Control in Electronic Devices*, edited by A. Balandin, American Scientific Publishers, Stevenson Ranch, CA, USA, 2002.
23. *Proceedings of the 17th Int . Conf. on Noise and Fluctuations Physical Systems*, edited by J. Sikula, CNRL s.r.o., Brno, Czech Republic, 2003
24. L. Varani, L. Reggiani, T. Kuhn, T. González, D. Pardo, *IEEE Trans. Electr. Dev.* **41**, 1916 (1994)
25. B. Pellegrini, *Phys. Rev. B* **34**, 5921 (1986).
26. A. Kim, A. S. Min, T. W. Tang and Y. J. Park, *Solid State Electronics* **34**, 1251 (1991)
27. T. González, D. Pardo, L. Varani and L Reggiani, *IEEE Trans. Electron. Dev.* **42**, 991 (1995)
28. A. Maestrini, J. Ward, J. Gill, H. Javadi, E. Schlecht, G. Chattopadhyay, F. Maiwald, N. R. Erickson, and I. Mehdi, *IEEE Microwave and Wireless Components Letters* **14**, 253 (2004).
29. W. Knap, D. Dur, A. Raymond, C. Meny, J. Leotin, S. Huant, B. Etienne, *Review of Scientific Instruments* **63**, 3293 (1992)
30. J. Mateos, T. González, D. Pardo, S. Bollaert, T. Parenty and A. Cappy, *IEEE Trans. Electron. Dev.* **51**, 521 (2004), and *IEEE Trans. Electron. Dev.* **51**, 1228 (2004)
31. G. M. Dunn, A. Phillips and P. J. Topham, *Semicond. Sci. Technol.* **16,** 562 (2001).

Unsolved Problems of Low Frequency Noise in GaN-Based HFETs

S. L. Rumyantsev[*†], A. P. Dmitriev[†], M. E. Levinshtein[†], and M. S. Shur[*]

[*] *Rensselaer Polytechnic Institute, Troy, NY 12180-3590*

[†] *The Ioffe Institute of Russian Academy of Sciences, 194021, St. Petersburg, Russia.*

Abstract. Measured temperature and bias dependences of the low frequency noise in AlGaN/GaN HFETs along with the measurements of correlation between gate and drain current fluctuations suggest that the noise is caused by the electrons tunneling from the channel into GaN or/and AlGaN layers. However, many features of the noise behavior remain unclear:

1. Tunneling models of noise in AlGaN/GaN HFETs (and in Si MOSFETs) predict linear or faster reduction of the *1/f* noise with the temperature decrease. However, this noise reduction had not been observed in either AlGaN/GaN HFETs or Si MOSFETs.
2. There is no direct proof that tunneling either into GaN or AlGaN layers is the main source of the 1/f noise.
3. Inhomogeneous distribution of carriers along the channel in nonlinear regime should lead to the sharp increase of the noise amplitude at the point of saturation. Such noise behavior was observed in AlGaN/GaN HFETs but not in Si MOSFETs where the main mechanism of the *1/f* noise is tunneling.
4. Unexplained decrease of the noise above saturation was found in AlGaN/GaN HFETs.

Keywords: low frequency noise, 1/f noise, GaN, gallium nitride, AlGaN/GaN, HFETs
PACS: 72.70.+m,

TEMPERATURE DEPENDENCE OF NOISE

Figure 1 shows a simplified band diagram of the GaN/$Al_xGa_{1-x}N$ heterostructure. Numerical values in Fig. 1 correspond to the typical doped HFET structure with the Al molar fraction $x = 0.3$ in the wide band gap barrier layer and with the GaN channel doped by shallow donors (Si) with the activation energy of $\varepsilon_c - \varepsilon_d = 0.02$ eV. Dashed line shows the band diagram for undoped HFETs. In the model [1], the 2D electrons in the channel can tunnel into the GaN, where they are captured by the tail states. The density of the tail states, $\rho(\varepsilon)$, decreases exponentially with the energy. Tunneling to the traps of the density of states tail is very similar to the McWhorter model of noise in Si MOSFET, which assumes homogeneous distribution of traps in space and energy. Both models give *1/f* noise with linear or close to linear temperature dependence of noise.

The expression for the noise spectra caused by tunneling on continuous levels of the tail states was obtained in ref. [1]:

CP800, *Unsolved Problems of Noise and Fluctuations: UPoN 2005*
edited by L. Reggiani, C. Pennetta, V. Akimov, E. Alfinito, and M. Rosini

$$\frac{S_n}{n_0^2} = \frac{2\pi\rho_0 x_o}{WLn_0^2\omega}\int_{-\infty}^{\varepsilon_C}[e^{\frac{\varepsilon-\varepsilon_C}{\varepsilon_0}} f(\varepsilon)(1-f(\varepsilon))]\,d\varepsilon =$$
$$= \frac{2\pi\rho_0 x_o}{WLn_0^2\omega}\int_{-\infty}^{\varepsilon_C} e^{\frac{\varepsilon-\varepsilon_C}{\varepsilon_0}} \frac{ge^{\frac{\varepsilon-\varepsilon_F}{kT}}}{\left(1+ge^{\frac{\varepsilon-\varepsilon_F}{kT}}\right)^2}\,d\varepsilon \qquad (1)$$

where $\rho_0 = \rho(\varepsilon_c)$ is the density of the tail states at the bottom of the conduction band, x_0 is the characteristic tunneling distance, n_o is the electron concentration in the channel, $f(\varepsilon)$ is the occupancy function, g is the degeneracy factor, ε_F is the position of the Fermi level, W and L are the channel width and length, respectively. In Fig.2 solid lines show results of calculations using Eq.(1) for different level of GaN layer compensation. As seen in accordance with this mechanism the noise amplitude tends to zero with the temperature decrease.

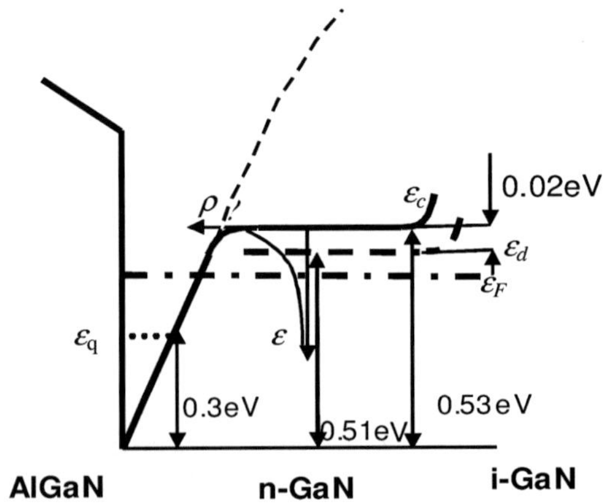

FIGURE 1. Simplified band diagram of the GaN/AlGaN quantum well Parameters are calculated for GaN/$Al_{0.3}Ga_{0.7}N$ heterojunction with doping level of GaN layer 2×10^{17} cm^{-3}. T = 300K, $n_0 = 1.3\times10^{13}$ cm^{-2}. Thin dashed line shows the band diagram for undoped HFETs.

However, measurements of noise in Si MOSFETs and AlGaN/GaN HFETs at cryogenic temperatures do not show a substantial decrease of noise in comparison with the room temperature.

Another mechanism of noise in doped channel AlGaN/GaN HFETs is tunneling to the local Si level. Since donor atoms are randomly distributed in space this tunneling process to the local level also results in the $1/f$ noise [2]. The spectral noise density in this case is given by [2]:

$$\frac{S_n}{n_0^2} \approx \frac{8N_d x_F f_d(\varepsilon_d)[1-f_d(\varepsilon_d)]}{3WLn_0^2\omega} \tag{2}$$

where $f_d(\varepsilon_d)$ is the occupancy function of the Si donor level.

Fig.2 shows results of calculations of the Hooge parameter α for the noise caused by electron tunneling from 2D channel to Si doping level (dashed lines). As seen, tunneling to the local level does not lead to the decrease of noise with the temperature decrease. The explanation is that, at low temperatures, the Fermi level is always close to the trap level and the impurity bandwidth in the model was assumed to be smaller than energy kT at all temperatures. By other words, if in the adjoining to the channel material (GaN or AlGaN in HFETs, SiO_2 in MOSFET) Fermi level is pinned to some local level, which energy bandwidth is smaller than energy kT, the temperature dependence of noise caused by the tunneling to this level is weak.

Can this mechanism explain the experimental weak temperature dependence of noise in doped channel AlGaN/GaN HFETs?

Can tunneling into a local level pinning the Fermi level be responsible for the weak temperature dependence of noise in Si MOSFETs?

In ref. [1,2] we analyzed tunneling of electrons from the channel to the GaN layer. However tunneling to the AlGaN layer also should produce 1/f noise.

Can we distinguish experimentally which layer gives the main contribution to noise?

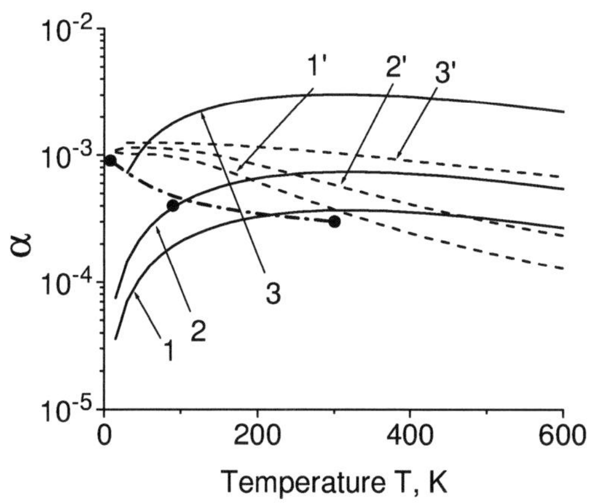

FIGURE 2. Calculated temperature dependencies of Hooge parameter α for different compensation levels N_a/N_d. Solid lines correspond to the noise caused by tunneling on the tail states. Dashed lines represent the noise caused by tunneling on the doping Si level. N_a/N_d: 1,1' – 0.8; 2,2' – 0.9; 3,3' – 0.98; Dots show experimental data.

NOISE IN THE CURRENT SATURATION REGIME

Analysis of noise in AlGaN/GaN HFETs in nonlinear regime predicts that, close to saturation, S_I increases faster than the square of the drain current I_d and diverges when $V_d=V_g-V_t$ [3]:

$$\frac{S_I}{I_d^2} \propto \frac{1}{I_d} \ln\left(\frac{V_g - V_t}{V_g - V_t - V_d}\right) \quad (3)$$

where V_g is the gate voltage and V_t is the threshold voltage.

This model predicts that Si MOSFETs should demonstrate the same behavior. This is because at the point of saturation the channel is pinched off near the drain. The small volume where the channel is pinched off gives the main contribution to the channel resistance and to the noise. As shown in [3], this noise increase close to saturation should be more pronounced for the tunneling mechanism than for the bulk mechanism of noise [4].The noise increase close to saturation was observed in AlGaN/GaN HFETs, see Fig. 3 [3].

However measurements of noise Si MOSFETs in saturation usually do not show much deviation from the $S_I \sim I_d^2$ *law [5,6]. We know only one exception from the relatively old publication [7].*

As seen in Fig.3 noise in AlGaN/GaN HFETs *decreases above saturation. This kind of behavior was not reported before and has no explanation yet.*

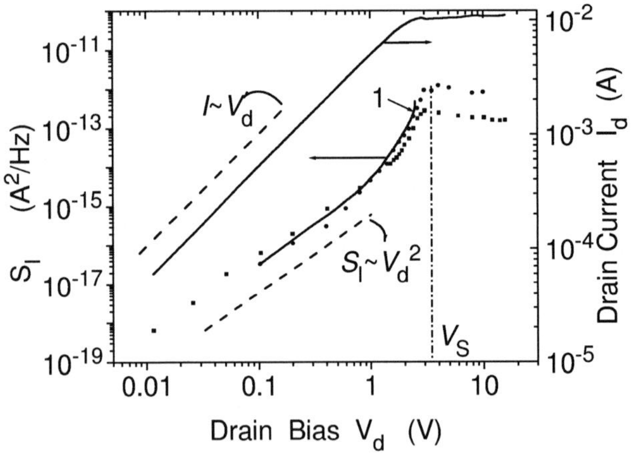

FIGURE 3. The dependencies of the drain current I_d and relative spectral noise density of the short circuit drain current fluctuations S_I/I_d^2 on drain voltage ($V_g = 0$). Symbols show the experimental data for two transistors with $L_g = 1$ μm. Curve 1 is theoretical prediction.

SUMMARY OF UNSOLVED PROBLEMS

Measurements of noise in AlGaN/GaN HFETs and Si MOSFETs at cryogenic temperatures do not exhibit decrease of noise compared to the room temperature measurements.

Tunneling into GaN or AlGaN layers should be proved or disproved to be the dominant mechanism of noise in HFETs

Measurements of noise Si MOSFETs in saturation usually do not show any deviation from the $S_I \sim I^2$ law, in contrast to GaN-based HFETs

Decrease of noise in AlGaN/GaN HFETs above saturation has no explanation

ACKNOWLEDGMENTS

At the Ioffe Institute this work was supported by Russian Foundation for Basic Research (grant 05-02-17774)

REFERENCES

1. A. P. Dmitriev and M. E. Levinshtein, S. L. Rumyantsev and M. S. Shur, accepted to *J. Appl. Phys.*
2. S. L. Rumyantsev, Y. Deng, E. Borovitskaya, A. Dmitriev, W. Knap, N. Pala, M. S. Shur, M. E. Levinshtein, M. Asif Khan, G. Simin, and J. Yang, *J. Appl. Phys.* **92**, 4726-4730 (2002).
3. S.L. Rumyantsev, N. Pala, M.S. Shur, M.E. Levinshtein, M. Asif Khan, G. Simin, J. Yang, *J. Appl. Phys.* **93**, 10030-10034 (2003).
4. Van der Ziel, *Proc. of the IEEE* **41**, 1670 (1963)
5. L.K.J. Vandamme, X.Li, and D. Rigaud, *IEEE Trans.*, **41**, 1936-1944 (1994).
6. G. Ghibaudo, Phys. Stat. Sol. (a) **113**, 223-240 (1989).
7. L.K.J. Vandamme, and H.M.M. de Werd, *Sol. St. El.* **23**, 325-329 (1980).

$1/f$ And RTS Noise In Submicron Devices: Faster Is Noisier

L.K.J. Vandamme[1] and M. Macucci[2]

[1] *Eindhoven University of Technology, Department of Electrical Engineering EH 9.13, 5600MB Eindhoven, The Netherlands, L.K.J.Vandamme@tue.nl*
[2] *Dipartimento di Ingegneria dell'Informazione, Università di Pisa, Via Caruso 16, I-56122 Pisa Italy macucci@mercurio.iet.unipi.it*

Abstract. The origin of $1/f$-like noise in devices is still under discussion. There is one school of thought explaining low-frequency noise as a surface effect due to trapping and suggesting number fluctuations as the origin. The number of (surface) traps is the key-parameter. According to another school, pure $1/f$ noise is considered as a bulk phenomenon due to mobility fluctuations. There the $1/f$ noise parameter α plays an important role. Low frequency noise is often due to a mixture of conductance fluctuations due to number fluctuations and due to mobility fluctuations. A good strategy to improve the analysis is as follows. First step: make a decomposition of the observed noise into pure $1/f$ noise ($1/f^{\gamma}$ with $0.9<\gamma<1.1$) and, if possible, a Lorentzian stemming from e.g., random telegraph signal noise (RTS). Second step: investigate how the different noise components depend on bias and device size. We explain why: i) for $N<1/\alpha$ in sub-micron devices we can expect RTS on top of $1/f$ noise, ii) faster devices are noisier; iii) $1/f$ and $1/f$-like noise are often confused. For silicon resistors there is a possibility of RTS noise on top of the $1/f$ noise if the number of carriers is less than $1/\alpha$. The RTS criterion is normally fulfilled for sub-micron MOSFETs with an area $LW<1\ \mu m^2$. With $N = C_{gate}V_G^{*}/q <1/\alpha$, the RTS noise on top of the $1/f$ noise can become dominant. RTS on top of the $1/f$ noise is to be expected in forward biased diodes at low currents if $I_0 < I < I_{RTS} \approx q/\alpha\tau$; where I_0 is the saturation current, q the elementary charge and τ the minority carrier life time.

Keywords: Fluctuations, 1/f noise, RTS-noise, low-frequency noise, MOSFET, diode, corner frequency, unit-current gain frequency.
PACS: 72.70+m; 74.40+k; 78.30.Fs; 85.40.Qx;

INTRODUCTION

The volume formalism and the mobility fluctuation origin for the $1/f$ noise in metals and semiconductors are well established by Hooge's empirical relation [1]. This idea is not adhered to by the majority of the noise community involved in low frequency noise in MOSFETs. The Mc Whorter model looks so simple and explains the $1/f$ shape of the spectrum, that many people accept it blindly. The McWhorter model explains the $1/f$ noise as a surface effect, due to trapping of carriers into the oxide and leading to number fluctuations. Hooge's empirical relation considers the $1/f$ noise as a bulk phenomenon due to mobility fluctuations. This paper is a contribution to the old surface-volume debate and to the unsolved problem of ΔN-$\Delta\mu$ as origin for the $1/f$ noise.

CP800, *Unsolved Problems of Noise and Fluctuations: UPoN 2005*
edited by L. Reggiani, C. Pennetta, V. Akimov, E. Alfinito, and M. Rosini

RTS ON TOP OF 1/F NOISE

There are strong indications that RTS and $1/f$ noise have different physical origins. RTS noise depends strongly on the position of the Fermi level, E_F, compared to the trap level, while $1/f$ noise does not. RTS noise can appear and disappear on top of the $1/f$ noise in a small temperature range, while $1/f$ noise cannot. RTS and $1/f$ noise often show a different dependence on bias, because RTS is very sensitive to E_F, while $1/f$ is not. Both show a trend toward an increase with decreasing number of carriers, but that is common for all types of conductance fluctuations. The $1/f$ noise can often be considered as a bulk phenomenon given by Hooge's empirical relation [1]. The relative resistance fluctuations are inversely proportional to the number of carriers N as in

$$\frac{S_I}{I^2}=\frac{S_V}{V^2}=\frac{S_R}{R^2}=\frac{\alpha}{Nf} \tag{1}$$

The RTS noise from one single trap is inversely proportional to N^2 as can be seen in the RTS relation for noise spectra by Machlup [2] as:

$$\frac{S_{I_{RTS}}}{I^2}=\frac{S_N}{N^2}=\frac{<(\Delta N)^2>}{N^2}\frac{4\tau_{RTS}}{1+(2\pi f\tau_{RTS})^2} \tag{2}$$

with $1/\tau_{RTS}$ = $1/\tau_e+1/\tau_c$ and τ_{RTS} being the characteristic time constant of the Lorentzian, S_N is the power spectrum of the carrier number (N) fluctuations and τ_e and τ_c are the emission and capture time constant, respectively. The maximum value for the variance which is a worst case condition, equals $\overline{(\Delta N)^2}$ = 1/4 for symmetric trapping, $\tau_e = \tau_c$. For a strategic trap located at a spot with a higher electric field than the average field, we found $\Delta I/I > \Delta N/N = 1/N$ [3]. Hence, S_I/I^2 =K S_N/N^2 with K>1. For symmetric trapping in non uniform fields $K\overline{(\Delta N)^2} = K/4$ with $(K \neq 1)$ appear the RTS equation (3) below:

$$\frac{S_{I_{RTS}}}{I^2}=K\frac{S_N}{N^2}=K\frac{<(\Delta N)^2>}{N^2}\frac{4\tau_{RTS}}{1+(2\pi f\tau_{RTS})^2}\Rightarrow S_{I_{RTS}}=\frac{KI^2}{N^2}\frac{\tau_{RTS}}{1+(2\pi f\tau_{RTS})^2} \tag{3}$$

with K = 1 for a uniform field situation or a trap located at an average electric field position $E_{loc} = E_{av}$, and $K > 1$ for a trap at a strategic high field location ($E_{loc}>E_{av}$) [3]. For K we have [3]:

$$K \approx \frac{E_{loc}^4}{E_{av}^4} \tag{4}$$

We compare $1/f$ noise and RTS noise at the strategic frequency $f_c = 1/(2\pi\tau_{RTS})$, because at frequencies much lower or higher than f_c the $1/f$ noise will always be dominant compared to the RTS noise. We are interested in a worst case condition. Then the condition to observe RTS on top of 1/f noise is $S_{I_{RTS}}(f_c) \geq S_{I_{1/f}}(f_c)$ and this results in the following inequality for an active symmetric trap at a slightly higher electric field than the average (E_{loc}~$2E$av and $K = 4\pi$):

$$N \leq \frac{K}{4\pi\alpha} \approx \frac{1}{\alpha} \tag{5}$$

1/F NOISE IN SILICON WIRES

The effect of different ratios of sample volume-to-surface was investigated in silicon wires, all having an effective length of about, $l = 5\mu m$, but different cross sections [4]. The suspended silicon wires were fabricated in the top layer of a commercially available SOI wafer by partially removing the buried silicon dioxide layer. The n-type silicon top layer was doped at 10^{15} cm^{-3}. The spectra at room temperature showed a $1/f$ spectrum as reported in Fig. 1 [4]. The experimental results obtained on a sample of 150 kΩ for the current range 0.3 μA < I < 1.1 μA can be summarized in a C-value, which is typical for Ohmic samples:

$$\frac{S_I}{I^2} = \frac{C}{f} \tag{6}$$

with $C = 4.4\text{x}10^{-8}$. The number of electrons in the sample can be estimated from $N = l^2/(q\mu R)$ and the mobility is estimated to be μ = 1500 cm^2/Vs which leads to $N = 6.9\text{x}10^3$. Applying the empirical relation (1) for $1/f$ noise, we get $\alpha = CN = 3\text{x}10^{-4}$. Such α-value is typical for silicon with some lattice damage [5]. For silicon without implantation damage values between 10^{-6} and $2\text{x}10^{-6}$ are observed at room temperature and 77 K [6]. With $N = 6.9\text{x}10^3 > (1/\alpha) = 3.3\text{x}10^3$ we have no reasons to expect RTS noise on top of the $1/f$ noise [3] as confirmed by the room temperature measurements reported in Fig. 1.

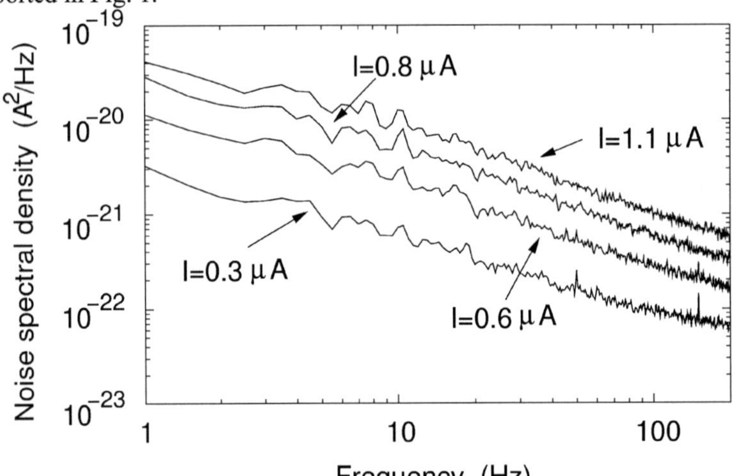

FIGURE 1. Noise current power spectral density as a function of frequency, for the suspended silicon wires, at room temperature.

1/ F AND RTS IN MOSFETS

Hooge's empirical relation (1) with α-values as found in bulk silicon is also applicable to $1/f$ noise in MOSFETs [7, 8]. For MOSFETs operated in strong inversion in their linear regimes with their sources and substrates grounded, it holds at $|V_G - V_T| \equiv V_G^*$ > 25 mV and $|V_{ds}| < V_G^*/10$ that:

$$N = \frac{WLC_{ox}V_G^*}{q} = \frac{L^2}{q\mu R} = \frac{L^2 g_m}{q\mu} \tag{7}$$

In saturation and sub-threshold with $|V_{ds}| \geq V_G^*$ it holds, respectively:

$$S_{I_S} = \frac{2\alpha I_S^2}{Nf} = \frac{\alpha q g_{ms}^2 V_G^*}{2fC_{ox}WL} = \frac{\alpha q \mu^2 C_{ox} W V_G^{*3}}{2fL^3} \tag{8}$$

$$S_{V_{eq}}(V_G^*) = \frac{S_{I_S}}{g_{ms}^2} = \frac{\alpha q V_G^*}{2fWLC_{ox}} = \frac{\alpha q}{fWLC_{ox}} \frac{I_S}{g_{ms}} \tag{9}$$

$$\frac{S_{I_{sub}}}{I_{sub}^2} = \frac{\alpha}{(N+N_0)f} \approx \frac{\alpha}{N_0 f} = \frac{\alpha q^2}{kTC_{ox}Wlf} \tag{10}$$

Studies on IC-graded n and p- type Si wafers with impurity concentrations $6\text{x}10^{14} < N_{imp} < 6\text{x}10^{15}$ cm^{-3} show $10^{-6} < \alpha < 2\text{x}10^{-6}$ at 300 K and 77 K. The α-values were obtained from constriction resistances on Si-wafers, where the $1/f$ noise is unambiguously a bulk phenomenon [6].The $1/f$ noise parameter α in poor quality silicon shows a large spreading, $10^{-5} < \alpha < 2\text{x}10^{-3}$. The results on implanted resistors in quality wafers showed $\alpha = 3\text{x}10^{-4}$ for poorly annealed (450 °C) samples and down to $\alpha = 10^{-6}$ for samples annealed at 900°C [5]. Such results are important to understand the noise levels in modern MOSTs where implantations and low thermal budgets are used for e.g., threshold voltage adjustment and reduction of short channel effects.

The $1/f$ noise in MOSTs over the last 35 years shows $4\text{x}10^{-7} < \alpha < 10^{-3}$ for n- and p-channels. An α-value of $5\text{x}10^{-4}$ has been observed in modern p-type MOSFETs with SiGe channels away from the notorious interface and with an atomic layer deposition of a high-K dielectric [9]. The lower values, in the range $4\text{x}10^{-7} < \alpha < 4\text{x}10^{-6}$, are more often found in bulk p-channel [7]. Surface channel n-MOSTs show in general higher values. Hence, all α-values in MOSTs are similar to the values observed in bulk resistors made in Si wafers.

A $1/f^\gamma$ noise on top of a $1/f$ noise with either $\gamma > 1.3$ or $\gamma < 0.8$ is a strong indication for ΔN-noise, stemming from several generation recombination contributions on top of the omnipresent $1/f$ noise, which is often due to mobility fluctuations.

RTS noise is a ΔN-noise and quite different from the often superimposed $1/f$ [10]. In MOSTs with $LW < 1$ μm^2 or a low number of carriers, $N < 1/\alpha$, RTS-noise often becomes the dominant low frequency noise source [11, 12].

The corner frequency, f_c where the $1/f$ noise equals the thermal noise is a figure of merit for the low frequency noise. The frequency for a unit current gain, f_T is used as a characterization of the high frequency behaviour of MOSTs. From the expressions for thermal and $1/f$ noise we obtain for MOSTs biased in saturation f_c (V_G^*) and f_c (I_s) [12].

$$f_c(V_G^*) = \frac{3\alpha q V_G^{*2} \mu}{16kTL^2} = \frac{3\alpha q V_G^* g_{ms}}{16kTWLC_{ox}} \tag{11}$$

$$f_c(I_S) = \frac{3\alpha q I_S}{8kTWLC_{ox}} \tag{12}$$

The unit current gain frequency f_T of a MOSFET is defined as the frequency where the current gain, $g_{ms}/(\omega C_{ox} WL) = 1$:

$$f_T = \frac{g_{ms}}{2\pi C_{ox} WL} = \frac{\mu V_G^*}{2\pi L^2} \tag{13}$$

Both frequencies f_c and f_T are proportional to $1/l^2$. Hence, the ratio f_c/f_T is independent of channel length as can be seen from:

$$\frac{f_c}{f_T} = \frac{3\pi\alpha q V_G^*}{8kT} \tag{14}$$

From the relation, $f_c/f_T \approx \alpha q V_G^*/kT$, it is clear that faster devices with a higher f_T are inherently noisier considering the $1/f$ noise expressed in f_c. The ratio f_T/f_c is in general larger than 10^3 for normal bias and α-values. The integrated noise from f_L =1Hz << f_h up to $f_h = f_T$ is then dominated by thermal noise. The variance in a bandwidth f_h-f_l is also a figure of merit. Smaller is noisier also holds for thermal noise, at least if we consider the integrated equivalent input noise voltage. We find the surprising result that $<\Delta V_{eq}^2>$ is independent of bias and mobility and inversely proportional to the total gate capacitance

$$<\Delta V_{eq}^2> \approx <\Delta V_{eq_{th}}^2> = \frac{4kT}{3\pi WlC_{ox}} \tag{15}$$

Smaller and faster is noisier holds for the $1/f$ noise and the integrated noise. Smaller devices with LW<1 μm^2 or, more specifically, $N<1/\alpha$ often result in an inevitable RTS-noise on top of the $1/f$ noise.

DIODE: FASTER IS NOISIER

The $1/f$ noise in semiconductor diodes was explained [1] in terms of mobility fluctuations. For a long diode the following reasoning can be proposed [13]. At a fixed forward biasing with $V > kT/q$, the following proportionalities hold: $I \propto I_0 \propto D/L \propto (\mu/\tau)^{1/2}$ with I_0 the saturation current, D the diffusion coefficient of the dominant carriers, L the diffusion recombination length defined by $(D\tau)^{1/2}$, τ the minority carrier lifetime, and μ the mobility. Hence $\ln I = 1/2 \ln \mu +$ constant, thus $\Delta I/I = ½\, \Delta\mu/\mu$ and $S_I/I^2 = ¼\, S_\mu/\mu^2$. Here the Einstein relation $D/\mu = kT/q$ was used, which means that $1/f$ mobility fluctuations cause $1/f$ fluctuations in the diffusion coefficient. We apply the empirical relation (1) with N being the number of excess minority carriers, which is given by the charge control approach as $I = qN/\tau$, and that results in:

$$\frac{S_{I_{1/f}}}{I^2} = \frac{1}{4}\frac{S_\mu}{\mu^2} = \frac{\alpha}{4Nf} = \frac{q\alpha}{4I\tau f} \therefore S_{I_{1/f}} = \frac{q\alpha I}{4\tau f} \tag{16}$$

For the shot noise holds $S_I^{shot} = 2qI$ and the so-called corner frequency f_c, where the shot noise is equal to the $1/f$ noise $S_I^{shot}(f_c) = S_{I_{1/f}}(f_c)$, becomes:

$$f_c = \frac{\alpha}{8\tau} \tag{17}$$

The quantity f_c is an excellent figure of merit, if it is independent of current. The lower f_c, the better the diode is, considering the low frequency noise, at least if there are no generation-recombination noise contributions. The relation for f_c shows that low $1/f$ noise goes hand in hand with slow devices (high values of τ). For low frequency applications, rather high values of τ can be chosen, leading to low leakage currents and hence low dark currents. Pure silicon with its indirect bandgap can exhibit high τ values and hence low $1/f$ noise. A gold doped Si switching diode has a reduced τ–value, hence an enhanced leakage current, higher $1/f$ noise and f_c values. In general, τ values from direct gap semiconductors (GaAs and other III-V compounds) are much smaller (typical 10^{-9} s) than for indirect bandgap pure semiconductors like Si (typical 10^{-6} s). Hence III-V (laser) diodes have higher f_c values than silicon diodes, and this has nothing to do with a better technology. The fastest switching frequency of a long diode, here called f_{Ts}, is inversely proportional to the minority carrier lifetime outside the depletion region, if recombination in the space charge region is ignored. The fastest switching frequency of a diode is given by $f_{Ts} \approx 1/4\tau$, which means faster switching diodes are noisier:

$$\frac{f_c}{f_{Ts}} = \frac{\alpha}{2} \tag{18}$$

CURRENT CRITERION TO OBSERVE RTS ON TOP OF 1/*F* NOISE IN DIODE: $I < I_{RTS}$

Diode type devices often show a Lorentzian contribution on top of the $1/f$ noise at low current levels. We can expect RTS on top of the $1/f$ noise if $I_0 < I < I_{RTS}$, if at least one active trap at the Fermi level position is present. Again, such a condition can be derived if the $1/f$ and RTS noise sources are considered as two independent phenomena, as was already proven for GaAs based quantum dot light emitting diodes [10, 14]. Both components of the low frequency noise were separated, showed a different dependence on current, and were uncorrelated. This is a strong indication for a different physical origin for RTS and $1/f$ noise. We apply the RTS relation (2) [2, 3] and take into account that for the current I in a forward biased diode holds: $I = qN/\tau$ with τ being the minority carrier life time. We find for the RTS spectrum in diodes with active symmetric traps

$$S_{I_{RTS}} = \frac{Kq^2\tau_{RTS}}{\tau^2\left(1+(2\pi f\tau_{RTS})^2\right)} \tag{19}$$

with $K = 1$ for a trap located at an average electric field position $E = E_{\text{loc}} = E_{\text{av}}$, and $K > 1$ for a trap at a strategic high field location ($E_{\text{loc}} > E_{\text{av}}$) [3]. Now again we compare the $1/f$ noise and the RTS noise at the strategic frequency $f_c = 1/(2\pi\tau_{RTS})$. Then the condition to observe RTS on top of $1/f$ noise in diodes is: $S_{I_{RTS}}(f_c) \geq S_{I_{1/f}}(f_c)$. This result in:

$$\frac{q^2 K\tau_{RTS}}{2\tau^2} \geq \frac{2\pi q\alpha I\tau_{RTS}}{4\tau} \therefore\ I_{RTS} \leq \frac{qK}{\pi\alpha\tau} \approx \frac{q}{\alpha\tau} \tag{20}$$

for $K \approx \pi$. For the experimentally obtained parameter values α and τ, for Si or III-V compound diodes we expect RTS noise on top of the $1/f$ noise for a current $I_{RTS} < 100$ nA ($\alpha\tau=1.6\text{x}10^{-12}$ s) and $I_{RTS} < 10$ μA ($\alpha\tau=1.6\text{x}10^{-14}$ s), respectively. This is consistent with the observation of two-level RTS in e.g., GaN LEDs [15].

DISCUSSION AND CONCLUSIONS

There is ample experimental evidence that proves that the $1/f$ noise in homogeneous Si is mobility noise [1,16]. Values of α obtained from constriction resistances on homogeneous Si wafers are confined in the narrow range $1\text{x}10^{-6}$ and $2\text{x}10^{-6}$ [6]. Boron implanted resistors in n-type Si wafers showed $\alpha \approx 3\text{x}10^{-4}$ if poorly annealed at 450°C. After the lattice damage had been restored by annealing at 900 °C α was down to 10^{-6} [5]. These were the wafers in which the MOSFET structures are made. The numerical values of α in MOSFETs do not prove or disprove on their own the mobility model or the McWhorter model [16]. However, the surprising result remains those numerical values of α obtained from bulk samples and from MOSFETs are in the same range.

The V_G^*-dependence of α in $\alpha = aV_G^{*\gamma}$ plays an important role in the ΔN-$\Delta\mu$ discussion in MOSFETs. A $\Delta\mu$-source explains the γ = 0 value of p-channel MOSFETs, but for the γ = -1 for n-channels one has to resort to complications. The McWhorter model gives the opposite result: n-channels are straightforward, whereas p-channels require complications. For different γ-values it still holds that faster devices are noisier. The relation f_c/f_T becomes V_G^* independent for γ = -1. The dependence of α on V_G does not rule out a $\Delta\mu$-model or the McWhorter ΔN-model [16].

The volume-surface controversy can be investigated with samples with different volume-to-surface ratios as in silicon wires. With MOSFETs this comparison is impossible because the inversion layer has for all WL-values a fixed thickness of about 10 to 20 nm.

A local change in conductivity at some spot can have different effects at the terminals of a biased device, depending on the sensitivity coefficient at that spot [3]. This concept leading to strategic and less strategic traps is used to explain the wide range of RTS amplitudes often observed. Hence, without invoking multi-electron capture or strong changes in carrier mobility by one single charged trap, the wide distribution of RTS amplitudes is explained. The observed transition between RTS and 1/f noise at $N<1/\alpha$ [10, 11, 14, 15] was also observed by Kirton and Uren [17]. However, they claim that there can be no doubt that the measurements of noise and RTS in microstructures have shown definitively that 1/f noise in MOSFETs and MIM tunnel diodes is generated through carrier trapping.

REFERENCES

1. F. N. Hooge, T. G. M. Kleinpenning, and L. K. J. Vandamme. Experimental studies on 1/f noise. Rep. Prog. Phys. 44 (5), pp. 479-532 (1981)
2. S. Machlup. Noise in Semiconductors: Spectrum of a Two-Parameter Random Signal. J. Appl. Phys. 25 (3), pp. 341-343, (1954).
3. L. K. J. Vandamme, D. Sodini, and Z. Gingl. On the anomalous behaviour of the relative amplitude of RTS noise. Solid-State Electronics 42 (6), pp. 901-905, (1998)
4. M. Macucci, B. Pellegrini, G. Penneli and M.Piotto. Microelectronic Eng 61-62, pp.701-705 (2002).
5. L. K. J. Vandamme and S. Oosterhoff. Annealing of ion-implanted resistors reduces the 1/f noise. J. Appl. Phys. 59 (9), pp. 3169-3174, (1986).
6. R. H. M. Clevers. Volume and temperature dependence of the 1/f noise parameter α in Si. Physica B 54, pp. 214-224, (1989).
7. L.K.J. Vandamme et al. IEEE Trans on Elec. Dev. **41** (11), pp. 1936-1945, (1994).
8. E. P. Vandamme and L. K. J. Vandamme. Critical discussion on unified 1/f noise models for MOSFETs. IEEE Trans. on Elec. Dev. 47 (11), pp. 2146-2152, (2000).
9. M. von Haartman, D. Wu, B.G. Malm, P.-E. Hellstom, S.-L. Zhang, M. Ostling . Low –frequency noise in $Si_{0.7}Ge_{0.3}$ surface channel pMOSFETs with ALD HfO_2/Al_2O_3 gate dielectrics. Solid-State Electronics 48, pp. 2271-2275 (2004).
10. A.V. Belyakov, L.K.J. Vandamme et al. Fluctuation and Noise Letters (FNL) **3** (3): L325-L339, (2003).
11. M. Toita, L. K. J. Vandamme, S. Sugawa, A. Teramoto, and T Ohmi. Geometry and Bias Dependency of Low-Frequency Random Telegraph Signal and 1/f Noise Amplitude in MOSFETs. FNL 2005 (submitted).
12. L. K. J. Vandamme. 1/f Noise in MOSTs: Faster is Noisier; Advanced Experimental Methods for Noise Research in Nanoscale Electronic Devices (pp 109-120) Eds. J.Sikula and M. Levinshtein (ARW, Brno, NATO), Kluwer, ISBN 1-4020-2169-0, (2004).
13. L. K. J. Vandamme, E. P. Vandamme, and J. J. Dobbelsteen. Impact of silicon substrate, iron contamination and perimeter on saturation current and noise in n^+p diodes. Solid-State Electronics 41 (6), pp. 901-908, (1997).
14. L. K. J. Vandamme, A. V. Belyakov, M. Y. Perov, and A. V. Yakimov. Difference in dependence of 1/f and RTS noise on current in quantum dots light emitting diodes. In: NOISE IN DEVICES AND CIRCUITS, edited by M. J. Deen, Z. CelikButler, and M. E. Levinshtein, Proc. of SPIE Vol. 5113, pp. 368-378, (2003).
15. S. Bychikhin, D. Pogany, L. K. J. Vandamme, G. Meneghesso and E. Zanoni. Low frequency noise sources in as-prepared and aged GaN-based light emitting diodes. J. Appl. Phys. 97, pp. 123714 -1 to 123714-7, (2005).
16. L. K. J. Vandamme and F. N. Hooge. On the additivity of generation recombination spectra Part 3: The McWhorter model for 1/f noise in MOSFETs. Physica B, Vol 357, pp 507-524 (2005).
17. M.J. Kirton and M.J. Uren. Noise in solid-state microstructures: a new perspective on individual defects, interface states and low-frequency(1/f) noise. Advances in Phys. Vol 38, pp 367-468 (1989).

Noise And Charge Transport In Carbon Nanotube Devices

Shahed Reza,[*] Quyen T. Huynh,[*] Gijs Bosman,[*] Jennifer Sippel[†] and Andrew G. Rinzler[†]

[*] *Department of Electrical and Computer Engineering, University of Florida, Gainesville, Florida*
[†] *Department of Physics, University of Florida, Gainesville, Florida*

Abstract. The charge transport and noise properties of three terminal, gated devices containing multiple, single wall, metallic and semiconductor carbon nanotubes have been measured as a function of gate and drain bias at 300K. Using pulsed bias the metallic tubes could be burned sequentially enabling the separation of measured conductance and low frequency excess noise into metallic and semiconductor contributions. The relative low frequency excess noise of the metallic tubes was about a factor 100 lower than that of the semiconductor tubes, whereas the conductance of the metallic tubes was significantly higher (10 to 50 times) than that of the semiconductor tubes.

Keywords: noise, carbon nanotubes.
PACS: 73.63Fg, 72.20Jv

INTRODUCTION

Based on their favorable electrical, mechanical, and growth properties, metallic and semiconductor carbon nanotubes (CNTs) have moved to the forefront of the research community as serious contenders for leading roles in future nano-scale device technology and circuitry[1]. Two terminal devices have potential uses as wires or sensors, whereas three terminal, gated devices may be used to replace MOSFETs. The charge transport and noise properties of these structures are presently not well understood. Specifically the role of the metal contacts, defects and the gate control mechanism are topics of current investigations for both the metallic and semiconductor tubes. Our research has focused on gated, single wall carbon nanotube devices with physically supported carbon nanotubes. Due to the random nature of our CNT growth process and the small feature size of the tubes, typically a large number of semiconductor and metallic tubes in parallel are contacted in an electrical experiment. The focus of this paper is to demonstrate a pulsed bias tube burning method to delineate the noise and charge transport properties of the metallic and semiconductor tubes.

CP800, *Unsolved Problems of Noise and Fluctuations: UPoN 2005*
edited by L. Reggiani, C. Pennetta, V. Akimov, E. Alfinito, and M. Rosini

EXPERIMENT

Nanotubes were grown by CVD on silicon chips with a 600 nm thermal oxide layer. The catalyst used was 10μg/ml $Fe(NO_3)_3{\cdot}9H_2O$ dissolved in IPA spun dried onto the oxide/silicon[2]. The nanotubes were grown at 900° C in a 1" tube furnace flowing 200 sccm hydrogen and 200 sccm methane[3]. Optical lithography was used to pattern Cr/Pd (5nm/45nm) electrodes spaced 1 μm apart (the Pd likely overcoats the edge of the Cr layer so that electrical contact to the nanotubes is via the Pd). The length of the electrodes was 500 μm and an estimated total of 21 and 54 nanotubes in parallel spanned the 1 μm spacing between adjacent source and drain electrodes in samples A and B respectively. Gate control was achieved via a back contact to the silicon-silicon dioxide substrate resulting in a gate to tube distance of 600nm. A positive gate bias of +14V cuts off the p-channel semiconductor tubes leaving only the metallic tubes to conduct. At a gate bias of -14V both the metallic and semiconductor tubes conduct with the semiconductor tubes fully turned-on at low drain bias.

The samples were placed at room temperature in a pulsed bias circuit that allowed for 1-10μs duration voltage bias pulses applied at low repetition rates as depicted in figure 1.

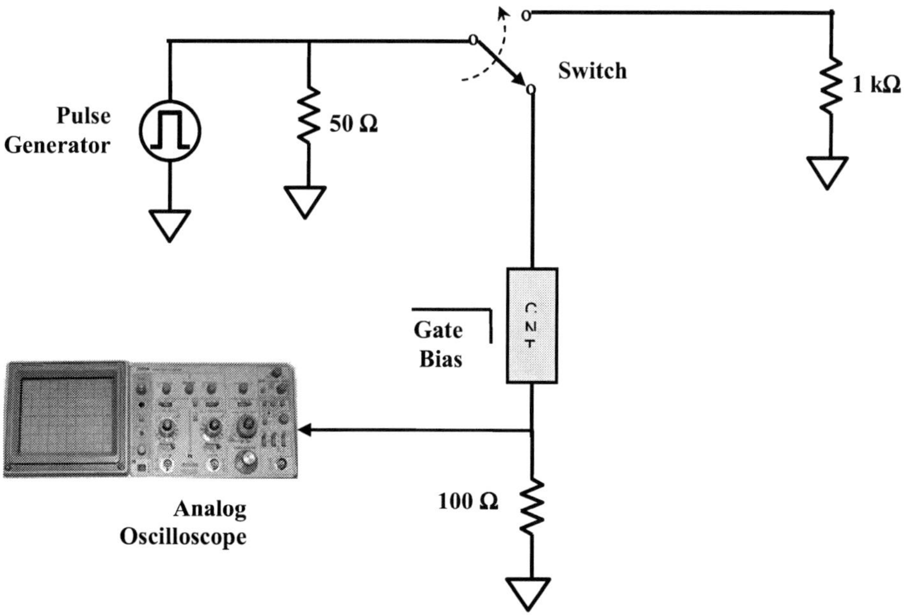

FIGURE 1. Pulsed bias experimental set-up for sequential burning of metallic CNTs.

A positive, continues gate bias of +14V was applied so that only the metallic tubes contribute to conduction. Upon slowly increasing the drain bias voltage across the sample and monitoring the device current on an analog oscilloscope, the onset of Joule heating in a metallic tube could be observed from small fluctuations in the current.

Upon further increase of the pulsed drain bias, the tube would burn and the pulsed bias was switched quickly to the 1kΩ dissipative resistor preventing damage to additional tubes. Current-voltage and noise measurements were carried out before and after each tube burning using a HP 4351 semiconductor parameter analyzer and a HP3561A low frequency spectrum analyzer operating between 10Hz and 100KHz.

The noise observed is a combination of low frequency 1/f-like excess noise and Nyquist noise. Shot noise levels were below our detection window.

EXPERIMENTAL RESULTS AND DISCUSSION

The linear conductance of sample A, biased at V_G=+14V, is plotted as a function of the number of tube burnings in figure 2.

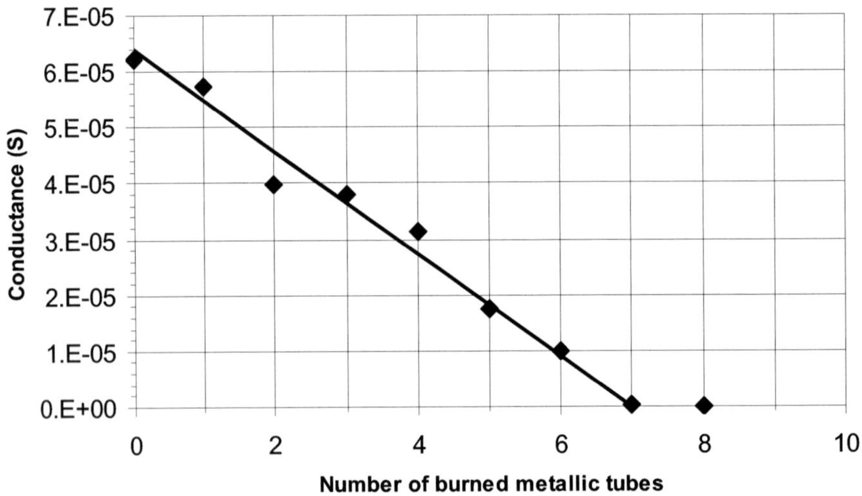

FIGURE 2. The measured conductance of sample A, biased at V_G=+14V, plotted as a function of the number of burned metallic tubes.

The observed linear dependence reveals the sequential burning of single metallic tubes. From the inverse slope of the data the resistance of a single metallic tube is calculated to be 113kΩ. Sample B shows a similar behavior but with a tube resistance of 73.5kΩ. Figure 2 indicates that 7 burns were needed to reduce the sample conductance to zero, hence the original sample contained 7 metallic tubes. Under random CNT growth conditions typically 1/3 of the tubes shows metallic properties while 2/3 becomes a p-type semiconductor for the Pd contact system used in our experiment[1]. Consequently sample A is expected to contain 14 semiconductor tubes for a total of 21 tubes originally. For sample B the numbers are 18 metallic, 36 semiconductor for a total of 54 tubes originally.

By measuring the linear conductance of the samples at V_G=-14V (semiconductor and metallic tubes conducting) and at V_G=+14V (metallic tubes only) the conductance of the semiconductor tubes at V_G=-14V can now be estimated. For sample A this resulted in 814kΩ per semiconductor tube, while sample B produced 3.4MΩ per tube. This spread in data is in line with values reported in the literature and may be due to small variations in the Pd metal contact-CNT interfaces from sample to sample[4].

Following Collins[5] example, the measured excess noise of carbon nanotubes can be expressed as

$$S_{1/f}(f) = \frac{A \cdot I^2}{f^\beta} \tag{1}$$

where the excess noise factor A models the relative magnitude of the excess noise and β is the frequency exponent. The value of A/R, where R is the ohmic resistance of the sample, has been shown to be close to 10^{-11} Ω^{-1} for single CNTs, bundles and films[5].

Assuming that the parallel CNTs in our samples produce noise independently and using the tube conductances measured from the current voltage characteristics, the excess noise factors A_m of the metallic tubes follow from noise measurements at V_G=+14V, where the semiconductor tubes are cut-off and presumably contribute negligible noise levels. For samples A and B A_m was measured at 2.5×10^{-5} resulting in $A_m/R_{mA} = 2.2 \times 10^{-10}$ and $A_m/R_{mB} = 3.4 \times 10^{-10}$ Ω^{-1}, respectively. Analyzing the noise data measured at V_G=-14V which includes semiconductor and metallic tube noise contributions resulted in $A_{scA} = 3 \times 10^{-3}$ and $A_{scB} = 10^{-3}$ or $A_{sc}/R_{scA} = 3.7 \times 10^{-9}$ Ω^{-1} and $A_{sc}/R_{scB} = 2.9 \times 10^{-10}$ Ω^{-1} for samples A and B, respectively. A summary of this data is presented in table 1.The values of β range between 0.8 and 1.2 for the samples studied.

Note that the measured A/R values all exceed the average 10^{-11} value observed by Collins. From this we conclude that our samples are relatively noisy which may be attributed to the sputtering deposition of the Pd/Cr contacts. Also it is clear that in contrast to conventional metal and semiconductor systems where the excess noise of metals can be typically ignored, the excess noise of metallic tubes needs to be accounted for.

TABLE 1. Noise And Conductance Data Of Samples A And B.

Device Data	Sample A	Sample B
Number of Metallic Tubes	7	18
Number of Semiconductor Tubes	14	36
Total Number of Tubes	21	54
Resistance of Metallic Tubes	113kΩ	74 kΩ
Resistance of Semiconductor Tubes	814 kΩ	3.4MΩ
A_m	2.5×10^{-5}	2.5×10^{-5}
A_m/R_m	2.2×10^{-10}	3.4×10^{-10}
A_{sc}	3×10^{-3}	10^{-3}
A_{sc}/R_{sc}	3.7×10^{-9}	2.9×10^{-10}

CONCLUSIONS

The pulsed bias experiments were successful in demonstrating the sequential burning of single metallic carbon nanotubes. The conductance and relative excess noise factors of semiconductor and metallic tubes were determined showing that both produce 1/f like excess noise but with the metallic tubes a factor 100 less, in relative measure, than semiconductor CNTs.

REFERENCES

1. Dresselhaus, M.S., Dresselhaus, G., and Avouris, P., *Carbon Nanotubes: Synthesis, Structure, Properties, and Application*, Berlin: Springer Verlag, 2001, 1st edition.
2. Hafner, J. Cheung, C., Oosterkamp, T., and Lieber, C., *J. Phys. Chem. B* **105**, 743 (2001).
3. Li, Y., Kim, W., Zhang, Y., Rolandi, M., Wang, D., and Dai, H., *J. Phys. Chem. B* **105**, 11424 (2001).
4. Javey, A., Guo, J.G., Wang, Q., Lundstrom, M., and Dai., H., *Nature,* **424**,654 (2003).
5. Collins, P. G, Fuhrer, M. S., and Zettle, A. *App. Phys. Letters* **76,** 894 (2000).

Key issues in trap-assisted low-frequency device noise simulation in nonlinear large-signal conditions

F. Bonani, F. Bertazzi, G. Conte, S. Donati Guerrieri, G. Ghione

Dipartimento di Elettronica, Politecnico di Torino,
Corso Duca degli Abruzzi, I-10129 Torino, ITALY

Abstract. The paper presents a detailed analysis of the conversion phenomena observed in trap-assisted low-frequency noise conversion in forced large-signal device operation. The discussion is based on a drift-diffusion physical model, with stationary and cyclostationary noise analysis capabilities. In particular, we show that in a semiconductor diode, traps located in a resistive area yield terminal fluctuations whose frequency conversion is milder than in the case of traps in the depletion region. This leads to a different frequency dependence for noise at the baseband or near the fundamental frequency.

Keywords: Semiconductor device noise, Circuit noise, Semiconductor device modeling, Microwave devices, Nonlinear systems
PACS: 85.40.Qx, 07.05.Tp

INTRODUCTION

Trap-assisted generation-recombination (GR) low frequency noise is an important issue in device modeling, in particular under large-signal nonlinear operation. In this regime, low-frequency noise is upconverted towards the transmitting/receiving (TX/RX) frequency bands of RF systems, thus impairing the overall performance of the TX/RX link. While trap-assisted GR noise has been often called for as the basic mechanism giving rise to low-frequency noise in semiconductor devices, and, in some cases also as a paradigm for the modelling of $1/f$ noise, the actual physical phenomena underlying noise upconversion in the nonlinear regime has not been fully addressed yet. In this work we exploit physics-based simulation of trap-assisted GR noise both in small- (stationary) and large-signal (cyclostationary) conditions to investigate the effect on device terminal noise of the superposition of noninteracting traps distributed in the bulk and at the surface of bipolar devices. Starting from the microscopic noise sources, we exploit a Green's function approach to recover terminal noise fluctuations, allowing for a detailed analysis of the regions of the device contributing to terminal noise. In particular we demonstrate that material differences arise, depending both on the presence of memory effects in the device, and on the location of the traps.

CP800, *Unsolved Problems of Noise and Fluctuations: UPoN 2005*
edited by L. Reggiani, C. Pennetta, V. Akimov, E. Alfinito, and M. Rosini

TRAP-ASSISTED GR NOISE MODEL

In this contribution we exploit a unified description of trap-assisted GR noise for physical, PDE-based simulators amenable for implementation both in small- (stationary) and large-signal (cyclostationary) device operation. This model is described in detail in [1]. The main advantage of the model is the use of stationary white microscopic noise sources [2, 3], which allows for a uniquely defined modulation scheme in cyclostationary conditions [4]. Such sources are added to the carrier continuity equations and to the rate equations of each trap level included in the simulation: the total PDE system is therefore made of the standard drift-diffusion model plus one rate equation for each trap. The propagation of the noise sources to the terminals is carried out, for small-signal and forced large-signal conditions, by means of Green's functions [2]. Following the discussion in [1], for trap level k, the stationary short-circuit noise current spectrum $S_{\delta i,k}(\omega_{SS})$ in the case of a two-terminal device is evaluated as:

$$S_{\delta i,k}(\omega_{SS}) = \sum_{\alpha,\beta=n,p,k} S_{\delta i,k}^{(\alpha,\beta)}(\omega_{SS}) = \sum_{\alpha,\beta=n,p,k} \int_{\Omega} G_{\alpha}(\mathbf{r},\omega_{SS}) K_{\gamma_\alpha,\gamma_\beta}(\mathbf{r}) G_{\beta}^{*}(\mathbf{r},\omega_{SS})\, d\mathbf{r}, \quad (1)$$

where ω_{SS} is the small-signal angular frequency, $S_{\delta i,k}^{(\alpha,\beta)}$ is the partial contribution to the overall noise due to the correlation of carrier/trap α and β, $K_{\gamma_\alpha,\gamma_\beta}(\mathbf{r})$ is the corresponding microscopic noise source and * denotes complex conjugate. If several non-interacting traps are present, the total noise spectrum at the device terminal is recovered by simply adding the contributions in (1) for all the traps. In the following we will examine a device with four trap levels: a bulk level ($k = \mathrm{B}$) and three surface levels ($k = 1,2,3$).

In the cyclostationary case, the above expression must be modified, in order to take into account all the possible correlations among the noise sideband components. The terminal diode noise is described by the noise current sideband correlation matrix (SCM), given by a generalization of (1) [1]:

$$\mathsf{S}_{\delta i,k}(\omega) = \sum_{\alpha,\beta=n,p,k} \mathsf{S}_{\delta i,k}^{(\alpha,\beta)}(\omega) = \sum_{\alpha,\beta=n,p,k} \int_{\Omega} \mathsf{G}_{\alpha}(\mathbf{r};\omega) \cdot \mathsf{K}_{\gamma_\alpha,\gamma_\beta}(\mathbf{r}) \cdot \mathsf{G}_{\beta}^{\dagger}(\mathbf{r};\omega)\, d\mathbf{r}, \quad (2)$$

where ω is the sideband frequency, $\mathsf{G}_{\alpha}(\mathbf{r};\omega)$ is the conversion Green's function (CGF) corresponding to unit impulsive injection in point $\mathbf{r}$ and in equation α, and $\dagger$ denotes hermitian conjugation.

The model is implemented in a bipolar, 2D drift-diffusion simulator developed in our group, where an arbitrary number of (position-dependent) trap levels can also be included, with stationary and cyclostationary noise analysis capabilities. A similar methodology was developed in [3] and applied to cyclostationary analysis in [5].

LOW-FREQUENCY NOISE CONVERSION IN A *PN* DIODE

In order to gain insight into the intricated physical phenomena giving rise to trap-assisted noise upconversion in bipolar devices, we will exploit a simple test device: a semiconductor n^+p junction diode, as shown in Fig. 1 (left), where we included a bulk trap, uniformily distributed in the device volume, and three surface traps modelled as present in

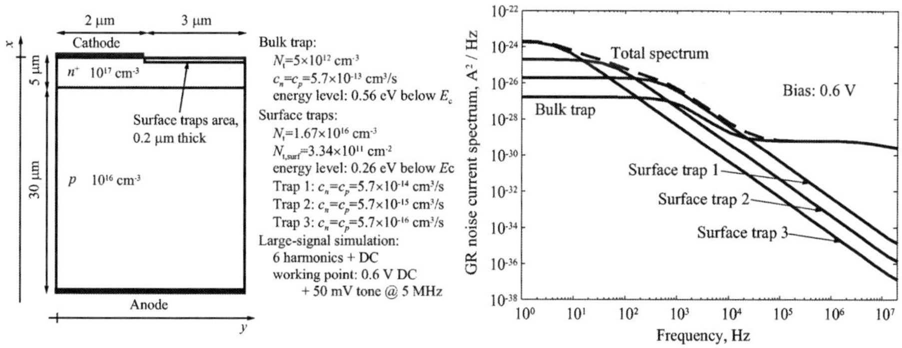

FIGURE 1. Structure and trap parameters for the n$^+$p Si junction (left). Small-signal stationary GR spectra for the n$^+$p abrupt junction (right). The surface traps yield a $1/f$-like spectrum.

a thin surface volume. The three surface traps were introduced in order to reproduce the possible degradation of the device surface during device fabrication [3]. The diode trap parameters are also shown in Fig. 1; in particular the surface traps c coefficents (product of the trap cross section and the thermal velocity) follow a logarithmic distribution since, in a uniformely doped sample and stationary conditions, they would give rise to $1/f$-like noise spectrum, due to the superposition of three lorentzian spectra with corner frequencies logarithmically spaced from 10 Hz to 1kHz. As a first test we investigate if the same $1/f$-like noise due to surface traps is also found for the nonhomogeneous, diode case. The results of the diode simulation in stationary conditions (with a forward 0.6 V bias) are shown in Fig. 1 (right): the three surface traps contributions dominate over the bulk trap one, yielding a $1/f$-like spectrum at low frequency. The bulk trap frequency dependency is not lorentzian, but is characterized by two plateaux, as already suggested in [6]. The surface traps are located in a resistive device region, and thus contribute to the terminal noise in a way similar to the one expected in a homogeneous sample. In particular, the stationary noise due to traps in a homogeneous sample [1] is directly proportional to the square of the current flowing through the same resistive region. The same also holds in the diode case for the surface traps. Concerning the bulk trap uniformely distributed in the device volume, its contribution to the overall terminal noise stems mainly from the depleted junction region, as also suggested in [6].

Let us now turn to cyclostationary noise analysis. The device has been driven by a 0.6 V bias and a 0.05 V tone at 5 MHz. Note that the fundamental frequency is higher than the device cut-off frequency, so that the diode shows a very strong capacitive behavior. Fig. 2 (left) shows the noise simulation results in the nonlinear case. The diode shows a strong nonlinear behavior with a noticeable noise conversion from the baseband to the upper sidebands. In order to gain more insight in the mechanisms originating this noise conversion, Fig. 2 (right) reports the diagonal elements of diode noise current SCM as a function of the sideband frequency. The understanding of the baseband noise behavior is quite straigthforward, as a detailed analysis, based on the right part of Fig. 2, shows that the element $(0,0)$ of the SCM is basically equal to the stationary spectrum evaluated for a bias point (0.62 V) with a DC current equal to the DC component of the large-

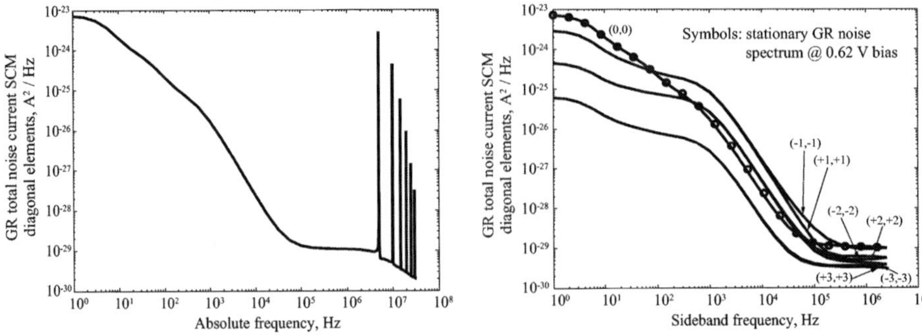

FIGURE 2. Absolute (left) and sideband (right) frequency dependence of the diagonal SCM elements for the total GR noise current. Symbols in the right figure are the stationary total GR noise current spectrum for a DC bias yielding the DC component of the LS current. On the left figure, 6 noise sidebands are shown, while 3 only are represented on the right one.

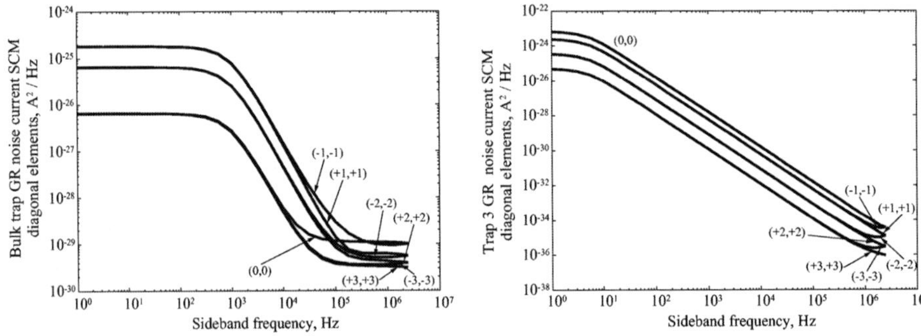

FIGURE 3. Sideband frequency dependence of the diagonal SCM elements for the GR noise current due to the bulk trap (left) and surface trap 3 (right).

signal device current. On the contrary, noise conversion shows an intricated behaviour: the most noticeable effect is that noise is not converted to the upper sidebands with the same frequency dependence as the baseband noise, i.e. $1/f$ noise is not upconverted to the upper sidebands. This phenomenon has been further investigated, finding that surface traps noise is converted towards the upper sidebands less strongly than the bulk trap noise. This is demonstrated in Fig. 3, where the contributions of the bulk trap and of surface trap 3 (the same holds for traps 1 and 2) are compared. While for the surface trap the frequency conversion exhibits a decreasing value of the low (sideband) frequency plateau as the sideband order increases, for the bulk trap the trend is different. This behavior suggests that the conversion effects for the bulk trap is stronger than for the surface traps, probably due to fact that the surface traps are located in a resistive region of the device where nonlinear effects are milder, while the bulk trap is also present in the depletion region where the nonlinear effects (and therefore the noise conversion) are stronger.

In order to validate this picture, we have investigated, for $k = \mathrm{B}, 1, 2, 3$ separately, the various terms contributing to the $(\mathsf{S}_{\delta i,k}(\omega))_{0,0}$ and $(\mathsf{S}_{\delta i,k}(\omega))_{1,1}$ elements of the diode SCM, for a sideband frequency of 1 Hz (subscripts here represent sideband indices). The table in Fig. 4 (left) shows that, for each trap, the main contribution to $\mathsf{S}_{\delta i,k}(\omega)$ in (2) is given by the $\mathsf{S}_{\delta i,k}^{(k,k)}(\omega)$ term (i.e. the auto-correlation of trap sources), and, within this, the dominant terms are $(\mathsf{D}^{(k)})_{0,0}$ and $(\mathsf{D}^{(k)})_{1,1}$ defined as follows:

$$(\mathsf{D}^{(k)})_{0,0} = \int_{\Omega} (\mathsf{G}_k(\mathbf{r};\omega))_{0,0}\,(\mathsf{K}_{\gamma_k,\gamma_k}(\mathbf{r}))_{0,0}\,(\mathsf{G}_k^{\dagger}(\mathbf{r};\omega))_{0,0}\,\mathbf{dr} \tag{3}$$

$$(\mathsf{D}^{(k)})_{1,1} = \int_{\Omega} (\mathsf{G}_k(\mathbf{r};\omega))_{1,0}\,(\mathsf{K}_{\gamma_k,\gamma_k}(\mathbf{r}))_{0,0}\,(\mathsf{G}_k^{\dagger}(\mathbf{r};\omega))_{1,0}\,\mathbf{dr} \tag{4}$$

$(\mathsf{D}^{(k)})_{0,0}$ is easily recognized as the term propagating the baseband microscopic noise source for trap k to the diode terminal, while the term $(\mathsf{D}^{(k)})_{1,1}$ corresponds to the upconversion of the same baseband noise to the upper sideband though the element (1,0) of the CGF for the same trap k. Terminal noise at the first sideband is therefore identified as the result of the upconversion from the baseband only. Once the dominant terms that give rise to the diode terminal noise are identified, we can investigate their behavior inside the device. Fig. 4 (right) shows $(\mathsf{K}_{\gamma_k,\gamma_k}(\mathbf{r}))_{0,0}$, i.e. the baseband term of the microscopic noise source for each trap present in the device. The microscopic noise source for the bulk trap is high in the diode depletion region, while the surface trap noise is only located in the surface area where they are present. Concerning the Green's functions, Fig. 5 shows the behavior of the (0,0) and (1,0) elements for the bulk trap and the surface trap 3 (for $k = 1,2$ similar results apply). We observe that both the real and imaginary parts of the element (1,0) of bulk trap CGF are much higher than the element (0,0), therefore explaining the strong conversion of noise due to the bulk trap. Furthermore it is observed that the imaginary part is dominant, therefore confirming that noise conversion is related to a capacitive effect in the depletion region. On the contrary, the elements (0,0) and (1,0) of the CGF for surface trap 3 are roughly of the same magnitude, therefore explaining that the surface trap noise in the resistive region is less strongly upconverted.

CONCLUSION

The effect on device terminal noise of the superposition of noninteracting traps distributed in the bulk and at the surface of a junction diode has been investigated. In the cyclostationary case, results show that the diode's terminal noise near the fundamental frequency is due to the upconversion of the baseband noise through to the off-diagonal terms of the CGF. Such upconversion is much stronger in the case of a bulk trap, while for surface traps located in a resistive area the noise conversion effects are milder. The conversion of noise related to the bulk trap is mainly capacitive and arises in the junction depletion region. The different behaviour with respect to a simple homogeneous sample (resistor) suggests that this device includes a filtering effect on noise mixed with a nonlinear behavior, i.e. the device can be interpreted as a nonlinear system with memory. These results explain the different frequency behavior of the terminal diode's noise at the baseband with respect to the one at the upper sidebands. The different upconversion

	k=B (A^2/Hz)	k=1 (A^2/Hz)	k=2 (A^2/Hz)	k=3 (A^2/Hz)
$(\mathbf{S}_{\delta l,k})_{0,0}$	6.412×10^{-27}	6.692×10^{-26}	6.689×10^{-25}	6.412×10^{-24}
$(\mathbf{S}_{\delta l,k}^{(k,k)})_{0,0}$	6.795×10^{-27}	6.041×10^{-26}	6.619×10^{-25}	6.405×10^{-24}
$(\mathbf{D}^{(k)})_{0,0}$	6.787×10^{-27}	6.041×10^{-26}	6.619×10^{-25}	6.405×10^{-24}
$(\mathbf{S}_{\delta l,k})_{1,1}$	1.783×10^{-25}	2.503×10^{-26}	2.502×10^{-25}	2.398×10^{-24}
$(\mathbf{S}_{\delta l,k}^{(k,k)})_{1,1}$	1.786×10^{-25}	2.560×10^{-26}	2.502×10^{-25}	2.398×10^{-24}
$(\mathbf{D}^{(k)})_{1,1}$	1.784×10^{-25}	2.503×10^{-26}	2.502×10^{-25}	2.398×10^{-24}

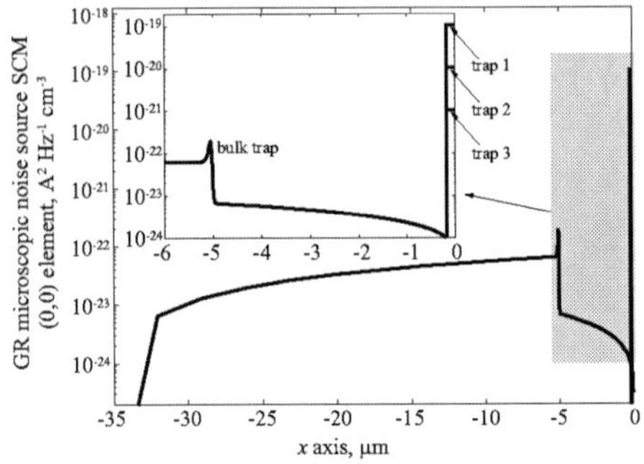

FIGURE 4. Relative contributions to (0,0) and (1,1) SCM elements (left). Element (0,0) of the microscopic noise source SCM due to trap-trap correlation. The behavior is shown on a vertical section along x axis for $y = 2.8\ \mu$m. Sideband frequency is 1 Hz.

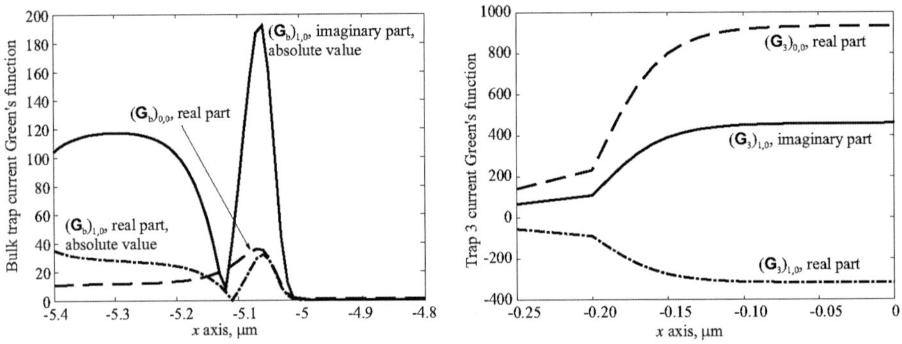

FIGURE 5. (0,0) and (1,0) elements of the CGF for the bulk trap (left) and surface trap 3 (right). The behavior is shown on a section along x axis for $y = 2.8\ \mu$m. Sideband frequency is 1 Hz.

mechanisms for bulk and surface traps have interesting implications. In fact this result highlights that identifying an analytical/compact model for the device terminal noise in small- and large-signal operation is still an open problem, at least for complex structures, and that a simple terminal level modulation of the stationary spectrum [7], often exploited in circuit-level cyclostationary noise simulations, is not in general accurate.

REFERENCES

1. S. Donati Guerrieri *et al.*, *FaN04* pp. 37–48, May 2004.
2. F. Bonani, G. Ghione, *Noise in semiconductor devices: modelling and simulation*, Springer: Heidelberg, 2001.
3. F. Hou, PhD Thesis, University of Florida, 2002.
4. F. Bonani *et al.*, *IEEE Trans. El. Dev.*, Vol. ED-49, No. 9, pp. 1640–1647, September 2002.
5. J. E. Sanchez *et al.*, *IEDM Technical Digest*, pp. 477–480, Washington, 2001. J. E. Sanchez, PhD Thesis, Univ. of Florida, 2002.
6. K. M. Van Vliet, *IEEE Trans. El. Dev.*, Vol. ED-23, No. 11, pp. 1236–1246, Nov. 1976.
7. F. Bonani *et al.*, *IEDM Technical Digest*, pp. 133–136, San Francisco, December 2002.

Mesoscopic Modeling Of Depletion Forces

P. Kotelenez[1], J. A. Mann[2], and M. J. Leitman[3]

[1]*Mathematics, Case Western Reserve University, Cleveland, OH 44106-7058, USA, peter.kotelenez@case.edu, 216 368 xxxx (analog), 216 368 5163 (fax).*
[2]*Chemical Engineering, Case Western Reserve University, Cleveland, OH 44106-7058, USA, j.mann@case.edu, 216 368 4122 (analog), 216 368 3016 (fax).*
[3]*Mathematics, Case Western Reserve University, Cleveland, OH 44106-7058, USA, marshall.leitman@case.edu, 216 368 2890 (analog), 216 368 5163 (fax)*

Abstract: Our object is to formulate and analyze a physically plausible and mathematically sound model to better understand the phenomenon of "clumping" in colloids. The term *depletion force* refers to those forces (in a Newtonian sense) which are associated with the clumping. Our model is stochastic but rigorously derived from a deterministic setup in a Newtonian setting. After a scaling limit, the stochastic motion of the large particles can then be described by a system of correlated Brownian motions. The scaling in the transition preserves a small parameter, the correlation length. From our limiting stochastic equations, we compute the probability fluxes for the difference of two large particles, provided the distribution is uniform. We show that, for short times, two particles tend to be attracted to each other when the distance between them is smaller than a critical value determined by the correlation length. Our result suggests that large particles sufficiently close together will tend to move even closer. This agrees with the observed *depletion phenomena* in colloids

Keywords: Depletion forces, colloid forces, correlated Brownian motion
PACS: 02.50Ey, 02.50Ez, 05.10Gg, 05.40.Jc

INTRODUCTION

A kinetic model for Brownian motion was introduced by Einstein [1] and von Smoluchowski [2]. A corresponding dynamical model was analyzed by Uhlenbeck and Ornstein [3]. The mathematical model, first formulated for one large particle suspended in a (stationary) liquid, could be easily generalized to models of several large particles under the assumption that their positions (and, respectively, their velocities) are spatially independent of one another. The requirement of independence has become the basic assumption in the theory of stochastic differential equations, mathematical particle systems and their macroscopic limits. However, the assumption of independence appears to be in conflict with the fact that the large particles are all suspended in the same fluid (cf. Spohn [4]). This begs the question of how to define correlations for physically relevant models. We address this question by considering systems of two types of interacting particles large particles and small particles. The terms *large* and *small* here refer to their different masses. After a suitable transition from micro-scales to meso-scales, the positions (and, respectively, the velocities) of the large particles will perform Brownian motions.

CP800, *Unsolved Problems of Noise and Fluctuations: UPoN 2005*
edited by L. Reggiani, C. Pennetta, V. Akimov, E. Alfinito, and M. Rosini

FIGURE 1: The Brownian particles are separated far enough that the small solvent particles are not depleted.

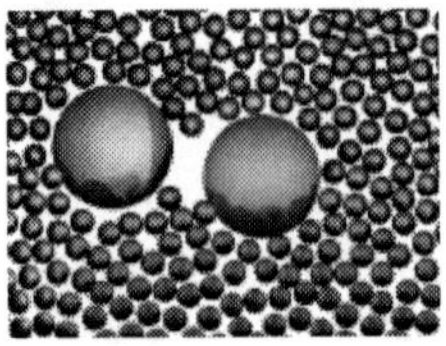

FIGURE 2: The depletion effect is observed when the Brownian particles are separated by a small distance.

The fluid around one large particle appears homogeneous and isotropic, Fig. 1, which leads to a. relatively simple statistical description of the displacement of that particle because of the collisions with the small particles. Under the assumption that the collisions of the large particle with the small particles is elastic, a Brownian motion as an approximation to the position (and, respectively, the velocity) of one large particle has been obtained by numerous authors (cf. the references in Kotelenez [5]).

Now suppose there are two large particles, very close together, Fig. 2. The fluid around each of them will no longer appear homogeneous and isotropic. In fact, the fluid between the two large particles will become *depleted* in the sense that fewer small particles per unit volume will be found between them (cf. references [6-10]). The colloid science of the depletion effect is reviewed in reference [9] and [10]. When the large particles are sufficiently close together, the collisions of the many small particles with two large particles cause their motions to become statistically correlated. We call the critical distance at which this occurs the *correlation length* and denote it by $\sqrt{\varepsilon}$.

To capture this effect, we analyze a model in which the elastic collisions between large and small particles are replaced by a mean-field interaction. The mean-field force is derived from the probability density of the velocity field of the small particles and the scaling preserves the correlation length. The positions and velocities of $N, N \geq 1,$ large particles and infinitely many small particles are given by a deterministic system of coupled oscillators with independent random initial conditions. The oscillators are coupled through the rescaled mean-field force, G_ε. Using coarse graining in space and time, we obtain N correlated diffusions in a scaling limit as an approximation to the positions of the N large particles, where the spatial correlations are computed from the probability density of the velocity field of the small particles (cf. Kotelenez [5] for details). Our analysis is carried out in space dimension $d \geq 2$. Beginning with the N limiting diffusions, we next outline the main features of our analysis.

ANALYSIS OF THE N DIFFUSIONS

Let $w(dq, ds)$ denote the standard Gaussian white noise on $\mathbb{R}^d \times \mathbb{R}_+$, defined on the same probability space as $(R^1(0),...,R^N(0))$, the initial positions of the large particles. We assume that $(R^1(0),...,R^N(0))$ and $w(dq, ds)$ are independent. Then, as the number of small particles approaches infinity, the limiting motion of the positions of the N large particles are the solutions of the following system of stochastic integral equations.

$$R^i(t) = R^i(0) + \int_0^t \int_{\mathbb{R}^d} G_\varepsilon(R^i(u) - q) w(dq, du), \quad i = 1, \dots, N, \tag{1}$$

where G_ε , the *forcing kernel,* embodies the rescaled mean-field force between large and small particles in the original system of coupled deterministic oscillators (cf. Kotelenez [5]). The system of Equations (1) has a unique solution with continuous sample paths in $\mathbb{R}^{d\times N}$, which is a Markov diffusion process in $\mathbb{R}^{d\times N}$. The system of Equations (1) may appear to be N -versions of the same equation, albeit with different initial values; in fact, they are coupled through the convolution of the kernel G_ε with the space-time white noise $w(dq, du)$. More precisely, if $G_{\varepsilon,j}$ denotes the jth component of the kernel G_ε, the diffusion coefficients for the system of Equations (1) are shown to be

$$D_{kl}^{ij}(\hat{r}) := \int G_{\varepsilon,k}(r^i - r^j - q) G_{\varepsilon,l}(-q) dq, \tag{2}$$

where the subscripts $k, l = 1, 2, \dots, d$ refer to spatial coordinates and the superscripts $i, j = 1, 2, \dots, N$ identify the large particles. See reference [8].

Paul Levy's theorem then implies that, for the ith large particle, the d -dimensional component $R^i(\cdot)$ of Equations (1) separately performs a standard Brownian motion, as required by Einstein and Smoluchowski. The significant point here is that *the joint distribution is not Brownian,* since the diffusion governed by the coefficients in Equation (2) clearly depends in a non-trivial way on the distance between distinct particles. However, for $|R^2 - R^1| >> \sqrt{\varepsilon}$ the particles are almost uncorrelated and are essentially indistinguishable from two independent Brownian motions..Accordingly, the solutions of Equations (1) are called *correlated Brownian motions.*

Henceforth we consider just two distinct particles whose motions in $\mathbb{R}^{2d}$, $(R^1(\cdot, R^1(0)), R^2(\cdot, R^2(0)))$, are governed by Equations (1). Recall that, $(R^1(0)^1, R^2(0))$ are their (random) initial positions. We are interested in local effects of the diffusion coefficient on the distance $|R^2 - R^1|$ between the two particles. For convenience, we define the differences the diffusion coefficient on the distance $|R^2 - R^1|$ between the two particles. For convenience, we define the differences

$$a := R^2(0) - R^1(0) \text{ and } b(t) := R^2(t, R^2(0))) - R^1(t, R^1(0))). \tag{3}$$

After an orthogonal transformation of coordinates in $\mathbb{R}^{2d}$, we obtain a function, $b \mapsto \sigma(b)$ on $\mathbb{R}^d$ whose values are nonnegative definite, $(d \times d)$ -matrices, for which the following assertion holds (cf. Kotelenez [8]):

There exists a process $\beta(\cdot)$ in $\mathbb{R}^d$ whose components are one-dimensional, i.i.d., standard Brownian motions, such that the difference process $b(\cdot)$, defined in Equations (3), can be represented as the unique solution to the following stochastic Itô integral equation

$$b(t) = a + \int_0^t \sigma(b(u)) \beta(du). \tag{4}$$

Using the diffusion coefficients determined by Equation (2), we next compute the probability flux associated with a vector b (cf. Van Kampen [11]). This flux will enable us to identify regions in $\mathbb{R}^d$ in which there is a bias towards attraction between the two particles or a bias towards repulsion. Assuming a uniform distribution, the flux is proportional to

$$-J(b) := \sum_{k,l=1}^{d} \frac{\partial}{\partial b_l}(D_{kl}^{11}(0) - D_{kl}^{12}(\sqrt{2}b)). \tag{5}$$

Define two complimentary sets, E_+, E_-, in $\mathbb{R}^d$ by

$$\begin{aligned} E_+ &:= \{b \in \mathbb{R}^d \ : \ \sum_{j=1}^{d} b_j \geq 0\}, \\ E_- &:= \{b \in \mathbb{R}^d \ : \ \sum_{j=1}^{d} b_j < 0\}. \end{aligned} \tag{6}$$

A Borel set B in $\mathbb{R}^d$ is said to be *attraction biased* if either $\{ B \subset E_+$ and $-J(b) > 0, \forall b \in B \}$ or $\{ B \subset E_-$ and $-J(b) < 0, \forall b \in B \}$. Conversely, a Borel set B in $\mathbb{R}^d$ is said to be *repulsion biased* if either $\{ B \subset E_+$ and $-J(b) < 0, \forall b \in B \}$ or $\{ B \subset E_-$ and $-J(b) > 0, \forall b \in B \}$.

If the forcing kernel, G_ε, is derived from a Maxwellian velocity field for the small particles, which seems a plausible assumption, $-J(b)$ can be computed explicitly. There is a positive constant c_ε such that

$$-J_u(b) = c_\varepsilon \left(d + 2 - \frac{|b|^2}{\varepsilon} \right) \left(\sum_{k=1}^{d} b_k \right) \exp\left(-\frac{|b|^2}{2\varepsilon} \right). \tag{7}$$

Hence, for the Maxwellian distribution, the region $\{b \in \mathbb{R}^d : |b| < \sqrt{\varepsilon(d+2)}\}$ is attraction-biased, and the region $\{b \in \mathbb{R}^d : |b| > \sqrt{\varepsilon(d+2)}\}$ is repulsion-biased. We should expect a similar result for any unimodular distribution.

REMARKS

We expect that for short times the distance between the two particles in attraction-biased regions would decrease and in repulsion-biased regions this distance would increase. In particular, for particles sufficiently close together, the attraction bias would be consistent with the depletion phenomena observed in colloids (cf. [9], [10]). However, the long-time behavior may be different. Indeed, a proof of the following result can be found in Skorohod [12]. Assume $d = 1$ and let $b(t, a)$ be the unique solution of Equation (4) with $b(0) = a \neq 0$. Then,

$$P[\lim_{t \to \infty} b(t, a) = 0] = 1. \tag{8}$$

This means that, on the average, the distance between two particles on the real line will eventually tend to zero. In other words, the whole positive and negative real lines are attractive regions; any repulsion-biased regions will have no long-term effect on the distances between the two particles. We expect the apparent difference between long-time behavior and short-time behavior in repulsive regions for dimension $d = 1$ to be the result of the recurrence of one-dimensional Brownian motions.

In dimension $d \geq 3$, Brownian motion is transient and, therefore, attraction-biased and repulsion-biased domains may play a different role in the long-time behavior of the distance between two particles than in the case $d = 1$. This open and interesting problem (also for $d = 2$) will be investigated in future research. Indeed, we expect to develop the theory, which will include more realistic Brownian particle interactions. These results will be compared to experimental and simulation data that are represented in reference [9].

ACKNOWLEDGMENTS

J. Mann gratefully acknowledges support in part by a DoD - MURI grant, DAAD 19-03-1-0169 to Case Western Reserve University.

REFERENCES

1. A. Einstein, *Ann.d.Phys.* 11 (1905).
2. M. von Smoluchowski, *Ann.d.Phys*, Ser. A, 21, 756-780 (1906).
3. Uhlenbeck and Ornstein *Physical Review*, 36, 823-841 (1930).
4. Spohn , *Large Scale Dynamics of Interacting Particles*, Springer-Verlag, Berlin Heidelberg, New York (1991).
5. P. Kotelenez, *Stochastics and Dynamics* (to appear) (2005).
6. S. Askura and F Oosawa, *J. Chem. Physics* 22,1255 (1954); A. Vrij, *Pure Appl. Chem.* 48, 471(1976)
7. B. Gotzelmann, R. Evans, S. Dietrich, *Physical Review E*, 57(6), 6785-6800 (1998); R. Roth, R. Evans and S. Dietrich, *Phys. Rev.* E **62** 5360 (2000).
8. P. Kotelenez, *Ukrainian Math. J. T* 57, No. 6, 757-769(2005).
9. S. Biggs, D. C. Prieve, R.R. Dagastine, *Langmuir*, 21(12), 5421-5428 (2005); Y. Mao, M.E. Cates, H.N.W.Lekkerkerker, *Physica A* 222, 10-24 (1995); A.P. Gast, C.K. Hall, W.B. Russell, *J. Colloid Interface Sci.* 96, 251 (1983); M. Dijkstra, R. van Roij, R. Evans, *Physical Review Letters*, 82(1), 117-120 (1999); M. Piech, P. Weronski, X. Wu, J. Walz, *J. Colloid and Intf Science*, 247,327-341(2002).
10. P. Kotelenez, M. Leitman, J.Adin Mann, Jr. *Manuscript in Preparation* (2005).
11. N. van Kampen, *Stochastic Processes in Physics and Chemistry*, North Holland Publ. Co., Amsterdam, New York (1983).
12. A. V. Skorohod, *Asymptotic Methods of the Theory of Stochastic Differential Equations*, Naukova Dumka, Kiev (in Russian) (1987).

Upconversion as a Universal Mechanism of Noise Transformation

E. Starikov*, P. Shiktorov* and V. Gružinskis*

*Semiconductor Physics Institute, A. Goštauto 11, LT 01108 Vilnius, Lithuania

Abstract. Upconversion of low-frequency noise to high frequency centered around the eigen frequency of a physical system is proposed as a universal mechanism of noise transformation of the system under test. This conjecture is supported by various analytical models and Monte Carlo simulation of the current noise spectrum in bulk materials, inhomogeneous resistors, and Schottky barrier diodes.

Keywords: High-frequency noise, Noise upconversion
PACS: 72.20.Ht, 72.30.+q, 72.70.+m

INTRODUCTION

In general, the noise spectrum at the output of a two terminal system can be represented as a linear transformation of some noise sources taken as input fluctuations. For example, in the framework of the Langevin approach, the input fluctuations are described by the Langevin force, while the output noise is represented through fluctuations of dynamical variables of the system under test (e.g. instantaneous velocity and energy of a single carrier, current flowing through the sample, etc.). It is well know that under thermal equilibrium the spectral density of velocity (current) fluctuations in bulk semiconductors (homogeneous diffusive resistors) exhibits a Lorentzian shape $\sim (1+\omega^2\tau_v^2)^{-1}$ with a high-frequency cutoff determined by the velocity relaxation time τ_v. Under nonequilibrium conditions there can appear additional contributions (the so called excess noise, e.g. generation-recombination noise, intervalley noise, etc.) which also exhibit a Lorentzian shape, but usually with much longer relaxation times and, hence, much lower cutoff frequencies. Moreover, in the presence of hot-carrier conditions there can appear low-frequency noise reduction, which can be explained as an additional contribution due to energy relaxation process described by a negative Lorentzian with a cut-off determined by the energy relaxation time, $\tau_\varepsilon >> \tau_v$ (see, e.g. [1] and referencies therein). Thus, various peculiarities of the current noise spectra under steady-state conditions in bulk semiconductors (or homogeneous resistors) can be described by a superposition of several Lorentzian-shape contributions.

Under cyclostationary conditions (when a periodic driving force at a given frequency is applied to the system under test) or in semiconductor structures with proper eigen frequencies (e.g. the plasma frequency, etc.) the transformation of an input noise source depends on the properties of the physical system under test and, in the general case, it can be of sufficiently complicate character. It is natural to pose the question: *can the noise spectrum of a physical system be decomposed into a small number of primitive transformations* which form the noise response of the physical system under test? An

CP800, *Unsolved Problems of Noise and Fluctuations: UPoN 2005*
edited by L. Reggiani, C. Pennetta, V. Akimov, E. Alfinito, and M. Rosini

upconversion process which redistributes some region of the noise spectrum from one frequency region to another without or with minima additional changes of the original spectrum can be considered as one of the possible candidates for such a primitive transformation. The aim of the present work is to provide arguments in favour of such a point of view.

MATHEMATICAL FORMULATION

The simplest mathematical model to describe the fluctuations of a given random process $\{x(t)\}$ whose spectrum contains the contribution of a noise upconversion phenomenon consists in the amplitude modulation of a periodic signal with frequency ω_0 by some stationary random process $\{u(t)\}$ as [2]:

$$\{x(t)\} = \{u(t)\}cos(\omega_0 t) \tag{1}$$

It is assumed that the random process $\{u(t)\}$, has zero average value and is characterized by the correlation function $C_{uu}(\tau) =< u(t-\tau/2)u(t+\tau/2) >$, with the corresponding spectral density $S_{uu}(\omega)$. The two-time symmetric correlation function of the random process $\{x(t)\}$ can then be expressed as [2]:

$$C_{xx}(t,\tau) = \frac{1}{2}[cos(\omega_0 \tau) + cos(2\omega_0 t)]C_{uu}(\tau) \tag{2}$$

According to this expression, the instantaneous spectral density of $x(t)$ is decomposed into two terms, $S_{xx}(t,\omega) \equiv 2\int_{-\infty}^{\infty} C_{xx}(t,\tau)cos(\omega\tau)d\tau = \overline{S}_{xx}(\omega) + S'_{xx}(t,\omega)$, corresponding, respectively, to the stationary component

$$\overline{S}_{xx}(\omega) = \frac{1}{4}[S_{uu}(\omega+\omega_0) + S_{uu}(\omega-\omega_0)] \tag{3}$$

and to the nonstationary (periodic) component

$$S'_{xx}(t,\omega) = \frac{1}{2}cos(2\omega_0 t)S_{uu}(\omega) \tag{4}$$

Due to the stochastic amplitude modulation of the harmonic signal, the noise spectrum of the stationary random process $\{u(t)\}$ given in Eq. (3) is shifted to the high-frequency region centered around the excitation frequency $\pm\omega_0$. As a signature of the upconversion process, Eq. (4) indicates that the low-frequency nonstationary component of the instantaneous spectral density, $S'_{xx}(t,\omega)$, should exhibit an harmonic behavior with frequency $2\omega_0$. Let us emphasize, that in accordance with Eqs. (3) and (4) only half of $S_{uu}(\omega)$ is upconverted to higher frequencies while the other half of the noise power remains in the low-frequency region, thus manifesting harmonic oscillations which reflect the temporal nonstationarity of the process.

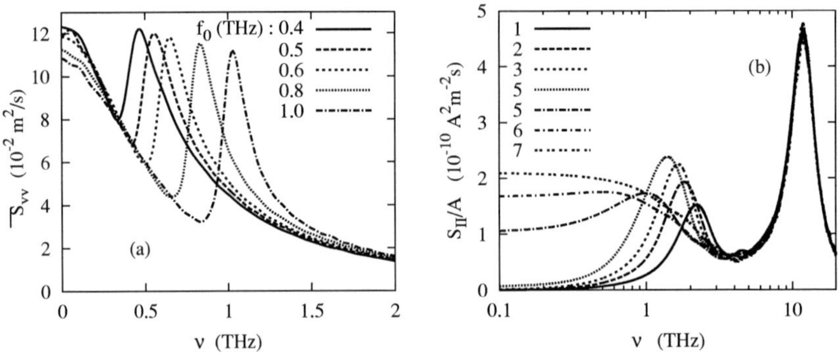

FIGURE 1. Resonant-like enhancement of current noise spectra as obtained by Monte Carlo simulations of: (a) velocity fluctuations in a bulk GaAs in the presence of a microwave electric field with the reported values of frequency f_0; (b) current fluctuations in an n^+n-metal GaAs SBD under constant applied voltages $U = 0.65, 0.75, 0.8, 0.85, 0.925, 0.95$ and 1 V (respectively curves from 1 to 7).

PHYSICAL EXAMPLES

Ideally, such a model of upconversion process can be realized only under the so called cyclostationary conditions, when the physical system undergoes the influence of an external regular harmonic signal with frequency $\omega_0 = 2\pi f_0$. Typical examples where such a model works well are the frequency shift of: (i) low-frequency 1/f noise [3], (ii) generation-recombination noise [4,5] and, (iii) intergroup (or intervalley) noise [6]. A characteristic feature of these examples is the presence of some force governed by an external harmonic signal which, by synchronizing the motions appearing in the system under test, avoids the appearance of phase noise. As a consequence a resonant-like behavior of the noise spectrum appears in the high-frequency region around the excitation frequency and its harmonics (see, e.g., [4-6]).

As an example, Fig. 1(a) reports the frequency dependence of the stationary component of $S_{xx}(t,\omega)$ calculated by the Monte Carlo method for a bulk n-GaAs with doping level $N_D = 10^{16}\ cm^{-3}$ at room temperature under the action of a microwave electric field $Ecos(2\pi f_0 t)$ with different values of frequency $f_0 = 0.4, 0.5, 0.6, 0.8, 1.0$ THz and amplitude $E = 7, 8, 9, 10, 12, 14$ kV/cm, respectively. The appearance of the resonant peak centered near the applied field frequency f_0 is related with the upconversion of the intervalley noise caused by random electron transitions between the lowest and upper valleys (see Ref. [6] for more details). The general behavior of the peaks agrees well with the upconversion model described above. The upconversion origin of the noise peaks is also confirmed by the near-harmonic behavior of the noise nonstationary component at the doubled frequency (not presented here).

It is well known that the noise spectrum of physical systems, which have intrinsic harmonic motions such as plasma oscillations, etc., also exhibits a resonant-like behaviour at the eigen frequencies of the system. As an example, Fig. 1(b) reports the results of MCP simulations of the current-noise spectra of a 0.2 μm GaAs n^+n-metal Schottky-barrier diode (SBD) with doping levels $N_+ = 2\times10^{18}\ cm^{-3}, N = 5\times10^{16}\ cm^{-3}$. Curves

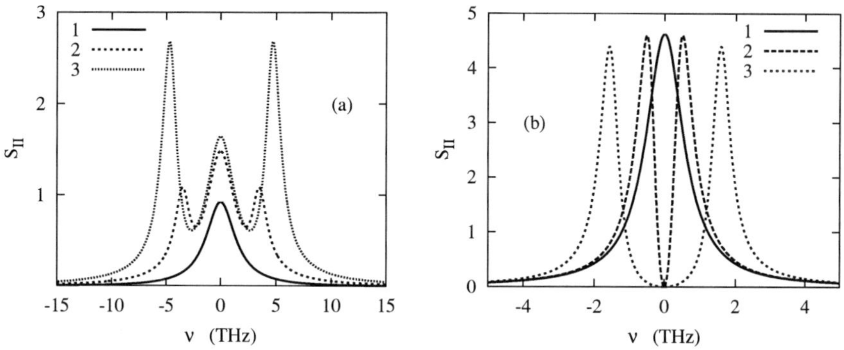

FIGURE 2. Noise spectra of current fluctuations calculated by the analytical approach of Ref. [8] for: (a) 0.1-0.1 μm n^+n-GaAs structures with variations of the n^+-region doping level N_+ (curves 1 to 3 refer, respectively, to $N_+ = N, 4N, 8N$, $N = 5 \times 10^{16}$ cm^{-3}), and (b) 0.2 μm (n-GaAs)-metal SBD with the doping level of $N = 5 \times 10^{17}$ cm^{-3} and depletion region lengths of 0, 0.001 and 0.01 μm (respectively curves from 1 to 3).

1 to 7 refer to variations of the applied voltage from barrier-limited ($U = 0.65$ V, curve 1) to flat-band condition ($U = 1$ V, curve 7), corresponding to a depletion region length l_d which varies from some finite value to zero. Two kind of resonant-like peaks can be detected in the noise spectrum . The peak at the highest frequency around 10 THz is caused by the plasma resonance at n^+n-homojunction [7] and is practically independent of U. The peak centered near 1 THz is related with the resonance of carriers reflected from the Schottky barrier. With the increase of U (and corresponding decrease of l_d) this peak shifts to lower frequencies and eventually disappears under flat-band conditions. The low-frequency plateau corresponds to full shot noise at the lowest voltages. Its value increases with U and transforms finally into a Lorentzian near thermal noise by approaching the flat-band conditions.

To consider the details of the resonant-like enhancement of noise under static conditions we use the analytical model of Ref. [8] based on Langevin sources of noise in inhomogeneous structures transformed by the collective motion of carriers in the self-consistent electric field. Figure 2 illustrates two typical cases of transformation of the spectrum of current fluctuations under constant voltage operation with increase of the eigen frequency of an intrinsic oscillation motion initiated by inhomogeneities inside the structure. Figure 2 (a) reports the noise spectrum transformation in an inhomogeneous resistor with n^+n-homojunction when the oscillation motion is originated by carrier reflection from the n^+n homobarrier. Figure 2 (b) shows the noise spectrum transformation in a homogeneous semiconductor with Schottky-barrier contact when the intrinsic oscillations appear due to reflection of carriers from the n-metal barrier.

It should be underlined that the characteristic feature of both the examples is that the total power of fluctuations integrated over the whole frequency range,

$$< \delta I^2 > \equiv \frac{1}{2\pi} \int_{-\infty}^{\infty} S_{II}(\omega) d\omega = \frac{kTe^2 N_0}{l^2 m} \tag{5}$$

is determined only by the temperature T, and the total number of carriers inside the structure N_0. However, as follows from the comparison between Fig. 2 (a) and (b), the scenarios of the spectrum transformation with the increase of the frequency of the eigen oscillations are different.

In the first case, the presence of the n^+n-homojunction does not destroy the current flow through the structure. Therefore, the low-frequency Lorentzian component of the noise spectrum (i.e. the thermal noise of the n and n^+ regions) always remains and even growth with N_+ due to growth of the n^+-region contribution. Simultaneously, when $N_+ > N$ there exist two symmetrical resonant peaks at the plasma frequency $\pm\omega_p$ which with the increase of N_+ are shifted to high frequencies following the increase of ω_p.

In the second case, with the appearance of the depletion region, the current flow through the SBD stops, so that $S_{II}(0) = 0$. As a result, the initial spectral density (solid curve at $l_d = 0$) is splitted into two parts so that $S_{II}(\omega) \sim \omega^2$ in the low-frequency region. With a further increase of l_d, two resonant peaks are shifted to the high frequency region in accordance with the increase of the eigen frequency.

However, as follows from Eq. (5), in both the cases considered above, the normalized power $< \delta I^2 > /N_0$ remains the same, thus indicating that there takes place only a redistribution of the noise power inside the whole spectrum. Therefore, under the variation of the eigen frequencies of an inhomogeneous resistor or SBD the transformation of the power spectrum of fluctuations can be reduced to a partial or a full upconversion of the low-frequency spectrum of thermal fluctuations when, respectively, a part or the whole low-frequency spectrum is shifted into the frequency regions of the SBD resonances.

DISCUSSION

As shown above (see Fig. 1 (a)), under cyclostationary conditions the mathematical model of noise upconversion based on the stochastic amplitude modulation (see Eq. (1)) works reasonably well. It is therefore natural to pose the question: *is it possible to use such a model to describe a resonant-like behavior of noise in the static case?* As follows from Eqs. (1) to (4), to apply the idea of a stochastic amplitude modulation to an intrinsic oscillation process under stationary conditions it is necessary to reformulate the model in order to exclude the nonstationary component (see Eq. (4)) which accounts for the presence of phase synchronization of fluctuations under cyclostationary conditions that is absent in the static case. Such a reformulation can be satisfied if the amplitude-modulated stochastic process can be represented as:

$$\{x(t)\} = \{u_1(t)\}cos(\omega_0 t) + \{u_2(t)\}sin(\omega_0 t) \tag{6}$$

The nonstationary component of the spectral density at the output of the system (see Eq. (4)) can be excluded from consideration by supposing that both the random processes $\{u_1(t)\}$ and $\{u_2(t)\}$, are mutually uncorrelated, but are characterized by the same spectral density. In such a model the nonstationary component of the spectral density is absent. However, similarly to the case considered above of a Langevin scheme for the current noise description (see Eq. (5)), the integral power of noise $< \Delta I^2 >$ does not depend on the intrinsic frequency ω_0. In essence, such a mathematical model, from one hand, is consistent with the Langevin approach, and, from another hand, it recovers the approximation scheme of the redistribution of the noise power in the frequency range, thus, conserving the total power of fluctuations. Here, the output spectrum, $S_{xx}(\omega)$, is related with the input spectrum $S_{uu}(\omega)$ in accordance with Eq. (3) where the factor $1/4$ must be replaced by $1/2$. Therefore, the random process which follows from Eq. (6) can be considered as a primitive process which forms a noise spectrum of a physical system with one or more eigen resonance frequencies.

It is worthwhile to underline that such a model allows us also to discuss an often posed unsolved question, namely: *what is the fluctuating quantity: the amplitude or the phase?* Indeed, Eq. (6) can be also rewritten as

$$\{x(t)\} = \{A(t)\}cos[\omega_0 t + \{\phi(t)\}] \tag{7}$$

where $\{A(t)\} = \sqrt{\{u_1(t)\}^2 + \{u_2(t)\}^2}$ and $\{\phi(t)\} = arctg[\{u_1(t)\}/[\{u_2(t)\}]$. In such a representation, one has two sources of fluctuations of the characteristics of the harmonic process at frequency ω_0, namely: random fluctuations of the amplitude $\{A(t)\}$, and random fluctuations of the phase $\{\phi(t)\}$. It is evident, that these sources cannot be treated as absolutely independent physical quantities. Indeed, to somewhat extent, fluctuations of the amplitude $\{A(t)\}$, can be interpreted as fluctuations of the phase $\{\phi(t)\}$. For example, sign-varied fluctuations of the amplitude, when its sign is randomly changed ($A \to -A$) are fully equivalent to the phase fluctuations $\phi \to \phi \pm \pi$. This allows us to formulate the following two open question:

(i) *what is the primary source of noise in physical systems under a near-oscillating behavior: amplitude or phase fluctuations*,

(ii) *can amplitude fluctuations be reduced to phase fluctuations and vice versa?*

One possible answer can be that here, in full analogy with quantum mechanics, one should satisfy the complementary principle: when the description of noise phenomena cannot be reduced to either the amplitude or the phase modulation only, but they must be present together complementing each other.

ACKNOWLEDGMENTS

This work is supported by NATO Grant PST.EAP.CLG 980629.

REFERENCES

1. L. Varani, et al., J. Appl. Phys. **80**, 5067 (1996).
2. J.S. Bendat and A.G. Piersol, *Random Data* (John Wiley and Sons, 1986).
3. A. Hajimiri and T. H. Lee, IEEE J. Solid-State Circuits **33**, 179 (1998).
4. S. Pérez, T. González, S.L. Delage, J. Obregon, J. Appl. Phys. **88**, 800 (2000).
5. F. Bonani and G.Ghione, *Noise in Semiconductor Devices. Modeling and Simulation* (Springer, Berlin, 2001).
6. P. Shiktorov et al., *Phys. Rev. B* **67**, 165201 (2003).
7. E. Starikov et al., *Appl. Phys. Lett.* **66**, 2361 (1995).
8. P. Shiktorov et al., *IEEE Electron Device Lett.* **26**, 2 (2005).

On A Semi-Classical Theory Of Noise In Quantum Well Infrared Photodetectors With A Discrete Set Of Emission-Capture Centers

A. Carbone*, R. Introzzi*, H.C. Liu† and L. Reggiani**

*Dipartimento di Fisica and CNR-INFM
Politecnico di Torino, Corso Duca degli Abruzzi 24, I-10129 Torino, Italy
†Institute for microstructural Science, National Research Council
Ottawa - K1A 0R6 - Ontario, Canada
**Dipartimento di Ingegneria dell'Innovazione and CNR-INFM National Nanotechnology Laboratory, Università di Lecce, via Arnesano s/n, 73100 Lecce, Italy

Abstract. Using a Langevin-transport approach, the current noise power spectrum of a multi Quantum Well Infrared Photodetector (QWIP) is obtained considering a discrete distribution of generation-recombination centers (quantum wells). The inhomogeneous charge distribution in the quantum well structure affects the intensity and the frequency dependence of the noise power spectrum, giving rise to features different than what expected on the basis of a model based on fully uncorrelated fluctuation processes in homogeneous structures. This model reproduces the behavior of the current noise power spectral density. In particular, its relation to the discrete structure of the device, i.e. to the number of quantum wells N, arises as a consequence of the imbalance between the current injected at the emitter and the stream of photoelectrons drifting through the structure. The $1/N$ scaling of noise is an issue whose interest exceeds the application area of QWIP and is under intense investigation in periodic substrates, in devices based on chain of organic molecules, in superlattice and others. Understanding the origins of the unavoidable deviations of noise from a simple $1/N$ dependence, exhibited by the aforementioned and many other devices in different operative conditions, represents one of the most intriguing and fascinating problem that the noise community is facing in the last decade.

Keywords: Heterostructures, periodic structure, noise experiment, noise theory, photodetectors
PACS: 85.35.Be, 85.60.Gz, 73.50.Td, 73.40.Kp, 07.57.Kp, 73.50.Pz

INTRODUCTION

Mesoscopic devices have attracted the noise community, interested by the possibility to exploit nonequilibrium fluctuations to provide additional information, not contained in the average transport characteristics. The noise level represents the ultimate limit to the sensitivity of any device. In the specific case of a photodetector, the background detectivity D^*, the specific detectivity D^*_λ and the Background Limited Infrared Performance Temperature T_{blip} are usually adopted to quantify its performances. At temperatures above $T > T_{blip}$, the fluctuations related to thermally activated processes and to background radiation prevail. Conversely, the *photoinduced noise* limits the detectivity of a photodetector operating at $T < T_{blip}$. A figure of merit currently used is the noise gain:

$$g \equiv S_I/4eI \tag{1}$$

CP800, *Unsolved Problems of Noise and Fluctuations: UPoN 2005*
edited by L. Reggiani, C. Pennetta, V. Akimov, E. Alfinito, and M. Rosini

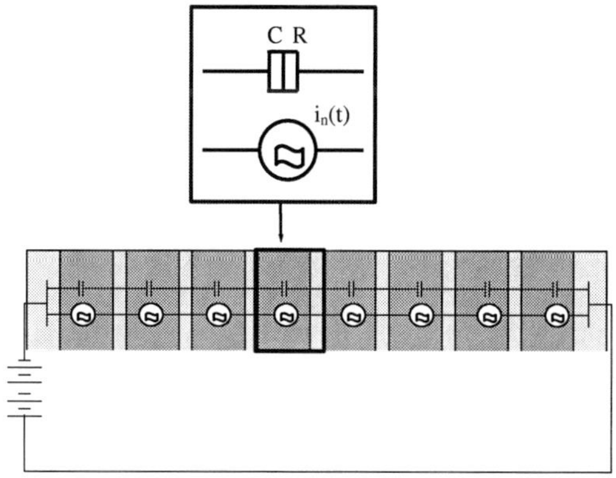

FIGURE 1. Scheme of the equivalent circuit of an array of N heterojunctions. In the specific case of a multi Quantum Well structure, the rectangle with the blue border outlines a period, constituted by one quantum well and an adjacent barrier. Each period contributes a noise generator S_I including a generation-recombination noise source S_{g-r}, related to stochastic capture-emission processes from each well, of a shot-noise source S_{shot} related to the injection of the charge carriers in each period and a thermal noise component S_{th}

where S_I is the low frequency density of current fluctuations, e is the electron charge and I is the average current. Before going into the details of our work, consider an array of N identical heterojunctions similar to that shown in Fig.(1).

The N heterojunctions are assumed to be under an external applied voltage large enough so that all the electrons emitted by each unit cross over to the next one. A device based on the repetition of identical elementay structures is the Multi Quantum Well Infrared Photodetectors. The N periods of the Quantum Well are assumed to have a voltage drop large enough so that all the electrons arriving from the previous period cross over to the Quantum Well, where they have a probability p_c to be captured and a probability $1 - p_c$ to go to the next period. Each heterojunction is characterized by the noise generators: $S_{g-r}(f)$ (*g-r* noise source), $S_{shot}(f)$ (*shot* noise source), $S_{th}(f)$ (*thermal* noise source) and finally a modulation noise source related to the dielectric relaxation of the interwell barriers. The *g-r* noise source $S_{g-r}(f)$ is related to the capture-emission processes from-and-to each Quantum Well. The *shot* noise source $S_{shot}(f)$ is related to the injection over the interwell barrier from each quantum well to the adjacent period. Under the assumption that each heterojunction exhibit independent and uncorrelated noise, the noise current flowing in the external circuit is given by:

$$S_{QWIP}(0) = S_p(0)/N \tag{2}$$

where $S_p(0)$ is the current noise generator in parallel to each QW period. This result can be simply obtained by observing that the noise current generator in each QW period is supplied to a parallel circuit. One branch consists of the impedance (admittance) Z_p

($G_p = Z = p^{-1}$) of the particular period, the other branch consists of the remaining periods connected in series. Thus if $\sqrt{S_p(0)}$ is the current noise of each period, the total noise is given is:

$$S(0) = S_p(0)\frac{G_p}{G_p + G_{N-1}} \tag{3}$$

A simple calculation shows that the factor g is approximately equal to $1/N$. The reason is that the noise originated in each period is strongly shunted by the junction impedance which is much smaller than the total impedance of other periods [1-6]

The most characteristic property of a Quantum Well Infrared Photodetectors is the discrete structure of the generation-recombination centers (QWs). The discreteness should be carefully taken into account if the number N of periods is small. To describe the photodetection process in Quantum Well Infrared Photodetectors, the standard model valid for homogeneous semiconductors including the simple generation-recombination noise mechanisms, have been adopted relying on a charge carriers relaxation dynamics occurring via a simple exponential process [16-27].

Let us first recall the main steps to calculate the $g-r$ noise in bulk semiconductors [29-32]. The transport and the recombination of carriers in bulk semiconductors are indeed described by a first-order continuity equation:

$$\frac{\partial \delta n}{\partial t} + v_d \frac{\partial \delta n}{\partial x} = -\frac{\delta n}{\tau_c} \tag{4}$$

where δn is the charge carrier density at the time t, at a distance x from the emitter. The dominant transport mechanism is *drift* with velocity v_d. The drift time between the electrodes is $\tau_d = L/v_d$, being L the interelectrode distance. The fluctuations $\delta n(t)$ is calculated by adding a white noise excitation term $\xi(x,t)$ to the right hand side of the Eq.(4). The power spectrum of the fluctuations is calculated by multiplying the Fourier transform $\delta n(f)$ by its complex conjugate $\delta n^*(f)$.

Under the assumption that $\tau_d \gg \tau_c$, (τ_c typically ranges from 0.01 to 1 μs for AlGaAs/GaAs QWIPs at 77K) the power spectral density $S_n(f)$ of the fluctuations in the photogenerated charge carrier density n becomes:

$$S_n(f) = 2\eta S_\phi(f)\frac{\tau_c^2}{1 + 4\pi^2 f^2 \tau_c^2} \tag{5}$$

where $S_\phi(f)$ is the power spectrum of the photon excitation noise source $\xi_\phi(x,t)$. $S_\phi(f)$ is *white* for all the frequencies of practical interest, η is the quantum efficiency, f is the frequency and τ_c is the recombination time. The quantity $\eta S_\phi(f)$ is the optical generation rate. When the carriers are thermally generated, the noise power spectrum is still given by the relationship (5) provided that $2\eta S_\phi(f)$ is replaced by $4g_0$, being g_0 the thermal generation-recombination rate.

The frequency-dependent response $R_n(f)$, defined as the number of excited carriers $\delta n(f)$ per incident optical power $\phi(h\nu)$, can be also deduced from Eq.(4), it is:

$$R_n(f) = \eta \frac{\tau_c}{1 + i2\pi f \tau_c}. \tag{6}$$

Since each generated carrier induces a current pulse of height eg/τ_c in the external circuit before recombination, the current noise power spectral density due to g-r processes $S_I^{gr}(f)$ can be calculated by:

$$S_I^{gr}(f) = \frac{I^2}{N_0^2} S_n(f) \tag{7}$$

where I is the total current, and N_0 is the total average number of free carriers given by $g_0\tau_c$ when the carriers are thermally generated, and by $\eta S_\phi(f)\tau_c$ when the carriers are optically generated. Using Eqs.(7) and (5) one obtains :

$$S_I^{gr}(f) = 4eIg\frac{1}{1+4\pi^2 f^2 \tau_c^2}. \tag{8}$$

Analogously, the current responsivity $R_I(f)$, defined as the current per the incident radiation power $\phi(h\nu)$, i.e. $\delta I(f)/\phi(h\nu)$, can be obtained multiplying Eq.(6) by the elementary current carried by each electron, i.e. eg/τ_c. Moreover, keeping in mind that the current responsivity is usually normalized to the photon energy $h\nu$, one obtains:

$$R_I(f) = \frac{e\eta}{h\nu} g \frac{1}{1+i2\pi f\tau_c} \tag{9}$$

The previous relationships have been used to describe the noise and the responsivity of QWIPs. The current noise power spectrum and the responsivity (Eqs.(8,9)) are written in terms of the capture probability p_c of the quantum wells:

$$p_c = \frac{\tau_d}{\tau_d + \tau_c} \tag{10}$$

in the following form:

$$S_I^{gr}(0) = \frac{4eI}{N}\left[\frac{1}{p_c} - \frac{1}{2}\right] \tag{11}$$

$$R_I(0) = \frac{e\eta_p}{h\nu}\frac{1}{p_c} \tag{12}$$

where the relationships $g = 1/(Np_c)$ and $\eta = \eta_p N$, with η_p the quantum efficiency of each period, have been also used. Moreover, it is worthy of note that in Eq. (11) the term $2eI/N$ has been subtracted. The term $2eI/N$ corresponds to the shot noise of a chain of N uncorrelated diodes in series. This subtraction is necessary since the shot noise component, of the chain of the N uncorrelated QW periods above considered, is already taken into account in the $g-r$ noise component (the first term in square brackets of Eq.(11)). In fact an electron emitted from a QW must recombine not in the same but in an adjacent QW layer to contribute to the $g-r$ noise of the chain of N uncorrelated QW periods. The need to subtract the term $2eI/N$ can be demonstrated by rigorous calculations for single and multi QWIPs [17, 21, 23, 24, 25].

According to the relationships (11,12), for a given value of the current I, the current noise power spectrum should therefore vary as $1/N$ and the responsivity should be independent of N. However, deviations of the behavior of the dynamic and static photoresponse [7-11] and of the noise power spectra [26, 27] have been reported.

The validity of the previous relationships for real QWIPs requires a number of limiting assumptions:

1. the modulation noise arising as a consequence of the stochastic dynamics of the dielectric relaxation processes occurring in the interwell *AlGaAs* regions, is neglected,
2. the device is treated as a medium with a continuous distribution of generation-recombination centers, i.e. as a bulk semiconductors whose noisy behavior is described by the standard *g-r noise* theory,
3. the carrier drift velocity, lifetime and generation rate are taken constant across the structure,
4. the statistical correlation among the noise sources in each QW is disregarded,
5. the diffusion noise is neglected.

This work is devoted to the development of a noise model based on a continuity equation of the free charge carrier including the discreteness of the quantum well layer, i.e. valid for a multi-quantum well infrared photodetector (MQWIP). The noise model is based on a transport continuity equation including the discrete geometry of the quantum well layers, whereas the Eq. (4) is valid for homogeneous semiconductors with a uniformly distributed set of generation-recombination centers [13, 14, 15].

This work has been motivated by an extensive experimental study that has evidenced characteristic features related to the discreteness of the structure in the noise power spectra of QWIPs [24, 26].

NOISE MODEL FOR A DISCRETE DISTRIBUTION OF QUANTUM WELLS

Let us consider the simplest quantum well structure constituted by N identical doped QW's, separated by identical relatively thick undoped barriers. In the absence of external photoexcitation, the processes of emission and capture of electrons are thermally activated, therefore the *linearized* electron balance in each QW can be written:

$$\frac{\partial \delta\Sigma_k}{\partial t}\frac{p_c}{e}\cdot \delta j|_{x=kL} \tag{13}$$

where Σ_k is the electron sheet concentration in the bound state with Σ_{0k} the steady state value, k is the QW index, L is the width of each period of the QW structure. $\delta\Sigma_k$ represents the fluctuation of the electron sheet concentration Σ_k around Σ_{0k}. Furthermore

$$\delta j = ev_s\delta n \tag{14}$$

is the current change due to a change $\delta n = n(t) - n_0$ in the concentration of the charge carriers, $n(t)$ being the electron concentration in the continuum states with n_0 the steady state value. In the previous equation, the bias voltage is assumed constant and high enough to provide the electron drift across the barriers with the saturation velocity v_s. Thus the electron diffusion can be disregarded.

Under the same assumptions, the continuity equation for the electrons in the continuum states can be written [13, 14, 15]:

$$\frac{\partial \delta n}{\partial t} + v_s \frac{\partial \delta n}{\partial x} = -\sum_{k=1}^{N} \frac{\partial \delta \Sigma_k}{\partial t} \delta(x - kL). \tag{15}$$

The carrier concentration n in the continuum states fluctuates due to the fluctuations of the electron sheet concentration Σ_k in the $k-th$ Quantum Well. The power spectrum of the fluctuating carrier concentration $\delta n(t) = n(t) - n_0$ can be calculated by means of Eq.(15) after introducing a white noise $\xi_k(x,t)$ excitation source in the Eq.(13):

$$\frac{\partial \delta n}{\partial t} + v_s \frac{\partial \delta n}{\partial x} = -\sum_{k=1}^{N} \left[\frac{p_c}{e} \cdot \delta j|_{x=kL} + \xi_k(x,t) \right] \delta(x - kL). \tag{16}$$

The previous equation is the Langevin-transport equation for a semiconductor having a discrete set of generation-recombination centers. The equation (16) has been solved numerically. The response $\delta n(t)$ and its Fourier transform $\delta n(\omega)$ are related to the current density by Eq.(14). The current noise power spectral density can be obtained by multiplying $\delta j(\omega)$ by its complex conjugate $\delta j^*(\omega)$ and summing over the ensemble of the quantum wells. The sum has been performed here by assuming the independence of the noise source in each period. [24, 26].

We are interested in comparing the behavior of power spectra measured in the dark and at different radiation intensities, at constant current I. The condition $\tau_c \gg \tau_d$, i.e. $p_c \rightarrow 0$ means that the rate of capture from the QW is low. The quantum wells are weakly depleted (dark or low radiation power). Conversely, when $p_c \rightarrow 1$, i.e. $\tau_c \ll \tau_d$, the rate of capture from the QW is high, thus the quantum wells are strongly depleted (high radiation power). The reason of this is that when the capture probability is low, the spectral dependence of the fluctuations is determined primarily by the electron transit time. the charge relaxation processes occur via simple exponential. Otherwise, when p_c is close to unity the relaxation processes are dominated by the capture-emission from the QWs. In such case the discrete structure of the QW layer plays a major role leading to the deviation from a simple exponential relaxation. The effect of noise saturation at high values of the capture probability $p_c \rightarrow 1$ at small number of wells N might have an origin similar to the reduction of noise in the two port devices: the so-called *partition noise*. The partition noise is due to the random distribution of the electrons between the base and the collector, in a transistor, or between the screen grid and the anode, in a vacuum pentode. In the QWIP, the strongly depleted quantum wells close to the emitter might act as the base or equivalently as the screen grid [29].

CONCLUSION

The current noise in Quantum Well Infrared Photodetectors has been modeled by means of the Eq. (16) including the discrete structure of the recombination centers in the device (QWs) instead of Eq.(4) valid for homogeneous semiconductors.

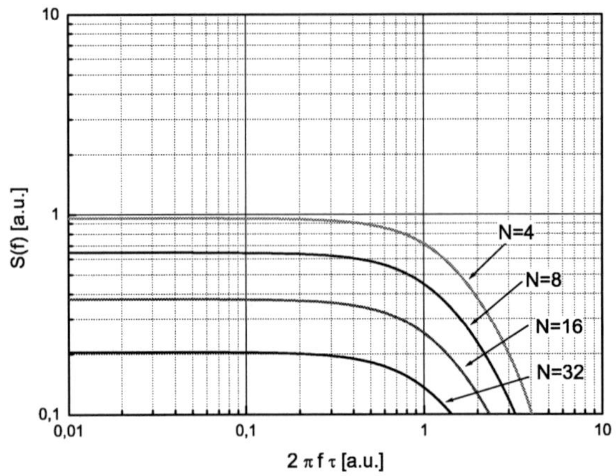

FIGURE 2. Power spectral densities of current noise obtained by numerical solution of Eq.(16). The curves refer to $N = 4, 8, 16, 32$ and to $p_c \rightarrow 0$.

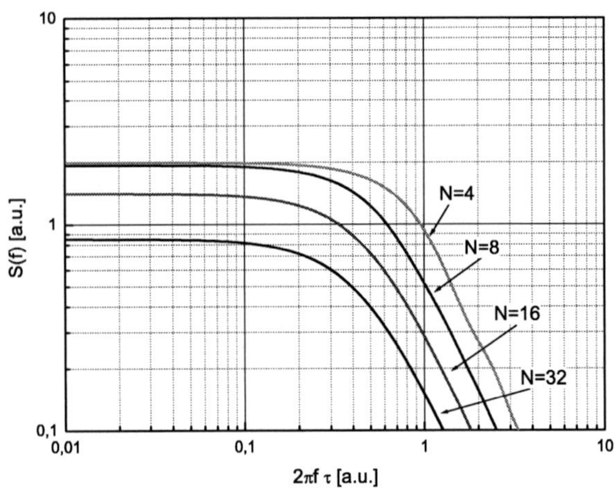

FIGURE 3. Power spectral densities of current noise obtained by numerical solution of Eq.(16). The curves refer to $N = 4, 8, 16, 32$ and to $p_c \rightarrow 1$.

The numerical results qualitatively reproduce the experimental ones [26, 27], particularly for what concerns the anomalies observed in the N-dependence of the noise intensity. The anomalous behavior of the photocurrent noise in excess to the dark noise is intrinsically related to the imbalance between the photoelectron emission and the electron capture processes in the quantum wells.

ACKNOWLEDGEMENT

This work has been supported by the Italian Ministry of University and Research under the contract PRIN 2003029008.

REFERENCES

1. Quantum Well Infrared Photodetectors, Proc. of the Int. Workshop, Torino, Italy, G. Sarusi, A. Carbone, S.D. Gunapala, H.C. Liu Eds., Elsevier. Infrared Physics and Technology, 44, (2003).
2. A. Rogalski, J. AppL. Phys., **93**, 4355, (2003).
3. P. Bhattacharya, S. Chakrabarti, A.D. Stiff-Roberts, S. Gunapala, S. Bandara, S.B. Rafol and S.W. Kennerly, IEEE Phot. Tech. Lett. **16**, 1361 (2004).
4. M.B. Johnston, M. Gal, N. Li, Z. Chen, X. Liu, N. Li, W. Lu, S.C. Shen, L. Fu, H.H. Tan and J. Jagadish, Appl. Phys. Lett. **75**, 923 (1999).
5. L. Fu, H H. Tan, C. Jagadish, N. Li, X.Q. Liu, W. Lu, S.C. Shen, Appl. Phys. Lett., **78**, 10, (2001).
6. L. Gendron, M. Carras, A. Huynh, V. Ortiz, C. Koeniguer and V. Berger, Appl. Phys. Lett., 2824, **85** (2004).
7. M. Ershov, V. Ryzhii, C. Hamaguchi, Appl. Phys. Lett., 3146, **67** (1995).
8. M. Ershov, H. C. Liu, M. Buchanam, Z. R. Wasileski and V. Ryzhii, Appl. Phys. Lett. **70**, 416 (1997).
9. V. Letov, M. Ershov, S. G. Matsik, A. G. U. Perera, H. C. Liu, Z. R. Wasilewski and M. Buchanan, Appl. Phys. Lett., **79**, 2094 (2001).
10. M. Ershov, H. C. Liu, M. Buchanan, Z. R. Wasileski, V. Ryzhii, Appl. Phys. Lett. **70**, 1828 (1997).
11. J. E. Hubbs, D. C. Arrington, M. E. Gramer, G. A. Dole, Opt. Eng. **39**, 2660, (2000).
12. L. Thibaudeau, P. Bois and J. Y. Duboz, J. Appl. Phys. **79**, 446 (1996).
13. V. Ryzhii J. Appl. Phys. **86**, 6596 (1997).
14. V. Ryzhii, I. Khmyrova, M. Ryzhii, IEEE Trans. on Electron Devices, **45**, 293 (1999).
15. M. Ryzhii, V. Ryzhii, R. Suris, C. Hamaguchi, Phys. Rev. **B61**, 2742 (2000).
16. H. C. Liu, Appl. Phys. Lett., **61**, 2703 (1992).
17. W. A. Beck, Appl. Phys. Lett., **63**, 3589 (1993).
18. K. K. Choi, Appl. Phys. Lett. **65**, 1266 (1993).
19. D. C. Wang, G. Bosman, S. S. Li, Appl. Phys. Lett. **65**, 183 (1994).
20. D. C. Wang, G. Bosman, S. S. Li, J. Appl. Phys. **79**, 1486 (1996).
21. M. Ershov and H. C. Liu, J. Appl. Phys., **86**, 6580 (1999).
22. M. Ershov, J. Appl. Phys., **89**, 6253 (2001).
23. A. Carbone and P. Mazzetti, Appl. Phys. Lett., **70**, 28 (1997).
24. A. Carbone and P. Mazzetti, Inf. Phys. Tech., **42**, 185 (2001).
25. M. Ershov and A. N. Korotkov, Appl. Phys. Lett., **71**, 1667 (1997).
26. A. Carbone, R. Introzzi and H. C. Liu, Appl. Phys. Lett. **82**, 4292 (2003).
27. A. Carbone, R. Introzzi and H. C. Liu, Inf. Phys. Tech., **44**, 363 (2003).
28. A. Carbone, R. Introzzi and H. C. Liu, Inf. Phys. Tech. **46**, 363 (2005).
29. A. van der Ziel, *Fluctuation Phenomena in Semiconductors*, Butterworth, London, (1963).
30. A. Rose, *Concept in Photoconductivity and Allied Problems*, Butterworth, London, (1963).
31. K.M. Van Vliet and J.R. Fassett, *Fluctuation Phenomena in Solids*, edited by R.E. Burgess, Academic, New York, (1965).
32. C. Van Vliet, IEEE Trans. on Electron Devices, **41**, 1902 (1994).
33. R. Landauer, Physica B, **227**, 156 (1996).

Is It Possible To Suppress Noise By Noise In Semiconductors?

L. Varani*, C. Palermo*, C. De Vasconcelos*, J.-F. Millithaler*, J.-C. Vaissiére*, J.P. Nougier*, E. Starikov†, P. Shiktorov† and V. Gruzhinskis†

*Centre d'Electronique et de Micro-optoélectronique de Montpellier (UMR CNRS 5507), Université Montpellier II, Place Bataillon, 34095 Montpellier Cedex 5, France
†Semiconductor Physics Institute, Goshtauto 11, 2600 Vilnius, Lithuania

Abstract. We investigate the possibility to suppress the diffusion noise in semiconductor bulk materials by adding to the constant electric field a fluctuating contribution characterized by a gaussian distribution and a characteristic time. The general theory is applied to the specific cases of Si and GaAs and investigated by Monte Carlo simulations. Numerical results confirm the possibility to suppress the intrinsic noise and the best conditions to realize this effect are analyzed in detail.

Keywords: Noise, suppression, semiconductors, Monte Carlo
PACS: 72.70.+m; 72.80.Cw; 05.40.Ca

INTRODUCTION

Since several years an increasing interest is devoted to constructive aspects of noise and fluctuations in different systems. In general, this kind of studies concerns the mutual interplay between a fluctuating random component (the noise) and a deterministic component (the signal). Accordingly, the addition of noise is able to enhance under some specific conditions the response to the signal. However, very few studies concern the mutual interplay between two fluctuating random signals. In this context, one of the most intriguing results was reported by Vilar and Rubí in 2001 [1]: they showed that adding an external noise to systems with an intrinsic noise may result in a less noisy system. More recent studies have extended the original theory (developed only for the low–frequency spectrum) to the full frequency spectrum of the output signal [2]. An experimental validation of the general theory applied to the case of a microscopic bead held in a noisy optical trap has been proposed in Ref. [3].

The original theory was developed using very general statistical properties of nonlinear systems without involving any microscopic detail of the system under study. Therefore the following question arises: is the general theory applicable to a realistic situation where a bulk semiconductor material is submitted to a fluctuating electric field? It is the purpose of this contribution to clarify these points using Monte Carlo simulations applied to bulk Si and GaAs.

CP800, *Unsolved Problems of Noise and Fluctuations: UPoN 2005*
edited by L. Reggiani, C. Pennetta, V. Akimov, E. Alfinito, and M. Rosini

ANALYTICAL THEORY

We recall here the main steps of the analytical theory [1] specialized to the case of diffusion noise in semiconductors. The system under consideration is a system relaxing fast to a stationary state and whose state depends on some external parameter. Let us consider as relevant variable the component v of the velocity of a single charge carrier along the direction of an applied electric field of modulus E which represents the external parameter:

$$v(t,E) = \langle v(E) \rangle + \delta v(t,E), \tag{1}$$

where the stochastic behavior of the variable velocity is given by the sum of the average value $\langle v(E) \rangle$ and the fluctuation $\delta v(t,E)$, this last being characterized by the correlation function

$$C(\tau,E) = \langle \delta v(t,E) \delta v(t+\tau,E) \rangle = \langle v(t,E) v(t+\tau,E) \rangle - \langle v(t,E) \rangle^2 . \tag{2}$$

Assuming that the characteristic time scale of the intrinsic fluctuations is much smaller than any other characteristic time of the system enables the correlation function to be approximated by a delta function as $C(\tau,E) = S(E)\delta(\tau)$, where $S(E)$ is the low frequency value of the spectral density of velocity fluctuations.

Let us consider now the case when an external noise is added to the system in the sense that also the electric field becomes time dependent according to $E(t) = \langle E \rangle + \delta E(t)$. The fluctuating term $\delta E(t)$ is assumed to be a Gaussian noise with zero mean and correlation function $\langle \delta E(t) \delta E(t+\tau) \rangle = \left\langle \delta E^2 \right\rangle \exp(-t/\tau_F)$, where τ_F is the characteristic time of the fluctuations of the electric field. Under the assumptions that the time τ_F is long enough so that the system can follow the variations of $E(t)$ and that $\left\langle \delta E^2 \right\rangle$ is small, the main result of the theory can be summarized in the following two formulas:

$$\langle v(E_0) \rangle_0 = \langle v(E_0) \rangle + \frac{1}{2} \frac{d^2 \langle v(E_0) \rangle}{dE^2} \left\langle \delta E^2 \right\rangle \tag{3}$$

and

$$S_0(E_0) = S(E_0) + \left\{ 2\tau_F \left[\frac{d \langle v(E_0) \rangle}{dE} \right]^2 + \frac{1}{2} \frac{d^2 S(E_0)}{dE^2} \right\} \left\langle \delta E^2 \right\rangle \tag{4}$$

where the subscript 0 means averaging in the presence of external noise. The former relation indicates that the average velocity is modified by the presence of the external noise, but the most interesting result is represented by the latter equation which demonstrates that the intrinsic noise of the system can be reduced if the term $\frac{d^2 S}{dE^2}$ is negative and the time τ_F is small enough.

If $\left\langle \delta E^2 \right\rangle$ is not small enough an alternative procedure can be employed: after having fitted the average velocity and the spectral density by analytical expressions, such as for instance $\langle v(E) \rangle \propto \frac{E}{\left(1+E^2\right)^{1/2}}$ and $S(E) \propto \frac{1}{\left(1+E^2\right)^{1/2}}$ for the case of covalent semiconductor materials, one can numerically calculate the average of the values of $\langle v(E) \rangle$ and $S(E)$ over a gaussian distribution for E. As a consequence, one can conclude that within this model: (i) since $\frac{d^2 \langle v \rangle}{dE^2}$ is always negative, the average velocity will decrease [see eq. (3)],

(ii) since $\frac{d^2S}{dE^2}$ is negative in the low field region, the new spectral density will decrease in this field region if τ_F is small enough [see eq. (4)].

CRITICAL ANALYSIS

Before investigating the problem of noise suppression at a kinetic level, it is the purpose of this section to critically analyse some points of the analytical theory.

First of all the concept itself of noise suppression needs some clarification. As a matter of fact, there are two possible scenarios:

1. The variance of velocity fluctuations $\langle \delta v^2(E) \rangle = \int S(\nu, E) d\nu$ is constant, thus implying that if the noise is suppressed in a certain frequency range it must be enhanced in another range in order to keep a constant value for the total power of the fluctuating signal;
2. The variance of velocity fluctuations $\langle \delta v^2(E) \rangle$ decreases, thus implying that the noise can be suppressed for every frequency or in some frequency range but the total power of the signal also decreases.

Of course it is impossible to discriminate between the two cases if the modification of the spectral density is considered only in a restricted frequency domain (for instance at zero frequency). Even if both scenarios are usually associated with noise suppression, we believe that the former one should be indicated rather as noise redistribution than noise suppression, analogously to what happens in the well known case of upconversion process [4], keeping the denomination of suppression only for the latter case. It can be easily shown that the model presented in Ref. [1] implies strictly $\langle \delta v^2 \rangle$=const., thus leading to the conclusion that this theory corresponds to the case of noise redistribution. However in a real semiconductor the addition of a fluctuating field may modify the average energy due to carriers heating, and this effect should modify the value of $\langle \delta v^2 \rangle$ since this quantity is strictly related to the electronic temperature [5].

Moreover, it is assumed that the external noise is adiabatically slow with respect to the processes responsible for the intrinsic noise, i.e. all average characteristics of the system (velocity, correlation functions, spectral densities) depend only parametrically on the external action described by $E(t)$. Such an assumption implies that the internal and external sources of noise are completely uncorrelated while in a real semiconductor the action of the external noise can in principle modify the internal noise spectrum.

Finally, another questionable point is related to the assumption that the internal noise is strictly white, i.e. its correlation function is approximated by a δ function. It is evident that this assumption is not correct in realistic cases where there always exists some correlation for the velocity during some relaxation time, and it is just this correlation which is at the origin of the standard Lorentzian cut-off for the diffusion noise spectrum.

MONTE CARLO RESULTS

To clarify all the above open problems and to overcome the simplifying assumptions inherent to the analytical theory we have employed a Monte Carlo simulation since it is

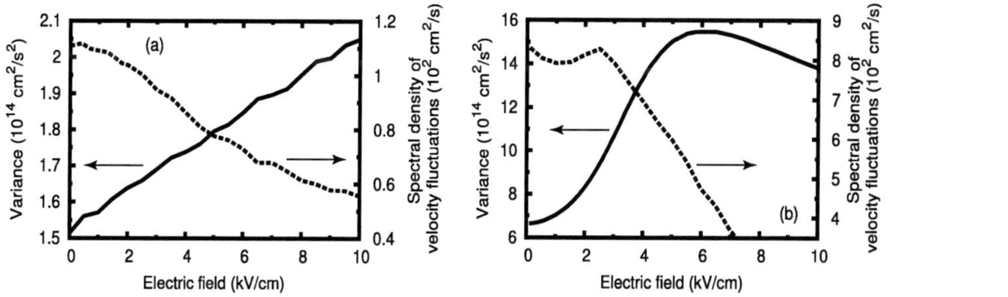

FIGURE 1. Variance of velocity fluctuations (continuous lines) and spectral density of velocity fluctuations at zero frequency (dashed lines) as functions of the electric field calculated by the Monte Carlo simulation at a temperature $T = 300$ K for the case of n-Si with a donor concentration $N_D = 0$ cm^{-3} (a) and n-GaAs with a donor concentration $N_D = 10^{16}$ cm^{-3}(b).

a method which is appropriate for the calculation of noise spectra in semiconductors at a kinetic level taking into account the main details of band structure, scattering processes as well as heating effects. For the sake of completeness we will compare the results obtained for the two cases of n-Si and n-GaAs at 300 K since they exhibit quite different behaviors under nonequilibrium conditions and they represent the two most popular prototypes of covalent and III–V semiconductor materials. We recall that while in the case of Si the average velocity saturates at high fields, thus corresponding to the behavior discussed in the analytical theory [1], in the case of GaAs the average velocity exhibits a maximum at a threshold electric field of about 3 kV/cm.

Figure 1 reports the results of Monte Carlo calculations of the variance of velocity fluctuations and of the spectral density of velocity fluctuations at zero frequency as functions of the applied electric field. In Si the spectral density exhibits a maximum at zero field thus implying a negative d^2S/dE^2 which is a necessary condition to obtain the noise suppression in the low field region [see eq. (4)] according to Ref. [1]. However, the quantity $\langle \delta v^2 \rangle$, representing the integrated spectral density in the whole frequency domain, increases monotonously due to the presence of hot carrier effects [5]. This enable us to interpret the decreasing of S as the result of the dominant effect associated with the decrease in the momentum relaxation time with respect to the increase in the variance of velocity fluctuations. As a consequence, while according to the theory of Ref. [1] the low field region is favourable to obtain the noise suppression, the Monte Carlo results, including energetic considerations in the theoretical analysis, contradicts this conclusion. In contrast, in GaAs due to the presence of intervalley transfers, both $S(E)$ and $\langle \delta v^2 \rangle$ exhibit a maximum at intermediate fields, therefore we conclude that the most favourable condition to obtain noise suppression (in the sense of decreasing of the total power of the fluctuating signal) should be found in GaAs and for a static electric field around 6 kV/cm corresponding to the maximum variance. Therefore, in the following we will present the results obtained only in GaAs.

The above expectation is confirmed by the results for the correlation functions of velocity fluctuations [Fig. 2(a)] and for the frequency dependent spectral density of velocity fluctuations [Fig. 2(b)] obtained for the case of a static electric field equal to 6.5

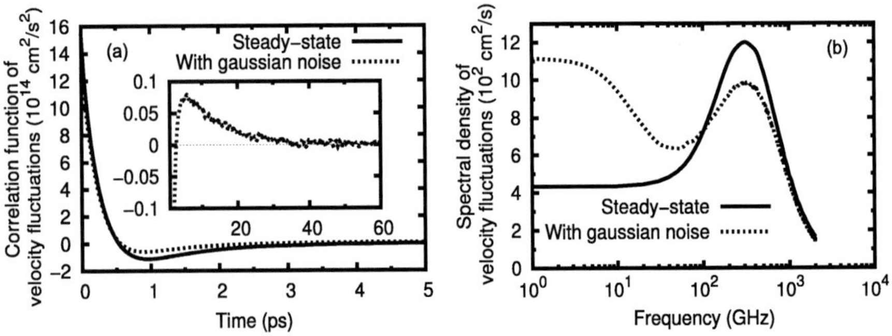

FIGURE 2. Correlation functions of single particle velocity fluctuations (a) and frequency dependent spectral density of velocity fluctuations (b) calculated by the Monte Carlo simulation for the case of n-GaAs. Continuous lines correspond to the calculation with a fixed electric field $E = 6.5$ kV/cm while dashed lines correspond to the calculation with a fluctuating electric field characterized by a standard deviation $\sqrt{\delta E^2} = 3$ kV/cm and a characteristic time $\tau_F = 10$ ps. The insert in (a) reports a zoom of the long-time tail of the correlation function.

kV/cm and when a fluctuating electric field (characterized by $\sqrt{\Delta E^2} = 3$ kV/cm and $\tau_F = 10$ ps) is superimposed to the static field. We notice that the negative part in $C(\tau)$, which is a typical feature of nonequilibrium conditions [5], is slightly reduced for the case of an externally added noise (we recall that the negative part of the correlation function in the picosecond time scale produces a peak in the spectral density at frequencies in the TeraHertz region). Additionally, we remark the appearance of a long-time positive tail of the correlation function which vanishes for a time corresponding nearly to $3\tau_F$ as one may conclude by treating τ_F as a relaxation time characteristic of field fluctuations. As a matter of fact, the presence of a fluctuating field introduces an additional correlation in the fluctuations of carrier velocity which gives a positive contribution to the low-frequency spectral density. This is confirmed by the calculation of the Fourier transform of the correlation functions which is shown in Fig. 2(b). We remark that, while the spectral density is enhanced in the low–frequency region, it is suppressed in the high–frequency region near the peak associated with intervalley transfers. Due to the logarithmic representation in the frequency domain, this suppression plays a dominant role in the total noise power spectrum.

This is confirmed by the results shown in Fig. 3 which report the integrated spectral density (i.e. $\langle \delta v^2 \rangle$). We remark in Fig. 3(a) that the variance decreases monotonously by increasing the amplitude of electric field fluctuations thus confirming the possibility of obtaining the suppression of diffusion noise by using an external noise. In some sense, the amplitude of field fluctuations plays the role of an *amplifying* factor for noise suppression in agreement with eq. (4). >From a microscopic point of view, this suppression arises from the fact that the fluctuating electric field obliges the carriers to visit regions of the momentum space characterized by a smaller variance (in full analogy with recent experiments performed in a noisy optical trap [3]). To complete the numerical analysis, Fig. 3(b) presents the results obtained for different values of τ_F but using a uniform instead of a gaussian distribution for the fluctuating electric

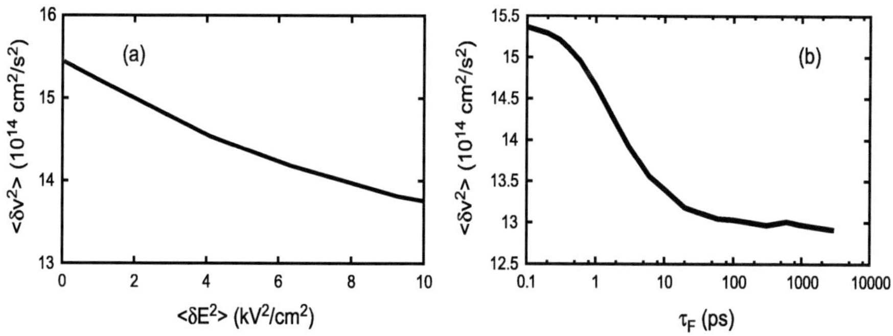

FIGURE 3. Variance of velocity fluctuations calculated by the Monte Carlo simulation for the case of n-GaAs when a fluctuating electric field centered around a constant value of 6.5 kV/cm is superimposed to the stationary value. Figure (a), obtained for a gaussian distribution, reports $\langle \delta v^2 \rangle$ as a function of the variance of electric field fluctuations and a characteristic time $\tau_F = 100$ ps. Figure (b) reports $\langle \delta v^2 \rangle$ as a function of the characteristic time τ_F for a fluctuating electric field uniformely distributed between 0.5 and 12.5 kV/cm.

field. We remark that the suppression is enhanced at increasing τ_F, finally saturating for characteristic times in the nanoseconds scale.

CONCLUSIONS

We have investigated the possibility to suppress the diffusion noise in covalent and III–V semiconductor materials by using a time dependent electric field fluctuating according to a gaussian distribution with a given characteristic time. A more precise definition of noise suppression is proposed which enables to discriminate the two cases corresponding to a suppression in a limited frequency domain and to a suppression of the total noise power of the fluctuating signal. The analytical theory is critically analyzed by evidencing some intrinsic limitation and, in particular, the assumption of complete independence of internal and external noises. The Monte Carlo results show that the suppression of diffusion noise is possible and that the most favourable conditions are found in GaAs for a fluctuating electric field around 6.5 kV/cm. Additional calculations are needed to extend the model to more complex one-port and two-port devices and eventually to other types of electrical noises.

REFERENCES

1. J.M.G. Vilar and J.M. Rubí, *Phys. Rev. Lett.*, **86**, 950–953 (2001).
2. D.B. Walton and K. Visscher, *Phys. Rev.*, E **69**, 051110-1–051110-8 (2004).
3. Y. Seol, K. Visscher, D.B. Walton, *Phys. Rev. Lett.*, **93**, 160602-1–160602-4 (2004).
4. P. Shiktorov, E. Starikov, V. Gruzhinskis, S. Pérez, T. González, L. Reggiani, L. Varani, J.C. Vaissière, *Phys. Rev.*, B **67**, 165201 (2003).
5. T. Kuhn, L. Reggiani, L. Varani, V. Mitin, *Phys. Rev.*, B **42**, 5702 (1990).

Simulating Noise Performance of Advanced Devices down to Cryogenic Temperatures

Alina Caddemi, Francesco Catalfamo, and Nicola Donato

Dept. Fisica della Materia e Tecnologie Fisiche Avanzate, University of Messina, Salita Sperone 31, S. Agata, Messina 98166, Italy

Abstract. The aim of the present work is the development of a suitable neural network structure to compute the microwave noise parameters of High Electron Mobility Transistors (HEMT). The noise parameters (NP) here considered are the magnitude ($|\Gamma_{opt}|$) and phase ($\angle\Gamma_{opt}$) of the optimum noise source reflection coefficient, the minimum noise figure (F_{min}) and the noise resistance (R_n). By this procedure, we are able to reproduce the above noise parameters of several device types from only one measured scattering parameter set, one frequency point and one generic noise figure value at different bias and/or temperature conditions. The employed noise figure is that measured in input matched conditions (i.e. 50 Ω source impedance), namely F_{50}. The trained network outputs were compared with data of noise measurements performed in our lab and the results have shown to be in a very good agreement. We observed that the difference occurring between the expected value of the noise parameters and the value calculated by the neural network is negligible. The procedure here presented is the most effective choice to avoid the need of any circuit simulations. Our neural network allows great time saving, thus leading to a rapid and complete characterization of the active microwave device under test with a remarkable generalization feature.

Keywords: Noise parameters; HEMT; neural networks; cryogenics temperatures.
PACS: 84.35.+i, 84.40.-x, 84.40.Dc, 84.40.Lj.

INTRODUCTION

The characterization of solid-state devices, in terms of scattering and noise parameters, is a topical aspect of modern advanced microwave electronics for low noise applications. More specifically, the noise characterization of microwave transistors is based on a set of time consuming measurement procedures performed with very expensive instrumentation [1]. Alternatively, a mixed approach can be used to derive the noise parameters by computer-aided analysis of a circuit model having noisy elements as it can be extracted from the scattering parameters and a reduced set of noise figure measurements [1, 2]. In the present work we develop a neural network structure to compute the microwave noise parameters of High Electron Mobility Transistor (HEMT) [3, 4] without the need of any circuit model extraction and parameter computation by CAD tools.

The noise parameters of microwave devices are expressed in the reflection coefficient form as follows:

CP800, *Unsolved Problems of Noise and Fluctuations: UPoN 2005*
edited by L. Reggiani, C. Pennetta, V. Akimov, E. Alfinito, and M. Rosini

$$F(\Gamma_S) = F_{\min} + \frac{4R_n}{Z_0} \cdot \frac{\left|\Gamma_S - \Gamma_{opt}\right|^2}{\left|1+\Gamma_{opt}\right|^2 \cdot \left(1-\left|\Gamma_S\right|^2\right)} \tag{1}$$

where F_{min} is the minimum noise figure, R_n is the noise resistance, and Γ_{opt} is the optimum noise source reflection coefficient, that is a complex quantity. Moreover, $F(\Gamma_S)$ and Γ_S are the noise figure and the relevant input termination reflection coefficient of the device, respectively. The normalization impedance is typically Z_0 = 50 ohm. The state-of-the-art in modeling the noise performance of active devices by artificial neural networks consists of simulating a specific component by giving its bias conditions, its operating temperature and key structural parameters (like gate length for FET's) as inputs and by processing the values of the equivalent circuit model and the noise parameters as output variables [5-8].

ARTIFICIAL NEURAL NETWORK MODELING

Artificial Neural Networks (ANN) have emerged as a powerful technique for modeling general input/output relationships, especially highly non-linear ones. Moreover, ANNs have been recently adopted as an interesting modeling tool for the active device characterization in the microwave frequency range. Our most recent works in this field have been focused upon GaAs lattice-matched and pseudomorphic HEMT's previously characterized in our laboratory [5, 9]. By implementing these procedures, we have been able to reproduce the small-signal circuit model and noise parameters of several device types from only one measured scattering parameter set, one frequency point and one noise figure value. The employed noise figure is that measured in input matched conditions (i.e. 50 ohm source impedance), namely F_{50}. The outputs consisted of one noise temperature (that associated to the drain-source resistance R_{DS}, according to Pospieszalski's model) and the typical 17-element small-signal circuit model of HEMTs. The trained network outputs were used in a commercial CAD tool to simulate measurements performed down to cryogenic temperatures with very good agreement. In the present work, we describe a new ANN procedure that directly gives as outputs the four noise parameters, thus avoiding the need of final circuit simulations. The block diagram of the developed neural network is shown in Figure 1.

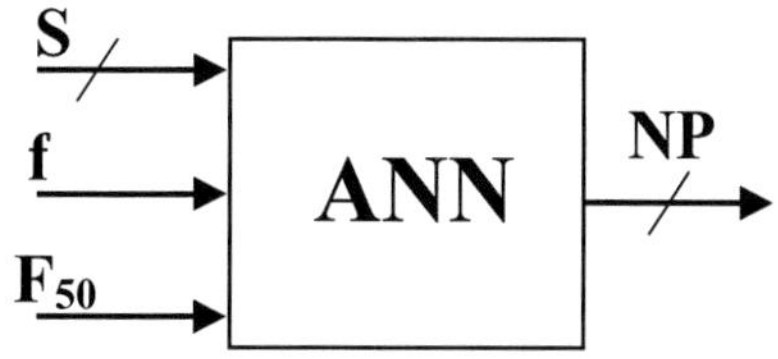

FIGURE 1. Schematic block diagram of the neural network.

The adopted ANN structure is a multi-layered system known as multilayer perceptron (MLP) with a single hidden layer such as that reported in Figure 2.

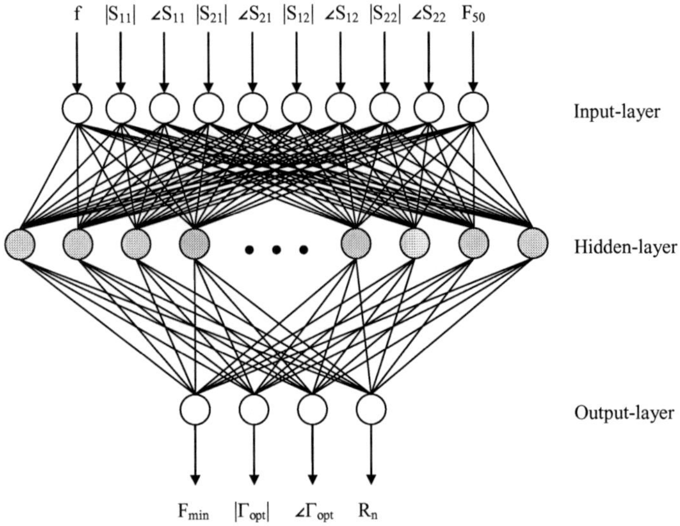

FIGURE 2. MLP (three-layers) artificial neural network.

Out optimized MLP structure is composed of three layers. The input layer consists of ten neurons and its inputs are the scattering parameters S_{11}, S_{12}, S_{21}, S_{22} (magnitude and phase), one frequency value and one noise figure value (i.e. F_{50}). The output layer consists of four neurons, each corresponding to the four noise parameters of the device under test (DUT). Finally, the hidden-layer employs twenty seven neurons.

TRAINING PROCEDURE

The training is certainly the most critical and time-consuming step in the development of a modeling procedure based on ANN. Therefore, a great care must be taken in choosing the best data training input/output set.

For the training step optimization of ANN structure, we used the back propagation algorithm of Levenberg-Marquardt. This algorithm develops an optimization process in the weight space.

We have acquired training data from measurement sets carried out in our lab on the following super low-noise pHEMTs:

1. NEC: NE20283 (vs temperature and bias conditions);
2. Mitsubishi: MGF4318 (vs temperature);
3. Celeritek: CF001 (vs temperature).

Several ANN implementations with different numbers of hidden neurons were used for training step. By means of a performance analysis we were able to choose the best ANN model with the right balance between training time and output error.

RESULTS

We here examine the performance of the implemented ANN procedure. We have tested the neural network performance with NEC and Mitsubishi devices. For these components, the measurements were performed at different values of operating temperatures (NEC: from 333K down to 233K, step 20K; Mitsubishi: from 290K down to 90K, step 50K). In the training step, we included all measured available data except those utilized for the validation test. So, we utilized the measured data of NE20283 at 233K, 273K and 293K and MGF4318 at 140K and 240K, for the testing step. Therefore, the generated data values have been compared with measured values of noise parameters. This comparison is shown in Figures 3 – 4 for NE20283 and in Figures 5 – 6 for MGF4318.

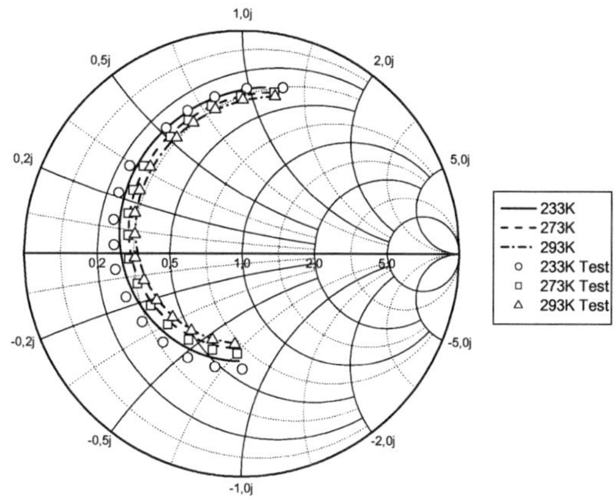

FIGURE 3. NE20283 testing: comparison between test data and measured data for the optimum noise source reflection coefficient vs. temperature, in the 6 – 18 GHz frequency range.

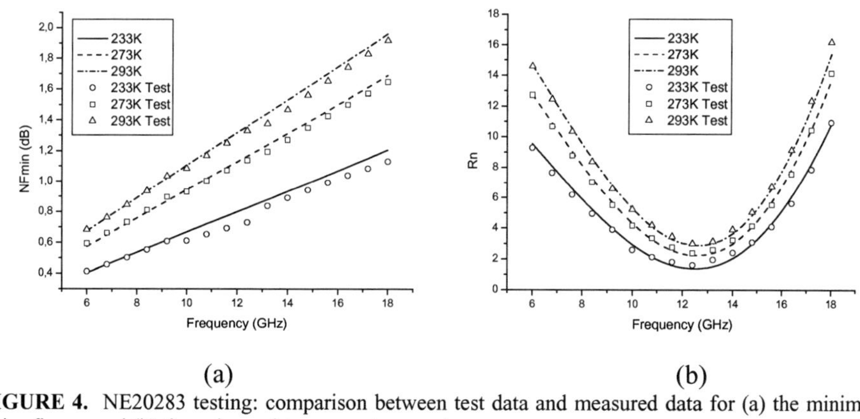

FIGURE 4. NE20283 testing: comparison between test data and measured data for (a) the minimum noise figure, and (b) the noise resistance, vs. temperature.

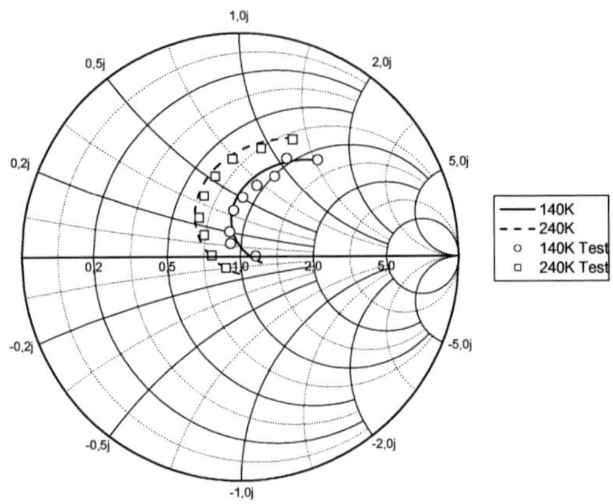

FIGURE 5. MGF4318 testing: comparison between test data and measured data for the optimum noise source reflection coefficient vs. temperature, in the 6 – 18 GHz frequency range.

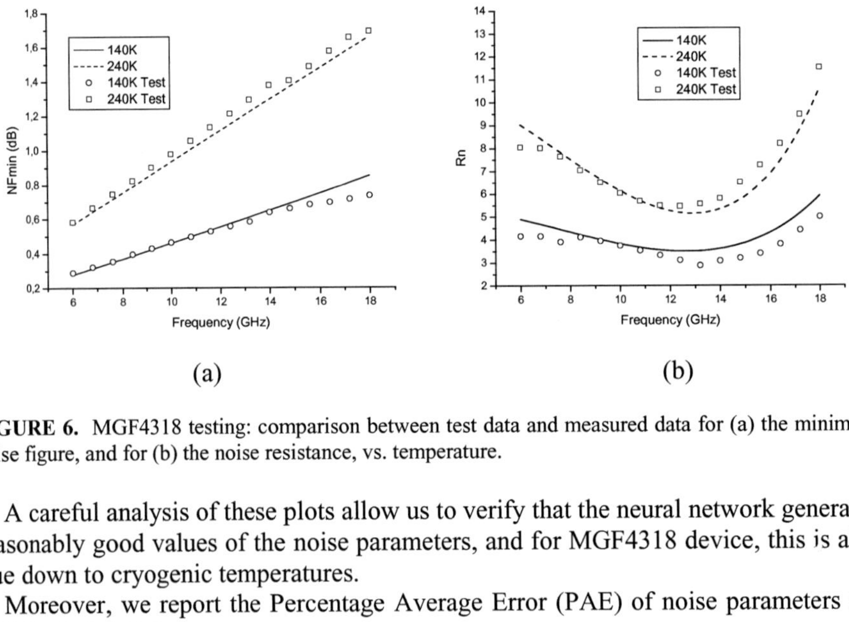

FIGURE 6. MGF4318 testing: comparison between test data and measured data for (a) the minimum noise figure, and for (b) the noise resistance, vs. temperature.

A careful analysis of these plots allow us to verify that the neural network generates reasonably good values of the noise parameters, and for MGF4318 device, this is also true down to cryogenic temperatures.

Moreover, we report the Percentage Average Error (PAE) of noise parameters for NE20283, in Table 1.

TABLE 1. PAE for NE20283.

Noise Parameters	233K	253K	293K
F_{min}	5	2.3	2.1
$\|\Gamma_{opt}\|$	4.5	1.6	0.7
$\angle\Gamma_{opt}$	1.3	0.4	0.6
R_n	8.3	3.8	3.1

CONCLUSION

In this paper we have presented a new noise modeling procedure for HEMT devices. By this ANN-based technique, we avoid the need of circuit model extraction and relevant CAD simulations for the determination of the noise parameters from the scattering parameters and a reduced set of noise figure measurements. As a direct consequence, there is great time saving in the characterization of different device types, thanks to the remarkable generalization feature of this optimized ANN structure.

It is worthy to remember here that the state-of-the-art ANN techniques applied to noise parameter computation work with a fixed device type, thus presenting a limited range of usability when an extensive activity of microwave FET characterization has to be carried out.

REFERENCES

1. A. Caddemi, A. Di Paola, and M. Sannino, *Full Characterization of Microwave Low-Noise Packaged HEMT's: Measurements Versus Modeling,* IEEE Trans. Instrumentation and Measurement, vol. 46, No. 2, April 1997, pp. 490-494.
2. G. Dambrine, H. Happy, F. Danneville, A. Cappy, *A new method for on wafer noise measurement*, IEEE Trans. Microwave Theory Tech., Vol. MTT-41, March 1993, pp.375-381.
3. V. K. Devabhaktuni, M. Yagoub, and Q. J. Zhang, *A Robust Algorithm for Automatic Development of Neural Network Models for Microwave Applications,* IEEE Trans. Microwave Theory Tech., vol. MTT-49, Dec. 2001, pp. 2282-2290.
4. Q. J. Zhang, and K. C. Gupta, *Neural Networks for RF and Microwave Design,* 2000 Artech House, Inc. 685 Canton Street, Norwood, MA 02062.
5. A. Caddemi, N. Donato, and M. G. Xibilia, *Advanced Simulation of Semiconductor Devices by Artificial Neural Networks,* Journal of Computational Electronics, vol. 3, No. 1, 2004, pp. 301-307.
6. Z. Marinkovic, V. Markovic, A. Caddemi, and B. Milovanovic, *Microwave Transistor Noise Models Including Temperature Dependence,* Proc. of TELSIKS 2003 – Serbia Montenegro, Nis, October 1-3, 2003, pp. 561-564.
7. Z. Marinkovic, V. Markovic, and B. Milovanovic, *Implementation of Temperature Dependence in Small Signal Models of Microwave Transistors Including Noise,* Proc. of MIEL 2004, May 2004, pp. 355-358.
8. V. Markovic, and Z. Marinkovic, *HEMT Noise Neural Model based on Bias Conditions,* COMPEL, vol. 23, No. 2, 2004, pp. 426-434.
9. A. Caddemi, F. Catalfamo, N. Donato, *Cryogenic HEMT Noise Modeling by Artificial Neural Networks*, to appear on Fluctuations and Noise Letters, vol. 5, No. 3, 2005.

Noise In Adaptive Systems

Miguel Angel Carrasco and Rogerio Enríquez

INAOE, Electronics Department, 72000 Puebla, Mexico

Abstract. This paper presents a statistic analysis for an adaptable filter's behavior configured in an Adaptive Line Enhancer (ALE) which is used in an ample range of applications at integrated circuit level, such as: instantaneous frequency estimation, spectrum analysis, voice codification and predictive convolution. Whilst this type of technologies become incorporated at integrated levels, it is necessary to know their noise characteristics, which much the same as the electronic noise (thermal, shot flicker, etc.) have an impact on the processing of the signals wherever they are applied. For the afore mentioned adaptable filter's configuration, the process of adaptation takes place through the Least Mean Square (LMS) algorithm, which by it's mere adjusting characteristic, generates adaptable internal noise. This noise is characterized by means of statistical tools, resulting in its probabilistic density function (PDF) by experimentation.

Experiments are performed and the resulting numeric values from this adaptable system in steady state are grouped together and acquiring a plot of the probability distribution function. Afterwards, a best fit is tried using both, a Laplacian and a Gaussian function. An inspection of the resulting graphs along with the least square error between the data and corresponding adjusted values indicate that the best representation corresponds to that of the Laplacian function.

To continue the study, the difference equation for a first and second order Digital TanLock Loop system with an adaptive filter is established and identified as a Chapman-Kolmogorov equation. Meanwhile we have found a solution for a first order Chapman-Kolmogorov (CH-K) equation with Laplacian noise, for a second order with Laplacian noise we have had some difficulties.

We may conclude that, whilst the adaptable technologies integrate, and in general, those which require an adaptation process (for example Kalman filters) it is necessary to characterize their operation with the purpose of knowing the impact on the associated circuits and in their totality. The adaptation process of this type of elements generates what we call adaptation noise and shown that is Laplacian in nature when using the LMS algorithm. Taking into account a response of such nature, it would be of great interest to study the total behavior of an application for a generalized nth order system and for all adaptation algorithms, this would complement the study of a system in which circuits of another kind would interact, which present different noises (electrical) characterized as Gaussians

Keywords: Adaptive Systems, Laplacian Noise, Chapman-Kolmogorov.
PACS: 84.30.Bv

INTRODUCTION

J. Ch. Lee and Ch. K. Un first reported the DTL as a digital sampling, non uniform PLL that uses a linear phase detector. This configuration is shown in Figure 1. From Figure 1 It can be easily seen the phase detector linear characteristic which allows easy system analysis without non-linear approximation.

CP800, *Unsolved Problems of Noise and Fluctuations: UPoN 2005*
edited by L. Reggiani, C. Pennetta, V. Akimov, E. Alfinito, and M. Rosini

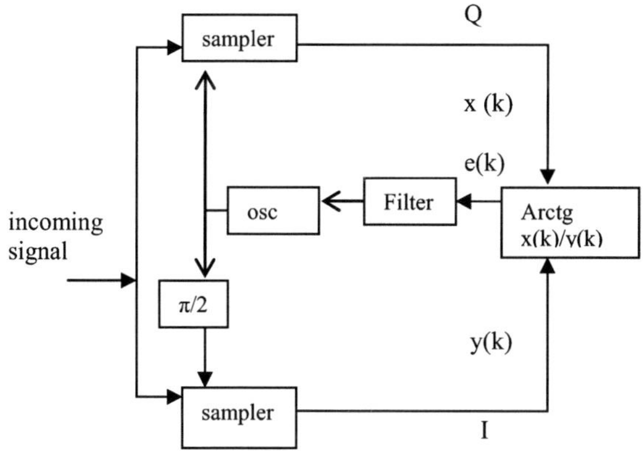

FIGURE 1. TanLock Loop.

Usually, the system's equations are derived under two conditions: in the absence of noise and in the presence of Gaussian noise and then obtaining the phase error's PDF which is the difference between the incoming phase and local generated phase.

For the above digital PLL, the filter may be implemented using an infinite impulse response (IIR) [5-8] configuration where the filter's coefficients are constant. An adaptive IIR configuration was proposed in [2] with the advantage of obtaining a greater initial locking range and also, allowing to compensate for phase changes of high orders at the input. Such phase perturbations are due to the dynamics of the communication system.

Due to the proposed adaptive filter configuration proposed in [2], the adaptive noise has a Laplacian type PDF distribution as shown in [3]. However, it is adequate to ask if there is a dependence relation between the entrance and the exit noises. By doing a scatter diagram of experimental data, it can be assured that noises are independent. This result allows studying the system under the Laplacian noise influence only.

The behaviour of the first order DTL when Laplacian noise is present, is obtained by finding the phase error PDF for the steady state. Based on the phase error equation the process is identified as a Markov process with an initial phase error. Because of this, the phase error PDF satisfies the C-K equation which solution finally defines the phase error PDF in steady state.

We obrtain the PDF for the first order DTL in the presence of Laplacian noise which is proven to be internally generated for the adaptive process and we intend to obtain the PDF solution for the second order DTL.

INPUT AND OUTPUT NOISE INDEPENDENCE

Frenquently, the statistic data reveal a relation between two or more variables, this can be expressed mathematically with a curve adjustment. A visual approximation of the shape of the curve is obtained when the data are presented in a scatter diagram as shown in Figure 2. Here we can observe the input and output noise independence, establishing a null correlation coeficient.

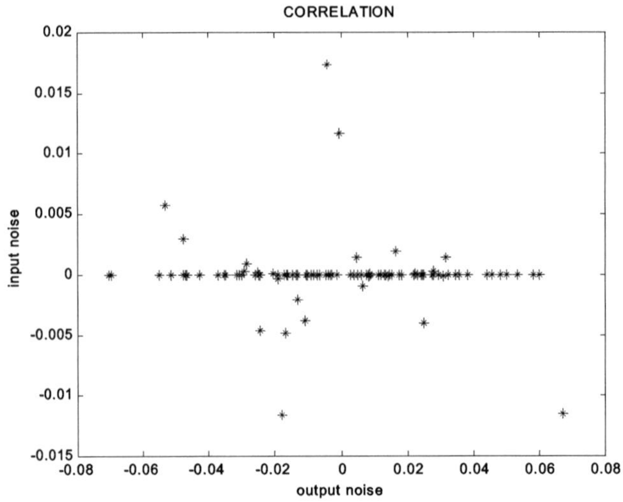

FIGURE 2. Input and output noise correlation

FIRST ORDER STEADY STATE PHASE ERROR PDF

When noise is present in first order DTL, the system equation for the phase error $\phi(k)$ may be written as [1]:

$$\phi(k+1) = \phi(k) - k_1\, f[\phi(k)] + \Lambda_0 - k_1\, \eta(k) \quad (1)$$

where $k_1 = G_1$ w and G_1 is the initial filter gain, $\Lambda_0 = 2\,\pi\,\Delta w/w_0$, w is the frequency and wo the initial frequency and $\eta(k)$ the random variable which represents the disturbance due to the filter's adaptation noise. $f[\phi(k)]$ is is the characterisitc function of the phase error detector.

The $\eta(k)$s are mutually independent for different k values.Therefore, the $\phi(k)$ error abeys a process which may be considered as a first order, time discrete Markov process. According to this, given a $\phi(0)$ initial phase error, the $\phi(k)$ PDF satisfies the following Chapman-Kolmogorov (C-K) equation [9,10]:

$$p_{k+1}(\phi|\phi_0) = \int_{-\infty}^{\infty} q_k(\phi|u) p_k(u|\phi_0)\, du \quad (2)$$

where $\phi_0 = \phi(0)$ is the initial phase error, $p_k(\phi|\phi_0)$ is $\phi(k)$*'s PDF given* ϕ_0 and, $q_k(\phi|u)$ is the transition PDF for $\phi(k+1)$ given $\phi(k) = u$. As the $\eta(k)$ range lies in the interval $(-\pi - f[\phi(k)], \pi - f[\phi(k)])$ then, It can be shown that the $q_k(\phi|u)$ range lies in the interval $(u+\Lambda_0 - k1\pi, u+\Lambda_0 + k1\,\pi)$. Consequentally $q_k(\phi|u)$ may be represented for the 0 to $\pi/2$ interval as:

$$q_k(\phi|u) = \sqrt{2} / (2\, k_1'\sigma\, [\exp(-\sqrt{2} / (k_1'\sigma\, |\phi - \phi p|\,] \quad (3)$$

where ϕp = u- k_1f[u]+ Λ_0 and σ is the noise standard deviation. The ϕ range, as the z range, in the C-K equation is (-∞,∞). Usually, the interest region for ϕ is in the (-π,π) range. A direct adjustment in the original C-K equation given an appropriate ecuation for $-\pi \leq \phi \leq \pi$:

$$P_{k+1}(\phi\,|\phi_o)=\int_{-\pi}^{\pi} r_k(\psi|z)p_k(z|\psi_o)dz$$

where

$$r_k(\psi|z) = \sum_{n=-\infty}^{\infty} q_k(\psi+2n\pi \mid z) \tag{4}$$

for ψ and z in the (-π,π) interval.

It can be seen of the last equation that r_k (ψ |z) is obtained as the infinite duration PDFs sum for the q_k (ψ | z) finite values that repeat every 2π. Finaly, the steady state PDF p(ψ) satisfies the following integral equation [10]:

$$p(\psi)= \int_{-\pi}^{\pi} r(\psi|z)p(z)dz \tag{5}$$

The p(ψ) PDF is obtained in steady state by solving equation (5) as indicated in Weinberg and Liu [9]. The results are shown in figure 3 for various parameters.

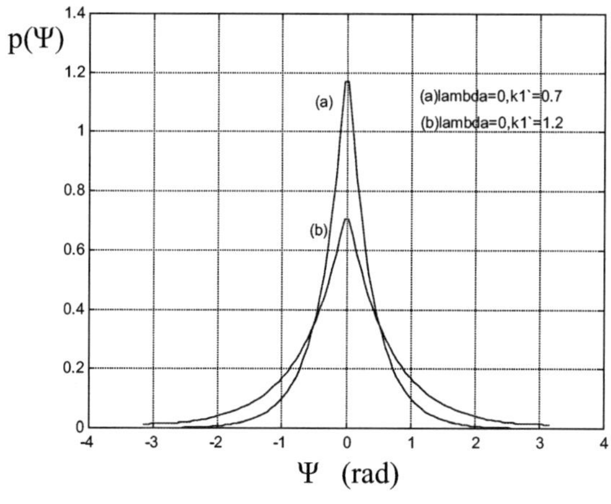

FIGURE 3. First order Steady-state pdf's for ψ for different values of k1' and $\Lambda_{0=0}$ (SNR=7 dB).

SECOND ORDER C-K WITH LAPLACIAN NOISE

For this case, the system now is described by:

$$\psi(k+2)= (2-r\,k_1)\,\psi(k+1) + (k_1-1)\,\psi(k) - r\,k_1\,\eta(k+1) + k_1\eta(k) \tag{6}$$

where $k_1 = G_1\,\omega$ y r = 1+ /G_1 and G_2 is the second initial coefficient for the filter.

To assure a Markov process for the phase error $\{\psi(k)\}$, equation (6) can be written as two equations of first order:

$$x_1(k+1) = x_2(k) \text{ and } x_2(k+1) = (k_1'-1)\, x_1(k) + (2-r\, k_1')\, x_2(k) + k_1'\eta(k) \quad (7)$$

Now $\psi(k) = x_1(k) - r\, x_2(k)$ and then, the vector $(x_1(k),x_2(k))$ can be considered as a two dimensional Markov process of first order.

To solve for the PDF $\psi(k)$, It is necessary to get the total PDF of $x_1(k)$ and $x_2(k)$ from the following C-K equation:

$$p_{k+1}(x_1,x_2 \mid x_{10},x_{20}) = \int_{-\pi}^{\pi}\int_{-\pi}^{\pi} r_k(x_1,x_2 \mid z_1,z_2)\, p_k(z_1,z_2 \mid x_{10},x_{20})\, dz_1\, dz_2 \quad (8)$$

where

$$r_k(x_1,x_2 \mid z_1,z_2) = \sum_{n=-\infty}^{\infty}\sum_{n=-\infty}^{\infty} q_k(x_1+2n\pi,\, x_2+2m\pi \mid z_1,z_2). \quad (9)$$

In equation (8), $p_k(x_1,x_2 \mid x_{10},x_{20})$ is the total PDF of $x_1(k)$ and $x_2(k)$ given $x_1(0)= x_{10}$ and $x_2(0)= x_{20}$, and $q_k(x_1,x_2 \mid z_1,z_2)$ is the total transition PDF of $x_1(k+1)$ and $x_2(k+1)$ given $x_1(k)= z_1$ y $x_2(k)= z_2$. Thus, from (7) the marginal PDF of ψ can be found.

The total transition PDF can be represented as:

$$q_k(x_1,x_2|z_1,z_2) = \delta(x_1-z_2)\sqrt{2} / (2\, k_1\sigma\, [\exp(-\sqrt{2} / (k_1\sigma^2\, x_2 - (k_1-1)\, z_1 - (2-r\, k_1)] \quad (10)$$

which is independent of k? With this, equation (8) can be represented as:

$$p_{k+1}(x_1,x_2 \mid x_{10},x_{20}) = \quad (11)$$

$$\int_{-\pi}^{\pi} \sqrt{2} / (2\, k_1\sigma\, [\exp(-\sqrt{2} / (k_1\sigma^2\, x_2 - (k_1-1)\, z-(2-r\, k_1)\, x_1|]\, p_k(z, x_1| x_{10},x_{20})\, dz$$

and then, C-K equation in steady state is:

$$P_{k+1}(x_1,x_2) = \int_{-\pi}^{\pi} K(x_1,x_2,z)\, P_k(z, x_1)\, dz \quad (12)$$

where

$$K(x_1,x_2,z) = \sum_{n=-\infty}^{\infty} \sqrt{2} / (2\, k_1'\sigma\, [\exp(-\sqrt{2} / (k_1\sigma^2\, x_2 - (k_1-1)\, z - (2-r\, k_1)\, x_1+2n\pi|] \quad (13)$$

Therefore, solving the previous equation will give the unique steady-state PDF.

Preliminary results with numerical methods for the given parameters are shown in Figure 4.

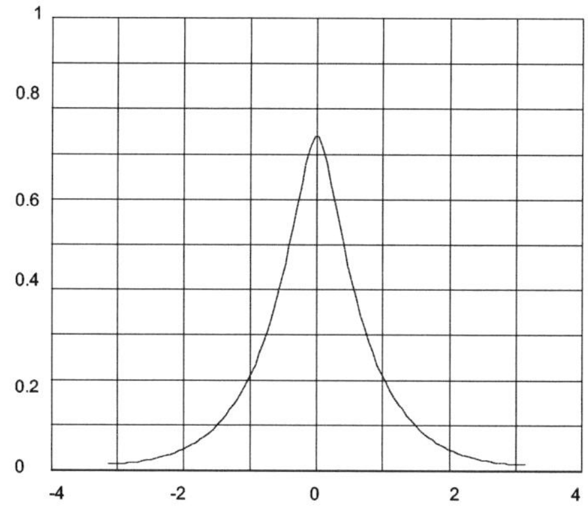

FIGURE 4. A second order Steady-state PDF for ψ (k_1= 0.77, $\Lambda_0 = \pi/5$, SNR=7 dB).

CONCLUSIONS

It was proven that the entrance and exit noises are not correlated. This permits to assure the independence between them and to perform separate noise analysis. By dividing in ranges, the absolute values, for the Laplacian noise, the desired PDF equation is obtaied. The steady state PDF for the phase error in first order DTL was obtained by numericaly solving the corresponding C-K eqution. It was also written the second order C-K equation as a two dimentional Markov process and a first numerical test was shown but convergence remains to be proven.

REFERENCES

1. Lee J. C., and Un, C. K.,"Performance análisis of digital tanlock loop", IEEE transactions on Communication, COM-30 ,pp 2398-2411. 1982.
2. Enríquez, R. A.,"Costas Tanglock Loop an adaptive filter for detection of Doppler shifted signals", Tesis doctoral, Universidad de New Brunswick Canada, Enero 1994
3. M.A. Carrasco A., Enriquez, R.A., "Ruido en sistemas adaptables", IV Encuentro de Investigación, INAOE, pags. 165-168, 2001.
4. Lasota Andrzej, Mackey C. Michael, C*haos, Fractals, and Noise. Stochastic Aspects of Dynamics*, Springer-Verlag New York, Second Edition.
5. Floyd Gardener, *Phaselock Techniques*, John Wiley & Sons, 2nd Edition.
6. Dan H. Wolaver, *Phase-Locked Loop Circuit Design*, Prentice Hall, 1st Edition.
7. Ronald E. Best, *Phase-Locked Loop: Design, Simulation, and Applications*, McGraw Hill, 4th Edition.
8. William Egan, *Phase-Lock Basics*, Wiley Inter-science, July 1998.
9. Weinberg A. and Liu B. ,"Discrete time analyses of nonuniform sampling first- and second-order digital phase lock loops." IEEE Trans. Commun., vol. COM-22, pp. 123-137, 1974.
10. Chie, C.M. , "Mathematical analogies between first-order digital and analog phase-locked loops," IEEE Trans. Commun., vol. COM-26, pp. 860-865, 1978.
11. F.N.H.Robinson, Noise and Fluctuations in Electronic Devices and Circuits. Clarendon Press, Oxford, 1974.

Transverse Velocity Fluctuations Of Hot Electrons In N-type GaAs In Crossed Electric And Magnetic Fields By Monte Carlo Methods

F. Ciccarello and M. Zarcone

Dipartimento di Fisica e Tecnologie Relative Viale delle Scienze, edificio 18, 90128 Palermo (Italy)

Abstract. In this work we investigate some steady-state stochastic properties of hot electron dynamics in bulk n-type GaAs in the presence of crossed, static electric and magnetic fields. To this aim, a single particle, three valleys-Monte Carlo method is adopted. In order to include the non-parabolicity of valleys a local parabolic approximation is performed allowing us to get an analytic expression for $\mathbf{k}(t)$ during free flights. It is shown how the room temperature-spectrum of longitudinal velocity fluctuations is markedly affected by the oscillations at the cyclotron frequency and is lowered or increased by the magnetic field depending on the considered frequency. On the contrary, the transverse spectral density turns out to be reduced in a suitable range of electric field strengths, as a consequence of the lowering of transverse velocity fluctuations's variance. We found that in the same electric field range the mean kinetic energy is lowered too.

Keywords: GaAs, Monte Carlo, hot electrons, electric and magnetic field
PACS: 72.20.Ht, 72.20.My

INTRODUCTION

Up to now the Monte Carlo (MC) determination of hot electron transport properties of *n*-type semiconductors in crossed electric and magnetic fields has been rarely addressed in literature. By using a two-valleys *parabolic* model for the conduction band, Boardman *et al.* used a single-particle MC simulation to deduce the fields-dependence of drift velocity, mean energy, Hall angle and Hall factor in GaAs at room temperature [1].

On the other hand, the MC method proved to be a powerful tool for deriving the stochastic properties of carriers motion in semiconductors in the presence of intense electric fields (see for example ref. [2, 3]). However, to our knowledge, no effort was done up to now to generalize this study to the case when a magnetic field is simultaneously applied.

RESULTS AND DISCUSSION

In a recent work [4], we derived some features of electron noise regarding the longitudinal velocity (that is the component along the applied electric field) in *n*-type GaAs. In the present paper, we try to investigate whether the electron velocity fluctuations under an applied electric field $\mathbf{E}_A$ are affected by the simultaneous presence of a magnetic field $\mathbf{B}$. With this aim in mind, we will finally focus on the analysis of transverse velocity fluctuations. Assuming that $\mathbf{E}_A$ is along $\hat{\mathbf{x}}$ and $\mathbf{B}$ along $\hat{\mathbf{z}}$, by "transverse velocity" we

CP800, *Unsolved Problems of Noise and Fluctuations: UPoN 2005*
edited by L. Reggiani, C. Pennetta, V. Akimov, E. Alfinito, and M. Rosini

mean the z-component of the electron velocity $\mathbf{v}$.

For the conduction band of bulk n-type GaAs we use a nonparabolic, three valleys-model taking into account all the relevant scattering mechanisms (scattering with polar optical and acoustic phonons and with impurities and phonon-assisted equivalent and non-equivalent intervalley processes). Since we are interested to the steady-state regime, a single-particle MC approach can be adopted. The treatment of the scattering events proceeds as usual [5]. The main difficulty comes from the wavevector-dynamical evolution $\mathbf{k}(t)$ between two next collisions when nonparabolicity effects are considered. Indeed, in the latter case the electron classical equations of motion during free flights are analytically unsolvable since the velocity $\mathbf{v}(\mathbf{k})$ (and thus the Lorentz force) is not a linear function of $\mathbf{k}$. A way to overcome this difficulty is to expand the Kane dispersion law $\varepsilon(k^2)$ up to the first order in k^2 around $k^2(0)$, that is the squared wavevector at the beginning of a free flight [4]. It may be easily shown that this local parabolic approximation (LPA) (following the idea suggested in ref. [6]) implies that at each free flight the electron effective mass m^* is renormalized to the higher value $m_R^* = m^* \sqrt{1+4\alpha \frac{\hbar^2 k^2(0)}{2m^*}}$, where α is the nonparabolicity coefficient of the considered valley. By denoting with $\mathbf{E_T} = \mathbf{E_A} + \mathbf{E_H}$ the total electric field ($\mathbf{E_H}$ being the Hall electric field) the solutions for $\mathbf{k}(t)$ obtained in this way are given by

$$
\begin{gathered}
k_x(t) = \left(\frac{m_R^* E}{\hbar B} + k_{y_0}\right) \sin(\omega_R t) + k_{x_0} \cos(\omega_R t) \\
k_y(t) = \left(\frac{m_R^* E}{\hbar B} + k_{y_0}\right) \cos(\omega_R t) - k_{x_0} \sin(\omega_R t) - \frac{m_R^* E}{\hbar B} \\
k_z(t) = k_{z_0}
\end{gathered} \tag{1}
$$

where k_{x_0}, k_{y_0} and k_{z_0} stand for the x, y and z components of $\mathbf{k}$ at the beginning of the free flight and where $\omega_R = \frac{eB}{m_R^*}$ is the renormalized cyclotron frequency.

Eqs. (1) differ from those used in ref. [1] (where nonparabolicity of valleys was fully neglected) just for the replacement of m^* with m_R^*. The LPA is physically reasonable in view of the short duration of free flights determining a small variation of k^2 between two next collisions. Indeed, we checked that the results obtained by a numerical integration of the equations of motion under assumption of the Kane dispersion law show an excellent agreement with those obtained by LPA. Of course, the latter approach is to be preferred to get shorter computational times and to facilitate the understanding of the involved physical processes. As suggested in ref.[1], once the total electric field $\mathbf{E_T}$ is chosen as an independent variable and the MC simulation is performed, $\mathbf{E_T}$ is decomposed into two components, parallel and perpendicular to the direction of the drift velocity $\bar{\mathbf{v}}$ (obtained from a temporal average of the electron velocity): the former is interpreted as the applied electric field $\mathbf{E_A}$ while the latter as the Hall field $\mathbf{E_H}$. Due to the exact solvability of the free flights equations of motion under LPA, simple accumulation formula can be derived to calculate $\bar{\mathbf{v}}$ (and thus $\mathbf{E_A}$) with no need of sampling $\mathbf{v}(t)$ at fixed time steps.

In view of eqs. (1), it turns out that in the presence of a magnetic field the amount of energy that an electron can absorb from the electric field during a free flight is upper

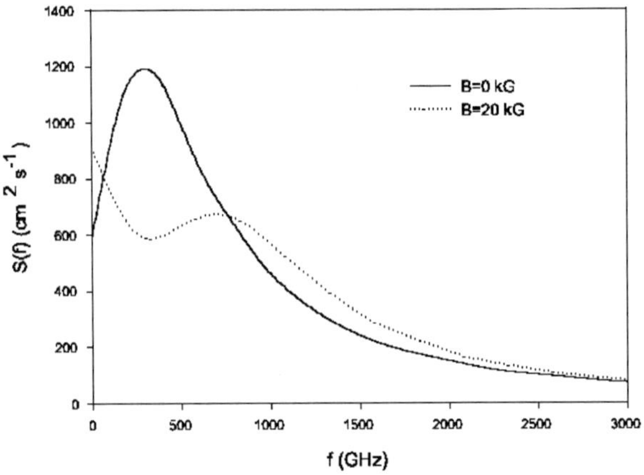

FIGURE 1. Spectral density of longitudinal velocity fluctuations $S(f)$ ($\mathrm{cm}^{-2}\,\mathrm{s}^{-1}$) at room temperature for B=0, 20 kG and E_A=6 kV/cm.

bounded. This means that at high electric fields (when the system exhibits a nonlinear response) the dynamics of the scattering processes can be generally affected by the presence of **B**. The latter expectation motivates the investigation of the auto-correlation function $C(t)$ and the spectral density $S(f)$ of velocity fluctuations. The approach used for computing $C(t)$ from a single-particle MC simulation follows standard methods [5, 7] while, in view of the Wiener-Kintchine theorem, $S(f)$ can be computed as the $C(t)$'s Fourier transform.

Fig. 1 shows the room temperature-behavior of $S(f)$ at $E_A = 6$ kV/cm and B =0, 20 kG for the *longitudinal velocity*[1]. Apart from a narrow region around $f = 0$ GHz where for B =20 kG the spectral density $S(f)$ shows a maximum (as it has been explained explained in ref. [4]), it can be appreciated that at frequencies lower than $\simeq$750 GHz, the spectral density of fluctuations is decreased by the presence of the magnetic field. However, for higher frequencies the magnetic field seems to determine the opposite effect (even if it is not so strong) and no definitive answer to the question whether *fluctuations* are lowered or not may be given. However, it may be argued that the longitudinal velocity-analysis is not a good test to answer the latter question. Indeed, in view of eqs. (1), the behavior of $v_x(t)$ and $v_y(t)$ (and thus of the longitudinal velocity) is strongly affected by the oscillations at the cyclotron frequency. A better probe to investigate the purely stochastic behavior of the system is to analyze the trend of $C(t)$ and $S(f)$ related to the *transverse* velocity $v_z(t)$. Of course $k_z(t) = k_{z_0}$ during free flights regardless the magnetic field is switched on or not. This implies that the overall trend of $v_z(t)$ is expected to be mainly determined by the occurrence of scattering events[2]. Fig.

[1] Of course, before computing $C(t)$, the longitudinal velocity's samples must be referred to the drift velocity $\bar{v}$

[2] Rigorously speaking, in view of the Kane dispersion law, the relationship between v_z and k_z does depend on the electron energy $\varepsilon(\mathbf{k})$ and it is generally influenced by the cyclotron frequency-oscillations of $\mathbf{k}(t)$.

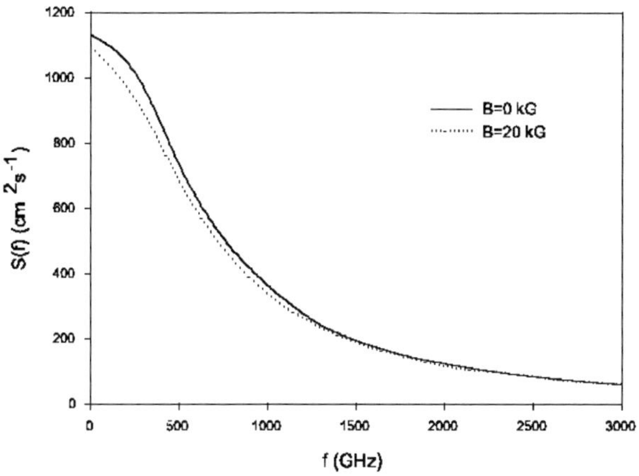

FIGURE 2. Spectral density of transverse velocity fluctuations $S(f)$ ($\mathrm{cm}^{-2}\,\mathrm{s}^{-1}$) at room temperature for B=0, 20 kG and E_A=6 kV/cm.

2 shows the room temperature-behavior of $S(f)$ at $E_A = 6$ kV/cm and B =0, 20 kG for the *transverse velocity*.

It may be appreciated how in this case the spectral density of fluctuations is lowered by the action of **B** in the whole range of frequencies, the effect getting weaker as f increases. An analysis of the $C(t)$'s in the latter range shows that, while $C(0)$ is decreased by the action of the magnetic field, the v_z-relaxation time is unchanged [3]. Since $C(0)$ represents $\langle v_z^2 \rangle$, it turns out that the observed decreasement of $S(f)$ is a consequence of the reduced variance of transverse velocity fluctuations. The above phenomenon is observed in a range of E_A between $\simeq$ 3 kV/cm and $\simeq$ 10 kV/cm.

Fig. 3 shows the mean kinetic energy $\bar{\varepsilon}$ as a function of the applied electric field at room temperature for B =0, 20 kG. It may be appreciated that $\bar{\varepsilon}$ is lowered by the presence of the magnetic field in a range of electric field strenghts between $\simeq$ 3 kV/cm and $\simeq$ 10 kV/cm. We interpret this phenomenon as due to the previously mentioned confinement of the wavevector-motion during free flights (see eqs. (1)). Interestingly, the range in which $\bar{\varepsilon}$ turns out to be decreased is approximately the same where a reduction of $\langle v_z^2 \rangle$ is observed.

CONCLUSIONS

In conclusion, in this work we have presented a preliminar room-temperature analysis of velocity fluctuations in bulk n-type GaAs in crossed electric and magnetic fields.

However, the latter effect is expected to be very weak in the regime here considered and it thus does not play any relevant role.

[3] Actually, it is even obtained that, at a fixed applied electric field strength, once the $C(t)$'s with and without **B** are normalized to their respective values at $t = 0$, they turn out to be identical.

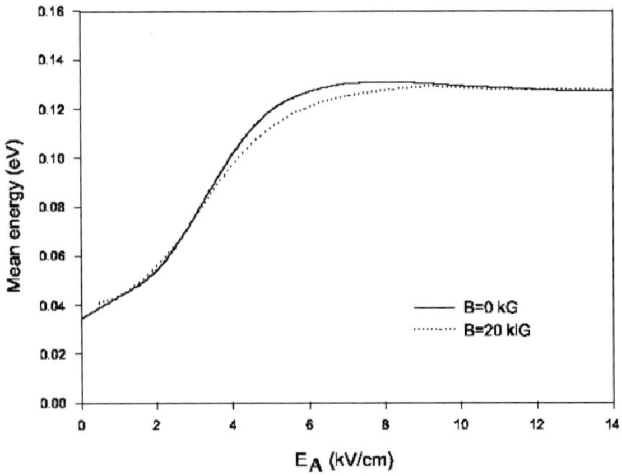

FIGURE 3. Mean kinetic energy (eV) as a function of E_A(kV/cm) at room temperature for B=0, 20 kG.

After showing the MC approach used for taking into account nonparabolicity effects, we have showed how the spectrum of longitudinal fluctuations is markedly changed by the presence of the magnetic field. Since the latter one is not negligibly affected by the oscillations at the cyclotron frequency, an analysis of tranverse velocity fluctuations has been performed. We have found the existence of a range of electric field strengths where the spectral density of these is lowered by the magnetic field as a consequence of the reduction of the transverse velocity variance. Remarkably, in the same range of electric field values the electron mean kinetic energy turns out to be reduced as well. Further investigations (possibly exploring the behavior of the distribution function) are required in order to accurately and definitively assess the physical mechanisms underlying the reduction of transverse fluctuations.

REFERENCES

1. A. D. Boardman, W. Fawcett, and J. G. Ruch, *Phys. Stat. Sol. (a)*, **4**, 133 (1971)
2. R. Fauquembuergue, J. Zimmerman, A. Kaszynski, E. Constant, and G. J. Microondes, *J. Appl. Phys.*, **51**, 1065 (1980)
3. D. K. Ferry, and J. R. Barker, *J. Appl. Phys.*, **52**, 818 (1980)
4. F. Ciccarello, and M. Zarcone, to be pubblished in *Proc. of the 18th Intl. Conf. of Noise and Fluctuations* (Salamanca 2005)
5. C. Jacoboni, and L. Reggiani, Rev. Mod. Phys., **55**, 645 (1983)
6. P. Warmenbol, F. M. Peeters, J. T. Devreese, G. E. Alberga, and R. G. van Welzenis, *Phys. Rev. B*, **31**, 5285 (1985)
7. C. Jacoboni, and P. Lugli, *The Monte Carlo Method for Semiconductor Device Simulation*, Springer-Verlag (1989)

Investigation of longitudinal velocity fluctuations in MOSFETs by means of ensemble Monte Carlo simulation

Raúl Rengel[1*], Javier Mateos[1], Tomás González[1], Daniel Pardo[1], Gilles Dambrine[2], François Danneville[2] and María J. Martín[1]

[1] *Departamento de Física Aplicada, Universidad de Salamanca, 37008 Salamanca, Spain*
[2] *Département Hyperfréquences et Semiconducteurs, I.E.M.N., F-59655, Villeneuve d'Ascq, France*
**e-mail : raulr@usal.es*

Abstract. A microscopic investigation of high-frequency noise in a deep-submicrometer MOSFET device is carried out by means of an ensemble Monte Carlo simulator. Together with the bias dependence of the spectral densities of drain current fluctuations at terminals (directly provided by the simulator) the noisy nature of carriers within the channel is analyzed. Results show that, under saturation bias conditions, carriers in the velocity overshoot section present a much stronger intensity of longitudinal velocity fluctuations than electrons in the diffusive section. However, the amplitude of fluctuations in the diffusive section of the channel is increased as the gate voltage is raised (in good agreement with the augmentation of electric field in that section). This, together with the increase of sheet carrier concentration, leads finally to a local current noise source higher than in the overshoot section.

Keywords: Fully-Depleted SOI MOSFET, Monte Carlo simulation, high-frequency noise.
PACS: 72.70.+m, 85.30.Tv, 85.30.De, 05.10.Ln

INTRODUCTION

To get an optimized development of analog MOS technologies a detailed knowledge of the noise properties of deep-submicrometer MOSFET transistors is necessary. As a consequence, the modeling of high-frequency noise in such devices is receiving a lot of attention in the last years, and the consideration of advanced simulation techniques that incorporate physically-based noise sources is becoming mandatory. One important issue when addressing the investigation of high-frequency noise in downscaled MOSFETs operating under saturation conditions is the appearance of an excessive drain current noise as compared to long-channel predictions: the origin of this behavior is still a subject of controversy, since some authors suggest that hot carriers near the drain are responsible for this phenomenon [1] while others affirm that the noise generated within the device is due only to carriers in the diffusive region (where the average longitudinal velocity is under the velocity saturation value in bulk material) [2]. Therefore, the noisy nature of carriers in the different sections of the channel and their final contribution to drain current noise remain an open question.

CP800, *Unsolved Problems of Noise and Fluctuations: UPoN 2005*
edited by L. Reggiani, C. Pennetta, V. Akimov, E. Alfinito, and M. Rosini

In this work, we have analyzed the spectra of electron velocity fluctuations at high-frequencies within different sections of the channel of a fabricated MOSFET by means of an ensemble Monte Carlo (MC) simulator. The stochastic character of this simulation technique, together with the important information provided about internal physical quantities of interest (concentration, velocity and energy of carriers, scattering mechanisms, etc.) makes it the ideal tool to carry out this research and to give a physical insight of noise sources within the channel.

DEVICE UNDER TEST AND MONTE CARLO PROCEDURE

The device under test in this work is a Fully-Depleted Silicon-on-Insulator (SOI) n-MOSFET with effective gate length of 160 nm. The thickness of the active layer is 30 nm (therefore quantum confinement effects are not significant) and the channel is doped. The buried oxide thickness is 400 nm and the gate oxide thickness is 5.5 nm, so leakage and tunneling effects are negligible. More details about the experimental device can be found in [3]. The simulations were carried out by means of a 2D bipolar ensemble MC simulator: the main experimental figures of merit have been correctly reproduced for the same devices under consideration in the present work [4], which confirms the reliability of the simulator.

For the calculation of current noise at terminals, we chose the typical noisy two-port device representation, the noiseless device represented as a black box (described by admittance parameters) together with two current noise generators, one at the input (gate) and the other at the output (drain), both correlated [5]. The values of the noise sources are determined by processing the instantaneous values of current at terminals recorded during long simulations: in this way, no assumptions are made about the physical origin or the value of the noise source. From the instantaneous current fluctuations, the correlation function is evaluated; for the drain current, i.e.:

$$C_{ID}(t) = \lim_{T\to\infty} \frac{1}{T} \int_0^T \delta I_D(t')\delta I_D(t'+t)dt' \qquad (1)$$

Fourier transformation of this quantity to the frequency domain gives the spectral density of current fluctuation at that terminal:

$$S_{ID}(f) = 2\int_{-\infty}^{\infty} C_{ID}(t)e^{j2\pi ft}dt \qquad (2)$$

To address the investigation of velocity fluctuations within the channel, a different procedure is considered. Within the framework of some deterministic approaches, the local current noise source can be expressed through [6]:

$$S_{in}(x,f) = q^2 N(x) S_v(x,f) W / dx \qquad (3)$$

q being the electron charge, $N(x)$ the local sheet carrier density, dx the length of the local section, W the device width, $S_v(x,f)$ the local noise spectrum of velocity fluctuations and $S_{in}(x,f)$ the local noise source. Under these approaches, to obtain the final spectrum of current noise at the drain terminal it is necessary to integrate this quantity along the channel weighted by the adequate transfer field. In our MC simulation, this is not needed since the procedure to evaluate S_{ID} is based on a different approach, as explained above [eq. (2) and (3)]. Moreover, the evaluation of the transfer field by means of Monte Carlo simulation is neither reliable nor affordable from a practical point of view due to the stochastic nature of the method.

Nevertheless, from the information provided by the MC simulation it is indeed possible to determine the values of $S_v(x,f)$ and $S_{in}(x,f)$ in different sections of the channel. To properly determine S_v (which refers to single-electron fluctuations), the contribution of number fluctuations was eliminated and the simulation was performed in the absence of fluctuations of the self-consistent electric field [7], which is "frozen" after the first 100 ps. Under the above described conditions, the instantaneous values of longitudinal velocity of every carrier present at each channel section are recorded, and the correlation functions are determined in an analogous way to equation (1). Afterwards, $S_v(x,f)$ is calculated including the adequate normalization by the average number of particles present in each section during the whole simulation time. In the results shown in this paper, two sections of the channel are considered: the diffusive section (where the average longitudinal velocity of carriers is below the value of velocity saturation in bulk silicon, v_{sat}) and the overshoot section (with average velocity over v_{sat}) that appears at the drain end of the channel in saturation conditions.

SPECTRAL DENSITIES OF CURRENT FLUCTUATIONS

The spectral densities of current fluctuations and their cross-correlation showed the typical frequency dependence for FETs in the RF and microwaves bias range: S_{IG} exhibits a f^2 dependence, S_{ID} has a white noise behavior and the imaginary part S_{IGID} has a linear dependence on f, its real part being negligible. Let's focus on the bias dependence of S_{ID}, since it is usually considered as the primary intrinsic noise source in a MOSFET device from a circuital point of view. In Figure 1(a), the MC results for S_{ID} as a function of V_{DS} for V_{GS} equal to 1.0 V are shown, together with the results predicted by the theoretical Klaassen/van der Ziel's model [6] ($S_{ID} = 4\gamma g_{d0} K_B T$, the value for g_{d0} extracted from the I_D-V_{DS} curve and γ varying from 1 for $V_{DS} = 0.0$ V to 2/3 in saturation). As V_{DS} is increased over 0.5 V (saturation condition) the numerical values for S_{ID} are progressively higher than those predicted by the long-channel calculation (that remains constant). This behavior has been attributed by other authors to channel length modulation effects [2]. The nonlinearity in the local channel conductance created by the appearance of the pinch off and the progressive reduction of the length of the diffusive section with V_{DS} (at fixed V_{GS}) yields to a stronger transfer of the noise generated in that region. However, it must be also taken into account that, for the device under consideration, as V_{DS} is raised in saturation the average energy of carriers near the drain is significantly increased [Figure 1(b)], as well as the longitudinal electric field. As discussed in the next section, S_{in} in the velocity overshoot area also increases with V_{DS} and its values are similar to those obtained in the diffusive section. The noise generated by hot carriers should be also taken into account as a contribution to the excess current noise obtained at the drain, as shown by other authors [8].

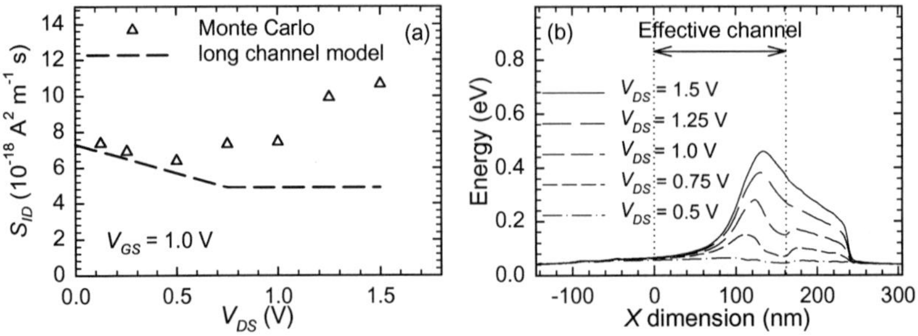

FIGURE 1. S_{ID} as a function of V_{DS} for V_{GS} = 1.0 V (a) and average energy within the channel (b).

In Figure 2(a) we plot the MC results obtained for S_{ID} as a function of V_{GS} for a fixed V_{DS} in saturation (1.5 V), together with the Klaasen/van der Ziel prediction. As in the case previously commented, the values obtained in the simulation are significantly higher than the theoretical prediction. Checking the energy [Figure 2(b)], the concentration and the average number of scattering mechanisms in the channel, it can be observed that as V_{GS} is raised the average energy of carriers is increased in the diffusive section of the channel and the peak value near the drain is decreased. At the same time, the concentration and the average electric field in the diffusive section are significantly raised. The combination of these facts produces an important increase of the local noise source, as we shall see in the next section, thus leading to the increase of S_{ID} vs. V_{GS}.

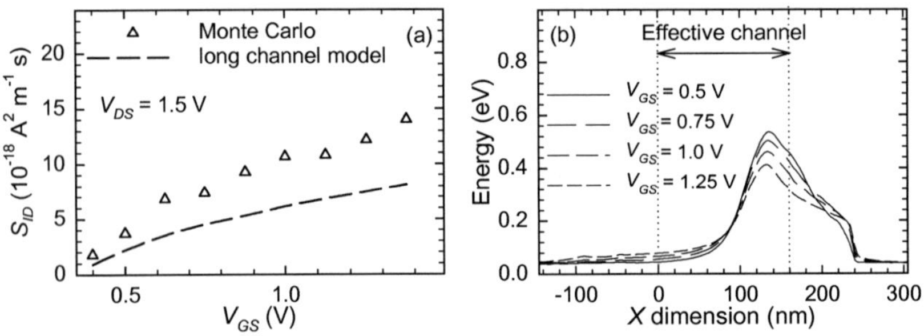

FIGURE 2. S_{ID} as a function of V_{GS} for V_{DS} = 1.5 V (a) and average energy within the channel (b).

ANALYSIS OF VELOCITY FLUCTUATIONS

Following the procedure indicated above, we performed the calculation of the spectral density of velocity fluctuations S_v in the diffusive and overshoot sections of the channel [an example of the single-particle normalized velocity fluctuations δv(t) obtained is shown in Figure 3(a),(b); as it can be observed, the amplitude of the fluctuations is significantly larger in the overshoot section, what yields a higher zero-time value of the correlation function]. Figure 3(c) shows the results for S_v as a function of V_{GS} for V_{DS} = 1.5 V (the same bias conditions as in Fig. 2). The dashed line shows the values for S_v obtained from the relationship S_v = 4D and considering the bulk diffusion coefficient corresponding to the channel doping concentration. The inset of Fig. 3(c) shows the results for S_v as a function of V_{DS} for V_{GS} = 1.0 V. S_v was found to be frequency independent in the frequency range of interest (1 to 10 GHz). As it can be observed, carriers in the velocity overshoot section present a stronger S_v in the whole V_{GS} range, specially for low V_{GS}, this is, in deep saturation and near the weak inversion regime. Transport in the overshoot section is much closer to ballistic regime: as an example, the Monte Carlo simulation showed that the average mean free path is 4.1 nm in the diffusive section and 7.3 nm in the overshoot section for V_{GS} = 0.5 V and V_{DS} = 1.5 V.

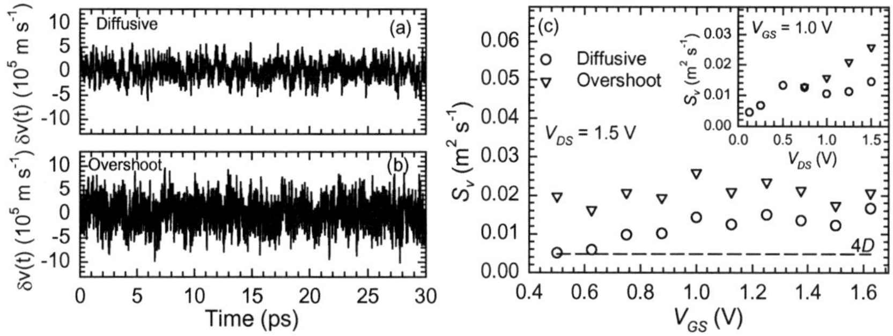

FIGURE 3. Velocity fluctuations as a function of time for V_{GS} = 1.0 V and V_{DS} = 1.5 V (a),(b) and S_v as a function of V_{GS} for V_{DS} = 1.5 V (b). Inset shows S_v as a function of V_{DS} for V_{GS} = 1.0 V.

As V_{GS} is increased, so do the average longitudinal electric field in the diffusive section of the channel and the mean free path (5.4 nm for V_{GS} = 1.25 V). As a consequence, the amplitude of the velocity fluctuations is augmented (which leads to higher zero-time values of the correlation function) and S_v is raised monotonically in this section, thus reducing the differences with the results for the overshoot section. No significant differences in the decay time of the correlation function were found, what indicates that the augmentation of phonon scattering is compensated by the reduction in the impurity scattering (it must be reminded that the channel is doped) as the average energy in the diffusive section is raised. It can be also observed that the values for S_v are in general well over the value predicted by considering the bulk diffusion coefficient corresponding to the active layer doping concentration. The values only agree reasonably for low V_{GS} and in the diffusive section, a situation in which the average energy of carriers is close to thermal equilibrium.

Considering the dependence on V_{DS} (inset in Figure 3), for the beginning of saturation (V_{DS} = 0.75 V) both sections show similar values for S_v. However, as V_{DS} is augmented, S_v is clearly increased in the overshoot section, while in the diffusive section remains quite constant. This behavior is in good qualitative agreement with the bias dependence found for the total S_{ID} by other authors [8], who attributed the increase of drain current noise S_{ID} with V_{DS} in saturation to a progressive higher contribution of hot carriers located at the drain end of the channel. The enhanced velocity fluctuations of hot carriers in the overshoot region could be at the origin of the behavior observed in Fig. 1(a), although within the framework considered in this work a stronger transfer of noise generated in the diffusive section due to the channel modulation length effect can not be discarded as the origin of excess noise.

To calculate the complete local current noise source (S_{in}), it is necessary to take into account the sheet carrier concentration and determine the values of q^2NS_v, as indicated in eq. (3). Figure 4 shows the results for this factor as a function of V_{GS} for the two sections of the channel under consideration. Once the local sheet concentration is taken into account, the values for this factor are slightly higher in the diffusive section than in the overshoot section due to the important augmentation of the inversion layer concentration as V_{GS} is raised. In any case, the results are comparable in both sections of the channel, especially in deep saturation and low current conditions. Regarding the dependence on V_{DS} (inset), the value of the q^2NS_v factor in the diffusive section is, as in the previous case, more important, although for the highest V_{DS} (1.5 V) where carriers reach more elevated energies in the overshoot section and the channel length modulation effect is quite relevant the results are rather similar for both sections.

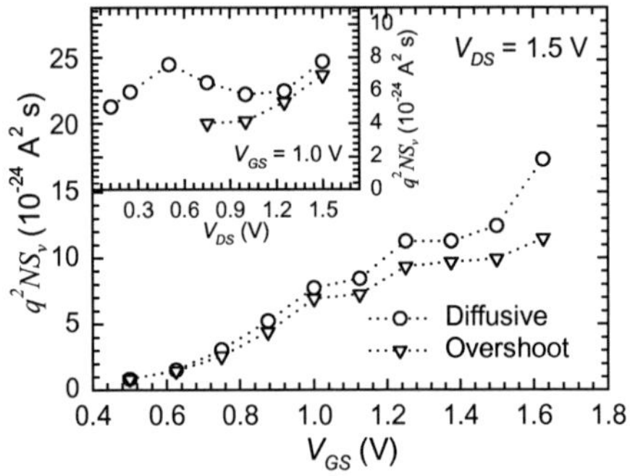

FIGURE 4. q^2NS_v as a function of V_{GS} for V_{DS} = 1.5 V.

4. CONCLUSIONS

An investigation of the spectral intensity of carrier velocity fluctuations within the channel of a MOSFET has been performed by means of a Monte Carlo simulator. The results obtained show that carriers in the overshoot section of the channel have stronger velocity fluctuations than carriers in the diffusive section, especially at low V_{GS} and high V_{DS}. As the average longitudinal electric field in the inversion layer is augmented, the amplitude of fluctuations in that area of the channel is increased, thus obtaining a higher S_v than that obtained by considering the bulk diffusion coefficient. Although it is not possible in our case to determine the separate contribution to S_{ID} of hot carriers at the drain end of the channel, the results obtained show that the local current noise source in that area is quite similar to that of the diffusive section.

REFERENCES

1. Knoblinger, G. *et al.*, *IEEE J. Solid-State Circ.* **36**, 831-837 (2001).
2. Chen, C.-H. and Deen M. J., *IEEE Trans. Electron Devices* **49**, 1484-1487 (2002).
3. Vanmackelberg, M. *et al.*, *Solid-State Electron.* **46**, 379-386 (2002).
4. Rengel, R. *et al.*, *Semicond. Sci. Technol.* **17**, 1149-1156 (2002).
5. van der Ziel, A., *Noise in Solid State Devices and Circuits*, New York: Wiley, 1986.
6. Cappy, A. and Heinrich, W., *IEEE Trans. Electron Devices* **36**, 403-409 (1989).
7. Shitkorov P. *et al.*, *Rivista Nuovo Cimento* **24**, 1-71 (2001).
8. Triantis, D. P. *et al.*, *IEEE Trans. Electron Devices* **43**, 1950-1955 (1996).

Why Noise Can Serve As Precursor Of Catastrophes

P. Shiktorov*, E. Starikov*, V. Gružinskis* and L. Reggiani†

*Semiconductor Physics Institute, A. Goštauto 11, LT 01108 Vilnius, Lithuania
†CNR-INFM - National Nanotechnology Laboratory, Dipartimento di Ingegneria dell' Innovazione, Università di Lecce, Via Arnesano s/n, 73100 Lecce, Italy

Abstract. We present a model where the notion of time-arrow describing the direction of time flow is formulated in the framework of the time-reversible Hamilton dynamics. The possibility to predict the future without contradicting the causality principle is discussed in the framework of such a model.

Keywords: Causality, noise, time irreversibility, time-arrow
PACS: 72.20.Ht, 72.30.+q, 72.70.+m

INTRODUCTION

Modern physical theories are based on two fundumental but conflicting principles: (i) total energy conservation, holding at each time moment and, (ii) the irreversibility of time flow (the latter is often formulated in terms of unidirectional time-arrow). Since now any violation of both principles has not been observed. Nevertheless, the mathematical formulation of these principles that is commonly adopted leads to contradictory results as we will detail in the following. Since the Hamilton dynamics is inherently invariant with respect to the direction of time flow, an additional constraint is required to single out one of the two possible directions. Usually such a constraint is formulated in terms of the causality principle as the requirement of the physical realization of a system, stating that a present state depends only on actions taken place in the past and independent of those which will occur in future [1]. In the framework of the Green-function (GF) formalism used to describe a response to some external action this requirement is formulated as: (i) the GF $G(t) = 0$ at all time moments $t < 0$ preceding the local in time action taking place at $t = 0$ (in the time-domain), and (ii) the formal replacement:

$$\frac{1}{\omega} \rightarrow P\{\frac{1}{\omega}\} + i\pi\delta(\omega) \tag{1}$$

when an integration over circular frequency ω is performed in the frequency-domain [2]. Here $P\{...\}$ is the principal value of the integral. The second term in r.h.s. of Eq. (1) is responsible for appearance of the dissipative part of the generalized susceptibility. This introduces an energy relaxation, i.e. an exponential decay at all time moments $t > 0$ after a delta like excitation has taken place at time $t = 0$.

From one hand, such an exponential decay is usually interpreted as the time-arrow direction, that is: *the physical time flows in the relaxation direction only*. From another hand, this decay corresponds to a violation of the energy conservation law, since the

CP800, *Unsolved Problems of Noise and Fluctuations: UPoN 2005*
edited by L. Reggiani, C. Pennetta, V. Akimov, E. Alfinito, and M. Rosini

energy is relaxed at time moments $t > 0$ when the interaction is already absent (that is without a reason at a given time $t > 0$).

To understand the origin of the conflict between unidirectional time-flow formulated by using the causality principle and energy conservation, and to find possible ways for its removal, in the present work we relax the general assumption that *the time flows in an absolute continuous way* and, by contrast, we suppose a *discrete time flow* even in the case of infinitesimal short steps. This conjecture allows us to consider the physical evolution of the system as an explicit sequence of primitive states: past, present, and future when the past and future states are, respectively, just before and just after the present state.

DISCRETE REPRESENTATION OF TIME

Time discretization can be introduced in two ways (see, e.g. [3] and references therein):
(i) by considering time as a continuum, where events can take place (discontinuously) only at discrete intervals of time Δ, and
(ii) by attributing to time a discrete structure, i.e. by regarding time not as a continuum, but as one-dimentional time-grid of equal elementary steps Δ.

In both approaches the time derivative is replaced by a discrete analog:

$$\frac{d}{dt}y \rightarrow \frac{1}{\Delta}\begin{cases} y(t)-y(t-\Delta) & \text{, retarded} \\ y(t+\Delta)-y(t) & \text{, advanced} \\ 0.5[y(t+\Delta)-y(t-\Delta)] & \text{, symmetric} \end{cases} \tag{2}$$

the choice of which determines entirely the character of dynamics. For example, the eigen solutions of the harmonic oscillator equation $dy/dt = i\omega_0 y$ (assuming that $y = y(t)$ in the right hand side) for the three discretization schemes given by Eq. (2) take the form:

$$y(t) = C\begin{cases} [1 \mp i\omega_0\Delta]^{\mp\frac{t}{\Delta}} & \text{, "-" retarded, "+" advanced} \\ exp[-i\frac{t}{\Delta}arcsin(\Delta\omega_0)] & \text{, symmetric} \end{cases} \tag{3}$$

As follows from Eq. (3), the amplitude of the oscillations (the energy) is conserved in time in the symmetric scheme only, while for the asymmetric ones the amplitude exponentially either grows or decreases.

Notions of past and future with respect to the present time moment have absolute physical meaning only in the presence of an arrow for the time flow. Therefore, in the framework of the commonly adopted definition of the time-arrow as that corresponding to the direction of exponential decay, both asymmetric representations of the time derivative describe the same dynamics occurring only in one direction of time flow, *thus satisfying inherently the causality principle*. Indeed, in the retarded (or advanced) scheme a state at time moment t is determined by the preceding state at $t-\Delta$ (or $t+\Delta$) which corresponds to a time-arrow directed from $-\infty$ to $+\infty$ (or vice versa). In doing so, both the asymmetric schemes defined in the framework of Hamilton formalism violate energy conservation.

In the framework of the symmetric scheme, which conserves energy, the usual definition of the time-arrow is impossible at all. Here, the present state of the system at time moment t mixes both the past and future states, $t \pm \Delta$, independently of the way used to determine the time-arrow direction. Therefore, one can pose the natural question: *by using the conjecture of the symmetric discrete time-derivative in the framework of the Hamilton dynamics is it possible to introduce the time arrow without violating the causality principle?* In the following we offer an answer to this question.

CENTRAL-DIFFERENCE SCHEME FOR TIME-DERIVATIVE

Let us consider two examples of constructing the GF on a time discrete grid in the framework of the central-difference scheme. *The first example corresponds to the free motion of a particle under the action of an arbitrary external force $f(t)$.* Accordingly, in the continuous and discrete representation the equations of motion can be written as (left- and right-hand equations, respectively): :

$$\frac{d}{dt}v = f(t) \quad <==> \quad v(i+1) - v(i-1) = \Delta f(i) \tag{4}$$

By using the time-domain GF formalism the inhomogeneous solution is given as the convolution integral:

$$v(t) = \int_{-\infty}^{\infty} G_v(t-t')f(t')dt' \quad <==> \quad v(i) = \Delta \sum_{j}^{\pm\infty} G_v(i-j)f(j) \tag{5}$$

with the corresponding GFs satisfying:

$$\frac{d}{dt}G_v(t-t') = \delta(t-t') \quad <==> \quad G_v(i+1) - G_v(i-1) = \delta_{ij} \tag{6}$$

Here δ_{ij} is the Kroneker symbol, j corresponds to time moment of the impulsive perturbation on the discrete time-grid $t_i = \Delta i$ with the index $i = 0, \pm 1, \pm 2, ...$ determining the present value of time.

The homogeneous solution of the discrete Eq. (4) is looked for in the form: $v(i) = C\lambda^i$ with C a dimensional constant. As a result, one obtains the characteristic equation $\lambda^2 = 1$ to determine two eigen solutions, which describe a different (but to a somewhat extent "conjugate") time behavior. The first solution with $\lambda_+ = 1$ corresponds to a regular function which behaves monotonously on the time grid, while the second one with $\lambda_- = -1$ changes of sign at each time step. Under the limit $\Delta \to 0$ the former transforms into a continuous function of the current time, while the latter solution cannot be characterized by any usual function since it must contain discontinuities at any point of the continuous time. In the following we shall refer to it as the "noise-like" solution since it corresponds to δ-correlated processes with zero average value at any finite time-interval, which are usually used in noise description [4].

In the asymmetric schemes there is only one solution with $\lambda_+ = 1$ since the motion on the time-grid evolves in one direction. The two time-conjugate solutions of the central-difference scheme reflect the possibility for the motion to evolve in both the time directions. Two GFs which satisfy Eq. (6) can be constructed. To construct them, both

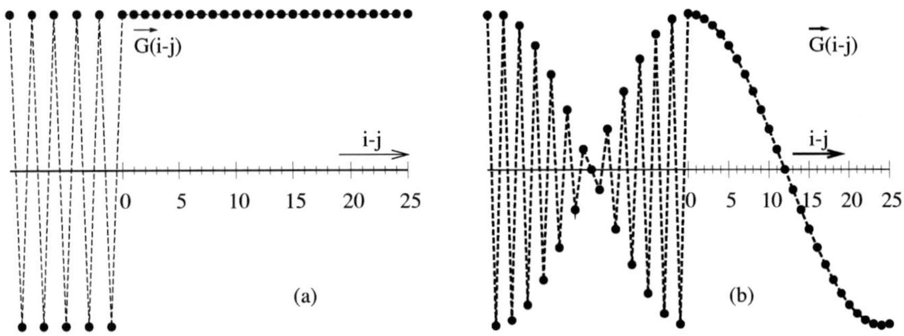

FIGURE 1. Green functions obtained in the framework of the symmetric scheme for the time derivatives on the discrete time grid for: (a) the free carrier and (b) the harmonic oscillator.

the regular and noise-like solutions, $v(i) = C_1\lambda_+^i + C_2\lambda_-^i$, must be used. Analogously with the asymmetric schemes we shall call them as the retarded and advanced GF. The retarded GF (see Fig 1(a)) is obtained when $C_2 = 0$, $C_1 = 0.5$ for $i \geq j$ and $C_2 = 0.5$, $C_1 = 0$ for $i < j$:

$$\overrightarrow{G}_v(i-j) = 0.5\begin{cases} \lambda_+^i & ,i \geq j \\ \lambda_-^i & ,i < j \end{cases} \equiv 0.5[sig(i-j)]^{(i-j)} \tag{7}$$

The advanced GF, $\overleftarrow{G}(i-j) = \overrightarrow{G}(j-i)$, is obtained by changing in Eq. (7) i with j.

The second example corresponds to the linear oscillator subjected to the influence of some external force $f(t)$. In the symmetric scheme it is:

$$\frac{1}{2\Delta}[x(i+1) - x(i-1)] = v(i) + f(i), \quad \frac{1}{2\Delta}[v(i+1) - v(i-1)] = -\omega_0^2 x(i) \tag{8}$$

In this case, the characteristic equation which determines the eigen solutions takes the form: $\lambda_{1,2}^2 = exp(\pm i2\Omega)$, where $\Omega = 0.5arctg[2\omega_0\Delta\sqrt{1-\Delta^2\omega_0^2}/(1-2\Delta^2\omega_0^2)]$. Analogously to the previous example, in parallel with each regular solution $\lambda_{1,2}$ there also exists a conjugate sign-changing solution with $-\lambda_{1,2}$. For example, the retarded GF for the velocity (see Fig. 1(b)) takes the form:

$$\overrightarrow{G}_v(i-j) = \frac{sin\Omega}{2\Delta sin2\Omega}[sig(i-j)]^{(i-j)}cos\Omega(i-j) \tag{9}$$

while the advanced GF, $\overleftarrow{G}(i-j)$, is obtained by interchanging i with j.

The main difference between the GFs of the symmetric and asymmetric schemes is that: in the former there exists the noise-like dynamics, which in the latter case is washed out by taking $G(i-j) \equiv 0$. This difference acts essentially on the evolution of the total energy of the system. In the asymmetric scheme, an energy jump takes place at the onset of the impulsive action. By contrast, in the symmetric scheme the energy does not change at the onset, and there it takes place merely a change of dynamics from the noise-like to the regular one (or vice versa) as shown in Fig. 1. Furthermore, in order to satisfy the law of energy conservation for a system subjected to an external action $f(i)$, the condition of energy balance can be properly formulated only by taking into account the noise-like solutions. Under the symmetric scheme, the instantaneous power related to the energy exchange can be represented as:

$$P(i) = \frac{1}{2} v(i)[f(i+1) + f(i-1)] \tag{10}$$

By substituting $v(i)$ given by Eq. (5) into Eq. (10) and performing the averaging of $P(i)$ over the time interval $[-N, N]$ one obtains the following expression for the average power

$$\overline{P} = \frac{\Delta}{2} \sum_{m}^{\pm\infty} G_v(m)[C_{ff}(m+1) + C_{ff}(m-1)] \tag{11}$$

where $C_{ff}(m) = \frac{1}{2N} \sum_i^{\pm\infty} f(i) f(i-m)$ is the correlation function of the external action which on the time grid has the same (regular or noise-like) behavior as $f(i)$. As follows from Eq. (11), the direction of the energy transfer originated by the external action is opposite for the regular and noise-like forces $f(i)$. In the former case, $\overline{P}_r > 0$ and the system absorbs energy from the outside. Here the main contribution to $\overline{P}_r$ comes from the summation at $m > 0$, that is over the GF regular part. In the latter case, $\overline{P}_n < 0$, i.e. the system loses energy to the outside. Here, the main contribution is due to the GF noise-like part at $m < 0$, which is not vanishing even at $\Delta \to 0$.

In the framework of the symmetric scheme, the presence of opposite directions for energy transfer allows us to formulate the stationary condition of the system under test subjected to an external action $f(t)$ in terms of the balance equation between the absorbed and lost powers, namely: $\overline{P}_r + \overline{P}_n = 0$, in full accordance with the energy conservation law.

By omitting noise-like solutions, only the process of energy absorption with $P_r > 0$ will remain. As a consequence, the system cannot achieve stationary conditions for any external action. The energy of the system will grow infinitely, what contradicts physical reality. To eliminate such a scenario of time-evolution, a phenomenological term responsible for energy relaxation is introduced into the equations of motion. It is the so called Langevin approach [4], which, however, contradicts the basic principles of the Hamilton formalism. Thus, in the framework of the Hamilton formalism, the implementation of energy conservation law is directly related to a noise-like dynamics. However, under the continuous-time limit, $\Delta \to 0$, the noise-like solutions correspond to a dynamics, which cannot be described by a function that is continuous in time due to the sign-varied discontinuities at each time moment. Therefore, such trajectories cannot be considered as observable ones from a physical point of view. Indeed, any physical measurement takes some finite time-interval. This is equivalent to a local-time averaging. Such an operation gives a zero value for the contribution of the noise-like solution.

MODEL

Eventually one comes to the dilemma: *what to do with noise-like solutions ?*

(i) If we simply omit them as unphysical solutions, then we must come back to the asymmetric schemes with all their consequences;

(ii) If we try to keep such a dynamics, which comes from a mathematical ground, then we need to find its place inside a physical framework.

Below we propose a logical model of speculations, which should satisfy simultaneously both the time-reversibility of Hamilton dynamics and the irreversibility of the direction of time-flow by taking into account both the regular and the noise-like dynamics. The model assumes that both the time-conjugate dynamics are physically realizable and there is no preference between them, namely:

(i) the conjugate dynamics are mutually unobservable, while the observable processes and the observer must pertain to the same dynamics;

(ii) time-inversion is the result of some action (since, as shown above, the energy is conserved it can be even some internal process) which initiates a transition from one conjugate dynamics to the other (see, e.g. Fig. 1 where the action occurs at t=0), and;

(iii) with respect to the inversion time moment the time-arrow always follows the direction of the observable dynamics (with reference to Fig. 1 the time-arrow is from left to right for the regular dynamics and viceversa for the noise-like dynamics).

Such a scheme implies the co-existence of two parallel worlds in which dynamics are conjugate and the directions of time arrows are opposite. The notions of observable (continuous function) and unobservable (discontinuous function everywhere) have no absolute meaning but they must be considered merely as antonyms with respect to the observer. Inside each world, the causality principle is satisfied for the description of the interaction and the physical time flows in one direction only. The time inversion in the system under test is perceived by an observer as the disappearance of the system from the observable world. In other words, the observer cannot detect any causal sequence of events describing the interaction on the system that have changed the dynamics to the conjugate one. He can only detect the presence of the energy related with the disappeared system. The mutual influence between the conjugate worlds takes place only at the energetic level through spontaneous phenomena (which by definition are not-causal). Without these phenomena the energetic balance cannot be fulfilled if one remains only in the framework of a causal dynamics which keeps the same time arrow.

OPEN QUESTIONS

Do in the physical world exist phenomena that can be explained by the proposed model?

Examples of such phenomena can be: the quantum-mechanical process of spontaneous emission without which the energy-balance formulation is impossible [5]; the Big Bang and the matter crossing the Schwarcschild sphere of a black hole [6] as the time-flow inversion; the hidden mass of the Universe as the coexistence of the two worlds with opposite time-arrows [7]; etc.

A rather intriguing application of the above model takes place when considering phenomena related with the prediction of future such as the occurrence of a catastrophic event. This subject, in essence, was never considered at the level of physical models, since the possibility of such a prediction in itself contradicts the causality scheme. In the model proposed here, the noise-like dynamics is out of the causality scheme, which can be applied only to the regular observable solutions, and thus the model offers the opportunity to open such a discussion. Indeed, in the framework of the symmetric scheme for the time derivative we have stressed that the GF can be constructed only by accounting for noise-like solutions. These solutions can provide the possibility to identify at the energetic level (directly or indirectly) events that will appear in the future, for example the step-like change of a regular motion ("catastrophe") as reported in Fig. 1. Since the transfer of information coming from the future is only provided by noise-like solutions, this transfer cannot be described in terms of the usually adopted causality scheme. Such an information can be obtained (detected) as some spontaneous physical process, since the physical phenomenon responsible for its origin is not observable.

REFERENCES

1. E. Merzbacher, Quatum Mechanics, Wiley, New York (1970); Paul C.W. Davies, Proc. Ot the 2nd International Conference on Unsolved Problems of Noise and Fluctuations, Edited by Derek Abbott, and Laszlo B. Kish, AIP Conference Proceedings 511, 16 (2000).
2. C. Kittel, Elements de Physique Statisque, (Dunod, Paris 1961) p.217
3. Ruy H.A. Farias and E. Recami, *Cond. Matt. arXiv:quant-ph/9706059*
4. N.G. Van Kampen: Stochastic Processes in Physics and Chemistry, (North Holland, Amsterdam, 1992).
5. http://en.wikipedia.org/wiki/Spontaneous emission
6. http://map.gsfc.nasa.gov/m uni.html
7. W.H. Tucker in: http://history.nasa.gov/SP-466/contents.htm

Effect Of Non-Equilibrium Term In Two-Particle Correlation Function On Electron-Phonon Collision Integrals

A.Ya. Shul'man

Institute of Radioengineering and Electronics of the RAS, 125009 Moscow, Russia.[1]

Abstract. The derivation of kinetic equation for one-particle distribution function $F(p;r;t)$ from BBGKY chain leads to the collision integral expressed in terms of the two-particle correlation function. The latter is in turn expressed by F using Boltzmann's Stosszahlansatz that implies the neglect of initial one-time correlation function $g2(p1;p2;r;t)$ which is non-diagonal in the momentum space. In nonequilibrium case, however, it has been established that pair collisions generate the non-diagonal two-particle correlations that give the essential contribution to the current fluctuations of hot electrons. These correlations have been shown giving also a contribution to the collision integrals, i.e., to kinetic properties of nonequilibrium gas The expression for electron energy loss rate P via phonons is re-derived in detail from this point of view. The order of value of the new term in P in the cases of acoustic- and optical-phonon scattering is evaluated in the electron temperature approximation using the experimental study of hot-electron energy relaxation time in $n-InSb$ at helium temperatures.

Keywords: Nonequilibrium fluctuations, Two-particle interaction
PACS: 72.70.+m, 72.20.Dp

INTRODUCTION

It was established that pair collisions generate in nonequilibrium case non-zero two-particle correlations that are non-diagonal in momentum space and give the essential contribution to the current fluctuations of hot electrons [1, 2]. These correlations have been shown giving also a contribution to the collision integrals, i.e., to kinetic properties of nonequilibrium gas [3, 4]. The expression for electron energy loss rate P via phonons is re-derived from this point of view. The formula for P in terms of electron density-density correlation function $\langle \rho_k \rho_k^+ \rangle_\omega$ was obtained by Kogan [5] using Fermi Golden Rule. However, in order to come to the well-known result for P the fluctuation-dissipation theorem was implicated to express $\langle \rho_k \rho_k^+ \rangle_\omega$ in terms of the dielectric function of the electron gas. On this way the non-equilibrium correction has been lost. Here the contribution of the non-diagonal part of the nonequilibrium pair correlator to phonon-electron collision integral and to P is obtained and explicitly calculated in the electron temperature approximation. The contribution obtained does not contain the extra powers of the small parameter of kinetic theory unlike the virial expansions in equilibrium thermodynamics. The order of value of the new term in P is evaluated us-

[1] E-mail: ash@cplire.ru

CP800, *Unsolved Problems of Noise and Fluctuations: UPoN 2005*
edited by L. Reggiani, C. Pennetta, V. Akimov, E. Alfinito, and M. Rosini

ing the experimental study of hot-electron energy relaxation time in $n-InSb$ at helium temperatures [6].

BASIC DESIGNATION

Hamiltonian

$$H = H_e + H_p + V_{ep} \tag{1}$$

Electron part

$$H_e = \sum_{\mathbf{k}} \hbar\Omega_{\mathbf{k}} a^{+}_{\mathbf{k}} a_{\mathbf{k}} + V_{ee} \tag{2}$$

where V_{ee} stands for the Coulomb electron-electron interaction.

Phonon part

$$H_p = \sum_{\mathbf{f}} \hbar\omega_{\mathbf{f}} b^{+}_{\mathbf{f}} b_{\mathbf{f}} \tag{3}$$

and electron-phonon interaction

$$V_{ep} = \frac{1}{\sqrt{V_0}} \sum_{\mathbf{f}} \left(c_f \rho^{+}_{\mathbf{f}} b_{\mathbf{f}} + c^{*}_{f} \rho_{\mathbf{f}} b^{+}_{\mathbf{f}} \right) \tag{4}$$

where V_0 is the normalizing volume.

Fourier components of the electron density operator

$$\rho_{\mathbf{f}} = \sum_{k} a^{+}_{\mathbf{k}+\frac{\mathbf{f}}{2}} a_{\mathbf{k}-\frac{\mathbf{f}}{2}}, \quad \rho^{+}_{\mathbf{f}} = \rho_{-\mathbf{f}} \tag{5}$$

Phonon distribution function

$$N_{\mathbf{f}}(t) = \left\langle \left(b^{+}_{\mathbf{f}} b_{\mathbf{f}} \right)_t \right\rangle \tag{6}$$

Electron distribution function

$$F_{\mathbf{k}}(t) = \left\langle \left(a^{+}_{\mathbf{k}} a_{\mathbf{k}} \right)_t \right\rangle \tag{7}$$

The angle brackets denote the averaging with the density matrix corresponding to the noninteracting electrons and phonons at initial instant of time. The subscript t indicates the Heisenberg representation for electron and phonon operators.

KINETIC EQUATION FOR PHONONS

We obtain the kinetic equation for slow variables $N_f(t)$, smoothing Heisenberg equations of motion over mesoscopic time interval. With accuracy to the second-order terms in $e-p$ interaction we have

$$\frac{\Delta N_{\mathbf{f}}(t)}{\Delta t} = -\frac{1}{\hbar^2 \Delta t} \left\langle S^{+}(t) \int_{t}^{t+\Delta t} dt\prime \int_{t}^{t\prime} dt\prime\prime \left[V^{0}_{ep}(t\prime\prime), \left[V^{0}_{ep}(t\prime), (b^{+}_{\mathbf{f}} b_{\mathbf{f}})_t \right] \right] S(t) \right\rangle. \tag{8}$$

The superscript 0 denotes operators in the interaction representation with respect to the electron-phonon interaction V_{ep} .

Calculations of the internal commutator lead to the desired result involving the electron density-density correlation function:

$$\frac{dN_{\mathbf{f}}}{dt} = \frac{|c_f|^2}{V_0\hbar^2}\left\{\left\langle \rho_{\mathbf{f}}^{+}\rho_{\mathbf{f}}\right\rangle_{-\omega_f}^{0}(N_{\mathbf{f}}+1) - \left\langle \rho_{\mathbf{f}}\rho_{\mathbf{f}}^{+}\right\rangle_{\omega_f}^{0} N_{\mathbf{f}}\right\}, \tag{9}$$

where the condition $\omega_{\mathbf{f}}\Delta t >> 1$ was used. The function

$$\left\langle \rho_{\mathbf{f}}\rho_{\mathbf{f}}^{+}\right\rangle_{\omega} = \int_{-\infty}^{\infty} d\tau \left\langle \rho_{\mathbf{f}}(t+\tau)\rho_{\mathbf{f}}^{+}(t)\right\rangle \exp(i\omega\tau) \tag{10}$$

is the spectral density of the correlation function of the electron density fluctuations.

SPECTRAL DENSITY OF THE HIGH-FREQUENCIES ELECTRON FLUCTUATIONS IN THE ELECTRON TEMPERATURE APPROXIMATIONS

The spectral density must be calculated without V_{ep} taken into account. From equations of motion for the electron operators we obtain in the random phase approximation:

$$\left\langle \rho_{\mathbf{f}}^{+}\rho_{\mathbf{f}}\right\rangle_{\omega}^{0} = \frac{\left\langle \rho_{\mathbf{f}}^{+}\rho_{\mathbf{f}}\right\rangle_{\omega}^{00}}{|\varepsilon(\omega,\mathbf{f})|^2}, \tag{11}$$

where $\varepsilon(\omega,\mathbf{f})$ is the dielectric function of the electron gas and $\left\langle \rho_{\mathbf{f}}^{+}\rho_{\mathbf{f}}\right\rangle_{\omega}^{00}$ is the spectral density of the density-density correlation function for non-interacting electrons. The latter is connected with one-time correlation function $\left\langle \rho_{\mathbf{f}}(\mathbf{k})\rho_{\mathbf{f}}^{+}(\mathbf{k}_1)\right\rangle_t$ by equations of motion for free electrons. The one-time correlation function is defined as

$$\left\langle \rho_{\mathbf{f}}(\mathbf{k})\rho_{\mathbf{f}}^{+}(\mathbf{k}_1)\right\rangle_t \equiv \left\langle a_{\mathbf{k}-\frac{\mathbf{f}}{2}}^{+} a_{\mathbf{k}+\frac{\mathbf{f}}{2}} a_{\mathbf{k}_1+\frac{\mathbf{f}}{2}}^{+} a_{\mathbf{k}_1-\frac{\mathbf{f}}{2}}\right\rangle_t . \tag{12}$$

The one-time correlation function is independent of time (in steady state) under following condition on the electron energies $\Omega_k = \hbar k^2/2m$:

$$\Omega_{\mathbf{k}-\frac{\mathbf{f}}{2}} - \Omega_{\mathbf{k}+\frac{\mathbf{f}}{2}} + \Omega_{\mathbf{k}_1+\frac{\mathbf{f}}{2}} - \Omega_{\mathbf{k}_1-\frac{\mathbf{f}}{2}} \equiv \frac{\hbar}{m}\mathbf{f}(\mathbf{k}_1-\mathbf{k}) = 0. \tag{13}$$

Because $\mathbf{f} \neq 0$ then it is necessary to be $\mathbf{k}_1 = \mathbf{k}$. Therefore, this condition provides the one-time correlator to be the integral of motion during the time interval Δt like the distribution functions $N_{\mathbf{f}}$ and $F_{\mathbf{k}}$.

KINETIC EQUATION WITH ACCOUNT FOR THE CONTRIBUTION OF THE NON-EQUILIBRIUM PAIR CORRELATIONS

The substitution of the explicit expression Eq.(11)for correlators to the kinetic equation Eq.(9) gives rise to

$$\begin{aligned}\frac{dN_{\mathbf{f}}}{dt} &= \frac{2\left|C_f\right|^2}{V_0\hbar^2\left|\varepsilon\left(\omega_{\mathbf{f}},\mathbf{f}\right)\right|^2}\sum_{\mathbf{k}}\left\langle a^+_{\mathbf{k}+\frac{\mathbf{f}}{2}}a_{\mathbf{k}-\frac{\mathbf{f}}{2}}a^+_{\mathbf{k}-\frac{\mathbf{f}}{2}}a_{\mathbf{k}+\frac{\mathbf{f}}{2}}\right\rangle_t\delta\left(\Omega_{\mathbf{k}+\frac{\mathbf{f}}{2}}-\Omega_{\mathbf{k}-\frac{\mathbf{f}}{2}}-\omega_{\mathbf{f}}\right)\left(N_{\mathbf{f}}+1\right)-\\ &-\frac{2\left|C_f\right|^2}{V_0\hbar^2\left|\varepsilon\left(\omega_{\mathbf{f}},\mathbf{f}\right)\right|^2}\sum_{\mathbf{k}}\left\langle a^+_{\mathbf{k}-\frac{\mathbf{f}}{2}}a_{\mathbf{k}+\frac{\mathbf{f}}{2}}a^+_{\mathbf{k}+\frac{\mathbf{f}}{2}}a_{\mathbf{k}-\frac{\mathbf{f}}{2}}\right\rangle_t\delta\left(\Omega_{\mathbf{k}+\frac{\mathbf{f}}{2}}-\Omega_{\mathbf{k}-\frac{\mathbf{f}}{2}}-\omega_{\mathbf{f}}\right)N_{\mathbf{f}}. \quad (14)\end{aligned}$$

Taking into account the commutation rules for $a^+_{\mathbf{k}}a_{\mathbf{k}_1}$ it is easy to obtain

$$\left\langle a^+_{\mathbf{k}}a_{\mathbf{k}_1}a^+_{\mathbf{k}_1}a_{\mathbf{k}}\right\rangle_t = F_{\mathbf{k}}(t)-\left\langle a^+_{\mathbf{k}}a^+_{\mathbf{k}_1}a_{\mathbf{k}_1}a_{\mathbf{k}}\right\rangle_t. \quad (15)$$

According to the results from [1]

$$\left\langle a^+_{\mathbf{k}}a^+_{\mathbf{k}_1}a_{\mathbf{k}_1}a_{\mathbf{k}}\right\rangle = F_{\mathbf{k}}F_{\mathbf{k}_1}+\phi\left(\mathbf{k},\mathbf{k}_1\right), \quad (16)$$

where $\phi\left(\mathbf{k},\mathbf{k}_1\right)$ is a non-equilibrium correction to the pair correlation function.

Finally, the kinetic equation for phonons with non-equilibrium corrections of electron pair correlators takes the form:

$$\begin{aligned}\frac{dN_{\mathbf{f}}}{dt} &= \frac{2\left|C_f\right|^2}{V_0\hbar^2\left|\varepsilon\left(\omega_f,\mathbf{f}\right)\right|^2}\sum_{\mathbf{k}}\left[F_{\mathbf{k}+\frac{\mathbf{f}}{2}}\left(1-F_{\mathbf{k}-\frac{\mathbf{f}}{2}}\right)-\phi\left(\mathbf{k}+\mathbf{f}/2,\mathbf{k}-\mathbf{f}/2\right)\right]\left(N_{\mathbf{f}}+1\right)\times\\ &\times\delta\left(\Omega_{\mathbf{k}+\frac{\mathbf{f}}{2}}-\Omega_{\mathbf{k}-\frac{\mathbf{f}}{2}}-\omega_{\mathbf{f}}\right)-\\ &-\frac{2\left|C_f\right|^2}{V_0\hbar^2\left|\varepsilon\left(\omega_f,\mathbf{f}\right)\right|^2}\sum_{\mathbf{k}}\left[F_{\mathbf{k}-\frac{\mathbf{f}}{2}}\left(1-F_{\mathbf{k}+\frac{\mathbf{f}}{2}}\right)-\phi\left(\mathbf{k}-\mathbf{f}/2,\mathbf{k}+\mathbf{f}/2\right)\right]N_{\mathbf{f}}\times\\ &\times\delta\left(\Omega_{\mathbf{k}+\frac{\mathbf{f}}{2}}-\Omega_{\mathbf{k}-\frac{\mathbf{f}}{2}}-\omega_{\mathbf{f}}\right) \quad (17)\end{aligned}$$

Certainly, it needs to keep in mind the Eq.(17) describes the part of full time dependence of the phonon occupation numbers that resulting from electron-phonon interaction only.

It should be stressed that for ordinary classical equilibrium electron gas there are no pair correlations between different states of momentum space. This fact can be easily seen from the form of the Gibbs distribution. In conclusion of these considerations it needs to underline that the contribution of non-equilibrium correlations to kinetics should be expected in the cases when the particles considered are scattered by non-equilibrium many-particle system.

ELECTRON ENERGY LOSS RATE VIA PHONONS

The energy loss rate of electrons due to phonons per unit volume can be obtained using the relation

$$P = \frac{1}{V_0} \sum_{\mathbf{f}} \hbar \omega_{\mathbf{f}} \frac{dN_{\mathbf{f}}}{dt}. \tag{18}$$

It is seen from the kinetic equation for phonons Eq.(17) that the loss rate can be broken up into two parts:

$$P = P_1 + P_2, \tag{19}$$

where the first term on the right hand side is related to the one-particle electron distribution function and it is described by the known expressions. The second one is due to the correlation contribution determined by the function ϕ .We assume during the calculations of P the electron-electron collisions are sufficiently frequent in order to provide the use of the electron temperature approximation for hot electrons. As a result of these assumptions the following formulae may be derived

$$\begin{aligned} P_1(T,T_0) \quad &= \quad \sum_{\mathbf{f}} \frac{2\left|C_f\right|^2}{(V_0\hbar)^2 \left|\varepsilon(\omega_f,\mathbf{f})\right|^2} \hbar\omega_f \left[N_{\mathbf{f}}^0(T) - N_{\mathbf{f}}\right] \\ &\times \sum_{\mathbf{k}} \left[F_{\mathbf{k}}^0(T) - F_{\mathbf{k+f}}^0(T)\right] \delta\left(\Omega_{\mathbf{k+f}} - \Omega_{\mathbf{k}} - \omega_{\mathbf{f}}\right) \end{aligned} \tag{20}$$

and

$$P_2(T,T_0) = -\sum_{\mathbf{f}} \frac{2\left|C_f\right|^2}{(V_0\hbar)^2 \left|\varepsilon(\omega_{\mathbf{f}},\mathbf{f}\right|^2} \hbar\omega_{\mathbf{f}} \sum_{\mathbf{k}} \phi\left(\mathbf{k}+\mathbf{f},\mathbf{k}\right) \delta\left(\Omega_{\mathbf{k+f}} - \Omega_{\mathbf{k}} - \omega_{\mathbf{f}}\right) \tag{21}$$

Here $N_{\mathbf{f}}^0(T) = \left[\exp\left(\hbar\omega_{\mathbf{f}}/T\right) - 1\right]^{-1}$ is the equilibrium Bose-Einstein distribution with electron temperature T and $N_{\mathbf{f}}$ is the true non-equilibrium phonon distribution function. Further, we neglect possible phonon heating and consider that the phonons remain in thermal equilibrium at temperature T_0, so $N_{\mathbf{f}} = N_{\mathbf{f}}^0\left(T_0\right)$.

The function $\phi\left(\mathbf{k},\mathbf{k}_1\right)$ was found in [2] under accepted assumptions for the case of nondegenerate electron gas and can be presented in the form:

$$\phi(\mathbf{k},\mathbf{k}_1) = Q(T,T_0) \frac{\partial F_{\mathbf{k}}^0}{\partial T} \frac{\partial F_{\mathbf{k}_1}^0}{\partial T} \tag{22}$$

where

$$Q(T,T_0) = \frac{T^2 N_c(T)}{nc_e} \left[\frac{P(T,T_0)/\left(T-T_0\right)}{dP/dT} - 1\right]. \tag{23}$$

Here $N_c(T)$ and $nc_e = 3n/2$ are the effective density of states and specific heat capacitance of electron gas with concentration n, respectively. The expression for ϕ in [2] was obtained without accounting for the contribution of ϕ itself in P .Therefore, the quantity P_1 was only implied as the total energy loss rate. Now evidently we have the differential equation for determination of $P_2(T,T_0)$ or full P since this quantity is defined by the right hand side of the equality Eq.(21) including $P_2(T,T_0)$ itself. The situation is quite resembling the self-consistent field approach in plasma theory. It needs to stress that the substitution of the total loss rate P instead of P_1 in the expression for $\phi\,(k,k_1)$ is justified because all symmetry properties of the electron-electron and electron-phonon integrals of collisions used in the derivation are not destroyed by introducing $\phi\,(\mathbf{k},\mathbf{k}_1)$ in collision integrals.

It is of interest to estimate the interrelation between P_1 and P_2. Instead carrying out the integrations in Eqs. (20)-(21) and following solution of a nonlinear differential equation for P (or P_2) we can use the experimental data from Ref. [6] where hot-electron energy relaxation time at helium temperatures in $n-InSb$ was investigated and all necessary information on n, P, dP/dT, and T can be extracted from. It was shown there that at low heating the electron energy loss is due to acoustic phonon scattering and next it is changed by longitudinal optical (LO) phonon scattering. So we can evaluate P_2/P_1 for both types of the electron-phonon scattering.

It is necessary to carry out preliminary the integrations over $\mathbf{k}$ in Eqs.(20)-(21) for all essential parameters be explicitly introduced. Taking into account nondegeneracy of electron gas the expressions for $P_1$and P_2 can be transform as following

$$P_1(T,T_0) = (T-T_0)\frac{2ms^2}{(2\pi)^2\hbar^3}\sum_{\mathbf{f}}\frac{2\left|C_f\right|^2}{\hbar V_0\left|\varepsilon(\omega_f,\mathbf{f})\right|^2}F^0_{\mathbf{f}/2}(T), \tag{24}$$

$$P_2(T,T_0) = \pm\frac{9Tm^2s}{4c_e\hbar^3}\left[\frac{P(T,T_0)/(T-T_0)}{dP/dT}-1\right]\sum_{\mathbf{f}}\frac{2\left|C_f\right|^2}{\hbar V_0\left|\varepsilon(\omega_f,\mathbf{f})\right|^2}F^0_{\mathbf{f}/2}(T/2)\times \times[1+O(T)] \tag{25}$$

$$F^0_{\mathbf{k}}(T) = \frac{n}{N_c(T)}\exp\left(-\frac{\hbar^2k^2}{2mT}\right). \tag{26}$$

for acoustical phonon scattering. Here s is the sound velocity and accounted for the near-elastic scattering of electrons by acoustic phonons. While the sign of expression 25 for P_2 leaves indeterminate since implicit $O(T)$ terms may redefine it,.the order of magnitude of P_2/P_1 can be found The expression Eq.(24) is just the Kogan's formula from [5] corrected for the screening of electron-phonon interaction.

In the case of LO phonon scattering ($\hbar\omega_{LO}/T << 1$) we obtain

$$P_1(T,T_0) = \frac{m^2T\hbar\omega_{LO}}{\pi^2\hbar^4}\frac{n}{N_c(T)}\left(1-\mathrm{e}^{-\hbar\omega_{LO}/T}\right)\left[N^0_{\mathbf{f}}(T)-N^0_{\mathbf{f}}(T_0)\right] \times\sum_{\mathbf{f}}\frac{2\left|C_f\right|^2}{\hbar V_0\left|\varepsilon(\omega_f,\mathbf{f})\right|^2 f}\exp\left(-\frac{\hbar^2}{2mT}\left[\frac{f^2-f^2_{LO}}{2f}\right]^2\right) \tag{27}$$

$$P_2(T,T_0) = \pm\frac{m^2\hbar^2\omega^2_{LO}}{2(2\pi)^2\hbar^4}\frac{n}{N_c(T)}\left[\frac{P(T,T_0)/(T-T_0)}{dP/dT}-1\right]\exp\left(-\frac{\hbar\omega_{LO}}{T}\right) \times\sum_{\mathbf{f}}\frac{2\left|C_f\right|^2}{\hbar V_0\left|\varepsilon(\omega_f,\mathbf{f})\right|^2 f}\exp(-\frac{2\hbar^2}{2mT}\left[\frac{f^2-f^2_{LO}}{2f}\right]^2[1+O(T)] \tag{28}$$

where $f_{LO}^2 = 2m\omega_{LO}/\hbar$.

The sums over **f** in corresponding expressions are of the same order of magnitude. Therefore, we have to evaluate following ratios

$$\frac{P_2}{P_1} = \frac{3\pi^2 T}{T - T_0}\left[\frac{P(T,T_0)/(T-T_0)}{dP/dT} - 1\right] \tag{29}$$

for the acoustic phonon scattering and

$$\frac{P_2}{P_1} = \frac{\hbar\omega_{LO}}{8T}\left[\frac{P(T,T_0)/(T-T_0)}{dP/dT} - 1\right] \tag{30}$$

for the low-temperature LO phonon scattering. Turning to experimental data [6] and using relation between dP/dT and measured electron energy relaxation time $\tau_P = nc_e/(dP/dT)$ we obtain for the acoustic scattering $P_2/P_1 \simeq 1.3$ at $T \simeq 7.8K$, $P \simeq 6\cdot 10^{-3}W/cm^3$, $\tau_P \simeq 4\cdot 10^{-7}s$ and $P_2/P_1 \simeq 0.6$ at $T \simeq 15.8K$, $P \simeq 3.8\cdot 10^{-2}W/cm^3$, $\tau_P \simeq 0.8\cdot 10^{-7}s$ for LO phonon scattering. All data are taken at $T_0 = 4.2K$. These estimations may be considered as an indication that the pair correlations contribute to hot electron energy loss via phonons comparable with common accepted expressions.

It is of importance to note in conclusion that the expression for $\phi\,(k,k_1)$ found in the limit of high frequency of the electron-electron collisions does contain coupling constants of neither electron-electron nor electron-phonon interactions. Therefore, the corresponding corrections to the integral of phonon-electron collisions do not contain additional powers of the small parameters of the kinetic theory. The interesting but still open question: are there analogous contributions to the hydrodynamic equations?

This work was partially supported by Russian Foundation for Basic Researches.

REFERENCES

1. S.V. Gantsevich, V.L. Gurevich, and R. Katilius. FTT 11, 308 (1969) [Sov. Phys. Sol. State **11**, 247 (1969)]
2. A.Ya. Shul'man. FTT 12, 1181 (1970) [Sov. Phys. Sol. State 12, 922 (1970)]
3. A.Ya. Shul'man. Cond-mat/0212595
4. A.Ya. Shul'man. In: Progress in Nonequilibrium Green's Functions-2, eds. M. Bonitz and D. Semkat. World Scientific Publishing, Singapore, p. 74 (2003)
5. Sh.M. Kogan. FTT 4, 2474 (1962) [Sov. Phys. Sol. State 4, 1813 (1963)]
6. T.M. Lifshits, A.Ya. Oleinikov, and A.Ya. Shul'man. Phys. Stat. Sol. 14, 511 (1966)

NOISE IN COMPLEX SYSTEMS AND NON-GAUSSIAN FLUCTUATIONS

Experimental results on the applications of Fluctuation Theorems in out of equilibrium systems

N. Garnier* and S. Ciliberto*

*Laboratoire de Physique de l'ENS Lyon — CNRS UMR 5672
46, Allée d'Italie — 69364 Lyon Cedex 07, France

Abstract. The fluctuations of global variables in out of equilibrium systems are a subject of current investigation ranging from dynamical systems theory to stochastic systems. There are only a few theoretical results which predict several properties of these fluctuations. This is the case, for example of the Fluctuation Theorem (FT). In this paper we mainly consider out of equilibrium stochastic systems. We summarize the theoretical results which can be applied both to a Brownian particle and a electric circuits driven out of equilibrium. As an example we describe an experiment where the fluctuations of the injected and dissipated powers have been measured. We find the theoretical predictions are verified. The applications of these results to more complex cases is discussed.

Keywords: fluctuation theorem, thermal noise, out of equilibrium, stochastic systems
PACS: 05.40.-a, 05.70.-a, 07.50.-e, 84.30.Bv m

INTRODUCTION

The second law of thermodynamics states that, on average, heat flows from hot to cold bodies or more generally that the entropy of a closed system tends to increase. However this principle does not make any statement regarding fluctuations; indeed over short time intervals the heat flux may be reversed. While fluctuations always play a negligible role in large systems, their influence may become extremely important in small systems (nano and micro devices, biological systems etc.) driven out of equilibrium. In small systems where relevant energies are comparable to thermal agitation, fluctuations are of the order of average values. In particular, large fluctuations will occur, with mostly unknown consequences. There are only a few theoretical results which may predict several statistical properties of these fluctuations. This is the case of the fluctuation theorem(FT) which has been first found in the context of dynamical systems[1, 2] and later extended to stochastic systems[3, 4, 5, 6]. The FT gives a relation between the probabilities to observe a positive value of the (time averaged) "entropy production rate" and a negative one. This relation is of the form

$$\frac{P(\sigma)}{P(-\sigma)} = \exp[\sigma\tau], \tag{1}$$

where σ and $-\sigma$ are equal but opposite values for the entropy production rate, $P(\sigma)$ and $P(-\sigma)$ give their probabilities and τ is the length of the interval over which σ is measured. In these systems, the above mentioned FT is derived for a

CP800, *Unsolved Problems of Noise and Fluctuations: UPoN 2005*
edited by L. Reggiani, C. Pennetta, V. Akimov, E. Alfinito, and M. Rosini

mathematical quantity σ, which has a form similar to that of the entropy production rate in Irreversible Thermodynamics.

The original proof of FT requires several restrictive hypothesis which have to be tested in experiments. However, there are not many experimental tests of FT. A few of them are performed in dynamical systems [7] in which the interpretation of the results is very difficult. A few other experiments are done on stochastic systems, one on a Brownian particle in a moving optical trap [8] and another on electrical circuits driven out of equilibrium by injecting in it a small current [9]. The latter is a very general and particularly interesting example because fluctuations of the steady current through a system in contact between two reservoirs is one of the simplest and most fundamental problems of non equilibrium physics[11]. Specifically in this talk we will focus on the properties of the fluctuations of the power injected and dissipated by an electrical resistance driven out of equilibrium by a small constant current I. This problem has been extensively studied both theoretically [4, 5] and experimentally [9] in its simplest configuration, but many problems remain open to extend these results to more complex cases, that we will discuss at the end of this article.

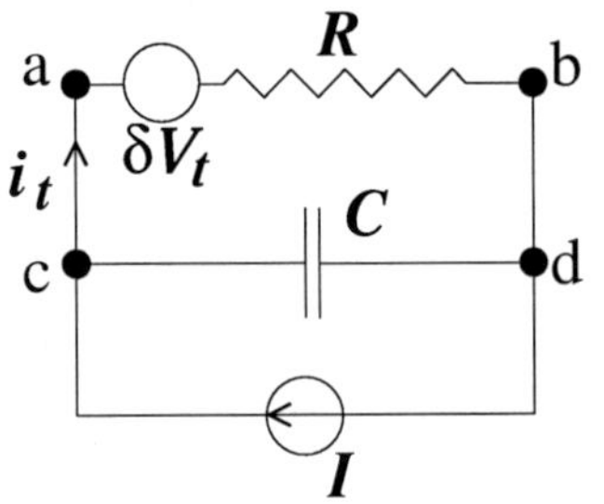

FIGURE 1. Model circuit : an electrical dipole is composed of a resistive part R and a capacitive part C. Due to thermal fluctuations of charges positions, a fluctuating voltage δV_t is observed. We drive the system away from equilibrium by imposing a constant flux of electrons, via a constant current I.

FLUCTUATION THEOREM IN AN ELECTRICAL CIRCUIT

Our system is an electrical dipole constituted of a resistance R in parallel with a capacitor C (Fig. 1). We drive it out of equilibrium by making a constant current I flow in it. Energy is being dissipated in the resistor and according to the fluctuation-dissipation theorem (FDT) this means there are fluctuations too. If one assumes that the constant current I does not modify the statistical properties of the resistance thermal noise, to a good approximation, one can use a Gaussian random noise term δV_t to describe these fluctuations[13], which is depicted in the figure by a voltage generator. In addition, we define q_t as the charge that has gone through the resistor, i_t as the current that is going through it (so $i_t = dq_t/dt$) and q'_t as the charge on the capacitor, all at time t. Using that $q'_t = \int_0^t dt'\,(I - i_{t'}) = It - q_t$ and that

$U = V_{ab} = q_t'/C$, standard calculations for electric circuits give

$$0 = -R\frac{dq_t}{dt} - \delta V_t - \frac{q_t - It}{C}, \tag{2}$$

in which δV_t satisfies[13]

$$\langle \delta V_t \rangle = 0; \quad \langle \delta V_t \delta V_{t'} \rangle = 2k_B T R\, \delta(t - t'). \tag{3}$$

where $\langle\rangle$ denotes an average over an ensemble of similar. The statistical properties of eq.2 for $I = 0$ can be found in text book (see for example [13]). When $I \neq 0$ and the system is out of equilibrium the statistical properties of the power injected and dissipated by the system deserve interesting features which merit to be studied as an instructive example for the analysis of more complex systems.

Before proceeding into this analysis we want to point out that the problem is a very general one as eq.2 can be mapped on equation of a small particle of negligible mass trapped by an opital tweezer moving at constant speed inside a viscous fluid. Indeed the form of the Langevin equation for the Brownian particle in the experiment of Wang *et al.*[4]:

$$m\frac{d^2x_t}{dt^2} = -\alpha\frac{dx_t}{dt} + \xi_t - k(x_t - v^* t). \tag{4}$$

Here, m is the mass of the Brownian particle, x_t is its position at time t, α is the (Stokes') friction coefficient, k is the strength of the harmonic potential induced by the optical tweezer, and v^* is the constant speed at which the tweezer is moved. ξ_t represents a Gaussian white noise, which satisfies

$$\langle \xi_t \rangle = 0; \quad \langle \xi_t \xi_{t'} \rangle = 2k_B T \alpha\, \delta(t - t'), \tag{5}$$

TABLE 1. The analogy of a circuit and a Brownian particle.

Brownian particle:	ξ_t	α	T	x_t	$\frac{dx_t}{dt}$	k	v^*
RC circuits:	$-\delta V_t$	R	T	q_t	i_t	$1/C$	I

We see that Eqs. (2) and (3) are of a very similar form as those for a one-dimensional massless Brownian particle dragged through a fluid by means of a harmonic potential in Eqs. (4) and (5). Table 1 gives not only the the well-known analogies between the Brownian motion and Nyquist noise (ξ_t, α, T,k vs. δV_t, R, T,$1/C$) but also the correspondence of the coefficients related to the external forces which drive the system out of equilibrium. Given this analogy, we turn to the heat and work fluctuations in this circuit. The power injected in the circuit is $\mathscr{P}_{\text{in}} = UI$ but power is dissipated in the resistive part only, so the dissipated power is $\mathscr{P}_{\text{diss}} = U i_t$ where i_t is the current flowing in the resistor (Fig. 1). As already noted [12], in average, one expects $\langle \mathscr{P}_{\text{in}} \rangle = \langle \mathscr{P}_{\text{diss}} \rangle \equiv \bar{\mathscr{P}}$.

The dissipated heat over a time τ is given by the time integral of $\mathscr{P}_{\text{diss}}$ i.e.,

$$Q_\tau = \int_0^\tau dt\, i_t [i_t R + \delta V_t] = -\int_0^\tau dt\, i_t \frac{q_t - It}{C}, \tag{6}$$

where we used Eq. (2). Instead, the total work done in the circuit is the time integral of the injected power $\mathscr{P}_{\rm in}$ so that

$$W_\tau = \int_0^\tau dt\, I \left[-\frac{q_t - It}{C} \right]. \tag{7}$$

As q_t,U and i_t are fluctuating quantities we are interested in studying the fluctuations of Q_τ and W_τ and whether a fluctuation theorem can be defined of the heat and the work for this electrical circuit. It can be shown [4, 5] that for a eq.2 a FT holds both for the $P(Q_\tau)$ and $P(W_\tau)$ (see also ref.[6]). Specifically defining the fluctuation function for heat by

$$f_Q(\tau, p_{\varrho}) \equiv \frac{k_B T}{\langle Q_\tau \rangle} \ln \left[\frac{P(+Q_\tau)}{P(-Q_\tau)} \right] \tag{8}$$

and a scaled heat fluctuation by $p_{\varrho} \equiv Q_\tau / \langle Q_\tau \rangle$, one has for large τ

$$f_Q(\tau, p_{\varrho}) = \begin{cases} p_{\varrho} + O(\tau^{-1}) & \text{if } 0 < p_{\varrho} < 1 \\ p_{\varrho} - \frac{1}{4}(p_{\varrho} - 1)^2 + O(\tau^{-1}) & \text{if } 1 < p_{\varrho} < 3 \\ 2 + O\left(\sqrt{(p_{\varrho} - 3)/\tau}\right) & \text{if } p_{\varrho} > 3. \end{cases} \tag{9}$$

Here, we only gave the orders of magnitude of the finite-τ correction terms. Their detailed forms can be found in Ref. [4]. We see that although a FT exists for Q_τ it differs from the conventional one given in eq.1. Instead it can be shown [4, 5] that the fluctuations of W_τ, satisfy the conventional FT for $\tau \to \infty$. Specifically defining the fluctuation function for work by

$$f_W(\tau, p_{w}) \equiv \frac{k_B T}{\langle W_\tau \rangle} \ln \left[\frac{P(+W_\tau)}{P(-W_\tau)} \right] \tag{10}$$

one finds

$$f_W(\tau, p_{w}) = \frac{p_{w}}{1 - \varepsilon(\tau)}, \tag{11}$$

where $p_{w} = W_\tau / \langle W_\tau \rangle$ and

$$\varepsilon(\tau) = 1 - \frac{\langle [W_\tau - \langle W_\tau \rangle]^2 \rangle}{2 \langle W_\tau \rangle} = \frac{\tau_o (1 - e^{-\tau/\tau_o})}{\tau}, \tag{12}$$

where $\tau_o = RC$. Note that $\varepsilon(\tau) \to 0$ for $\tau \to \infty$. We also note that the behavior of the heat fluctuations in Eq. (9) differs from that of the work fluctuations in Eq. (11). As we will see in the next section this difference is due to exponential tails of the distribution of heat fluctuations[4, 5], while those of $P(W_\tau)$ are Gaussian. Indeed if one assumes that fluctuations of U have a Gaussian distribution, which is the case at equilibrium when $I = 0$, then $\mathscr{P}_{\rm in}$ has also a Gaussian distribution, because I is constant. On the contrary, i_t fluctuates, as we see from Kirchoff's laws:

$$I = i_t + C \frac{\mathrm{d}U}{\mathrm{d}t}, \qquad \text{so} \qquad \mathscr{P}_{\rm in} = \mathscr{P}_{\rm diss} + C \frac{\mathrm{d}U^2}{\mathrm{d}t}, \tag{13}$$

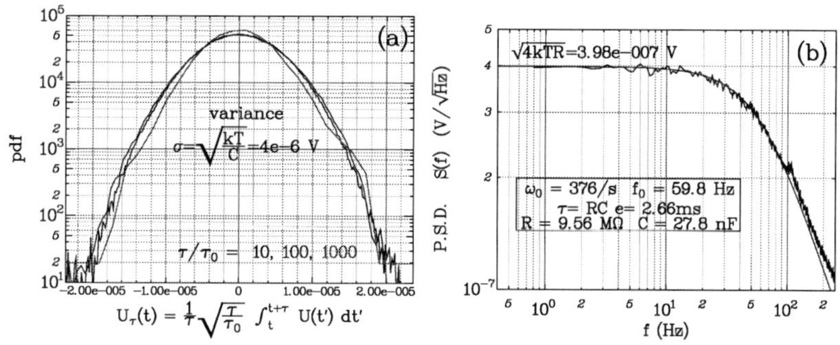

FIGURE 2. Thermal noise at $I = 0$. a) Johnson-Nyquist noise has a Gaussian distribution. b) Spectral density of U. the continuous line is a fit of the noise spectrum obatined using FDT.(Nyquist-Johnson noise)

and therefore, the probability distribution of $\mathscr{P}_{\text{diss}}$ is not Gaussian [5]. It is worth noting that for large current I, some orders of magnitude larger than the one we use, $\mathscr{P}_{\text{in}}$ and $\mathscr{P}_{\text{diss}}$ will be much larger than the conservative part $C\frac{dU^2}{dt}$ and therefore the probability distributions of both the injected and dissipated power will be Gaussian, as it is usually expected in macroscopic systems.

EXPERIMENTAL TEST OF THE THEORETICAL PREDICTIONS ON FLUCTUATION THEOREMS

In order to check whether the theoretical results summarized in the previous section can be seen experimentally we have performed a series of measurements of the injected and dissipated power in a circuit as the one depicted in fig.1. The used resistance is a standard metallic one of nominal value $R = 9.52$ MΩ. In parallel, we have an equivalent capacitor of value $C = 280$ pF. This accounts for the capacitance of the all set of coaxial connectors and cables that we used. The time constant of the circuit is $\tau_0 \equiv RC = 2.67$ms. Using a 50 GΩ resistance, we inject in the circuit a constant current I ranging from 0 to 6×10^{-13} A. This current corresponds to an injected power I^2R ranging from 0 to $1000 k_B T/\text{s}$. Experiments are conducted at room temperature $T = 300$ K. The typical values of the injected energy for, e.g., $I = 1.4 \times 10^{-13}$ A and $\tau = 10\tau_0$ are of the order of a few hundreds of $k_B T$. The fluctuating voltage U across the dipole is measured with a resolution of 10^{-11} V sampled at 819.2 Hz. More details on the experimental set-up can be found in ref.[9]. As it can be seen in fig.2, we find that, at $I = 0$, the spectral density of U is the one predicted by the FDT (Nyquist-Johnson noise) [10]. The circuit is driven out of equilibrium by letting a current $I \neq 0$ flowing into the circuit. We first checked that $\langle \mathscr{P}_{\text{in}} \rangle = \langle \mathscr{P}_{\text{diss}} \rangle \equiv \bar{\mathscr{P}} = R\, I^2$. Next we measured $P(W_\tau)$ and $P(Q_\tau)$. We used values of τ spanning from τ_0 up to hundreds of τ_0. From the experimentally measured $U(t)$ for a given value of I, we compute $\mathscr{P}_{\text{in}}$ and $\mathscr{P}_{\text{diss}}$. We then build the probability density functions of cumulated variables W_τ and Q_τ using 10^6 points;

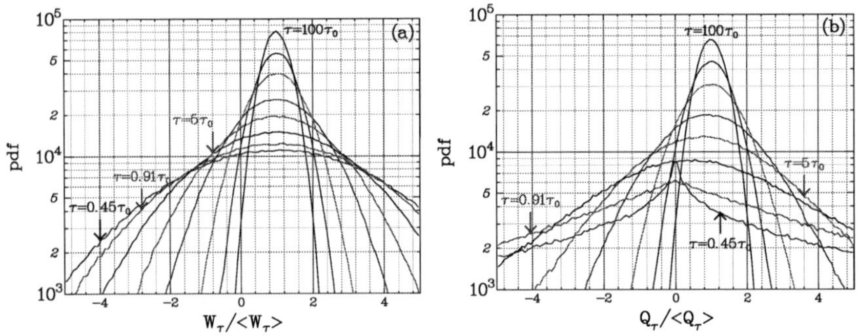

FIGURE 3. Histograms of W_τ and Q_τ when a current $I = 1.4 \times 10^{-13}$A is flowing in the dipole ($\bar{\mathscr{P}} = RI^2 = 45k_BT/s$).

their typical distributions are plotted on Fig. 3. As noticed above, fluctuations of W_τ are Gaussian for any τ; on the contrary, heat fluctuations are not Gaussian for small values of τ. Furthermore we see that, although $\langle \mathscr{P}_{\text{in}} \rangle$ is positive, there are negative fluctuations both of W_τ and of Q_τ. This means that instantaneously heat may flow from the cold to the hot source. Of course this heat flow reversal is not a violation of the second principle, which makes a statement only on the mean values, it shows only the importance of FT to know the probabilities of these flux reversals. Thus, in order to check FT, we construct the symmetry functions defined in eq.8 and eq.10. First, for any given τ we checked that the symmetry function $f_W(\tau, p_w)$ is linear in p_W (Fig. 4a). We measured the corresponding proportionality coefficient $\sigma_W(\tau) = f_W(\tau, p_w)/p_w$. This coefficient $\sigma_W(\tau)$ tends to 1 when τ is increased as it can be seen in Fig. 5a). For the heat fluctuations we find that $f_Q(\tau, p_Q)$ is linear in p_Q only for $p_Q < 1$, as expected by eq.9. Again, as $\tau \to \infty$, the limit slope of the symmetry function is 1 whereas for $p_Q > 3$, $f_Q(\tau, p_Q)$ tends to two.

In [5, 4], expressions for the convergence towards the asymptotic limits $f_W^\infty(w)$ and $f_Q^\infty(p_Q)$ are given in terms of τ. We can check these predictions with our data. The hyperbolic convergence for the work W_τ is very well reproduced by these predictions, as can be seen on Fig. 4a: continuous straight lines are theoretical predictions for small values of τ, using eqs.11 and eq.12, with no adjustable parameters. On Fig. 5a the function $\sigma_W(\tau) - 1$ of the experimental symmetry function is plotted. The solid line in the figure is $\varepsilon(\tau)/[1 - \varepsilon(\tau)]$ defined in eq.12. The agreement is perfect. For the heat, we distinguish two regimes for the convergence towards the asymptotic symmetry function. For $p_Q < 3$, we find that when τ is increased, symmetry functions are converging to the asymptotic function from below (Fig. 4), which is the opposite of what is observed for the work. On the contrary, for $p_Q > 3$, convergence to the asymptotic function is from above, thus enhancing the peculiarity of the point $p_Q = 3$. On Fig. 5, we have plotted the evolution of $f_Q(\tau, a) - f_Q^\infty(a)$ versus τ for several fixed values of $a = p_Q$. The convergence from above for $a \geq 3$ and from below for $a < 3$ is clear. For increasing values of τ, only smaller and smaller values of a are accessible because of the averaging process. Therefore the accessible values of a are quickly lower than 1, and the linear part of $f_Q(\tau, a)$ can be

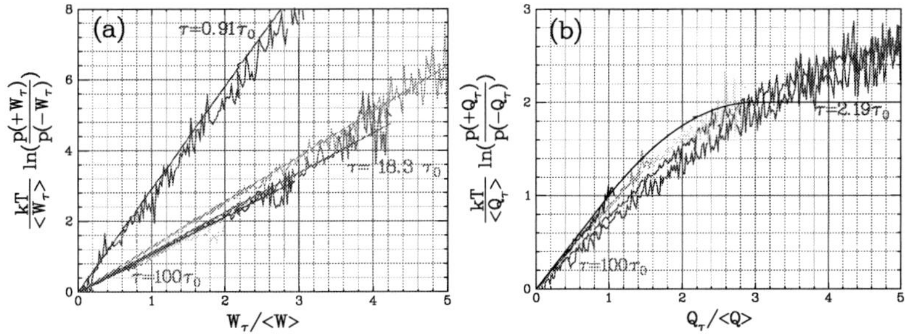

FIGURE 4. Normalized symmetry functions $f_W(\tau, p_w)$ and $f_Q(\tau, p_Q)$ for W_τ and Q_τ when a current $I = 1.4 \times 10^{-13}$A is flowing in the dipole. For any τ, $f_W(\tau, a)$ is a linear function of $a = p_w$. For $\tau \to \infty$, $f_W(\tau, p_w)$ tends to have a slope of 1, whereas $f_Q(\tau, a)$ tends to a limit function (black curve) which is a straight line of slope 1 for $a = p_Q < 1$ only.

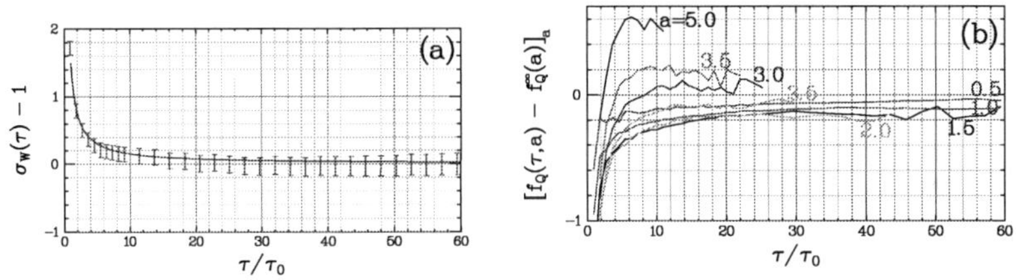

FIGURE 5. (a): dependence on τ of the slope of S_W in Fig. 4. Slope is converging from above. (b): distance between $f_Q(\tau, p_Q)$ for finite τ and theoretical prediction of $f_Q^\infty(p_Q)$ for infinite τ, for several values of $a = p_Q$. for $a < 3$, convergence is from below, whereas it is from above for $a > 3$.

experimentally tested. Nevertheless, for intermediate time scales τ we see in Fig. 4 that the data converge towards the theoretical asymptotic nonlinear symmetry function [5, 4] (smooth curve in Fig. 4b).

We observed that the convergence reproduced on Fig. 5 depends on the injected current I, as pointed out in [5]. Other experiments with a larger current ($\bar{\mathscr{P}} = 186 k_B T/\mathrm{s}$) give a faster convergence.

CONCLUSIONS

We have discussed the general results of the Fluctuation Theorem for the injected and dissipated power in out of equilibrium systems which are described by Langevin equations. These results can be applied either to a Brownian particle or to electric circuits. As an example we have shown that the theoretical predictions can be observed in a simple electrical circuit driven out of equilibrium by a fixed current I. We also show that the asymmetry of the probability distribution functions of work and heat is linked to the averaged dissipated power in the system. The first simple application of this result is that FT (eq.9, eq.11) can be used to measure an unknown averaged dissipated power $\bar{\mathscr{P}} = \lim_{\tau \to \infty} \frac{Q_\tau}{\tau}$ by using only the symmetries of the fluctuations, *i.e.* computing f_τ^W or f_τ^Q and measuring their asymptotic slope. We operated with energies of order of $k_B T$ in order to have strong fluctuations compared to the averaged values. It is worth noting that as the driving current is increased up to macroscopic values, the fluctuations become more and more negligible, therefore Fluctuation Theorems become harder and harder to use, and therefore less relevant. As a consequence FT are especially useful for all of the systems, such as nano et and micro systems, where thermal fluctuations play an important role. The example given in this talk is a very simple one but many complex applications of these results are possible. For example one can study the FT in a aging materials [14] where the statistical properties of the noise are strongly affected by the aging and FDT does not hold necessarily [15, 16]. Furthermore one can also study the motion of Brownian particle in an out of equilibrium liquid [17]. Work is progress in this direction.

REFERENCES

1. D. J. Evans, E. G. D. Cohen, and G. P. Morriss, Phys. Rev. Lett. **71**, 2401 (1993); D. J. Evans and D. J. Searles, Phys. Rev. E **50**, 1645 (1994).
2. G. Gallavotti and E. G. D. Cohen, Phys. Rev. Lett. **74**, 2694 (1995); E. G. D. Cohen, Physica **240**, 43 (1997); E. G. D. Cohen and G. Gallavotti, J. Stat. Phys. **96**, 1343 (1999).
3. J. Kurchan, J. Phys. A, Math. Gen. **31**, 3719 (1998); J. L. Lebowitz and H. Spohn, J. Stat. Phys. **95**, 333 (1999).
4. van Zon, R., Cohen, E.G.D., *Phys. Rev. Lett.* **91** (11) 110601 (2003), *Phys. Rev. E* **67** 046102 (2003)
5. van Zon, R., Ciliberto, S., Cohen, E.G.D., *Phys. Rev. Lett.* **92** (13) 313601 (2004)
6. J. Farago, J. Stat. Phys. **107**, 781 (2002). See also these proceedings.
7. S. Ciliberto and C. Laroche, J. Phys. IV, France **8**, 215 (1998); , S. Ciliberto, N. Garnier, S. Hernandez, C. Lacpatia J.-F. Pinton, G. Ruiz Chavarria, Physica A340, 240 (2004).
8. G. M. Wang *et al.*, Phys. Rev. Lett. **89**, 050601 (2002).
9. N. Garnier, S. Ciliberto, Phys. Rev. E 71, 060101 (2005)
10. Nyquist, H., *Phys. Rev.* **32** pp. 110-113 (1928); Johnson, J.B., *Phys. Rev.* **32** pp. 97-109 (1928).
11. Bodineau, T., Derrida, B. *Phys. Rev. Lett.* **92** 1800601 (2004)
12. Aumaitre, S., Fauve, S., McNamara, S., Poggi, P., *Eur. Phys. J. B* **19** pp. 449-460 (2001)
13. See e.g. N. G. van Kampen, *Stochastic Processes in Physics and Chemistry* (North Holland, Amsterdam, 1992).
14. Crisanti A. and Ritort F., Europhys. Lett., **66**, 253 (2004).
15. L. Bellon, S. Ciliberto, Physica D **168**, 169 (2002).
16. L. Buisson and S. Ciliberto,Physica D **204**,1 (2005).
17. F.Zamponi, L.F.Cugliandolo, J.Kurchan, *Fluctuation theorem for non-equilibrium relaxational systems driven by external forces* cond-mat/0504750.

Non-equilibrium stationary states in dissipative systems

Jean Farago

LCP-UMR 8000, Bat. 349, Université Paris-Sud, 91405 Orsay cedex, France

Abstract. We consider different models of stochastic dissipative equations and compute theoretically the probability distribution functions (actually the associated large deviation functions) of the *time averaged* injected power required to sustain a nontrivial stationary state. We discuss the results and in particular draw from our results some general features shared by these distributions in realistic dissipative systems.

Keywords: Fluctuation phenomena, random processes, noise, and Brownian motion ; Probability theory, stochastic processes, and statistics.
PACS: PACS: 05.40.-a; 02.50.-r

Recently, non equilibrium stationary states (NESS) received renewed attention, due to some works which showed that large deviation techniques are tools well adapted to the description of NESS ([1, 2, 3] and references therein). These studies were however restricted to systems where the dynamics are conservative in-the-bulk : a quantity (typically the energy or the number of particles) is conserved by the inner dynamics and can change only through an interaction with the boundaries.

However, the class of so-called NESS is larger, since dissipative systems form a large class of statistical systems where non trivial NESS can be sustained [5]. Granular materials and hydrodynamical flows are typical examples of such dissipative media where NESS are ubiquitous (Of course, the boundaries are often periodically varying, but on a coarse-grained timescale, the stationarity is recovered). Actually, any system where some degrees of freedom have a "macroscopical" status, whereas the others can be treated as a perfect sink for the conserved quantities are eligible for the denomination "dissipative systems".

The two classes are very different : the natural reference for the conservative in-the-bulk dynamics is the thermal equilibrium (example; a rod heated at the ends with two different thermostats examplifies this class. On the contrary, the natural fixed point for a dissipative system is the full rest (when the external forcing ceases), and the thermal equilibrium is irrelevant.

The absence of reference to a true thermal equilibrium makes the dissipative NESS particularly difficult to study, and generally the techniques and the concepts are specific to each context.

We have addressed in preceding papers [6, 7, 8] the question of dissipative NESS considered from a general point of view, having remarked that all such systems share the same balance equation for the global energy (or any quantity of interest) :

$$\dot{E} = \mathscr{I} - \mathscr{D}$$

CP800, *Unsolved Problems of Noise and Fluctuations: UPoN 2005*
edited by L. Reggiani, C. Pennetta, V. Akimov, E. Alfinito, and M. Rosini

where $\mathscr{I}$ is an injection term, usually a surface term, and $\mathscr{D}$ is a dissipation term, always a volumic one, proportional to a dissipation coefficient (viscosity...), and always *positive*. This imposes that $\langle \mathscr{I} \rangle > 0$, but does not prevent occasional occurences of negative excursions of $\mathscr{I}(t)$.

To fade out short-time dynamical details, and consider the system on a coarse-grained time scale, it is natural to partially integrate the observable :

$$\varepsilon = \frac{1}{\tau}\int_0^\tau \mathscr{I}(t)dt, \quad \delta = \frac{1}{\tau}\int_0^\tau \mathscr{D}(t)dt$$

If the process is stationary (as in a NESS), and has a finite correlation time, the large deviation theorem ensures that there exists a fonction $f(\varepsilon)$ such that

$$\log(P(\varepsilon)) \underset{\tau\to\infty}{\sim} \tau f(\varepsilon)$$

This function codes the way the pdf of ε goes to a Dirac δ function when τ is increased, beyond the Gaussian approximation. It is analogous to a thermodynamical potential in the context of signal processing, except that in thermodynamics the potentials are large deviation functions (ldf) associated with a spatial averaging.

STOCHASTIC MODELS OF DISSIPATIVE NESS

In [6] and [7], I studied a general class of minimal models of dissipative NESS. It consists in a single coordinate experiencing a dissipative dynamics :

$$\ddot{x} + \varphi(\dot{x}) + V'(x) = 0 \tag{1}$$

φ is a odd increasing function (the friction) and $V(x)$ is a potential. The natural definition for the energy of the particle is $E = \frac{1}{2}\dot{x}^2 + V(x)$, and in the absence of external driving, this energy recedes in course of time : $\dot{E} = -\dot{x}\varphi(\dot{x})$. To drive the system out of equilibrium, we could use a periodic forcing, and actually it is an interesting perspective (as far as a chaotic region of the parameters phase space is selected), but the simplest way is probably to use a stochastic forcing, say $\psi(t)$. Thus, the complete dynamical equation reads

$$\ddot{x} + \varphi(\dot{x}) + V'(x) = \psi(t) \tag{2}$$

where $\psi(t)$ is a noise.

The reader is invited not to consider he faces a Langevin equation, for although it resembles formally a Langevin equation (especially if the friction is linear), the underlying physics are entirely different (in a Langevin equation, the friction term and the fluctuation term are two faces of the *same* cause, what is here absolutely excluded a priori, owing to the modelization itself). In particular, the noise is not necessarily white, and a differential equation driven out-of-equilibrium by a non-white noise cannot describe a situation *à la Langevin* [4].

I shew in [6] that the whole distribution of $\varepsilon = \frac{1}{\tau}\int_0^\tau dt\,\psi(t)v(t)$ is computable for the case of a linear friction ($\varphi = \gamma\dot{x}$) and white noise ($\langle\psi(t)\psi(t')\rangle = 2\Gamma\delta(t-t')$). It was

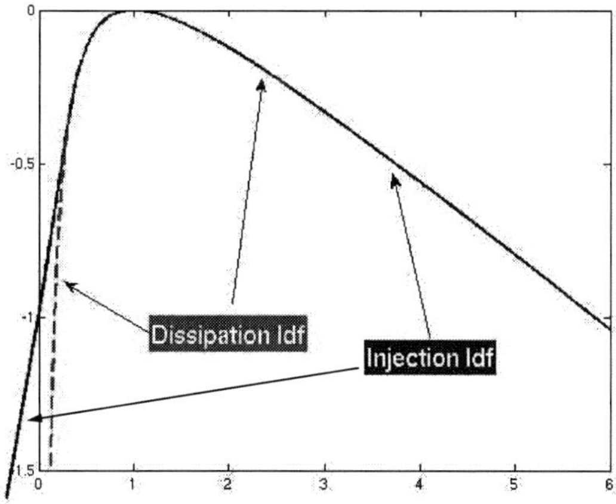

FIGURE 1. Large deviation functions of injected (ε) and dissipated (δ) energy

thus possible to highlight a curious phenomenon affecting the large deviation function : in the permanent regime, the rare energetic initial conditions (by "initial conditions" I term the state of the system at the beginning of the time averaging in the formula : $\varepsilon = \frac{1}{\tau}\int_0^\tau \mathscr{I}$) are responsible for a breaking of analyticity in $f(\varepsilon)$, giving rise to a second order singularity and a negative branch in the $\varepsilon < 0$ region:

$$f(\varepsilon) = \begin{cases} -\dfrac{\gamma}{4\varepsilon/\Gamma}(\varepsilon/\Gamma - 1)^2 & \text{if } \varepsilon/\Gamma \geqslant \frac{1}{3} \\ -\gamma(1 - 2\varepsilon/\Gamma) & \text{if } \varepsilon/\Gamma \leqslant \frac{1}{3} \end{cases} \tag{3}$$

It is worth pointing out that this splitting is a kind of extremal fluctuation/dissipation relation, in that the negative branch witnesses to large fluctuations of energy, whereas the other branch is closely related to the dissipation : actually, the large deviation function of the dissipation is exactly $f(\delta) = -(\gamma\Gamma/4\delta)(\delta/\Gamma - 1)^2$ also for $\delta \in]0, 1/3]$ ($f(\delta < 0)$ is obviously irrelevant since $\mathscr{D} > 0$) : see fig. 1 This simple model exhibited also a very interesting "reduction" property : if one adds a potential $V(x)$ to the dynamics, i.e. $\ddot{x} + \gamma\dot{x} + V'(x) =$white noise, the ldf $f(\varepsilon)$ is *independent* of $V(x)$! This illustrates the relevance (for this model at least) of the intuitive statement, that the time averaging can fade out specific dynamical details and reveal a common behaviour of different models as regards the energy flow.

Exponentially correlated noise

In [7], I generalized the previous model to a more complicated situation where the noise is a Ornstein Uhlenbeck process : $\psi(t)$ is Gaussian with zero mean and $\langle\psi(t)\psi(t')\rangle = \frac{\Gamma}{\tau_0}\exp(-|t-t'|/\tau_0)$. The aim was to test whether the observations made in [6] were robust against a slight modification of the external input, and/or the "nonlinearization" of the friction.

For a linear friction, the computations appeared to be tractable for a quadratic potential $V(x) = \frac{1}{2}\omega^2 x^2$ only, and the result ($\alpha = \gamma\tau_0$, $\beta = \omega\tau_0$) (*we do not write the left "fluctuative" part of the distribution, even it is present, see [7] for details*),

$$f(\varepsilon) = g(\lambda) + \lambda\varepsilon \;\text{ with } \lambda \text{ such that } g'(\lambda) = -\varepsilon \tag{4}$$

$$g(\lambda) = \frac{1}{2\tau_0}(\alpha + 1 - \theta(\lambda)) \tag{5}$$

$$(\theta^2 - \alpha^2 - 1 + 2\beta^2)^2 - 8\beta^2\theta - 4(\beta^4 + \alpha^2 - 2\beta^2 + 4\Gamma\lambda\gamma\tau_0^2 \qquad = 0 \tag{6}$$

(θ is the largest positive root of the polynomial), shows the complexity of the quantity we are looking of (due to its time-averaged nature). It can be seen immediately that now the potential $V(x)$ (represented by β) plays a role at the ldf level; thus, the previous irrelevance is just a peculiarity of the previous class only (a pity!). On the contrary, the singularity is still here, and actually is *generic* for systems where the stationary distribution of the energy is *unbounded*: for instance, such a singularity would be absent in a chaotic model where the injection is a deterministic (periodic for instance) function —in that case, the distributions are as a rule bounded. It stresses also the difficulty one would have to find a numerical signature of such a singularity, owing to the boundedness *de facto* of any distribution numerically generated.

Nonlinear frictions were also investigated, and it was shown that the asymptotics of the ldf are closely related to the functional form of the dissipation. Actually, the linear behaviour of $f(\varepsilon)$ for large ε is entirely due to the linear form of the dissipation in the previous examples.

CHARACTERISTIC ENERGY, CHARACTERISTIC TIME

Another interesting point emerging from these studies is the possibility to define a characteristic time associated with the energy flow as

$$\tau_e = \frac{1}{\langle\mathscr{I}\rangle^2}\int_0^\infty dt[\langle\mathscr{I}(t)\mathscr{I}(0)\rangle_{st} - \langle\mathscr{I}\rangle_{st}^2] = \frac{1}{\langle\mathscr{D}\rangle^2}\int_0^\infty dt[\langle\mathscr{D}(t)\mathscr{D}(0)\rangle_{st} - \langle\mathscr{D}\rangle_{st}^2] \tag{7}$$

or similarly a characteristic energy (by dividing by $\langle\mathscr{I}\rangle$ instead of $\langle\mathscr{I}\rangle^2$). This time can be termed "characteristic", for it is the same for the injection or the dissipation. Usually, one defines the correlation time of an observable as $\tau_X = \left[\int_0^\infty dt\langle X(t)X(0) - \langle X\rangle^2\rangle\right]/\langle X^2 - \langle X\rangle^2\rangle$. Thus, τ_e is more or less a bare time whence τ_I and τ_D are constructed as

$$\tau_I = \frac{\langle\mathscr{I}\rangle^2}{\langle\mathscr{I}^2\rangle - \langle\mathscr{I}\rangle^2}\tau_e \tag{8}$$

(and a similar one with $\mathscr{D}$). These relations are interesting, since if one imagines that τ_e is fixed (or taken as the natural time unit), one sees that at the injection level, the

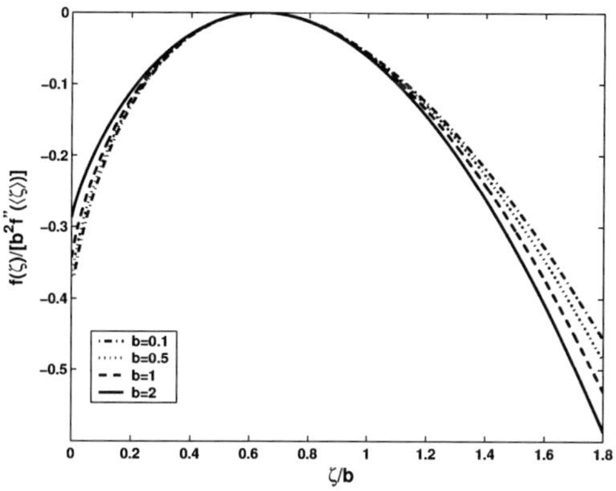

FIGURE 2. Large deviation function of the energy stored in the N first sites.

fluctuations level is enslaved by the characteristic time of the injection mechanism. At the other side, this balance exists with a dissipation time which is possibly quite different of the former; in that case the fluctuations are again constrained to adjust correspondingly. This simple reasoning of course does not take into account the fact that the "reference" time τ_e is itself a property of the established regime.

SPATIAL LDF AND DISSIPATIVE NESS

For extended systems, dissipative NESS can be highly inhomogeneous. Thus, it is not sure that an analysis of dissipative NESS in terms of spatial LDF, as it is done for example (for conservative dynamics) in [1] be even feasible. In [8], I studied a one-dimensional system, a semi-infinite chain of Potts spins, with dissipative inner dynamics (the zero-T Glauber dynamics). The system is driven out of rest by random flipping of the spin at the edge, and a nontrivial NESS appears.

Using random walk counting techniques, it was possible to show that the distribution of the energy located in the N first spins has a large deviation property with respect to $\log N$ (as the diverging parameter), and with the ldf $f(x)$ given by the inverse Legendre transform of

$$g(\mu) = \frac{1}{8} - \left[\frac{\sqrt{2}}{\pi}\text{Arccos}\left(\frac{be^{-2\mu}+1-b}{\sqrt{2}}\right)\right]^2 \quad \text{where } b = 2\times\left(1-\sum_{\ell=1}^{q} P_0(\ell)^2\right) \tag{9}$$

Figure 2 shows the result for various values of b. The ldf look vaguely parabolic, but they are not exactly such, and have the curious property to stop abruptly at 0 without divergence.

An essential point highlighted by this study —beyond the fact that dissipative nonequilibrium profile can have a ldf property for its fluctuations— is that the fluctuations are in some way decorrelated. We refer to [8] for details, since it is rather technical, but to summarize, the probability to observe simultaneously a certain energy in the N

first spins, and another in the N^2 following is just the product of the two probabilities, *at the level of the large deviation function*. The need to take N^2 spins for the second block (for the second block fluctuations to be relevant at the ldf level) reflects the strong inhomogeneity of the dissipative profile. Thus, in dissipative systems, no special weak-and-long-range correlations take place (as in [9, 1] for instance), but we have to deal with inhomogeneities, which can be so strong that it could certainly destroy the ldf structure : for instance, in our model, the introduction of a drift of the domain wall motion toward the boundary would make the spatial ldf disappear, since the inner part of the chain would be in that case statistically almost never visited by any domain wall (we would have an effective zero-dimensional model). Thus, even if in our case the fluctuations of the profile obeys a large deviation theorem, it does not guaranty that such a favourable case always occurs.

Conversely, the time averaging of the injected/dissipated power could be studied in this model, and it would be interesting to get the corresponding ldf, and to test its possible (partial or not) irrelevance with respect to the injection mechanism. This question remained to be answered.

CONCLUSION : UNSOLVED PROBLEMS

The main question concerning the physics of far-from-equilibrium stationary states is the question of universality. Presently the known results are restricted to stochastic one-dimensional models, and the validity of the conclusions one can draw from these models is not ensured for other systems. Actually, the non-equilibrium world seems to be the realm of exceptions and details and even the beautiful universality found in [2] seems to being partially amended [3].

A particular simple question which is yet not fully answered is the existence of a criterion for the direction of the energy flow (the "temperature gradient" for non thermal systems) : if one couples extremely weakly two dynamical systems, which gives energy to the other ? Does exist only a criterion for answer this question, or is the answer specific to each particular situation, depending on the coupling form, the nature of the two systems, etc...?

REFERENCES

1. B. Derrida, J.L. Lebowitz, E. R. Speer, *Phys. Rev. Lett.*, **87**, 150601 (2001) ; *J. Stat Phys.*,**107**, 599 (2002); *Phys. Rev. Lett.*, **89**, 030601 (2002); *J. Stat. Phys.*,**110**, 775 (2003).
2. T. Bodineau, B. Derrida, *Phys.Rev.Lett.*,**92**,180601 (2004).
3. L.Bertini, A. de Sole, D. Gabrielli, G. Jona-Lasinio and C. Landim, *Phys.Rev.Lett.*,**94**,030601 (2005).
4. R.Kubo, M.Toda, N.Hashitsume, *Statistical Physics II: Nonequilibrium Statistical Mechanics*, Springer Series in Solid-State Sciences, Vol 31.
5. S.Aumaître, S.Fauve, S.McNamara, P.Poggi, *Eur. Phys. J. B*, **19**, 449 (2001).
6. J. Farago, *J. Stat. Phys.*, **107**, 781 (2002).
7. J. Farago, *Physica A*, **331**, 69 (2004).
8. J. Farago, *J. Stat. Phys.*, **118**, 373 (2005).
9. H. Spohn, *J. Phys. A*, **16**, 4275 (1983).

Ensemble Inequivalence In A XY Model With Long-Range Interactions

Pierre de Buyl*, David Mukamel† and Stefano Ruffo**

*Center for Nonlinear Phenomena and Complex Systems,
Faculté des Sciences, Université Libre de Bruxelles
Campus de la Plaine CP231, boulevard du Triomphe, 1050 Bruxelles, Belgium
†Department for Physics of Complex Systems,
Weizmann Institute of Science, Rehovot, Israel 76100
**Dipartimento di Energetica "S. Stecco", Università di Firenze,
CSDC and INFN, via s. Marta, 3 50139 Firenze, Italy

Abstract. We introduce a new XY mean-field spin system whose canonical and microcanonical thermodynamics can be solved exactly. Microcanonical entropy is obtained by a novel use of large deviation techniques. The model desplays ensemble inequivalence and the presence of negative specific heat and temperature jumps in the microcanonical ensemble.

Keywords: Long-range interactions, Large deviation techniques, XY model, Ensemble Inequivalence
PACS: 05.20.Gg, 05.70.Fh, 02.50.Cw

INTRODUCTION

When the thermodynamic limit is taken in statistical mechanics, the system is implicitly assumed to be *additive*. This hypothesis is not respected by long-range interactions, for which the interparticle potential decreases as $V(r) \sim r^{-\alpha}$, with $\alpha < d$ in d space dimensions [1]. Indeed, for this type of interactions, the contribution of the "surface" energy is not necessarily negligible and the system is *non additive*. A consequence of non additivity is the presence of "convex intruders" in microcanonical entropy, which can give rise to *negative specific heat* [2] and *temperature jumps* [3]. In this paper we demonstrate the presence of such features in an infinite-range (mean-field) XY model, which generalizes the so-called Hamiltonian mean field (HMF) model [4].

We first introduce and briefly justify the model. We then discuss *ensemble inequivalence* and the presence of negative specific heat in the microcanonical ensemble. The solution of the model is obtained by the use of large deviations, a technique of probability theory which has nowadays received a wide range of applications [5, 6, 7] and which has recently been employed to solve a large class of mean-field models [8]. Finally, we draw some conclusions.

THE MODEL

The spins of the XY model are two-dimensional unit vectors $\vec{s}_i = (\cos\theta_i, \sin\theta_i)$. Each pair (i, j) of spins contribute to the energy with a term $-J\vec{s}_i \cdot \vec{s}_j$. With each spin we

CP800, *Unsolved Problems of Noise and Fluctuations: UPoN 2005*
edited by L. Reggiani, C. Pennetta, V. Akimov, E. Alfinito, and M. Rosini

associate an angular momentum $p_i = \dot{\theta}_i$. Here we generalize this model and consider the Hamiltonian

$$H = \sum_{i=1}^{N} \frac{p_i^2}{2} - \frac{J}{2N}(\sum_{i=1}^{N} \vec{s}_i)^2 - \frac{K}{4N^3}[(\sum_{i=1}^{N} \vec{s}_i)^2]^2 , \tag{1}$$

where N is the total number of spins. The coupling constants J and K are both positive and, hence, tend to align the spins in a ferromagnetic order. Introducing the order parameter

$$\vec{m} = (m_x, m_y) = \frac{1}{N} \sum_{i=1}^{N} \vec{s}_i = \left(\frac{1}{N} \sum_i \cos\theta_i, \frac{1}{N} \sum_i \sin\theta_i \right) \tag{2}$$

and the kinetic energy per particle

$$\mathscr{E}_K = \frac{1}{N} \sum_{i=1}^{N} \frac{p_i^2}{2} , \tag{3}$$

it is possible to rewrite the energy in an intensive form

$$\mathscr{E} = \frac{H}{N} = \mathscr{E}_K + \mathscr{E}_I = \mathscr{E}_K - \frac{J}{2} m^2 - \frac{K}{4} m^4, \tag{4}$$

where m is the modulus of $\vec{m}$. The variables $\mathscr{E}_K$ and m have been called *global* variables in Ref. [8], because they result from an average over the whole system. Large deviation techniques allow one to estimate the number of microscopic states compatible with a given set of values of global variables.

Model (1) reduces to the HMF model for $K = 0$. In this limit, the model displays a second order phase transition and the canonical and microcanonical ensembles yield equivalent predictions. This is expected on the basis of the criterion for ensemble inequivalence formulated in Ref. [3], which states that ensembles are inequivalent if a first order phase transition is present in the canonical ensemble. We will see that the ensembles become inequivalent for $K/J > 1/2$.

LARGE DEVIATIONS AND ENSEMBLE INEQUIVALENCE

Large deviations

The method of large deviations allows to characterize the statistical behavior of a sum of random variables. We describe here the method for independent identically distributed *i.i.d.* vector random variables $\vec{X}$, forming a sample of N elements $\vec{X}_i$, $i = 1, \ldots, N$. The sample mean is defined as

$$\vec{S}_N = \frac{1}{N} \sum_{i=1}^{N} \vec{X}_i . \tag{5}$$

The sample mean is said to obey a *large deviation principle* if its probability density scales as

$$P(\vec{S}_N = \vec{x}) \simeq \exp -N I(\vec{x}), \tag{6}$$

where $I(\vec{x})$ is the *rate function*.

Cramér's theorem [6] allows one to derive this probability distribution in the large N-limit, providing a method to compute the rate function. One first computes the cumulant generating function

$$\Psi(\vec{\lambda}) = \langle e^{\vec{\lambda}\cdot\vec{X}} \rangle \,, \tag{7}$$

where the average is performed over the probability distribution of $\vec{X}$. Then, under quite general assumptions, the rate function is given by the Legendre-Fenchel transform of $\ln(\Psi(\vec{\lambda}))$

$$I(\vec{x}) = \sup_{\vec{\lambda}}(\vec{\lambda}\cdot\vec{x} - \ln\Psi(\vec{\lambda})). \tag{8}$$

We will use this theorem to determine the entropy-energy function of model (1).

Microcanonical ensemble

The model has three global variables m_x,m_y and $\mathscr{E}_K$, which are sample means of the *i.i.d.* vector random variable

$$\vec{X} = \left(\cos\theta, \sin\theta, p^2/2\right) \,. \tag{9}$$

Following the same strategy of Ref. [8], we determine the cumulant generating function

$$\Psi(\vec{\lambda}) = \langle e^{\vec{\lambda}\cdot\vec{X}} \rangle \simeq \frac{I_0(\sqrt{\lambda_x^2+\lambda_y^2})}{\sqrt{-\lambda_K}} \tag{10}$$

where $\vec{\lambda} = (\lambda_x, \lambda_y, \lambda_K)$ and I_0 is the modified Bessel function of zero order. In formula (10) we have neglected irrelevant multiplicative constants and we have performed the average at leading order in λ_K. As a consequence, the rate function, up to irrelevant additive constants, is given by

$$I(\vec{x}) = \sup_{\vec{\lambda}}\left(\frac{\lambda_K p^2}{2} + \lambda_x m_x + \lambda_y m_y + \frac{\ln(-\lambda_K)}{2} - \ln(I_0(\sqrt{\lambda_x^2+\lambda_y^2}))\right) \,, \tag{11}$$

where $\vec{x} = (m_x, m_y, \mathscr{E}_K)$. Let $(\bar{\lambda}_x, \bar{\lambda}_y, \bar{\lambda}_K)$ be the vector which maximizes (11). The optimization problem in (11) can be explicitly solved to yield

$$I(\vec{x}) = \bar{\lambda}m - \frac{\ln(2\mathscr{E}_K)}{2} - \ln I_0(\bar{\lambda}) \,, \tag{12}$$

again apart from constants, where $\bar{\lambda} = \sqrt{\bar{\lambda}_x^2+\bar{\lambda}_y^2}$ is obtained by inverting the consistency equation

$$m = \frac{I_1(\bar{\lambda})}{I_0(\bar{\lambda})}, \tag{13}$$

in which I_1 is the modified Bessel function of first order. The entropy is then found by solving the following variational problem

$$s(\mathscr{E}) = \sup_{\vec{x}}\{-I(\vec{x}) \text{ with } \mathscr{E}_K = \mathscr{E} + \frac{Jm^2}{2} + \frac{Km^4}{4}\}\ , \tag{14}$$

which amounts to solve in m the equation

$$\bar{\lambda}(m) = \frac{Jm + Km^3}{2\mathscr{E} + Jm^2 + K/(2m^4)}. \tag{15}$$

By solving analytically (for second order transition lines) and numerically (for first order lines) these variational problems, and taking into account the standard relation between inverse temperature and energy

$$\beta = \frac{1}{T} = \frac{ds}{d\mathscr{E}}\ , \tag{16}$$

with $k_B = 1$, one gets both the phase diagram and the "caloric curves" discussed below[1].

Canonical ensemble

The derivation of canonical free energy of model (1) is more standard and will be discussed elsewhere. We give here only the final result of the Landau canonical free energy as a function of β and m

$$f = \frac{Jm^2}{2} + \frac{3Km^4}{4} - \frac{\ln(2\pi I_0(\beta(Jm + Km^3)))}{\beta} \tag{17}$$

Phase diagram and caloric curves

In Fig. 1 we report the phase diagram of model (1) in the parameter space $(T/J, K/J)$ in both the canonical and the microcanonical ensemble. As expected [3], the two phase diagrams differ in the region where the canonical transition line is first order. The second order phase transition lines at $T/J = 1/2$ coincide up to the canonical tricritical point, located at $K/J = 1/2$, but then the predictions of the two ensembles differ: while the canonical ensemble gives a first order phase transition, in the microcanonical ensemble the critical line at $T/J = 1/2$ persists up to the microcanonical tricritical point at $K/J = 5/2$. Even more striking is the difference between the two ensembles for $K/J > 5/2$. Here, two different temperatures coexist at the transition energy in the microcanonical ensemble, showing what we have called a *temperature jump* [3]. This corresponds to a microcanonical first order phase transition, whose "phase coexistence"

[1] A full derivation will be reported in a future publication.

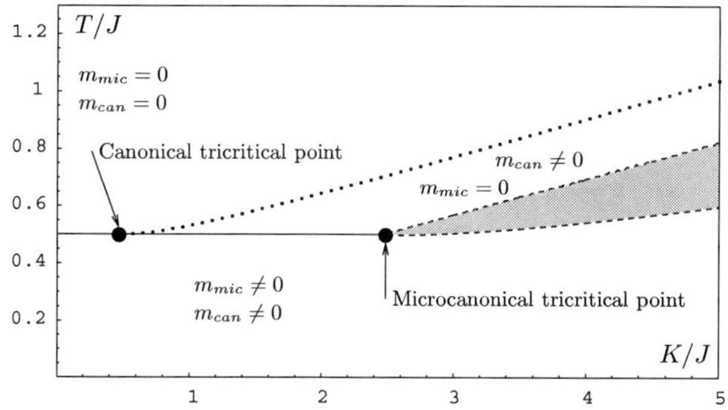

FIGURE 1. Phase diagram of model (1). The canonical second order transition line (full line at $T/J = 1/2$) becomes first order (dotted line, determined numerically) at the canonical tricritical point. The microcanonical second order transition line coincides with the canonical one below $K/J = 1/2$ but extends further right up to the microcanonical tricritical point at $K/J = 5/2$. At this latter point, the transition line bifurcates in two first order microcanonical lines, corresponding to a temperature jump. The behavior of the order parameter in the two ensembles is also shown in the figure, to highlight the striking difference in the predictions of the two ensembles.

region is shown by the shaded area in Fig. 1. As for the order parameter m, the most impressive difference between the two ensembles is in the intermediate region between the canonical first order phase transition line and the upper microcanonical first order line, where the canonical ensemble predicts a non vanishing order parameter $m_{can} \neq 0$, while the microcanonical ensemble gives $m_{mic} = 0$.

Fig. 2 displays temperature vs. energy, so-called "caloric curve", in the region where both the canonical and the microcanonical ensemble predict a first order phase transition. However, the microcanonical ensemble gives two temperatures at the transition energy, both below the canonical phase transition temperature (but the order of the transitions may change for other parameter values). Both a region of negative specific heat and a temperature jump are present in the microcanonical ensemble. The specific heat is always positive in the canonical ensemble and latent heat is released. In a microcanonical simulation performed with $N = 100$ spins (shown by the points in Fig. 2), the temperature jump is smoothed by finite size effects and negative specific heat is well reproduced. The agreement with microcanonical thermodynamics predictions is good.

CONCLUSIONS

We have shown how a simple generalization of the HMF model (mean-field XY system) [4] can give rise to all the features of ensemble inequivalence displayed by the Blume-Emery-Griffiths model [3], *i.e. negative specific heat* and *temperature jumps* in the microcanonical ensemble. The solution of the model also gives the opportunity to illustrate the power of *large deviation techniques* in the calculation of microcanonical

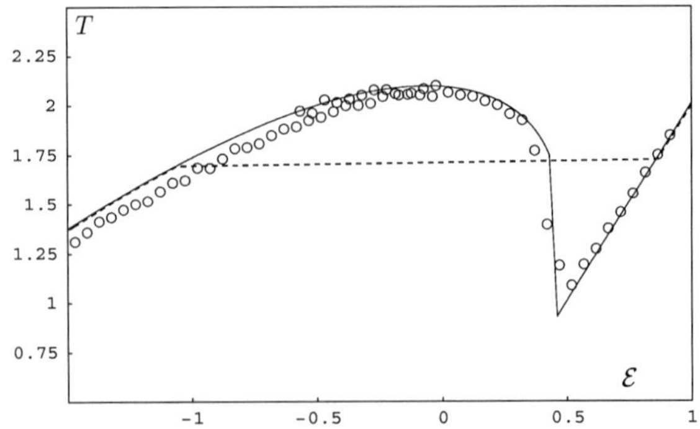

FIGURE 2. Caloric curve $T(\mathscr{E})$ at $K/J = 10$. The microcanonical ensemble (full line) predicts a region of negative specific heat, where temperature T decreases as the energy is increased. Moreover, a temperature jump is present at the transition energy. In the canonical ensemble we have a first order phase transition (dashed line). The points are the result of a molecular dynamics simulation performed by solving numerically the equations of motion given by Hamiltonian (1) with $N = 100$.

ACKNOWLEDGMENTS

We thank A. Campa, A. Giansanti and A. Torcini for useful discussions. S.R. acknowledges financial support of the PRIN03 project *Order and chaos in nonlinear extended systems*. D.M. was partially supported by INFN-Florence. P.d.B. thanks P.Gaspard for suggesting him the stage at Florence University which gave rise to this work.

REFERENCES

1. T. Dauxois, S. Ruffo, E. Arimondo, M. Wilkens (Eds.), *Dynamics and Thermodynamics of Systems with Long-Range Interactions*, Lecture Notes in Physics **602**, Springer-Verlag, New York, 2002.
2. V.A. Antonov, *Vest. Leningrad Univ.*, **7**, 135 (1962); Translation in *IAU Symposium*, **113**, 525 (1995); D. Lynden-Bell and R. Wood, *Mon. Not. R. Astron. Soc.*, **138**, 495 (1968); P. Hertel and W. Thirring, *Ann. of Phys.*, **63**, 520 (1971).
3. J. Barré, D. Mukamel, S. Ruffo, *Phys. Rev. Lett.*, **87**, 030601 (2001).
4. M. Antoni, S. Ruffo, *Phys. Rev. E*, **52**, 2361 (1995).
5. R. S. Ellis, *Entropy, Large Deviations, and Statistical Mechanics*', Springer-Verlag, New York, 1985.
6. A. Dembo, O. Zeitouni, *Large Deviations Techniques and their Applications*, Springer-Verlag, New York, 1998.
7. R. S. Ellis, *Physica D*, **133**, 106 (1999).
8. J. Barré, F. Bouchet, T. Dauxois and S. Ruffo, *J. Stat. Phys.*, **119**, 677 (2005).

Stochastic Model Reconstruction From Incomplete Noisy Measurements

D. G. Luchinsky*,**, V.N. Smelyanskiy* and J. Smith†

*Newstead Mission Critical Technologies Inc., 9100 Wilshire Blvd. Suite 540 East Beverly Hills, CA 90212-3437
†NASA Ames Research Center, Mail Stop 269-2, Moffett Field, CA 94035, USA
**Department of Physics, Lancaster University, Lancaster LA1 4YB, UK

Abstract. A technique of reconstruction of both unknown state and unknown vector-field of stochastic nonlinear dynamical system is introduced. It is based on the application of the path-integral theory to the full Bayesian inference and extended Kalman filter theory. We illustrate the application of this technique to the reconstruction of the model of FitzHugh-Nagumo oscillator from the corrupted by noise measurements. A number of important unsolved problems is identified.

Keywords: <Enter Keywords here>
PACS: <Replace this text with PACS

INTRODUCTION

We consider the following general problem of applied stochastic nonlinear dynamics (see e.g.[1]). We observe a L-dimensional time series of signals $\theta(t) = \theta(t_0 + n\tau_s) = \theta_n$ measured at each τ_s seconds starting at some time t_0. Signal $\theta(t)$ includes measurement errors. The actual state of the dynamical system is described by N-dimensional ($N \ll L$) state vector $\mathbf{x}(t)$ which follows stochastic dynamics driven by white Gaussian vector process $\xi(t)$ in the form

$$\dot{\mathbf{x}} = \mathbf{f}(\mathbf{x}) + \sqrt{\hat{\mathbf{D}}}\xi(t), \qquad \theta(t) = \hat{\Gamma}\mathbf{x}(t) + \sqrt{\mathcal{M}}\eta(t), \tag{1}$$

where $\hat{\Gamma}$ is a measurement matrix, $\mathcal{M}$ is the intensity matrix of the measurement noise. The actual state of the dynamical system $\mathbf{x}$ is not known. In a great number of important problems the vector field $\mathbf{f}(\mathbf{x})$ is not usually known exactly from "first principles" and one is faced with a rather broad range of possible parametric models to consider. Furthermore, the experimental data can be extremely skewed, whereby important "hidden" features of a model such as coupling coefficients between the dynamical degrees of freedom can be very difficult to extract due to the intricate interplay between noise and nonlinearity.

The question is how and with what accuracy can we determine $\mathbf{x}(t)$ and functional form of $\mathbf{f}(\mathbf{x})$. Examples are numerous and include molecular motors [2], semiconductor lasers [3], and epidemiology [4] to mention a few. In many areas of research the problem formulated above is so complicated (see e.g. discussion in application to ecological

---

[1] d.luchinsky@lancaster.ac.uk

CP800, *Unsolved Problems of Noise and Fluctuations: UPoN 2005*
edited by L. Reggiani, C. Pennetta, V. Akimov, E. Alfinito, and M. Rosini

systems in [5]) that the statement: "The population went up and down, and so did the model" is not totally unjustified as a caricature of the way in which e.g. ecological models are often compared to data. In this talk we discuss a novel approach to the solution of this problem based on the application of the path-integral approach to the full Bayesian inference. We demonstrate a reconstruction of stochastic dynamics of a limit cycle system from the corrupted by noise measurements of one dynamical coordinate. The emphasis are on the theoretical analysis. The results are compared with the results of earlier research. Unsolved problems are identified.

STATE AND MODEL INFERENCE

From a Bayesian prospective, which is the most general approach to the model reconstruction (see e.g. [6]), the time series θ_n is a vector sample picked at random from a probability density. This probability density conditioned on a given state $\mathbf{x}$ and model parameters $\mathbf{c}$ and (known as *likelihood* $\ell(\theta|\mathbf{x},\mathbf{c})$) is the only source of the information about stochastic dynamical model underlying observed time series data. Accordingly, the derivation of the likelihood function and its optimization with respect to $\mathbf{x}$ and $\mathbf{c}$ constitute the main subject of research in this area.

In the path-integral approach to the solution of this problems the sampling distribution can be factorized as a product of two main factors (cf. [7, 8])

$$\ell(\theta|\mathbf{x},\mathbf{c}) = p_{\rm o}(\theta|\mathbf{x})\,\mathscr{F}_{\mathbf{c}}[\mathbf{x}(t)], \tag{2}$$

where $p_{\rm o}v$ represents the conditional probability density of observations and $\mathscr{F}_{\mathbf{c}}[\mathbf{x}(t)]$ is the probability density functional to have given dynamical trajectory conditioned on fixed values of the model parameters. Using the fact that the distribution of the measurement noise is known and applying results of [9] to the evaluation of the $\mathscr{F}_{\mathbf{c}}[\mathbf{x}(t)]$ we obtain the following expression for the negative log-likelihood to have a given realization of the dynamical variable $\mathbf{x}(t)$ for given model parameters $\mathbf{c}$

$$\begin{aligned} S\left[\mathbf{x}(t),\mathbf{c}\right] &= \frac{1}{2h\varepsilon}\int_0^t d\tau \left(\theta(t)-\hat{\mathbf{\Gamma}}\mathbf{x}(t)\right)^T \hat{\mathbf{M}}^{-1}\left(\theta(t)-\hat{\mathbf{\Gamma}}\mathbf{x}(t)\right) + \\ &\quad + \frac{1}{2}\int_0^t d\tau \frac{\partial \mathbf{f}}{\partial \mathbf{x}} + \frac{1}{2}\int_0^t d\tau \left(\dot{\mathbf{x}}-\mathbf{f}\right)^T \hat{\mathbf{D}}^{-1}\left(\dot{\mathbf{x}}-\mathbf{f}\right). \end{aligned} \tag{3}$$

EXTENDED KALMAN FILTER

We assume first that the nonlinearity of the minus log-likelihood function with respect to the dynamical variables is weak and an extended Kalman theory [8] can be used. We note, however, that the commonly used in Bayesian inference likelihood function derived for the maps [7, 8] does not include the prefactor. Accordingly, the theory of the extended Kalman filter derived for maps [8] is not applicable for the flows. Here we presented the corrected analytical theory of the extended Kalman filter for the flows. The calculations are local. One has to choose initial distribution of a dynamical variable $\mathbf{x}_0$

at time t_0 in the form of Gaussian

$$\rho_0(\vec{x}_0) = \left[(2\pi)^L \det\left(\hat{\mathbf{A}}_0^{-1}\right)\right]^{-1/2} \exp\left\{-\frac{1}{2}\left[\mathbf{x}_0 - \mu_0\right]^T \hat{\mathbf{A}}_0 \left[\mathbf{x}_0 - \mu_0\right]\right\}, \tag{4}$$

with mean μ and inverse dispersion matrix $\hat{\mathbf{A}}$. Next the probability for the system to be in the state $\mathbf{x}_1$ is predicted using result (3) for small time step $h = t_1 - t_0$

$$\begin{aligned} \rho(\mathbf{x}_1|\mathbf{x}_0,\mathbf{c}) &= \left[(2\pi)^L \det\left(\hat{\mathbf{D}}\right)\right]^{-1/2} \times \\ &\times \exp\left\{-\frac{h}{2}\frac{\partial \mathbf{f}(\mathbf{x}_0,\mathbf{c})}{\partial \mathbf{x}} - \frac{h}{2}\left[\dot{\mathbf{x}}_0 - \mathbf{f}(\mathbf{x}_0,\mathbf{c})\right]^T \hat{\mathbf{D}}^{-1}\left[\dot{\mathbf{x}}_0 - \mathbf{f}(\mathbf{x}_0,\mathbf{c})\right]\right\}. \end{aligned} \tag{5}$$

Substituting (5) and the likelihood of observation of θ_1

$$\rho(\theta_1|\mathbf{x}_1) = \left[(2\pi)^L \det\left(\hat{\mathbf{M}}\right)\right]^{-1/2} \exp\left\{-\frac{1}{2}\left(\theta_1 - \hat{\Gamma}\mathbf{x}_1\right)^T \hat{\mathbf{M}}^{-1}\left(\theta_1 - \hat{\Gamma}\mathbf{x}_1\right)\right\} \tag{6}$$

into Bayesian rule

$$\rho_{post}(\mathbf{x}_1|\theta_1\mathbf{c}) = \frac{\rho(\theta_1|\mathbf{x}_1)\rho_0(\mathbf{x}_1|c)}{\int \rho(\theta_1|\mathbf{x}_1\mathbf{c})\rho_0(\mathbf{x}_1|\mathbf{c})d\mathbf{x}_1} \tag{7}$$

one can update the predicted value of $\mathbf{x}_1$. Here $\rho_0(\mathbf{x}_1|\mathbf{c}) = \int d\mathbf{x}_0 \rho(\mathbf{x}_1|\mathbf{x}_0,\mathbf{c})$ plays the role of a prior distribution. Expanding $\mathbf{f}(\mathbf{x}_0,\mathbf{c})$ in the Taylor series about μ_0, collecting terms up to quadratic powers in $\mathbf{x}_0$, and taking resulting integrals in (6) and (7) one to find final distribution for $\mathbf{x}_1$ in the form

$$\rho_1(\vec{x}_1) = \sqrt{\frac{\det(\hat{\mathbf{M}}\hat{\mathbf{A}}_1\hat{\mathbf{T}})}{(2\pi)^{3L}\det\left(\hat{\mathbf{G}}\hat{\mathbf{D}}\right)}} \exp\left\{-\frac{1}{2}\left[\mathbf{x}_1 - \mu_1\right]^T \hat{\mathbf{A}}_1 \left[\mathbf{x}_1 - \mu_1\right]\right\}, \tag{8}$$

where

$$\begin{aligned} \mu_1 &= \hat{\mathbf{A}}^{-1}\mathbf{q}, \quad \mathbf{g} = \frac{\mu_0}{\mathbf{h}} + \mathbf{f}(\mu_0), \quad \hat{\mathbf{P}} = \hat{\mathbf{I}}/\mathbf{h} + \frac{\partial \mathbf{f}}{\partial \mathbf{x_n}}, \\ \hat{\mathbf{A}}_1 &= \hat{\Gamma}^T \hat{\mathbf{M}}\hat{\Gamma} + \frac{\mathbf{1}}{\mathbf{h}}\hat{\mathbf{Q}} + (\hat{\mathbf{Q}}\hat{\mathbf{P}})\hat{\mathbf{G}}^{-1}(\hat{\mathbf{Q}}\hat{\mathbf{P}})^{\mathbf{T}}, \\ \mathbf{q}^{\mathbf{T}} &= \theta_1^T \hat{\mathbf{M}}\hat{\Gamma} + \mathbf{g}^{\mathbf{T}}\hat{\mathbf{Q}} - \mathbf{p}^{\mathbf{T}}\hat{\mathbf{G}}^{-1}(\hat{\mathbf{Q}}\hat{\mathbf{P}})^{\mathbf{T}}, \\ \mathbf{p}^{\mathbf{T}} &= h\mathbf{g}^{\mathbf{T}}(\hat{\mathbf{Q}}\hat{\mathbf{P}}) + \frac{\mathbf{h}}{\mathbf{2}}\frac{\partial^2 \mathbf{f_i}(\mu_0)}{\partial \mathbf{x_i}\partial \mathbf{x}}, \\ \hat{\mathbf{G}} &= \hat{\mathbf{A}}_0 + \frac{h}{2}\frac{\partial^3 f_i(\mu_0)}{\partial x_i \partial \mathbf{x}^2} + h\hat{\mathbf{P}}^{\mathbf{T}}\hat{\mathbf{Q}}\hat{\mathbf{P}}. \end{aligned}$$

These procedure should be repeated now at each time step to.

GLOBAL OPTIMIZATION IN THE SPACE OF DYNAMICAL TRAJCTORIES

To find the global minimum of the minus likelihood function (3) derived above in the space of dynamical trajectories we employ a conjugate gradient method [10]. The computational efficiency is achieved by analytical calculations of the gradient and the Hessian matrix of $S = -\ln \ell(\theta|\mathbf{c})$.

The gradient function is given by the following expression

$$\begin{aligned} \frac{\partial S}{\partial \mathbf{x}_n} &= -(\theta_n - \Gamma \mathbf{x}_n)^T \hat{\mathbf{M}}^{-1}\Gamma + \frac{h}{2}\frac{\partial}{\partial \mathbf{x}_n}\left[\operatorname{tr}\frac{\partial \mathbf{f}}{\partial \mathbf{x}}\right] \\ &\quad + [\dot{\mathbf{x}}_{n-1} - \mathbf{f}(\mathbf{x}^*_{n-1}|\mathbf{c})]^{\mathrm{T}}\hat{\mathbf{D}}^{-1} - h[\dot{\mathbf{x}}_n - \mathbf{f}(\mathbf{x}^*_n|\mathbf{c})]^{\mathrm{T}}\hat{\mathbf{D}}^{-1}\mathbf{F}_n, \end{aligned} \tag{9}$$

where $\mathbf{x}^*_n = \frac{1}{2}(\mathbf{x}_n + \mathbf{x}_{n+1})$ and $\mathbf{F}_n = \hat{\mathbf{I}}/h + \frac{\partial \mathbf{f}}{\partial \mathbf{x}_n}$.

The Hessian matrix has a tridiagonal block structure. In particular case of two dimensional dynamical model it takes the form

$$\hat{\mathbf{H}} = \begin{pmatrix} \hat{\mathbf{B}}_{00} & \hat{\mathbf{B}}_{01} & \hat{\mathbf{0}} & \dots & \hat{\mathbf{0}} & \hat{\mathbf{0}} \\ \hat{\mathbf{B}}_{10} & \hat{\mathbf{B}}_{11} & \hat{\mathbf{B}}_{12} & \dots & \hat{\mathbf{0}} & \hat{\mathbf{0}} \\ \dots & \dots & \dots & \dots & \dots & \dots \\ \hat{\mathbf{0}} & \hat{\mathbf{0}} & \hat{\mathbf{0}} & \dots & \hat{\mathbf{B}}_{n-1,n} & \hat{\mathbf{B}}_{nn} \end{pmatrix}, \tag{10}$$

where matrices $\hat{\mathbf{B}}_{nm}$ are given by the following expression

$$\begin{aligned} \hat{\mathbf{B}}_{nm} &= \frac{\partial S}{\partial \mathbf{x}_n \partial \mathbf{x}_m} = \Gamma^T \hat{\mathbf{M}}^{-1}\Gamma\delta_{nm} + \frac{h}{2}\frac{\partial}{\partial \mathbf{x}_n \partial \mathbf{x}_m}\left[\operatorname{tr}\frac{\partial \mathbf{f}}{\partial \mathbf{x}}\right]\delta_{nm} + \frac{1}{h}\hat{\mathbf{D}}^{-1}\delta_{nm} + h^T\hat{\mathbf{D}}^{-1}\mathbf{F}_n\delta_{nm} \\ &\quad - h\left[\dot{\mathbf{x}}_n - \mathbf{f}(\mathbf{x}^*_n|\mathbf{c})\right]^T \hat{\mathbf{D}}^{-1}\left[\frac{\partial^2 \mathbf{f}}{\partial \mathbf{x}_n^2}\right]\delta_{nm} - h\mathbf{F}^T_{n-1}\hat{\mathbf{D}}^{-1}\delta_{m,n-1} - h\hat{\mathbf{D}}^{-1}\mathbf{F}_n\delta_{m,n+1}. \end{aligned} \tag{11}$$

NUMERICAL EXAMPLE

Example of corrupted by noise measurements of the nonlinear limit cycle system

$$\dot{x} = y - x^2 y + \xi_x(t) \qquad \dot{y} = -x + 0.1(1-x^2)y + \xi_y(t), \tag{12}$$

is shown in the Fig. 1(a). Our technique allows to recover stochastic dynamics of the system (12) as shown in the Fig. 1(b) and to estimate model parameters (correct values are shown in the brackets): 1.07 (1); -1.065 (-1); 0.17 (0.1). The results of convergence are summarized in the table 1.

Our second example shows that stochastic dynamics of system (12) can be recovered even in the case of incomplete noise measurements. In this case we measure only $y(t)$ with intensity of the measurement error 0.4 as shown in Fig. 2(b) by black line. As an initial guess for the unknown dynamical variable $x(t)$ we choose white Gaussian noise shown by black line in the Fig. 2(a). Applying our technique we can recover actual stochastic dynamics of both variables. Note how the learning of the actual dynamics is improved with time.

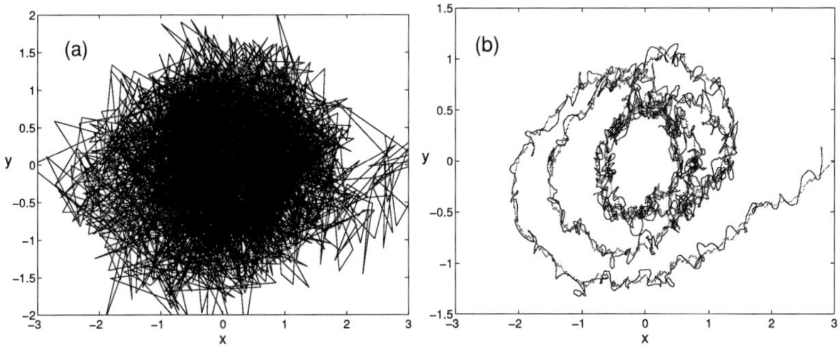

FIGURE 1. (a) Example of corrupted by noise measurements (12) with intensity of dynamical noise $\langle\xi_x^2(t)\rangle = 0.1$ and $\langle\xi_y^2(t)\rangle = 0.2$ and intensity of measurements noise 0.4 in both coordinates. (b) Recovered stochastic dynamics of the system (12) (blue line) is shown in comparison with the actual dynamical trajectory (green line).

TABLE 1. Convergence of some coefficients of the system (12). We have used one block of data with 40000 points.

coefficients	true values	inferred values	updated values
c_3	1	1.46	1.08
c_4	0	-42.64	0.02
c_{10}	-1	-10.93	-1.07
c_{11}	0.1	-35.68	0.258
c_{15}	0	1.82	0.005
c_{16}	-0.1	-27.96	-0.17
D_{11}	0.04	325	0.045
D_{12}	0	6	0.006
D_{22}	0.04	318	0.03

DISCUSSION

A technique of reconstruction of both unknown state and unknown vector field from corrupted by noise measurements is introduced. The probability density for the reconstructed dynamical trajectories and for the model parameters are derived from the probability density of the observations using path-integral approach. In the limit of small nonlinearities we obtain equations for the extended Kalman filter which are compensated for dynamical noise-induced bias by including correct prefactor term into the likelihood function. In the limit of arbitrary large nonlinearities and measurement noise we obtain efficient global optimization by calculating Jacobian and the Hessian of the cost function. We illustrate the reconstruction using as an example model of the FitsHugh-Nagumo oscillator.

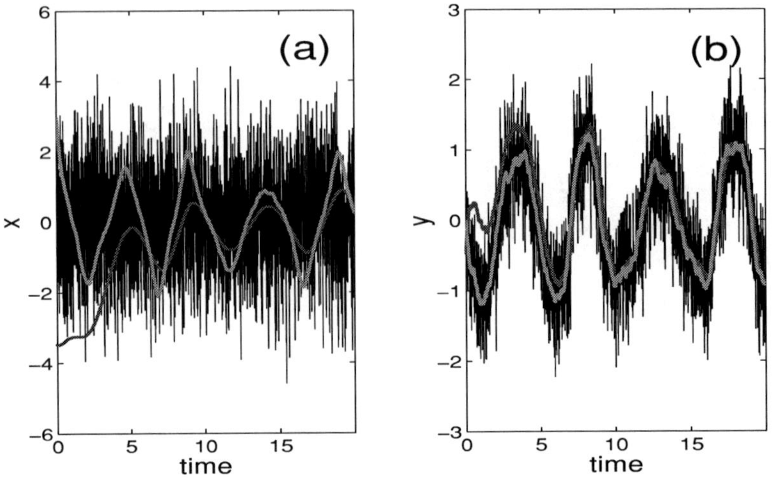

FIGURE 2. (a) Initial guess for the unknown dynamical variable $x(t)$ of system (12) (black line). Recovered dynamical variable is shown by red line in comparison with the actual dynamical variable (green line). (b) Measured dynamical variable $y(t)$ of system (12) (black line) with intensity of dynamical noise $\langle \xi_x^2(t) \rangle = 0.2$ and intensity of measurements noise 0.4 in both coordinates. Recovered stochastic variable $y(t)$ (red line) is shown in comparison with the actual dynamical trajectory (green line).

The main direction of the further research will be outlined. In particular, we note the following major problems that has yet to be tackled

- In most practical situations, the model is not known exactly *a priori*: one is faced with many possibilities. However, in the earlier research the functional form of the equations was assumed known and the problem of model identification as such was not addressed.
- Systems of interest are usually multidimensional, and an extension of the method to multidimensional case is a formidable task.
- In general, dynamical noise in the system is coloured and multiplicative – circumstances that were not usually considered in earlier investigations.
- Another major challenge is how to encompass the (widespread) case of time-varying parameters.

Solution of these problems will open a new venue for extracting dynamical and statistical information from experimental observations in e.g. physics, chemistry, biology, physiology, engineering, econometrics.

ACKNOWLEDGMENT

This work was supported by the Engineering and Physical Sciences Research Council (UK), NASA CICT–IS–IDU Project (USA), the Russian Foundation for Fundamental Science, and INTAS.

REFERENCES

1. H. D. I. Abarbanel, in *Nonlinear dynamics and statistics*, edited by A. I. Mees (Birkhäuser, Boston, 2001), Chap. I, pp. 1–30.
2. K. Visscher, M. J. Schnitzer, and S. M. Block, Nature **400**, 184 (1999).
3. M. Willemsen, M. P. van Exter, and J. P. Woerdman, Phys. Rev. Lett. **84**, 4337 (2000).
4. D. J. D. Earn, S. A. Levin, and P. Rohani, Science **290**, 1360 (2000).
5. S. N. Wood, Ecological Monographs **71**, 1 (2001).
6. P. Congdon, *Bayesian statistical modelling, Wiley series in probability and statistics* (Wiley, Chichester, 2001).
7. R. Meyer and N. Christensen, Physical Review E **62**, 3535 (2000).
8. R. Meyer and N. Christensen, Phys. Rev. E **65**, 016206 (2001).
9. V. N. Smelyanskiy, D. G. Luchinsky, D. A. Timucin, and A. Bandrivskyy, to appear in Phys. Rev. E (2005).
10. W. H. Press, B. P. Flannery, S. A. Teukolsky, and W. T. Vetterling, *Numerical Recipes in FORTRAN: The Art of Scientific Computing*, 2 ed. (Cambridge University Press, Cambridge, England, 1992).

Distribution of Return Periods of Rare Events in Correlated Time Series

C. Pennetta and E. Alfinito

Dipartimento di Ingegneria dell'Innovazione, Università di Lecce, Via Arnesano, 73100 Lecce, Italy, and National Nanotechnology Laboratory, CNR - INFM, Via Arnesano, 73100 Lecce, Italy
cecilia.pennetta@unile.it

Abstract. We study the effect on the distribution of return periods of rare events of the presence in a time series of finite-term correlations with non-exponential decay. Precisely, we analyze the auto-correlation function and the statistics of the return intervals of extreme values of the resistance fluctuations displayed by a resistor with granular structure in a nonequilibrium stationary state. The resistance fluctuations, δR, are calculated by Monte Carlo simulations using the SBRN model introduced some years ago by Pennetta, Trefán and Reggiani and based on a resistor network approach. A rare event occurs when δR overcomes a threshold value q significantly higher than the average value of the resistance. We have found that for highly disordered networks, when the auto-correlation function displays a non-exponential decay but yet the resistance fluctuations are characterized by a finite correlation time, the distribution of return intervals of the extreme values is well described by a stretched exponential, with exponent largely independent of the threshold q. We discuss this result and some of the main open questions related to it, also in connection with very recent findings by other authors concerning the observation of stretched exponential distributions of return intervals of extreme events in long-term correlated time series.

Keywords: Extreme values in random processes, Fluctuation phenomena, Time series analysis.
PACS: 05.40.-a, 05.45.Tp, 02.50.-r

INTRODUCTION

It is well known that in a Poisson process the return periods between extreme events associated with the exceedance of a threshold are exponentially distributed [1, 2]. In other terms, by considering a time series $x(t)$ made by uncorrelated records, the return periods r_q of events above a threshold q (i.e. the time intervals between two consecutive occurences of the condition $x(t) > q$) are distributed according to the probability density function (PDF):

$$P_q(r) = (1/R_q)\exp(-r/R_q) \quad (1)$$

where R_q is the mean return interval of these events [1, 2]. Of course, the higher the value of q (quantile), the rarer are the events over the threshold and bigger is the value of R_q. Fluctuations of prices in financial markets, wind speed data or daily precipitations in a given location for a same time windows are typically described by uncorrelated records [1, 3]. However, particularly in the last ten years, it has become clear that several other important examples of time series display long-term correlations [3, 4, 5]. This is the case of physiological data, like heartbeats [6, 7] and neuron spikes [8], hydro-meteorological records, like daily temperatures [3, 4, 9], geophysical or astrophysical data, concerning for example the occurrence of earthquakes [10, 11] or solar flares [12], stock market volatility [3, 13] and internet traffic records [4]. Long-term correlated series

CP800, *Unsolved Problems of Noise and Fluctuations: UPoN 2005*
edited by L. Reggiani, C. Pennetta, V. Akimov, E. Alfinito, and M. Rosini

are characterized by an auto-correlation function which decays as a power-law:

$$C_x(s) = < x_i x_{i+s} > = \frac{1}{N-s} \sum_{i=1}^{N-s} x_i x_{i+s} \sim s^{-\gamma_x} \quad , \quad 0 < \gamma_x < 1 \tag{2}$$

In this case, since the mean correlation time τ is the integral of $C_x(s)$ over s, it is easy to see that this time diverges when the correlation exponent γ_x is between 0 and 1.

Recently, Bunde et al. [4] investigated the effect of long-terms correlations on the statistics of the return periods of extreme events r_q. These authors found that: i) long-terms correlations leave unchanged the mean return interval R_q; ii) they significantly modify the distribution of return intervals whose PDF becomes a stretched exponential

$$P_q(r) = a \, \exp[-(b \; r/R_q)^{\gamma}] \tag{3}$$

where the values of the exponents γ and γ_x are found the same; and that iii) the return intervals themself are long-term correlated, with an exponent γ' close to the exponent γ of the original records [4]. The result i) was obtained by Bunde et al. [4] by statistical arguments and, as clarified by Altmann and Kantz in a very recent paper [3], it can identified be with Kac's lemma [14] in the context of area preserving dynamical systems. The results ii) and iii) in Ref. [4] were obtained by numerical analysis of long-term correlated Gaussian time series generated by the Fourier transform technique and by imposing a power-law decay of the power spectrum [15]. Very recently, it has been realized [3, 5] that result ii) only applies to linear time series (i.e. to series whose properties are completely defined by the power spectrum and by the probability distribution, regardless of the Fourier phases). However, a part from this restriction, the stretched exponential distribution of the return intervals of extreme events seems to be a general feature in presence of long-term correlations in a time series [3, 5]. It must be noted that this feature has important consequences on the observation of extreme events: in fact it implies a strong enhancement of the probability of having return periods well below R_q and well above R_q, in comparison with the occurrence of extreme events in an uncorrelated time series. Finally, it must be underlined that the distribution in Eq. (3) is a single parameter function, being a and b in Eq. (3) functions only of the exponent γ, as shown in Ref. [3]. Here, we study the effect on the distribution of return periods of rare events of the presence in a time series of finite-term correlations with non-exponential decay. Precisely, in the following section we will analyze the auto-correlation function and the statistics of the return intervals of extreme values of the resistance fluctuations displayed by a resistor with granular structure in a nonequilibrium stationary state [16, 17, 18, 19, 20]. In particular we will consider time series characterized by a stretched exponential decay of the auto-correlation function, thus a behavior intermediate between an exponential and a power law decay. A situation which can occur in systems which are approaching criticality [2, 21].

METHOD AND RESULTS

We analyze time series of resistance fluctuations of a thin resistor with granular structure in nonequilibrium stationary states in contact with a thermal bath at temperature T_0 and biased by an external current I. More precisely, we indicate with R the resistance of the resistor, with $< R >$ its average value and with σ the root-mean-square deviation from the average and we analyze the normalized signal: $x(t) \equiv (R(t) - < R >)/\sigma$, with zero average and unit variance. The $R(t)$ series is calculated by using the stationary and biased resistor network (SBRN) model introduced in 1999 by Pennetta, Trefán and Reggiani [22] and developed in Refs. [16, 17, 18, 19]. This model describes a thin film with granular structure as a resistor network in a stationary state determined by the competition between two stochastic processes, breaking and recovery of the elementary resistors. Both processes are thermally activated and biased by the external current. The network resistance and its fluctuations are calculated by Monte Carlo simulations. All the details about the model and its results are reported in Refs. [16, 17, 18, 19, 20]. Within this model, the level of intrinsic disorder in the network (average fraction of broken resistors in the vanishing current limit [23]) is controlled by a characteristic parameter: $\lambda \equiv (E_D - E_R)/k_B T_0$, where E_D and E_R are the activation energies respectively of the breaking and recovery processes. It holds the relationship: $\lambda_{min} < \lambda < \lambda_{max}$, where λ_{max} corresponds to an homogeneous resistor (perfect network) and $\lambda = \lambda_{min} \approx 0$ to the maximum level of intrinsic disorder compatible with a stationary state of the network (stationary resistance fluctuations) [17, 18, 19, 20, 23]. In addition to this intrinsic disorder, a disorder biased by the current I is also present in the network. As a consequence, for a given value of λ, and for a network of given size, nonequilibrium stationary states of the resistor exist only for $I \leq I_B$ (breakdown threshold). When $I > I_B$ the resistor undergoes an electrical breakdown, associated with an irreversible divergence of its resistance [16, 17, 18, 20]. For a generic value of λ this breakdown corresponds to a first order transition [17, 18, 19]. However, for decreasing λ values, when $\lambda \rightarrow \lambda_{min}$, the system becomes more and more close to its critical point [17, 18, 19]. We note that the SBRN model provides a good description of many features associated with nonequilibrium stationary states and with the electrical instability of composite and granular materials [16, 17, 18], including the electromigration damage of metallic lines [20].

Long $x(t)$ time series (typically made of $1 \div 2 \times 10^6$ records) have been generated and analyzed for different values of λ, of the external current and of the network size. The analysis has been performed by calculating the auto-correlation function C_x and the PDF of the x records, the return intervals r_q of the extreme values for different threshold q and their distribution $P_q(r)$. The values of q are expressed in root-mean-square deviation units. At small λ values (high level of intrinsic disorder), we have found that the auto-correlation function displays a non-exponential (but non-power-law) decay. This behavior is different from that displayed at high λ values (low level of intrinsic disorder), where C_x exhibits an exponential decay (consistent with the Lorentzian power spectrum reported in previous works [16, 17, 19]). Here we will focus on the case of non-exponential and non-power-law decay of correlations, a situation typical of systems which are approaching criticality, and we will show the results obtained by taking $\lambda = 0.33$ for a network of size 125×125, biased by a current $I = 0.011$ A. Figure 1(a) displays a portion of the $x(t)$ time series (the total number of records in $x(t)$ is 1.6×10^6). The figure clearly shows a strong non-Gaussianity of the resistance fluctuations. Actually, with the above specified choice of the parameters, the PDF of the resistance fluctuations is well described by the Bramwell-Holdworth-Pinton [24] distribution, as discussed in Refs. [18, 19]. Figure 1(b) reports the auto-correlation function which significantly deviates from a single exponential or a power-law behavior.

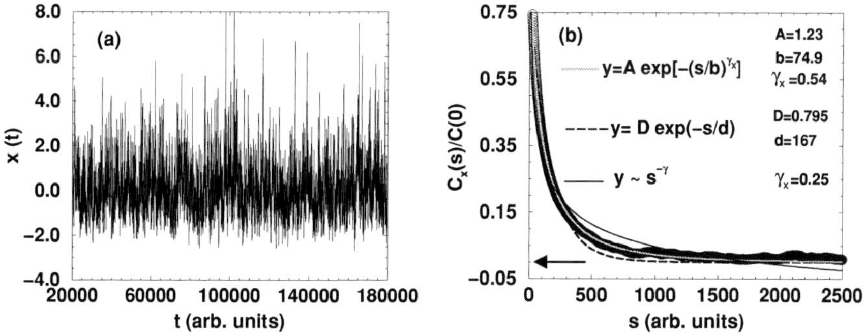

FIGURE 1. (a): The $x(t)$ signal (resistance fluctuations calculated by the SBRN model) as a function of the time expressed in simulation time steps. Only a small portion of the records are shown. The data have been normalized to provide a zero average and a unit variance for $x(t)$. (b): The black circles display the auto-correlation function, $C_x(s)$, of the signal in Fig. 1(a). C_x has been normalized to $C_x(s=0)$. The black arrow at the bottom points out the origin on the y-axis. The grey curve represents the best-fit with a stretched exponential. The black curves, dashed and solid, show respectively the best-fit with an exponential and a power-law. The figure also reports the values of the best-fit parameters.

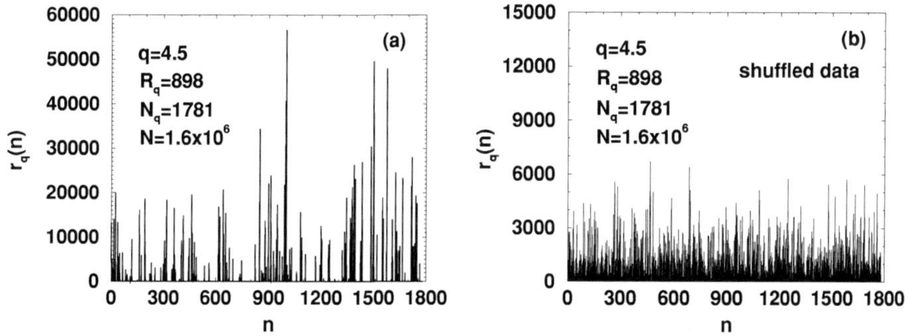

FIGURE 2. (a): Return intervals of an extreme value above the threshold $q = 4.5$ (in units of root-mean square deviation) for the time series in Fig. 1(a). The series consists of 1.6×10^6 records. The values over the threshold are only $N_q = 1781$ and the mean return period is $R_q = 898$ (in simulation time-step units). (b): The same as Fig. 2(a) but obtained by shuffling the data (see the text).

We have found that C_x is well fitted by a stretched exponential function: $C_x(s) = A \exp[(-s/b)^\gamma)]$, with the following values of the fitting parameters: $A = 1.23$, $b = 74.9$ and $\gamma = 0.54$. We have also considered many other functions for the best-fit of the C_x data. However, we have found that the stretched exponential function optimizes the best-fit procedure with the minimum numbers of fitting parameters. We note that this expression of $C_x(s)$ recovers the simple exponential dependence for $\gamma = 1$ while it provides a constant behavior for $\gamma = 0$. Moreover, it provides a finite value of the average correlation time τ. Figure 2(a) shows the return intervals of the extreme values above the threshold $q = 4.5$. For comparison, in Fig. 2(b) we have reported the return

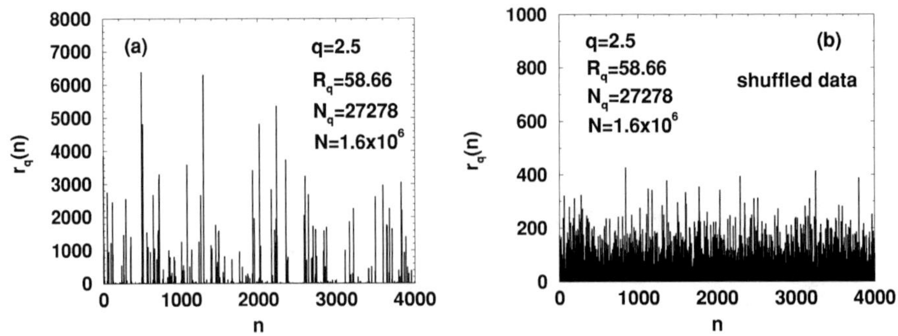

FIGURE 3. (a): Return intervals of an extreme value above the threshold $q = 2.5$ (in units of root-mean square deviation) for the same time series considered in the previous figures. The values over the threshold now are $N_q = 27268$ and the mean return period is $R_q = 58.66$ (in simulation time-step units). (b): The same as Fig. 3(a) but obtained by shuffling the data (see the text).

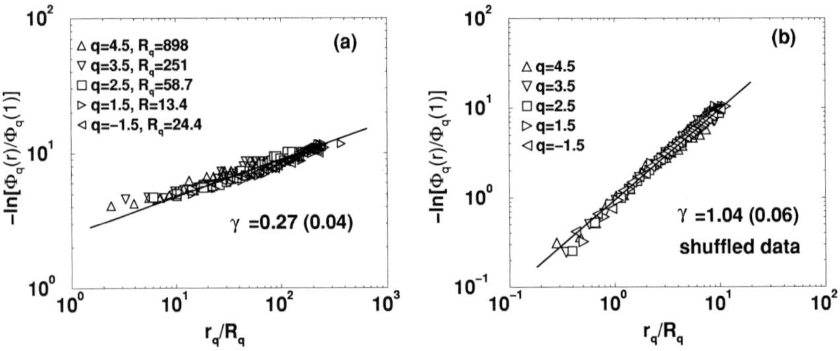

FIGURE 4. (a): Double-logarithmic plot of the probability density $P_q(r)$ of the return intervals distribution as a function of r/R_q for different quantiles q. The probability densities have been normalized to $P_q(1)$. The slope of the straight line shown in the figure gives the value of the exponent γ of a stretched exponential function fitting $P_q(r)$. (b): The same as Fig. 4(a) but obtained by shuffling the data.

intervals obtained for the same quantile by random shuffling the records of $x(t)$ (an operation which leaves unchanged the PDF of the x-data). The different scales of Fig. 2(a) and 2(b) must be underlined. Similarly to the results of Bunde et al. [5] concerning long-term correlated records, Fig. 2(a) shows a strong clustering of the extreme events: sequences of very short return intervals follow sequences of typically long intervals. This strong clustering is present even if the x records are not long-term correlated while they are characterized by a finite correlation time. A situation completely different from that reported in Fig. 2(b), which corresponds to fully uncorrelated records. The clustering of extreme values persists also by lowering the threshold. This is displayed in Fig. 3(a) which shows the return intervals obtained for $q = 2.5$. Again, Fig. 3(b) reports the returns intervals calculated by shuffling the x-data. The probability density function $P_q(r)$ of the return intervals of extreme values is plotted in Fig. 4(a) in a double-logarithmic scale as a function of r/R_q and for different quantiles q (the probability density has been normalized to $P_q(1)$). This representation has been adopted for convenience: in fact in this plot a stretched exponential function with exponent γ is represented by a straight line, whose slope just provides the value of γ. For check and comparison we have reported in Fig. 4(b) the probability density function $P_q(r)$ of the return intervals calculated after random shuffling the x-records: in this case $\gamma = 1$, i.e. the distribution of the r_q is exponential, as it must be for a Poisson process. Thus, Fig. 4(a) shows that the distribution of return intervals of extreme values of $x(t)$ is well described by a stretched exponential and that the value of the exponent γ is independent of the threshold q in a large range of q-values. Moreover, at least in the case of the data reported in Figs. 1-4, the value of the exponent γ of the return interval distribution is just $1/2$ of the value of γ_x, the exponent of the auto-correlation function. An analysis performed on other sets of data, obtained for high disordered networks of different size L and/or biased by different current density I/L, seems to confirm this last result. However, further investigations are necessary to establish the possible relationship between the two exponents.

CONCLUSIONS AND OPEN QUESTIONS

We have studied the distribution of return intervals of extreme events in time series with finite-term correlations. Precisely, we have analyzed the auto-correlation function and the distribution of the return intervals of extreme values of the resistance fluctuations displayed by a resistor with granular structure in nonequilibrium stationary states. The resistance fluctuations were calculated by Monte Carlo simulations using the SBRN model [16, 17, 18, 19, 20]. We have found that for highly disordered networks, when the auto-correlation function displays a non-exponential (precisely, a stretched exponential) decay and the resistance fluctuations are characterized by a finite correlation time, the distribution of return intervals of the extreme values is well described by a stretched exponential with exponent γ largely independent of the threshold q. Therefore our results show that the stretched exponential distribution [21] describes the distribution of return intervals of extreme events not only in the case of long-term correlated time series [3, 4, 5] but also when the records are characterized by finite-term correlations, with non-exponential decay of the auto-correlation function. Many open questions arise from this study and we limit to mention the following ones: i) why is the distribution of return intervals of extreme values a stretched exponential? what is the basic reason for this behavior which, as we have found, is not only limited to long-term correlated records [3, 4, 5] but is present also when non-exponential, finite term correlations are present? ii) what is the relationship between the exponents γ and γ_x? The correlations among the r_q have been studied in Refs. [3, 4, 5] in the case of long-term correlated processes, thus we can ask iii) what kind of correlations exist among the r_q in the case of a finite-term, non-exponentially correlated processes ?

ACKNOWLEDGMENTS

This work has been partially supported by the SPOT NOSED project IST-2001-38899 of E.C. and by the cofin-03 project "Modelli e misure in nanostrutture" financed by Italian MIUR. The authors thank S. Ruffo (University of Florence, Italy), P. Olla (ISAC-CNR, Lecce, Italy) and G. Salvadori (University of Lecce, Italy) for helpful discussions.

REFERENCES

1. H. von Storch and F. W. Zwiers. *Statistical Analysis in Climate Research.* Cambridge University Press, Cambridge, 2001.
2. S. Kotz and S. Nadarajah. *Extreme Value Distributions, Theory and Applications.* Imperial College Press, London, 2002.
3. E. G. Altmann and H. Kantz. Recurrence time analysis, long-term correlations, and extreme events. *Phys. Rev. E*, 71:056106–1–9, 2005.
4. A. Bunde, J. F. Eichner, S. Havlin, and J. W. Kantelhardt. The effect of long-term correlations on the return periods of rare events. *Physica A*, 330:1–7, 2003.
5. A. Bunde, J. F. Eichner, J. W. Kantelhardt, and S. Havlin. Long-term memory: a natural mechanism for the clustering of extreme events and anomalous residual times in climate records. *Phys. Rev. Let.*, 94:048701–1–4, 2005.
6. A. Bunde, S. Havlin, J. W. Kantelhardt, T. Penzel, J.H. peter, and K. Voigt. Correlated and uncorrelated regions in heart-rate fluctuations during sleep. *Phys. Rev. Lett.*, 85:3736–3739, 2000.
7. Y. Ashkenazy, P. C. Ivanov, S. Havlin, C. K. Peng, A. L. Goldberger, and H. E. Stanley. Magnitude and sign correlations in heartbeat fluctuations. *Phys. Rev. Let.*, 86:1900, 2001.
8. J. Davidsen and H. G. Schuster. Simple model for $1/f^{\alpha}$ noise. *Phys. Rev. E*, 65:026120–1–4, 2002.
9. E. Koscielny-Bunde, A. Bunde, S. Havlin, H. E. Roman, Y. Goldreich, and H.J. Schellnhuber. Indication of a universal persistance law governing atmospheric variability. *Phys. Rev. Lett.*, 81:729–732, 1998.
10. P. Bak, K. Christensen, L. danon, and T. Scanlon. Unified scaling law for earthquakes. *Phys. Rev. Lett.*, 88:178501–1–4, 2002.
11. A. Corral. Long-term clustering, scaling, and universality in the temporal occurrence of earthquakes. *Phys. Rev. Lett.*, 92:108501–1–4, 2004.
12. G. Boffetta, V. Carbone, P. Giuliani, P. Veltri, and A. Vulpiani. Power-laws in solar flares: self-organized criticality or turbulence? *Phys. Rev. Lett.*, 83:4662–4665, 1999.
13. Y. Liu, P. Cizeau, M. Meyer, C. K. Peng, and H. E. Stanley. Correlations in economic time series. *Physica A*, 245:437–440, 1997.
14. M. Kac. -. *Bull. of the Am. Math. Soc.*, 53:1002, 1947.
15. H. A. Makse, S. Havlin, M. Schwartz, and H. E. Stanley. Method for generating long-range correlations for large systems. *Phys. Rev. E*, 53:5445–5449, 1996.
16. C. Pennetta, L. Reggiani, G. Trefán, and E. Alfinito. Resistance and resistance fluctuations in random resistor networks under biased percolation. *Phys. Rev. E*, 65:066119–1–10, 2002.
17. C. Pennetta. Resistance noise near to electrical breakdown: steady state of random networks as a function of the bias. *Fluct. and Noise Let.*, 2:R29–R49, 2000.
18. C. Pennetta, E.Alfinito, L. Reggiani, and S. Ruffo. Non-gaussian resistance noise near breakdown, in granular material. *Physica A*, 340:380–387, 2004.
19. C. Pennetta, E. Alfinito, L. Reggiani, and S. Ruffo. Non-gaussian resistance flutuations in disordered materials. In Z. Gingl, J. M. Sancho, L. Schimansky-Geier, and J. Kertesz, editors, *Noise in Complex Systems and Stochastic Dynamics II*, number 5471 in Proceedings of SPIE, pages 38–47, Bellingham, 2004. Int. Soc. Opt. Eng.
20. C. Pennetta, E. Alfinito, L. Reggiani, F. Fantini, I. de Munari, and A. Scorzoni. Biased resistor network model for electromigration failure and related phenomena in metallic lines. *Phys. Rev. B*, 70:174305–1–15, 2004.
21. D. Sornette. *Critical Phenomena in Natural Sciences, Chaos, Fractals, Selforganization and Disorder: Concepts and Tools.* Springer, Berlin, 2004.
22. C. Pennetta, G. Trefán, and L. Reggiani. A percolative approach to resistance fluctuations. In D. Abbott and L. B. Kish, editors, *Unsolved Problems of Noise and Fluctuations*, number 511 in AIP Conf. Proceedings, pages 447–452, New York, 1999. AIP.
23. C. Pennetta, L. Reggiani, and G. Trefán. Scaling law of resistance fluctuations in stationary random resistor networks. *Phys. Rev. Let.*, 85:5238–5241, 2000.
24. S.T. Bramwell, P.C.W. Holdsworth, and J. F. Pinton. Universality of rare fluctuations in turbulence and critical phenomena. *Nature*, 396:552–554, 1998.

Stochastic Dynamics of Anæsthesia

A. Bandrivskyy*, M. Entwistle†, P.V.E. McClintock*, B. Musizza**,‡, M. Paluš§, J. Petrovčič**, S. Ribarič¶, A.F. Smith† and A. Stefanovska‡,*

*Department of Physics, Lancaster University, Lancaster LA1 4YB, UK
†Department of Anæsthesia, Royal Lancaster Infirmary, Ashton Road, Lancaster LA1 4RP, UK
**Jožef Stefan Institute, Jamova 39, Ljubljana, Slovenia
‡Group of Nonlinear Dynamics and Synergetics, Faculty of Electrical Engineering, University of Ljubljana, Tržaška 25, Ljubljana, Slovenia
§Institute of Computer Science, Academy of Sciences of the Czech Republic, Pod vodárenskou věží 2, 182 07 Prague 8, Czech Republic
¶Institute of Pathophysiology, Faculty of Medicine, University of Ljubljana, Zaloška 7, Ljubljana, Slovenia

Abstract.
Recent developments in the analysis of synchronization and directionality of couplings for noisy nonlinear oscillators are being applied to study the complex interactions between cardiac and respiratory oscillations, and brain waves (especially delta and gamma), during anæsthesia. It is found that marked changes occur in the inter-oscillator interactions during anæsthesia in both rats and humans. These could form a new basis for measurement of depth of anæsthesia. The new EC programme BRACCIA will explore and quantify causal relationships between the oscillatory processes for the different stages of anæsthesia and consciousness.

Keywords: Synchronization, noise, nonlinear oscillator, directionality, anæsthesia, cardiovascular system, phase transition

INTRODUCTION

There is abundant evidence favouring the perception of the cardiovacular system as a *noisy* dynamical system that behaves much like a collection of coupled oscillators [1]. The inter-oscillator couplings give rise to many of the complex and emergent phenomena seen in classical oscillator systems, including chaos [2, 3], modulation [1] and episodes of synchronization [4, 5, 6]. Numerical modelling of the cardiovascular system [7] has shown that the role of noise in facilitating these phenomena can be crucial. Yet the physiological origin of the oscillators is not yet fully established, nor the nature of the interactions between them, nor the origins or character of the the noise itself. The nonlinear dynamics of anæsthesia therefore represents *par excellence* an unsolved problem of noise.

It has become clear that the oscillatory processes – their amplitudes, frequencies and especially their mutual couplings – can be used to characterise the state of the system. It is known, for example, that that marked changes in cardio-respiratory synchronisation occur in anæsthesia [8]; furthermore, recent work by two of us [9] has shown that cerebral (cortical) oscillations interact with those of the cardiac and respiratory systems and that these interactions change in anæsthesia. Although the work is in its infancy, and is extremely challenging, there is a clear opportunity for applying techniques from

CP800, *Unsolved Problems of Noise and Fluctuations: UPoN 2005*
edited by L. Reggiani, C. Pennetta, V. Akimov, E. Alfinito, and M. Rosini

stochastic nonlinear dynamics, information theory and control theory to illuminate the underlying physiological processes.

Application of the techniques of nonlinear science to rhythmic signals of biological origin [10] has already illuminated many fundamental processes, especially those connected with cardiac function. These include fractal dynamics [11], deterministic chaos [3] in heart rate variability (HRV) data and the fact (see above) that the blood circulatory system behaves in many respects like a collection of five coupled oscillators [1] with frequencies ranging from 1 Hz (cardiac) down to 0.01 Hz (endothelial/metabolic [12]). Episodes of cardio-respiritory synchronization are known [4, 8] to occur in humans; synchronization phenomena involving a 0.1 Hz blood flow component, believed to be of myogenic origin, can also occur [13, 14]. Pomfrett suggested [15] that HRV or a bispectral index could be used to quantify depth of anæsthesia but that, in practice, it will probably be better to use several indicators rather than to rely on any particular technology; Larson and Galletly have commented [16] that the standard methods used for HRV analysis are limited in their ability to detect the presence of cardioventilatory coupling in heart rate time series, although there are some graphical methods [17, 18] that seem promising. A different approach, based on synchronization indices and synchrograms, has recently been used to reveal an unexpected and very marked increase of cardio-respiratory synchronization during anæsthesia [8] in rats.

ANÆSTHESIA

In anæsthesia a chemical perturbation of the organism leads to a temporary loss of consciousness, but just how this happens remains to a large extent a mystery. Perhaps on account of this lack of deep understanding, there are not yet any reliable markers for depth of anæsthesia. Consequently, as Pomfrett [15] puts it –

> "...the anæsthetist is still unable to measure the depth of anæsthesia in order to prevent inadvertent awakening during anæsthesia."

It is perhaps not surprising, therefore, that the incidence of awareness among patients undergoing surgery is nonzero: see Table 1. Note that even the *lowest* estimate of 0.2% is unacceptably high. The problem can be particularly severe in cases where for clinical reasons the anæsthesia is kept as light as possible (e.g. Cæsarian section), and where a mechanically ventilated curarised patient is unable to give any voluntary indication of awareness.

A well advertised approach to assessment of depth of anæsthesia is the bispectral index technique known as BIS [20]. Its use is controversial, however, and it is certainly not 100% accurate [21, 19]. Other promising approaches involves measurements of changes in heart rate variability (HRV) [15] and cardio-respiratory coupling [8, 17]. What we propose in this project takes such ideas much further, partly by adopting a more sophisticated mathematical analysis of the data, and partly by including consideration of the neural (cortical) oscillations.

TABLE 1. Incidence of awareness during surgery (after Pomfrett [15] and Myles at al [19]).

Author	Date	Sample	Awareness %
Hutchinson	1960	656	1.2
Harris	1971	120	1.6
McKenna	1973	200	1.5
Wilson	1975	490	0.8
Liu et al	1990	1000	0.2
Myles et al	2004	1238	0.9

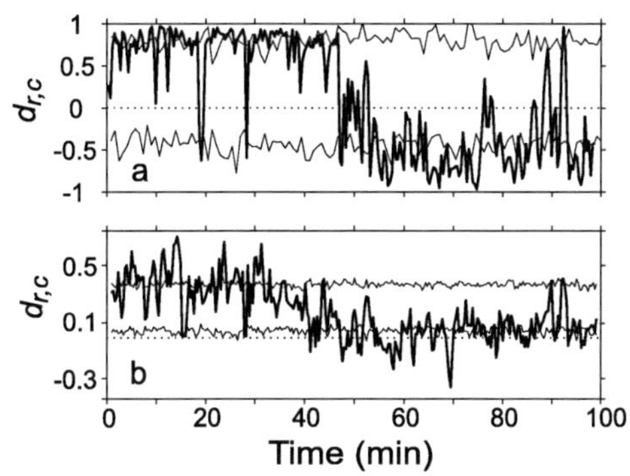

FIGURE 1. Direction of the cardio-respiratory interaction measured for a rat during anæsthesia. Top: calculated from the data according to the method proposed by Rosenblum et al [22]. Bottom: calculated from the same data using mutual information as proposed by Paluš et al [23]. The nearly horizontal thin lines are plotted ± 2 standard deviations outside a set of surrogate data. If $d_{r,c} > 0$ it means that respiration drives cardiac function, and *vice versa* if $d_{r,c} < 0$. In each case the directionality index exhibits a change of sign at a time that coincides approximately with the end of deep anæsthesia ([24]).

Anæsthesia in rats

The first simultaneous studies of the neural and cardio-respiratory phenomena associated with anæsthesia were carried out on rats, revealing a number of interesting effects. Rats exhibit similar cardiovascular dynamics to humans, except that the oscillatory components are 3–4 times faster [25], with heart beat at $\sim$3 Hz and respiration at $\sim$1 Hz. Analyses based on techniques drawn from nonlinear science [22] and information theory [23, 26] are being used to reveal and quantify *causal* relationships between the cardiac, respiratory and neural oscillations, and how these change during anæsthesia. Several important results have emerged, and show that the more systematic study now proposed is likely to yield important results. As one example, we show in Fig. 1 how the cardio-respiratory directionality index evolves as deep anæsthesia gives way to lighter anæsthesia. There is a dramatic change of sign showing that, during the lighter anæsthesia, the heart is predominantly driving respiration rather than *vice versa* as is observed in the conscious state or in deep anæsthesia.

Computation of the cardio-respiratory directionality index and how it evolves with

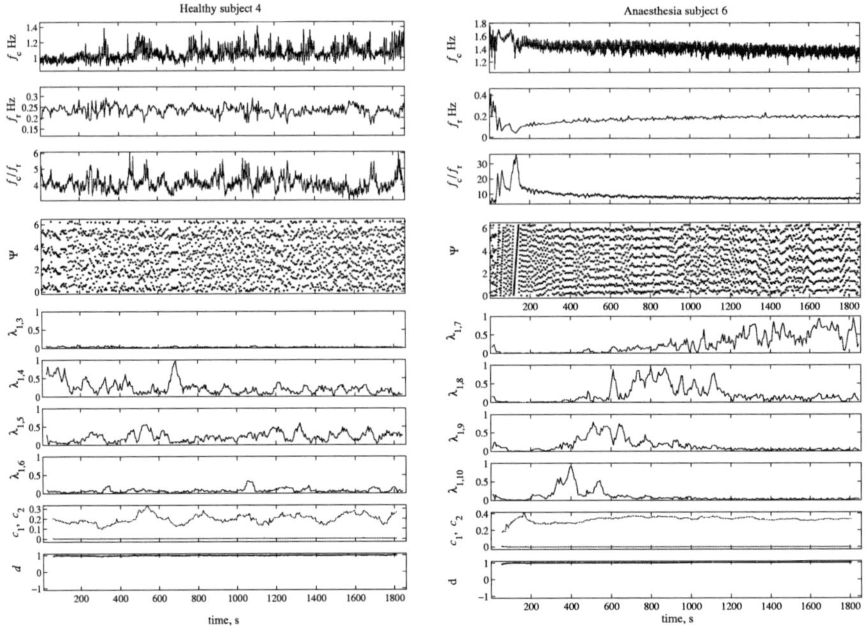

FIGURE 2. Typical analysis of cardiorespiratory data from human volunteers in the wakeful state (left) and under anæsthesia (right). From the top in each case, the plots show as functions of time: cardiac f_c and respiratory f_r frequencies and their ratio f_r/f_c; synchrogramme Ψ; synchronization indices $\lambda_{1,n}$; coupling coefficients c_1 and c_2; and directionality index d.

time from measured data could therefore provide one way of assessing depth of anæsthesia. There are several others: from the original research [8] it appears that strength of synchronization, synchronization index, and respiratory rate variability are good indicators; the latest results show that, in addition to the cardio-respiratory directionality index (Fig. 1), characteristic changes occur in the brain, especially δ and γ waves, and in the interactions and causalities with the cardiac and respiratory oscillations.

An initial modelling of coupled oscillators within both the micro and macroscopic worlds and has been undertaken [27]. A large number of oscillators acting on a similar time scale has been be represented as an ensemble that, on the macroscopic scale, may be taken as a single oscillator. Furthermore, their interaction on the macroscopic scale with other similar type of oscillators has been described, but acting on widely different time scales.

Anæsthesia in humans

The crucially important question of "Does what applies to rats also apply to humans?", i.e. whether anæsthetised humans display similar phenomena to those seen earlier in rats, is being addressed. The preliminary investigation [28] has involved studies of ten healthy men aged between 20 and 40 undergoing elective surgery. Standard clinical monitoring was supplemented by forehead BIS Quatro electrodes (Aspect Medical Systems, Leiden) for EEG capture and a TSD201 respiratory effort transducer (BIOPAC Systems Inc, Goleta, CA) for respiratory frequency (f). A standard anæsthetic (midazolam 2mg, fentanyl $1.5\mu\text{gkg}^{-1}$, sleep-dose propofol and isoflurane 1% in 40% oxygen in air) was administered by the same anæsthetist. Recordings were made for 30 minutes in the anæsthetic room prior to surgery. Waveforms were digitised at 1000 Hz and stored on a PC for later analysis.

A typical set of results is shown in Fig. 2 for healthy subjects in the waking (left) and anæsthetised (right) states. The synchrogramme Ψ plots points of fixed phase of the cardiac oscillator as a function of the phase of the respiratory oscillator, and therefore yields n horizontal lines where there is $1{:}n$ synchronisation; the synchronisation index $\lambda_{1,n}$, calculated from the data, is equal to unity for perfect $1{:}n$ synchronisation, but is zero in its absence; the coupling coefficients c_1, c_2 provide measures of the cardiac-to-respiratory (lower lines) and respiratory-to-cardiac (upper lines) couplings; and the directionality index d specifies the direction of coupling, being $+1$ for exclusively respiratory-to-cardiac coupling, and -1 for the opposite extreme. In all cases the idealised behaviour is to some extent modified by noise. In the waking state, the synchrogramme reveals two *brief* episodes of 1:4 synchronisation starting at 0 and 650 s into the measurements. Under anæsthesia, in marked contrast, there is a virtually continuous state of synchronization as shown by the horizontal lines in the synchrogramme and finite values of the synchronization indices. As time evolves, the system passes through 1:10, 1:9, 1:8 and 1:7 synchronized states – very similar to the behaviour seen earlier in rats [8]. The directionality index d, calculated from the two unidirectional coupling coefficients c_1, c_2, shows that during deep anæsthesia, as in the waking state, respiration tends to drive the cardiac oscillator. In rats, d changes sign during emergence from anæsthesia (Fig. 1); it is not yet known what happens for humans.

We now plan to monitor patients throughout surgery and into recovery to explore how synchronisation and directionality alter with changing depth of anaesthesia.

CONCLUSION

Many problems remain to be solved. In particular it will be necessary to develop algorithms for calculation of the directionality indices that are robust in the presence of physiological noise. It is also highly desirable to reach an understanding of the underlying processes giving rise to interactions between the oscillatory processes. If such problems can be solved, the potential impact of the work is high, in both specific and general terms. Specifically in relation to anæsthesia, we note that: anæsthesia is universally employed throughout the world for all but the simplest surgery; awareness during anæsthesia is not as uncommon as it should be (Table 1) and, as already mentioned, a better way of measuring depth of anæsthesia would be widely welcomed. More generally, we note that: coupled oscillators are widespread in nature, at all levels; and causal relationships between such oscillators are of great interest because they shed light on the underlying mechanisms of coupling. Improved techniques for establishing causalities based on measured noisy time series will be applicable in many areas.

REFERENCES

1. A. Stefanovska, and M. Bračič, *Contemporary Phys.*, **40**, 31–55 (1999).
2. A. Babloyantz, and A. Destexhe, *Biol. Cybern.*, **58**, 203–211 (1988).
3. C. S. Poon, and C. K. Merrill, *Nature*, **389**, 492–495 (1997).
4. C. Schäfer, M. G. Rosenblum, J. Kurths, and H. H. Abel, *Nature*, **392**, 239–240 (1998).
5. C. Schäfer, M. G. Rosenblum, H. H. Abel, and J. Kurths, *Phys. Rev. E*, **60**, 857–870 (1999).
6. A. Pikovsky, M. Rosenblum, and J. Kurths, *Synchronization – A Universal Concept in Nonlinear Sciences*, Cambridge University Press, Cambridge, 2001.
7. A. Stefanovska, D. G. Luchinsky, and P. V. E. McClintock, *Physiol. Meas.*, **22**, 551–564 (2001).
8. A. Stefanovska, H. Haken, P. V. E. McClintock, M. Hožič, F. Bajrović, and S. Ribarič, *Phys. Rev. Lett.*, **85**, 4831–4834 (2000).
9. B. Musizza, and A. Stefanovska, "Interactions between cardiac, respiratory and brain activity in humans," in *Fluctuations and Noise in Biological, Biophysical and Biomedical Systems III*, edited by N. G. S. et al, SPIE, Bellingham, 2005, vol. 5841, pp. 139–149.
10. L. Glass, *Nature*, **410**, 277–284 (2001).
11. A. L. Goldberger, L. A. N. Amaral, J. M. Hausdorff, P. C. Ivanov, C.-K. Peng, and H. E. Stanley, *Proc. National Acad. of Sciences* Suppl. 1, **99**, 2466–2472 (2002).
12. H. D. Kvernmo, A. Stefanovska, K.-A. Kirkebøen, and K. Kvernebo, *Microvasc. Res.*, **57**, 298–309 (1999).
13. A. Stefanovska, and M. Hožič, *Progress of Theoretical Physics* Supplement, **139**, 270–282 (2000).
14. N. B. Janson, A. G. Balanov, V. S. Anishchenko, and P. V. E. McClintock, *Phys. Rev. E*, **65**, 036212 (2002).
15. C. J. D. Pomfrett, *Brit. J. Anæsth.*, **82**, 659–662 (1999).
16. P. D. Larson, and D. C. Galletly, *Br. J. of Anæsth.*, **87**, 819–826 (2001).
17. D. C. Galletly, and P. D. Larsen, *Brit. J. Anæsth.*, **87**, 827–833 (2001).
18. D. C. Galletly, and P. D. Larsen, *Brit. J. Anæsth.*, **86**, 777–788 (2001).
19. P. S. Myles, K. Leslie, J. McNeil, A. Forbes, and M. T. V. Chan, *The Lancet*, **363**, 1757–1763 (2004).
20. J. M. Vernon, E. Lang, P. S. Sebel, and P. Manberg, *Anesthesia and Analgesia*, **80**, 780–785 (1995).
21. C. Lennmarken, and R. Sandin, *The Lancet*, **363**, 1747–1748 (2004).
22. M. G. Rosenblum, L. Cimponeriu, A. Bezerianos, A. Patzak, and R. Mrowka, *Phys. Rev. E.*, **65**, 041909 (2002).
23. M. Paluš, V. Komárek, Z. Hrnčíř, and K. Štěrbová, *Phys. Rev. E*, **63**, 046211 (2001).
24. B. Musizza, A. Stefanovska, P. V. E. McClintock, M. Paluš, J. Petrovčič, and S. Ribarič, (in preparation) (2005).
25. F. Bajrović, M. Čenčur, M. Hožič, S. Ribarič, and A. Stefanovska, *Pflüg. Arch.: Europ. J. Physiol. Supl.*, **439**, R158–R160 (2000).
26. M. Paluš, and A. Stefanovska, *Phys. Rev. E*, **67**, 055201(R) (2003).
27. B. Musizza, and A. Stefanovska, "Noise and synchronization on micro and macroscopic scales," in *Fluctuations and Noise in Biological, Biophysical and Biomedical Systems II*, edited by D. A. et al, SPIE, Bellingham, 2004, pp. 273–282.
28. M. Entwistle, A. Bandvivskyy, B. Musizza, A. Stefanovska, P. V. E. McClintock, and A. Smith, *Br. J. of Anæsthesia*, **93**, 608P–609P (2004).

Absolute Stability Margin Of Nonequilibrium Stationary Systems With I/f Noise.

Andrey A. Snarskii* and Igor V. Bezsudnov†

*National Technical University of Ukraine "KPI", Kiev, UKRAINE
† "Nauka-Servis" JSC, Moscow, RUSSIA

Abstract. In the article one of the base problems for 1/f investigations are discussed: what parameters do manage the amplitude of 1/f noise? What is 1/f noise in much more complex systems - biological, sociological, etc., i.e. those systems in which the self-organized structures supported by a constant flow of energy are observed.

Keywords: 1/f noise, polycrystal, effective conductivity, heartbeat time series.
PACS: 72.70.+m, 72.80.Ng.

Here we will discuss behaviour of 1/f noise in different systems and under different conditions: which factors govern the amplitude of 1/f noise; what does it mean, that 1/f noise is present in the system; which properties of the system will be affected if energy or information flows through the system is modulated by 1/f noise.

As the instance of the systems which illustrate to us the answer to first question about governing factors we consider macroscopically inhomogeneous media. The 1/f noise will be characterized by noise intensity

$$C(r) = V \cdot S \qquad S = \frac{S_R}{R^2}, \tag{1}$$

where V - volume, S - relative power spectral density, S_R - power spectral density of resistor fluctuations $S_R = \{\delta R \delta R\}$.

In inhomogeneous media 1/f noise is described by effective noise intensity C_e [1,2]

$$C_e = \left\langle C(\mathbf{r}) \frac{\mathbf{E}(\mathbf{r})^2}{\langle \mathbf{E}(\mathbf{r}) \rangle^2} \frac{\mathbf{j}(\mathbf{r})^2}{\langle \mathbf{j}(\mathbf{r}) \rangle^2} \right\rangle, \quad \langle ... \rangle = \frac{1}{V} \int ... dV, \tag{2}$$

In two phase composites for common case when $\sigma_1 / \sigma_2 \neq 0$ above percolation threshold according [3] effective noise intensity will be

$$C_e^+ = C_1 |\tau|^{-k} + C_2 \left(\frac{\sigma_2}{\sigma_1} \right)^2 |\tau|^{-w}, \quad p > p_c, \quad w_3 = 6.49, \quad w_3 > \quad k_3 = 1.52, \tag{3}$$

below Percolation threshold [3],[4]

$$C_e^- = C_2 |\tau|^{-k'} + C_1 \left(\frac{\sigma_2}{\sigma_1} \right)^2 |\tau|^{-w'}, \quad p < p_c, \quad w_3' = 6.98, \quad w_3' > k_3' = 1.03, \tag{4}$$

and finally on the percolation threshold in smearing region [3]

CP800, *Unsolved Problems of Noise and Fluctuations: UPoN 2005*
edited by L. Reggiani, C. Pennetta, V. Akimov, E. Alfinito, and M. Rosini

$$C_e^- = C_2|\tau|^{-k'} + C_1\left(\frac{\sigma_2}{\sigma_1}\right)^2 |\tau|^{-w'}, \quad p < p_c, \quad w_3' = 6.98 \quad w_3' > k_3' = 1.03 \quad , \tag{5}$$

In ideal simple case when $\sigma_1 / \sigma_2 = 0$ [5], [6] from (3) and (4) one can find that C_e depends only on closeness to percolation threshold $\tau, \tau = (p - p_c)/p_c$, where p - concentration of σ_1 phase. The common case (3), (4) is really differs from simple ideal case [5], [6] in all regions – above, below and especially on the percolation threshold, where C_e depends on ratio of phase conductance σ_1 / σ_2.

Even more surprising behaviour demonstrates 1/f noise in selfduality media [7]. Main heat dissipation in such media is space irregular and concentrated in a very small part of the media [8]. Detailed investigation [9] shows that there is specific limit in the system $(\sigma_1 / \sigma_2)_C$. If $(\sigma_1 / \sigma_2) > (\sigma_1 / \sigma_2)_C$ then C_e increased extremely high, which means in reality that C_e makes step up to anomalous high values.

The same behaviour has two dimensional polycrystalline media. In such systems in case of high anisotropy $\sigma_\parallel >> \sigma_\perp$, where $\sigma_\parallel, \sigma_\perp$ major components of local conductivity tensor of monocrystal. The current and Joule heat dissipation situated in so called current traps [10]. Investigations based in current trap model [11] shows that the system has anisotropy limit $(\sigma_\parallel / \sigma_\perp)_C$ that if $(\sigma_\parallel / \sigma_\perp) > (\sigma_\parallel / \sigma_\perp)_C$ C_e rising extremely high.

Therefore in last two considered cases despite of first one C_e has not monotone increase but in stepwise way. Even more variety of behaviour has 1/f noise in media with thermoelectric phenomena, in this case 1/f noise is generated not only by resistance fluctuations but with fluctuations of Seebeck's coefficient [12]. Even we will not take into consideration Seebeck's source of 1/f noise but only use thermoelectric phenomena to find currents in random inhomogeneous media we will find [13] that C_e is proportional to temperature difference applied to system and inversally to square of voltage applied.

Let us emphasize that above results was not devoted to properties or states of media that initiate 1/f noise presence in the system.

It is well known that energy flow through the system is able to create non-equilibrum stationary structures. Particularly any living creature is an example of such complex structure. Parameters described such structures are fluctuating and fluctuations have 1/f noise present.

We can show a lot of examples of such behavour

- 1/f noise in heartbeat time series, Long range power correlations exist in such systems (heartbeat time series. [14-17] …
- 1/f noise in traffic flow[18,19]., …
- 1/f noise in biological macroevolution [20].

In this article we put together a number of hypothesis (questions). Each thesis is not absolute new but we not only put them together but believe that they are to be handled together.

Is it correct that:

A. Non equilibrium systems with stable structures always have long-range power correlations? –
B. Long-range power correlations in the system indicate the existence of stable structures in non equilibrium system?
C. The amplitude of long-range power correlations shows reserve (resourse) of system stability? (For example, for certain person, if the amplitude of 1/f noise in heartbeat time series is higher then the period of error free job is longer.).
D. The resource of system stability will be higher if the energy or signal flows through the system has long range power correlations? (For example, listening to Bach's music will increase the period of error free job for person [21]).

To test above set of hypothesis experiments can be proposed. Below you find tree examples of many thousands of possible tasks, experiments and observation.

Experiment #1

Let we have two or more operators and they have to execute error free very important and complex job (say, flight traffic dispatcher in airport). How to select the person which has higher stability resource i.e. longer error free period of job? Authors believe that selection criteria (if all others tests are passed equally positive) can be based on the value of 1/f noise amplitude of heartbeat time series or Herst parameter of those series etc.

Let us emphasize that we do not divide healthy and ill persons but we perform selection among healthy persons to find maximum stability margin of each.

Experiment #2

The flow of mercury vapor (Hg) through surface of the earth is one of earthquake prediction parameter:[22] a few days before an event of earthquake the mercury flow is getting up to 2-3 times higher. [Tatsy, Bezsudnov,Patent 1999]. Authors believe that prediction parameter in mercury measurement series is not only concentration but the value of 1/f noise amplitude of mercury concentration series and such amplitude can predict earthquake earlier than concentration itself.

Experiment #3

How to evaluate parameters of society behaviour or rather change of such parameters. Authors believe that the value of 1/f noise amplitude of fluctuation of human behaviour (time series of underground step in – step out control points, auto traffic in cities, voice of crowd in mass events) is real society parameter that correlates with society stability, labor productivity, common life conflict possibility etc.

Authors believe that such or similar experiments will be executed in the near future and they will answer our questions.

ACKNOWLEDGMENTS

Authors would like to thank A.M. Dykhne, S.Lukyanets, V. Sevrukov for fruitful discussions and interest to this problem.

REFERENCES

1. H.Butterweck, Philips Res.Rep., 30, 316, 1975.
2. M.Wolf, K.-H.Muller, Phys.Stat.Sol.(a), 92, 151, 1985.
3. A.E.Morozovsky, A.A.Snarskii, Sov.Phys. JETP, 68, 1066, 1989
4. A.-M. S. Tremblay, B.Fourcade, P.Breton, Physica A, 157, 89 1989.
5. R. Rammal, C Tannous, P.Brenton, A.-M. S.Tramblay, Phys Rev. Lett. 54, 1718, 1985.
6. D.C.Wright, D.J. Bergman, Y.Kantor Phys rev.B 33, 396, 1986 .
7. A.M.Dykhne, Sov.Phys.JETP 32, 63, 1971.
8. M.Soderberg, G.Grimvall, J.Phys. C16, 1085, 1983.
9. A.M.Dykhne, Preprint TPI-MINN-90/59-T (Theor.Phys.Inst., Univ. of Minnesota, Minneapolis), 1990.
10. A.Dreizin, A.M.Dykhne, Sov.Phys. JETP, 57, 1024, 1983.
11. A.Snarskii, I.Bezsudnov, in print, 2005.
12. J. Brophy, Rhys Rev, 27, 148, 1958.
13. A.Snarskii, I.Bezsudnov, in print, 2005.
14. P. Bak, C.Tang, K.Wiesenfeld PRL 1987.
15. C.-K.Peng, H.Stanley, A.L.Goldberger, 1995.
16. A.Goldberger, B.West 1987
17. B.West, Deering Phys.Rep. 1994.
18. Musha, Higuchi Jp.J.Appl.Phys. 1976
19. Choi, Lee PRE 1995.
20. P.A.Rikvold, arXiv:q-bio.PE/0502046 v1 28 Feb2005.
21. R.Voss, J.Clark, 1/f noise misic, J.Acoust.Soc.Amer.,1978.
22. Yu.Tstsy, Yu.Stacheev, I.Bezsoudnov, V.Sevrukov Patent, 1999.

Small Josephson Junction As Detector Of Non-Gaussian Noise

Rene K. Lindell*, Mika A. Sillanpää*, Tero T. Heikkilä*, Julien Delahaye†,*, Takahide Yamaguchi**, Edouard B. Sonin‡,* and Pertti J. Hakonen*

*Low Temperature Laboratory, Helsinki University of Technology, P.O. Box 2200, 02015 TKK, Finland
†LEPES-CNRS, BP 166, 38042 Grenoble Cedex 9, France
**National Institute for Materials Science, Tsukuba 305-0003, Japan
‡The Racah Institute of Physics, The Hebrew University of Jerusalem, Jerusalem 91904 Israel

Abstract. The current in a a Coulomb blockaded Josephson junction is sensitive to voltage or phase fluctuations in its electromagnetical environment. This property can be used to detect noise also from non-Gaussian noise sources. We have experimentally studied IV-curves of as a function of the shot noise current in a nearby tunnel junction. We observe a linear enhancement of zero-bias conductance of the Josephson junction as a function of the shot noise power. Furthermore, with increasing shot noise power, the IV-curves becomes asymmetric with a shift of minimum conductance and the appearance of a local conductance maximum. A quantitative explanation for the asymmetric features due to the non-Gaussian noise sources is still an open question. There is also motivation to pursue modifications of the current scheme in order to detect individual statistical moments of noise and to connect different noise sources to the Josephson junction.

Keywords: Josephson junction, non-Gaussian, shot noise
PACS: 73.23.-b, 73.23.Hk, 85.25.Cp

Josephson junction devices has proven to be strong candidates for realization of solid state qubits [1]. The coherence in these devices is destroyed by interactions with the environment and outside noise. Therefore, currently a lot of theoretical [2, 3, 4, 5] and experimental [6, 7, 8, 9] research on noise in mesoscopic systems is carried out. Especially, there is an interest in non-Gaussian noise sources, such as shot noise in tunnel junctions, where the odd moments of the noise statistics are non-zero. Measuring higher moments with standard spectrum analyzer methods [6] can be difficult and slow due to the long averaging times required to capture the rare events. In this paper we discuss the scheme of using an on-chip detector [10, 11], a Coulomb blockaded Josephson junction, which is sensitive to the asymmetry of the noise source. For other methods suggesting using Josephson junction as a switching detector for obtaining the full counting statistics see [12, 13]. Deblock et al. used photon-assisted quasiparticle tunnelling in a Josephson junction as the basis of their noise measurement [7].

The Coulomb blockade (CB) of a Josephson junction (JJ) is sensitive to the phase fluctuations caused by the junction's environment. This fact, can be used to probe an on-chip noise source, which in our case, is a single normal-superconducting-insulator tunnel junction. Therefore, we work in the Coulomb blockade regime with voltages from zero to about the value of the charging energy $V_C = e/2C$, while Deblock et al. worked at the

CP800, *Unsolved Problems of Noise and Fluctuations: UPoN 2005*
edited by L. Reggiani, C. Pennetta, V. Akimov, E. Alfinito, and M. Rosini

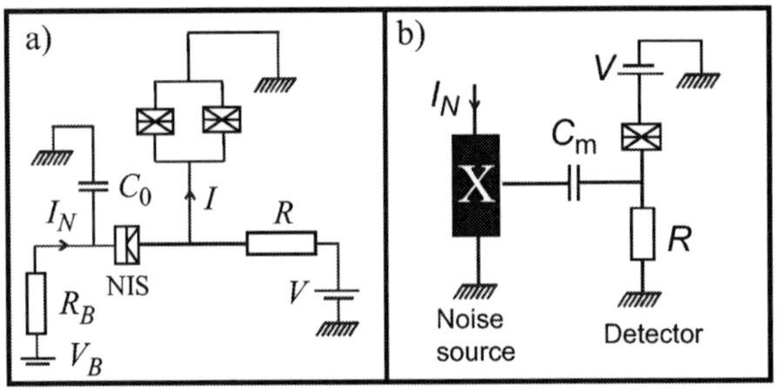

FIGURE 1. Circuit diagram for the directly coupled experiment considered here (a) and an alternative capacitive coupling with some noise source X (b).

quasiparticle threshold, or the superconducting gap voltage of 2Δ.

Our circuits were fabricated using electron beam lithography and four-angle evaporation. We place the Josephson junction (smaller than 100×100 nm area) in a resistive environment made out of a thin film, on-chip, Cr resistor ($R \sim 100$ kΩ) located within a few μm from the Josephson junction (for the circuit diagram, see Fig. 1).

We generate shot noise by a separately biased normal-insulator-superconductor tunnel junction (NIS TJ). The resistor R_B is located outside the cryostat, and it is taken as large as 10 GΩ for good current biasing of the NIS TJ, which, effectively, will be voltage biased (a necessary condition for shot noise generation) by the large capacitance C_0 of the measurement lead.

In earlier measurements [10, 11]with samples that had chromium resistors in the order of $20 - 67$ kΩ, the response of the zero-bias conductance of the Josehpson junction to the shot-noise current was treated as an effective noise tempearature. Hence, the response could be characterized by a power-law dependence on the shot noise current, $G_0 \propto (T + \gamma T_N)^{2\rho-2}$, where $\rho = R/R_Q$ and T is the sample base temperature and T_N could be related to the shot noise current by $T_N = eI_N R/2k_B$. However, this treatment was only approximate and does not hold when $\rho \gg 1$. In fact, for a large $\rho = 27.8$ as in the case of the sample described here (see Table 1), the dependence of G_0 on I_N turns out to be linear (see Fig. 4).

The IV characteristic of a Coulomb blockaded Josephson junction in a linear, Gaussian, electromagnetic environment can be described by the standard $P(E)$-theory [14, 15, 16]. However, in the case of non-Gaussian phase fluctuations the theory has to be modified, as was done in Ref. [17]. The main modification lies in the treatment of the phase fluctuations, which govern the inelastic Cooper pair tunneling in the Josephson junction. The current through the Josephson junction can be given by the expression

$$I(V) = -\frac{2eE_J^2}{\hbar^2}\mathrm{Im}\left\{\int_0^{\infty} \mathrm{d}t\, e^{J_0(t)} \times \left\langle \sin\left(\frac{2eVt}{\hbar} + \Delta\varphi_s\right)\right\rangle_{t_0}\right\}, \tag{1}$$

where $\Delta\varphi_s = \varphi_s(t_0) - \varphi_s(t_0 - t)$ is the phase fluctuation due to the non-Gaussian sources and, the averaging $\langle ... \rangle$ is done with respect to all times t_0. The normal Gaussian electromagnetic environment is contained in the function $J_0(t)$, which follows from the

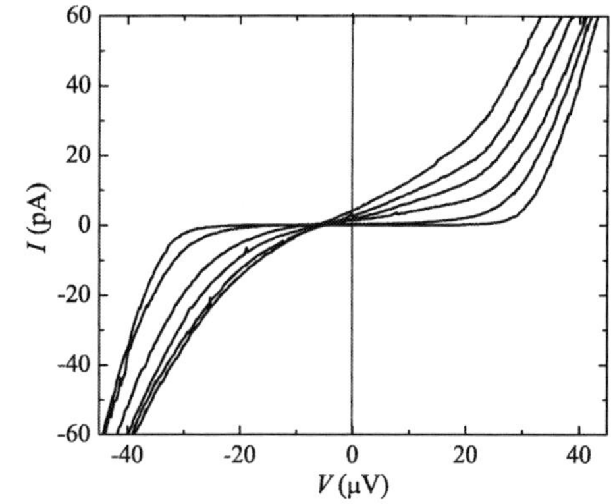

FIGURE 2. *IV* curves for the detector junction in the Coulomb blockade region for noise currents $I_N = 0 - 10$ pA, with 2 pA steps per curve from bottom upwards. The voltage is measured across the Josephson junction and chromium resistor but subtracted by RI_N.

the fluctuation-dissipation theorem [14]. However, now the problem is to evaluate the above expression for non-Gaussian sources. When the extra noise source is made of quasiparticle current from another tunnel junction directly coupled to the island, the phase fluctuations can be found by looking at the voltage fluctuations caused by the tunnelling quasiparticles

$$V_s(t) = \frac{I_N e}{|I_N|C} \sum_i \Theta(t - t_i) e^{-(t-t_i)/\tau}, \tag{2}$$

where I_N is the current through the noise junction and $\tau = RC$ is the time-constant for the decay of the charge through the external resistor. The sum goes over the tunnelling events at times t_i, which depend on the statistics of the tunneling events. The corresponding phase fluctuations can be found by using the Josephson relation and, hence, integrating Eq. (2). If one expands Eq. (1) around zero bias voltage [17]

$$I(V) = I_0 + GV + aV^2 + bV^3. \tag{3}$$

and assumes that the current is low enough, $I_N \ll e/RC$, so that current pulses are not overlapping (thus ignoring higher order interactions), one obtains analytic results that show a shift of the conductance minimum, a ratchet effect (defined here as a non-zero current at zero voltage, see [8, 17]) and a linear dependence of the minimum conductance as a function of shot noise current (power). In the case of large currents, *i.e.* $I_N \sim e/RC$, the voltage pulses or phase steps become overlapped, meaning that the simple treatment in the above references do not apply. Numerically, one can do the sum (2) for any pulse sequence, but the calculation quickly becomes time consuming and the question of convergence becomes important.

In Fig. 2 the *IV* curves for the detector junction at the blockade region is shown for different values of the noise current, while in Fig. 3, the corresponding differential conductance curves measured with lock-in amplifiers are shown. The behavior reveals a

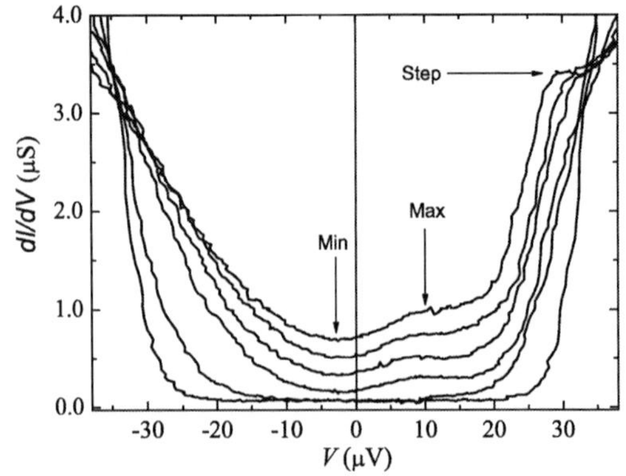

FIGURE 3. Differential conductance for the Coulomb blockade region for noise currents $I_N = 0-10$ pA, with 2 pA steps per curve from bottom upwards. The voltage is measured across the Josephson junction and chromium resistor but subtracted by RI_N.

large sensitivity to shot noise. The theory is strictly relevant for the zero bias case only, and it predicts the observed small shift of the minimum conductance. The competing shift is the trivial shift of RI_N, due to the fact that the noise current goes through the chromium resistor. When we subtract this trivial shift we find that the extrema are located at constant points of voltage. The dependence of the minimum conductance as a function of the noise current is found to be linear as expected (see Fig 4) and the slope was shown in Ref. [8] to correspond to the theoretical prediction in [17]. When the sign of the noise current is changed, the conductance curves of Fig. 3 are mirrored around the y-axis, showing that the effects are indeed due to an asymmetric noise source (see e.g. [8]). In the case of the thermal noise, the conductance curve would be totally symmetric and change of sign of the noise current would not have an effect. However, a quantitative explanation for the observed maximum point and the conductance step at +30 μV seen in Fig. 3 still remains to be found. However, preliminary numerical solution of Eq. (1) does indeed show a conductance maximum. In simple terms, this could be seen as the convolution of two $P(E)$-functions [16], one due to the chromium resistor and the other due to the shot noise source, and the resulting lifting of the blockade.

The directly connected noise source set-up presented here has a large sensitivity to the noise power, or second moment, but with the drawback of not being able to discriminate between higher moments. The observed asymmetric features in the IVs are, however, clear indications of the influence of odd moments. The present theory, however, does not show how the odd moments affect the asymmetry observed in the conductance curves.

An open question is how to find the complete statistics (full counting statistics) of the noise source. Also, one needs to keep in mind that the detector has a back-action on the noise source and this will affect the statistics. The source and detector are thus always interacting in this scheme, and a refined theoretical treatment to take into account this interaction might be necessary. An alternative set-up with a capacitive coupling of the noise source has been proposed [18, 19] (see also Fig. 1a). In this way, one can indirectly study the effect of the second moment or noise power in the excess

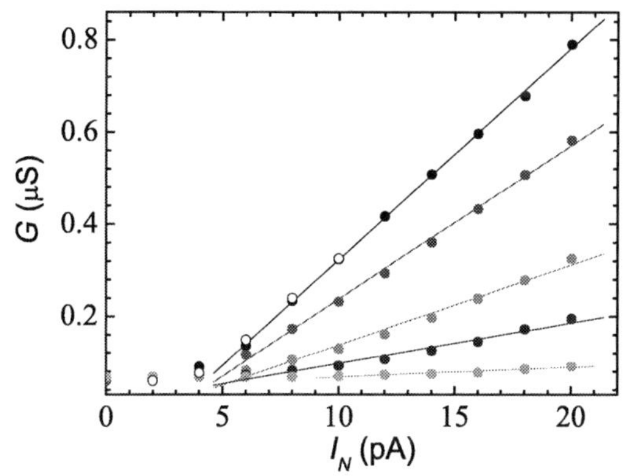

FIGURE 4. Zero-bias conductance as a function of the noise current I_N for different values of the ratio $E_J/E_C = 0.06, 0.11, 0.16, 0.24$ and 0.3 (from lowest to highest curve). The dependence is strongest for $E_J/E_C = 0.3$ and linear for all cases when the shot noise power exceeds the thermal noise from the chromium resistor at $I_N \sim 5$ pA. The open circles shows the points for negative values of I_N.

TABLE 1. Sample parameters for the directly coupled noise source measurement presented in this paper (see setup in Fig. 1a). The energies are given in units of μeV.

R_T^{JJ} (kΩ)	R_T^{NIS} (kΩ)	R (kΩ)	E_C	E_J^{min} / E_J^{max}
24	73	179	80	2.4/24

antisymmety $\delta I_A = [I(V) - I(-V)]/2 - I_D(V)$ of the IV curve (to avoid confusion, recall that the IV curve is intrinsically an odd function because $I(-V) = -I(V)$). Here $I_D(V)$ is the IV without the noise source on, meaning in the case of shot noise that zero current is fed through the noise source. Furthermore, the effect of the third moment is seen in the symmetric part $I_S = [I(V) + I(-V)]/2$, which would be identically zero for a Gaussian noise source. The method also gives a way to find the Fano factor of the noise source depending on the detection band and the shape of the noise spectral density. If the noise spectral density for order n is flat and linear in noise current, $S_{nI} = e^{n-1} F_n I_N$, then the symmetric and antisymmetric components of the detector current also respond linearly to the noise current and the Fano factor is proportional to I/I_N. The constant of proportion should be found from a calibration measurement which can be carried out for example at the linear branch of a tunnel junction IV curve, where the Fano factor can be assumed to be known. In this manner, one could connect different noise sources, for example a superconducting SET, or an array of tunnel junctions, to the detector, and measure the Fano factor as a function of the noise source bias point, in either voltage or current biased modes. A problem arises if the S_{nI} are not flat but the frequency dependence has to be taken into account, due to the fact that the response of the detector is a convolution between a kernel, which contains the detector IV for zero noise current, and the $S_{nI}(\omega)$ (for details, see Ref. [19]).

In summary, we found that a Coulomb blockaded Josephson junction in a resistive environment is an interesting device for measuring both Gaussian and non-Gaussian noise. Many open questions remain, regarding the analytical and numerical methods of calculating *IV* curves for high current when the pulses overlap, especially for voltages far from zero bias, where a series expansion is no longer valid. In the detection of higher order moments, methods for comparison between theory and experiment still remains to be found in the case when the spectral density of cumulants is not flat in the frequency regions of interest.

ACKNOWLEDGMENTS

We acknowledge fruitful discussions with G. Johansson and P. Virtanen. This work was supported by the Academy of Finland and by the Large Scale Installation Program ULTI-4 of the European Union (Contract No. RITA-CT-2003-5D5313).

REFERENCES

1. Y. Makhlin, G. Schön, and A. Shnirman, *Rev. Mod. Phys.*, **73**, 357 (2001).
2. L. S. Levitov, H. Lee, and G. B. Lesovik, *J. Math. Phys.*, **37**, 4845 (1996).
3. R. Aguado, and L. Kouwenhoven, *Phys. Rev. Lett.*, **84**, 1986 (2000).
4. L. S. Levitov, and M. Reznikov, *Phys. Rev. B*, **70**, 115305 (2004).
5. C. W. J. Beenakker, M. Kindermann, and Y. V. Nazarov, *Phys. Rev. Lett.*, **90**, 176802 (2003).
6. B. Reulet, J. Senzier, and D. Prober, *Phys. Rev.Lett.*, **91**, 196601 (2003).
7. R. Deblock, E. Onac, L. Gurevich, and L. P. Kouwenhoven, *Science*, **301**, 203 (2003).
8. R. K. Lindell, J. Delahaye, M. A. Sillanpää, T. T. Heikkilä, E. B. Sonin, and P. J. Hakonen, *Phys. Rev. Lett.*, **93**, 197002 (2004).
9. J. Pekola, T. Nieminen, M. Meschke, J. Kivioja, A. Niskanen, and J. Vartiainen, *cond-mat/0502446* (2005).
10. J. Delahaye, R. Lindell, M. S. Sillanpää, M. A. Paalanen, E. B. Sonin, and P. Hakonen, *cond-mat/0209076* (2002).
11. J. Delahaye, R. Lindell, M. Sillanpää, M. Paalanen, E. Sonin, and P. Hakonen, *J. Phys. Soc. Jpn.*, **72**, 187 (2002).
12. J. Tobiska, and Y. V. Nazarov, *Phys. Rev. Lett.*, **93**, 106801 (2004).
13. J. P. Pekola, *Phys. Rev. Lett.*, **93**, 206601 (2004).
14. G.-L. Ingold, and Y. V. Nazarov, "Charge Tunneling Rates in Ultrasmall Junctions," in *Single Charge Tunneling*, edited by H. Grabert, and M. H. Devoret, Plenum Press, New York, 1992.
15. G. Ingold, H. Grabert, and U. Eberhardt, *Phys. Rev. B*, **50**, 395 (1994).
16. R. Lindell, J. Delahaye, M. Sillanpää, M. Paalanen, E. Sonin, and P. Hakonen, "Mesoscopic Josephson junction as a Noise Detector," in *Noise and Information in Nanoelectronics, Sensors and Standards II, Proceedings of SPIE*, edited by J. M. Smulko, Y. Blanter, M. I. Dykman, and L. B. Kish, SPIE, Bellingham, 2004.
17. E. B. Sonin, *Phys. Rev. B.*, **70**, 140506(R) (2004).
18. J. Delahaye, T. T. Heikkilä, R. Lindell, M. Sillanpää, T. Yamaguchi, and P. Hakonen, "Ultrasensitive noise measurement scheme for mesoscopic circuits using a Coulomb blockaded Josehpson junction," in *17th International Conference on Noise and Fluctuations*, edited by J. Sikula, CNRL s. r. o, Brno, 2003.
19. T. T. Heikkilä, P. Virtanen, G. Johansson, and F. K. Wilhelm, *Phys. Rev. Lett.*, **93**, 247005 (2004).

How To Measure A Subdiffusion Coefficient And A Subdiffusion Parameter

Tadeusz Kosztołowicz

Institute of Physics, Świętokrzyska Academy,
ul. Świętokrzyska 15, 25-406 Kielce, Poland

Abstract. We derive theoretical formulas which allows one to extract the subdiffusion parameter α and subdiffusion coefficient D_α form measured concentration profiles of transported substance in a subdiffusive medium. The parameters occur in the relation defining the subdiffusion $\langle \Delta x^2 \rangle = \frac{2D_\alpha t^\alpha}{\Gamma(1+\alpha)}$ with $\alpha < 1$, where $\langle \Delta x^2 \rangle$ denotes the mean-square displacement of a Brownian particle. The analysis is pe.formed separately for the medium, where optical measurements of concentrations are possible (the medium is transparent for a laser beam), as well as for the non-transparent medium.

Keywords: Subdiffusion, gels, measurement
PACS: 05.40.-a, 66.10.-x, 06.30.-k, 82.70.-y

INTRODUCTION

Subdiffusion occurs in many systems such as porous media or gel solvents. It is defined by the relation [1]

$$\langle \Delta x^2 \rangle = \frac{2D_\alpha t^\alpha}{\Gamma(1+\alpha)}, \tag{1}$$

for $\alpha < 1$, where $\langle \Delta x^2 \rangle$ denotes the mean-square displacement of a transported particle, α is the subdiffusion parameter and D_α denotes the subdiffusion coefficient measured in the units m^2/s^α. While the subdiffusion is theoretically rather well understood, there are only a few papers concerning the problem of measurement of the subdiffusive parameters [2, 3]. In the pioneering study [3] the interdiffusion of heavy and light water in a porous medium was observed by means of nuclear magnetic resonance. The subdiffusion coefficient D_α was determined using the special case $\alpha = 2/3$ solution of the subdiffusive equation. The procedure is neither very accurate nor of general use. In our very recent papers [2] we have proposed a method to measure the parameters α and D_α, which exploits a membrane system where a substance of interest is transported in a solvent from one vessel to another across a thin membrane which plays here only an auxiliary role. Comparing the experimental and theoretical results we found the values of subdiffusion parameter and subdiffusion coefficient for the glucose and sucrose transported in agarose gel. The concentration profiles of transported substances were measured by means of the interferometric method. However, the method can be used only when transported substance and a medium are transparent for a laser light. In this paper we extend the method to non-transparent media. Specifically, we propose to use a system where a subdiffusive non-transparent material is in contact with the transparent diffusive one. We derive formulas which allows one to infer the subdiffusion parameters

CP800, *Unsolved Problems of Noise and Fluctuations: UPoN 2005*
edited by L. Reggiani, C. Pennetta, V. Akimov, E. Alfinito, and M. Rosini

of non-transparent medium from the concentration profiles measured in the transparent diffusive one.

MEASUREMENT IN TRANSPARENT MEDIUM

Let us consider the system containing two semi-infinite vessels with a thin membrane in between which separates the initially homogeneous solute of the substance of interest from the pure solvent. A schematic view of the system is presented in Fig. 1. Instead of the mean-square displacement (1), our method refers to the temporal evolution of the so-called near-membrane layers (NML). The thickness of NML δ is defined as the distance from the membrane where the substance concentration $C(x,t)$ drops κ times with respect to the membrane surface

$$C(\delta,t) = \kappa C(0^+,t), \tag{2}$$

where $x = 0$ is the location of the membrane and κ is an arbitrary number smaller than unity. In the following we consider the concentrations in the right vessel ($x > 0$) only.

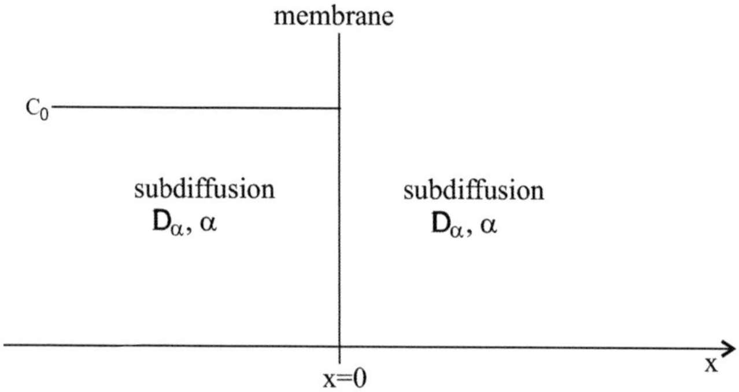

FIGURE 1. Schematic view of the membrane system, C_0 denotes a initial concentration.

The subdiffusion is described by the equation [1]

$$\frac{\partial^\alpha C(x,t)}{\partial t^\alpha} = D_\alpha \frac{\partial^{1-\alpha}}{\partial t^{1-\alpha}} \frac{\partial^2 C(x,t)}{\partial x^2} \tag{3}$$

where $\frac{\partial^{1-\alpha}}{\partial t^{1-\alpha}}$ is the Riemann-Liouville fractional derivative [1, 4]. To obtain the theoretical concentrations we solve the eq.(3) in the region occupied by right vessel for the initial concentration

$$C(x,0) = \begin{cases} C_0, & x < 0 \\ 0, & x > 0 \end{cases} . \tag{4}$$

Since the subdiffusion equation is of the second order with respect to the space variable, it requires two boundary condition at a thin membrane. The first one simply assumes the continuity of the flux J flowing through the membrane

$$J(0^-,t) = J(0^+,t). \tag{5}$$

The flux J gets a simple form after the Laplace transformation $L[f(t)] \equiv \hat{f}(s) = \int_0^\infty dt f(t)e^{-st}$, namely $\hat{J}(x,s) = -D_\alpha s^{1-\alpha} d\hat{C}(x,s)/dx$. There is no obvious choice of the second boundary condition (BC). We choose the BC which is formulated as follows [5, 6]: *If during a given time interval N particles reach the membrane, the fraction σ of them will be stopped while $1-\sigma$ will go through.* This condition leads to a BC of the form

$$C(0^+,t) = \lambda C(0^-,t), \tag{6}$$

where $\lambda = (1-\sigma)/(1+\sigma)$. For $\lambda = 1$, one deals with continuous concentration profiles.

For $x > 0$ the solution of subdiffusive equation is [2]

$$C(x,t) = C_0 \frac{2\lambda}{(1+\lambda)\alpha} H_{11}^{10}\left(\left(\frac{x^2}{D_\alpha t^\alpha}\right)^{1/\alpha} \middle| \begin{matrix} 1 & 1 \\ 0 & 2/\alpha \end{matrix} \right), \tag{7}$$

where H denotes Fox function, which can be given as follows [7]

$$H_{11}^{10}\left(\frac{a^{1/\beta}}{t} \middle| \begin{matrix} 1 & 1 \\ (1+\nu)/\beta & 1/\beta \end{matrix} \right) = \beta \left(\frac{a^{1/\beta}}{t}\right)^{1+\nu} \sum_{k=0}^{\infty} \frac{1}{\Gamma(-k\beta-\nu)k!} \left(-\frac{a}{t^\beta}\right)^k. \tag{8}$$

Putting eq. (7) into (6) and using following relation

$$H_{11}^{10}\left(0 \middle| \begin{matrix} 1 & 1 \\ 0 & 2/\alpha \end{matrix} \right) = \frac{\alpha}{2}, \tag{9}$$

which is easy derived from eq. (8), we get

$$H_{11}^{10}\left(\left(\frac{\delta}{\sqrt{D_\alpha t^\alpha}}\right)^{2/\alpha} \middle| \begin{matrix} 1 & 1 \\ 0 & 2/\alpha \end{matrix} \right) = \frac{\kappa\alpha}{2}. \tag{10}$$

From eq. (10) we derive the relation

$$\delta(t) = A(\alpha, D_\alpha, \kappa) t^{\alpha/2}, \tag{11}$$

where the coefficient A equals

$$A(\alpha, D_\alpha, \kappa) = \sqrt{D_\alpha} \left[\left(H_{11}^{10}\right)^{-1} \left(\frac{\alpha\kappa}{2} \middle| \begin{matrix} 1 & 1 \\ 0 & 2/\alpha \end{matrix} \right) \right]^{\alpha/2}. \tag{12}$$

Comparing the temporal evolution of NML obtained from experimental data with the formula (11) one can extract the parameters A and α. Indeed, after simple calculations we obtain

$$D_\alpha = \frac{A^2}{\left[\left(H_{11}^{10}\right)^{-1} \left(\frac{\alpha\kappa}{2} \middle| \begin{matrix} 1 & 1 \\ 0 & 2/\alpha \end{matrix} \right) \right]^{\alpha}}. \tag{13}$$

Thus, knowing experimental values of A, α and κ, one can deduce D_α. At the end of this section we note that the time evolution of NML was also used to determine the normal diffusion coefficient [6].

MEASUREMENT IN NON-TRANSPARENT MEDIUM

To derive the formula for subdiffusive parameters in the case of non-transparent medium, we consider the system presented on Fig. 2. The subdiffusive medium under study is located between two thin membranes placed at $x=0$ and $x=d$; the membranes separate subdiffusive medium form the ones with normal diffusion (where $\alpha=1$). Such a system seems to be appropriate for experimental investigations; the concentration profiles are recorded in right vessel $x>d$. In the following we assume that the temporal evolution of concentration is measured at $x=d^+$ only. At an initial moment the homogeneous solvent with non-zero initial concentration C_0 is present in the left vessel $x<0$.

To solve the subdiffusion and diffusion equations describing the transport process in the system under considerations we adopt the following boundary conditions

$$J(0^-,t)=J(0^+,t)\ ,\ C(0^-,t)=\lambda C(0^+,t), \tag{14}$$

$$J(d^-,t)=J(d^+,t)\ ,\ \lambda C(d^-,t)=C(d^+,t), \tag{15}$$

where the dimensionless parameter λ controls, as previously, the concentration discontinuities at the boundaries. In terms of Laplace transform the concentration profiles for $x>d$ are [8]

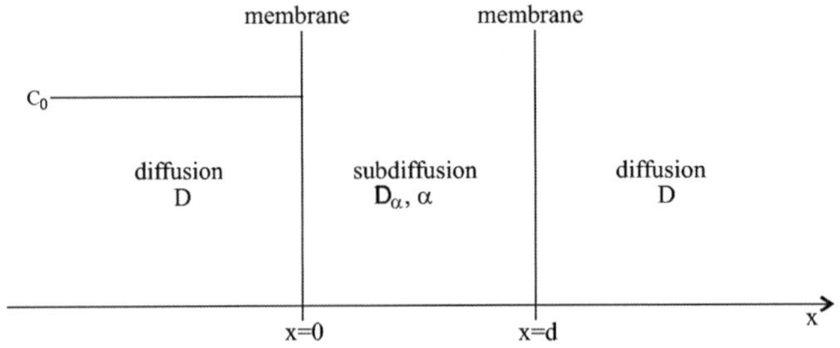

FIGURE 2. The slab of non-transparent subdiffusive medium is located between the two vessels filled with diffusive one. The diffusive medium is assumed to be transparent for laser light.

$$\hat{C}(x,s)=\frac{2\lambda\gamma}{s}\sqrt{\frac{D}{D_\alpha s^{1-\alpha}}}e^{-\sqrt{s/D}(x-d)-\sqrt{s^\alpha/D_\alpha}d}, \tag{16}$$

where

$$\gamma=\left[(1+\lambda\sqrt{D/D_\alpha s^{1-\alpha}})^2-(1-\lambda\sqrt{D/D_\alpha s^{1-\alpha}})^2e^{-2\sqrt{s^\alpha/D_\alpha}d}\right]^{-1}. \tag{17}$$

In the long time limit, which corresponds to the limit of small parameter s, the Laplace transformed concentration at the the membrane surface $x=d^+$ equals

$$\hat{C}(d,s)=\frac{D_\alpha}{\sqrt{D}d\lambda s^{\alpha+1/2}}. \tag{18}$$

Using the formula $L^{-1}(1/s^{\mu}) = \Gamma(\mu)t^{\mu-1}$, $\mu > 0$, we obtain a very simple equation of the time evolution of the concentration at the membrane surface

$$C(d,t) = \frac{D_\alpha \Gamma(\alpha + 1/2)}{\sqrt{D} d \lambda} t^{\alpha - 1/2}. \tag{19}$$

Comparing experimental results for $C = C(d,t)$ with the prediction of eq. (19), we can extract the subdiffusive parameter α. We note that extracting of the subdiffusion coefficient D_α from eq. (19) is rather difficult as one then needs a value of the parameter λ which controls the membrane permeability. To determine D_α one should rather compare the experimental and theoretical concentration profiles in a wider range of the space variable.

FINAL REMARKS

Using the method based on time evolution of NML in transparent medium we have already obtained [2] $\alpha = 0.90 \pm 0.01$ for the transport of glucose and sucrose in agarose gel. Thus, we have shown that the sugar transport in agarose is subdiffusive, in contrast to some papers (see for example [9]) where the transport has been assumed to be diffusive. We have found the subdiffusion coefficients as: $D_{0.90} = (9.8 \pm 1.0) \times 10^{-4}$ mm^2/s$^{0.90}$ for glucose and $D_{0.90} = (6.3 \pm 0.9) \times 10^{-4}$ mm^2/s$^{0.90}$ for sucrose. As we have also shown [2], the time evolution of the non-membrane layers does not depend on the membrane permeability in the long time limit. Unfortunately, the concentration (19) at the membrane surface does not obey this property. As seen the parameter λ enters the equation.

ACKNOWLEDGMENTS

The author wishes to express his thanks to Stanisław Mrówczyński for fruitful discussions and critical comments on the manuscript.

REFERENCES

1. R. Metzler and J. Klafter, *Phys. Rep.* **339**, 1 (2000); *J. Phys. A* **37**, R161 (2004).
2. T. Kosztołowicz, K. Dworecki and S. Mrówczyński, *Phys. Rev. Lett.* **94**, 170602, (2005); *Phys. Rev. E* **71**, 041105, (2005).
3. A. Klemm, R. Metzler and R. Kimmich, *Phys. Rev. E* **65**, 021112 (2002).
4. K.B. Oldham and J. Spanier, *The fractional calculus*, Academic Press, New York, 1974.
5. T. Kosztołowicz, *Physica A* **248**, 44 (1998); *J. Phys. A* **31**, 1943, (1998).
6. K. Dworecki, T. Kosztołowicz, S. Wąsik and S. Mrówczyński, *Eur. J. Phys. E* **3**, 389, (2000).
7. T. Kosztołowicz, *J. Phys. A* **37**, 10779, (2004).
8. T. Kosztołowicz, in preparation.
9. J. Gutenwik, B. Nilsson and A. Axelsson, *Bioch. Eng. J.* **19**, 1, (2004); Ch. Mattison et al., *J. Chrom. B* **743**, 151, (2000); I. Septi et al., *J. Food Eng.* **63**, 185, (2004).

LIST OF PARTICIPANTS

NOMI	e-mail	Address
Abbott Derek	dabbott@eleceng.adelaide.edu.au	CBME UNIVERSITY OF ADELAIDE, SA 5005 ADELAIDE - - Australia
Akimov Vladimir	vladimir.akinov@unile.it	DIPARTIMENTO DI INGEGNERIA DELL' INNOVAZIONE, VIA ARNESANO, 73100 LECCE, Italy
Aleshkin Vladimir	aleshkin@ipm.sci-nnov.ru	DEPARTMENT SEMICONDUCTOR PHYSICS IPM, 603950, NIZHNY NOVGOROD, GSP-105 Russia
Alfinito Eleonora	eleonora.alfinito@unile.it	DIPARTIMENTO DI INGEGNERIA DELL' INNOVAZIONE, VIA ARNESANO, 73100 LECCE, Italy
Allison Andrew	aallison@eleceng.adelaide.edu.au	CBME UNIVERSITY OF ADELAIDE, SA 5005 ADELAIDE - - Australia
Andreucci Daniele	andreucci@dmmm.uniroma1.it	DIPARTIMENTO DI METODI E MODELLI MATEMATICI VIA A. SCARPA 16 00161 ROMA - - Italy
Bardhan Kamal	bardhan@cmp.saha.ernet.in	I/AF BIDHANNAGAR, 700 064 KOLKATA - WEST BENGAL - India
Barone Carlo	barone@sa.infn.it	VIA S. ALLENDE, 84081 BARONISSI - SALERNO - Italy
Bellini Eleonora	bellini@le.infn.it	DIPARTIMENTO DI FISICA, VIA ARNESANO, 73100 LECCE, Italy
Belzig Wolfgang	Wolfgang.Belzig@unibas.ch	DEPARTMENT OF PHYSICS AND ASTRONOMY, KLINGELBERGSTR. 82, 4056 BASEL - - Switzerland
Bergou Janos	jbergou@hunter.cuny.edu	DEPARTMENT OF PHYSICS AND ASTRONOMY, 695 PARK AVENUE, 10021 NEW YORK - - U.S.A.
Bertoni Andrea	bertoni.andrea@unimo.it	DIPRTIMENTO DI FISICA, VIA CAMPI 213/A, 41100 MODENA - - Italy
Bezrukov Sergey	bezrukos@mail.nih.gov	NATIONAL INSTITUTE OF HEALTH, BETHESDA, MD 20892- - USA
Bezsudnov Igor		NAUKA-SERVIS- JSC, MOSCOW, Russia
Bonani Fabrizio	fabrizio.bonani@polito.it	DIPARTIMENTO DI ELETTRONICA, CORSO DUCA DEGLI ABRUZZI 24, 10100 TORINO - - Italy
Borondo Florentino Rodriguez	f.borondo@uam.es	DEPT DE QUIMICA C-IX 28049 CANTOBLANCO - MADRID - Spain
Bosman Gijs	gbosm@ece.ufl.edu	NEB 565 CAMPUS UF, 32611 GAINESVILLE - FLORIDA - USA
Brunetti Rossella	brunetti@unimo.it	DIPRTIMENTO DI FISICA, VIA CAMPI 213/A, 41100 MODENA - - Italy
Bulsara Adi	bulsara@spawar.navy.mil	SPACE AND NAVAL CENTER, SAN DIEGO CODE 2363, 53560 HULL STREET , CA 92152-5001 - - U.S.A.
Buttiker Markus	Markus.Buttiker@physics.unige.ch	DEPARTMENT PHYSICS THEOR. 24 QUAI E. ANSERMET, 1211 GENEVA - - Switzerland
Capobianco Enrico	ecapob@bu.edu	DPT OF BIOMEDICAL ENGINEERING CUMMINGTON STREET, BOSTON, MA 02215 - - USA
Carbone Anna	anna.carbone@polito.it	DIPARTIMENTO DI FISICA , CORSO DUCA DEGLI ABRUZZI 24, 10129 TORINO - - Italy
Catalfamo Francesco	fcatalfamo@ingegneria.unime.it	DIPARTIMENTO FISICA DELLA MATERIA, SALITA SPERONE 31, 98166 MESSINA - - Italy
Chan Ho Bu	hochan@phys.ufl.edu	DEPATMENT OF PHYSICS, UNIVERSIY OF FLORIDA, GAINESVILLE, FL 32608, USA.
Cherubini Maria	cherubin@pc-cher.unile.it	DIPARTIMENTO DI FISICA, VIA ARNESANO, 73100 LECCE, Italy
Ciccarello Francesco	ciccarello@difter.unipa.it	DIPARTIMENTO DI FISICA E TECNOLOGIE RELATIVE, VIALE DELLE SCIENZE , 90128 - PALERMO - Italy
Ciliberto Sergio	sergio.ciliberto@ens-lyon.fr	LABORATOIRE DE PHYSIQUE 46 ALLEE D'ITALIE, 69364 LYON - - France
Conte Simona	simona.conte@virgilio.it	DIPARTIMENTO DI FISICA, VIA ARNESANO, 73100 LECCE, Italy
Conte Gabriele	gabriele.conte@polito.it	DIPARTIMENTO DI FISICA, CORSO DUCA DEGLI ABRUZZI 24, 10129 TORINO - - Italy
D'Agostino Stefania		DIPARTIMENTO DI FISICA, VIA ARNESANO, 73100 LECCE, Italy

Di Jeso Bruno	bdijeso@ilenic.unile.it	DIPARTIMENTO DI SCIENZE E TECNOLOGIE BIOLOGICHE E AMBIENTALI, VIA ARNESANO, 73100 LECCE, Italy
Doering Charlie	doering@umich.edu	DEPT OF MATHEMATICS, 48109-1043, ANN ARBOR - MICHIGAN - U.S.A.
Dykman Mark	dykman@pa.msu.edu	DEPARTMENT OF PHYSICS AND ASTRONOMY, 48823 EAST LANSING - MICHIGAN - U.S.A.
Erdman Udo	udo.erdmann@physik.hu-berlin.de	HUMBOLDT UNIVERSITAT, INSTITUT FUR PHYSIK, NEWTONSTR. 15 , 12489 BERLIN - - Germany
Fanciulli Marco	marco.fanciulli@mdm.infm.it	LABORATORIO NAZIONALE MDM, VIA OLIVETTI 2, 20041 AGRATE BRIANZA - - Italy
Farago Jean	farago@lcp.u-psud.fr	LCP BAT 349 UNIVERSITE PARIS XI, 91405 ORSAY - FRANCE - France
Ferrari Giorgio	ferrari@elet.polimi.it	DIPARTIMENTO DI ELETTRONICA E INFORMAZIONE, P.ZA L. DA VINCI 32, 20133 MILANO, Italy
Fortuna Luigi	lfortuna@diees.unict.it	DIPARTIMENTO ELETTRICO ELETTRONICO E DEI SISTEMI , VIALE A. DORIA 6, 95125 CATANIA , Italy
Frasca Mattia	mfrasca@diees.unict.it	DIPARTIMENTO ELETTRICO ELETTRONICO E DEI SISTEMI , VIALE A. DORIA 6, 95125 CATANIA , Italy
Fronzoni Leone	leone.fronzoni@df.unipi.it	DIPARTIMENTO DI FISICA, LARGO PONTECORVO 2, 56127 PISA - - Italy
Galdi Alice	aliceg@mclink.it	VIA S. ALLENDE, 84081 BARONISSI - SALERNO - Italy
Galleani Lorenzo	galleani@polito.it	DIPARTIMENTO DI FISICA , CORSO DUCA DEGLI ABRUZZI 24, 10129 TORINO - - Italy
Gingl Zoltan	gingl@sol.cc.u-szeged.hu	DEPARTMENT OF EXPERIMENTAL PHYSICS, DOM TER 9, H 6720 SZEGED - Hungary
Glattli Christian	glattli@lpa.ens.fr	SERVICE DE PHYSIQUE DE L'ETAT CONDENSE CEA SACLAY F-91191 GIF-SUR-YVETTE - - France
Goldobin Denis	denis.goldobin@mail.ru	DEPARTMENT OF PHYSICS, POSTFACH 601553, D-14415 POTSDAM - Germany
Gonnella Giuseppe	Giuseppe.Gonnella@ba.infn.it	DIPARTIMENTO DI FISICA, VIA AMENDOLA 173, 70126 BARI - Italy
Hashiguchi Sumihisa	has@es.yamanashi.ac.jp	TAKEDA 4, 400 8511 KOFU - Japan
Heiblum Mordehai	moty.heiblum@weizmann.ac.il	WEIZMANN INSTITUTE, 76100 REHOVOT - I- Israel
Jacquod Philippe	Philippe.Jacquod@physics.unige.ch	DPT. DE PHIISIQUE THEORIQUE, UNIVERSITY OF GENEVA, 1211 GENEVA - - Switzerland
Jauho Antti-Pekka	antti@mic.dtu.dk	NANO DTU, MIC, BUILDING 345 EAST OERSTEDS PLADS, DK-2800 LYNGBY - - Denmark
King Maria	mdking@neo.tamu.edu	TEXAS A&M UNIVERSITY, MS 2128, 77843 COLLEGE STATION - TX 77843-2128, TEXAS - USA
Kish Laszlo	Laszlo.Kish@ee.tamu.edu	DEPARTMENT OF ELECTRICAL ENGINEERING 77843-3128 COLLEGE STATION - TEXAS - U.S.A.
Kosinska Ilona	kosinska@th.if.uj.edu.pl	REYMONTA 4, 30-059 KRAKOW - MALOPOLSKA - Poland
Kosztolowicz Tadeusz	tkoszt@pu.kielce.pl	INSTITUTE OF PHYICS, UL. SWIETOKRZYSKA 15, 25-406 KIELCE - KIELCE - Poland
Kotelenez Peter	pxk4@po.cwru.edu	MATHEMATICS CASE WESTERN RESEARCH UNIVERSITY, CLEVELAND, OH 44106-7058 - OHIO - U.S.A.
Kou Samuel	kou@stat.harvard.edu	DEPT. OF STATISTICS, I OXFORD STREET, 02138 CAMBRIDGE - MASSACHUSETTS - U.S.A.
Kullman Lisen	Lisen.Kullman@medcellbiol.uu.se	BIOMEDICAL CENTER, UPPSALA UNIVERSITY, BOX 571, SE-751 23 UPPSALA - - Sweden
Lamprecht Anita		DIPARTIMENTO DI FISICA, UNIVERSITA' DI LECCE -- Italy
Lefloch Francois	francois.lefloch@cea.fr	CEA/GRENOBLE 17 RUE DES MARTYRS, 38000 GRENOBLE - France
Leuzzi Maria	maria.leuzzi@unile.it	DIPARTIMENTO DI INGEGNERIA DELL' INNOVAZIONE, VIA ARNESANO, 73100 LECCE, Italy
Liljenstrom Hans	hans.liljenstrom@bi.slu.se	DEPT. OF BIOMETRY AND ENGINEERING, SLU , P.O. BOX 7032, 750 07 UPPSALA, Sweden
Lindell Rene	Rene.Lindell@hut.fi	LOW TEMPERATURE LAB , OTAKAARI 3 A, P.O. BOX 2220, 02015-HUT ESPOO , HELSINKI - - Finland
Lindenberg Katja	kl@hypatia.ucsd.edu	, UNIVERSITY OF CALIFORNIA, SAN DIEGO, 9500 GILMAN DRIVE LA JOLLA, CA 92093-0340 - USA
Lindgard Per-Anker	P.A.Lindgard@risoe.dk	RISOE NAT. LAB., DK4000 ROSKILDE - - Denmark

Lindner Benjamin	Lindner.Benjamin@science.uottawa.ca	MAX PLANCK INSTITUTE, NOETHNITZER STR. 38, D-01187 DRESDEN - - Germany
Luczka Jurek Jerzey	luczka@us.edu.p	UNIWERSYTECKA 4, 40-007 KATOWICE - POLAND - Poland
Macucci Massimo	macucci@mercurio.iet.unipi.it	DIPARTIMENTO DI INGEGNERIA DELL' INFORMAZIONE, VIA CARUSO 16, 56122 PISA - Italy
Mannella Riccardo	riccardo.mannella@df.unipi.it	DIPARTIMENTO DI FISICA, LARGO B. PONTECORVO 2, 56127 PISA - - Italy
Mateos Javier	javierm@usal.es	DEPARTAMENTEO DE FISICA APICADA, PLAZA DE LA MERCED, 37008 SALAMANCA - -- Spain
McClintock Peter	p.v.e.mcclintock@lancaster.ac.uk	DEPARTMENT OF PHYSICS, LAI 4YB LANCASTER - - U.K.
Olla Piero	p.olla@isac.cnr.it	ISAC-CNR, STRADA PROVINCIALE LECCE-MONTERONI, KM 1,200, 73100 LECCE, Italy
Oriols Xavier	xavier.oriols@uab.es	DEPT. D'ENGINIERIA ELECTRONICA , 08193 BELLATERRA - BARCELONA – Spain
Pagliara Stefano		DIPARTIMENTO DI FISICA, VIA ARNESANO, 73100 LECCE, Italy
Paparella Francesco	francesco.paparella@unile.it	DIPARTIMENTO DI MATEMATICA , VIA ARNESANO, 73100 LECCE, Italy
Paradisi Paolo	paradisi@le.isac.cnr.it	ISAC-CNR, STRADA PROVINCIALE LECCE-MONTERONI KM 1,200, 73100 LECCE - - Italy
Pennetta Cecilia	cecilia.pennetta@unile.it	DIPARTIMENTO DI INGEGNERIA DELL' INNOVAZIONE, VIA ARNESANO, 73100 LECCE - Italy
Prati Enrico	enrico.prati@mdm.infm.it	LABORATORIO NAZIONALE MDM, VIA OLIVETTI 2, 20041 AGRATE RIANZA ,Italy
Reggiani Lino	lino.reggiani@unile.it	DIPARTIMENTO DI INGEGNERIA DELL' INNOVAZIONE, VIA ARNESANO, 73100 LECCE, Italy
Rogerio Enriquez Caldera	rogerio@inaoep.mx	INAOE, ELECTRONICS DEPARTMENT, 72000 PUEBLA - Mexico
Rosini Marcello	marcello.rosini@unile.it	DIPARTIMENTO DI INGEGNERIA DELL' INNOVAZIONE, VIA ARNESANO, 73100 LECCE, Italy
Rudan Massimo	mrudan@deis.unibo.it	DEIS, VIALE RISORGIMENTO, 2, 40126 BOLOGNA, Italy
Ruffo Stefano	stefano.ruffo@unifi.it	DIPARTIMENTO DI ENERGETICA, VIA S. MARTA 3, 50139 FIRENZE, Italy
Rumyantsev Sergey	roumis2@rpi.edu	POLYTECHNIC INSTITUTE 110 8-TH STREET, 12180-3590 TROY - NEW YORK - U.S.A.
Sampietro Marco	sampietr@elet.polimi.it	DIPARTIMENTO DI ELETTRONICA E INFORMAZIONE, P.ZA L. DA VINCI 32, 20133 MILANO, Italy
Safonov Leonid	safonov@p.u-tokyo.ac.jp	THE UNIVERSITY OF TOKYO 7-31 HONGO BUNKYO - KU, 113-0033 TOKYO - - Japan
Schneider Raul		DIPARTIMENTO DI FISICA, VIA ARNESANO, 73100 LECCE, Italy
Shiktorov Pavel	pavel@pav.pfi.lt	SEMICONDUCTOR PHYSICS INSTITUTE, A. GOSTAUTO 11, LT 01108 VILNIUS - - Lithuania
Shulman Alexander	ash@east.ru	LAVRUSHINSKII PER. 17-24, 119017 MOSCOW - - Russia
Sikula Josef	sikula@feec.vutbr.cz	CZECH NOISE RESEARCH LABORATORY, TECHNIKA 8, 616 00 BRNO - - Czech Republic
Singer Amit	amit_sin@zahav.net.il	5 BUKI BEN YOGLI ST., 63296 TEL AVIV - - Israel
Smelianvskiy Vadim	vadim@email.arc.nasa.gov	NASA AMES RESEARCH CENTER MAIL-STOP 269-2, 94035-1000 MOFETT FIELD - CALIFORNIA - U.S.A.
Snarskii Andrew	phys@ln.com.ua	NATIONAL TECHNICAL UNIVERSITY OF UKRAINE, KPI, KIEV - Ukraine
Solovieva Anna	anna@polymer.chph.ras.ru	4 KOSYGINA ST., 119991 MOSCOW - RUSSIA - Russia
Soskin Slava	slava@soskin.semicond.kiev.ua	INSTITUTE OF SEMICONDUCTOR PHYSICS , PROSPECT NAUKI 45, 03028 KIEV, Ukraine

Starikov Jevgenij	pavel@pav.pfi.lt	SEMICONDUCTOR PHYSICS INSTITUTE A. GOSTAUTO 11, LT 01108 VILNIUS - - Lithuania
Stefanovska Aneta	aneta@osc.fe.uni-lj.si	UNIVERSITY OF LJUBIJANA, TRZASKA 25, 1000 LJUBLJANA - Slovenia
Sverdlov Viktor	sverdlov@iue.tuwien.ac.at	INSTITUTE FUR MICROELECTRONICS, GUSSHAUSSTRASSE 27-29/E360, 1040 VIENNA - - Austria
Tacano Munecazu	tacano@ee.meisei-u.ac.jp	MEISEI UNIVERSITY, HODOKUBO HINO, 191-8506 TOKYO - - Japan
Timashev Serge	serget@mail.ru	INSTITUTE OF PHYSICAL CHEMISTRY, 10 VORONTSOVO POLE ST., 105064 MOSCOW - - Russia
Trifone Schettino	schettin@ilenic.unile.it	DIPARTIMENTO DI BIOLOGIA VIA MONTERONI, 73100 LECCE, Italy
van Beijeren Henk	H.vanBeijeren@phys.uu.nl	LEUVENLAAN 4, 3584 CE UTRECHT - - The Netherlands
Van Kampen Nico	N.G.vanKampen@phys.uu.nl	OUDEDIJK 38, 3612 AB TIENHOVEN - - The Netherlands
Vandamme Lode	L.K.J.Vandamme@tue.nl	DEPARTMENT OF ELECTRICAL ENGINEERING, EH 9.13, 5600MB EINDHOVEN, The Netherlands
Varani Luca	varani@cem2.univ-montp2.fr	UNIVERSITY OF MONTPELLIER II CEM 2 (CC084) PLACE E. BATAILLON, 34095 MONTPELLIER - France
Vitusevich Svetlana	s.vitusevich@fz-juelich.de	ISG2 FORSCHUNGSZENTRUM JUELICH, 52425 JUELICH - Germany
Ward Lawrence	lward@psych.ubc.ca	UNIVERSITY OF BRITISH COLUMBIA, 2136 WEST MALL, V6T 1Z4 VANCOUVER - BC - Canada
Whitney Robert	Robert.Whitney@physics.unige.ch	DEPT. DE PHYSIQUE THEORIQUE, 24 QUAI ERNEST-ANSERMET, CH-1211 GENEVE - - Switzerland
Wilkinson Michael	m.wilkinson@open.ac.uk	APPLIED MATHEMATICS, MK7 6AA ENGLAND - BUCKINGHAMSHIRE - U.K.
Winters-Hilt Stephen	swinters@chnola-research.org	COMPUTER SCIENCE DEPT, 2000 LAKESHORE DRIVE, 70148 NEW ORLEANS - LOUISIANA - U.S.A.
Yamamoto Yoshiharu,	yamamoto@p.u-tokyo.ac.jp	UNIVERSITY OF TOKYO, 7-3-1 HONGO, 113-0033 BUNKYO-KU - TOKYO - Japan

Author Index

G

H

I

J

K

L

M

N

O

P

R

S

T

U

V

W

Y

Z